STRENGTH OF MATERIALS

STRENGTH OF MATERIALS

AN UNDERGRADUATE TEXT

Graham M. Seed

SAXE-COBURG
PUBLICATIONS

published 2000 by
Saxe-Coburg Publications
10 Saxe-Coburg Place
Edinburgh, EH3 5BR, UK

Saxe-Coburg Publications is an imprint of Civil-Comp Ltd

ISBN 1-874672-12-1

British Library Cataloguing in Publication Data
A catalogue record for this book is available from the British Library

Printed in Great Britain by Bell & Bain Ltd, Glasgow

Contents

Part III Advanced Topics 269

List of Examples

Preface

Numerous texts exist which more than adequately cover the subject matter of the *Strength of Materials* discipline for an undergraduate degree in Mechanical, Civil and Offshore Engineering. However, there are few texts which students can use throughout their entire undergraduate studies. This book aims to provide that single text. The whole book is divided into three main sections (Parts I, II and III) with the subject matter chosen to reflect the core syllabus of a typical three year Mechanical/Civil/Offshore Engineering Honours degree. Each part is chosen to correspond to a typical year's material, with the final part covering more advanced material via *Further* chapters such as *Further Elasticity* and *Further Finite Elements*. Several of the latter Chapters in Part III present an overview of selected topics in the strength of materials such as further finite elements, composite materials and contact mechanics. These topics will not generally form part of the core syllabus of the majority of degree courses and may be optional courses in either final year or Masters courses. Each Chapter concludes with *Further Reading* and *Exercises* sections. The recommended reading chosen is predominantly text books and not research papers. Solutions to all Chapter exercises are provided and can be downloaded via the Saxe-Coburg Publications web site. It is acknowledged that the scope of the book will not cover all degree courses but hopefully the majority of material is covered, particularly in the first two parts.

The structure of the book is as follows. Part I *Introductory Topics* contains six chapters which provide an introduction to several of the key topics in the discipline of *Strength of Materials*. Chapter 1 provides an introduction to the notions of stress and strain and typical stress-strain curves for engineering materials. The Chapter also includes a discussion of the constitutive laws for linear elastic materials and material constants such as Young's modulus, shear modulus and Poisson's ratio. Chapter 2 introduces the first and second moments of area with numerous examples illustrating their evaluation. In Chapter 3 the general torsion formulae are introduced and applied to the torsion of circular cross-section bars and cater for both solid, tubular and thin-walled cross-sections. The fundamental stress-strain formulae for the pressurisation of thin-walled cylindrical and spherical vessels are considered in Chapter 4. The topic of simple beam theory plays a key role in the majority of strength of materials texts and lectures for introducing the key concepts and techniques of the subject. Chapter 5 covers all of the main properties of beam theory from their classification, shear force and bending moment diagrams, shear stress and bending stress formulae and on to the deflection of beams. Part I concludes with Chapter 6 which provides an examination of the compression of struts. The Chapter begins by introducing Euler's formula for the four main types of end-conditions. Eccentric loading is then addressed, and the Chapter concludes with a discussion of semi-empirical formulae and why they are required.

Part II *Intermediate Topics* begins with taking more of a continuum approach to the strength of materials. Chapter 7 begins by deriving the transformation equations of stress and their associated principal and maximum shear stress equations. The graphical approach of Mohr's circle is then covered. The Chapter concludes by re-visiting the second moments of area discussed in Chapter 2 but now subject to a pure rotation of coordinate axes, with the analogy between the stress and moments of area tensors being made. Chapter 8 is analogous to that of Chapter 7 except that we now discuss strain rather than stress. In addition, the topic of strain gauge rosettes is covered. The concept of strain energy is central to a variety of different analyses in the field of strength of materials, and Chapter 9 spends some time introducing this important concept in detail. The key formulae for determining the strain energy under conditions of pure torsion and pure bending are then derived. These are then applied to the problems of torsion, bending of beams, helical springs and proving rings. The Chapter concludes by discussing Castigliano's theorems and providing an insight into strain energy due to impact loading. In Chapter 10 the principles of virtual displacement, virtual work and total potential energy are discussed. This Chapter aims to provide a theoretical basis for later Chapters such as Chapter 13 which discusses the finite element method. Chapter 11 extends the discussion of Chapter 4 to thick-walled cylindrical and spherical pressure vessels. The key formulae due to Lame are derived and applied to vessels with different combinations of boundary conditions and end conditions. The Chapter also covers the shrink fit of a collar onto a shaft and compound cylinders.

Chapter 12 provides a detailed introduction to the plasticity of materials since the majority of material covered in previous Chapters deals with linear elastic materials. The Chapter begins by discussing what is meant by plasticity and why we need to consider it. Typical elastic-plastic constitutive relationships are examined emphasising the tensile macro approach. The five key criteria of yielding are then introduced. The latter half of the Chapter takes a more applied approach by examining the torsion of circular components and bending of beams. The finite element method now forms the most popular and available tool used by engineers for the analysis of engineering components. Chapter 13 introduces the finite element method assuming no previous experience. The mathematical theory then proceeds to build from considering the one-dimensional bar element in some detail. The vector-matrix formulations of the constitutive relations and the variational formulation of the stiffness matrix are then discussed. The Chapter then extends the analysis to two-dimensional pin-jointed structures and concludes by drawing analogies between the elasticity formulation and that of one-dimensional heat transfer and fluid flow finite element formulations.

The final part of the book, Part III *Advanced Topics*, begins with Chapter 14 which takes a closer look at stress and strain and the general theory of elasticity. The first seven sections cover key topics such as the equations of equilibrium and the classical approach of Airy stress functions. The stress analysis of Chapter 7 is then extended to three dimensions by deriving the cubic equation of stress. As a supplement to Chapter 13 a matrix representation of the stress-strain constitutive relations is presented. The complex-variable representation of elasticity is then covered. The bending of plates and general theory of torsion of prismatic bars are then introduced. The Chapter concludes by examining the problems of a hole in a plate and a point-loaded wedge in preparation for later Chapters on fracture mechanics and contact mechanics. Chapter 15 extends the previous discussion of Chapter 12 on plasticity. The notion of equivalent stress and strain are introduced. The two key rules of flow are then covered and applied to the two problems of an elastic-plastic thick-walled pressure vessel and the plastic instability of a thin-walled pressure vessel. Chapter 16 extends the discussion of Chapter 13 by presenting the finite element formulations of two-dimensional continuum elements with emphasis on the constant strain triangle. Higher order elements are then examined with emphasis on the bar and triangle elements. Chapter 17 provides a comprehensive examination of the fracture and fatigue of materials. The governing crack tip parameters of stress intensity, T-stress and Griffith's energy release rate are introduced. Several stress intensity factors and T-stresses are presented for several popular configurations. Three methods for determining crack tip stress intensity factors are then presented. The discussion is extended to take account of crack tip plasticity and the determination of the fracture toughness of materials. An important application of fracture mechanics is in the design of pressure vessels in order to prevent catastrophic failure. Chapter 17 includes a discussion on the damage assessment of defects in pressure vessels. The subject of fatigue crack growth law is then discussed with reference to both long and short cracks. The Chapter concludes with a discussion of the J, L and M elastic-plastic fracture mechanics parameters.

The creep of engineering materials is examined in Chapter 18. Following a discussion on the mechanisms of creep, the time, stress and temperature dependencies of creep are examined. The three problems of the tension of a bar, torsion of a tube and pure bending of a beam are analysed. The reader learns about the viscoelastic response of materials in Chapter 19. The Maxwell, Voigt and the Standard Linear Solid models are examined with respect to both creep and stress relaxation. Chapter 20 presents a brief introduction to the subject of continuum damage mechanics. The Paris long fatigue crack growth law presented in Chapter 17 is re-examined from a damage mechanics perspective. The material and mechanical properties of composite materials are examined in Chapter 21 and their advantage over metallic engineering materials. The elastic properties and strength of unidirectional laminae are derived. The discipline of contact mechanics is introduced in Chapter 22. The Chapter begins by analysing an elastic half-plane subject to concentrated normal and tangential forces. These concentrated forces are then distributed over a finite interval with particular emphasis given to the case of uniformly distributed surface tractions. In addition, the problems of indentation by a flat rigid punch and Herzian contact are also considered. Chapter 22 concludes with a discussion on indentation fracture in which cracks are present on the surface of the half-plane. The final Chapter of the book presents a statistical examination of the strength of brittle materials such as ceramics. The first part of the Chapter includes discussions on necessary statistical measures such as relative and cumulative frequencies, probability and the normal distribution. The latter half of the Chapter focuses on the Weibull distribution which is frequently used when determining the strength of brittle materials from both tensile and three-point bend specimens.

Chapter Exercises

At the end of each Chapter there are seven exercises which are chosen to reflect the key subject matter. The solutions, `seed_sol.pdf`, to the Chapter exercises, in Acrobat version 4 format, can be downloaded from the Saxe-Coburg Publications web page: `http://www.saxe-coburg.co.uk`.

Acknowledgements

The author acknowledges students of the Department of Mechanical and Chemical Engineering at Heriot-Watt University who have, unknowingly, both influenced and contributed to the present text. Thanks also to my colleagues at the department, in particular to Norman Loch for proof reading a draft version of the book. Thanks to everyone at Saxe-Coburg Publications for their support and help in bringing this book to the shelf! In particular, Barry Topping for championing the publication process, Rosemary Brodie for proof reading and Jelle Muylle, Roman Putanowicz and Alasdair Closier for all their hard work in typesetting, layout and graphical design. Finally, a huge thanks to my partner Melanie for her endless support and encouragement throughout the completion of this book.

Correspondence

The author welcomes communication and comments on the book and can be contacted at the following virtual addresses:

Internet: `g.m.seed@hw.ac.uk`
World-Wide Web: `http://www.hw.ac.uk/mecWWW/staff/gms.htm`

Graham M. Seed
December 2000

Part I

Introductory Topics

Fundamental to the subject of *Strength of Materials* are the concepts of stress and strain and these are introduced in Chapter 1. This Chapter also covers the elastic constitutive equations of Hooke's law and provides definitions of several material parameters such as Young's modulus and Poisson's ratio. In Chapter 2 we will learn about the first and second moments of area. Chapter 3 examines the torsion of bars of circular cross-section and caters for bars of solid, tubular and thin-walled cross-sections. Both cylindrical and spherical thin-walled pressure vessels are examined in Chapter 4. One of the major Chapters of Part I is Chapter 5 which presents an overview of the loading of beams. We begin by classifying beams and types of loading and then move on to the determination of shear force and bending moment and their associated diagrams, the bending stress formulae and finally to the deflection of beams. Part I concludes with a discussion on struts in Chapter 6.

Chapter 1

Stress and Strain

1.1 Introduction

This Chapter introduces the concepts of stress and strain. The linear elastic constitutive relation, Hooke's law, which relates stress and strain is then introduced. This is then followed by an overview of the stress-strain diagram. Shear strain, shear modulus, Poisson's ratio, volumetric strain, bulk modulus and one-dimensional thermal strain are introduced. Stress and strain are then re-examined by introducing a tensor notation for stress and strain for general three-dimensional systems. The Chapter concludes with a discussion on the ideal tensile and shear strength of materials.

1.2 Stress

Consider Figure 1.1 which illustrates a three dimensional body in a state of equilibrium under the action of external forces $F_1, F_2, \ldots, F_n$. If the body is sliced at the cross-section C then the forces which were acting through surface C are maintained in order to preserve equilibrium. Considering a small area, δA, of C there will be the associated proportion δF of the total force F acting on δA. The force δF can be resolved into components perpendicular to the cut surface, δP, and parallel to the cut surface, δQ, as shown in Figure 1.1. The intensity of these two forces (that is, force per unit area) at a point is termed the *stress*. Thus, the stresses associated with the perpendicular or *direct*, σ, and tangential or *shear*, τ, components of force δF are given by

$$\sigma = \lim_{\delta A \to 0} \frac{\delta P}{\delta A}; \qquad \tau = \lim_{\delta A \to 0} \frac{\delta Q}{\delta A} \tag{1.1}$$

It is worth noting that stress is a scalar quantity and not a vectorial quantity as are the resolved forces. It is a common mistake to confuse stress and force when first introduced to the definition of stress. In an analogous manner to pressure, stress acts at a point and is frequently referred to as the *stress at a point*.

In Figure 1.2 a bar of uniform cross-sectional area A is subject to an axial force P in the x direction. Since P is acting perpendicularly to the cross-section then from the above definition of stress the direct stress is

$$\sigma_{xx} = \frac{P}{A} \tag{1.2}$$

If the bar is elongated then σ_{xx} is referred to as tensile stress whereas a compressive stress is induced if the bar is compressed. If the units of P are N and the units of cross-sectional area A are m^2 then the units of σ_{xx} are N/m^2 or Pa.

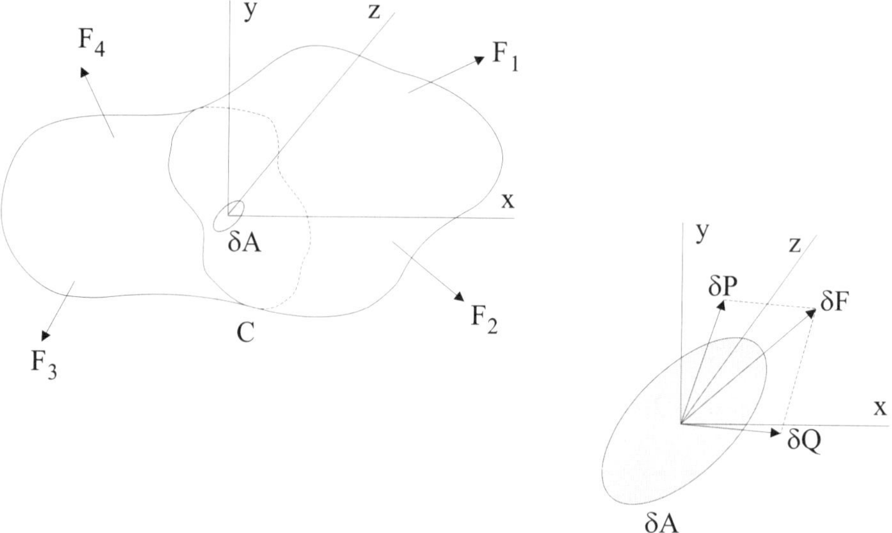

Figure 1.1: Three dimensional body under arbitrary loading. Also shown are the resolved differential forces δP and δQ on the differential element δA.

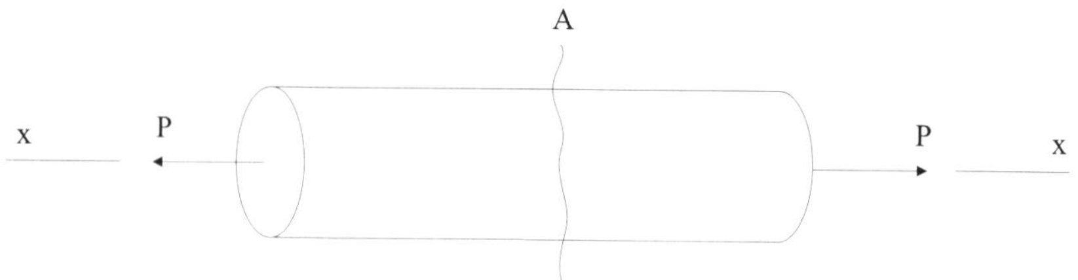

Figure 1.2: A bar of uniform cross-sectional area A subject to an axial load P.

1.3 Strain

Staying with the bar subject to tensile axial loading P in Figure 1.2, then if the initial length of the bar is L_0 and the action of loading extends the bar to L_1 then the elongation, δ, is $L_1 - L_0$. Provided that the elongation remains small and that the bar remains elastic throughout then the *engineering strain*, e, is defined as the elongation normalised with respect to the original length

$$e = \frac{\delta}{L} = \frac{L_1 - L_0}{L_0} \tag{1.3}$$

Thus, strain is a dimensionless quantity. For example, if a steel bar of initial length 2m is loaded in tension and extends 0.25mm then the engineering strain is equal to $0.25 \times 10^{-3}/2 = 125 \times 10^{-6}$. Strain is typically expressed in micro-strain; for example 125×10^{-6} or $125\mu\varepsilon$.

An alternative definition of strain is the *natural* or *true* strain, ε, which in terms of a strain increment is

$$\mathrm{d}\varepsilon = \frac{\mathrm{d}L}{L} \tag{1.4}$$

Integrating from L_0 to L_1 then ε is given by

$$\varepsilon = \int_{L_0}^{L_1} \frac{\mathrm{d}L}{L} = \ln L_1 - \ln L_0 = \ln\left(\frac{L_1}{L_0}\right) \tag{1.5}$$

When the extension is small then the nominal and true strains differ only by terms of order $(\delta/L_0)^2$. In general, for small-displacement elasticity use of the nominal strain suffices, but when displacements are large or in the case of plasticity the true strain is generally used.

1.4 Young's Modulus

In an analogous manner to the linear relationship between load and extension of a spring, the linear relationship between stress and strain in either tension or compression are related by *Young's modulus*, E, or the *modulus of elasticity*

$$E = \frac{\sigma}{\varepsilon} \tag{1.6}$$

Young's modulus is a material property and for a variety of engineering materials E is typically of the order 1×10^9Pa or GPa. Table 1.1 lists several values of E.

Material	**E(GPa)**	**G(GPa)**	ν	**K(GPa)**
aluminium	70	26	0.33	69
cast iron	80-170	30-65	0.2-0.3	118
copper	120	47	0.33	175
steel	210	80	0.3	67-142

Table 1.1: Typical values for Young's modulus (E), Poisson's ratio (ν), shear modulus (G) and bulk modulus (K) for several engineering materials.

Young's modulus is the slope of the stress-strain diagram which is discussed in the next section.

1.5 Stress-Strain Diagram

When a bar of cylindrical cross-section is subject to an increasing tensile load then a characteristic stress-strain diagram results. Figure 1.3 illustrates the stress-strain diagram of a typical engineering steel. The key regions of the diagram are as follows:

O-A: there is a proportional relationship between stress and strain with the material responding in a linear manner. If the load is removed at any stage between O and A then the strain will return to O and no residual strain will remain in the test specimen. Beyond point A the material no longer responds in a linear manner because the *proportional limit* of the material has been exceeded. Up to the proportional limit the slope of the straight-line A-B is characterised by Young's modulus, E.

A-B: in the region between A and B the material responds in a non-linear elastic manner. Point B is referred to as the *yield point* of a material, after which the material responds in a non-linear elastic-plastic manner. The yield point is generally denoted by the *yield stress*, σ_Y. The yield stress for a typical machine steel is in the range 340-700MPa.

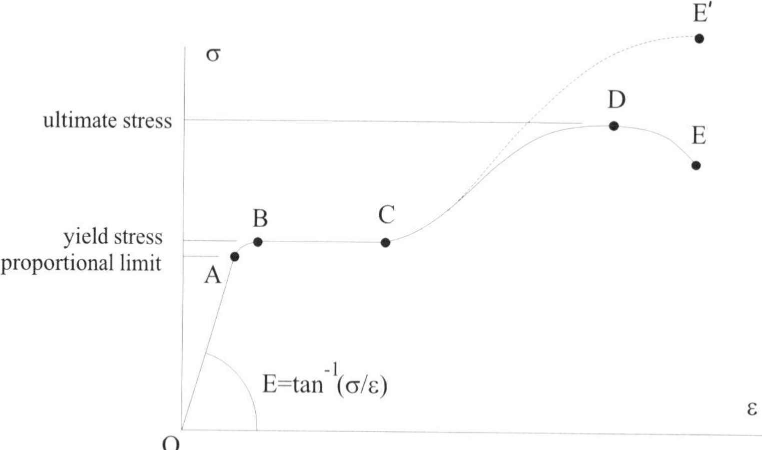

Figure 1.3: Stress-strain diagram for a typical engineering steel.

B-C: in the region between B and C (which is not experienced by all engineering materials) the material is responding in a *perfectly plastic* manner which is characterised by a significant increase in strain for a small increase in stress.

C-D: from point B the material has been experiencing *strain hardening* which increases significantly throughout the region C to D. Strain hardening is a result of the crystalline structure of a material readjusting to the increasing applied load and also due to the formation and propagation of dislocations. Essentially, in region C to D the applied load is increasing while the cross-sectional area is significantly reducing. At point D the load attains a maximum value, the *ultimate stress* σ_{UTS}. The ultimate stress for a typical machine steel is in the range 550-860MPa.

D-E: from point D the cross-sectional area of the bar has reduced to such an extent that the material can no longer sustain the level of load and further extension of the bar results in a reduction of load until the point of failure at point E.

In the above discussion the stress at any point was calculated with reference to the original cross-sectional area of the test specimen and is referred to as the *nominal stress*. If, however, at any stage throughout the test the actual cross-sectional area is used then the stress is referred to as the *true stress*. Similarly, if the original length of a specimen is used then the resulting strain is the *nominal strain* whereas if the actual length of the specimen is used then we obtain the *true strain*.

The stress-strain diagram discussed above for a typical engineering steel is a *ductile* response due to the large strains before failure occurred. A measure of the ductility of a material is *the percent elongation*

$$\text{percent elongation} = \frac{L_f - L_0}{L_0}(100) \tag{1.7}$$

where L_o and L_f denote the original and final lengths of a specimen. Because the elongation is not uniform throughout the entire length of a specimen but concentrated at a region of localised necking the percent elongation is a function of the original *gauge length* which must always be stated when quoting a percent elongation value. For a gauge length of 50mm pure aluminium experiences a 60% elongation while aluminium alloys experience 1-45% elongation.

An alternative measure of ductility is the *percent reduction in area* which directly accounts for the amount of localised necking in a specimen

$$\text{percent reduction in area} = \frac{A_0 - A_f}{A_0}(100) \tag{1.8}$$

where A_o and A_f denote the original and final cross-sectional areas of a specimen. A typical ductile engineering steel will experience approximately 50% reduction in area.

For a detailed discussion on the mechanical testing of engineering materials refer to Dowling (1999).

1.6 Shear Strain

Figure 1.4 shows a rectangle $ABCD$ subject to a force F at point B parallel to edge BC. In the case of irrotational simple shear $ABCD$ displaces to the rhombus $AB'C'D$ with B and C displaced by distance δ. The angle of shear, γ, (in radians) measures the shearing strain

$$\tan\gamma = \frac{\delta}{h} \tag{1.9}$$

where h is the length of edge AB. For small γ with $\tan\gamma \approx \gamma$ then $\gamma = \delta/h$.

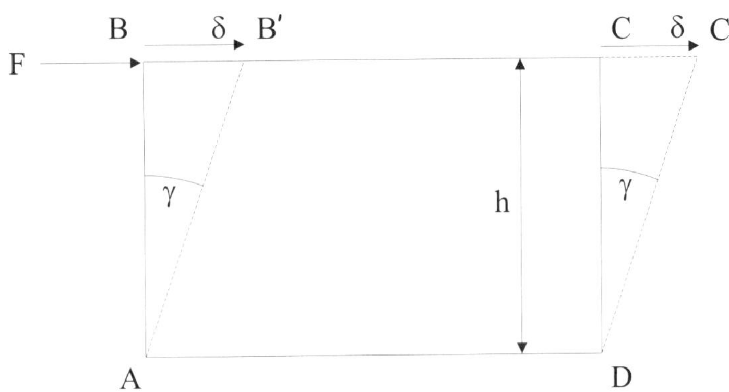

Figure 1.4: Simple shear of rectangle $ABCD$ displaced to the rhombus $AB'C'D$.

Now consider element $ABCD$ subject to a shear stress, τ, on edges BC and AD, Figure 1.5. The shearing strain, γ, is given by

$$\gamma \approx \tan\gamma = \frac{BB'}{AB} = \frac{CC'}{DC} = \frac{\tau}{G} \tag{1.10}$$

where G is the *shear modulus* or *modulus of rigidity* and relates the shearing stress to shearing strain (τ/γ) just as Young's modulus relates direct stress to direct strain (σ/ε). Young's modulus and the shear modulus are related by

$$E = 2G(1+\nu) \tag{1.11}$$

where ν is Poisson's ratio, to be discussed shortly in the next section. This relationship is derived in §7.7.

In order that element $ABCD$ is in a state of rotational equilibrium the complementary shearing stress, τ', must act on edges AB and CD. To determine τ' we can take moments about point D

$$\tau \text{Area}(AB)\, AD = \tau \text{Area}(BC)\, CD \tag{1.12}$$

If $ABCD$ is of unit thickness with edges AB and CD of length h and edges AD and BC of length l then $\tau' hl = \tau hl$, or

$$\tau = \tau' \tag{1.13}$$

As expected, for the element to be in a state of equilibrium the shearing strain and complementary shearing strain are equivalent.

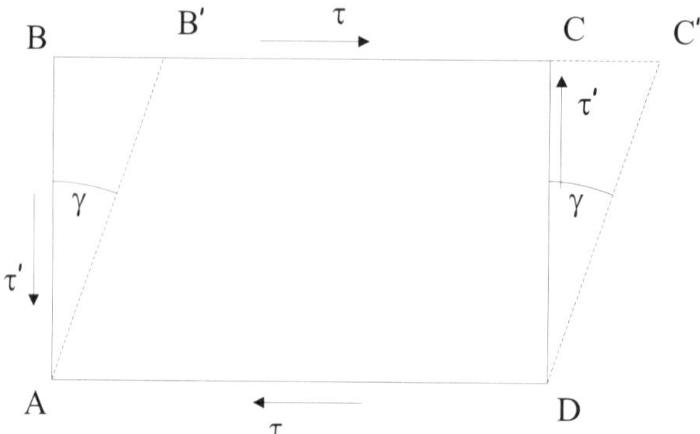

Figure 1.5: Rectangle $ABCD$ subject to shear stress τ on edges BC and AD.

1.7 Poisson's Ratio

When we considered the elongation of a bar in §1.2 we neglected to mention that the axial extension induces an associated lateral contraction, Figure 1.6. The ratio of lateral, ε_{yy}, to axial, ε_{xx}, strains is referred to as *Poisson's ratio*, ν

$$\nu = -\frac{\text{lateral strain}}{\text{axial strain}} = -\frac{\varepsilon_{yy}}{\varepsilon_{xx}} \tag{1.14}$$

For the majority of engineering materials ν is in the range $0 < \nu \leq \frac{1}{2}$ and typically $\nu = 0.33$; Table 1.1 lists values of ν for several materials. In the case of incompressible plastic materials $\nu = \frac{1}{2}$.

In the light of Poisson's ratio we can now extend the one-dimensional Hooke's law of (1.6) for a complete three-dimensional stress system

$$\varepsilon_{xx} = \frac{\sigma_{xx}}{E} - \nu\frac{\sigma_{yy}}{E} - \nu\frac{\sigma_{zz}}{E} = \frac{1}{E}\left[\sigma_{xx} - \nu\left(\sigma_{yy} + \sigma_{zz}\right)\right] \tag{1.15}$$

and similarly for ε_{yy} and ε_{zz}. Thus, Hooke's law in three dimensions is

$$\begin{aligned}
\varepsilon_{xx} &= \frac{1}{E}\left[\sigma_{xx} - \nu\left(\sigma_{yy} + \sigma_{zz}\right)\right] \\
\varepsilon_{yy} &= \frac{1}{E}\left[\sigma_{yy} - \nu\left(\sigma_{xx} + \sigma_{zz}\right)\right] \\
\varepsilon_{zz} &= \frac{1}{E}\left[\sigma_{zz} - \nu\left(\sigma_{xx} + \sigma_{yy}\right)\right]
\end{aligned} \tag{1.16}$$

These equations are the fundamental constitutive equations of linear elastic materials and will be used extensively throughout the entire text.

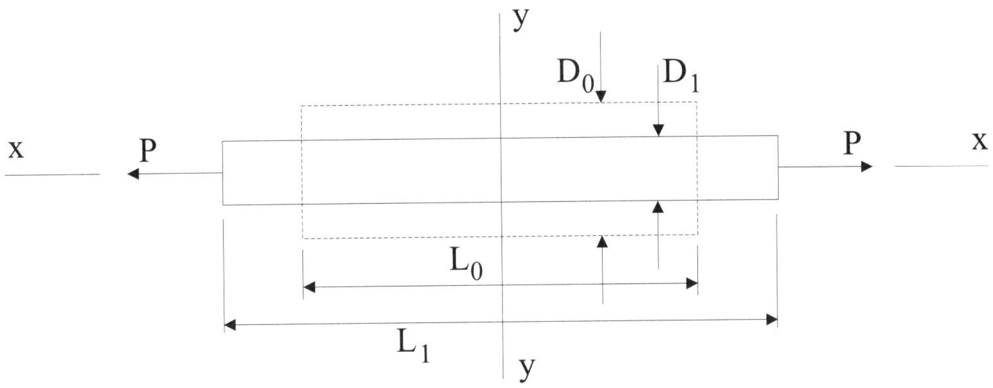

Figure 1.6: Axial extension and lateral contraction of a bar.

1.8 Volumetric Strain

Consider a three-dimensional element with initial edges a_1, a_2 and a_3 that deforms to edges a_1', a_2' and a_3' under the action of the stresses σ_{xx}, σ_{yy} and σ_{zz} acting in the x, y and z directions respectively, Figure 1.7.

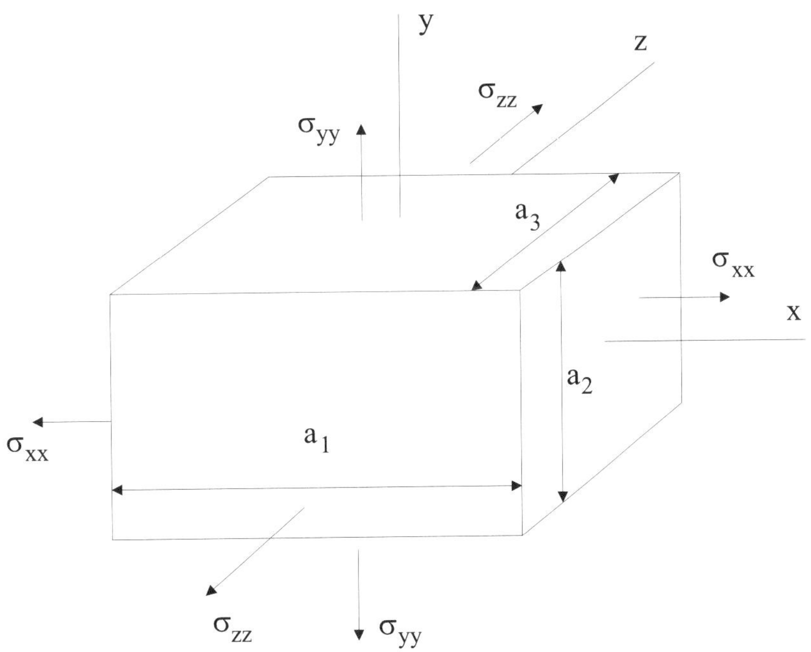

Figure 1.7: An element with initial edges a_1, a_2 and a_3 subject to stresses σ_{xx}, σ_{yy} and σ_{zz}.

Considering edge a_1, if σ_{xx} induces a strain of ε_{xx} in the x direction then $\varepsilon_{xx} = (a_1' - a_1)/a_1$ or $a_1' = (1 + \varepsilon_{xx})\, a_1$. Similarly, for edges a_2 and a_3 we have $a_2' = (1 + \varepsilon_{yy})\, a_2$ and $a_3' = (1 + \varepsilon_{zz})\, a_3$.

Defining a volumetric strain, ε_V, as the change in volume, ΔV, normalised with respect to the original volume, V, then

$$\varepsilon_V = \frac{\Delta V}{V} = \frac{a_1' a_2' a_3'}{a_1 a_2 a_3} = \frac{(1+\varepsilon_{xx})\, a_1 \,(1+\varepsilon_{yy})\, a_2 \,(1+\varepsilon_{zz})\, a_3 - a_1 a_2 a_3}{a_1 a_2 a_3} \tag{1.17}$$

which upon expanding is

$$\varepsilon_V = \varepsilon_{xx} + \varepsilon_{yy} + \varepsilon_{zz} + \varepsilon_{xx}\varepsilon_{yy} + \varepsilon_{yy}\varepsilon_{zz} + \varepsilon_{xx}\varepsilon_{zz} + \varepsilon_{xx}\varepsilon_{yy}\varepsilon_{zz} \tag{1.18}$$

For small strains, neglecting infinitesimal products, ε_V reduces to

$$\varepsilon_V = \varepsilon_{xx} + \varepsilon_{yy} + \varepsilon_{zz} \tag{1.19}$$

noting that if the element experiences no change in volume, $\varepsilon_V = 0$, under the action of the applied stresses then we have the *constancy of volume* condition

$$\varepsilon_{xx} + \varepsilon_{y} + \varepsilon_{zz} = 0 \tag{1.20}$$

which will prove useful when we discuss incompressible plasticity in Chapter 12.

1.9 Bulk Modulus

The volume modulus or *bulk modulus*, K, of a material relates the volumetric stress, σ_V, to the volumetric strain, ε_V

$$K = \frac{\sigma_V}{\varepsilon_V} \tag{1.21}$$

If a stress element is subject to a purely hydrostatic state of stress ($\sigma_{xx} = \sigma_{yy} = \sigma_{zz} = \sigma$ and $\varepsilon_{xx} = \varepsilon_{yy} = \varepsilon_{zz} = \varepsilon$), Figure 1.8, then $\sigma_V = \sigma$ and from (1.19) $\varepsilon_V = 3\varepsilon$. Using either of (1.16), say ε_{xx}

$$\varepsilon = \frac{\varepsilon_V}{3} = \varepsilon_{xx} = \frac{1}{E}\left[\sigma_{xx} - \nu\left(\sigma_{yy} + \sigma_{zz}\right)\right] = \frac{\sigma}{E}\left(1 - 2\nu\right) \tag{1.22}$$

and substituting into 1.21 gives

$$K = \frac{E}{3\left(1 - 2\nu\right)} \tag{1.23}$$

and illustrates that $K \to \infty$ as $\nu \to \frac{1}{2}$ and the material becomes incompressible, that is as $\varepsilon_V \to 0$.

1.10 One-Dimensional Thermal Strain

Thermal strain, ε_T, is defined by

$$\varepsilon_T = \alpha\left(\Delta T\right) \tag{1.24}$$

where α is the coefficient of thermal expansion (1/°C) and ΔT is the increase in temperature (°C). Table 1.2 lists typical values of α for four materials.

Consider the elongation of a simple bar solely due to an increase in temperature as shown in Figure 1.9. If the original length of the bar is L_0 and the elongation due to thermal strain is δ_T then the thermal strain in the x direction is

$$\varepsilon_{xx} = \frac{\delta_T}{L_0} \tag{1.25}$$

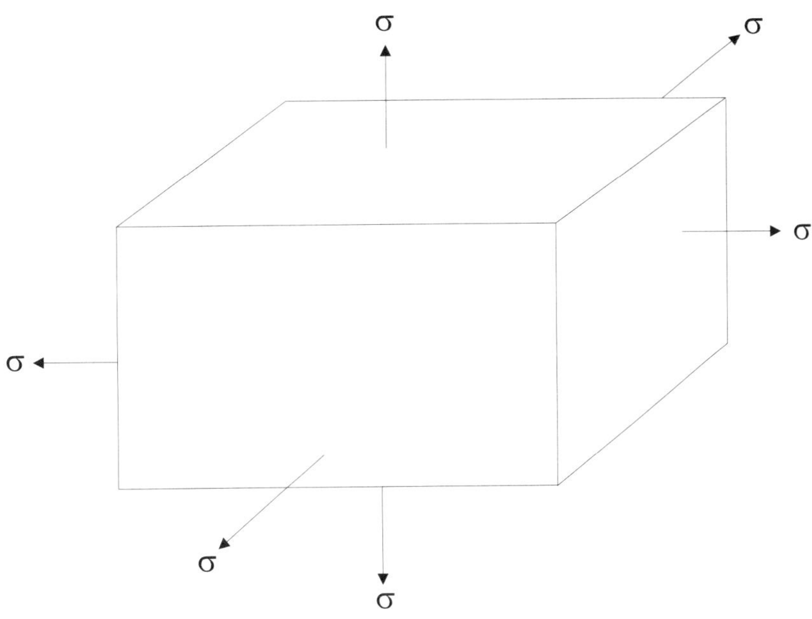

Figure 1.8: An element subject to a purely hydrostatic stress σ.

Material	$\boldsymbol{\alpha(\times 10^{-6}/^\circ\mathrm{C})}$
aluminium	23
cast iron	10
copper	17
steel	10-18

Table 1.2: Coefficient of thermal expansion, α, for aluminium, cast iron, copper and steel.

From (1.24) the elongation is

$$\delta_T = \alpha\,(\Delta T)\,L_0 \tag{1.26}$$

If L_1 denotes the new length of the bar then

$$L_1 = L_0 + \delta_T = L_0\,[1 + \alpha\,(\Delta T)] \tag{1.27}$$

As an example, a steel bar ($\alpha = 12 \times 10^{-6}/^\circ$C) of length 1m subject to an increase in temperature of 25°C produces an elongation of 0.3mm.

We will revisit thermal strains in §9.3 when we discuss strain energy.

1.11 Stress and Strain Revisited

Let us now revisit our definitions of stress and strain by defining a general three-dimensional notation for stress or the *stress tensor* and express the strain components in terms of displacement.

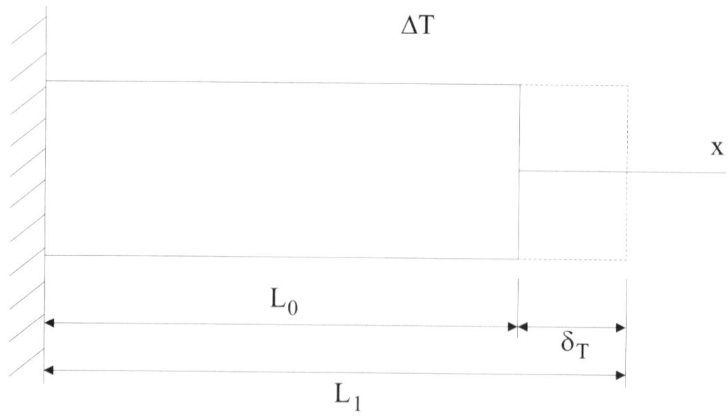

Figure 1.9: Elongation of a bar due to an increase in temperature.

In general, a hexahedral element within a three dimensional body will have both direct and shear stresses acting on each of its six faces. To form a coherent reference system it is necessary to resolve the stress on each face in accordance with the coordinate system (using a *face-direction* rule), see Figure 1.10. For example, the shear stress τ_{zy} acts on the face perpendicular to the z-axis and in a direction parallel to the y-axis whereas the normal stress σ_{xx} acts on face x and in direction x. As discussed above, §1.6, to preserve moment equilibrium in the stress element we have

$$\tau_{yx} = \tau_{xy}; \quad \tau_{zy} = \tau_{yz}; \quad \tau_{zx} = \tau_{xz} \tag{1.28}$$

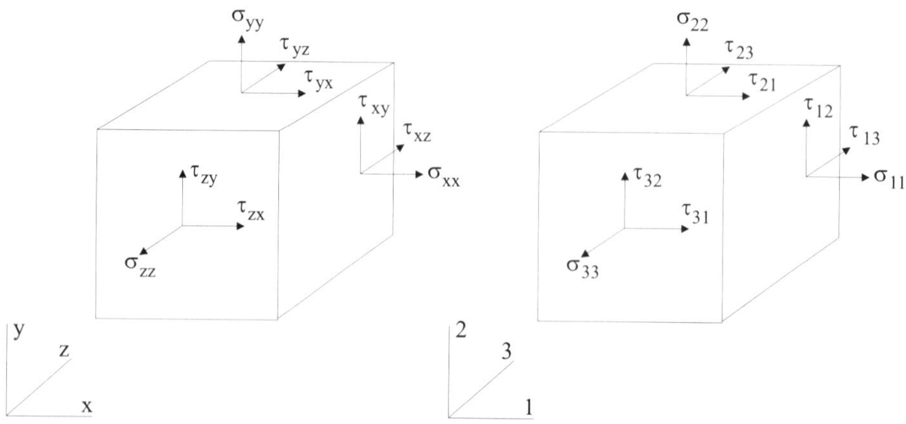

Figure 1.10: Cartesian stress element. For clarity only the stress components on the visible faces are shown.

Alternatively, in some circumstances (e.g. when modelling cylindrical pressure vessels) it may be more advantageous to use cylindrical coordinates rather than Cartesian coordinates, see Figure 1.11. In this instance moment equilibrium is ensured if

$$\tau_{r\theta} = \tau_{\theta r}; \quad \tau_{\theta z} = \tau_{z\theta}; \quad \tau_{zr} = \tau_{rz} \tag{1.29}$$

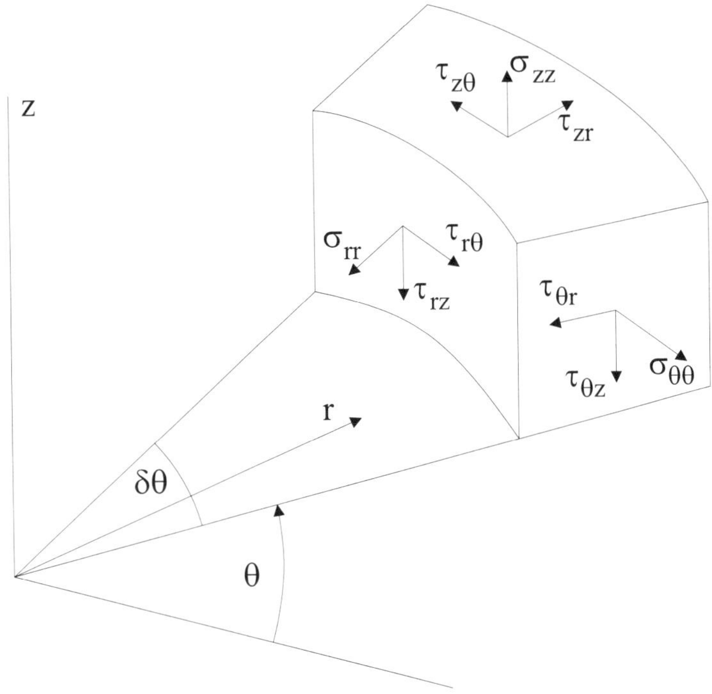

Figure 1.11: Cylindrical stress element. For clarity only the stress components on the visible faces are shown.

Other coordinate systems, such as Spherical, could equally be used.

Strain components can be expressed in terms of the displacement field of a body. Letting the displacement field (u, v, w) be associated with the coordinate system (x, y, z) then Figure 1.12 illustrates the deformation of edge ab to $a'b'$ of the quadrilateral $abcd$. If a and b are initially separated by δx with ab lying parallel to the x-axis then the strain ε_{xx} is

$$\varepsilon_{xx} = \lim_{ab \to 0} \frac{a'b' - ab}{ab} \tag{1.30}$$

Provided that $\delta u \ll \delta x$ then, to first order, $a'b' = \delta x + \delta u$ so that

$$\varepsilon_{xx} = \lim_{\delta x \to 0} \frac{(\delta x + \delta u) - \delta x}{\delta x} = \lim_{\delta x \to 0} \frac{\delta u}{\delta x} = \frac{\partial u}{\partial x} \tag{1.31}$$

Similarly

$$\varepsilon_{yy} = \frac{\partial v}{\partial y}; \quad \varepsilon_{zz} = \frac{\partial w}{\partial z} \tag{1.32}$$

To express the shear strain components in terms of displacement consider the orthogonal edges ab and ad which initially both lie parallel to the x and y axes respectively. Shear strain is defined as half the decrease in angle (measured in terms of dimensionless radians) between the two orthogonal axes. Therefore, from Figure 1.12

$$\tan \alpha = \frac{\delta v}{\delta x + \delta u} \tag{1.33}$$

Again, assuming that $\delta u \ll \delta x$ and $\tan\alpha \approx \alpha$ then

$$\alpha \approx \tan\alpha \approx \frac{\delta v}{\delta x} = \frac{\partial v}{\partial x} \tag{1.34}$$

Similarly, β can be shown to be equal to

$$\beta = \frac{\partial u}{\partial y} \tag{1.35}$$

Therefore, the shear strain γ_{xy} is

$$\gamma_{xy} = \frac{1}{2}(\alpha+\beta) = \frac{1}{2}\left(\frac{\partial u}{\partial y} + \frac{\partial v}{\partial x}\right) \tag{1.36}$$

and is often referred to as the mathematical shear strain whereas the engineering shear strain is $e_{xy} = 2\gamma_{xy}$. Similarly γ_{xz} and γ_{yz} are given by

$$\gamma_{xz} = \frac{1}{2}\left(\frac{\partial u}{\partial z} + \frac{\partial w}{\partial x}\right); \quad \gamma_{yz} = \frac{1}{2}\left(\frac{\partial v}{\partial z} + \frac{\partial w}{\partial y}\right) \tag{1.37}$$

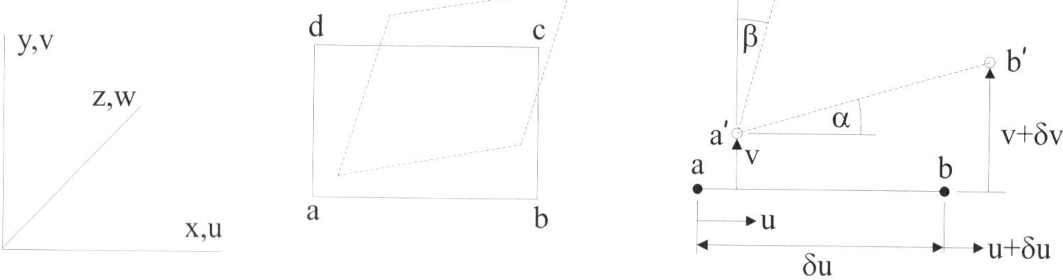

Figure 1.12: Deformation of a two-dimensional stress element.

Example 1.1 Stress, strain and displacement for a bar hanging under its own weight

Figure 1.13 shows a bar of length L, cross-sectional area A, Young's modulus E, Poisson's ratio ν and density ρ hanging under its own weight. The Cartesian displacements (u, v) can be expressed as follows

$$u = \frac{\rho g}{2E}\left(2xL - x^2 - \nu^2 y^2\right); \quad v = -\frac{\nu\rho g}{E}\left[(L-x)\,y\right]$$

where g is the gravitational constant. From (1.31), (1.32) and (1.36) the direct and shear strains are

$$\varepsilon_{xx} = \frac{\partial u}{\partial x} = \frac{\rho g}{E}(L-x); \quad \varepsilon_{yy} = \frac{\partial v}{\partial y} = -\frac{\nu\rho g}{E}(L-x); \quad \gamma_{xy} = \frac{1}{2}\left(\frac{\partial u}{\partial y} + \frac{\partial v}{\partial x}\right) = \frac{\nu\rho g y}{E}(1-\nu)$$

Let us now determine the total extension of the bar. At the free end of the bar there will be no stresses induced due to the weight of a given section of the bar. Gravity applies a force of magnitude $F = mg$ on the bar where m is the mass of the bar and F is the force due to gravity. With x measured from the fixed end then for the arbitrary element shown in Figure 1.13 the force on element $\mathrm{d}x$ is due to the weight of

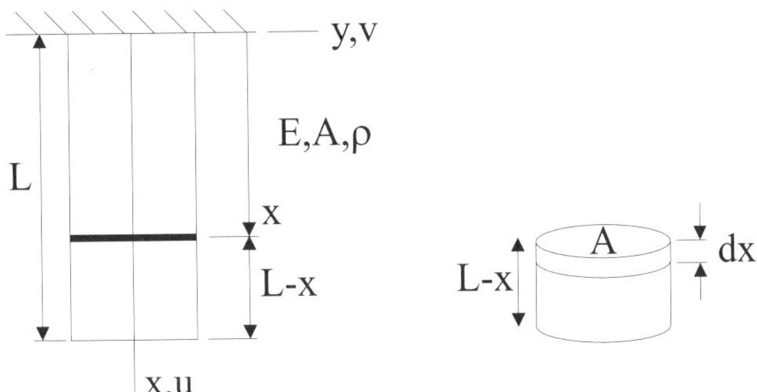

Figure 1.13: Example: A bar hanging under its own weight.

the bar section below. With mass(m) = density(ρ) × volume($A(L-x)$) then $F = \rho g A(L-x)$. The stress and strain at x are therefore

$$\sigma(x) = \frac{\rho g A(L-x)}{A} = \rho g(L-x)\,; \quad \varepsilon(x) = \frac{\sigma}{E} = \frac{\rho g(L-x)}{E}$$

From $\varepsilon(x) = \delta u/\delta x$ the total displacement at $x = L$ is

$$u = \frac{\rho g}{E}\int_0^L (L-x)\,\mathrm{d}x = \frac{\rho g L^2}{2E}$$

which agrees with the expression above for $x = L$ and noting that u is independent of A. For a steel bar of length 1m, ρ=7,850kg/m^3, E=210GPa the total elongation is 0.183mm.

1.12 Ideal Strength of Materials

This section examines the ideal strength of materials that are perfectly homogeneous, that is they are free of defects such as dislocations, voids, inclusions and cracks. We will consider separately the cases of pure cleavage (tensile strength) and pure shear (shear strength) of a body. When a body cleaves we are interested in the force required to separate two adjacent atomic planes whereas shear is based on the force required to shear one atomic plane relative to an adjacent plane.

Consider a pair of atoms sufficiently far apart that they do not influence each other. Let the total energy of each atom be W_a with the total energy of the pair equal to $2W_a$. We now bring the pair of atoms together to form a stable molecule of total energy W_0. The value of W_0 is less than $2W_a$ and the difference is referred to as the bond strength; $W_b = 2W_a - W_0$. Figure 1.14a) illustrates the variation of total energy as a function of separation distance d between the pair of atoms. At the equilibrium separation distance of d_0 the energy is seen to attain a minimum which is equal to the total energy or potential energy of the stable molecule, W_0. If the pair of atoms are pushed together closer than d_0 then the repulsive forces of the atoms result in a sharp increase in the total energy. The exact value of W_0 depends on the type of atoms, the type of bond and temperature. If the pair of atoms moves by a small amount δx with corresponding change in W of δW then the force is $\delta W/\delta x$ (work = force × distance). Thus, the force-separation curve is obtained from the energy-separation curve and is shown in Figure 1.14b).

At the equilibrium separation distance d_0 Figure 1.14b) illustrates that the attractive-repulsive force between the pair of atoms is zero and with the attractive force increasing to a maximum value of σ_{th}. This

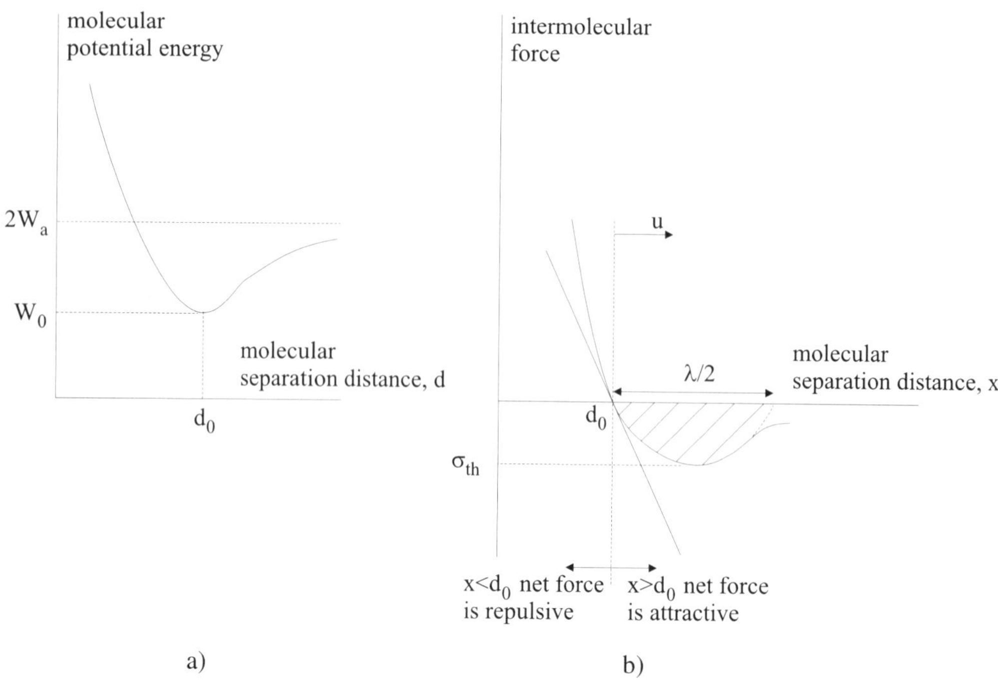

Figure 1.14: Ideal material strength: a) Variation of energy of a pair of atoms as a function of separation distance, d. b) Variation of force against separation distance.

maximum σ_{th} denotes the theoretical strength of the material. With the bond between the material atoms broken once σ_{th} is exceeded. Orowan (1955) was one of the first to estimate the theoretical cleavage or tensile strength of a perfectly homogeneous material. Let the shape of the force-separation curve in the vicinity of d_0 be approximated by a sine curve, see Figure 1.14b). The variation of σ as a function of x is then given by

$$\sigma = \sigma_{th} \sin\left(\frac{2\pi\,(x - d_0)}{\lambda}\right) = \sigma_{th} \sin\left(\frac{2\pi u}{\lambda}\right) \tag{1.38}$$

where λ is the wavelength and $u = x - d_0$ is measured from the equilibrium position d_0.

The Young's modulus of an elastic material is the ratio of stress to strain as defined earlier in (1.6). Thus, the slope of the force-separation curve at the equilibrium position is equivalent to the Young's modulus. Figure 1.14b) illustrates that the force-separation curve at d_0 is approximately a straight-line and therefore independent of displacement. For small displacements $\sin x \approx x$ so that σ is approximately

$$\sigma = \sigma_{th} \frac{2\pi u}{\lambda} \tag{1.39}$$

The strain, ε, is equal to u/d_0 so that Young's modulus is

$$E = \frac{\sigma}{\varepsilon} = \sigma_{th} \frac{2\pi d_0}{\lambda} \tag{1.40}$$

and re-arranging for σ_{th}

$$\sigma_{th} = \left(\frac{\lambda}{2\pi d_0}\right) E \tag{1.41}$$

Material	E**(GPa)**	$d_0(1 \times 10^{-10}\text{m})$	σ_{th}**(GPa)**	E/σ_{th}	σ_{UTS}**(MPa)**	E/σ_{UTS}
aluminium	70	4.049			60-169	438
copper	124	3.6153	39	4.9	200-350	354
zinc	97	2.664	3.8	9.2	110-200	485
steel(mild)	210				480	438
concrete (28 day)	10-17				27-55	309
glass	50-80				30-90	889
timber (along grain)	8-13				20-110	118

Table 1.3: Comparison of theoretical tensile strength, σ_{th}, and ultimate tensile strength, σ_{UTS}, of several materials. Compiled from Jayatilaka (1979), Howatson *et al.* (1972) and McClintock and Argon (1966).

The value of d_0 depends on the particular solid but is typically $\sim 3 \times 10^{-10}$m and λ can be taken to be approximately equal to d_0. Table 1.3 lists d_0 and σ_{th} from (1.41) for several materials with E/σ_{th} in the range 4 to 10. As a rule the theoretical strength is approximately given by

$$\sigma_{th} \approx \frac{E}{10} \tag{1.42}$$

Table 1.3 also summarises observed values for the ultimate tensile strength, σ_{UTS}, for several engineering materials. It is seen that E/σ_{UTS} is approximately one or two orders of magnitude higher than E/σ_{th}.

A second estimate of the ideal tensile strength of a material can be obtained by considering the work done to create new surface due to fracture. A more detailed discussion of the energy released due to fracture will be discussed in §17.4.1 when we discuss Griffith's energy release rate. The shaded area of Figure 1.14b) represents the work done by the applied load and is approximately given by, from (1.38)

$$\int_0^{\lambda/2} \sigma_{th} \sin\left(\frac{2\pi u}{\lambda}\right) du A = \frac{\sigma_{th}\lambda A}{\pi} \tag{1.43}$$

where A is the cross-sectional area. The surface energy, G, required to create an increase in area of $2A$ (the factor of 2 accounts for the two surfaces created) is

$$G = 2A\gamma_0 \tag{1.44}$$

where γ_0 (Jm^{-2}) is the surface energy of the material. Equating (1.43) and (1.44) and re-arranging for σ_{th}

$$\sigma_{th} = \frac{2\pi\gamma_0}{\lambda} \tag{1.45}$$

Substituting for λ from (1.41)

$$\sigma_{th} = \sqrt{\frac{\gamma_0 E}{d_0}} \tag{1.46}$$

Typically for several engineering materials $\gamma_0 \approx Ed_0/40$ so that $\sigma_{th} \approx E/6$ and is in approximate agreement with (1.42).

We will now estimate the theoretical strength of a material in pure shear following the method of Frenkel (1926). Figure 1.15 illustrates an idealised array of atoms subject to a shear stress τ. The distance between atom centres in the direction of slip is denoted by b while the distance between atom centres in the direction perpendicular to the slip direction is denoted by a_0.

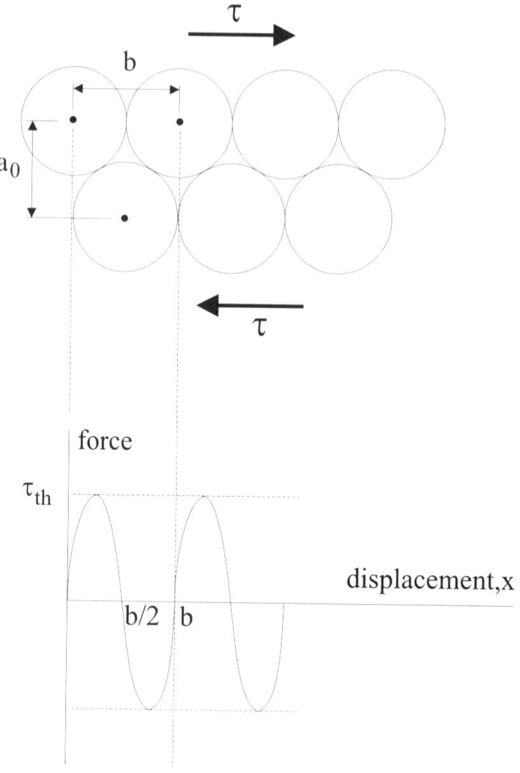

Figure 1.15: Variation of shear stress for an ideal array of atoms.

If the displacement of the upper layer of atoms with respect to the lower layer is denoted by x then the first two equilibrium positions are $x = 0$ and $x = b/2$. The shear stress attains a maximum, τ_{th}, at $x = b/4$ and for ease of calculation will be assumed to vary sinusoidally

$$\tau = \tau_{th} \sin\left(\frac{2\pi x}{b}\right) \approx \tau_{th} \frac{2\pi x}{b} \tag{1.47}$$

assuming small displacements. The shear modulus, G, is defined as the ratio of shear stress to shear strain, γ, with $\gamma \approx \tan\gamma = x/a_0$

$$G = \frac{\tau}{\gamma} = \frac{\tau a_0}{x} \tag{1.48}$$

Re-arranging for τ and equating to (1.47) we find that τ_{th} is given by

$$\tau_{th} = \left(\frac{b}{2\pi a_0}\right) G \tag{1.49}$$

For cubic materials $a_0 \approx b$ so that

$$\tau_{th} \approx \frac{G}{2\pi} \tag{1.50}$$

Material	G(GPa)	τ_{th}(GPa)	G/τ_{th}	τ_{act}(GPa)	G/τ_{act}
aluminium	23	2.62	8.8	0.76	30,263
copper	30.8	0.74	25.7	0.93	33,118
zinc	38	2.3	16.5	0.29	131,034

Table 1.4: Comparison of theoretical shear strength, τ_{th}, and actual shear strength, τ_{act}, of several materials. Compiled from Jayatilaka (1979).

Table 1.4 compares τ_{th} with the actual shear strength, τ_{act}, of several materials with τ_{th} seen to be orders of magnitude higher that τ_{act}.

We will conclude by referring the reader to Kelly and MacMillan (1986) for a more detailed discussion on the theoretical strength of solids.

1.13 Conclusion

The present Chapter has introduced the fundamentals of stress and strain which will be used throughout the entire book. We have also introduced Young's modulus which relates stress and strain, and Poisson's ratio which provides a measure of lateral contraction. A typical stress-strain diagram for a variety of engineering steels has been discussed. In addition we have introduced shear strain, volumetric strain (and associated bulk modulus) and strain due to thermal expansion. Section 1.11 extended our examination of strain by introducing a tensorial notation for stress and strain and expressing the strain components in terms of displacements. Finally, we examined the ideal cleavage and shear strengths of a material.

1.14 References and Further Reading

○ Dowling, N. E. (1999) *Mechanical Behaviour of Materials: Engineering Methods for Deformation, Fracture, and Fatigue*, 2nd Ed., Prentice-Hall, New Jersey.

○ Frenkel, J. (1926) Z. Phy., 37, 572.

○ Gere, J. M. and Timohenko, S. P. (1997) *Mechanics of Materials*, 4th Ed., PWS Publishing, Boston.

○ Howatson, A. M., Lund, P. G. and Todd, J. D. (1972) *Engineering Tables and Data*, Chapman and Hall, London.

○ Jayatilaka, A. de S. (1979) *Fracture of Engineering Brittle Materials*, Applied Science Publishers, London.

○ Kelly, A. and MacMillan, N. H. (1986) *Strong Solids*, 3rd Ed., Clarendon Press, Oxford.

○ McClintock, F. A. and Argon, A. S. (1966) *Mechanical Behaviour of Materials*, Addison-Wesley, Canada.

○ Orowan, E. (1955) Rep. Prog. Phys., 12, 185.

○ Rees, D. W. A. (1997) *Basic Solid Mechanics*, Macmillan, London.

○ Riley, W. F. and Zachary, L. (1989) *Introduction to Mechanics of Materials*, John Wiley and Sons.

1.15 Exercises

1.1 A bar of circular cross-section of radius 50mm is subject to an axial load of 30kN. Determine the axial stress.

1.2 The bar of Exercise 1.1 is of original length 0.5m and Young's modulus E=210GPa. Determine the extension of the bar.

1.3 The direct strains at a point are equal to $\varepsilon_{xx} = 40 \times 10^{-6}$ and $\varepsilon_{yy} = 60 \times 10^{-6}$ for a material with Young's modulus $E = 70$GPa and Poisson's ratio $\nu = 0.28$. Determine the direct stresses σ_{xx} and σ_{yy}.

1.4 Determine the bulk modulus of a steel with Young's modulus of 210GPa and Poisson's ratio of 0.3.

1.5 A titanium alloy bar ($E = 120$GPa and $\alpha = 11 \times 10^{-6}/°$C) of circular cross-section (r=15mm) and length L=0.75m is subject to a temperature increase of $\Delta T = 100°$C. Determine the value of the applied compressive load required to cancel out the increase in length of the bar due to thermal expansion.

1.6 Figure 1.16 illustrates a composite steel/copper solid bar subject to a compressive axial load of 150kN. The copper bar is 50mm in diameter and of length 0.5m whereas the steel bar is 75mm in diameter and of length 0.6m. Assuming the Young's modulii of steel and copper to be 200GPa and 120GPa respectively, determine the compressive stress in each bar and the total contraction of the composite bar.

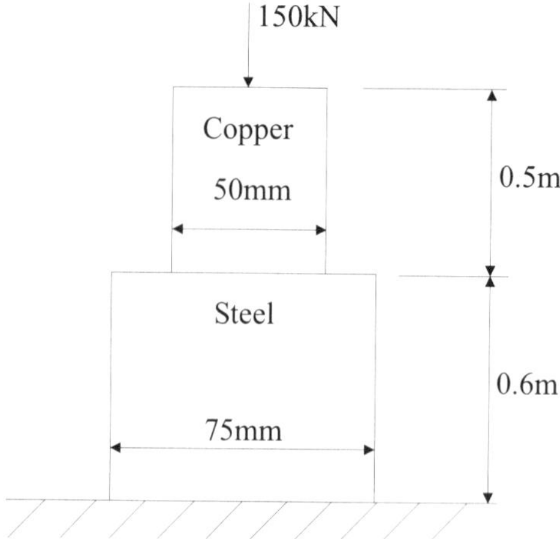

Figure 1.16: Exercise: A composite bar subject to a compressive axial load.

1.7 The in-plane displacements u and v are given by

$$u = \frac{1}{E}\left(\sigma_{xx} - \nu\sigma_{yy}\right)x; \quad v = \frac{1}{E}\left(\sigma_{yy} - \nu\sigma_{xx}\right)y$$

where E and ν are Young's modulus and Poisson's ratio respectively. Determine the in-plane strains ε_{xx}, ε_{yy} and γ_{xy}.

Chapter 2

Moments of Area

2.1 Introduction

This Chapter examines moments of area. Initially, we consider the first moments of area and the related centroid of a plane area. Next, the second moments of area are discussed and the related parallel-axis theorem for determining the second moments of area for a pair of coordinate axes which are not the centroidal axes of a plane area. This is then followed by the products of area and polar moment of area. The Chapter concludes by introducing the radius of gyration which provides a measure of the distribution of area from a given axis.

Allthough the cencept of moments of area is mathematical in origin, the properties evaluated find great application when we consider the action of bending and twisting of structural elements. This is covered in Chapters 3 and 5.

Throughout the literature the terminology *moments of inertia* is frequently used when referring to the moments of area. We will refrain from using moments of inertia because of the incorrect association made with dynamical modelling.

2.2 First Moments of Area and Centroid

Consider the arbitrarily bounded plane area of Figure 2.1 which lies in the plane of the (x, y) coordinate system and has total area A

$$A = \iint_A \mathrm{d}A \tag{2.1}$$

The *first moments of area*, Q_x and Q_y, about the x and y axes respectively are defined as

$$Q_x = \iint_A y\,\mathrm{d}A; \quad Q_y = \iint_A x\,\mathrm{d}A \tag{2.2}$$

The centroid $c(x_c, y_c)$ is simply the first moment divided by the area

$$x_c = \frac{\iint_A x\,\mathrm{d}A}{\iint_A \mathrm{d}A} = \frac{Q_y}{A}; \quad y_c = \frac{\iint_A y\,\mathrm{d}A}{\iint_A \mathrm{d}A} = \frac{Q_x}{A} \tag{2.3}$$

Equally, $Q_x = Ay_c$ and $Q_y = Ax_c$.

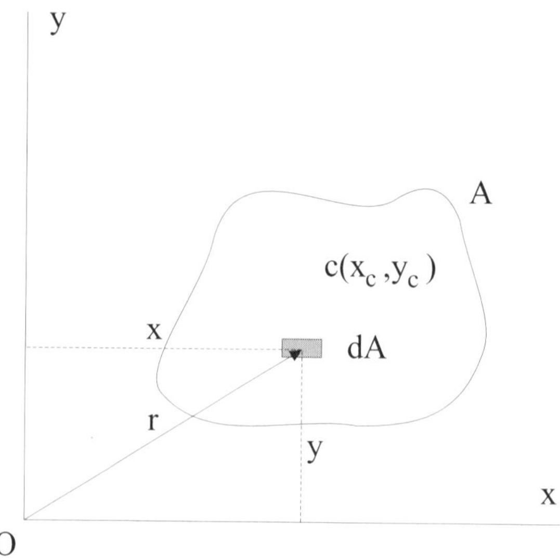

Figure 2.1: A plane area A within the (x, y) coordinate system with centroid c.

Example 2.1 First moments of area and centroid of a right-angled triangle

Figure 2.2 illustrates a right-angled triangle with vertices $v_1(x_1, y_1)$, $v_2(x_2, y_2)$ and $v_3(x_3, y_3)$. Vertex v_1 is at the origin of coordinates, edge (v_1, v_2) coincides with the x-axis and is of length b and edge (v_1, v_3) coincides with the y-axis and is of length h. Considering an elemental strip of width $\mathrm{d}y$ at height y then we observe that

$$y = h - \frac{h}{b}x; \quad x = b - \frac{b}{h}y$$

Determining Q_x with $\mathrm{d}A = x\,\mathrm{d}y$ then

$$Q_x = \iint_A y\,\mathrm{d}A = \int_0^h xy\,\mathrm{d}y = \frac{b}{h}\int_0^h (h - y)\,y\,\mathrm{d}y = \frac{bh^2}{6}$$

Similarly, letting $\mathrm{d}A = y\,\mathrm{d}x$ then Q_y is

$$Q_y = \iint_A x\,\mathrm{d}A = \int_0^b xy\,\mathrm{d}x = \frac{h}{b}\int_0^b (b - x)\,x\,\mathrm{d}x = \frac{hb^2}{6}$$

Thus, from (2.3) dividing through by $A(= bh/2)$ the triangle centroid is

$$(x_c, y_c) = \left(\frac{Q_y}{A}, \frac{Q_x}{A}\right) = \left(\frac{b}{3}, \frac{h}{3}\right)$$

This result can be confirmed by noting that the centroid of a general planar triangle with vertices (x_1, y_1), (x_2, y_2) and (x_3, y_3) is

$$(x_c, y_c) = \left(\frac{x_1 + x_2 + x_3}{3}, \frac{y_1 + y_2 + y_3}{3}\right)$$

with the centroid of coordinates seen to be the average of the three vertex coordinates. With $x_1 = 0$, $x_2 = b$, $x_3 = 0$, $y_1 = 0$, $y_2 = 0$ and $y_3 = h$ then $(x_c, y_c) = (b/3, h/3)$ as above.

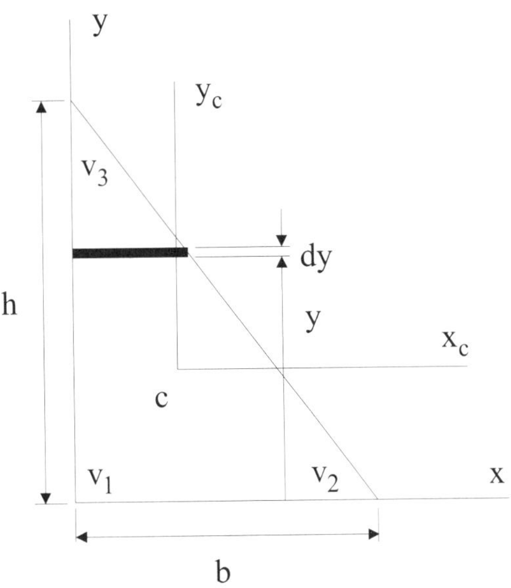

Figure 2.2: Example: A right-angled triangle.

2.3 Second Moments of Area

The *second moments of area*, I_x and I_y, with respect to the (x, y) coordinate system are defined as

$$I_x = \iint_A y^2 \,\mathrm{d}A; \quad I_y = \iint_A x^2 \,\mathrm{d}A \tag{2.4}$$

and are always positive. As an illustration of the second moments of area let us determine I_x and I_y for the rectangle shown in Figure 2.3. The origin of coordinates is at the centroid of the rectangle which is of width b and height h. Considering the elemental strip $\mathrm{d}y$ at position y then $\mathrm{d}A = bdy$ and I_x is given by

$$I_x = \iint_A y^2 \,\mathrm{d}A = \int_{-h/2}^{h/2} y^2 b \,\mathrm{d}y = \frac{bh^3}{12} \tag{2.5}$$

Similarly, by considering the elemental strip $\mathrm{d}x$ at position x then $\mathrm{d}A = h\,\mathrm{d}x$ and I_y is

$$I_y = \iint_A x^2 \,\mathrm{d}A = \int_{-b/2}^{b/2} x^2 h \,\mathrm{d}x = \frac{hb^3}{12} \tag{2.6}$$

Both of the second moments of area in (2.5) and (2.6) were determined with the origin of coordinates at the centroid of the rectangle. If we now determine I_x and I_y from the axes BB and HH respectively then

$$I_x\,(BB) = \int_0^h y^2 b \,\mathrm{d}y = \frac{bh^3}{3}; \quad I_y\,(BB) = \int_0^b x^2 h \,\mathrm{d}x = \frac{hb^3}{3} \tag{2.7}$$

illustrating that both I_x and I_y are larger in value when calculated on axes other than the centroidal axes. Using the parallel axis theorem, the next section will demonstrate the general result that the the moments of area increase as the reference axes move away from the centroid.

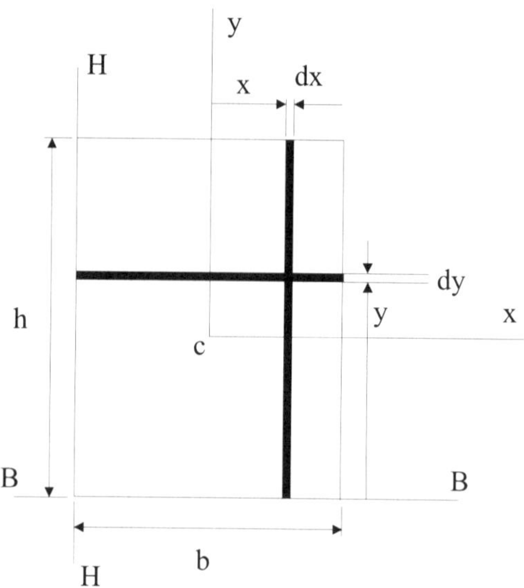

Figure 2.3: A rectangle of width b and height h.

Example 2.2 Second moments of area for a circular cross-section

The second moments of area are frequently required for a circular cross-section. Figure 2.4 illustrates the upper-right quadrant of a circle of radius r and diameter d. Considering I_x with $\mathrm{d}A = x\,\mathrm{d}y$ and $r^2 = x^2 + y^2$ then

$$I_x = \iint_A y^2\,\mathrm{d}A = 4\int_0^r xy^2\,\mathrm{d}y = 4\int_0^r y^2\sqrt{r^2 - y^2}\,\mathrm{d}y$$

where the value of 4 accounts for four equivalent quadrants due to symmetry. The integral can be evaluated by substituting $y = r\sin\theta$, $\mathrm{d}y = r\cos\theta\,\mathrm{d}\theta$ and noting the trignometric identity $\cos^2\theta + \sin^2\theta = 1$

$$I_x = 4r^4\int_0^{\pi/2}\sin^2\theta\cos^2\theta\,\mathrm{d}\theta = 4r^4\left[\frac{\theta}{8} - \frac{\sin 4\theta}{32}\right]_0^{\pi/2} = \frac{\pi r^4}{4} = \frac{\pi\,\mathrm{d}^4}{64}$$

where use of the following general indefinite integral has been made

$$\int \sin^2 ax\cos^2 ax\,\mathrm{d}x = \frac{x}{8} - \frac{\sin 4ax}{32a}$$

Due to symmetry $I_x = I_y$ for a circle.

Example 2.3 Second moments of area for a right-angled triangle

The second moments of area for the right-angled triangle in Example 2.1 above are given by

$$I_x = \iint_A y^2\,\mathrm{d}A = \int_0^h xy^2\,\mathrm{d}y = \frac{b}{h}\int_0^h (h - y)\,y^2\,\mathrm{d}y = \frac{bh^3}{12}$$

$$I_y = \iint_A x^2\,\mathrm{d}A = \int_0^b x^2 y\,\mathrm{d}x = h\int_0^b \left(1 - \frac{h}{b}x\right)x^2\,\mathrm{d}x = \frac{b^3 h}{12}$$

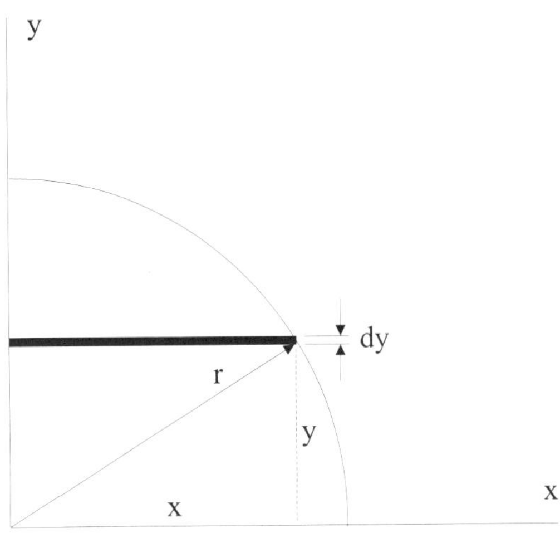

Figure 2.4: Example: Quadrant of a circle.

2.4 Parallel-Axis Theorem for the Second Moments of Area

The *parallel-axis theorem* enables the second moments of area with respect to an axis which is parallel to a centroidal axis to be determined from the centroidal second moments of area. Consider the arbitrary plane area of Figure 2.5. The centroidal axes (x_c, y_c) have their origin at the centroid c of the area and are parallel to the (x, y) coordinate system. Axes x and x_c are separated by distance d_1 and axes y and y_c are separated by distance d_2. The second moment of area I_x with respect to the x axis is

$$\begin{aligned} I_x &= \iint_A (y + d_1)^2 \,\mathrm{d}A = \iint_A y^2 \,\mathrm{d}A + 2d_1 \iint_A y \,\mathrm{d}A + d_1^2 \iint_A \mathrm{d}A \\ &= I_{x_c} + A d_1^2 \end{aligned} \tag{2.8}$$

The second integral on the right hand side of (2.8) vanishes because the x_c axis passes through the centroid for which $Q_x = 0$. The second moment of area I_y is similarly found

$$I_y = I_{y_c} + A d_2^2 \tag{2.9}$$

The parallel-axis theorem illustrates that the second moments of area vary according to the coordinate axes with the minimum attained about the centroidal axes.

Example 2.4 Application of the parallel-axis theorem to a right-angled triangle

The centroidal second moments of area for the right-angled triangle of Figure 2.2 are given by

$$I_{x_c} = \frac{bh^3}{36}; \quad I_{y_c} = \frac{b^3 h}{36}$$

With $d_1 = h/3$, $d_2 = b/3$ and $A = bh/2$ then the second moments of area with respect to the (x, y) axes are from (2.8) and (2.9)

$$I_x = I_{x_c} + A d_1^2 = \frac{bh^3}{36} + \left(\frac{bh}{2}\right)\left(\frac{h}{3}\right)^2 = \frac{bh^3}{12}$$

$$I_y = I_{y_c} + A d_2^2 = \frac{b^3 h}{36} + \left(\frac{bh}{2}\right)\left(\frac{b}{3}\right)^2 = \frac{b^3 h}{12}$$

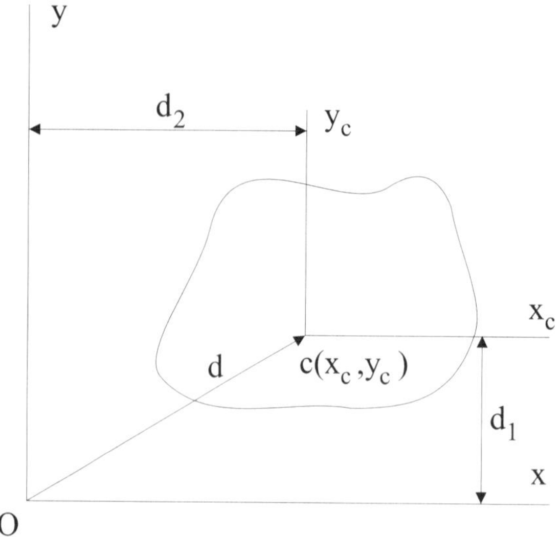

Figure 2.5: A plane area A with respect to the (x, y) and centroidal (x_c, y_c) coordinate systems.

which agree with those found in Example 2.3.

Example 2.5 Second moments of area for an I-section

Figure 2.6 illustrates an I-section which can be considered as composed of three rectangular sections. Evaluating I_x for the I-section with respect to axis AB in which the centroidal value of I_x for each of the three rectangles is $bh^3/12$ then

$$I_{x,AB} = I_{x,AB}\,(1) + I_{x,AB}\,(2) + I_{x,AB}\,(3)$$

and is the sum of the three section moments. The value of $I_{x,AB}$ can be determined from the parallel-axis theorem so that

$$I_{x,AB} = \left[8b\frac{(2h)^3}{12} + (16bh)h^2\right] + \left[\frac{b}{12}(4h)^3 + (4bh)(3h)^2\right] + \left[\frac{4b}{12}h^3 + (4bh)\left(\frac{13h}{2}\right)^2\right] = 260bh^3$$

2.5 Product of Areas

The *product of areas*, I_{xy}, with respect to the (x, y) coordinate system is defined as

$$I_{xy} = \iint_A xy\,\mathrm{d}A \tag{2.10}$$

and can be positive, negative or zero depending on the position of the coordinate axes relative to the area. The product of areas is zero when one of the coordinate axes is an axis of symmetry; for example the y centroidal axis of an isosceles triangle and both x and y centroidal axes of a circle.

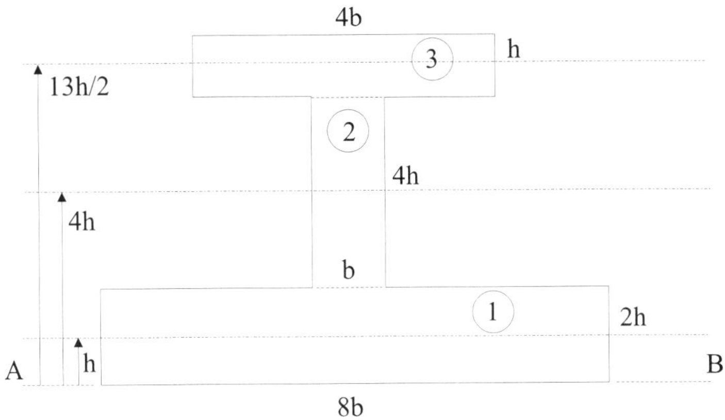

Figure 2.6: Example: An I-section.

Applying the parallel-axis theorem to the products of area we have, Figure 2.5

$$
\begin{aligned}
I_{xy} &= \iint_A (x+d_2)(y+d_1)\,\mathrm{d}A = \iint_A xy\,\mathrm{d}A + d_1 \iint_A x\,\mathrm{d}A + d_2 \iint_A y\,\mathrm{d}A + d_1 d_2 \iint_A \mathrm{d}A \\
&= I_{x_c y_c} + A\,\mathrm{d}_1\,\mathrm{d}_2
\end{aligned}
\tag{2.11}
$$

The second and third integrals on the right hand side of (2.11) vanish since they are the first moments of area with respect to the centroidal axes (x_c, y_c).

Example 2.6 Product of areas for a circular cross-section

Example 2.2 above derived I_x and I_y for a circle of radius r. Let us now derive I_{xy} for the circle. For the upper half of the circle shown in Figure 2.4 then $r^2 = x^2 + y^2$ and $\mathrm{d}A = x\,\mathrm{d}y = 2\sqrt{r^2 - y^2}\,\mathrm{d}y$ so that I_{xy} is given by

$$
\begin{aligned}
I_{xy} &= \iint_A xy\,\mathrm{d}A \\
&= \int_{-r}^{r} \sqrt{r^2 - y^2}\sqrt{r^2 - x^2}\left(2\sqrt{r^2 - y^2}\right)\mathrm{d}y = 2\int_{-r}^{r} (r^2 - y^2)y\,\mathrm{d}y = 2\left[\frac{r^2 y^2}{2} - \frac{y^4}{4}\right]_{-r}^{r} = 0
\end{aligned}
$$

illustrating that I_{xy} is equal to zero for the circle.

Example 2.7 Product of area for a right-angled triangle

Using the parallel-axis theorem (2.11) the products of area for the right-angled triangle of Figure 2.2 with respect to the axes (x, y) given that $I_{x_c y_c} = -b^2h^2/72$, $d_1 = h/3$ and $d_2 = b/3$, is equal to

$$
I_{xy} = I_{x_c y_c} + A d_1 d_2 = -\frac{b^2h^2}{72} + \left(\frac{bh}{2}\right)\left(\frac{h}{3}\right)\left(\frac{b}{3}\right) = \frac{b^2h^2}{24}
$$

2.6 Polar Moment of Area

The previous sections have considered the second moments of area with respect to the Cartesian coordinate axes (x, y). Let us now consider the polar moment of area, I_p, defined by

$$I_p = \iint_A r^2 \, \mathrm{d}A \tag{2.12}$$

where $r = (x^2 + y^2)^{1/2}$ is the distance from the origin of coordinates, O, to an elemental area, $\mathrm{d}A$, as shown in Figure 2.1. In terms of Cartesian coordinates I_p can be expressed

$$I_p = \iint_A r^2 \, \mathrm{d}A = \iint_A (x^2 + y^2) \, \mathrm{d}A = I_x + I_y \tag{2.13}$$

illustrating that I_p is the sum of I_x and I_y. Since I_p is the second moment of area about the z-axis then $I_z = I_p$.

As with the products of areas we can apply the parallel-axis theorem to the polar moment of area. Referring to Figure 2.5 and from (2.8) and (2.9)

$$\begin{aligned} I_p &= I_x + I_y = (I_{x_c} + Ad_1^2) + (I_{y_c} + Ad_2^2) \\ &= I_{pc} + Ad^2 \end{aligned} \tag{2.14}$$

where $I_{p_c} (= I_{x_c} + I_{y_c})$ and $d^2 (= d_1^2 + d_2^2)$.

Example 2.8 Polar moment of area for a circular cross section

The polar moment of area for a circle of radius R and diameter D can be found from $(I_x + I_y)$, with $I_x = I_y = \pi R^4/4 = \pi D^4/64$ from Example 2.2 then $I_p = \pi R^4/2 = \pi D^4/32$. Alternatively and equivalently, with $\mathrm{d}A = \mathrm{d}r \times r \, \mathrm{d}\theta$

$$I_p = \iint_A r^2 \, \mathrm{d}A = \int_0^R \int_0^{2\pi} r^3 \, \mathrm{d}r \, \mathrm{d}\theta = 2\pi \int_0^R r^3 \, \mathrm{d}r = \frac{\pi R^4}{2} = \frac{\pi D^4}{32}$$

2.7 Radius of Gyration

Figure 2.7a) illustrates an arbitrary plane figure of area A with second moments of area I_x, I_y and $I_z = I_p$. If the area of the figure is now concentrated in a narrow strip at a distance r_x from the x-axis, Figure 2.7b), then the second moment of area about the x-axis will be the same as that of the original area if $r_x^2 A = I_x$. Similarly, area A could be concentrated into a narrow strip at a distance r_y from the y-axis, Figure 2.4c) and concentrated into a narrow ring of radius r_z, Figure 2.7d). Thus, the *radii of gyration* for the three axes are

$$r_x = \sqrt{\frac{I_x}{A}}; \quad r_y = \sqrt{\frac{I_y}{A}}; \quad r_z = \sqrt{\frac{I_z}{A}} \tag{2.15}$$

The radius of gyration for a given section provides a measure of the distribution of the area from the associated axis. For the I-section in Example 2.5 we find that $r_x = \sqrt{65/6h}$. Comparing this value of $r_x(I)$ to the radius of gyration of a rectangular section, $r_x(R) = 4h/3$, of the same area with width $8b$ (same as the I-section) and height $3h$ then $r_x(I)/r_x(R) \approx 2.469$. This comparison demonstrates that I-sections have the property of larger radii of gyration for a given area than their rectangular counterparts.

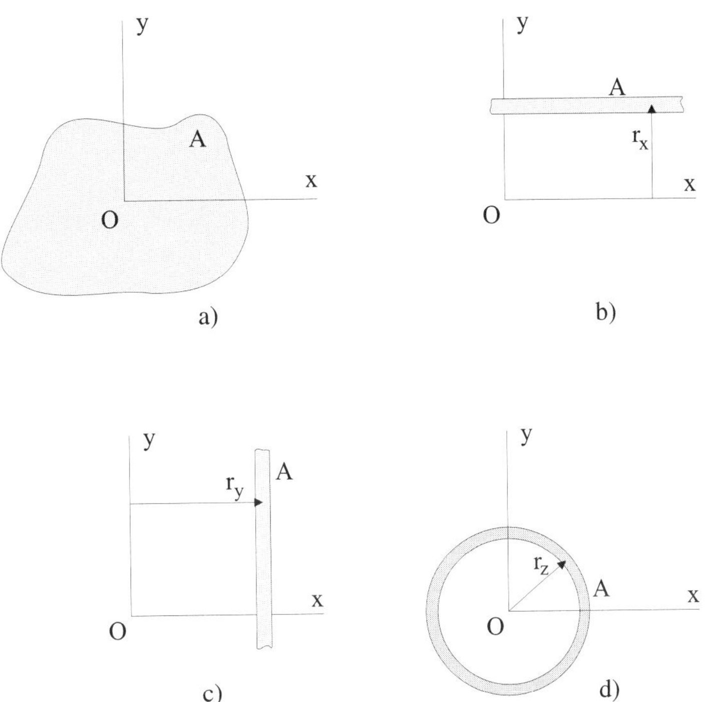

Figure 2.7: Radius of gyration. a) A plane area A and equivalent areas about the b) x-axis, c) y-axis and d) z-axis.

2.8 Conclusion

This Chapter has examined the centroid, and the first and second moments of area of plane sections. We also examined the parallel-axis theorem which assists in determining the second moments of area relative to a set of axes that have undergone a pure translation. In Chapter 7 we will also examine the case of a pure rotation of coordinate axes and their effects on the second moments of area. The importance of second moments of area will become evident in Chapter 5 when we examine simple beam theory.

2.9 References and Further Reading

Tables of second moments of area are given in the following text for both plane and three-dimensional bodies.

∘ Howatson, A. M., Lund, P. G. and Todd, J. D. (1972) *Engineering Tables and Data*, Chapman and Hall, London.

2.10 Exercises

2.1 Determine the area, first moments of area and centroid of a circle of radius R with respect to axes (x, y) centred at the origin of the circle.

2.2 Determine the area, first moments of area and centroid of the semicircle of radius R shown in Figure 2.8 with respect to the (x, y) axes centred at the origin of the circle.

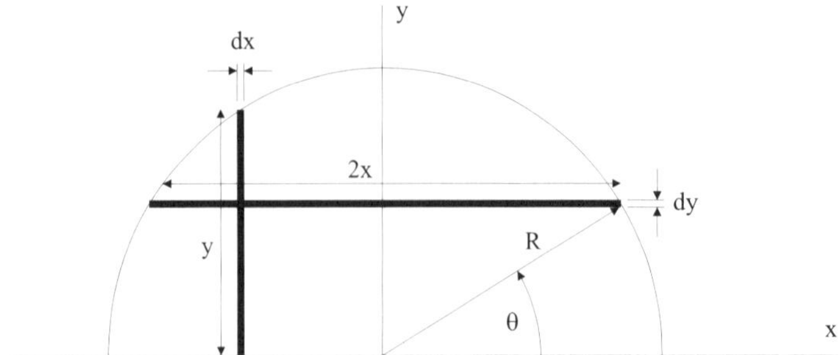

Figure 2.8: Exercise: A semicircle of radius R and elemental strips dx and dy.

2.3 Determine the area and second moments of area of an ellipse with half major and minor axes a and b respectively.

2.4 For the angle bracket shown in Figure 2.9 determine the centroid, and both the first and second moments of area with respect to both the global (x, y) and centroidal (x_c, y_c) sets of axes.

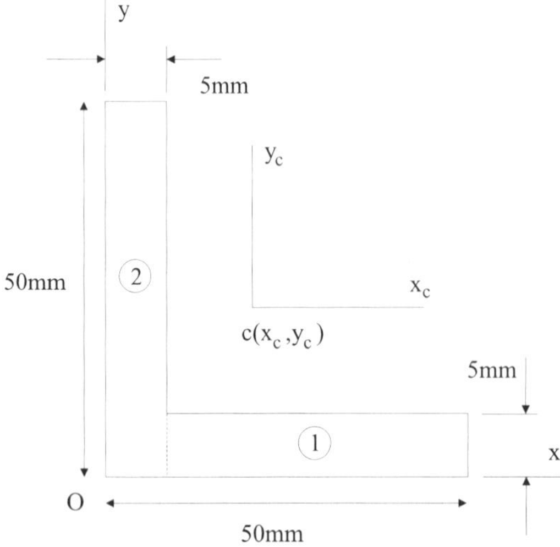

Figure 2.9: Exercise: Angle bracket cross-section.

2.5 Determine the product of areas I_{xy} for the rectangle shown in Figure 2.3.

2.6 Determine the polar moment of area for the elliptical cross-section of Exercise 2.3.

2.7 Determine the radii of gyration r_x and r_y for the elliptical cross-section of Exercise 2.3

Chapter 3

Torsion

3.1 Introduction

From a design point of view considerable interest is attached to the effects of pure torsion on machine elements, particularly shafts in both the elastic and plastic ranges of deformation. Although we will only consider the circular cross-section, it must be noted that in many cases of practical importance this geometry is no longer acceptable and more complicated shapes have to be adopted. For instance, non-circular solid cross-sections, hollow tubes and multiple cell tubes.

The following sections examine the torsion of shafts of circular cross-section in which the shaft material is assumed to respond elastically throughout. The elastic-plastic torsion of circular cross-section shafts will be considered in Chapter 12. The next section examines a circular shaft of solid cross-section. The following section then examines thin-walled shafts. The final section examines composite shafts in which a solid bar is bonded inside a tube.

3.2 Torsion of Circular Cross-Sectional Components

From an engineering point of view we are interested in relating the shear stress, τ, and the angular twist, θ, that are produced due to an externally applied torque, T, to a solid circular shaft of length L with material rigidity modulus G, section radius R or diameter D and the polar second moment of area J. Consider the shaft shown in the Figure 3.1. The polar moment of area, $J = I_z$, is, from (2.12)

$$J = \iint_A r^2 \, \mathrm{d}A; \quad J = \iint_A \left(x^2 + y^2\right) \mathrm{d}A = I_x + I_y = \frac{\pi D^4}{32} = \frac{\pi R^4}{2} \tag{3.1}$$

For a tubular section of inner and outer radii R_i and R_o respectively, then

$$J = 2\pi \int_{R_i}^{R_o} r^3 \, \mathrm{d}r = 2\pi \left[\frac{r^4}{4}\right]_{R_i}^{R_o} = \frac{\pi}{2}\left(R_o^4 - R_i^4\right) \tag{3.2}$$

The shaft is fixed at one end and free at the other. When the point A moves through to point B then the shear strain γ is related to the angular distortion ϕ by, (1.10)

$$\gamma = \frac{\tau}{G} = \tan\phi \tag{3.3}$$

with $\tan\phi \approx \phi$ for small ϕ. Since, for a given radius r, $AB = r\theta$ and from Figure 3.1a)

$$\gamma = \frac{\tau}{G} = \tan\phi \approx \phi = \frac{AB}{L} = \frac{r\theta}{L} \tag{3.4}$$

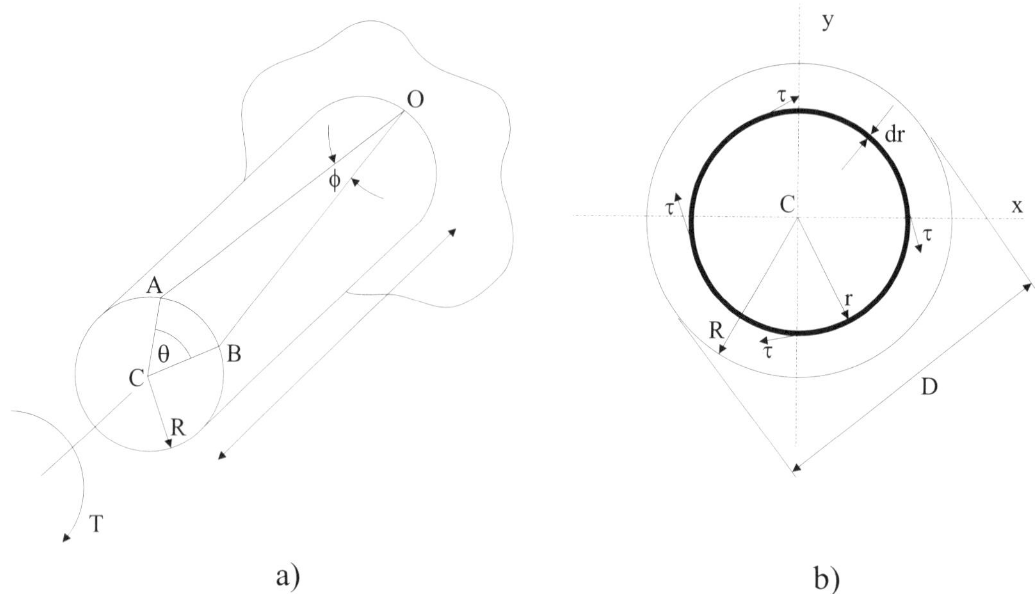

Figure 3.1: A shaft of length L and circular cross-section, radius R, subject to a torque T. a) General configuration. b) Shear stress τ acting on an annular ring of thickness dr at radius r from the centre C.

Re-arranging for τ we see that the shear stress is proportional to r and therefore attains a maximum value at the surface and is zero at the centre, see Figure 3.2.

$$\tau = \frac{Gr\theta}{L} \tag{3.5}$$

The resisting torque exerted by the cross-section must equilibrate the applied torque T. For a shear stress τ, acting on an elemental annular area of radius r and thickness $\mathrm{d}r$ the equilibrium condition results in the following elemental force $\mathrm{d}F$ as, Figure 3.1b)

$$\text{Elemental force } \mathrm{d}F = \tau\,(2\pi r dr) \tag{3.6}$$

Therefore, the elemental torque, $\mathrm{d}T$, is simply $\mathrm{d}F$ multiplied by the radius r

$$\mathrm{d}T = \mathrm{d}F.r = 2\pi r^2 \tau\,\mathrm{d}r \tag{3.7}$$

Integrating $\mathrm{d}T$ from the centre to the outer radius we find the total torque

$$T = \int_0^R 2\pi r^2 \tau\,\mathrm{d}r \tag{3.8}$$

Substituting τ from (3.5) we have

$$\begin{aligned} T &= \int_0^R 2\pi r^2 \left(\frac{Gr\theta}{L}\right) \mathrm{d}r \\ &= 2\pi \left(\frac{G\theta}{L}\right) \int_0^R r^3\,\mathrm{d}r \\ T &= \frac{G\theta}{L} J \end{aligned} \tag{3.9}$$

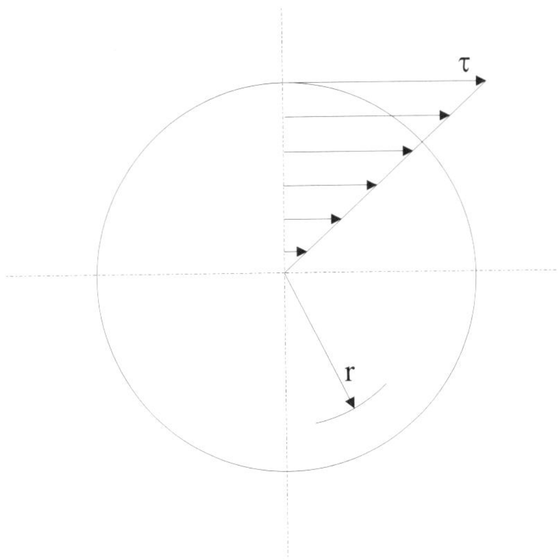

Figure 3.2: Variation of shear stress τ with radius r.

noting that J is given by

$$J = \iint_A r^2 \, \mathrm{d}A = \int_0^R r^2 2\pi r \, \mathrm{d}r = 2\pi \int_0^R r^3 \, \mathrm{d}r \tag{3.10}$$

Combining the above results leads to the following relationship which relates all parameters

$$\frac{T}{J} = \frac{G\theta}{L} = \frac{\tau}{r} \tag{3.11}$$

For a solid shaft the maximum shear stress, τ_{max}, occurs in the outer fibres, $r = R$

$$\tau_{max} = \frac{TR}{J} = \frac{16T}{\pi D^3} \tag{3.12}$$

Example 3.1 Torsion of a tapered bar

We will use this example to examine a solid bar with continuously varying cross-section subject to a constant torque T, Figure 3.3. When the torque is constant then from (3.12) we observe that the maximum shear stress will occur at the cross-section with the smallest diameter, that is d_1

$$\tau_{max} = \frac{16T}{\pi d_1^3}$$

By considering an elemental strip of thickness $\mathrm{d}x$ then the angle of twist for the entire bar is, from (3.11)

$$\theta = \int_0^L \frac{T \, \mathrm{d}x}{GJ(x)}$$

where the polar moment of area, $J(x)$, is now a function of x. To determine $J(x)$ we first determine the diameter d at a distance x from d_1 by linearly interpolating between the two ends

$$d = d_1 + \left(\frac{d_2 - d_1}{L} \right) x$$

with $J(x)$ now given by, from (3.1)

$$J(x) = \frac{\pi d^4}{32} = \frac{\pi}{32}\left[d_1 + \left(\frac{d_2 - d_1}{L}\right)x\right]^4$$

Substituting for $J(x)$ into θ

$$\theta = \frac{32T}{\pi G}\int_0^L \frac{\mathrm{d}x}{\left[d_1 + \left(\frac{d_2-d_1}{L}\right)x\right]^4}$$

To evaluate the integral we make use of the following standard indefinite integral

$$\int \frac{\mathrm{d}x}{(ax+b)^n} = -\frac{1}{(n-1)\,a\,(ax+b)^{n-1}}; \quad n \neq 1$$

Performing the integration θ is found to equal

$$\theta = \frac{32TL}{3\pi G\,(d_2 - d_1)}\left[\frac{1}{d_1^3} - \frac{1}{d_2^3}\right]$$

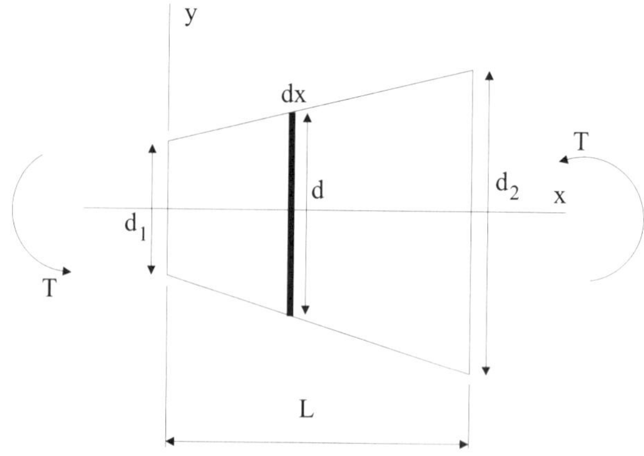

Figure 3.3: Example: A tapered bar subject to a constant torque T.

3.3 Torsion of Thin-Walled Circular Cross-Sectional Components

In addition to the previous section let us briefly consider the torsion of a thin-walled tube which has an inner radius R_i and outer radius R_o. The torque is, using an equivalent procedure to that used in the derivation of (3.8)

$$T = 2\pi \int_{R_i}^{R_o} \tau r^2 \,\mathrm{d}r \tag{3.13}$$

Assuming that the shear stress does not vary across the wall and that the average shear stress is τ_m then (3.13) is given by

$$T = 2\pi\tau_m \int_{R_i}^{R_o} r^2 \,\mathrm{d}r = \frac{2}{3}\pi\tau_m\left(R_0^3 - R_i^3\right) \tag{3.14}$$

and re-arranging for τ_m

$$\tau_m = \frac{3T}{2\pi\left(R_o^3 - R_i^3\right)} \tag{3.15}$$

If the wall thickness is denoted by $t = R_o - R_i$ and the mean radius of the tube is R_m then $R_i = R_m - t/2$ and $R_o = R_m + t/2$. Therefore, in terms of R_m and t the mean shear stress is

$$\tau_m = \frac{T}{2\pi R_m^2 t} \tag{3.16}$$

The exact shear stress, τ_e, at the mean radius $R_m = (R_i + R_o)/2$ can be found from $\tau = Tr/J$ of (3.11) with J given by (3.2)

$$\begin{aligned}\tau_e &= \frac{2TR_m}{\pi\left(R_o^4 - R_i^4\right)} = \frac{(R_i + R_o)\,T}{\pi\left(R_o^4 - R_i^4\right)} \\ &= \frac{(R_i + R_o)\,T}{\pi\left(R_o^2 + R_i^2\right)(R_o - R_i)(R_o + R_i)} = \frac{T}{\pi\left(R_o^2 + R_i^2\right)(R_o - R_i)}\end{aligned} \tag{3.17}$$

Comparing (3.16) and (3.17) for the case of $R_i = 10\text{mm}$ and $R_o = 10.25\text{mm}$ we find $\tau_e/\tau_m = 1.0001$ illustrating that (3.16) is an accurate measure of τ_e at the mean radius.

3.4 Torsion of Composite Circular Bars

Consider the composite bar of Figure 3.4 in which a solid bar A is bonded inside the tube B. It is assumed that the bond between A and B is perfect and no slipping occurs. If T_A and T_B denote the torques in A

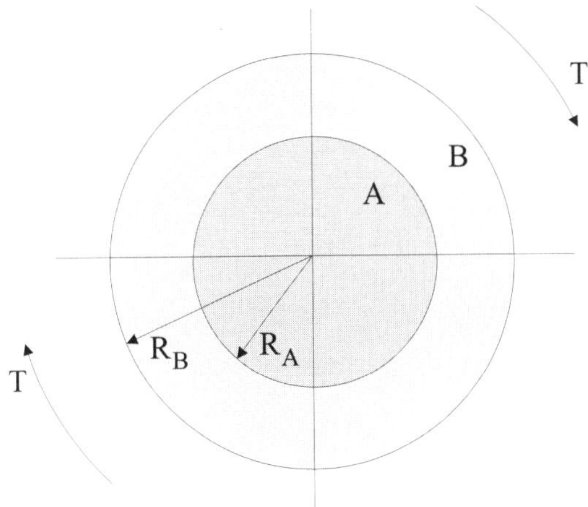

Figure 3.4: A composite bar of solid bar A and tube B subject to torque T.

and B respectively then the total torque T for the composite bar to be in equilibrium is

$$T = T_A + T_B \tag{3.18}$$

Assuming A and B are perfectly bonded then the angle of twist, θ, of A and B must be equivalent

$$\theta = \frac{T_A L}{G_A J_A} = \frac{T_B L}{G_B J_B} \tag{3.19}$$

where L is the length of the composite bar, G denotes the shear modulus and J is the polar moment of area. Solving for T_A and T_B from (3.18) and (3.19)

$$T_A = \left(\frac{G_A J_A}{G_A J_A + G_B J_B}\right) T; \quad T_B = \left(\frac{G_B J_B}{G_A J_A + G_B J_B}\right) T \tag{3.20}$$

and hence, from (3.19), the angle of rotation is

$$\theta = \frac{TL}{G_A J_A + G_B J_B} \tag{3.21}$$

Finally, the maximum shear stress in A and B are, from (3.12)

$$\tau_{A,max} = \frac{T_A R_A}{J_A}; \quad \tau_{B,max} = \frac{T_B R_B}{J_B} \tag{3.22}$$

It is worth noting that the shear strains in A and B at the common interface are equal whereas from (1.10) the ratio of the shear stresses is equal to the ratio of the shear moduli

$$\frac{\tau_A}{\tau_B} = \frac{G_A}{G_B} \tag{3.23}$$

3.5 Conclusion

This Chapter has examined the torsion of shafts of circular cross-section in which the shaft material is assumed to respond elastically. We considered the three cases of solid and thin-walled tube cross-sections and composite shafts. In Chapter 12 we will revisit the torsion of shafts of circular cross-section for elastic-plastic materials.

3.6 References and Further Reading

∘ Gere, J. M. and Timoshenko, S. P. (1997) *Mechanics of Materials*, PWS Publishers, Boston, MA.

3.7 Exercises

3.1 A solid bar of circular cross-section is of diameter 50mm and subject to an applied torque of 10kNm. Determine the maximum shear stress in the bar.

3.2 If the bar of Exercise 3.1 is of length 1.25m then determine the angle of twist.

3.3 A solid tapered bar of circular cross-section is of length 1m and subject to an applied torque of 12kNm. The diameter of the bar linearly varies from 50mm to 75mm. Determine both the maximum shear stress and the angle of twist of the bar.

3.4 A thin-walled tube of circular cross-section is subject to a torque of 100Nm. If the mean radius is 31.25mm and the wall-thickness is 0.1mm then determine the mean shear stress.

3.5 A thin-walled circular cross-section tube of 80mm mean diameter is subject to a torque of 1.5kNm. If the maximum permissible shear stress in the tube is 65MPa then determine the minimum wall thickness required.

3.6 A composite bar of length 0.75m consists of an inner solid bar of radius 12.5mm which is perfectly bonded inside a tube of outer radius 25mm. The shear moduli of the inner solid bar and outer tube are 45MPa and 30MPa respectively. Determine the angle of twist of the composite bar if it is subject to an applied torque of 5kNm.

3.7 For the composite bar of Exercise 3.6 determine the maximum shear stresses in both the inner solid bar and outer tube.

Chapter 4

Thin-Walled Pressure Vessels

4.1 Introduction

When a thin-walled vessel of mean diameter d and wall-thickness t is subject to an internal pressure then the induced stress distribution does not significantly vary throughout the wall thickness. As a result thin-walled pressure vessel theory assumes that the stress distribution is approximately constant and equal to the internal pressure across the wall. Generally, the through-wall or radial stress can be neglected when $d/t \gg 1$ because it is negligible in comparison with the maximum circumferential stress. In the following sections we will examine the two cases of cylindrical and spherical pressure vessels with the vessel material behaving in a linear elastic manner. The Chapter concludes by determining the second moments of area for thin-walled tubular sections.

4.2 Cylindrical Pressure Vessel

Consider the cylindrical vessel with mean radius r and thickness t shown in Figure 4.1a). For a stress element which is sufficiently removed from the ends of the vessel then the circumferential[1] stress, $\sigma_{\theta\theta}$, is denoted by the principal stress σ_1 and the axial stress, σ_{zz}, is denoted by σ_2. To determine σ_1 we cut the vessel with a plane that lies parallel to the z-axis as shown in Figure 4.1b). If the length of the vessel is l then the force acting on the two cut walls is $2(\sigma_1 lt)$. For the vessel to be in a state of equilibrium under the action of the internal pressure then this force is resisted by force $2lrp$ acting over the area $A_1 = 2rl$

$$2lt\sigma_1 - 2lrp = 0 \tag{4.1}$$

and re-arranging for σ_1

$$\sigma_1 = \frac{pr}{t} \tag{4.2}$$

To determine the axial stress σ_2 consider the vessel cut by a plane which is perpendicular to the z-axis as shown in Figure 4.1c). The force on the cut section is now $2\pi rt\sigma_2$ and is resisted by a force due to p acting on area $A_2 = \pi r^2 p$

$$2\pi rt\sigma_2 - \pi r^2 p = 0 \tag{4.3}$$

[1] In the literature, the circumferential stress is also referred to as the *hoop* stress.

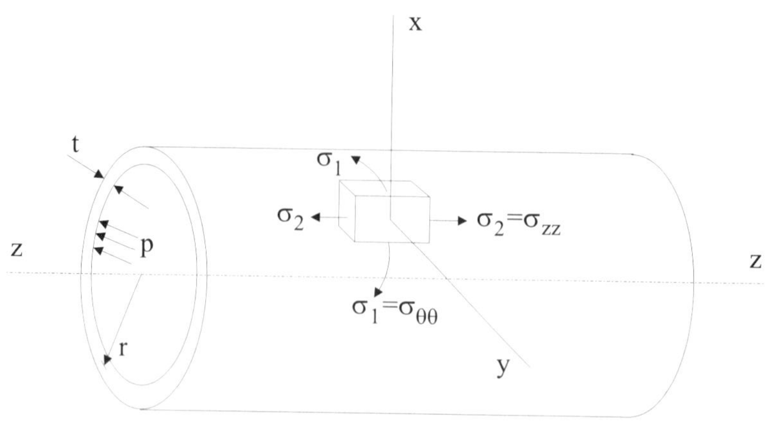

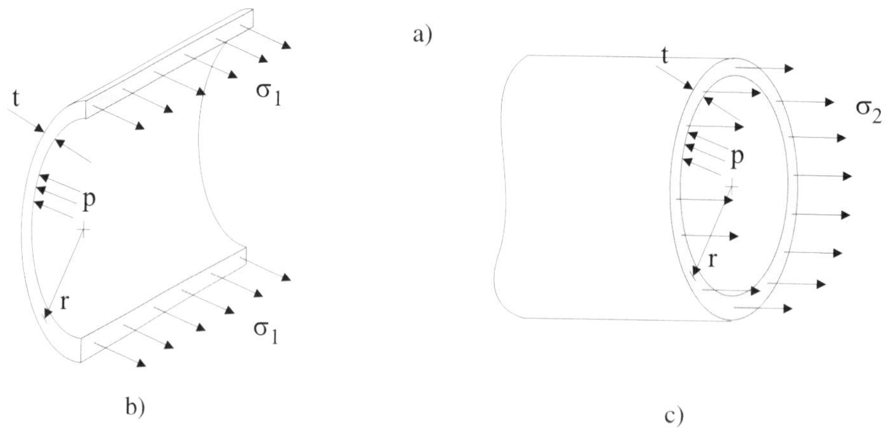

Figure 4.1: Cylindrical vessel of mean radius r and wall thickness t subject to an internal pressure p. a) A stress element with the two principal stress components σ_1 and σ_2. b) A diametrical cross-section through the vessel parallel to the z-axis. c) A cross-section through the vessel perpendicular to the z-axis.

and re-arranging for σ_2

$$\sigma_2 = \frac{pr}{2t} = \frac{\sigma_1}{2} \tag{4.4}$$

and illustrates that σ_2 is half the value of σ_1. Note that shear stresses are not present in the vessel because of the polar symmetry of the vessel and hence the reason why $\sigma_{\theta\theta}$ and σ_{zz} are in fact the principal stresses σ_1 and σ_2 respectively. A more detailed discussion of principal stresses is delayed until Chapter 7.

The circumferential and axial strains follow from the Hookian equations (1.16)

$$\begin{aligned}\varepsilon_1 &= \frac{1}{E}\left[\sigma_1 - \nu(\sigma_2+\sigma_3)\right] = \frac{1}{E}\left[\sigma_1 - \nu\frac{\sigma_1}{2}\right] = \frac{pr}{Et}\left(1-\frac{\nu}{2}\right)\\ \varepsilon_2 &= \frac{1}{E}\left[\sigma_2 - \nu(\sigma_1+\sigma_3)\right] = \frac{1}{E}\left[\sigma_2 - \nu(2\sigma_2)\right] = \frac{pr}{Et}\left(\frac{1}{2}-\nu\right)\\ \varepsilon_3 &= \frac{1}{E}\left[\sigma_3 - \nu(\sigma_1+\sigma_2)\right] = \frac{1}{E}\left[-\nu\left(\sigma_1+\frac{\sigma_2}{2}\right)\right] = -\frac{3pr}{Et}\nu\end{aligned} \tag{4.5}$$

with $\sigma_3 = \sigma_{rr} = 0$ and $\sigma_2 = \sigma_1/2$ and noting that $\varepsilon_3 \neq 0$. The volumetric strain can be found by considering the change in volume or from (1.19) in terms of principal strains

$$\varepsilon_V = \varepsilon_1 + \varepsilon_2 + \varepsilon_3 = \frac{3pr}{Et}\left(\frac{1}{2}-\nu\right) \tag{4.6}$$

Although the shear stresses are not present the maximum in-plane shear stress, τ_{max}, is non-zero and can be determined from (7.26)

$$\tau_{max,y} = \frac{\sigma_1-\sigma_2}{2} = \frac{pr}{4t} \tag{4.7}$$

and occurs on a plane rotated at 45° to the y-axis. The maximum out-of-plane shear stresses occur at 45° to the x and z axes

$$\tau_{max,x} = \frac{\sigma_2}{2} = \frac{pr}{4t}; \quad \tau_{max,z} = \frac{\sigma_1}{2} = \frac{pr}{2t} \tag{4.8}$$

Thus, the absolute maximum shear stress is

$$\tau_{max} = \frac{\sigma_1}{2} = \sigma_2 = \frac{pr}{2t} \tag{4.9}$$

Example 4.1 Stresses and strains of a thin-walled cylindrical pressure vessel

A soft drinks can in the form of a cylindrical vessel has a mean diameter, d, of 65mm and wall thickness, t, of 0.12mm and is subject to an internal pressure, p, of 90psi or 0.622MPa and atmospheric external pressure. Young's modulus and Poisson's ratio of the vessel material are 61GPa and 0.334 respectively. Let us begin by establishing that the vessel can be modelled in accordance with thin-walled pressure vessel theory and then we will determine the radial, circumferential and axial stresses and strains.

Since $d/t = 65/0.12 = 542 \gg 1$ then thin-walled pressure vessel theory will be assumed to be representative. From (4.2) and (4.4) the circumferential, σ_1, and axial, σ_2, stresses are

$$\sigma_1 = \frac{pr}{t} = 168\text{MPa}; \quad \sigma_2 = \frac{pr}{2t} = \frac{\sigma_1}{2} = 84\text{MPa}$$

Since the wall thickness is negligible and neglecting the external atmospheric pressure then the radial stress is equal and opposite to the internal pressure, $\sigma_{rr} \approx -p =$-0.622MPa, and can be taken as zero when compared to the circumferential and axial stresses. From (4.5) the strains are

$$\begin{aligned}\varepsilon_1 &= \frac{\sigma_1}{E}\left(1-\frac{\nu}{2}\right) = 2,300\times 10^{-6}\\ \varepsilon_2 &= \frac{\sigma_2}{E}(1-2\nu) = 458\times 10^{-6}\\ \varepsilon_3 &= -\frac{3\nu\sigma_1}{2E} = 1,380\times 10^{-6}\end{aligned}$$

4.3 Spherical Pressure Vessel

Figure 4.2a) illustrates a spherical pressure vessel of mean radius r and wall-thickness t. Due to the spherical symmetry of the vessel then just the σ_2 principal stress is present. Making a diametrical cut through the sphere, Figure 4.2b), from equilibrium considerations we have

$$2\pi r t\sigma_2 - \pi r^2 p = 0 \tag{4.10}$$

which reduces to

$$\sigma_2 = \frac{pr}{2t} \tag{4.11}$$

and is observed to be the same as the axial stress in the case of a cylindrical pressure vessel. As in the previous section, the strains for the spherical vessel are

$$\varepsilon_1 = \varepsilon_2 = \frac{pr}{2Et}(1-\nu); \quad \varepsilon_3 = -\frac{pr}{Et}\nu \tag{4.12}$$

The maximum in-plane shear stress is zero since $\sigma_1 = \sigma_2$. However, the maximum out-of-plane shear stresses are non-zero and given by

$$\tau_{max,x} = \tau_{max,z} = \frac{\sigma_2}{2} = \frac{pr}{4t} \tag{4.13}$$

so that τ_{max} for a thin-walled spherical pressure vessel is

$$\tau_{max} = \frac{\sigma_2}{2} = \frac{pr}{4t} \tag{4.14}$$

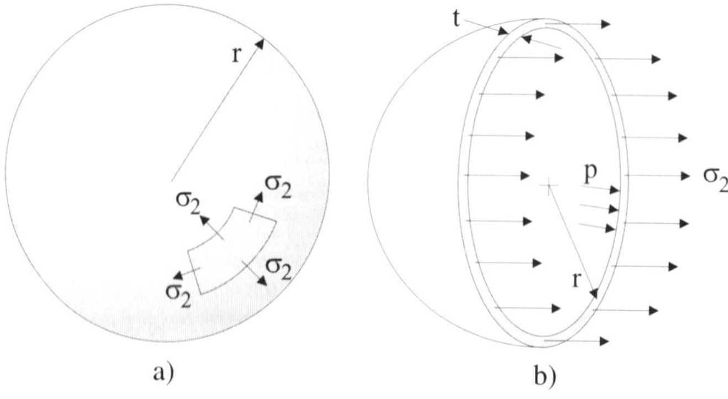

Figure 4.2: Spherical vessel of mean radius r and wall thickness t subject to an internal pressure p. a) A stress element with the principal stress component σ_2. b) A diametrical cross-section through the vessel.

Example 4.2 Thin-walled cylindrical pressure vessel with hemi-spherical end caps

Figure 4.3 illustrates a cylindrical pressure vessel with hemi-spherical end caps subject to an internal pressure p and of mean radius r. If the wall thickness of the cylindrical and spherical parts of the vessel are t_c and t_s respectively then let us determine the wall thickness for both parts of the vessel so that the radial expansion for each part is equivalent. Equating the circumferential strains (4.5) and (4.12)

$$\frac{pr}{Et_c}\left(1-\frac{\nu}{2}\right)=\frac{pr}{2Et_s}(1-\nu)$$

which reduces to

$$\frac{t_c}{t_s}=\frac{2-\nu}{1-\nu}$$

For example, if $\nu = 1/3$ then $t_c = 2.5t_s$.

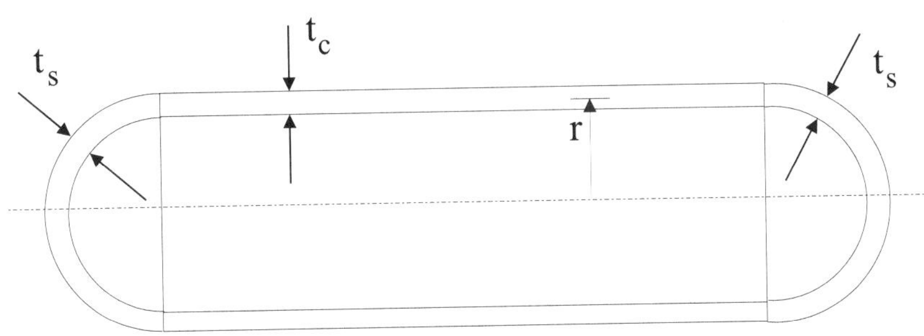

Figure 4.3: Example: Cylindrical pressure vessel with hemi-spherical end caps.

4.4 Calculation of the Second Moments of Area for Thin-Walled Tubular Sections

In Chapter 2 it was shown that the second moments of area I_x and I_y for a solid circular cross-section of radius r are, see Example 2.2

$$I_x = I_y = \frac{\pi r^4}{4} \tag{4.15}$$

with the origin of the (x, y) axes at the circle centre. Alternatively, in the case of a tubular cross-section with inner and outer radii a and b then

$$I_x = I_y = \frac{\pi}{4}(b^4 - a^4) \tag{4.16}$$

However, in the limit of the wall-thickness $t(= b-a) \to 0$ these values for I_x and I_y become increasingly inaccurate.

Alternative approximate expressions for I_x and I_y can be derived for a thin-walled tubular section by considering Figure 4.4. From the definition of I_x, (2.4), we have

$$I_x = \iint_A y^2\,\mathrm{d}A = 2\int_0^{\pi/2}(r\cos\theta)^2 t\,\mathrm{d}s = 2\int_0^{\pi/2}(r\cos\theta)^2 tr\,\mathrm{d}\theta = 2tr^3\int_0^{\pi/2}\cos^2\theta\,\mathrm{d}\theta \tag{4.17}$$

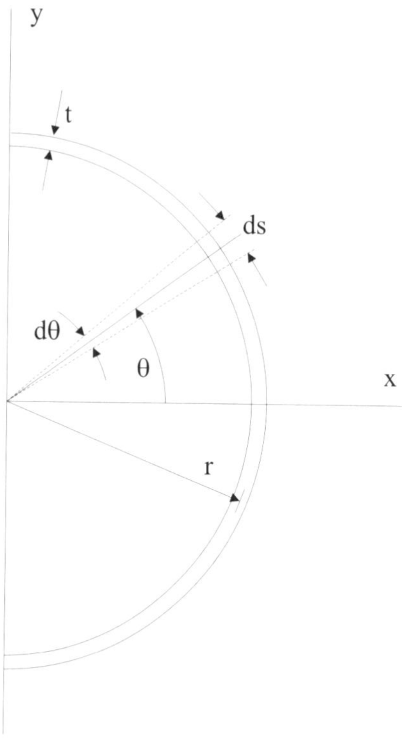

Figure 4.4: Half of a thin-walled tubular section of mean radius r and wall-thickness t.

where the factor of 2 accounts for both the upper and lower quadrants. Evaluating the integral

$$I_x = \frac{\pi r^3 t}{2} \tag{4.18}$$

and since there are two halves to the entire tubular section and noting that $I_x = I_y$ then we have finally

$$I_x = I_y = \pi r^3 t \tag{4.19}$$

and with polar second moment of area

$$I_p = I_z = I_x + I_y = 2\pi r^3 t \tag{4.20}$$

As a comparison of equations (4.16) and (4.20) with $r = 32.5$mm and $t = 0.1$mm then (4.16) predicts $I_x = 10,834\text{mm}^4$ whereas (4.20) gives $I_x = 10,784\text{mm}^4$.

4.5 Conclusion

This Chapter has examined thin-walled cylindrical and spherical pressure vessels. Although the formulae for the principal stresses and maximum shear stress are relatively simple, nevertheless they are applicable to a range of pressure vessels. In addition, these formulae provide a first estimate of the state of stress which then can be refined by more advanced analyses. The present discussion will be continued in Chapter 11 when we discuss thick-walled pressure vessels in which the stress distribution does vary across the wall thickness.

4.6 References and Further Reading

∘ Hibbeler, R. C. (1994) *Mechanics of Materials*, 2nd Ed., Macmillan, New York.

4.7 Exercises

4.1 A pipe of mean diameter 0.8m and thickness 15mm carries a liquid at a pressure of 50psi. Determine the state of stress in the walls of the pipe. 1lb/in^2(psi) = 6.895kN/m^2.

4.2 By the application of a strain gauge to the surface of a thin-walled cylinder the circumferential and axial strains are measured to be $\varepsilon_{\theta\theta} = 1,821 \times 10^{-6}$ and $\varepsilon_{zz} = 429 \times 10^{-6}$. Determine the corresponding stresses $\sigma_{\theta\theta}$ and σ_{zz} and the maximum in-plane shear stress. Assume that Young's modulus and Poisson's ratio are 70GPa and 0.3 respectively.

4.3 A spherical vessel of mean diameter 2m and thickness 25mm is submerged in water (density $\rho = 1,000$ kg/m^3). Using a safety factor of 10 determine the maximum depth of water into which the vessel can be lowered so that the circumferential stress in the sphere does not exceed 300MPa.

4.4 A thin-walled cylindrical tank with hemi-spherical ends is of mean diameter 0.25m and subject to an internal pressure of 0.75MPa, see Figure 4.3. If the wall thickness of the cylinder is 1mm then determine the wall thickness, t_s, of the hemi-spherical ends so that the radial expansion is constant throughout the entire vessel. Assume that Young's modulus and Poisson's ratio for the vessel material are equal to 210GPa and 1/3 respectively.

4.5 Figure 4.5 illustrates a force of 50kN acting on a piston of diameter 80mm. If the yield stress in pure shear of the cylinder material is 150MPa and a safety factor of 2 is used then determine the required thickness of the cylinder.

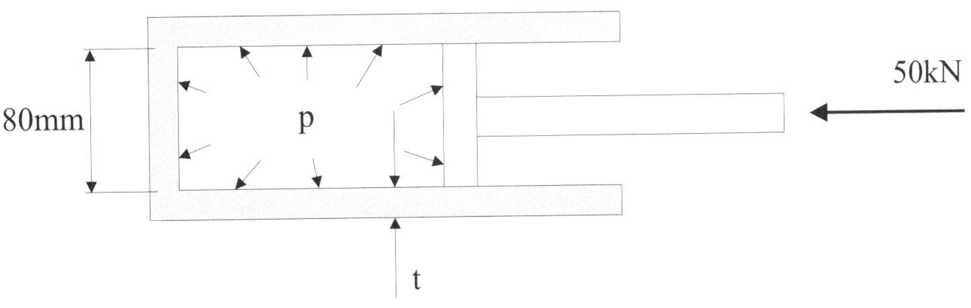

Figure 4.5: Exercise: Piston and cylinder.

4.6 A spherical pressure vessel has a mean radius 65mm and wall thickness 10mm. If the ultimate tensile stress of the vessel material is 737MPa then determine the maximum permissible internal pressure.

4.7 A tubular cylinder of circular cross-section has a mean radius 25mm and wall thickness 1mm. Determine the second moments of area I_x and I_y with origin of coordinates (x, y) centred at the cylinder centre.

Chapter 5

Beams

5.1 Introduction

This Chapter presents an overview of the extensive subject of beams. We begin by classifying beams and the different types of applied loading. Beams are generally categorised as either statically determinate or statically indeterminate and §5.2.3 provides a definition of the static determinacy of beams. Subsequent sections are concerned primarily with statically determinate beams until §5.7 when we discuss statically indeterminate beams. Section 5.3 provides a comprehensive discussion on shear force and bending moment at given points on a beam and their graphical illustration via the shear force and bending moment diagrams. This is followed by derivations of the key deflection, strain and stress formulae for a beam subject to pure bending. The beam shear formula is then derived for a beam that is subject to shear forces. Section 6 discusses the determination of the deflections of a beam by performing successive integrations of the flexure formula and includes a discussion of Macaulay's method which caters for discontinuities in the bending moment expression. We then examine two different methods for solving statically indeterminate beams in which all of the unknown reactions cannot be determined from the equations of equilibrium alone. Finally, we present a discussion on the position of the shear centre for beams with thin-walled open sections.

5.2 Classification of Beams

The following three sub-sections classify beams, the types of loads and statically determinate beams.

5.2.1 Types of Support

Beams are generally classified according to the manner in which they are supported. Figure 5.1 illustrates three beams with differing supports. The beam in case a) of Figure 5.1 illustrates a beam with a *pin-jointed* support at end A and a roller support at end B. Such beams with pin-jointed supports are referred to as *simply supported*. The pin-jointed support prevents translation of the beam but allows rotation about the support. Thus, a pin-jointed support is capable of inducing a force reaction with both horizontal and vertical components but it cannot support a moment reaction. End B has a roller support which prevents a vertical translation but allows a horizontal translation. As a consequence, a *roller support* can support a vertical force reaction but not a horizontal force reaction and is free to rotate like a pin-jointed support. The beam in case b) of Figure 5.1 illustrates a cantilever beam in which end A is fully built-in and end B is free. At the built-in end the beam is prevented from both translating and rotating and is therefore capable of sustaining both force and moment reactions. Case c) of Figure 5.1 illustrates a beam with an overhang. The overhang BC is similar to a cantilever beam except that the beam is able to rotate at the roller support.

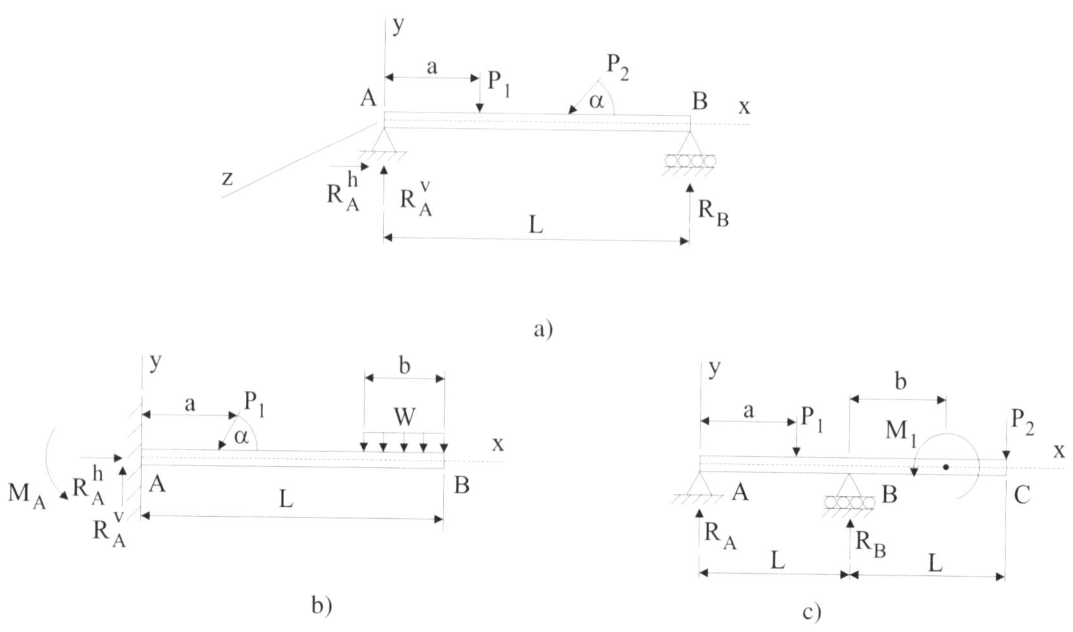

Figure 5.1: Different types of beams. a) Simply supported beam. b) Cantilever beam. c) Beam with overhang.

Case a) of Figure 5.1 illustrates the Cartesian coordinate system (x, y, z) centred at the left hand end A of the beam with associated displacements (u, v, w). Axis x is directed along the length of the beam with the y-axis directed vertically and with the z-axis acting out of the paper. All of the beams in Figure 5.1 (as with all beams examined in this Chapter) lie within a single plane; namely the xy-plane and are assumed to be symmetric about the plane. Furthermore, it will be assumed that all loads act in the plane of the beam and all couples have their moment vector perpendicular to the plane of the beam along the z-axis. Supports of a beam will be denoted by uppercase letters such as A, B and C and the length between supports generally denoted by L. Force and moment reactions will be denoted by R and M with a subscript to indicate the associated support and possibly a horizontal (h) or vertical (v) superscript to indicate a horizontal or vertical direction; for example R_A, R_B, R_A^h, R_A^v and M_A. Alternatively, a popular convention used by several texts (see for example Gere and Timoshenko (1997)) is to indicate a reaction by the use of a slash across the arrow.

5.2.2 Types of Loads

Figure 5.1 illustrated the three main types of loading that can act on beams. Loads P_1 and P_2 in Figure 5.1 are examples of *concentrated* or *point loads*. Case b) of Figure 5.1 illustrates a *distributed load* acting over a length b from the free end of the cantilever beam. Distributed loads can be of arbitrary intensity with the most common form being uniformly distributed as shown in Figure 5.1b). Case c) of Figure 5.1 illustrates an applied *couple* load, M_1, acting at a distance b from support B.

5.2.3 Statically Determinate Beams

The foregoing sections classified beams in terms of their supports and loading. In addition, beams are also categorised as either *statically determinate* or *statically indeterminate*. When a beam is statically determinate we can obtain all reactions, shear forces and bending moments from the free body diagram and equations of equilibrium. However, a statically indeterminate beam is one in which the number of reactions exceeds the number of independent equations of equilibrium. As a result, the reactions of such beams cannot be determined from statics alone and are therefore referred to as statically indeterminate. Statically indeterminate beams will be examined in more detail in §5.7.

Consider now the statically determinate cantilever beam of Figure 5.1b). Because the beam is fully built-in at end A, we require to determine the three unknown reactions R_A^h, R_A^v and M_A at end A. The equation of equilibrium for the horizontal forces is

$$\sum F^h = 0 : R_A^h - P_1 \cos\alpha = 0 \Rightarrow R_A^h = P_1 \cos\alpha \tag{5.1}$$

Similarly, the equation of equilibrium for the vertical forces gives

$$\sum F^v = 0 : R_A^v - P_1 \cos\alpha - Wb = 0 \Rightarrow R_A^v = P_1 \sin\alpha + Wb \tag{5.2}$$

where resultant force (Wb) for the uniformly distributed force W is equal to the area of the rectangular load diagram and acts through the centroid of the rectangle, that is $x = L - b/2$. Finally, the equation of moment equilibrium about end A gives

$$\sum M = 0 : M_A - P_1 a \sin\alpha - Wb\left(L - \frac{b}{2}\right) = 0 \Rightarrow M_A = P_1 a \sin\alpha + Wb\left(L - \frac{b}{2}\right) \tag{5.3}$$

5.3 Shear Force and Bending Moment

Figure 5.2a) illustrates a cantilever beam subject to concentrated loads P_1 and P_2. The applied loads induce stresses and strains within the beam which can be determined by considering the internal forces and couples that act at arbitrary cross-sections of the beam. Case b) of Figure 5.2 illustrates the free body diagram of the cantilever beam with a cut made at a distance $x(0 \leq x \leq a)$ from the free end. The free body is kept in a state of equilibrium due to the action of the stresses on the cut. These stresses are reduced to stress resultants in the form of a resultant *shear force* V_x and resultant *bending moment* M_x at x with both acting in the plane of the beam.

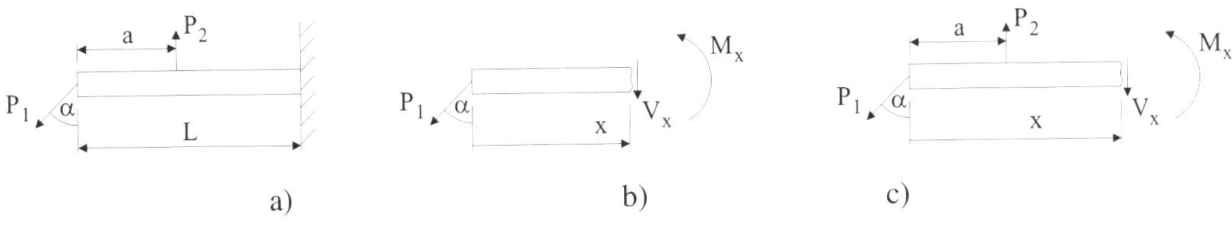

Figure 5.2: A cantilever beam subject to concentrated loads. a) Cantilever beam. b) Beam cut at left hand side of P_2 for $0 < x < a$. c) Beam cut at right hand side of P_2 for $a < x < L$.

Summing forces in the vertical direction and taking moments about the cut for the free body diagram of Figure 5.2b) we have

$$\begin{aligned} \sum F^h &= 0 : V_x + P_1 x \cos\alpha = 0 \Rightarrow V_x = -P_1 \cos\alpha \\ \sum M &= 0 : M_x + P_1 x \cos\alpha = 0 \Rightarrow M_x = -P_1 x \cos\alpha \end{aligned} \tag{5.4}$$

Similarly, for the free body diagram of Figure 5.2c)

$$\begin{aligned} \sum F^v &= 0 : V_x + P_1 x \cos\alpha - P_2 = 0 \Rightarrow V_x = P_2 - P_1 \cos\alpha \\ \sum M &= 0 : M_x + P_1 x \cos\alpha - P_2(x-a) = 0 \Rightarrow M_x = P_2(x-a) - P_1 x \cos\alpha \end{aligned} \tag{5.5}$$

Thus, by taking appropriate cuts through the beam both the shear force and bending moment can be determined via the use of a free body diagram and the equations of equilibrium.

Before proceeding further it is worth considering the sign conventions used for both shear force and bending moment. The sign convention is arbitrary but it is customary to assume that a positive shear force acts to rotate the material clockwise and a positive bending moment acts to compress the material on the upper part of the beam and elongate the material on the lower part. Figure 5.3a) illustrates the two possible sign conventions for both positive and negative shear force and bending moment acting on a small element cut out between two cross-sections. Figure 5.3b) illustrates the exaggerated deformations of the element for both positive and negative shear forces and bending moments.

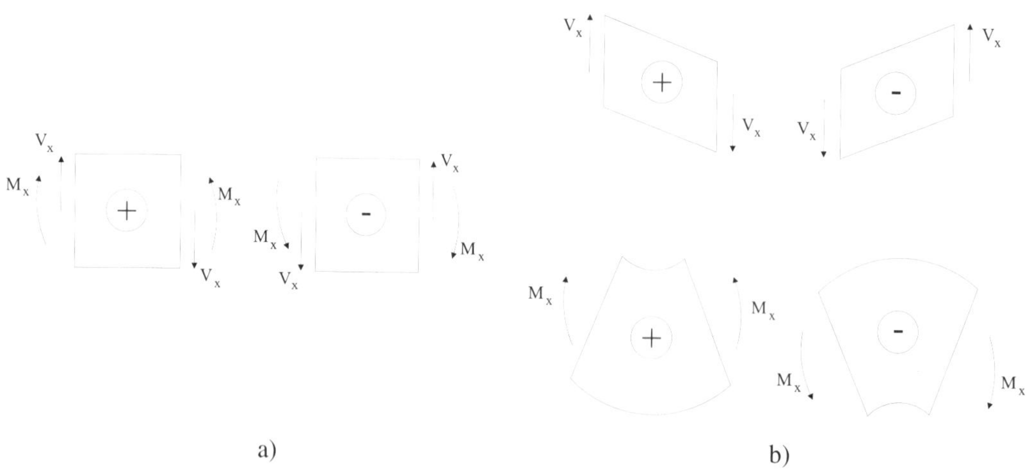

Figure 5.3: Sign conventions for shear force and bending moment. a) Positive and negative shear forces and bending moments. b) Deformations for positive and negative shear forces and bending moments.

We will now derive formulae which relate the applied loading, shear force and bending moment for a differentially small element subject to no applied load, uniformly distributed loading, concentrated loading and couple loading. Figure 5.4a) illustrates a section of a beam with a differentially small element e of length $\mathrm{d}x$ along the x-axis. Considering case b) first in which no external loads are applied along $\mathrm{d}x$. Taking moments about an axis at the left hand side of e we have

$$\sum M = 0 : (M_x + \mathrm{d}M_x) - M_x - (V_x + \mathrm{d}V_x)\,\mathrm{d}x = 0 \Rightarrow \frac{\mathrm{d}M_x}{\mathrm{d}x} = V_x \tag{5.6}$$

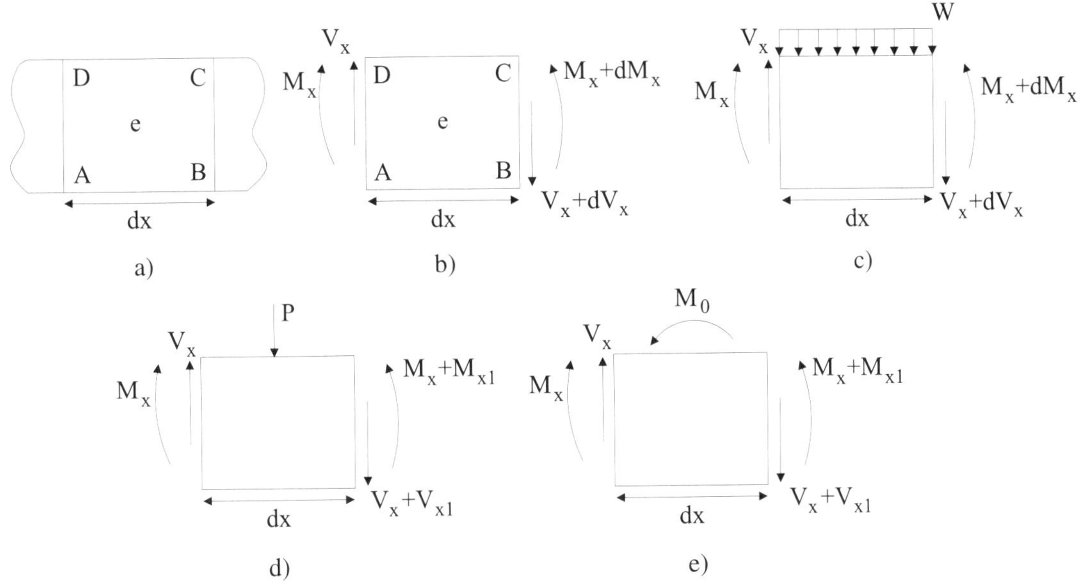

Figure 5.4: Element of a beam. a) Element $e(ABCD)$. b) No external loads along $\mathrm{d}x$. c) Uniformly distributed load W along $\mathrm{d}x$. d) Concentrated load P acting on $\mathrm{d}x$. e) Couple M_0 acting on $\mathrm{d}x$.

where the product of differentials $\mathrm{d}V_x\,\mathrm{d}x$ is assumed to be negligible compared to other terms. Equation (5.6) illustrates that the rate of change of bending moment at a point x is equal to the shear force. Next, consider case c) of Figure 5.4 where a distributed load W acts along $\mathrm{d}x$. From the equilibrium of forces we find

$$\sum F^v = 0 : (V_x + \mathrm{d}V_x) - V_x + W\,\mathrm{d}x = 0 \Rightarrow \frac{\mathrm{d}V_x}{\mathrm{d}x} = -W \tag{5.7}$$

and illustrates that the rate of change of V_x with respect to x is equal to the negative of the uniformly distributed load W. It follows that when $W = 0$ (that is, no loading over $\mathrm{d}x$) then $\mathrm{d}V_x/\,\mathrm{d}x = 0$. Taking moments about the left hand side of e once more

$$\sum M = 0 : (M_x + \mathrm{d}M_x) - M_x - (V_x + \mathrm{d}V_x)\,\mathrm{d}x - W\,\mathrm{d}x\left(\frac{\mathrm{d}x}{2}\right) = 0 \Rightarrow \frac{\mathrm{d}M_x}{\mathrm{d}x} = V_x \tag{5.8}$$

and is seen to be equivalent to (5.6).

Consider now case d) of Figure 5.4 where a concentrated load P acts on $\mathrm{d}x$; noting that the increments in shear force and bending moment may be finite and are therefore denoted by V_{x1} and M_{x1}. From the equilibrium of forces we have

$$\sum F^v = 0 : (V_x + V_{x1}) - V_x + P = 0 \Rightarrow V_{x1} = P \tag{5.9}$$

and illustrates that the shear force experiences an abrupt change as we move from left to right where a concentrated load acts. Taking moments about the left hand side of e we find

$$\begin{aligned} &\sum M = 0 : (M_x + M_{x1}) - M_x - (V_x + V_{x1})\,\mathrm{d}x - P\left(\frac{\mathrm{d}x}{2}\right) = 0 \Rightarrow \\ &M_{x1} = P\left(\frac{\mathrm{d}x}{2}\right) + V_x\,\mathrm{d}x + V_{x1}\,\mathrm{d}x \approx 0 \end{aligned} \tag{5.10}$$

The value of $M_{x1} \approx 0$ since $\mathrm{d}x$ is infinitesimally small and therefore the bending moment does not alter as we move from left to right where a concentrated load acts.

The final case to consider is that of a couple M_0 acting on $\mathrm{d}x$ as shown in Figure 5.4e). Equilibrium of forces in the y-direction illustrates that $V_{x1} = 0$ so that the shear force does not alter at the point of application of a couple. Taking moments about the left hand side of e we find

$$\sum M = 0 : (M_x + M_{x1}) + M_0 - M_x - (V_x + V_{x1})\,\mathrm{d}n = 0 \Rightarrow M_{x1} = -M_0 \qquad (5.11)$$

where the term $(V_x + V_{x1})\,\mathrm{d}x \approx 0$ since $\mathrm{d}x$ is infinitesimally small. Thus, the bending moment decreases by a value of M_0 as we move from left to right where a couple acts.

5.3.1 Shear Force and Bending Moment Diagrams

The cases of Figure 5.4 discussed above will now be illustrated graphically in the form of shear force and bending moment diagrams. By way of illustration consider the simply supported beam of Figure 5.5a). Let us first determine the two unknown reactions R_A and R_B from the equilibrium of forces for the entire beam

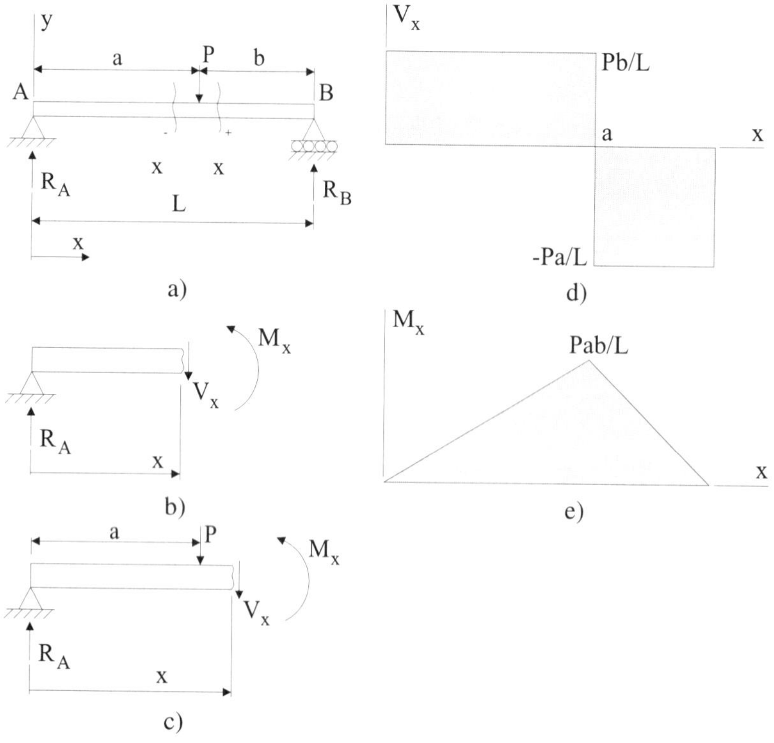

Figure 5.5: Shear force and bending moment diagrams for a simply supported beam. a) Beam subject to a concentrated load P. b) Beam cut to the left of $P(0 < x < a)$. c) Beam cut to right of $P(a < x < L)$. d) Shear force diagram. e) Bending moment diagram.

$$\sum M_A = 0 : R_B L - Pa = 0 \Rightarrow R_B = \frac{Pa}{L}$$
$$\sum M_B = 0 : Pb - R_A L = 0 \Rightarrow R_A = \frac{Pb}{L} \tag{5.12}$$

Insert a cut through the beam to the left of load P and from the force and moment equations of equilibrium for V_x and M_x, Figure 5.5b)

$$V_x = R_A = \frac{Pb}{L}; \quad M_x = R_A x = \frac{Pbx}{L}; \quad 0 < x < a \tag{5.13}$$

If we now cut the beam to the right of load P then V_x and M_x are given by Figure 5.5c)

$$V_x - R_A - P = \frac{-Pa}{L}; \quad M_x = R_A x - P(x-a) = \frac{Pbx}{L} - P(x-a); \quad a < x < L \tag{5.14}$$

noting that $L = a + b$.

Equations (5.13) and (5.14) are plotted in cases d) and e) of Figure 5.5. The shear force is equal to $R_A (= Pb/L)$ at $x = 0$ and remains constant to $x = a$. This follows from (5.7) with $W = 0$ so that $\mathrm{d}V_x/\,\mathrm{d}x = 0$ and V_x is therefore constant. At $x = a$ the shear force V_x changes abruptly from a value of Pb/L to $-Pa/L$ and agrees with (5.9). Along the remaining segment $a < x < L$ the shear force is constant ($\mathrm{d}V_x/\,\mathrm{d}x = 0$) and equal in magnitude to R_B. The maximum absolute value of V_x is equal to $R_A (= Pa/L)$ when $a > b$. Figure 5.5e) illustrates that the bending moment M_x increases from zero at $x = 0$ to a maximum value of Pab/L at $x = a$ in accordance with (5.13). This is in agreement with (5.6) in which the slope $\mathrm{d}M_x/\,\mathrm{d}x$ is constant and equal to V_x. At $x = a$ there is a change in slope of the bending moment diagram since V_x is constant in the interval $(a < x < L)$ but no abrupt change in M_x occurs at the load P. Along the right hand side segment $a < x < L$ the bending moment is again linear and decreases from the maximum value at $x = a$ to zero at $x = L$.

Example 5.1 Shear Force and bending moment for a simply supported beam

Figure 5.6 illustrates a simply supported beam with a uniformly distributed load W acting over the entire length of the beam. Determine the shear force and bending moment in the beam at a distance x from the left hand end A. To determine the unknown reactions R_A and R_B we take moments about both A and B

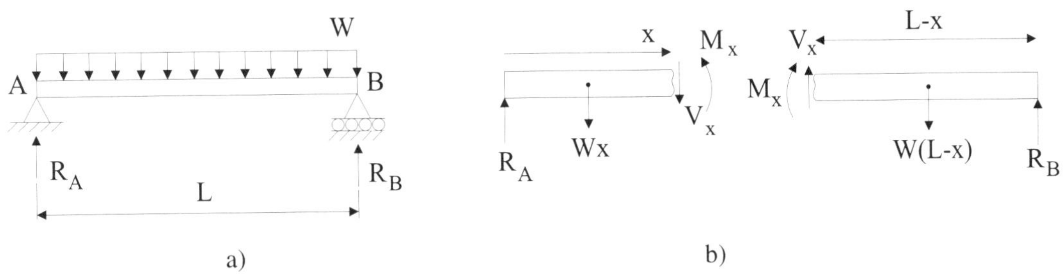

Figure 5.6: Example: Shear Force and bending moment for a simply supported beam. a) Beam subject to a uniformly distributed load W. b) Sections of the beam resulting from a cut at a distance x from A.

for the entire beam of Figure 5.6a)

$$\sum M_A = 0 : R_B L - WL\left(\frac{L}{2}\right) = 0 \Rightarrow R_B = \frac{WL}{2}$$
$$\sum M_B = 0 : WL\left(\frac{L}{2}\right) - R_A L = 0 \Rightarrow R_A = \frac{WL}{2}$$

with $R_A = R_B$ due to the symmetry of loading. Assuming that the unknown shear force V_x and bending moment M_x at the cut for the left hand section are assumed to be positive then the equations of equilibrium give

$$\sum F^v = 0 : V_x + Wx - \frac{WL}{2} = 0 \Rightarrow V_x = W\left(\frac{L}{2} - x\right)$$
$$\sum M = 0 : M_x + Wx\left(\frac{x}{2}\right) - \frac{WL}{2}(x) = 0 \Rightarrow M_x = \frac{Wx}{2}(L - x)$$

Alternatively, the equations of equilibrium for the right hand section give; again adopting a positive convention for V_x and M_x

$$\sum F^v = 0 : V_x + \frac{WL}{2} - W(L - x) = 0 \Rightarrow V_x = W\left(\frac{L}{2} - x\right)$$
$$\sum M = 0 : M_x + W(L - x)\left(\frac{L - x}{2}\right) - \frac{WL}{2}(L - x) = 0 \Rightarrow M_x = \frac{Wx}{2}(L - x)$$

Comparing V_x and M_x for both the left and right hand sections above we observe that they are equivalent since we were consistent in choosing the sign convention for each section.

Example 5.2 Shear Force and bending moment diagrams for a simply supported beam

Draw the shear force and bending moment diagrams for the simply supported beam shown in Figure 5.7a) subject to a uniformly distributed load W acting over the entire length L. Taking moments at A and B the

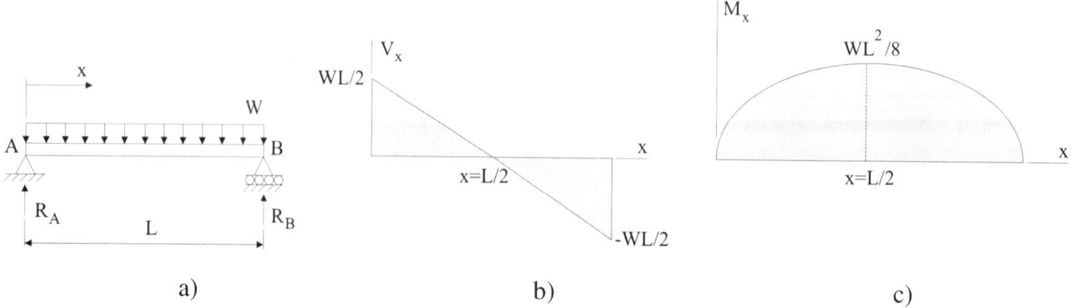

Figure 5.7: Example: Shear Force and bending moment diagrams for a simply supported beam. a) Beam subject to a uniformly distributed load. b) Shear force diagram. c) Bending moment diagram.

reactions R_A and R_B are both found to be equal to $WL/2$. The shear force, V_x, and bending moment,

M_x, at a distance x from A are

$$\sum F^v = 0 : V_x + Wx - R_A = 0 \Rightarrow V_x = R_A - Wx = W\left(\frac{L}{2} - x\right)$$

$$\sum M = 0 : M_x + Wx\left(\frac{x}{2}\right) - R_A x = 0 \Rightarrow M_x = R_A x - \frac{Wx^2}{2} = \frac{Wx}{2}(L - x)$$

with V_x and M_x illustrated graphically in Figure 5.7b) and Figure 5.7c) respectively. The shear force decreases from $R_A (= WL/2)$ at $x = 0$ to zero at $x = L/2$ and to $-R_B(= -WL/2)$ at $x = L$. The bending moment diagram is a parabolic curve and is zero at both $x = 0$ and $x = L$ and attains a maximum at $x = L/2$. From (5.8) $\mathrm{d}M_x/\mathrm{d}x = V_x$ at point x and can be confirmed by differentiating the above M_x

$$\frac{\mathrm{d}M_x}{\mathrm{d}x} = W\left(\frac{L}{2} - x\right)$$

At the maximum value of M_x then $\mathrm{d}M_x/\mathrm{d}x = 0$ and therefore $x = L/2$. Substituting $x = L/2$ into M_x then the maximum value of M_x is $WL^2/8$.

Example 5.3 Shear Force and bending diagrams for a cantilever beam

Draw the shear force and bending moment diagrams for the cantilever beam shown in Figure 5.8a). From

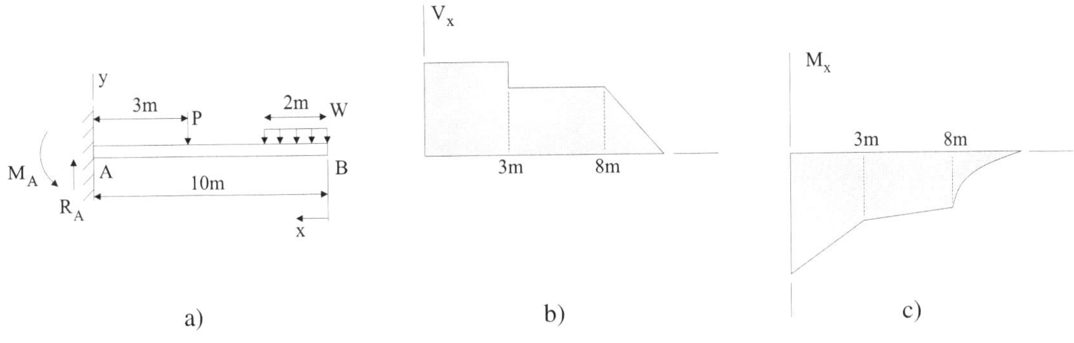

Figure 5.8: Example: Shear Force and bending diagrams for a cantilever beam. a) Beam subject to concentrated and distributed loads. b) Shear force diagram. c) Bending moment diagram.

the equations of equilibrium the shear force, V_x, and bending moment, M_x, for the three intervals i) $0 < x < 2$, ii) $2 < x < 7$ and iii) $7 < x < 10$ are

i) $0 < x < 2$

$$V_x = Wx$$
$$M_x = -\frac{Wx^2}{2}$$

ii) $2 < x < 7$

$$V_x = 2W$$
$$M_x = -2W(x - 1)$$

iii) $7 < x < 10$

$$V_x = P + 2W$$
$$M_x = -2W(x-1) - P(x-7)$$

with V_x and M_x illustrated in Figure 5.8.

5.4 Deflection, Strain and Stress for a Beam Subject to Pure Bending

This section derives two key formulae for the deflection and axial stress of a beam under conditions of plane bending. A state of *pure bending* refers to a beam under a constant bending moment, that is regions in which the shear force is zero, (5.6). Alternatively, *non-uniform bending* refers to the flexure of a beam in the presence of shear forces such that the bending moment varies along the axis of the beam. Firstly, it is necessary to define the so-called *neutral axis*. Examining the exaggerated deformed beam of Figure 5.9a) we observe that the lower section of the beam is in a state of tension whereas the upper section is in a state of compression. Somewhere between the lower and upper sections lies a surface which must experience no stress and is referred to as the neutral surface. The intersection of the xy-plane with the neutral surface defines the neutral axis of the beam and is denoted by the curve ss in Figure 5.9a). The exact location of the neutral axis will be discussed later in §5.4.1 and will be shown to be coincident with the centroidal axis of the beam cross-section.

Consider now the deformation of a differentially small element $\mathrm{d}x$ on the neutral axis of the deformed beam in Figure 5.9a). Since the initial distance $\mathrm{d}x$ remains unchanged between the two lines Oa and Ob, the deformed arc length $\mathrm{d}x$ of the circular arc of radius R is $\mathrm{d}x = R\,\mathrm{d}\theta$ or

$$\frac{1}{R} = \frac{\mathrm{d}\theta}{\mathrm{d}x} \tag{5.15}$$

where $\mathrm{d}\theta/\,\mathrm{d}x$ is the curvature or the reciprocal of the *radius of curvature*, R.

If the original length of fibre ab is L and is a distance y from the neutral axis then

$$\tan\frac{\mathrm{d}\theta}{2} = \frac{L/2}{R-y} \tag{5.16}$$

and re-arranging for L, assuming small displacements and $\tan\,\mathrm{d}\theta \approx \mathrm{d}\theta$ and in view of (5.15)

$$L = (R-y)\,\mathrm{d}\theta = \mathrm{d}x - \frac{y}{R}\,\mathrm{d}x \tag{5.17}$$

If fibre ab is in a state of compression then after deformation the elongation (strictly speaking the contraction) of ab is $L - \mathrm{d}x$ or $-y\,\mathrm{d}x/R$. The axial strain, ε_{xx}, of ab is therefore

$$\varepsilon_{xx} = \frac{\text{elongation}}{\text{original length}} = \frac{-y\,\mathrm{d}x/R}{\mathrm{d}x} = -\frac{y}{R} = y\frac{\mathrm{d}\theta}{\mathrm{d}x} \tag{5.18}$$

and the axial stress, σ_{xx}, is

$$\sigma_{xx} = E\varepsilon_{xx} = -E\frac{y}{R} \tag{5.19}$$

where E is Young's modulus.

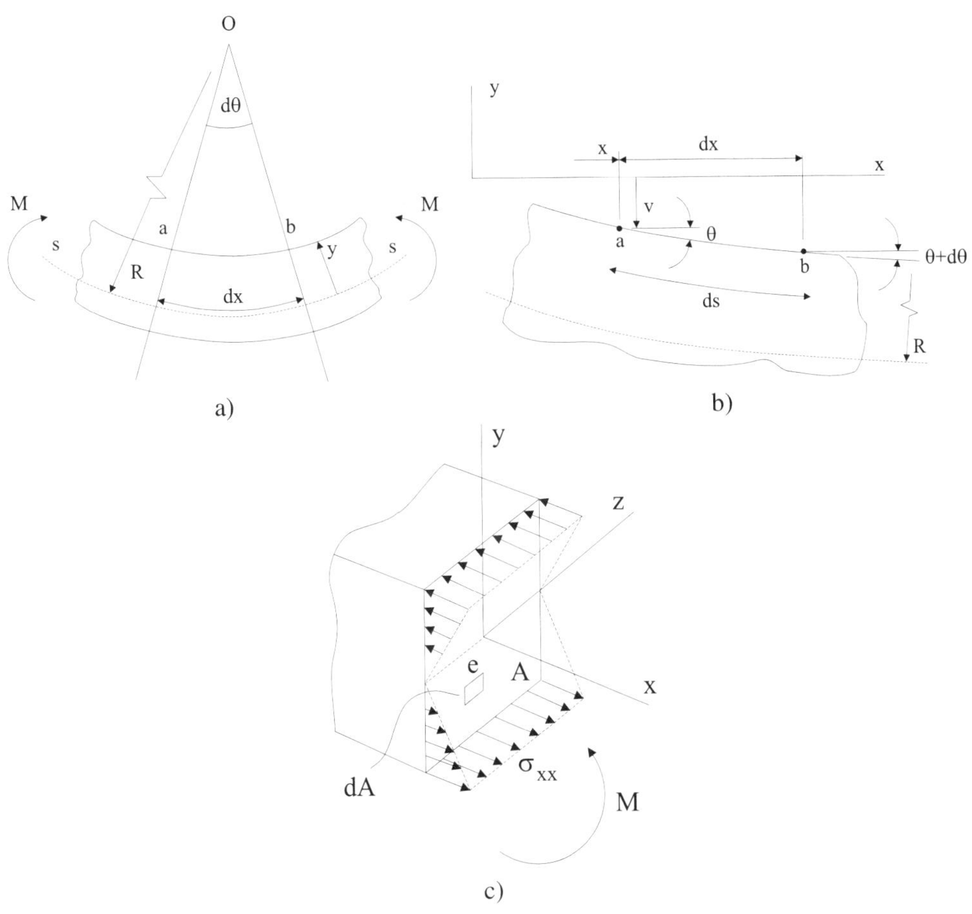

Figure 5.9: Bending of a beam. a) A beam subject to a bending moment M. b) Exaggerated deformation of points a and b on the upper surface of the beam. c) Differentially small element e on a cross-section of the beam.

Consider now the displacement of points a and b in more detail, Figure 5.9b). For small angles $d\theta$ then

$$\tan\theta \approx \tan\, \mathrm{d}\theta = \frac{\mathrm{d}v}{\mathrm{d}x} \tag{5.20}$$

where v is the vertical displacement associated with the y-axis and $\mathrm{d}v/\,\mathrm{d}x$ is therefore the slope of the deflection curve at point x. Similarly, from geometry we find

$$\frac{1}{R} = \frac{\mathrm{d}\theta}{\mathrm{d}x} \tag{5.21}$$

With $\mathrm{d}s = \mathrm{d}x/\cos\theta \approx \mathrm{d}x$ since $\cos\theta \approx 1$ for small θ we again arrive at (5.15). Once more using the fact that $\tan\theta \approx \theta$ for small θ we have from (5.15)

$$\theta = \frac{\mathrm{d}v}{\mathrm{d}x} \tag{5.22}$$

which, upon differentiating with respect to θ gives

$$\frac{\mathrm{d}\theta}{\mathrm{d}x} = \frac{d^2 v}{\mathrm{d}x^2} = \frac{1}{R} \tag{5.23}$$

Let us now derive a relationship between the reciprocal of the radius of curvature and the bending moment. For the differential element e of area $\mathrm{d}A$ at a distance $\mathrm{d}y$ from the neutral axis in Figure 5.9c) then the associated force $\mathrm{d}F_e$ is

$$\mathrm{d}F_e = -\sigma_{xx}\,\mathrm{d}A \tag{5.24}$$

The negative sign is necessary because element e has a positive y value for which the σ_{xx} stress is negative for a positive bending moment. Taking moments about the neutral axis then the moment of element e is

$$\mathrm{d}M_e = \mathrm{d}F_e y = -\sigma_{xx} y\,\mathrm{d}A \tag{5.25}$$

Integrating over the entire cross-section the total bending moment is

$$M = \iint_A \mathrm{d}M_e = -\iint_A \sigma_{xx} y\,\mathrm{d}A \tag{5.26}$$

Substituting σ_{xx} from (5.19)

$$M = \frac{E}{R}\iint_A y^2\,\mathrm{d}A = \frac{EI}{R} \tag{5.27}$$

where the integral over area A is seen to be the second moment of area $I = I_x$, §2.3. Re-arranging (5.27) for R we finally have the *moment-curvature equation*

$$\frac{1}{R} = \frac{M}{EI} \tag{5.28}$$

Collecting the above results (5.23) and (5.28) we have the simple beam formulae which relate R, θ, v and M

$$\frac{1}{R} = \frac{\mathrm{d}\theta}{\mathrm{d}x} = \frac{d^2 v}{\mathrm{d}x^2} = \frac{M}{EI} \tag{5.29}$$

Furthermore, substituting $E/R = M/I$ from (5.28) into (5.19) we find the bending stress formula

$$\sigma_{xx} = -\frac{My}{I} \tag{5.30}$$

The maximum tensile and compressive stresses occur at sections farthest away from the neutral axis. Let c_1 and c_2 denote the distances from the neutral axis to the beam extremities in the positive and negative y-directions respectively. From (5.30) the corresponding maximum stresses on c_1 and c_2 are

$$\sigma_1 = -\frac{Mc_1}{I} = -\frac{M}{S_1}; \quad \sigma_2 = -\frac{Mc_2}{I} = -\frac{M}{S_2} \tag{5.31}$$

where $S_1 (= I/c_1)$ and $S_2 (= I/c_2)$ are referred to as the *section moduli* and are frequently tabulated in designer handbooks of beam properties.

A note should be made of equations (5.19), (5.29) and (5.30) since they are the main equations of the theory of beams subject to pure bending.

5.4.1 Position of the Neutral Axis

Earlier we introduced the definition of the neutral axis but did not define exactly its location. Consider the cross-section of the beam shown in Figure 5.9c) with element e at a distance y from the neutral axis which is assumed to be coincident with the x-axis. The force acting on e of area $\mathrm{d}A$ is $\sigma_{xx}\,\mathrm{d}A$ and since the resultant force acting on the cross-section is zero then for equilibrium to be satisfied we have

$$\iint_A \sigma_{xx}\,\mathrm{d}A = 0 \tag{5.32}$$

Upon substituting into (5.19) for σ_{xx} then

$$-\iint_A \frac{Ey}{R}\,\mathrm{d}A = 0 \tag{5.33}$$

and since E and $1/R$ are non-zero constants for an arbitrary cross-section of a deformed beam then we have

$$\iint_A y\,\mathrm{d}A = 0 \tag{5.34}$$

Equation (5.34) informs us that the first moment of area with respect to the x-axis, Q_x, is equal to zero; see (2.2). From (2.3) the y component of the centroid, y_c, is equal to $y_c = Q_x/A$ and is therefore equal to zero since $Q_x = 0$. Thus, we conclude that the neutral axis passes through the centroid of the beam cross-section.

5.5 Shear Stresses in Beams

The previous section examined the bending of beams subject to pure bending only in which shear forces are not present. However, §5.3 illustrated that the majority of beams are subject to loads that induce both bending moments and shear forces, that is non-uniform bending. These non-zero shear forces have associated shear stresses which we will now derive. Consider the differential element e of the beam shown in Figure 5.10 with arbitrary cross-section. From (5.30) the normal stresses on sections AD and BC are, assuming tensile stresses

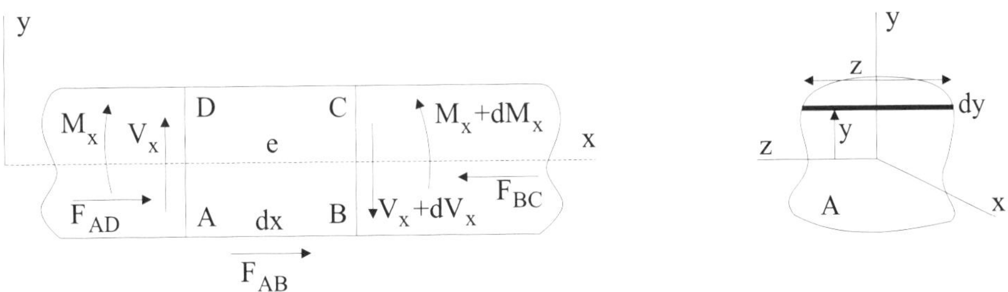

Figure 5.10: Section of a beam with differential element e.

$$\sigma_{AD} = \frac{M_x y}{I}; \quad \sigma_{BC} = \frac{(M_x + \mathrm{d}M_x)y}{I} \tag{5.35}$$

with corresponding forces

$$F_{AD} = \iint_A \sigma_{AD}\,\mathrm{d}A = \iint_A \frac{M_x y}{I}\,\mathrm{d}A; \quad F_{BC} = \iint_A \sigma_{BC}\,\mathrm{d}A = \iint \frac{(M_x + \mathrm{d}M_x)y}{I}\,\mathrm{d}A \quad (5.36)$$

To maintain element e in a state of equilibrium then the force on the lower face of e is now given by

$$F_{AB} = F_{BC} - F_{AD} = \iint_A \frac{(\mathrm{d}M_x)y}{I}\,\mathrm{d}A = \frac{\mathrm{d}M_x}{I}\iint_A y\,\mathrm{d}A \quad (5.37)$$

where the right hand side is obtained by noting that $\mathrm{d}M$ and I are constant at an arbitrary cross-section. Assuming that the shear stresses, τ, are uniformly distributed across the width of the beam then F_{AB} is

$$F_{AB} = \tau z\,\mathrm{d}x \quad (5.38)$$

where z is the width of the beam in the yz-plane at section AB. Equating (5.37) and (5.38) and re-arranging for τ

$$\tau = \frac{\mathrm{d}M_x}{\mathrm{d}x}\left(\frac{1}{Iz}\right)\iint_A y\,\mathrm{d}A = \frac{V_x}{Iz}\iint_A y\,\mathrm{d}A \quad (5.39)$$

noting that $V_x = \mathrm{d}M_x/\mathrm{d}x$ from (5.6). The integral in (5.39) is equal to the first moment of area Q_x, (2.2), so that τ reduces to

$$\tau = \frac{V_x Q_x}{Iz} \quad (5.40)$$

and is referred to as the *beam shear formula*.

Example 5.4 Determination of the shear stress for a beam of rectangular cross-section

Determine the shear stress for the beam of rectangular cross-section shown in Figure 5.11a).

The first moment of area is given by (2.2) with $\mathrm{d}A = b\,\mathrm{d}y$

$$Q_x = \iint_A y\,\mathrm{d}A = b\int_{y_1}^{h/2} y\,\mathrm{d}y = b\left[\frac{y^2}{2}\right]_{y_1}^{h/2} = \frac{b}{2}\left(\frac{h^2}{4} - y_1^2\right)$$

Substituting Q_x into (5.40) then τ at distance y_1 from the neutral axis is given by

$$\tau = \frac{V_x}{2I}\left(\frac{h^2}{4} - y_1^2\right)$$

and is seen to attain a maximum value at the neutral axis $(y = 0)$ equal to

$$\tau_{max} = \frac{V_x h^2}{8I} = \frac{3V_x}{2A}$$

where $A(= bh)$ is the cross-sectional area and $I(= bh^3/12)$ is from (2.5). The maximum shear stress τ_{max} is 1.5 times the average shear stress (V_x/A). The parabolic variation of τ across the rectangular beam section is shown in Figure 5.11b).

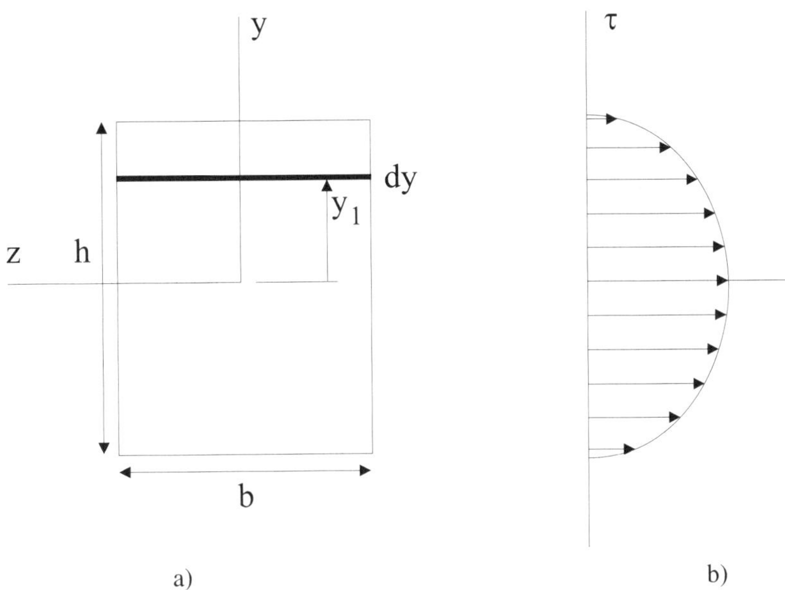

Figure 5.11: Example: Shear stresses in a beam. a) A beam of rectangular cross-section. b) Parabolic distribution of shear stress across the beam section.

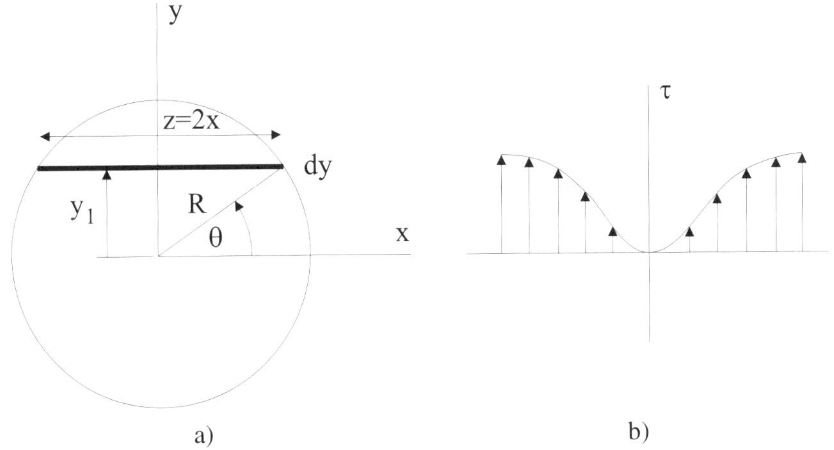

Figure 5.12: Example: Shear stresses in a beam. a) A beam of circular cross-section. b) Variation of shear stress across the beam section.

Example 5.5 Determination of the shear stress for a beam of circular cross-section

Determine the shear stress for the beam of circular cross-section shown in Figure 5.12a). The first moment of area is given by (2.2) with $\mathrm{d}A = 2x\,\mathrm{d}y = 2\sqrt{R^2 - y^2}\,\mathrm{d}y$

$$Q_x = \iint_A y\,\mathrm{d}A = 2\int_{y_1}^{R} y\sqrt{R^2 - y^2}\,\mathrm{d}y = -\frac{2}{3}\left[\left(R^2 - y^2\right)^{3/2}\right]_{y_1}^{R} = \frac{2}{3}(R^2 - y_1^2)^{3/2} = \frac{2x}{3} = \frac{z^3}{12}$$

Substituting Q_x into (5.40) then τ at distance y_1 from the neutral axis is, with $I = \pi R^4/4$ and $z = 2x = 2R\cos\theta$

$$\tau = \frac{V_x z^2}{3\pi R^4} = \frac{4V_x \cos^2\theta}{3\pi R^2}$$

and increases from zero at $\theta = \pi/2$ to a maximum at $\theta = 0°$ on the neutral axis of

$$\tau_{max} = \frac{4V_x}{3\pi R^2} = \frac{4V_x}{3A}$$

and is seen to equal 4/3 times the average shear stress.

5.6 Deflections of Beams

In §5.4 we derived the second order differential equation of the deflection of a beam (5.29)

$$v'' = \frac{M}{EI} \tag{5.41}$$

where a prime denotes differentiation with respect to x. The first integration of (5.41) produces the slope $v(= \mathrm{d}v/\mathrm{d}x)$ and a second integration of (5.41) produces the deflection v. To illustrate determining the deflection and slope (angle of rotation) from (5.41) consider the cantilever beam shown in Figure 5.13a) subject to a uniformly distributed load. From the equations of equilibrium applied to the beam of Figure 5.13a) the unknown reactions R_A and M_A are found to be

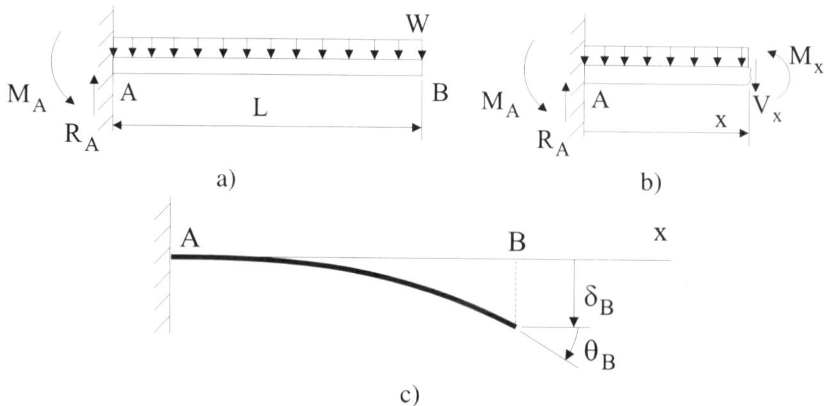

Figure 5.13: Deflection of a beam. a) Cantilever beam subject to a uniformly distributed load. b) Free body diagram of the beam cut at a distance x. c) Deflection of the beam.

$$R_A = WL; \quad M_A = \frac{WL^2}{2} \tag{5.42}$$

Considering the free body diagram of Figure 5.13b) the shear force, V_x, and bending moment, M_x, at a distance x from the built-in support A are

$$\begin{aligned} &\sum F^v = 0 : V_x + Wx - R_A = 0 \Rightarrow V_x = W(L-x) \\ &\sum M = 0 : M_x + M_A + Wx\left(\frac{x}{2}\right) - R_A x \Rightarrow M_x = WLx - \frac{WL^2}{2} - \frac{Wx^2}{2} \end{aligned} \tag{5.43}$$

Substituting M_x into (5.41) we have

$$EIv'' = WLx - \frac{WL^2}{2} - \frac{Wx^2}{2} \tag{5.44}$$

Integrating with respect to x then the slope of the beam is

$$EIv' = \frac{WLx^2}{2} - \frac{WL^2x}{2} - \frac{Wx^3}{6} + C_1 \tag{5.45}$$

The constant of integration C_1 is determined from the boundary condition that the slope is zero at end A

$$v'(0) = 0 \tag{5.46}$$

Applying the boundary condition to (5.45) we find that $C_1 = 0$

$$v' = -\frac{Wx}{6EI}(3L^2 - 3Lx + x^2) \tag{5.47}$$

The slope is zero at $x = 0$ and negative (that is, clockwise) over the entire length of the beam. The maximum slope occurs at the free end ($x = L$) of the beam and is given by, Figure 5.13c)

$$\theta_B = -v'(L) = \frac{WL^3}{6EI} \tag{5.48}$$

Integrating the slope (5.45) yields the deflection v

$$EIv = \frac{WLx^3}{6} - \frac{WL^2x^2}{4} - \frac{Wx^4}{24} + C_2 \tag{5.49}$$

The second constant of integration C_2 is determined from the boundary condition that the deflection is zero at end A

$$v(0) = 0 \tag{5.50}$$

Applying this boundary condition to (5.49) we obtain $C_2 = 0$ so that the deflection is given by

$$v = -\frac{Wx^2}{24EI}(6L^2 - 4Lx + x^2) \tag{5.51}$$

The deflection is zero at $x = 0$ and negative elsewhere. The maximum deflection occurs at the free end of the beam and is given by, Figure 5.13c)

$$\delta_B = -v(L) = \frac{WL^4}{8EI} \tag{5.52}$$

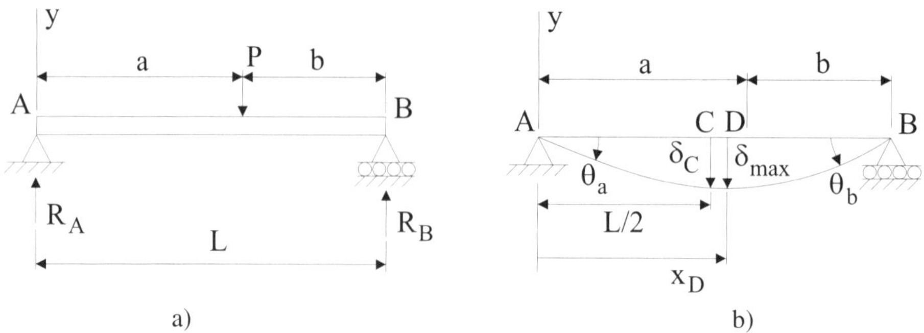

Figure 5.14: Example: a) A simply supported beam subject to a concentrated load. b) Deflection of the beam.

Example 5.6 Deflection of a simply supported beam

Figure 5.14a) illustrates a simply supported beam subject to a concentrated load P. Determine expressions for the deflection and slope throughout the beam.

The reactions have previously been determined in (5.12) to be $R_A = Pb/L$ and $R_B = Pa/L$ and the bending moments for the left and right sections of the beam are given by (5.13) and (5.14)

$$M = \frac{Pbx}{L} \quad (0 \le x \le a)$$
$$M = \frac{Pbx}{L} - P(x-a) \quad (a \le x \le L)$$

Substituting M into (5.41) then

$$EIv'' = \frac{Pbx}{L} \quad (0 \le x \le a)$$
$$EIv'' = \frac{Pbx}{L} - P(x-a) \quad (a \le x \le L)$$

Integrating these two equations then the slopes are

$$\begin{aligned} EIv' &= \frac{Pbx^2}{2L} + C_1 \quad (0 \le x \le a) \\ EIv' &= \frac{Pbx^2}{2L} - \frac{P(x-a)^2}{2} + C_2 \quad (a \le x \le L) \end{aligned} \tag{5.53}$$

Performing a second integration then the deflections are

$$\begin{aligned} EIv &= \frac{Pbx^3}{6L} + C_1 x + C_3 \quad (0 \le x \le a) \\ EIv &= \frac{Pbx^3}{6L} - \frac{P(x-a)^3}{6} + C_2 x + C_4 \quad (a \le x \le L) \end{aligned} \tag{5.54}$$

The four constants of integration C_1, C_2, C_3 and C_4 will now be determined from the two boundary conditions of $v = 0$ at $x = 0$ and $v = 0$ at $x = L$ and the two continuity conditions that the deflections and slopes for both the left and right sections of the beam at $x = a$ must be equivalent. Continuity of slopes at $x = a$ yields, (5.53)

$$\frac{Pba^2}{2} + C_1 = \frac{Pba^2}{2L} + C_2 \Rightarrow C_1 = C_2$$

Continuity of deflections at $x = a$ yields, (5.54)

$$\frac{Pba^3}{6L} + C_1 a + C_3 = \frac{Pba^3}{6L} + C_2 a + C_4 \Rightarrow C_3 = C_4$$

Applying the boundary condition $v = 0$ at $x = 0$ to $(5.54)_1$ leads to $C_3 = 0$ and $C_4 = 0$. Similarly, applying the boundary condition $v = 0$ at $x = L$ to $(5.54)_2$ leads to

$$C_1 = C_2 = -\frac{Pb(L^2 - b^2)}{6L}$$

Substituting C_1 and C_2 into (5.53) gives the final expressions for the slope of the beam

$$\begin{aligned} v' &= -\frac{Pb}{6LEI}(L^2 - b^2 - 3x^2) \quad (0 \le x \le a) \\ v' &= -\frac{Pb}{6LEI}(L^2 - b^2 - 3x^2) - \frac{P(x-a)^2}{2LEI} \quad (a \le x \le L) \end{aligned} \tag{5.55}$$

The slopes at ends A and B are

$$\theta_A = -v'(0) = \frac{Pab(L+b)}{6LEI}; \quad \theta_B = v'(L) = \frac{Pab(L+a)}{6LEI}$$

with θ_A and θ_B acting clockwise and anticlockwise, respectively, as shown in Figure 5.14b).

Substituting all constants of integration into (5.54) gives the final expressions for the deflection of the beam

$$\begin{aligned} v &= -\frac{Pbx}{6LEI}(L^2 - b^2 - x^2) \quad (0 \le x \le a) \\ v &= -\frac{Pbx}{6LEI}(L^2 - b^2 - 3x^2) - \frac{P(x-a)^3}{6EI} \quad (a \le x \le L) \end{aligned} \tag{5.56}$$

When $a > b$ the maximum deflection, δ_{max}, occurs at point D in the range $(0 \le x \le a)$ where the slope is horizontal. Substituting $v' = 0$ into $(5.55)_1$ and solving for x we obtain

$$x_D = \sqrt{\frac{L^2 - b^2}{3}} \quad (a \ge b) \tag{5.57}$$

noting that δ_{max} only occurs at $x = L/2$ when $a = b$ and P acts at the mid-span. Substituting x_D from (5.57) into $(5.56)_1$ gives δ_{max}

$$\delta_{max} = -v(x_D) = \frac{Pb}{9\sqrt{3}LEI}(L^2 - b^2)^{3/2} \quad (a \ge b)$$

The deflection at the mid-span is obtained by substituting $x = L/2$ into $(5.56)_1$

$$\delta_C = -v\left(\frac{L}{2}\right) = \frac{Pb(3L^2 - 4b^2)}{48EI} \quad (a \ge b)$$

Finally, for the special case of P acting at the mid-span then $a = b = L/2$ and the above formulae are found to reduce to

$$\begin{aligned} v' &= -\frac{P}{16EI}(L^2 - 4x^2) \quad (0 \le x \le L/2) \\ v &= -\frac{Px}{48EI}(3L^2 - 4x^2) \quad (0 \le x \le L/2) \\ \theta_A = \theta_B &= \frac{PL^2}{16EI}; \quad \delta_{max} = \delta_C = \frac{PL^3}{48EI} \end{aligned}$$

5.6.1 Macaulay's Method for the Deflections of Beams

The previous section has demonstrated how the deflections of a beam can be determined by repeated integration of the second order flexure formula, (5.41). Example 5.6 illustrated that discontinuities occur in the bending moment expression as x passes a concentrated load. A similar problem is encountered for couples and distributed loads which do not extend across the entire length of the beam. The solution of such problems can be lengthy since each discontinuity in loading introduces a new particular integral and requires the calculation of a new set of constants of integration. The simply supported beam subject to a single concentrated load of Example 5.6 required two constants of integration for the interval $0 \leq x \leq a$ and a further two constants for the interval $a \leq x \leq L$.

The most powerful method for such problems is the use of the Laplace transform. A simpler method, specific to beams, is *Macaulay's method* which replaces the buckling moment expression, M, on the right hand side of the flexure formula by the step function $M\{x - a\}$ in which a is a point at which a discontinuity occurs. Macaulay's method can be applied to both simply supported and built-in beams subject to the following conventions:

i) When forming the bending moment expression at a cut $x - x$ take the origin of coordinates at the left hand side of the beam and insert cut $x - x$ between the last load and the right hand end of the beam.

ii) Establish the expression for the bending moment at $x - x$ writing each discontinuity as $M\{x - a\}$ without expanding any brackets. Concentrated moments appear in the form $M\{x - a\}^0$.

iii) During integration bracket terms such as $\{x - a\}$ must not be broken so that the integration of $\{x - a\}$ is of the form $(1/2)\{x - a\}^2$.

iv) In the case of uniformly distributed loadings it may be necessary to extend and counterbalance loadings to the right hand end of the beam.

For example, Figure 5.15 illustrates a simply supported beam subject to n concentrated loads $P_1, P_2, \ldots,$ P_n acting at $a_1, a_2, \ldots, a_n$ from the left hand end of the beam. The unknown reactions are determined as usual from the equations of equilibrium and the flexure formula is of the form

$$EIv'' = R_A x - P_1\{x - a\} - \cdots - P_n\{x - a_n\} \tag{5.58}$$

Integrating with respect to x and satisfying condition iii) above we have the slope

$$EIv' = \frac{1}{2}R_A x^2 - \frac{1}{2}P_1\{x - a\}^2 - \cdots - \frac{1}{2}P_n\{x - a_n\}^2 + C_1 \tag{5.59}$$

followed by a further integration to give the displacement

$$EIv = \frac{1}{6}R_A x^3 - \frac{1}{6}P_1\{x - a\}^3 - \cdots - \frac{1}{6}P_n\{x - a\}^3 + C_1 x + C_2 \tag{5.60}$$

Expressions (5.59) and (5.60) define the slope and displacement along the entire length of the beam and the two constants C_1 and C_2 are determined from the boundary conditions at the ends of the beam. Note that Macaulay's method reduces the number of constants of integration to just two irrespective of the number of discontinuous loads.

Example 5.7 Deflection of a simply supported beam using Macaulay's method

In the present example we will use Macaulay's method to determine an expression for the deflection of the simply supported beam illustrated in Figure 5.14a). From the equations of equilibrium the two

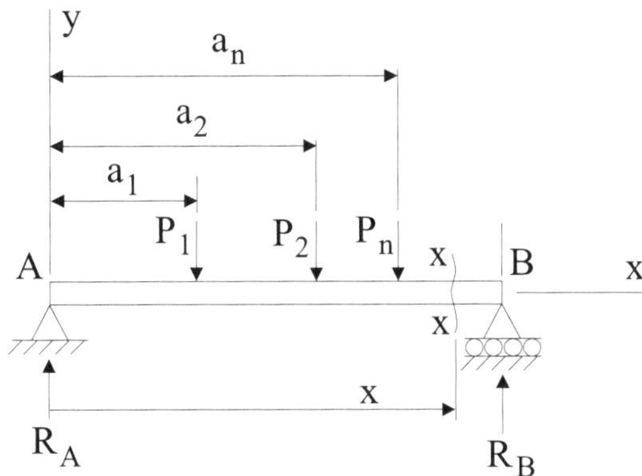

Figure 5.15: A simply supported beam subject to n concentrated loads $P_1, P_2, \ldots, P_n$.

unknown reactions are found to be $R_A = Pb/L$ and $R_B = Pa/L$. Inserting a cut between load P and the right hand end of the beam then the bending moment is found and upon substitution into the flexure formula (5.41), or from (5.58) we have

$$EIv'' = \frac{Pbx}{L} - P\{x-a\}$$

Integrating twice

$$\begin{aligned} EIv' &= \frac{Pbx^2}{2L} - \frac{P}{2}\{x-a\}^2 + C_1 \\ EIv &= \frac{Pbx^3}{6L} - \frac{P}{6}\{x-a\}^3 + C_1x + C_2 \end{aligned}$$

To determine the unknown constants of integration C_1 and C_2 we apply the two boundary conditions that $v = 0$ at $x = 0$ and $x = L$. Applying the first of these

$$0 = -\frac{P}{6}\{x-a\}^3 + C_2$$

From condition iii) of Macaulay's method the term in brackets is ignored since it is negative so that $C_2 = 0$. Applying the second boundary condition gives

$$0 = \frac{PbL^2}{6} - \frac{P}{6}\{L-a\}^3 + C_1L \Rightarrow C_1 = \frac{P}{6L}\{L-a\}^3 - \frac{PbL}{6}$$

Substituting C_1 and C_2 into v then the deflection of the beam is given by

$$EIv = \frac{Pbx^3}{6L} - \frac{P}{6}\{x-a\}^3 + \frac{Px}{6L}\{x-a\}^3 - \frac{PbLx}{6}$$

Example 5.8 Deflection of a simply supported beam with distributed loading using Macaulay's method

We will use the present example to illustrate the use of Macaulay's method to simply supported beams with distributed loads. The following illustrates the four possible cases which are correspondingly illustrated in Figure 5.16:

i) **W extends from $x = a$ to the right hand end of the beam.** Because the distributed load W extends over the entire length of the beam and to the right hand end of the beam then the bending moment at cut $x - x$ is

$$M_{xx} + Wx\frac{x}{2} - R_A x = 0 \Rightarrow M_{xx} = R_A x - \frac{Wx^2}{2}$$

ii) **W extends from $x = a$ to the right hand end of the beam.** As in the previous case, the distributed load extends to the right hand end of the beam so that M_{xx} follows immediately

$$M_{xx} + W\{x - a\}\frac{\{x - a\}}{2} - R_A x = 0 \Rightarrow M_{xx} = R_A x + \frac{W\{x - a\}^2}{2}$$

iii) **W extends from the left hand end of the beam to $x = a$.** When the distributed load does not extend to the right hand end of the beam then it is necessary to extend and counterbalance W to the right hand end. The bending moment at cut $x - x$ is

$$M_{xx} + Wx\frac{x}{2} - R_A x - W\{x - a\}\frac{\{x - a\}}{2} = 0 \Rightarrow M_{xx} = R_A x + \frac{W\{x - a\}^2}{2} - \frac{Wx^2}{2}$$

iv) **W extends from $x = a$ to $x = b$.** As in the previous case it is necessary to extend and counterbalance W to the right hand end of the beam

$$M_{xx} + W\{x - a\}\frac{\{x - a\}}{2} - W\{x - b\}\frac{\{x - b\}}{2} - R_A x = 0$$
$$\Rightarrow M_{xx} = R_A x + \frac{\{x - b\}^2}{2} - \frac{W\{x - a\}^2}{2}$$

5.7 Statically Indeterminate Beams

This section examines *statically indeterminate* beams in which the number of reactions exceeds the number of independent equations of equilibrium. Thus, it is not possible to determine all reactions from equilibrium considerations alone. Figure 5.17 illustrates two statically indeterminate beams. Case a) of Figure 5.17 illustrates a propped cantilever beam with the three unknown reactions R_A, R_B and M_A with no unknown horizontal reaction at A since the concentrated load acts vertically. Since only two of the three reactions can be determined from the equations of equilibrium then the beam is statically indeterminate. The built-in beam of case b) is similarly indeterminate because it has four unknown reactions at supports A and B but only two independent equations of equilibrium can be formed. The number of unknown reactions in excess of the number of independent equations of equilibrium is referred to as the *degree of static indeterminacy* with the excess reactions called the *static redundants*. Thus, the propped cantilever in Figure 5.17a) is statically indeterminate to the first degree while the built-in beam of Figure 5.17b) is statically indeterminate to the second degree.

We will see shortly that a suitable solution of redundant reactions leads to a solution of statically indeterminate beams. For example, reaction R_B of the propped cantilever of Figure 5.17a) can be chosen as the redundant reaction. Figure 5.18a) illustrates the cantilever beam with the support at B removed. Since the reaction R_B was in excess of those required to maintain equilibrium then removal of R_B leaves the beam in a state of equilibrium. The beam remaining after the removal of a redundant reaction is referred to as the *released beam* and must be both statically determinate and capable of carrying the prescribed loading. Alternatively, the moment reaction M_A could have been selected as the redundant reaction as shown in Figure 5.18b).

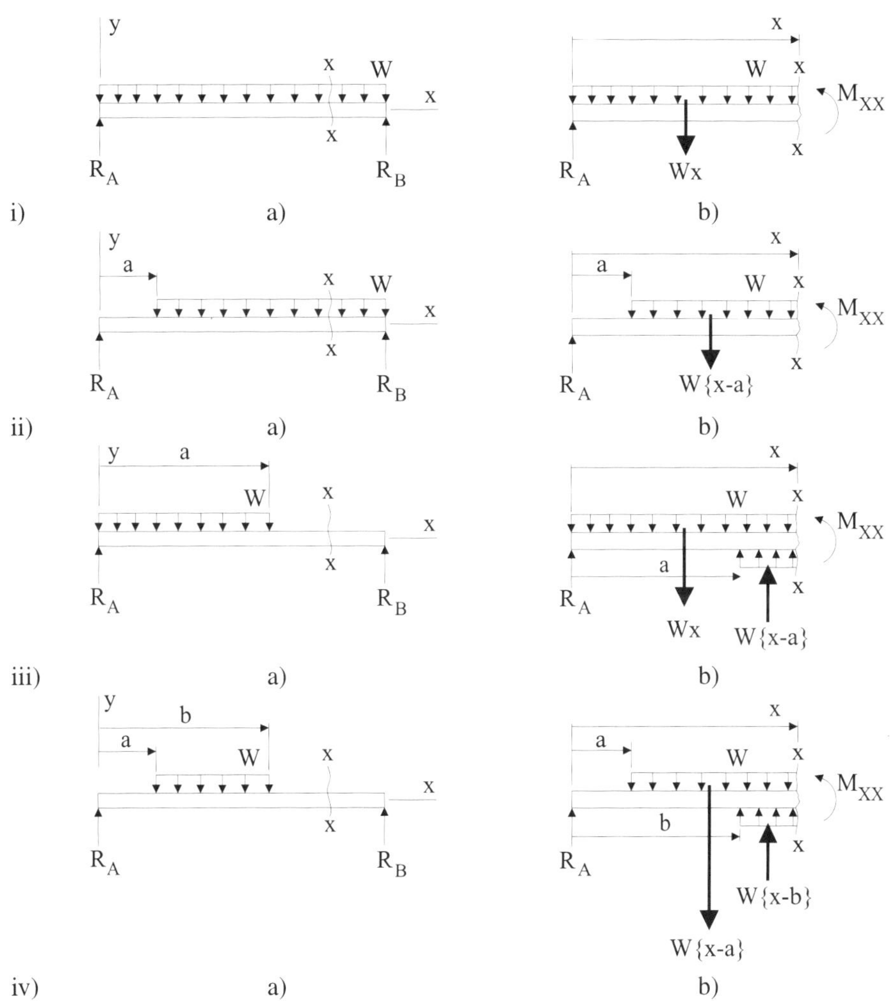

Figure 5.16: Example: Simply supported beams with distributed loads. i) W extends over the entire length of the beam. ii) W extends from $x = a$ to the right hand end of the beam. iii) W extends from the left hand end of the beam to $x = a$. iv) W extends from $x = a$ to $x = b$. Each of the beams in case b) illustrate the corresponding beam in case a) cut at section $x - x$.

In the following two sub-sections we will examine two methods for analysing statically indeterminate beams. The first method makes use of the differential equations of the deflection curve. The second method of superposition is a more general method in which the equations of equilibrium are supplemented with the compatibility equations and force-displacement equations.

5.7.1 Analysis of Statically Indeterminate Beams via the Differential Equations of the Deflection Curve

Let us illustrate the use of the differential equations of the deflection curve by considering the propped cantilever beam of Figure 5.17a). It was noted earlier that this beam is statically indeterminate to the first

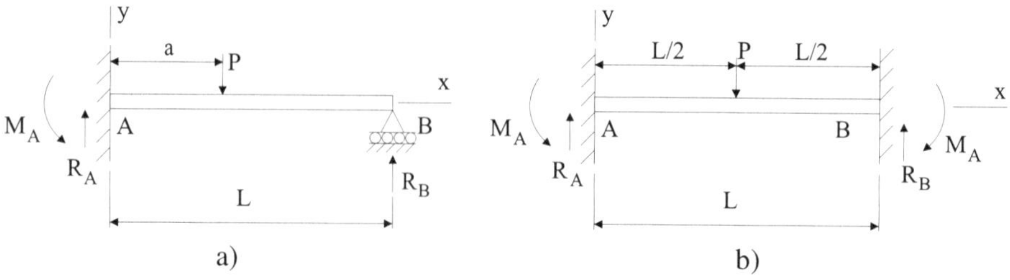

Figure 5.17: Statically indeterminate beams. a) Propped cantilever beam. b) Built-in beam.

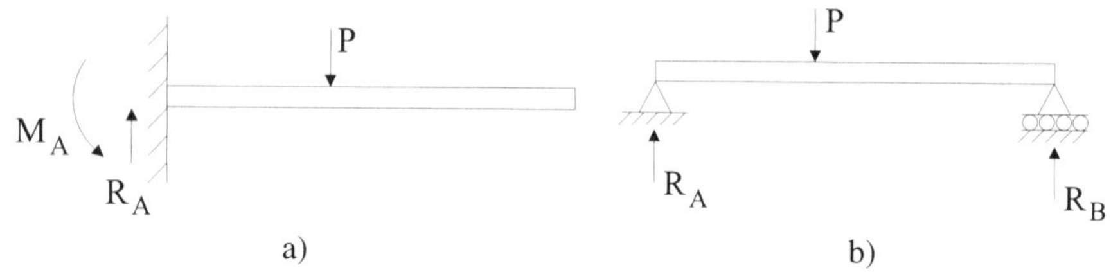

Figure 5.18: Released beams for the propped cantilever of Figure 5.17a). a) Released beam with R_B chosen as the redundant reaction. b) Released beam with M_A chosen as the redundant reaction.

degree and we will select the reaction R_B as the redundant reaction. From the equations of equilibrium we have

$$\begin{aligned} R_A + R_B - P = 0 &\Rightarrow R_A = P - R_B \\ M_A + R_B L - Pa = 0 &\Rightarrow M_A = Pa - R_B L \end{aligned} \tag{5.61}$$

where reactions R_A and M_A are expressed in terms of the redundant reaction R_B. Inserting a cut at a distance $x(a < x < L)$ from the left hand end then the bending moment is, using Macaulay's method

$$M_{xx} + M_A + P\{x-a\} - R_A x = 0 \Rightarrow M_{xx} = R_A x - M_A - P\{x-a\} \tag{5.62}$$

Substituting R_A and M_A from (5.61) into (5.62) then we can express M_{xx} in terms of the redundant reaction

$$M_{xx} = Px - R_B x - Pa + R_B L - P\{x-a\} \tag{5.63}$$

From the second order differential equation of the deflection curve, (5.41), we have

$$EIv'' = Px - R_B x - Pa + R_B L - P\{x-a\} \tag{5.64}$$

and performing two integrations

$$\begin{aligned} EIv' &= \frac{Px^2}{2} - \frac{R_B x^2}{2} - Pax + R_B Lx - \frac{P\{x-a\}^2}{2} + C_1 \\ EIv &= \frac{Px^3}{6} - \frac{R_B x^3}{6} - \frac{Pax^2}{2} + \frac{R_B L x^2}{2} - \frac{P\{x-a\}^3}{6} + C_1 x + C_2 \end{aligned} \tag{5.65}$$

where C_1 and C_2 are constants of integration. Using the boundary conditions $v' = 0$ at $x = 0$ and $v = 0$ at $x = 0$ reveals that $C_1 = 0$ and $C_2 = 0$. The boundary condition $v = 0$ at $x = L$ gives the redundant reaction

$$R_B = \frac{P\{x-a\}^3}{2L^3} - \frac{3Pa}{2L} - \frac{P}{2} \tag{5.66}$$

Substituting R_B into (5.61) gives the remaining reactions R_A and M_A

$$R_A = \frac{3P}{2} - \frac{P\{x-a\}^3}{2L^3} + \frac{3Pa}{2L}; \quad M_A = \frac{5Pa}{2} - \frac{P\{x-a\}^3}{2L^2} + \frac{PL}{2} \tag{5.67}$$

Having found all unknown reactions and constants of integration then the beam problem has been solved.

Example 5.9 Deflection of a statically indeterminate beam using the differential equations of the deflection curve

In this example we will examine the built-in beam of Figure 5.17b) subject to a concentrated load P at the mid-span of the beam. The beam has four unknown reactions but has only two independent equations of equilibrium and is therefore statically indeterminate to the second degree. From the equations of equilibrium we have

$$R_A + R_B - P = 0 \Rightarrow R_A + R_B = P$$

and due to the symmetry of the loading then

$$R_A = R_B = \frac{P}{2}; \quad M_A = M_B$$

Thus, the remaining unknown reactions are the moments M_A and M_B and we will select M_A as the redundant reaction. Inserting a cut at a distance $x(L/2 < x < L)$ from the left hand end of the beam then the bending moment is

$$M_{xx} + M_A + P\{x - L/2\} - R_A x = 0 \Rightarrow M_{xx} = \frac{Px}{2} - M_A - P\{x - L/2\}$$

Substituting M_{xx} into the flexure formula (5.41)

$$EIv'' = \frac{Px}{2} - M_A - P\{x - L/2\}$$

and integrating twice

$$EIv' = \frac{Px^2}{4} - M_A x - \frac{P}{2}\{x - L/2\}^2 + C_1$$

$$EIv = \frac{Px^3}{12} - \frac{M_A x^2}{2} - \frac{P}{6}\{x - L/2\}^3 + C_1 x + C_2$$

where C_1 and C_2 are constants of integration. From the boundary conditions $v' = 0$ at $x = 0$ and $v = 0$ at $x = 0$ we find that $C_1 = 0$ and $C_2 = 0$. The boundary condition $v = 0$ at $x = L$ gives the redundant reaction

$$M_A = \frac{PL}{8}$$

Substituting M_A into v then we have finally

$$EIv = \frac{Px^3}{12} - \frac{PLx^2}{16} - \frac{P\{x - L/2\}^3}{6}$$

Noting that the term in curly brackets is ignored for x in the range $0 \leq x \leq L/2$ then v reduces to

$$v = -\frac{Px^2}{48EI}(3L - 4x); \quad 0 \leq x \leq \frac{L}{2}$$

5.7.2 Analysis of Statically Indeterminate Beams via the Superposition Method

In this section we examine the application of the *method of superposition* to statically indeterminate beams. The method begins by determining the degree of static indeterminacy and selecting suitable redundant reactions. In an analogous manner to the use of the deflection curve, this is then followed by writing the equations of equilibrium that express the unknown reactions in terms of the redundant reactions. The original loading and redundant reactions are then applied separately to the released beam and the associated deflections are superimposed. The deflections of the released beam are found from tabulated values or by using previously discussed techniques for statically determinate beams since the released beam is required to be statically determinate. These load-deflection relations are referred to as the *force-displacement relations*. The superimposed deflections must agree with the deflections of the original beam which will be either zero or have prescribed values at the points of restraint. This is satisfied via the *equations of compatibility* which ensure that the deflections of the released structure at the points where restraints were removed agree with the deflections of the original beam.

An example will help illustrate the above outlined method. Consider the built-in beam shown in Figure 5.19a) which was previously examined in Example 5.9. The beam has four unknown reactions but only two independent equations of equilibrium, so that the beam is statically indeterminate to the second degree. From the equations of equilibrium for the entire beam and due to the symmetry of loading we have

$$R_A = R_B = \frac{P}{2}; \quad M_A = M_B \tag{5.68}$$

We will select M_A and M_B to be the redundant reactions with the released beam shown in Figure 5.19b). The concentrated load P and redundant reactions M_A and M_B are now applied to the released beam as shown in cases b), c) and d) of Figure 5.19 with corresponding deflections and slopes denoted by subscripts 1, 2 and 3. Since the slopes at the ends A and B of the original beam are zero then we have the following two equations of compatibility

$$\theta_A = \theta_{A1} - \theta_{A2} - \theta_{A3} = 0; \quad \theta_B = \theta_{B1} - \theta_{B2} - \theta_{B3} = 0; \tag{5.69}$$

where the minus sign indicates an upward deflection of the beam for cases c) and d).

The force-displacement relations give the slopes at the ends of the beam

$$\theta_{A1} = \theta_{B1} = \frac{PL^2}{16EI} \tag{5.70}$$

and can be determined from tabulated values or from an analysis of the statically determinate released beam of Figure 5.19b). Similarly, the slopes due to the redundant reaction M_A are, Figure 5.19c)

$$\theta_{A2} = \frac{M_A L}{3EI}; \quad \theta_{B2} = \frac{M_A L}{6EI} \tag{5.71}$$

and for the redundant reaction M_B, Figure 5.19d)

$$\theta_{A3} = \frac{M_B L}{6EI}; \quad \theta_{B3} = \frac{M_B L}{3EI} \tag{5.72}$$

Substituting the slopes (5.70), (5.71) and (5.72) into the compatibility equation (5.69) we arrive at the following two simultaneous equations

$$\begin{aligned} \frac{M_A L}{3EI} + \frac{M_B L}{6EI} &= \frac{PL^2}{16EI} \\ \frac{M_A L}{6EI} + \frac{M_B L}{3EI} &= \frac{PL^2}{16EI} \end{aligned} \tag{5.73}$$

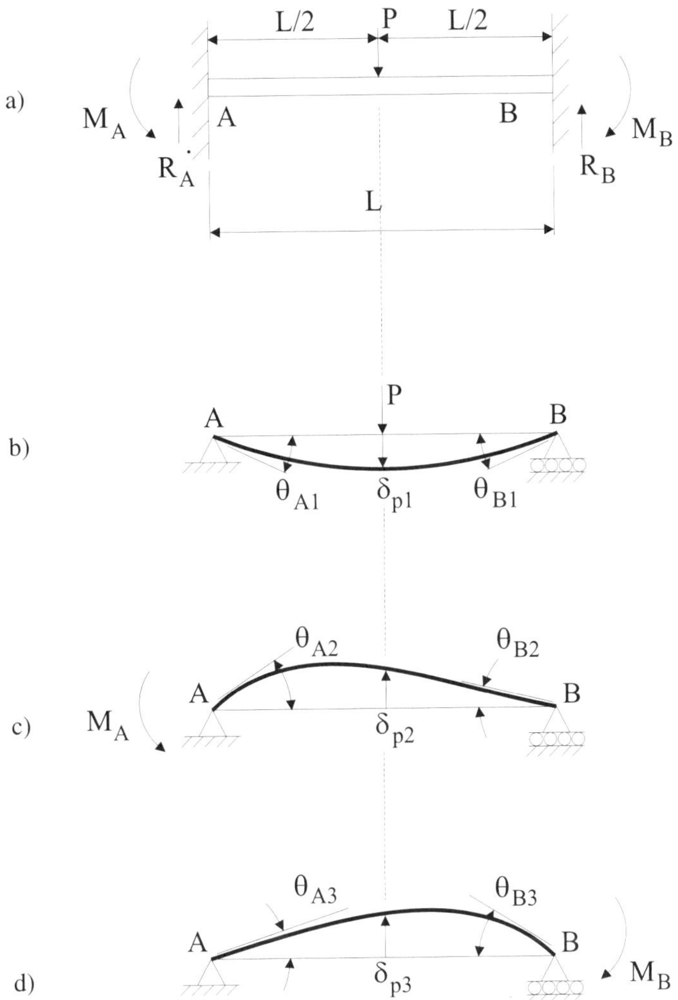

Figure 5.19: Analysis of a built-in beam using the method of superposition. a) The built-in beam subject to a mid-span load. b) The released beam with redundant reactions M_A and M_B and load P applied. c) The released beam with M_A applied. d) The released beam with M_B applied.

Solving these equations for the redundants we have

$$M_A = M_B = \frac{PL}{8} \tag{5.74}$$

All the reactions have now been determined.

To determine the deflection at the point of the concentrated load P we use the superposition of deflections to give the compatibility equation

$$\delta_p = \delta_{p1} - \delta_{p2} - \delta_{p3} \tag{5.75}$$

From tabulated values

$$\delta_{p1} = \frac{PL^3}{48EI}; \quad \delta_{p2} = \frac{M_A L^2}{16EI}; \quad \delta_{p3} = \frac{M_B L^2}{16EI} \tag{5.76}$$

Substituting for M_A and M_B from (5.74) then δ_{p2} and δ_{p3} are

$$\delta_{p2} = \delta_{p3} = \frac{PL^3}{128EI} \tag{5.77}$$

Finally, the compatibility equation (5.75) gives the deflection at load P

$$\delta_p = \frac{PL^3}{192EI} \tag{5.78}$$

5.8 Shear Centre

In this section we examine the shear centre of the cross-section of beams, with particular emphasis on thin-walled open cross-sections. The shear centre is a point in the plane of the cross-section through which the resultant of the transverse shearing stresses passes. A transverse load applied on a beam will produce bending only if the load acts through the shear centre. However, if the applied load acts at a point other than the shear centre then both bending and twisting of the beam will occur. Thus, the location of the shear centre is particularly relevant to thin-walled open sections which offer little resistance to twisting. The shear centre for a cross-section with one axis of symmetry (singly symmetric) lies on the axis of symmetry, Figure 5.20a). If a beam has two axes of symmetry (doubly symmetric) then the shear centre is coincident with the centroid, Figure 5.20b). It follows that doubly symmetric beams have high torsional rigidity and the effects of twisting can generally be neglected for such sections. If a beam has no axes of symmetry (unsymmetric) then the shear centre will not be coincident with the centroid, Figure 5.20c).

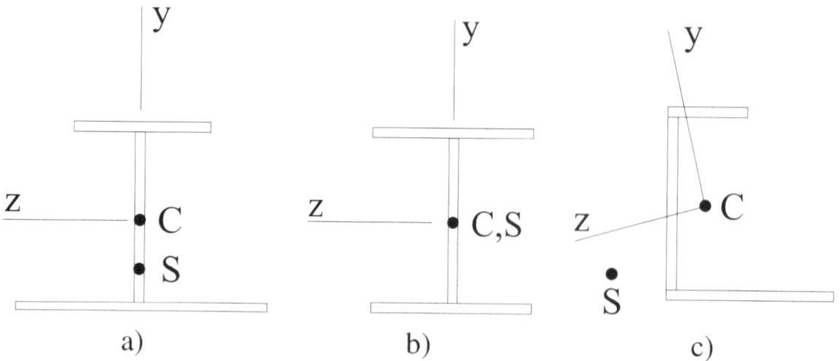

Figure 5.20: Open cross-section beams with centroid C and shear centre S. a) Singly symmetric. b) Doubly symmetric beam. c) Unsymmetric beam.

Figure 5.21 illustrates a thin-walled open section with the y and z axes denoting the principal centroidal axes of the cross-section and centroid C. To determine the shear centre, S, we begin by determining the shear stresses acting throughout the wall thickness using the beam shear formula (5.40)

$$\tau_{xy} = \frac{V_y Q_z}{I_z t} \tag{5.79}$$

and are assumed to be uniformly distributed over the wall thickness. The shear force acting on the cross-section in the y-direction is denoted by V_y, I_y is the second moment of area with respect to the neutral

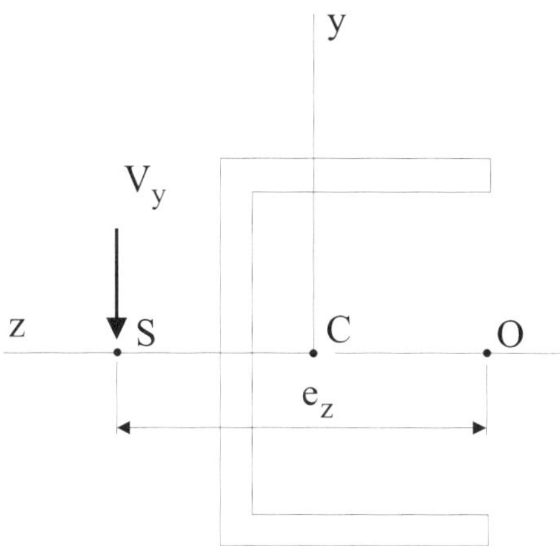

Figure 5.21: Open cross-section with centroid C and shear centre S.

axis, t is the thickness of the beam at the position where the shear stress is to be determined and Q_z is the first moment of area of the cross-sectional area outside the location where the shear stress is to be determined. The moment M_x of the shear stresses with respect to a point O is then obtained. The external moment due to the applied shear force V_y about O is equal to $V_y e_z$ so that e_z is therefore given by

$$e_z = \frac{M_x}{V_y} \tag{5.80}$$

We will illustrate the determination of the shear centre for thin-walled channel and semi-circular cross-sections in the following two examples.

Example 5.10 Shear centre for a thin-walled open channel cross-section

Figure 5.22 illustrates a singly symmetric open channel cross-section. The section is subject to a shear force V_y acting parallel to the y-axis. The force V_y gives rise to shear stresses acting in the web and flanges of the channel section as shown in Figure 5.22b). To determine the shear stress in the flanges from (5.79) we first need to determine Q_z. By considering an elemental strip of thickness $\mathrm{d}s$ at a mean distance of $h/2$ from the z-axis and distance s from the end of the flange then Q_z is, (2.2)

$$Q_z = \iint_A z\,\mathrm{d}A = \int_0^b \frac{h}{2}(t_f\,\mathrm{d}s) = \frac{sht_f}{2}$$

Substituting Q_z into (5.79) then the shear stress in the flanges is

$$\tau_{xy} = \frac{V_y Q_z}{I_z t_f} = \frac{V_y}{I_z t_f}\left(\frac{sht_f}{2}\right) = \frac{shV_y}{2I_z}$$

and is seen to vary from zero at the end of the flange to a maximum value at the flange-web intersection of

$$\tau_{xy}(max) = \frac{bhV_y}{2I_z}$$

The shear stress τ_{xy} at the top of the web is obtained in an analogous manner but with t_f replaced by t_w

$$\tau_{xz} = \frac{V_y Q_z}{I_z t_w} = \frac{bht_f V_y}{2I_z t_w}$$

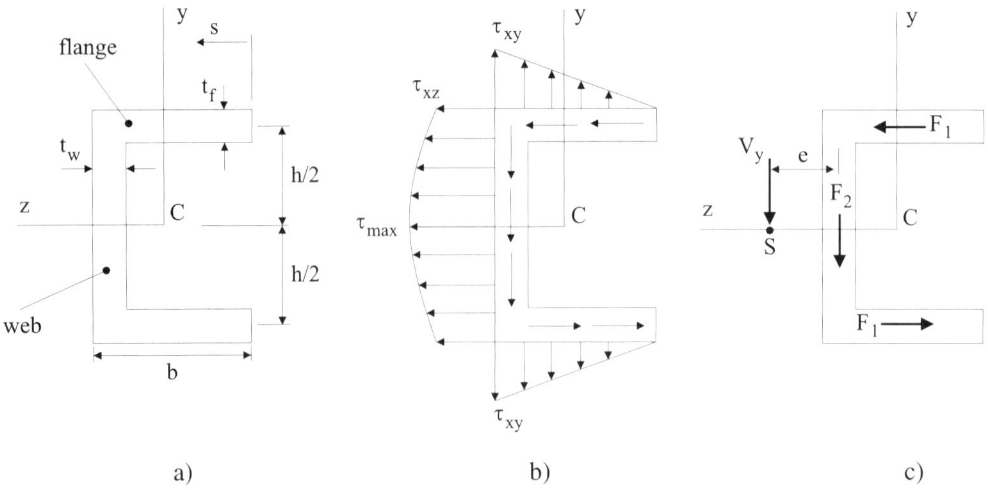

Figure 5.22: Example: Shear centre for a thin-walled channel section. a) Beam cross-section. b) Variation of shear stresses throughout the cross-section. c) Resultant cross-sectional forces and the applied shear force V_y acting at a distance e from the channel web.

To determine the maximum shear stress in the web assume that the maximum shear stress, $\tau_{xz}(max)$, occurs on a plane at a distance r from the $z-z$ neutral axis. The first moment of area is

$$Q_z = \frac{bht_f}{2} + \left(\frac{h}{2} - r\right) t_w \left(\frac{h/2 + r}{2}\right)$$

However, since the section is symmetric with respect to the z-axis then $r = 0$ and Q_z reduces to

$$Q_z = \frac{bht_f}{2} + \frac{h^2 t_w}{8}$$

Thus, the maximum shear stress in the web is

$$\tau_{xy}(max) = \frac{V_y Q_z}{I_z t_w} = \frac{hV_y}{2I_z}\left(\frac{bt_f}{t_w} + \frac{h}{4}\right)$$

The variation of τ_{xy} and τ_{xz} throughout the cross-section are shown in Figure 5.22b).

The resultant shear force, F_1, in each flange is found by summing τ_{xy} across b

$$F_1 = \left(\frac{\tau_{xy} h}{2}\right) t_f = \frac{hb^2 t_f V_y}{4I_z}$$

Since the vertical shear force transmitted by each flange is negligible and can be ignored then the resultant force in the web, F_2, must be equal to the applied shear force V_y

$$F_2 = V_y$$

The resultant forces acting in the section must be statically equivalent to the applied force so that the section is in a state of equilibrium. Therefore, the moment of V_y is equivalent to the moment of the flange and web resultant forces about an arbitrary point O. Selecting the shear centre as the centre of moment then for equilibrium to be satisfied

$$2F_1\frac{h}{2} - F_2 e = 0$$

and substituting F_1 and F_2 and re-arranging for e we find that the position of the shear centre is

$$e = \frac{b^2h^2t_f}{4I_z}$$

With I_z given by

$$I_z = \frac{bh^2t_f}{2} + \frac{h^3t_w}{12}$$

then e is

$$e = \frac{3b^2t_f}{ht_w + 6bt_f}$$

When the wall-thickness of the channel section is constant $(t_f = t_w = t)$ then e reduces to

$$e = \frac{3b^2}{h + 6b} = \frac{b}{2 + (h/3b)}$$

and is seen to be independent of t and varies from 0 to $b/2$ depending on the value of $h/3b$.

When the cross-section is subject to an applied load which acts through the shear centre then the beam will be subject to bending only. However, if the applied load acts through some point other than the shear centre then the beam will be subject to both bending and twisting.

Example 5.11 Shear centre for a thin-walled open semi-circular cross-section

In this second example we will determine the shear centre for the thin-walled semi-circular section shown in Figure 5.23. The section is of mean radius r and wall-thickness t. Since the section has one axis of symmetry then it follows that the shear centre, S, lies on the z-axis. Measuring θ from line OA the first moment of area, Q_z, between point A and a point at an angle θ is

$$Q_z = \iint_A y\,\mathrm{d}A = \int_0^\theta (r\cos\phi)(tr\,\mathrm{d}\phi) = r^2t\sin\theta$$

From (5.79) the shear stress in the section is, with $I_z(= \pi r^3t/2)$ given by (4.18)

$$\tau = \frac{V_yQ_z}{I_zt} = \frac{V_y(r^2t\sin\theta)}{(\pi r^3t/2)t} = \frac{2\sin\theta V_y}{\pi rt}$$

illustrating that $\tau = 0$ at $\theta = 0°$ and $\theta = \pi$ and $\tau = 2V_y/\pi rt$ at $\theta = \pi/2$.

To determine the shear centre, e, from (5.80) we need to determine the moment M_x of shear stresses about an arbitrary point, which we choose to be the centre O. The elemental moment M_x at radius r and angle ϕ is

$$\mathrm{d}M_x = r(\tau\,\mathrm{d}A)$$

and with τ and $\mathrm{d}A$ given by

$$\tau = \frac{2\sin\phi V_y}{\pi r t}; \qquad \mathrm{d}A = rt\,\mathrm{d}\phi$$

then M_x is

$$M_x = \int_0^\pi \mathrm{d}M_x = \frac{2rV_y}{\pi}\int_0^\pi \sin\phi\,\mathrm{d}\phi = \frac{4rV_y}{\pi}$$

and from (5.80) we find the shear centre to be located at

$$e = \frac{M_x}{V_y} = \frac{4r}{\pi}$$

and is equal to twice the distance of the centroid C from the centre O, that is $OS = 2OC$.

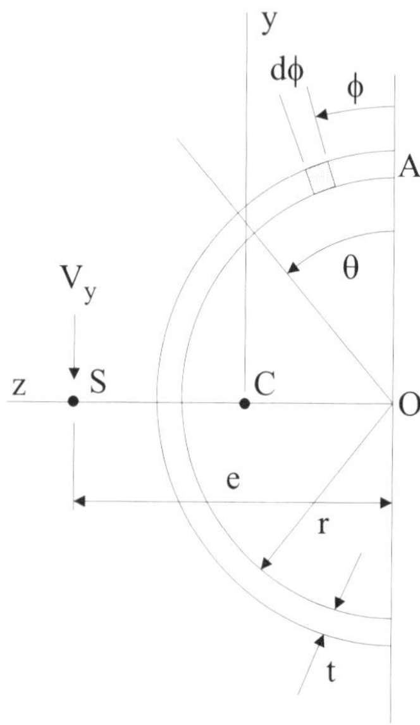

Figure 5.23: Example: Shear centre for a thin-walled semi-circular section.

5.9 Conclusion

The theory of beams plays a key role in the disciplines of mechanical and civil engineering. It is important to develop a good understanding of beam theory for the following two reasons: i) beams are widely used in structural engineering and ii) beam theory is often considered as a core topic in several *Strengths of Materials* courses to help familiarise engineers with the use of the fundamental parameters such as force, bending moment, stress and strain. This Chapter has introduced the different types of beams, their associated loading and static determinacy. The determination of both the shear force and bending moment at given points on a beam and their graphical illustration via the shear force and bending moment diagrams have been examined. Next we derived all of the key deflection, strain and stress formulae for a beam subject to pure bending and the beam shear formula for a beam that is subject to shear forces. Section 6 examined the determination of the deflection curve of a beam by performing successive integrations of the flexure formula. Statically indeterminate beams were then examined and two different methods of solution discussed. The Chapter concluded with a discussion on the determination of the position of the shear centre for beams with thin-walled open sections.

5.10 References and Further Reading

◦ Case, J., Chilver, A. H. and Ross, C. T. F. (1999) *Strength of Materials and Structures*, 4th Ed., Arnold, London.

◦ Gere, J. M. and Timoshenko, S. P. (1997) *Mechanics of Materials*, PWS Publishers, Boston, MA.

5.11 Exercises

5.1 Draw the shear force and bending moment diagrams for the simply supported beam shown in Figure 5.24.

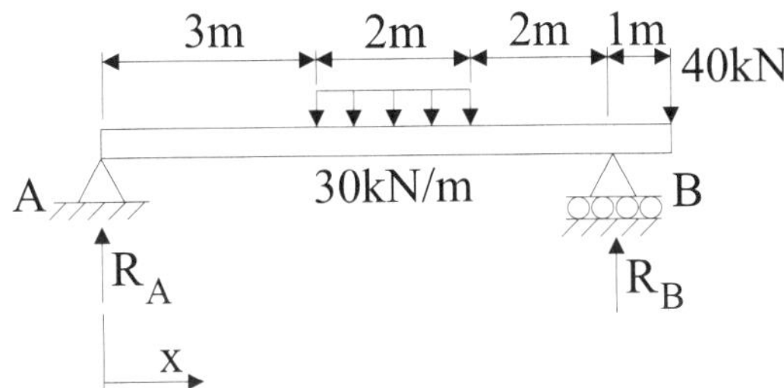

Figure 5.24: Exercise: A simply supported beam subject to concentrated and distributed loadings.

5.2 A simply supported beam, Figure 5.25, is subject to a uniformly distributed load of 5kN/m over the entire length of the beam. Determine the maximum tensile and compressive stresses due to bending.

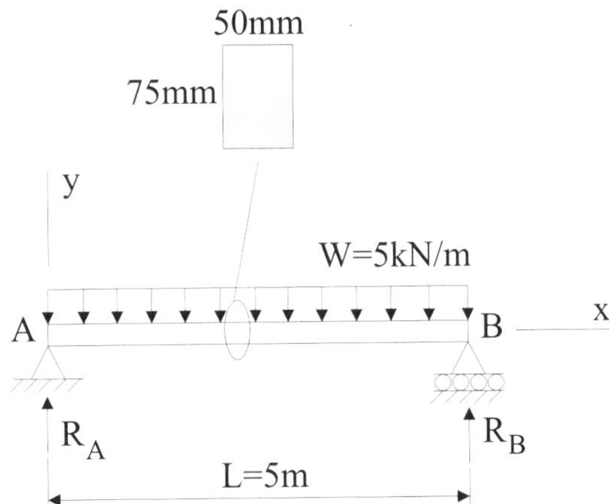

Figure 5.25: Exercise: A simply supported beam subject to a uniformly distributed load of 5kN/m.

5.3 For the simply supported beam of Exercise 5.2 determine the maximum and minimum values of the shear stress.

5.4 Determine expressions for both the slope and deflection of the simply supported beam shown in Figure 5.26 which is subject to a linearly varying distributed load of maximum value W at support B. In addition, determine the maximum deflection and the point at which it acts.

5.5 Determine an expression for the deflection of the beam shown in Figure 5.27a) using Macaulay's method.

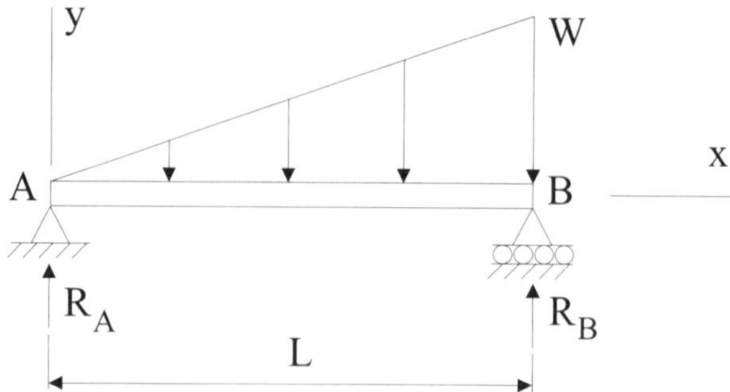

Figure 5.26: Exercise: Simply supported beam subject to a linearly varying distributed load.

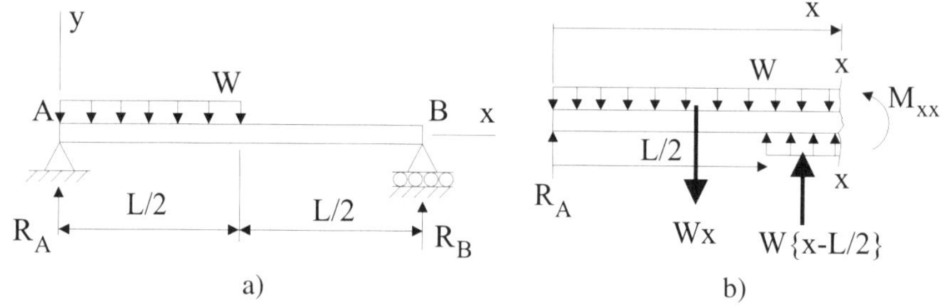

Figure 5.27: Exercise: A simply supported beam. a) Beam subject to a uniformly distributed load acting over one half the length of the beam. b) The beam cut at a distance x $(L/2 < x < L)$ from the left hand end of the beam with the uniformly distributed loading extended and counterbalanced to the right hand end of the beam.

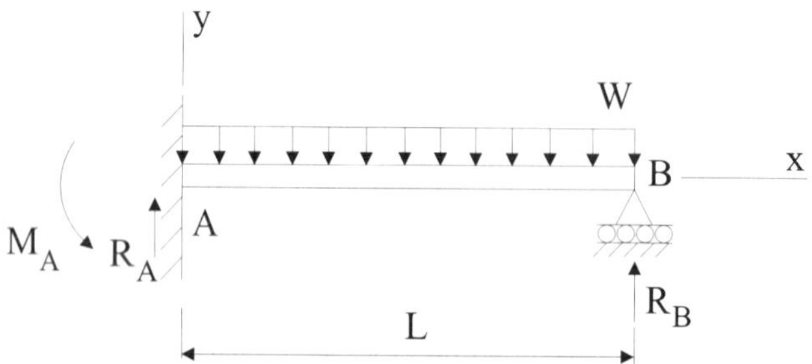

Figure 5.28: Exercise: Propped cantilever subject to a uniformly distributed load.

5.6 Determine an expression for the deflection of the propped cantilever beam shown in Figure 5.28 using the differential equations of the deflection curve.

5.7 Determine the unknown reactions R_A, R_B and M_A for the propped cantilever beam shown in Figure 5.28 using the superposition method.

Chapter 6

Struts

6.1 Introduction

The present Chapter examines the buckling of *struts* or *columns*. A strut is a long slender member which is loaded axially in compression. If a strut is sufficiently *slender* then it may fail due to buckling rather than failing due to the material of the strut. Struts occur frequently in both structural and mechanical engineering applications. For example, struts are used as vertical supports for construction, as spreaders in hoisting equipment and piston rods can be idealised as struts pin-jointed at the crankshaft and built-in and constrained at the piston.

The Chapter begins by considering struts with pin-jointed ends. Expressions for both the critical load and buckling shape of a strut are derived for the fundamental buckling mode and higher buckling modes. The radius of gyration, slenderness ratio and critical stress are then discussed. Section 6.4 examines struts with different end conditions and derives expressions for both the fundamental critical load and buckling shape for each set of end conditions. Pin-jointed struts are then examined further with eccentric loading. The Chapter concludes by discussing semi-empirical formulae which have been proposed for short struts in which Euler's curve is invalid.

6.2 Struts with Pin-Jointed Ends

To begin our examination of struts let us consider the slender strut with pin-jointed ends shown in Figure 6.1a). The column is assumed initially to be straight and of length L, of constant cross-sectional area A and of linear elastic material. The two ends, O and N, of the strut are pin-jointed such that the strut can only experience axial loading. End O is fully restrained whereas end N is free to move along the x-axis. The origin of coordinates is centred at end O with the x-axis taken to coincide with the initial position of the strut, the y-axis perpendicular to the strut and with the z-axis out of the paper. Thus, the coordinate system is equivalent to that used for beams previously discussed in Chapter 5.

The axial load P is applied at end N and acts along the length of the strut. For small values of P below the *critical load*, P_{cr}, the strut remains straight and undergoes axial compression only. The strut is in a state of *stable equilibrium* such that the strut will return to its original straight position following the removal of a small laterally applied force. When P reaches the critical load then we reach a state of neutral equilibrium with the strut able to adopt a bent shape. A strut in a state of neutral equilibrium is able to undergo small lateral displacements without a change in the applied axial force. Finally, when P exceeds P_{cr} then the column becomes unstable and fails due to excessive bending or buckling. An unstable strut will deflect laterally due to the slightest lateral force, resulting in complete failure.

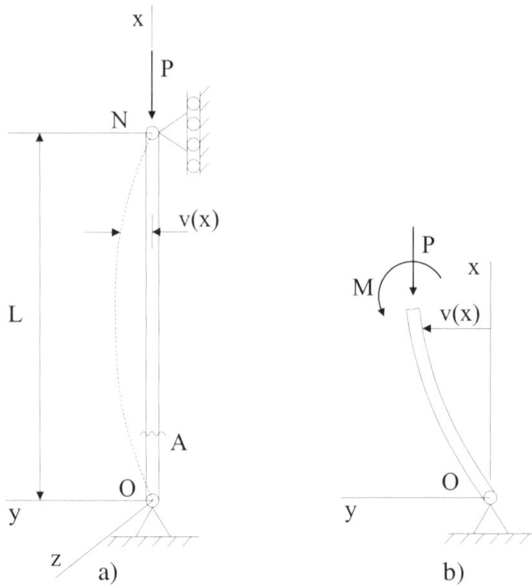

Figure 6.1: Strut with pin-jointed ends. a) Idealised strut. b) Axial force P and bending moment M at a given cut at a distance x from O.

Since a strut is simply a beam subject to axial compressive loading we will use the bending moment equation (5.41) to determine the buckling shape and critical load

$$EI\frac{d^2v}{dx^2} = M \tag{6.1}$$

where M is the bending moment, v is the lateral deflection in the y direction, E is the Young's modulus, I is the second moment of area and EI is the flexural rigidity. Since the ends of the strut are pin-jointed then equilibrium of moments about point O gives, Figure 6.1b)

$$M + Pv = 0 \tag{6.2}$$

Re-arranging for M and substituting into (6.1) gives the following homogeneous linear second order differential equation

$$EI\frac{d^2v}{dx^2} + Pv = 0 \tag{6.3}$$

which can be solved for the deflection v. Equation (6.3) can be more conveniently written by letting $k^2 = P/EI$ and denoting differentiation by a prime

$$v'' + k^2v = 0 \tag{6.4}$$

the solution of which is known to be, Stroud (1982)

$$v = C_1 \sin kx + C_2 \cos kx \tag{6.5}$$

where C_1 and C_2 are constants of integration to be evaluated from the prescribed boundary conditions. In the present case of pin-jointed ends the boundary conditions are $v = 0$ at both $x = 0$ and $x = L$

$$v(0) = 0 \quad \text{and} \quad v(L) = 0 \tag{6.6}$$

The first of these conditions gives $C_2 = 0$ from (6.5) and the second condition gives, from (6.5) with $C_2 = 0$

$$C_1 \sin kL = 0 \tag{6.7}$$

Thus, the two possible cases of $C_1 = 0$ and $\sin kL = 0$ need to be considered separately. The first case, $C_1 = 0$, is a trivial solution since from (6.5) with $C_2 = 0$ then v is zero irrespective of x or P. The second case, $\sin kL = 0$, is non-trivial and is referred to as the *buckling equation* and is satisfied when $kL = n\pi$ for $n = 0, 1, 2, \ldots$ For $n = 0$ then $kL = 0$ and $P = 0$ so that the solution $n = 0$ is not of interest. Thus, the solutions of importance are

$$kL = n\pi; \quad n = 1, 2, \cdots \tag{6.8}$$

With v given by (6.5) and $k^2 = P/EI$ then v and P can now be written

$$v = C_1 \sin kx = C_1 \sin \frac{n\pi x}{L}; \quad P = k^2 EI = \frac{n^2 \pi^2 EI}{L^2}; \quad n = 1, 2, \cdots \tag{6.9}$$

The expression for P in (6.9) in fact gives the *critical load* since (6.9) is based on the buckling equation and are the only values of P which result in a bent shape. The lowest value of critical load is attained when $n = 1$; with corresponding deflection or *mode shape*

$$v = C_1 \sin \frac{\pi x}{L}; \quad P_{cr} = \frac{\pi^2 EI}{L^2} \tag{6.10}$$

and is referred to as the *fundamental buckling* case. The constant C_1 represents the deflection at the mid-point of the strut ($x = L/2, \sin \pi x/L = \sin \pi/2 = 1$) as shown in Figure 6.2a). Considering higher values of n we obtain more complex buckling modes and higher values of critical load. For example, for $n = 2$ then v and P_{cr} are given by

$$v = C_1 \sin \frac{2\pi x}{L}; \quad P_{cr} = \frac{4\pi^2 EI}{L^2} \tag{6.11}$$

with v illustrated in Figure 6.2b). The critical load for $n = 2$ is four times the value of critical load for the fundamental case, $n = 1$, since P is proportional to n^2. The higher buckling modes can only be attained by providing lateral support of the strut, such as the mid-point for the case $n = 2$, and in practice it is the fundamental critical buckling load that is of most importance.

Let us conclude this section by noting that the fundamental critical load (6.10) is proportional to the flexural rigidity EI and inversely proportional to L^2. Thus, P_{cr} can only be increased by increasing the flexural rigidity (either by using a stiffer material or by redistributing the material to increase the second moment of area), reducing the length of the strut or by providing lateral support.

Example 6.1 Euler critical load for a strut with pin-jointed ends

A strut is of length $L = 1.5$m and of tubular circular cross-section of outside diameter $d_0 = 50$mm and wall thickness $t = 2$mm. Determine the Euler critical load, P_{cr}, for pin-jointed ends, assuming Young's modulus is equal to 210GPa.

With the internal diameter, d_i, equal to $d_i = d_0 - 2t = 46$mm then the second moment of area is, Exercise 2.2

$$I = \frac{\pi}{64}\left(d_0^4 - d_i^4\right) = \frac{\pi}{64}\left[\left(50 \times 10^{-3}\right)^4 - \left(46 \times 10^{-3}\right)^4\right] = 8.7 \times 10^{-8}\text{m}^4$$

From (6.10) P_{cr} is

$$P_{cr} = \frac{\pi^2 EI}{L^2} = \frac{\pi^2 \times 210 \times 10^9 \times 8.7 \times 10^{-8}}{1.5^2} = 80.14\text{kN}$$

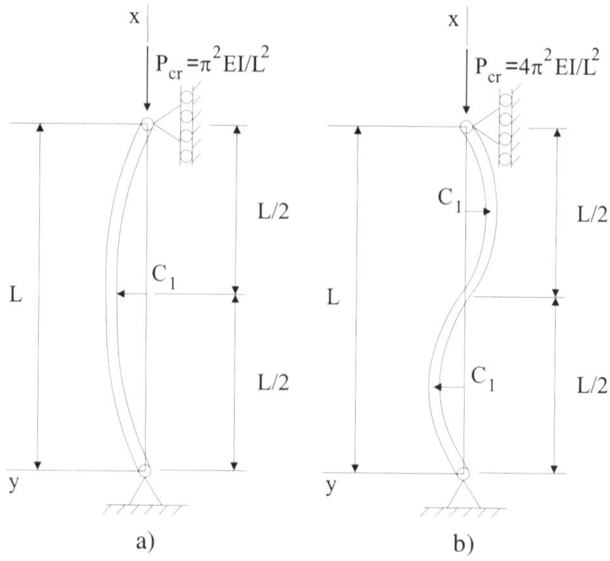

Figure 6.2: Buckling modes for a strut with pin-jointed ends. a) Fundamental buckling when $n = 1$. b) Buckling mode when $n = 2$.

6.3 Radius of Gyration, Slenderness Ratio and Critical Stress

In §2.7 we introduced the radius of gyration, which with respect to the x-axis is defined as

$$r_x = \sqrt{\frac{I_x}{A}} \tag{6.12}$$

and provides a useful measure of the distribution of area from the x-axis. When I is the second moment of area with respect to the principal axis about which buckling occurs then the radius of gyration will be denoted by $r = \sqrt{I/A}$. The non-dimensional *slenderness ratio*, l, is defined by

$$l = \frac{L}{r} \tag{6.13}$$

and provides a measure of the slenderness of a strut.

The fundamental critical load, P_{cr}, has a corresponding critical stress, σ_{cr}, which is conveniently expressed in terms of the slenderness ratio

$$\sigma_{cr} = \frac{P_{cr}}{A} = \frac{\pi^2 EI}{AL^2} = \frac{\pi^2 E}{l^2} \tag{6.14}$$

illustrating that σ_{cr} is inversely proportional to l^2, Figure 6.3. The graph of σ_{cr} against l is generally referred to as Euler's curve after Leonhard Euler who was the first person to investigate the buckling of struts. Euler's curve is only valid provided that the yield stress, σ_Y, of the material has not been exceeded. As a result, Euler's curve is generally truncated by σ_Y as shown in Figure 6.3.

Example 6.2 Radius of gyration, slenderness ratio and critical stress for a strut with pin-jointed ends

Determine the radius of gyration, slenderness ratio and critical stress for the strut specified in Example 6.1.

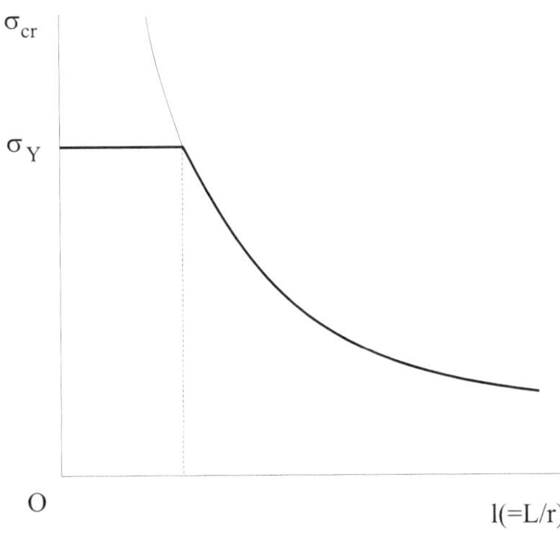

Figure 6.3: Euler's curve of σ_{cr} against l for a strut with pin-jointed ends and yield stress of σ_Y.

From (6.12) and (6.13) the radius of gyration and slenderness ratio are

$$r = \sqrt{\frac{I}{A}} = \sqrt{\frac{8.7 \times 10^{-8}}{3.0159 \times 10^{-4}}} = 16.98\text{mm}; \quad l = \frac{L}{r} = \frac{1.5}{16.98 \times 10^{-3}} = 88$$

From (6.14) the critical stress is

$$\sigma_{cr} = \frac{\pi^2 E}{l^2} = \frac{\pi^2 \times 210 \times 10^9}{88^2} = 268\text{MPa}$$

Example 6.3 Application of Euler's theory to a frame structure

Figure 6.4a) illustrates a frame consisting of two steel bars with pin-jointed ends. Using Euler's theory determine if the frame can safely support the load of 50kN at joint A without bar AC failing about both the xx-axis and yy-axis. Young's modulus is 210GPa. Application of the equations of equilibrium to the free body diagram of Figure 6.4b) gives

$$\sum M_B = 0 : -50 \times 10^3(5/\tan 40°) + F_C(5) = 0 \Rightarrow F_C = 60\text{kN}$$
$$\sum M_C = 0 : -50 \times 10^3(10) + F_B(10 \sin 40°) = 0 \Rightarrow F_B = 78\text{kN}$$

The second moments of area for the $x - x$ and $y - y$ axes are

$$I_{xx} = \frac{1}{12} 50 \times 10^{-3}(100 \times 10^{-3})^3 = 4.1\dot{6} \times 10^{-6}\text{m}^4$$
$$I_{yy} = \frac{1}{12} 100 \times 10^{-3}(50 \times 10^{-3})^3 = 1.041\dot{6} \times 10^{-6}\text{m}^4$$

Therefore, from (6.10) the critical load for both $x - x$ and $y - y$ axes are

$$P_{cr}(xx - \text{axis}) = \frac{\pi^2 EI_{xx}}{L^2} = \frac{\pi^2 \times 210 \times 10^9 \times 4.1\dot{6} \times 10^{-6}}{10^2} = 86\text{kN}$$
$$P_{cr}(yy - \text{axis}) = \frac{\pi^2 EI_{yy}}{L^2} = \frac{\pi^2 \times 210 \times 10^9 \times 1.041\dot{6} \times 10^{-6}}{10^2} = 22\text{kN}$$

With $P_{cr}(xx\text{-axis}) > F_C$ and $P_{cr}(yy\text{-axis}) < F_C$ then bar AC fails about the yy-axis only.

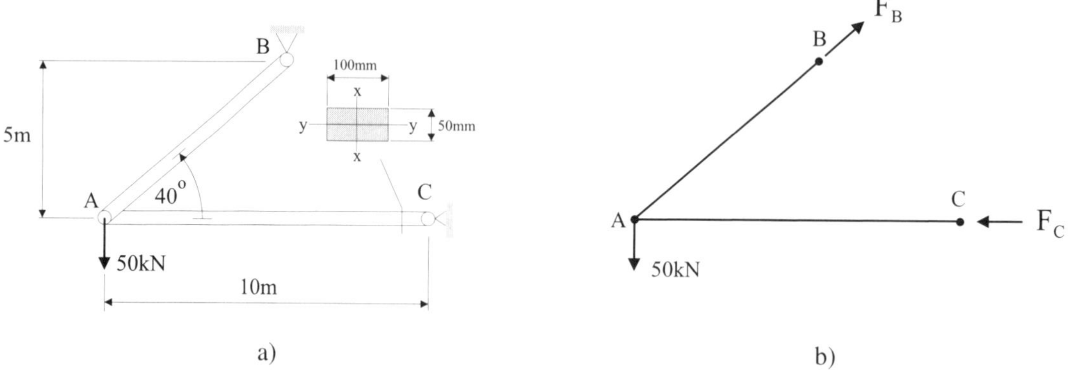

Figure 6.4: Example: Application of Euler's theory to a frame structure. a) Frame. b) Free body diagram.

6.4 Struts with other End Conditions

In the following sub-sections we examine struts with end conditions different to the preceding case of pin-jointed at both ends. The three cases considered are i) built-in at one end and free at the other end, ii) built-in at both ends and iii) built-in at one end and pin-jointed at the other end.

6.4.1 Struts Built-In at One End and Free at the Other End

Figure 6.5a) illustrates a strut built-in at end O and free at the other end N. Taking moments at a distance x from O then the bending moment is

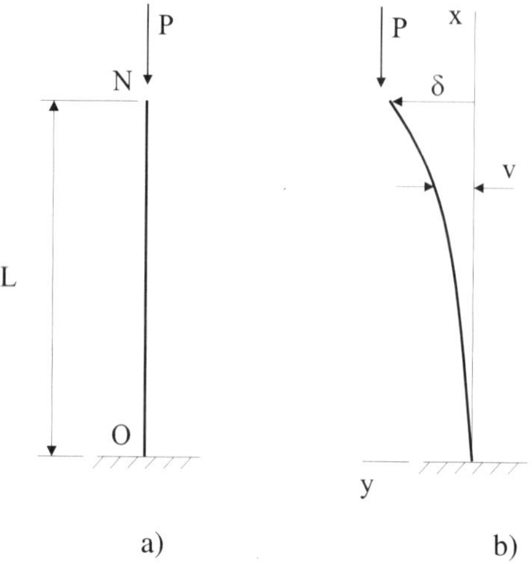

Figure 6.5: Strut built-in at one end and free at the other end. a) Idealised strut. b) Buckled shape of the strut.

$$M = P(\delta - v) \tag{6.15}$$

where δ is the deflection at the free end. From (6.1) and with $k^2 = P/EI$ then the differential equation of the deflection v is

$$v'' + k^2 v = k^2 \delta \tag{6.16}$$

Equation (6.16) is more difficult to solve than (6.4) because it is inhomogeneous, that is it consists of a non-zero right hand side. It is known that the general solution of an inhomogeneous differential equation consists of a complementary function and a particular integral. The complementary function is obtained by solving the equation with the right hand side equal to zero, as in (6.4). The particular integral is obtained by assuming the general form of the function on the right hand side, substituting this into the differential equation and equating coefficients; refer to Stroud (1982) for details. The complementary function is obtained from (6.5)

$$v_{CF} = C_1 \sin kx + C_2 \cos kx \tag{6.17}$$

where C_1 and C_2 are constants of integration determined from the prescribed boundary conditions. To find the particular integral we assume the general form of the right hand side $v = C$ (C =constant) since the right hand side of (6.16) is a constant. Therefore

$$\frac{\mathrm{d}v}{\mathrm{d}x} = 0; \quad \frac{d^2 v}{\mathrm{d}x^2} = 0 \tag{6.18}$$

Substituting these into (6.16) we have

$$0 + k^2 C = k^2 \delta \tag{6.19}$$

from which we find that $C = \delta$ and hence

$$v_{PI} = \delta \tag{6.20}$$

The complete solution consists of the complementary function plus the particular integral

$$v = v_{CF} + v_{PI} = C_1 \sin kx + C_2 \cos kx + \delta \tag{6.21}$$

To determine the three unknown constants (C_1, C_2 and δ) we require the following three boundary conditions

$$v(0) = 0; \quad v'(0) = 0; \quad v(L) = \delta \tag{6.22}$$

The first of these three boundary conditions gives $C_2 = -\delta$ from (6.21). To apply the second of (6.22) we need to firstly differentiate (6.21) with respect to x

$$v' = kC_1 \cos kx - kC_2 \sin kx \tag{6.23}$$

with the second of (6.22) giving $C_1 = 0$. With C_1 and C_2 now known, (6.21) can be written

$$v = \delta(1 - \cos kx) \tag{6.24}$$

Applying the third boundary condition of (6.22) to (6.24) we find

$$\delta \cos kL = 0 \tag{6.25}$$

which is analogous to condition (6.7) for the pin-jointed strut. The two cases to consider for (6.25) are $\delta = 0$ and $\cos kL = 0$. The first case, $\delta = 0$, is trivial since there is no deflection of the strut. The second case, $\cos kL = 0$, is the buckling equation and satisfied by the following

$$kL = n\frac{\pi}{2}; \quad n = 1, 3, 5, \cdots \tag{6.26}$$

With v given by (6.24) and $k^2 = P/EI$ then v and P_{cr} are

$$v = \delta\left(1 - \cos\frac{n\pi x}{L}\right); \quad P_{cr} = \frac{n^2\pi^2 EI}{4L^2}; \quad n = 1, 2, 5, \cdots \tag{6.27}$$

As discussed in §6.2 the fundamental buckling mode occurs when $n = 1$ in which case v and P_{cr} are given by

$$v = \delta\left(1 - \cos\frac{\pi x}{2L}\right); \quad P_{cr} = \frac{\pi^2 EI}{4L^2} \tag{6.28}$$

For the next higher buckling mode in which $n = 3$ then v and P_{cr} are

$$v = \delta\left(1 - \cos\frac{3\pi x}{2L}\right); \quad P_{cr} = \frac{9\pi^2 EI}{4L^2} \tag{6.29}$$

The buckling modes for $n = 1$ and $n = 3$ are illustrated in Figure 6.6.

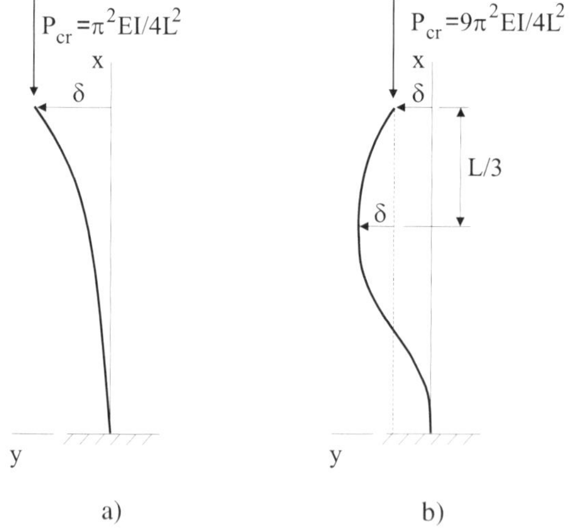

Figure 6.6: Buckling modes for a strut which is built-in at one end and free at the other end. a) Fundamental buckling when $n = 1$. b) Buckling mode when $n = 3$.

Before discussing other end conditions let us introduce the notion of an *effective length*. Figure 6.7 illustrates the deflection of a strut built-in at one end and free at the other end. The deflection curve has been reflected about the y-axis shown by the dashed curve. The effective length, L_e, is seen to be equal to twice the length of the strut so that the effective length is the length of a pin-jointed strut which has a deflection curve that matches all or part of the deflection curve of the original strut. In addition, an *effective length factor*, K, is introduced

$$L_e = KL \tag{6.30}$$

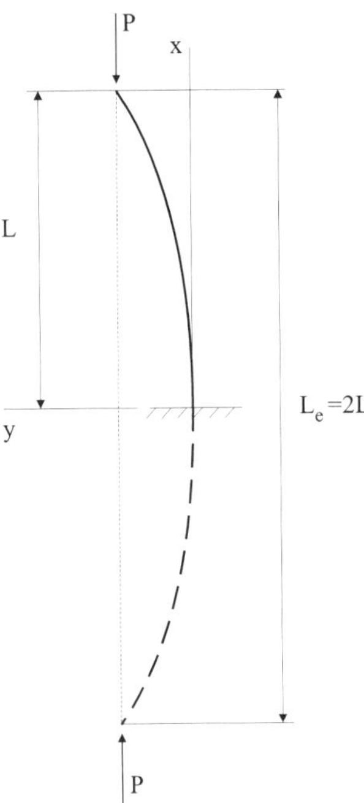

Figure 6.7: Deflection of a strut which is built-in at one end and free at the other end. The length of the strut is L and the effective length is $L_e = 2L$.

The factor K is equal to 2 for the built-in strut of Figure 6.5 and equal to 1 for the pin-jointed strut of Figure 6.1. Comparing (6.10) and (6.28) it follows that the fundamental critical load for all struts can be written in a general form in terms of L_e

$$P_{cr} = \frac{\pi^2 EI}{L_e^2} \tag{6.31}$$

6.4.2 Struts Built-In at Both Ends

Figure 6.8a) illustrates a strut which is built-in at both ends with the understanding that the strut is free to shorten in length. The buckled shape of the strut is illustrated in Figure 6.8b) and is seen to be symmetrical about the mid-point of the strut. There is zero slope at the mid-point and both ends of the strut. Rather than deriving a differential equation the critical load is easily determined by noting that the effective length is equal to $L/2$. Therefore, substituting $L_e = L/2$ into (6.31) the critical load is

$$P_{cr} = \frac{4\pi^2 EI}{L^2} \tag{6.32}$$

6.4.3 Struts Built-In at One End and Pin-Jointed at the Other End

Figure 6.9a) illustrates a strut which is built-in at one end and pin-jointed at the other end. When the strut deforms, a moment M_0 will be induced at the built-in end since there is no rotation at this end. From

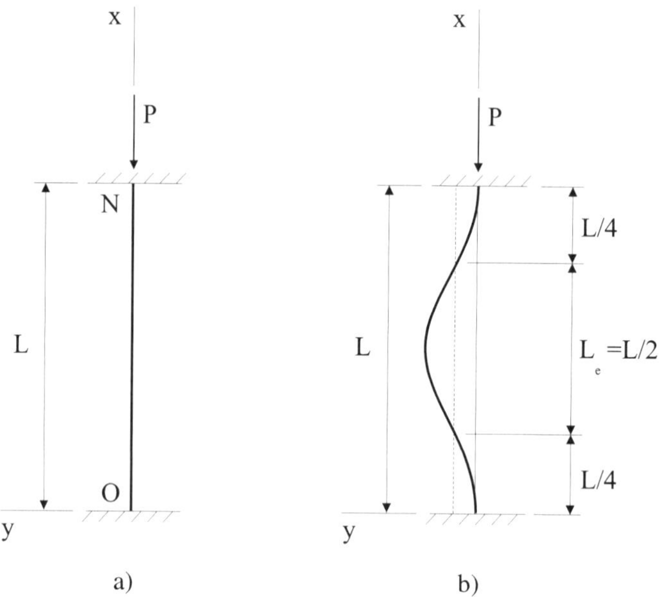

Figure 6.8: Strut built-in at both ends. a) Idealised strut. b) Buckled shape of the strut.

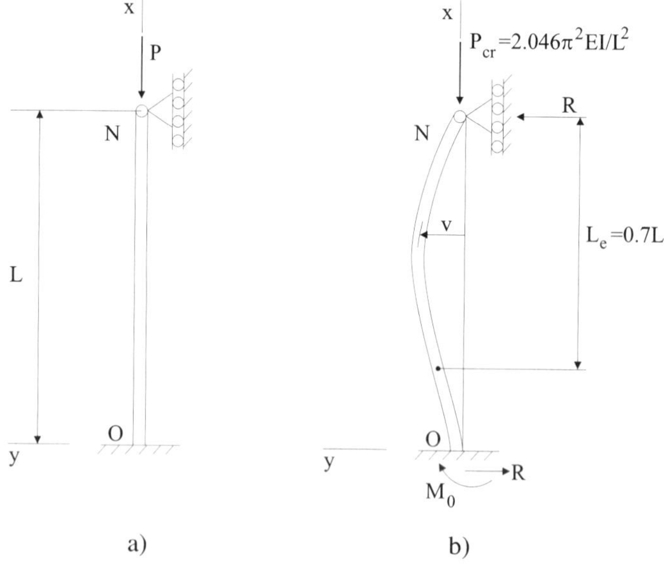

Figure 6.9: Strut built-in at one end and pin-jointed at the other end. a) Idealised strut. b) Buckled shape of the strut.

equilibrium $M_0 = RL$ where R is the horizontal reaction at each end. Taking moments at a distance x from the built-in end

$$M = M_0 - Pv - Rx = R(L - x) - Pv \tag{6.33}$$

where $v(x)$ is the deflection at a distance x from O. From the differential equation (6.1) and with $k^2 = P/EI$ then

$$v'' + k^2 v = \frac{R}{EI}(L - x) \tag{6.34}$$

Solving for the complementary function and particular integral as in §6.4.1 the general solution of (6.34) is

$$v = C_1 \sin kx + C_2 \cos kx + \frac{R}{P}(L - x) \tag{6.35}$$

To determine the three unknown constants (C_1, C_2 and R) we require the following three boundary conditions

$$v(0) = 0; \quad v'(0) = 0; \quad v(L) = 0 \tag{6.36}$$

Applying each of these boundary conditions to (6.35) we have the following equations

$$C_2 + \frac{RL}{P} = 0; \quad C_1 k - \frac{R}{P} = 0; \quad C_1 \tan kL + C_2 = 0 \tag{6.37}$$

Ignoring the trivial solution ($C_1 = C_2 = R = 0$) we first eliminate R from the first two equations of (6.37) which gives $C_2 = -kLC_1$ and substituting into the third equation of (6.37) gives the buckling equation

$$\tan kL = kL \tag{6.38}$$

To determine the lowest non-zero solution of (6.38) requires a numerical solution and is found to be

$$kL = 1.4303\pi \tag{6.39}$$

With $k^2 = P/EI$ then the critical load is

$$P_{cr} = \frac{2.0457\pi^2 EI}{L^2} \tag{6.40}$$

With $C_2 = -kLC_1$ and $R/P = kC_1$ from (6.37) then v is given by, (6.35)

$$v = C_1 \left[\sin kx - L \cos kx + k(L - x)\right] \tag{6.41}$$

Finally, comparing (6.40) and (6.31) the effective length is found to be

$$L_e = \frac{L}{\sqrt{2.0457}} = 0.7L \tag{6.42}$$

Let us conclude by tabulating P_{cr} for each of the four end conditions considered in the previous sections, Table 6.1.

	K			
	1	**2**	**1/2**	**0.7**
$\mathbf{P_{cr}/P_{cr}(K = 1)}$	1	1/4	4	2

Table 6.1: P_{cr} for four different end conditions.

Example 6.4 A strut built-in at both ends

A thin-walled tubular strut of circular cross-section is built-in at both ends and subject to a compressive load of 300kN. The length of the strut is 10m, the mean radius of the tube is 50mm and Young's modulus is 210GPa. Determine the minimum value of wall-thickness of the tube required in order to prevent failure due to buckling.

From §4.4 the second moment of area $I = I_x = I_y$ for a thin-walled tube of circular cross-section of mean radius r and wall-thickness t is $I = \pi r^3 t$. Substituting I into (6.31) with $L_e = L/2$ for a built-in strut and re-arranging for t we have

$$t = \frac{P_{cr}L^2}{4\pi^3 E r^3} = \frac{300 \times 10^3 \times 10^2}{4\pi^3 \times 210 \times 10^9 (50 \times 10^{-3})^3} = 9.2\text{mm}$$

so that the minimum required wall-thickness is 9.2mm.

6.5 Pin-Jointed Struts with Eccentric Axial Loading and the Secant Formula

Previous sections have examined the axial loading of struts in which it was assumed that the loading acted through the centroid of the cross-section. In the present section we will examine the case where the applied axial load acts at a small eccentricity e from the axis of the strut, Figure 6.10a). The eccentrically applied load is equivalent to a centric load P and a moment $M_0 = Pe$ as shown in Figure 6.10b). The buckled shape of the strut is also shown in Figure 6.10b) noting that the deflections are negative for a positive eccentricity measured along the positive y-axis.

Taking moments at a distance x from O we have

$$M = M_0 + P(-v) = Pe - Pv \tag{6.43}$$

From the differential equation (6.1) and with $k^2 = P/EI$ then

$$v'' + k^2 v = k^2 e \tag{6.44}$$

with general solution (= complementary function + particular integral)

$$v = C_1 \sin kx + C_2 \cos kx + e \tag{6.45}$$

The unknown constants C_1 and C_2 are determined from the following two boundary conditions

$$v(0) = 0; \quad v(L) = 0 \tag{6.46}$$

Applying these boundary conditions to (6.45) we find

$$C_2 = -e; \quad C_1 = -e \tan\frac{kL}{2} \tag{6.47}$$

The deflection v follows immediately from (6.45)

$$v = -e\left(\tan\frac{kL}{2}\sin kx + \cos kx - 1\right) \tag{6.48}$$

and illustrates that v is zero when e is zero. The maximum deflection, δ, occurs at the mid-point of the strut

$$\delta = -v\left(\frac{L}{2}\right) = e\left(\tan\frac{kL}{2}\sin\frac{kL}{2} + \cos\frac{kL}{2} - 1\right) = e\left(\sec\frac{kL}{2} - 1\right) \tag{6.49}$$

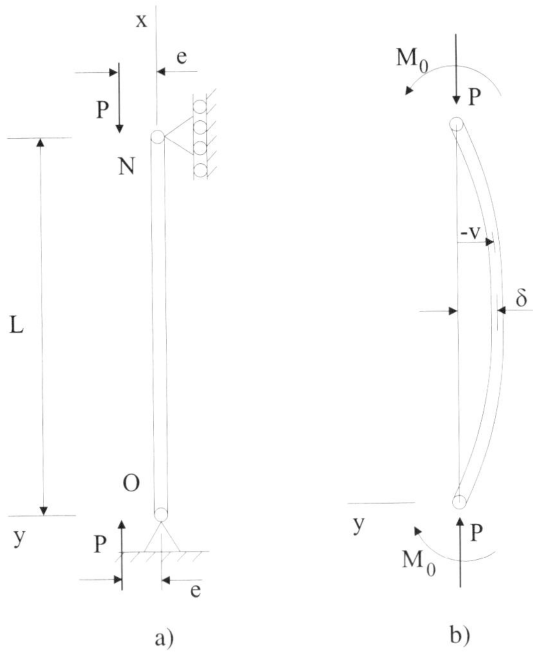

Figure 6.10: Strut with pin-jointed ends subject to eccentric axial loading. a) Idealised strut. b) Buckled shape of the strut with moments M_0 at the ends.

The critical load occurs when $e = 0$ and is therefore equivalent to (6.10)

$$P_{cr} = \frac{\pi^2 EI}{L^2} \tag{6.50}$$

and further noting that the eccentric load does not affect the magnitude of the critical load. The maximum deflection δ can alternatively be written by first expressing k in terms of P_{cr}. From (6.50) $EI = P_{cr}L^2/\pi^2$ and substituting into $k^2 = P/EI$

$$k = \sqrt{\frac{P}{EI}} = \sqrt{\frac{P\pi^2}{P_{cr}L^2}} = \frac{\pi}{L}\sqrt{\frac{P}{P_{cr}}} \tag{6.51}$$

and substituting into (6.49) then δ is now given by

$$\delta = e\left[\sec\left(\frac{\pi}{2}\sqrt{\frac{P}{P_{cr}}}\right) - 1\right] \tag{6.52}$$

and illustrates that $\delta \to \infty$ as $P \to P_{cr}$.

Let us conclude the present section by deriving an expression for the maximum stress in a strut subject to eccentric axial loading. The maximum stress occurs at the mid-point where the deflection and hence the bending moment are largest. Combining both the axial and bending stresses the maximum compressive stress, σ_{max}, is

$$\sigma_{max} = \frac{P}{A} + \frac{M_{max}y}{I} \tag{6.53}$$

where y is the distance from the centroidal axis to the extreme point on the concave side of the buckled strut. The maximum bending moment occurs at the mid-point and is given by

$$M_{max} = P(e + \delta) \tag{6.54}$$

Substituting for δ from (6.52)

$$M_{max} = Pe \sec\left(\frac{\pi}{2}\sqrt{\frac{P}{P_{cr}}}\right) \tag{6.55}$$

With P_{cr} given by (6.10) and $I = Ar^2$ from (6.12) then M_{max} can be written

$$M_{max} = Pe \sec\left(\frac{L}{2r}\sqrt{\frac{P}{EA}}\right) \tag{6.56}$$

Substituting M_{max} into (6.53) then σ_{max} is given by

$$\sigma_{max} = \frac{P}{A}\left[1 + \frac{ey}{r^2}\sec\left(\frac{L}{2r}\sqrt{\frac{P}{EA}}\right)\right] \tag{6.57}$$

and is referred to as the *secant formula*. We noted earlier in equation (6.13) that the ratio L/r is the slenderness ratio. In addition, the ratio ey/r^2 is referred to as the *eccentricity ratio* and is equivalently equal to eA/S since $r = I/A$ and the section modulus is $S = I/y$. Figure 6.11 illustrates the variation of σ_{max} against the slenderness ratio as a function of the eccentricity ratio. As the eccentricity ratio tends to zero then the curve tends to Euler's curve, Figure 6.3, again with the curve truncated by the material yield stress σ_Y.

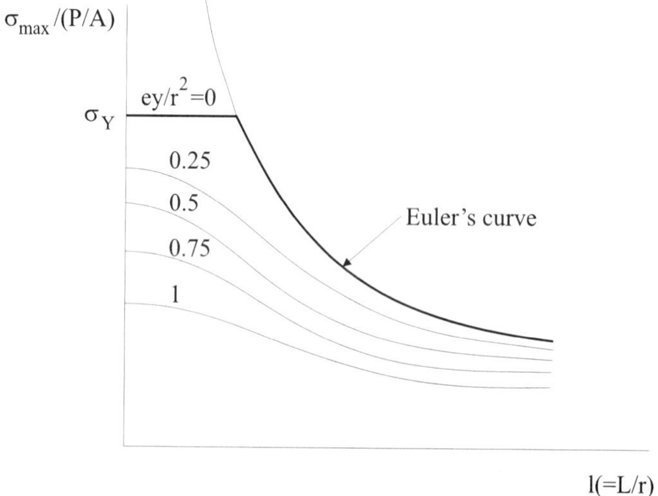

Figure 6.11: Variation of the maximum compressive stress, σ_{max}, against slenderness ratio, l, as a function of eccentricity ratio, ey/r^2.

Example 6.5 Maximum compression stress for a pin-jointed strut with eccentric loading

An I-section steel beam is compressed by a load of 100kN acting with an eccentricity of 4mm. The column is simply supported at its ends and has a length of 5m, cross-sectional area of 38×10^{-4}m^2, second moment of area of 354×10^{-8}m^4, flange thickness of $T = 9.6$mm and Young's modulus of 210GPa. Determine the maximum compressive stress in the strut assuming that $y = T/2$.

With radius of gyration $r = 0.0305$m and eccentricity ratio $ey/r^2 = 0.0206$ then from the secant formula, (6.57), the maximum compressive stress, σ_{max}, is

$$\sigma_{max} = 26.32 \times 10^6 \left[1 + 0.0206 \sec(0.9176)\right] = 69.6\text{MPa}$$

6.6 Semi-Empirical Formula

Let us now return to Euler's curve of Figure 6.3 which is redrawn in Figure 6.12. Euler's curve, (6.14), is applicable for long struts, that is struts which have a large slenderness ratio, $l(= L/r)$, with buckling occurring elastically at a stress below the yield stress, σ_Y, of the material. As the slenderness ratio decreases then Euler's curve diverges from observed behaviour. Struts having low slenderness ratios (e.g. $L/r < 30$ for steel) exhibit minimal buckling and failure is due to yielding of the material. The critical value of the slenderness ratio can be obtained by equating σ_{cr} to σ_Y

$$l_c = \left(\frac{L}{r}\right)_c = \sqrt{\frac{\pi^2 E}{\sigma_Y}} \tag{6.58}$$

noting that σ_{UTS} could alternatively be used instead of σ_Y. For example, for a steel with $E = 210$GPa and $\sigma_Y = 300$MPa then $l_c = 83$.

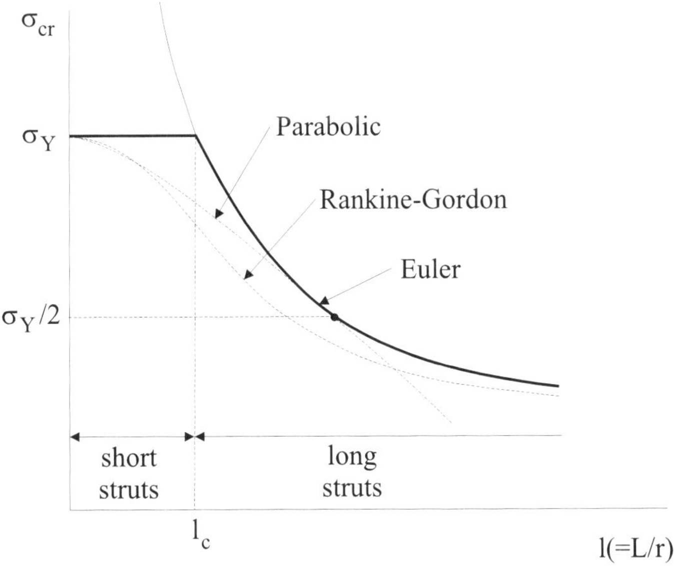

Figure 6.12: Euler's curve and the semi-empirical curves of Rankine-Gordon and Johnson's parabolic formula.

Several semi-empirical formulae have been proposed to account for the increasing influence of compressive yielding as l decreases below l_c in which Euler's curve is invalid. In the following two subsections we will examine the semi-empirical formulae of Rankine-Gordon and Johnson's parabolic formula.

6.6.1 Rankine-Gordon

In the case of pin-jointed struts the Euler buckling load, P_E, and compressive yield load, P_Y, are combined through the use of a reciprocal equation to give the Rankine-Gordon load, P_R

$$\frac{1}{P_R} = \frac{1}{P_E} + \frac{1}{P_Y} \tag{6.59}$$

from which P_R is given by

$$P_R = \frac{P_Y}{1 + (P_Y/P_E)} \tag{6.60}$$

Substituting P_E from (6.10) and $P_Y = A\sigma_Y$ then the Rankine-Gordon stress, σ_R, is given by

$$\sigma_R = \frac{\sigma_Y}{1 + al^2}; \quad a = \frac{\sigma_Y}{\pi^2 E} \tag{6.61}$$

The constant $a(= \sigma_Y/\pi^2 E)$ is replaced by a constant value which is found by equating (6.61) to experimental values. Table 6.2 lists typical values for three different materials. Table 6.2 also lists values of

Material	σ_Y (MPa)	a^{-1} (pin-jointed ends)	a^{-1} (built-in ends)
mild steel	300	7,500	30,000
cast iron	510	1,600	6,400
timber	32	3,000	1,200

Table 6.2: Rankine-Gordon constant a for different materials.

a for built-in end conditions in which the strut length, L, should be replaced by the effective length, L_e. The Rankine-Gordon stress σ_R against l is illustrated in Figure 6.12 and illustrates that σ_R is applicable throughout the entire range of l and $\sigma_R \leq \sigma_Y$.

6.6.2 Johnson's Parabolic Formula

Johnson's semi-empirical formula replaces Euler's curve for short struts by the following parabolic stress

$$\sigma_J = \sigma_Y(1 - bl^2) \tag{6.62}$$

where b is a constant and σ_Y is taken to be less than the actual compressive yield stress of the material in order to offset the need for a safety factor. The American Institute of Steel Construction (AISC) propose the following formula, that is $b = 1/2C_c^2$

$$\sigma_J = \sigma_Y\left(1 - \frac{l^2}{2C_c^2}\right) \tag{6.63}$$

where C_c is determined from the intersection between Johnson's curve and the Euler curve at $\sigma_E = \sigma_Y/2$; namely $C_c = l = (2\pi^2 E/\sigma Y)^{\frac{1}{2}}$ from (6.14). The variation of the Parabolic curve against l is illustrated in Figure 6.12.

Example 6.6 Application of Euler's critical load theory for a strut with pin-jointed ends

For the strut given in Examples 6.1 and 6.2, confirm that Euler's critical load theory is applicable for a strut having a length of 1.5m.

For Euler's theory to be applicable then the slenderness ratio l must be greater than or equal to the critical slenderness ratio l_c or from (6.58) L must satisfy

$$L \geq \sqrt{\frac{\pi^2 E r^2}{\sigma_Y}} = \sqrt{\frac{\pi^2 \times 210 \times 10^9 \times 2.8847 \times 10^{-4}}{300 \times 10^{-6}}} = 1.4117\text{m}$$

It follows that Euler's theory is valid for struts of length 1.5m.

Example 6.7 Johnson's parabolic formula

Determine the constant b in Johnson's parabolic formula when σ_J is to make a tangent with the Euler curve at the slenderness ratio $l = 100$.

Differentiating (6.14) with respect to l then the gradient of Euler's curve is

$$\frac{\mathrm{d}\sigma}{\mathrm{d}l} = -\frac{2\pi^2 E}{l^3}$$

and similarly differentiating Johnson's formula, (6.62)

$$\frac{\mathrm{d}\sigma}{\mathrm{d}l} = -2\sigma_Y bl$$

Equating these two gradients and re-arranging for σ_Y

$$\sigma_Y = \frac{\pi^2 E}{bl^4}$$

and substituting into (6.62) and re-arranging for b we find

$$b = \frac{1}{2l^2}$$

Hence, when $l = 100$ then $b = 5 \times 10^{-5}$.

6.7 Conclusion

This Chapter has examined the buckling of struts which are long slender members subject to axial compressive loading. Solving the bending moment equation for deflection leads to the buckling equation, which in turn leads to the Euler critical load for the fundamental buckling case. Higher order modes of buckling can be examined but it is the fundamental buckling mode which is of most importance. The Chapter examined several different combinations of pin-jointed, free and built-in end conditions. By introducing the notion of an effective length then a general expression for the critical buckling load can be written, with the effective length being a function of the end conditions. The case of eccentric axial loading was considered and the secant formula derived. The Chapter concluded by discussing the two semi-empirical formulae of Rankine-Gordon and Johnson which can both be used for the analysis of short struts in which Euler's formula becomes increasingly invalid.

6.8 References and Further Reading

○ Stroud, K. A. (1982) *Engineering Mathematics: Programmes and Problems*, MacMillan Publishers, London.

6.9 Exercises

6.1 A strut of solid circular cross-section is made of a material with Young's modulus equal to 210GPa and is 4m in length. If the strut is pin-jointed at both ends then determine the diameter of the strut so that it can carry a load of 200kN.

6.2 If three struts, each having the same length and Young's modulus, are of square, circular and equilateral triangle cross-section then determine which strut has the largest value of critical buckling stress according to Euler's theory.

6.3 Show that radius of gyration of a circular tubular section of mean radius a and thickness t is given by $a/\sqrt{2}$.

6.4 Figure 6.13 illustrates a structure supporting a load of W by the use of a strut ($50\text{mm} \times 50\text{mm}$ cross-section) which lies perpendicular to the support and a cable which is at an angle of $60°$ to the support. If the Young's modulus of the strut is 210GPa then determine the maximum value of W so that the strut does not fail due to buckling.

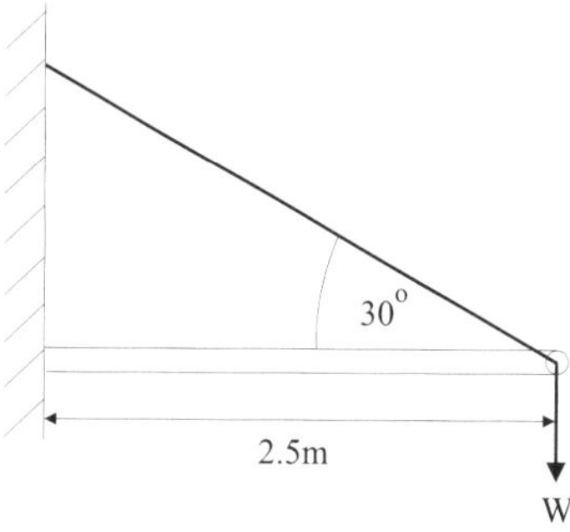

Figure 6.13: Exercise: Structure for Exercise 6.4.

6.5 A rectangular aluminium tube of constant thickness 10mm, Figure 6.14, is used as a 5m long strut which is built-in at both ends. Determine the critical stress in the strut assuming that the strut is made of aluminium alloy with Young's modulus of 70GPa.

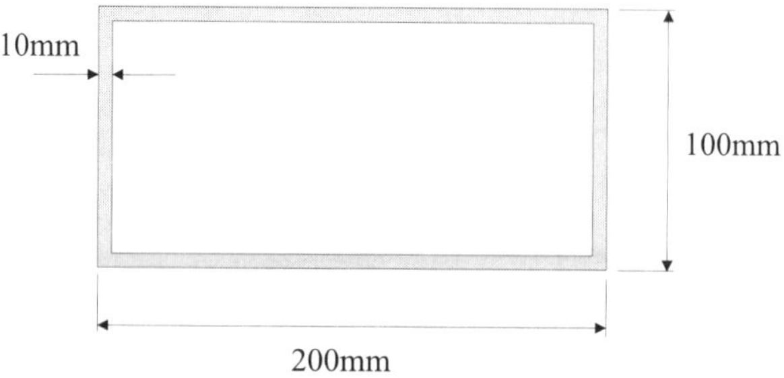

Figure 6.14: Exercise: Rectangular tube cross-section.

6.6 A 5m long pin-jointed steel strut is made from a 150×26 I-section with area $A = 3.27 \times 10^3\text{mm}^2$, second moment of area $I = 11 \times 10^6\text{mm}^4$ and section modulus $S = 144 \times 10^3$. If a load of $P = 350\text{kN}$ is applied to the strut with an eccentricity of $e = 25\text{mm}$ then determine both the maximum deflection and stress in the strut.

6.7 Determine the critical stress according to the Rankine-Gordon semi-empirical formula for a strut of 2m in length which is pin-jointed at both ends and made of mild steel. Assume that the radius of gyration of the strut cross-section is 39.6mm and the yield stress is 300MPa.

Part II

Intermediate Topics

In Part II, Chapters 7 and 8, we begin by examining stress and strain further via the *transformation equations* which enable the state of stress-strain at a point to be determined on a set of local axes with respect to a set of global axes. The transformation equations enable the principal stresses and maximum shear stress and their associated planes to be determined. In addition, Chapter 7 extends the discussion on second moments of area of Chapter 2 by determining the second moments of area with respect to a set of axes which have undergone a pure rotation relative to the global axes. Thereafter, Chapter 8 examines the application of the transformation equations to strain gauge rosettes. Chapter 9 is concerned with strain energy and its application. In Chapter 10 we will look at the principles of virtual displacement and work and potential energy. Chapter 11 extends the examination of thin-walled pressure vessels presented in Chapter 4 to thick-walled pressure vessels. An introduction to plasticity is given in Chapter 12 and we begin by defining the plastic behaviour of engineering materials. The key yield criteria of Tresca and Huber-von Mises are then derived and the Chapter concludes by examining the problems of the torsion of a circular bar and the bending of a beam of rectangular cross-section. In Chapter 13 we will study the finite element method. By way of introduction the Chapter focuses on the one- and two-dimensional pin-jointed bar elements. The Chapter concludes by illustrating the analogies between the elasticity, heat transfer and fluid flow finite element formulations.

Chapter 7

Stress Analysis

7.1 Introduction

Previous Chapters have presented methods for determining both the normal and shear stresses in beams, shafts and struts acting over the cross-section of a body. For example, the flexure and shear formulae for a beam of rectangular cross-section, see Figure 7.1, are given by (5.30) and (5.40). In the present Chapter we will be interested in determining the stresses that act not only on the cross-section of a body but the stresses on any given plane and point throughout a body. Once the state of stress is known for an arbitrary set of local axes then the maximum and minimum *principal stresses* follow immediately. Following the development of the *transformation equations* (which transform the state of stress between two axes) in the next section the principal stresses and their associated planes are developed. Similarly, the maximum and minimum shear stresses and their associated planes are developed. This is followed by expressing the transformation equations graphically with the aid of Mohr's stress circle. Section 7.6 develops the transformation equations for the second moments of area between two axes which have undergone a pure rotation with respect to each other. It will be shown that there is a direct analogy between the transformation of stress between two axes and the second moments of area. The Chapter concludes with an application of the transformation equations of stress to develop the well-known relationship between Young's modulus and the shear modulus, namely $E = 2G(1 + \nu)$.

7.2 Transformation Equations

To represent the state of stress within a given body we will consider the state of stress in an element, an example of which is shown in Figure 7.2. Since the element could have been chosen arbitrarily we need to associate a *local* set of coordinate axes to the element. Consider the plane elements in Figure 7.3 with (x, y) and (x', y') denoting the global and local sets of orthogonal axes respectively. The local axes are rotated through the angle θ measured positive in the anti-clockwise direction.

The problem to be addressed is to determine the local stress components $(\sigma_{x'x'}, \sigma_{y'y'}, \tau_{x'y'}, \dots)$ on an arbitrarily oriented element in terms of the global stress components $(\sigma_{xx}, \sigma_{yy}, \tau_{xy}, \dots)$. It is worth noting that we are dealing with one particular state of stress and it is only the stresses acting on different elements that differ. This problem is analogous to that of finding the coordinates of a given point, p, relative to two coordinate axes, Figure 7.4. In general, if $z = (x + iy)$ and $z' = (x' + iy')$ denote a complex point in the coordinate systems (x, y) and (x', y') respectively then $z' = ze^{-i\theta}$ or $(x' + iy') = (x + iy)(\cos\theta + i\sin\theta)$ where θ is the angle between the x and x' axes. Comparing real and imaginary

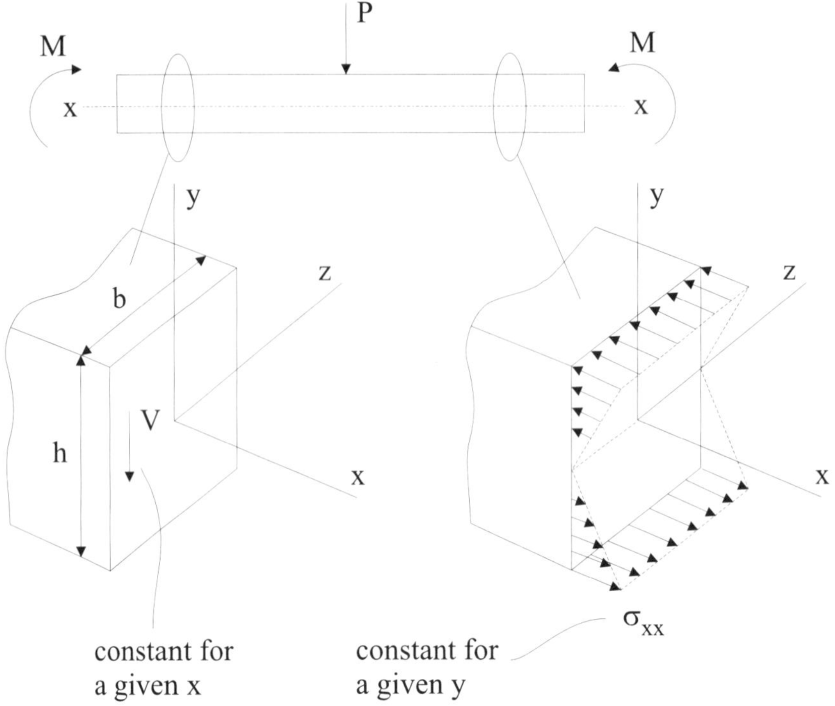

Figure 7.1: A beam subject to point, P, and bending moment, M, loadings with sections illustrating the shear force V and bending stress σ_{xx}.

parts and expressing the transformation in vector-matrix format

$$\left\{ \begin{array}{c} x' \\ y' \end{array} \right\} = \left[\begin{array}{cc} \cos\theta & \sin\theta \\ -\sin\theta & \cos\theta \end{array} \right] \left\{ \begin{array}{c} x \\ y \end{array} \right\} = \mathbf{T} \left\{ \begin{array}{c} x \\ y \end{array} \right\} \tag{7.1}$$

where $\mathbf{T}$ is the *transformation matrix*. Thus, for point p we have

$$\begin{aligned} p_{x'} &= p_x \cos\theta + p_y \sin\theta \\ p_{y'} &= -p_x \sin\theta + p_y \cos\theta \end{aligned} \tag{7.2}$$

However, for stresses the transformation relations are slightly more complicated!

Returning to the stress element of Figure 7.3 consider a wedge-shaped stress element that is chosen so that the wedge face coincides with the (x', y') axes, Figure 7.5a). To determine the equations of equilibrium consider the free-body diagram of Figure 7.5b) which illustrates the force acting on each face. The free-body diagram is determined with the aid of Figure 7.5c) in which the area of each face is proportional to the lengths of the sides since the thickness of the element, l, is constant.

Resolving forces along the x'-axis we have

$$\sigma_{x'x'} A \sec\theta - \sigma_{xx} A \cos\theta - \tau_{xy} A \sin\theta - \sigma_{yy} A \tan\theta \sin\theta - \tau_{xy} A \tan\theta \cos\theta = 0 \tag{7.3}$$

Resolving forces along the y'-axis

$$\tau_{x'y'} A \sec\theta + \sigma_{xx} A \sin\theta + \tau_{xy} A \tan\theta \sin\theta - \tau_{xy} A \cos\theta - \sigma_{yy} A \tan\theta \cos\theta = 0 \tag{7.4}$$

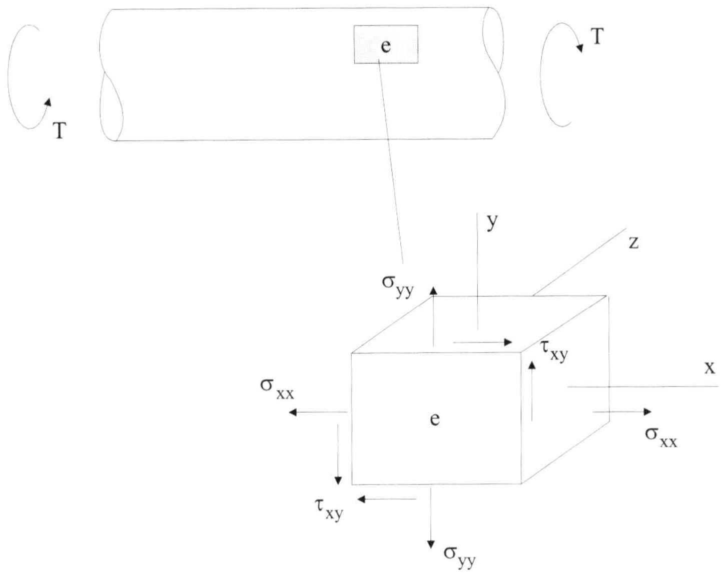

Figure 7.2: An arbitrary stress element in a body. The z-components of stress are not shown for the sake of clarity.

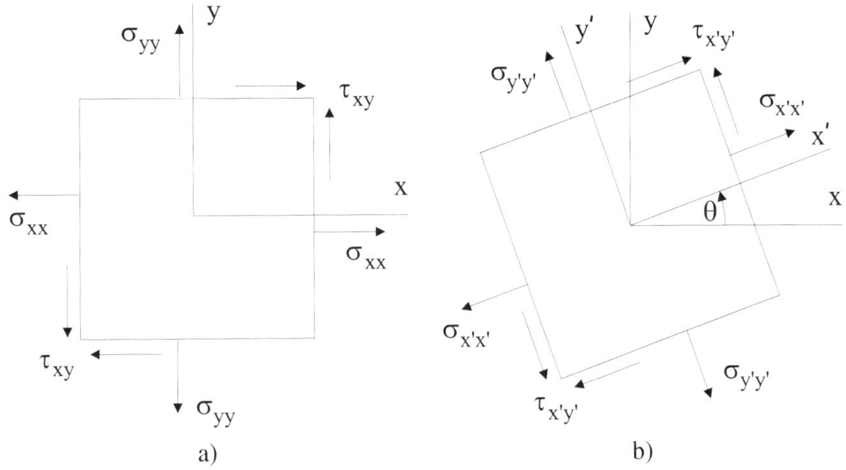

Figure 7.3: A plane stress element. a) Global axes (x, y) and global stresses σ_{xx}, σ_{yy} and τ_{xy}. b) Local axes (x', y') and local stresses $\sigma_{x'x'}$, $\sigma_{y'y'}$ and $\tau_{x'y'}$.

and dividing (7.3) and (7.4) by A and $\sec\theta$ and re-arranging for $\sigma_{x'x'}$ and $\tau_{x'y'}$

$$
\begin{aligned}
\sigma_{x'x'} &= \sigma_{xx}\frac{\cos\theta}{\sec\theta} + \sigma_{yy}\tan\theta\frac{\sin\theta}{\cos\theta} + \tau_{xy}\frac{\sin\theta}{\sec\theta} + \tau_{xy}\tan\theta\frac{\cos\theta}{\sec\theta} \\
\tau_{x'y'} &= -\sigma_{xx}\frac{\sin\theta}{\sec\theta} + \sigma_{yy}\tan\theta\frac{\cos\theta}{\sec\theta} + \tau_{xy}\frac{\cos\theta}{\sec\theta} - \tau_{xy}\frac{\tan\theta}{\sec\theta}\sin\theta
\end{aligned}
\tag{7.5}
$$

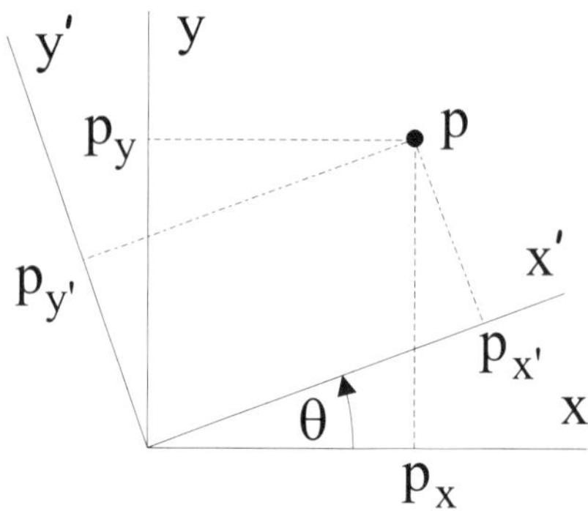

Figure 7.4: A point with respect to coordinate axes (x, y) and (x', y').

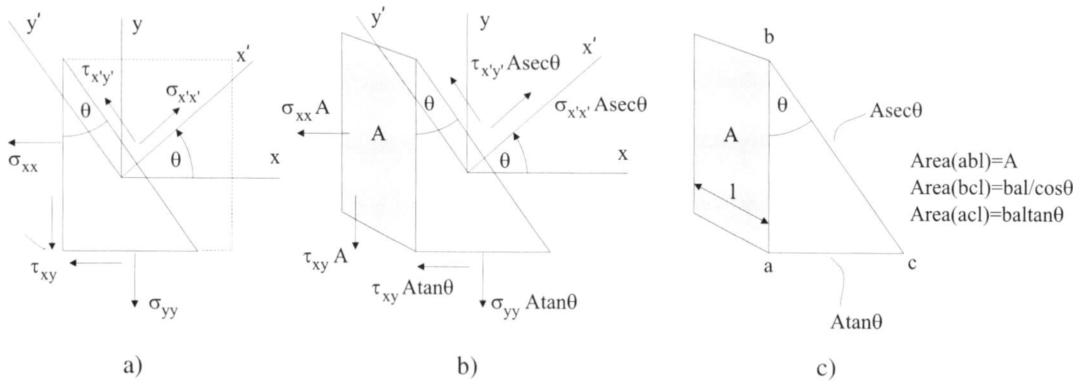

Figure 7.5: Wedge-shaped stress element. a) Stresses on each face. b) Free-body diagram. c) Areas of wedge faces.

from which

$$\begin{aligned} \sigma_{x'x'} &= \sigma_{xx} \cos^2 \theta + \sigma_{yy} \sin^2 \theta + 2\tau_{xy} \sin \theta \cos \theta \\ \tau_{x'y'} &= -(\sigma_{xx} - \sigma_{yy}) \sin \theta \cos \theta + \tau_{xy} (\cos^2 \theta - \sin^2 \theta) \end{aligned} \tag{7.6}$$

and using the trigonometric identities

$$\cos^2 \theta = \frac{1}{2}(1 + \cos 2\theta); \quad \sin^2 \theta = \frac{1}{2}(1 - \cos 2\theta); \quad 2 \sin \theta \cos \theta = \sin 2\theta \tag{7.7}$$

we have, finally

$$\begin{aligned} \sigma_{x'x'} &= \left(\frac{\sigma_{xx} + \sigma_{yy}}{2} \right) + \left(\frac{\sigma_{xx} - \sigma_{yy}}{2} \right) \cos 2\theta + \tau_{xy} \sin 2\theta \\ \tau_{x'y'} &= -\left(\frac{\sigma_{xx} - \sigma_{yy}}{2} \right) \sin 2\theta + \tau_{xy} \cos 2\theta \end{aligned} \tag{7.8}$$

and are referred to as the *stress transformation equations*. Let us note several points about the transformation equations:

- The transformation equations show that the local stresses $\sigma_{x'x'}$ and $\tau_{x'y'}$ acting normal and parallel to a plane vary due to the angle of the plane θ.
- The transformation equations transform the stress components from one set of axes to another.
- The transformation equations are derived purely from equilibrium considerations and as a result are applicable to stresses in any kind of material such as elastic and elastic-plastic.
- The 2θ-version of the transformation equations, (7.8), is generally preferred to the θ-version of (7.6) due to the similarity between stresses $\sigma_{x'x'}$ and $\tau_{x'y'}$ and the use of angle 2θ in Mohr's stress circle, to be discussed later in §7.5.
- For $\theta = 0°$ then $\sigma_{x'x'} = \sigma_{xx}$ and $\tau_{x'y'} = \tau_{xy}$ as expected since the two sets of axes coincide, Figure 7.6a).
- For $\theta = 90°$ then $\sigma_{x'x'} = \sigma_{yy}$ and $\tau_{x'y'} = -\tau_{xy}$, Figure 7.6b).

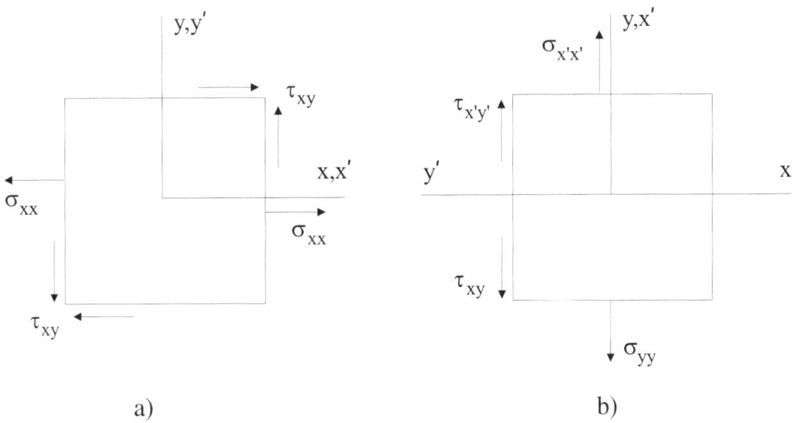

Figure 7.6: Global and local axes for a) $\theta = 0°$ and b) $\theta = 90°$.

The last two points above will prove useful when we discuss the construction of Mohr's stress circle in §7.5.1.

To obtain $\sigma_{y'y'}$ we simply replace θ by $(\theta + 90°)$ in $\sigma_{x'x'}$ in (7.8)

$$\sigma_{y'y'} = \left(\frac{\sigma_{xx} + \sigma_{yy}}{2}\right) + \left(\frac{\sigma_{xx} - \sigma_{yy}}{2}\right) \cos 2(\theta + 90°) + \tau_{xy} \sin 2(\theta + 90°) \tag{7.9}$$

and use of the trigonometric identities

$$\begin{aligned} \cos(\alpha + \beta) &= \cos\alpha \cos\beta - \sin\alpha \sin\beta \\ \sin(\alpha + \beta) &= \sin\alpha \cos\beta + \cos\alpha \sin\beta \end{aligned} \tag{7.10}$$

we find

$$\sigma_{y'y'} = \left(\frac{\sigma_{xx} + \sigma_{yy}}{2}\right) - \left(\frac{\sigma_{xx} - \sigma_{yy}}{2}\right) \cos 2\theta - \tau_{xy} \sin 2\theta \tag{7.11}$$

Combining $\sigma_{x'x'}$ and $\sigma_{y'y'}$ into a single expression

$$\sigma_{x'x',y'y'} = \left(\frac{\sigma_{xx} + \sigma_{yy}}{2}\right) \pm \left(\frac{\sigma_{xx} - \sigma_{yy}}{2}\right) \cos 2\theta \pm \tau_{xy} \sin 2\theta \tag{7.12}$$

where $\sigma_{x'x'}$ and $\sigma_{y'y'}$ adopt the $+$ and $-$ signs respectively. Hence, on inspection of (7.12) we see that

$$\sigma_{x'x'} + \sigma_{y'y'} = \sigma_{xx} + \sigma_{yy} \tag{7.13}$$

and indicates that the sum of the normal stress components in the global and local axes are equivalent and independent of θ.

Figure 7.7 illustrates $\sigma_{x'x'}$ and $\tau_{x'y'}$ plotted against θ for $\sigma_{yy} = 0.25\sigma_{xx}$ and $\tau_{xy} = 0.75\sigma_{xx}$ and simply illustrates that both $\sigma_{x'x'}$ and $\tau_{x'y'}$ have maxima and minima for various values of θ. It is these maximum and minimum values of stress that are of great importance to engineers and are considered in the next section.

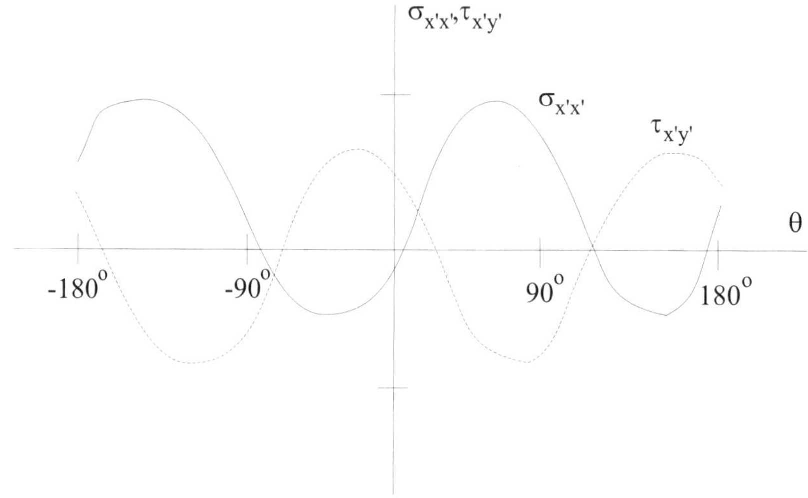

Figure 7.7: Variation of local stresses $\sigma_{x'x'}$ and $\tau_{x'y'}$ against θ for $\sigma_{yy} = 0.25\sigma_{xx}$ and $\tau_{xy} = 0.75\sigma_{xx}$.

7.3 Principal Stresses

To determine the angle, θ_p, at which the local stresses achieve maximum and minimum values we can differentiate $\sigma_{x'x'}$ with respect to θ and then set $\partial\sigma_{x'x'}/\partial\theta = 0$ to attain either a maxima or a minima

$$\begin{aligned}\frac{\partial\sigma_{x'x'}}{\partial\theta} &= \frac{\partial}{\partial\theta}\left[\left(\frac{\sigma_{xx} + \sigma_{yy}}{2}\right) + \left(\frac{\sigma_{xx} - \sigma_{yy}}{2}\right) \cos 2\theta + \tau_{xy} \sin 2\theta\right] \\ &= -(\sigma_{xx} - \sigma_{yy}) \sin 2\theta_p + 2\tau_{xy} \cos 2\theta_p = 0\end{aligned} \tag{7.14}$$

Re-arranging (7.14) in terms of θ_p

$$\tan 2\theta_p = \frac{2\tau_{xy}}{\sigma_{xx} - \sigma_{yy}} \tag{7.15}$$

Angle θ_p denotes the angle of the *principal planes* at which the maximum or minimum or *principal stresses* act. Due to the nature of the tangent function, equation (7.15) gives two values for θ_p in the interval $0 \leq \theta_p \leq 180°$.

To determine the principal stresses acting at θ_p we note from (7.15) that, Figure 7.8

$$\cos 2\theta_p = \frac{\sigma_{xx} - \sigma_{yy}}{R}; \quad \sin 2\theta_p = \frac{\tau_{xy}}{R}; \quad R = \sqrt{\left(\frac{\sigma_{xx} - \sigma_{yy}}{2}\right)^2 + \tau_{xy}^2} \tag{7.16}$$

where the positive square root of R is taken and the trivial solution ignored. Substituting (7.16) into $\sigma_{x'x'}$

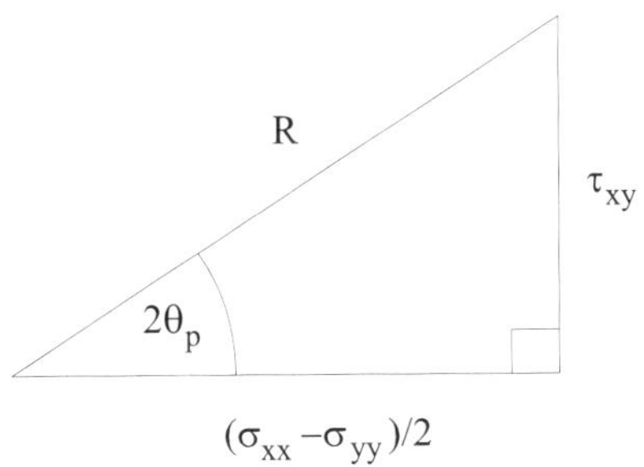

Figure 7.8: Principal plane angle, θ_p.

now denoted by σ_1 for $\theta = \theta_p$ we have

$$\sigma_1 = \left(\frac{\sigma_{xx} + \sigma_{yy}}{2}\right) + \sqrt{\left(\frac{\sigma_{xx} - \sigma_{yy}}{2}\right)^2 + \tau_{xy}^2} \tag{7.17}$$

and from (7.13)

$$\sigma_1 + \sigma_2 = \sigma_{xx} + \sigma_{yy} \tag{7.18}$$

Solving for σ_2 and combining σ_1 and σ_2 we have

$$\sigma_{1,2} = \left(\frac{\sigma_{xx} + \sigma_{yy}}{2}\right) \pm \sqrt{\left(\frac{\sigma_{xx} - \sigma_{yy}}{2}\right)^2 + \tau_{xy}^2} \tag{7.19}$$

with the positive sign resulting in the larger of the two principal stresses, that is $\sigma_1 > \sigma_2$. The principal stresses σ_1 and σ_2 occur at θ_{p1} and θ_{p2} respectively which are both determined from (7.15).

7.4 Maximum Shear Stress

To find the angle that the maximum shear stress acts we adopt a similar procedure to that in the previous section by differentiating $\tau_{x'y'}$ with respect to θ and then set $\partial\tau_{x'y'}/\partial\theta = 0$

$$\frac{\partial\tau_{x'y'}}{\partial\theta} = \frac{\partial}{\partial\theta}\left[-\left(\frac{\sigma_{xx} - \sigma_{yy}}{2}\right)\sin 2\theta + \tau_{xy}\cos 2\theta\right] = 0 \tag{7.20}$$

which leads to

$$\tan 2\theta_s = -\frac{\sigma_{xx} - \sigma_{yy}}{2\tau_{xy}} \tag{7.21}$$

where θ_s denotes the angle of the planes of the maximum shear stress.

Comparing (7.15) and (7.21) we observe that

$$\tan 2\theta_s = -\frac{1}{\tan 2\theta_p} = -\cot 2\theta_p \tag{7.22}$$

Noting, that for a general angle ϕ, $\tan(\phi \pm 90°) = -\cot\phi$ then

$$2\theta_s = 2\theta_p \pm 90° \quad \text{or} \quad \theta_s = \theta_p \pm 45° \tag{7.23}$$

which demonstrates that the planes of maximum shear stress occur at $45°$ to the principal planes.

To determine the *maximum shear stress*, τ_{max}, we follow a similar procedure to that for the principal stresses discussed in the previous section. If τ_{max} occurs at principal plane θ_{s1} then, with reference to Figure 7.9

$$\cos 2\theta_s = \frac{\tau_{xy}}{R}; \quad \sin 2\theta_s = \frac{-(\sigma_{xx} - \sigma_{yy})/2}{R}; \quad R = \sqrt{\left(\frac{\sigma_{xx} - \sigma_{yy}}{2}\right)^2 + \tau_{xy}^2} \tag{7.24}$$

which, upon substitution into (7.8), leads to τ_{max}

$$\tau_{max} = R = \sqrt{\left(\frac{\sigma_{xx} - \sigma_{yy}}{2}\right)^2 + \tau_{xy}^2} \tag{7.25}$$

The minimum shear stress, τ_{min}, has the same magnitude as τ_{max} but is opposite in sign. By subtracting σ_1 and σ_2 from (7.19) we observe that τ_{max} can be alternatively expressed as

$$\tau_{max} = \frac{\sigma_1 - \sigma_2}{2} \tag{7.26}$$

showing that the maximum shear stress is equal to one-half the difference of the principal stresses.

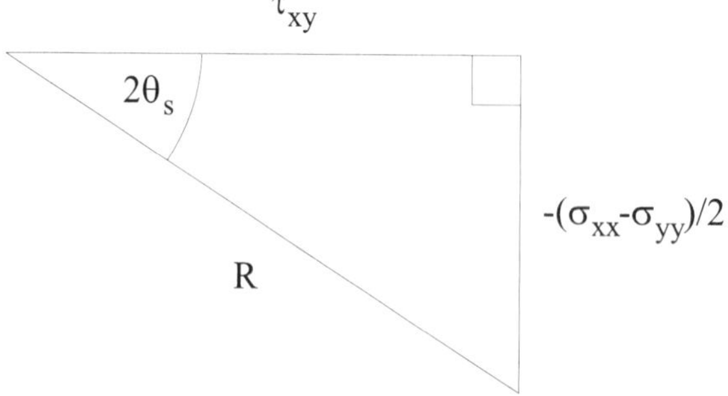

Figure 7.9: Maximum shear stress angle, θ_s.

Example 7.1 Uniaxial, biaxial and pure shear loading

In this example we will consider the state of stress under uniaxial, biaxial and pure shear loading.

Uniaxial Loading

Uniaxial loading results when a single loading is applied to a plate, Figure 7.10a). If the applied stress is σ_{xx} only then from the transformation equations (7.8)

$$\sigma_{x'x'} = \frac{\sigma_{xx}}{2}(1 + \cos 2\theta) = \sigma_{xx}\cos^2\theta$$
$$\tau_{x'y'} = -\frac{\sigma_{xx}}{2}\sin 2\theta = -\sigma_{xx}\sin\theta\cos\theta$$

with the variation in these stresses against θ shown in Figure 7.10b). The maximum value of $\sigma_{x'x'}$ occurs

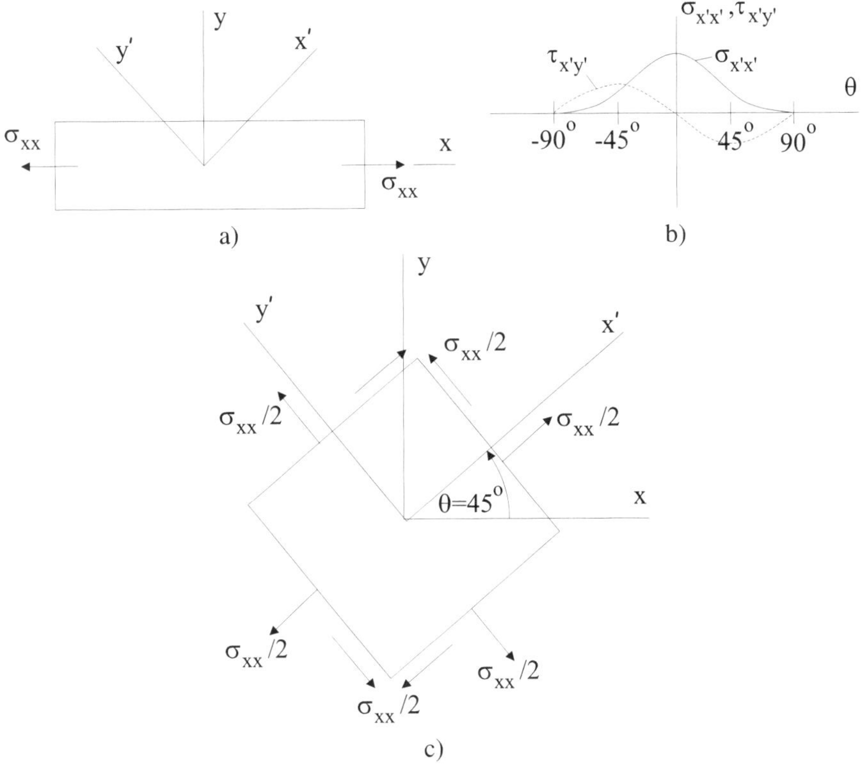

Figure 7.10: Example: Uniaxial loading. a) Stress σ_{xx} applied to a plate. b) The variation of local stresses $\sigma_{x'x'}$ and $\tau_{x'y'}$ against θ. c) Stress element for $\theta = 45°$.

at $\theta = 0°$ and is equal to σ_{xx} whereas at $\theta = 45°$ then $\sigma_{x'x'} = \sigma_{xx}/2$. The shear stress $\tau_{x'y'}$ is equal to zero for $\theta = 0°$ and $\theta = \pm 90°$ while the maximum value of $\tau_{x'y'}$ occurs at $\theta = \pm 45°$ and is equal to $\sigma_{xx}/2$. Thus, the maximum shear stress occurs on planes at $45°$ to the applied loading. The case of $\theta = 45°$ is illustrated in Figure 7.10c).

Biaxial Loading

Biaxial loading occurs when two mutually perpendicular loadings are applied to a plate, Figure 7.11. If the two applied loadings are equivalent then the loading is referred to as *equi-biaxial*. From the transformation equations (7.8)

$$\sigma_{x'x'} = \left(\frac{\sigma_{xx} + \sigma_{yy}}{2}\right) + \left(\frac{\sigma_{xx} - \sigma_{yy}}{2}\right)\cos 2\theta$$

$$\tau_{x'y'} = -\left(\frac{\sigma_{xx} - \sigma_{yy}}{2}\right)\sin 2\theta$$

which in the case of $\theta = 45°$ reduce to

$$\sigma_{x'x'} = \left(\frac{\sigma_{xx} + \sigma_{yy}}{2}\right); \quad \tau_{x'y'} = -\left(\frac{\sigma_{xx} - \sigma_{yy}}{2}\right)$$

In the case of equi-biaxial loading ($\sigma_{xx} = \sigma_{yy}$) then $\sigma_{x'x'} = \sigma_{xx}$, $\sigma_{y'y'} = \sigma_{yy}$ and $\tau_{x'y'} = 0$ irrespective of θ.

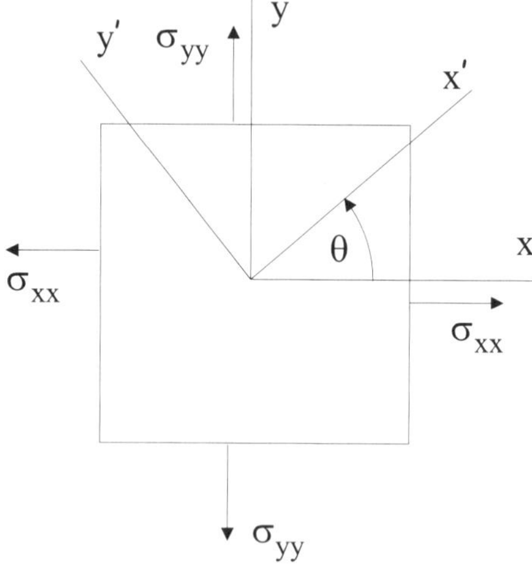

Figure 7.11: Example: Biaxial loading.

Pure Shear Loading

Figure 7.12 illustrates a plate subject to pure shear loading $\sigma_{xx} = \sigma_{yy} = 0$ and $\tau_{xy} \neq 0$. From the transformation equations (7.8)

$$\sigma_{x'x'} = 2\tau_{xy}\sin\theta\cos\theta$$

$$\tau_{x'y'} = \tau_{xy}(\cos^2\theta - \sin^2\theta)$$

which, in the case of $\theta = 45°$ reduce to $\sigma_{x'x'} = \tau_{xy}$ and $\tau_{x'y'} = 0$.

Example 7.2 Principal stresses and maximum shear stress for a shaft

In this example we will determine the principal stresses and maximum shear stress for a solid shaft of radius R and diameter D subject to pure bending moment M and torsion T, Figure 7.13. In the design of shafts subject to both bending and torsion it is the maximum bending and shear stresses that are of

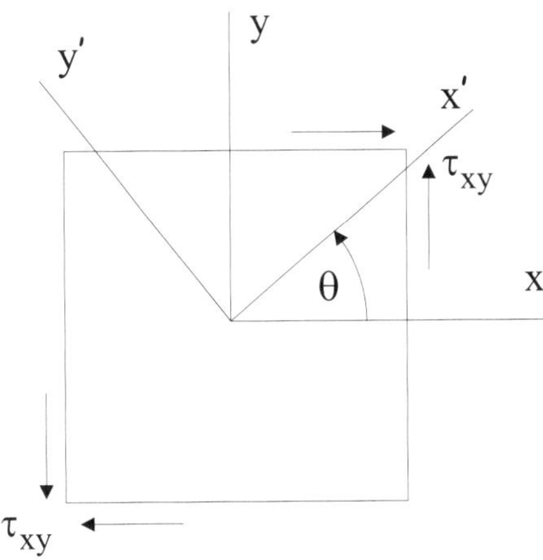

Figure 7.12: Example: Pure shear loading.

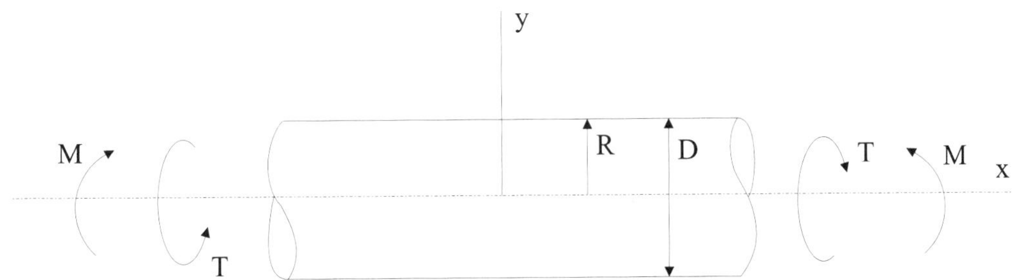

Figure 7.13: Example: A solid shaft subject to a pure bending moment M and torsion T.

greatest importance. These stresses achieve their maximum values at the surface of the shaft. The shear stress is given by (3.11) and with $J = \pi D^4/32$

$$\tau = \frac{TR}{J} = \frac{2T}{\pi R^3}$$

The direct stress as a result of bending is given by (5.30) and with $I = \pi R^4/4$ and $y = R$

$$\sigma = \frac{MR}{I} = \frac{4M}{\pi R^3}$$

An application of (7.19) with $\sigma_{xx} = \sigma$, $\sigma_{yy} = 0$ and $\tau_{xy} = \tau$ gives

$$\sigma_{1,2} = \frac{4M}{2\pi R^3} \pm \sqrt{\left(\frac{2M}{\pi R^3}\right)^2 + \left(\frac{2T}{\pi R^3}\right)^2} = \frac{2}{\pi R^3}\left[M \pm \sqrt{M^2 + T^2}\right]$$

The maximum shear stress is, from (7.26)

$$\tau_{max} = \frac{\sigma_1 - \sigma_2}{2} = \frac{2}{\pi R^3}\sqrt{M^2 + T^2}$$

Although these expressions are for elastic materials they are useful nevertheless when designing shafts.

If a hollow shaft is used instead of a solid shaft then, taking account of the different expressions for I and J, the above analysis is identical with τ and σ replaced by

$$\tau = \frac{2TR_o}{\pi(R_o^4 - R_i^4)}; \quad \sigma = \frac{4MR_o}{\pi(R_o^4 - R_i^4)}$$

where R_i and R_o are the inner and outer radii of the hollow shaft respectively.

7.5 Mohr's Circle for Plane Stress States

Re-arranging $\sigma_{x'x'}$ of (7.8)

$$\sigma_{x'x'} - \left(\frac{\sigma_{xx} + \sigma_{yy}}{2}\right) = \left(\frac{\sigma_{xx} - \sigma_{yy}}{2}\right) \cos 2\theta + \tau_{xy} \sin 2\theta \tag{7.27}$$

Squaring both (7.27) and $\tau_{x'y'}$ of (7.8) and adding together and noting that $\sin^2 2\theta + \cos^2 2\theta = 1$ then we arrive at

$$\left[\sigma_{x'x'} - \left(\frac{\sigma_{xx} + \sigma_{yy}}{2}\right)\right]^2 + \tau_{x'y'}^2 = \left[\left(\frac{\sigma_{xx} - \sigma_{yy}}{2}\right)^2 + \tau_{x'y'}^2\right] \tag{7.28}$$

which is of the form $(x-a)^2+(y-b)^2 = R^2$ for the general equation of a plane circle with centre at $[a, b]$ and radius R. Thus, with abscissa and ordinate axes $\sigma_{x'x'}$ and $\tau_{x'y'}$ respectively then (7.28) represents a circle with radius R and centre C

$$R = \tau_{max} = \sqrt{\left(\frac{\sigma_{xx} - \sigma_{yy}}{2}\right)^2 + \tau_{xy}^2}; \quad C = \left[\left(\frac{\sigma_{xx} + \sigma_{yy}}{2}\right), 0\right] \tag{7.29}$$

noting that the circle centre lies on the abscissa axis.

7.5.1 Construction of Mohr's Circle

In this section we will discuss a typical construction of Mohr's circle. Firstly, however, it is worth noting that throughout the literature there are generally two adopted conventions regarding the orientation of the axes, Figure 7.14. Referring back to Figure 7.3b), since we assumed θ to be positive when measured in the anti-clockwise direction then we will adopt the convention of Figure 7.14a) with the local axis $\tau_{x'y'}$ acting downwards. The construction of Mohr's circle can be broken down into the following four key steps, with reference to Figure 7.15:

1. Locate the centre of the circle $C[(\sigma_{xx} + \sigma_{yy})/2, 0]$.

2. Locate a point A on the circle that represents the stress state on the x-face of the stress element $(\theta = 0°)$ in which case $\sigma_{x'x'} = \sigma_{xx}$ and $\tau_{x'y'} = \tau_{xy}$.

3. Locate a point B on the circle that is diametrically opposite A that represents the stress state on the y-face of the stress element $(\theta = 90°)$ in which case $\sigma_{x'x'} = \sigma_{yy}$ and $\tau_{x'y'} = -\tau_{xy}$.

4. Draw a circle through points A and B with centre C.

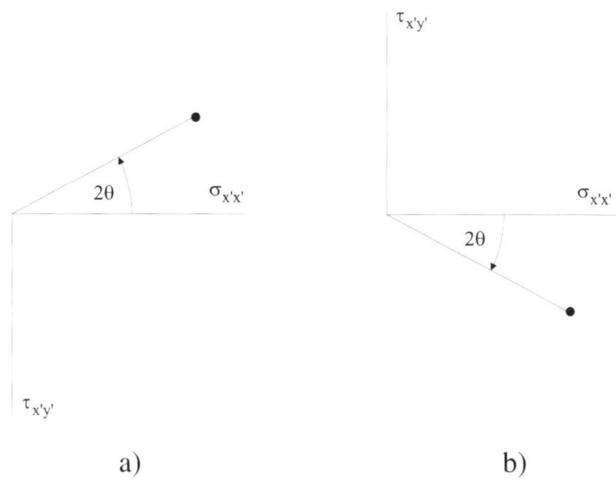

Figure 7.14: Local axes $\sigma_{x'x'}$ and $\tau_{x'y'}$ with a) θ measured in the anti-clockwise direction and b) θ measured in the clockwise direction.

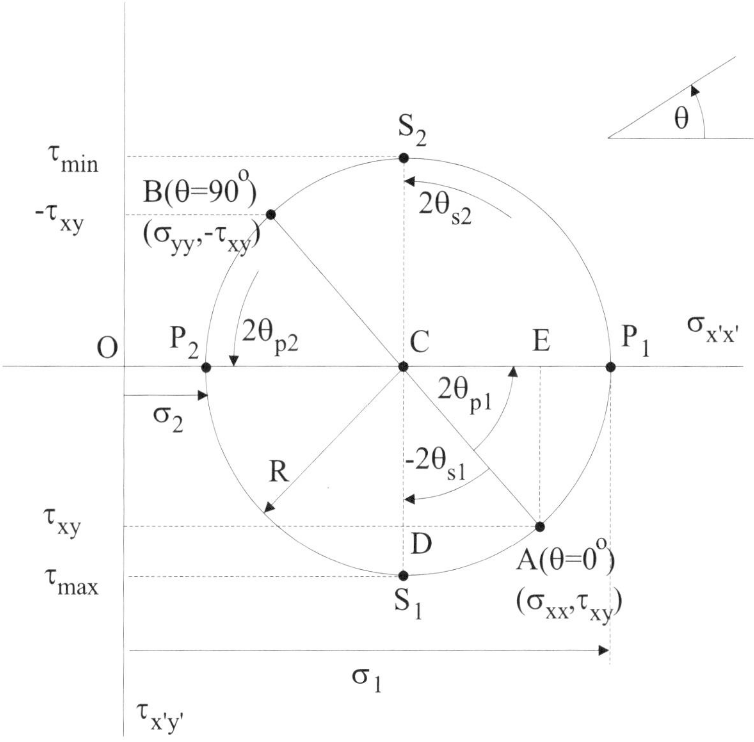

Figure 7.15: Mohr's stress circle.

Since A and B represent the stresses on planes $90°$ to each other then they are $180°$ (that is, 2θ) apart on the circle. It is worth emphasising that angle 2θ is measured in Mohr's circle and not θ. Also, when constructing Mohr's circle for the first time it is tempting to place the circle at the origin of coordinates, O, and not C. The principal stresses, σ_1 and σ_2, are the maximum and minimum values of $\sigma_{x'x'}$ respectively and by inspection of the circle are

$$\sigma_1 = OC + R; \quad \sigma_2 = OC - R \tag{7.30}$$

Angle ACP_1 is equal to $2\theta_{p1}$ and from the right-angled triangle AEC is given by

$$2\theta_p = \tan^{-1}\left(\frac{2\tau_{xy}}{\sigma_{xx} - \sigma_{yy}}\right) \tag{7.31}$$

which agrees with (7.15). Angle $2\theta_{p2}$ is seen to be equal to $180° + 2\theta_{p1}$. The radius of the circle is R and is equal to the maximum shear stress τ_{max} and can be found from, say, triangle ADC.

7.5.2 Calculation of Stresses on an Arbitrary Plane

In the previous sub-section we discussed the determination of the important stresses and angles which form Mohr's circle given the global plane stresses σ_{xx}, σ_{yy} and τ_{xy}. Let us now determine the stresses on a plane inclined at an angle θ from the x-axis using Mohr's circle, Figure 7.16a). Clearly, the transformation equations (7.8) can be used to determine the state of local stress but having previously constructed Mohr's circle then these local stresses quickly follow.

With reference to Figure 7.16b), since line CA represents $\theta = 0°$ if we want to determine the local stresses on a plane θ then we rotate 2θ in an anti-clockwise direction from CA to locate a point D on the circle. Point D has coordinates $(\sigma_{x'x'}, \tau_{x'y'})$ and represents the stresses acting on face x'. If angle DCP_1 is denoted by β then from point D we have

$$\sigma_{x'x'} = OC + R\cos\beta; \quad \tau_{x'y'} = R\sin\beta \tag{7.32}$$

and from point D' we have

$$\sigma_{y'y'} = OC - R\cos\beta \tag{7.33}$$

The $\sigma_{x'x'}$ stress at D' is equal to $\sigma_{y'y'}$ since D' is diametrically opposite D. Depending on whether $(2\theta_{p1} > 2\theta)$ or $(2\theta_{p1} < 2\theta)$ affects the sign of $\tau_{x'y'}$; see cases b) and c) of Figure 7.16. Therefore, if we always measure β from the plane 2θ to the plane $2\theta_{p1}$ the sign of $\tau_{x'y'}$ will be consistent.

7.5.3 Maximum Shear Stresses

Since the planes of maximum and minimum shear stress are at $45°$ to the principal planes then the planes of maxium and minimum shear stress are at right angles to σ_1 and σ_2 in Mohr's circle, Figure 7.15

$$\tau_{max} = R = CS_1; \quad -2\theta_{s1} = \angle ACS_1 = 90° - 2\theta_{p1} \tag{7.34}$$

and with $\tau_{min} = -\tau_{max}$ and $2\theta_{s2} = 180° + 2\theta_{s1}$ for the particular case shown in Figure 7.15.

Example 7.3 Construction of Mohr's circle

The state of global in-plane stress at a point is $\sigma_{xx} = 300$MPa, $\sigma_{yy} = 100$MPa and $\tau_{xy} = 200$MPa. To construct Mohr's circle we begin by determining the circle centre, C, radius R, and points A and B

$$C = \left[\left(\frac{\sigma_{xx} + \sigma_{yy}}{2}\right), 0\right] = [200, 0]$$

$$A(\theta = 0°) = (\sigma_{xx}, \tau_{xy}) = (300, 200)$$

$$B(\theta = 90°) = (\sigma_{yy}, -\tau_{xy}) = (100, -200)$$

$$R = \sqrt{\left(\frac{\sigma_{xx} - \sigma_{yy}}{2}\right)^2 + \tau_{xy}^2} = 223.6$$

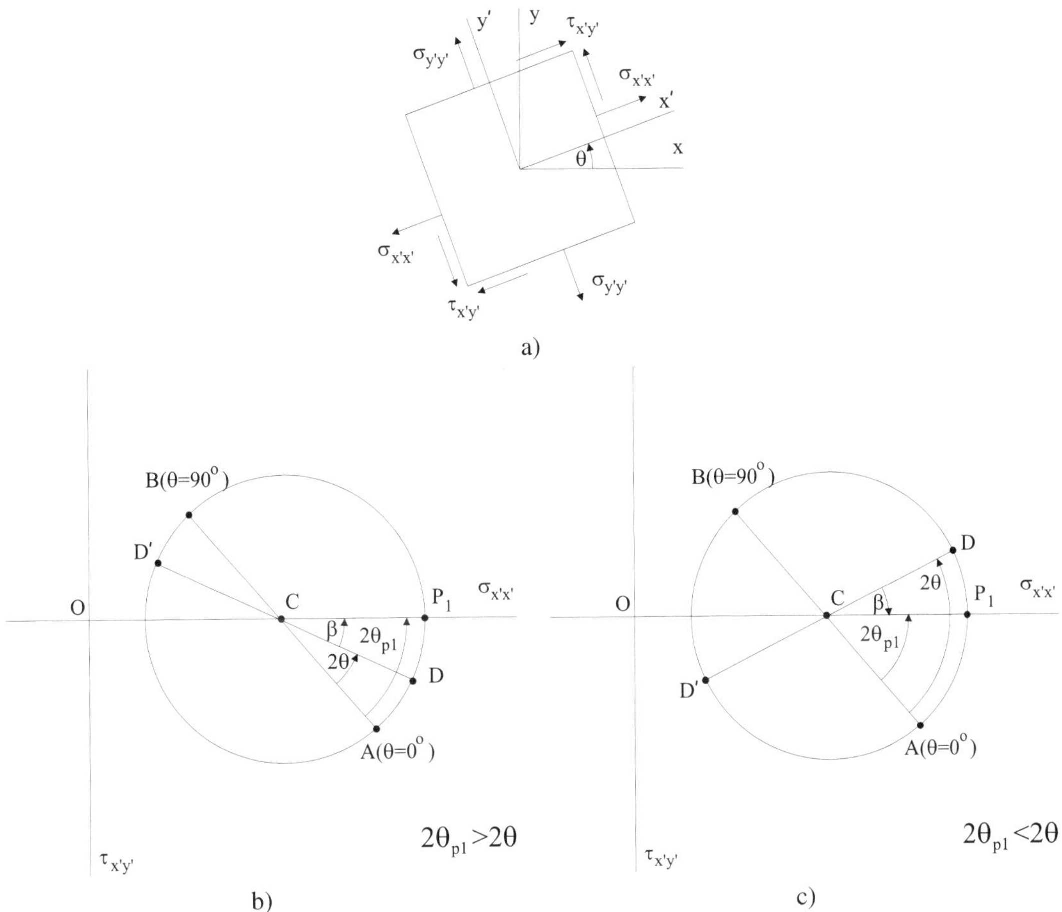

Figure 7.16: State of a stress on an arbitrary plane θ. a) Local stress element, b) Mohr's circle for $2\theta_{p1} > 2\theta$ and c) Mohr's circle for $2\theta_{p1} < 2\theta$.

The principal stresses from (7.30) are

$$\sigma_1 = OC + R = 423.6\text{MPa}; \quad \sigma_2 = OC - R = -23.6\text{MPa}$$

where O is the origin of the $(\sigma_{x'x'}, \tau_{x'y'})$ axes. Figure 7.17 illustrates the Mohr's circle for the present problem. The principal angle θ_{p1} associated with σ_1 is determined from the right-angled triangle AEC in Figure 7.17a)

$$\theta_{p1} = \frac{1}{2}\tan^{-1}\left(\frac{200}{100}\right) = 31.72^\circ$$

The principal angle θ_{p2} associated with σ_2 follows immediately from $2\theta_{p2} = 180^\circ + 2\theta_{p1}$ and hence $\theta_{p2} = 121.72^\circ$. The maximum shear stress is equal to R and so $\tau_{max} = 223.6$MPa and acts on the plane, from $\angle ACS_1$, $2\theta_{s1} = 90^\circ - 2\theta_{p1}$ and hence $\theta_{s1} = -13.29^\circ$

Let us now determine the state of stress for a stress element that is rotated through an angle of $\theta = 40^\circ$

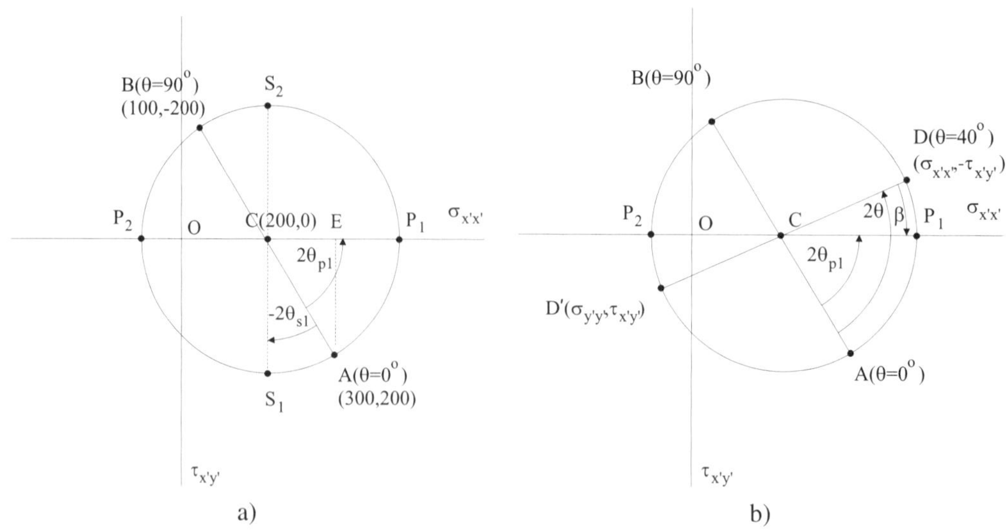

Figure 7.17: Example: Construction of Mohr's circle. a) Mohr's circle for the global stresses $\sigma_{xx} = 300$MPa, $\sigma_{yy} = 100$MPa and $\tau_{xy} = 200$MPa. b) Mohr's circle for the determination of the local stresses at an angle of $\theta = 40°$.

to the x-axis, Figure 7.17b). With $\beta = \angle DCP_1 = -(2\theta - 2\theta_{p1}) = -(80° - 63.43°) = -16.57°$ then

$$\begin{aligned}
\sigma_{x'x'}(D) &= OC + R\cos\beta = 200 + 223.6\cos(-16.57°) = 414.3 \\
\tau_{x'y'}(D) &= R\sin\beta = 200 - 223.6\sin(-16.57°) = -63.77 \\
\sigma_{x'x'}(D') &= OC - R\cos\beta = 200 - 223.6\cos(-16.57°) = -14.31 \\
\tau_{x'y'}(D') &= -R\sin\beta = -223.6\sin(-16.57°) = 63.77
\end{aligned}$$

and therefore the local stresses at $\theta = 40°$ are $\sigma_{x'x'} = 414.3$MPa, $\sigma_{y'y'} = -14.31$MPa and $\tau_{x'y'} = -63.77$MPa.

7.6 Second Moments of Area

In Chapter 2 we examined the second moments of area. In this section we will determine the second moments of area for a set of local axes in terms of the second moments of area for a global set of axes. The analogy between the transformation equations of stress will become self-evident.

The second moments of area I_x and I_y, and the product of area I_{xy} are dependent on the coordinate system used. For the xy axes

$$I_x = \iint_A y^2\,\mathrm{d}A; \quad I_y = \iint_A x^2\,\mathrm{d}A; \quad I_{xy} = \iint_A xy^2\,\mathrm{d}A; \tag{7.35}$$

To determine the second moments of area components for the $x'y'$-axes we begin by noting that a point (x, y) is transformed to the (x', y') axes by, Figure 7.18a)

$$\begin{aligned}
x' &= x\cos\theta + y\sin\theta \\
y' &= y\cos\theta - x\sin\theta
\end{aligned} \tag{7.36}$$

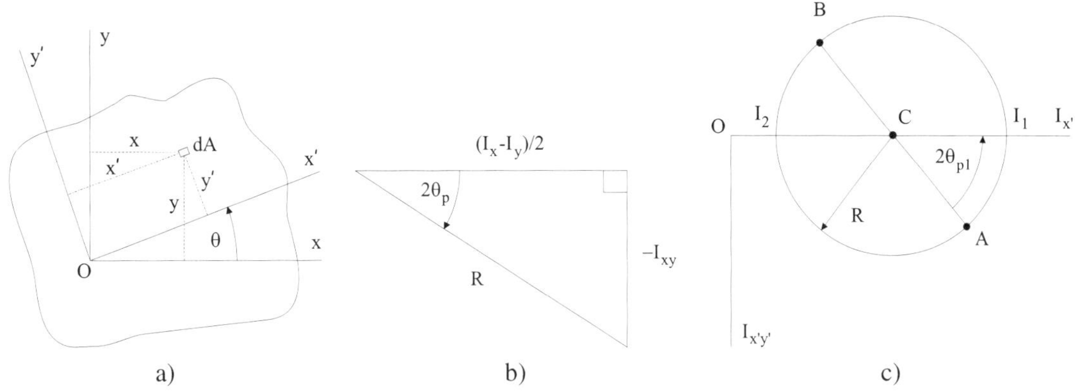

Figure 7.18: Second moments of area. a) Local axes (x', y') rotated through angle θ with respect to the global axes (x, y). b) Principal plane angle, θ_p. c) Mohr's circle for the second moments of area.

so that

$$\begin{aligned} I_{x'} &= \iint_A y'^2 \, \mathrm{d}A = \iint_A (y\cos\theta - x\sin\theta)^2 \, \mathrm{d}A \\ &= \cos^2\theta \iint_A y^2 \, \mathrm{d}A + \sin^2\theta \iint_A x^2 \, \mathrm{d}A - 2\sin\theta\cos\theta \iint_A xy \, \mathrm{d}A \\ &= I_x \cos^2\theta + I_y \sin^2\theta - 2I_{xy}\sin\theta\cos\theta \end{aligned} \tag{7.37}$$

Using the trigonometric identities $\cos^2\theta = (1+\cos^2\theta)/2$, $\sin^2\theta = (1-\cos^2\theta)/2$ and $2\sin\theta\cos\theta = \sin 2\theta$ then (7.37) can be written

$$I_{x'} = \left(\frac{I_x + I_y}{2}\right) + \left(\frac{I_x - I_y}{2}\right)\cos 2\theta - I_{xy}\sin 2\theta \tag{7.38}$$

A similar procedure can also be adopted for $I_{y'}$ so that the combined result is

$$I_{x',y'} = \left(\frac{I_x + I_y}{2}\right) \pm \left(\frac{I_x - I_y}{2}\right)\cos 2\theta \mp I_{xy}\sin 2\theta \tag{7.39}$$

Similarly, the product of area with respect to the (x', y') axes is

$$I_{x'y'} = \iint_A x'y' \, \mathrm{d}A = \iint_A (x\cos\theta + y\sin\theta)(y\cos\theta - x\sin\theta)\, \mathrm{d}A \tag{7.40}$$

and can be expressed in terms of 2θ with the aid of trigonometric identities

$$I_{x'y'} = \left(\frac{I_x - I_y}{2}\right)\sin 2\theta + I_{xy}\cos 2\theta \tag{7.41}$$

Equations (7.39) and (7.41) are referred to as *the transformation equations for the second moments and products of area* and are analogous to the transformation equations of stress (7.12) and (7.8). Taking the sum $I_{x'}$ and $I_{y'}$ we find from (7.39) that the sum of the second moments of area with respect to a set of axes remains unaltered if the axes are rotated about the origin

$$I_{x'} + I_{y'} = I_x + I_y \tag{7.42}$$

The transformation equations of the second moments of area illustrate how the moments of area vary according to the angle of rotation of the coordinate axes (and the planes on which they occur). In order to determine the planes of the minimum and maximum values of the moments of area we apply the condition $\mathrm{d}I_{x'} / \mathrm{d}\theta = 0$ to (7.39)

$$(I_{x'} - I_{y'}) \sin 2\theta + 2I_{xy} \cos 2\theta = 0 \tag{7.43}$$

and solving for θ gives

$$\tan 2\theta_p = -\frac{2I_{xy}}{I_x - I_y} \tag{7.44}$$

which has the two roots θ_{p1} and θ_{p2} which are referred to as the *principal planes*. Equation (7.44) illustrates that when $I_{xy} = 0$ then $\theta_{p1} = \theta_{p2} = 0$ and that the product of the moments of areas is equal to zero for the principal planes. Furthermore, it was noted in §2.5 and Example 2.6 that the product of areas is zero when one of the coordinate axes is an axis of symmetry of the plane area.

To determine the *principal moments of area* which act on the principal planes we note from (7.44) and with reference to Figure 7.18b) that

$$\cos 2\theta_p = \frac{I_x - I_y}{2R}; \quad \sin 2\theta_p = -\frac{I_{xy}}{R}; \quad R = \sqrt{\left(\frac{I_x - I_y}{2}\right)^2 + I_{xy}^2} \tag{7.45}$$

Substituting $\cos 2\theta_p$ and $\sin 2\theta_p$ into (7.39) we find the principal moments of area, I_1 and I_2

$$I_{1,2} = \frac{I_x + I_y}{2} \pm \sqrt{\left(\frac{I_x - I_y}{2}\right)^2 + I_{xy}^2} \tag{7.46}$$

with the algebraically larger of the two principal moments of area denoted by I_1.

Let us conclude by noting that Mohr's circle, radius R and centre $C(= (I_x + I_y)/2)$, can also be used to analyse the global-local moments of area as illustrated in Figure 7.18c).

Example 7.4 Principal second moments of area for a right-angled triangle

For the right-angled triangle shown in Figure 7.19 the second moments of area are given by

$$I_x = \frac{ba^3}{36}; \quad I_y = \frac{ab^3}{36}; \quad I_xy = -\frac{a^2b^2}{72};$$

with the origin of coordinates (x, y) centred at the triangle centroid G. Show that the principal second moments of area, I_1 and I_2, and their respective planes, θ_p, are given by

$$I_{1,2} = \frac{ab}{72}\left[(a^2 + b^2) \pm \sqrt{a^4 - a^2b^2 + b^4}\right]; \quad \tan 2\theta_p = \frac{ab}{b^2 - a^2}$$

From (7.46) I_1 and I_2 are

$$\begin{aligned} I_{1,2} &= \frac{1}{2}\left(\frac{ba^3}{36} + \frac{ab^3}{36}\right)^2 \pm \sqrt{\left(\frac{ba^3}{72} - \frac{ab^3}{72}\right)^2 + \left(\frac{a^2b^2}{72}\right)^2} \\ &= \frac{ab(a^2 + b^2)}{72} \pm \frac{1}{72}\sqrt{a^2b^2(a^4 - a^2b^2 + b^4)} \\ &= \frac{ab}{72}\left[(a^2 + b^2) \pm \sqrt{a^4 - a^2b^2 + b^4}\right] \end{aligned}$$

as required. The principal planes are found from (7.44)

$$\tan 2\theta_p = -\frac{2I_{xy}}{I_x - I_y} = -\frac{2\left(\frac{a^2b^2}{72}\right)}{\frac{ba^3}{36} - \frac{ab^3}{36}} = -\frac{a^2b^2}{ba^3 - ab^3} = \frac{ab}{b^2 - a^2}$$

Note that if $a = b$ then

$$I_1 = \frac{a^4}{24}; \quad I_2 = \frac{a^4}{72}; \quad \theta_p = 45^\circ$$

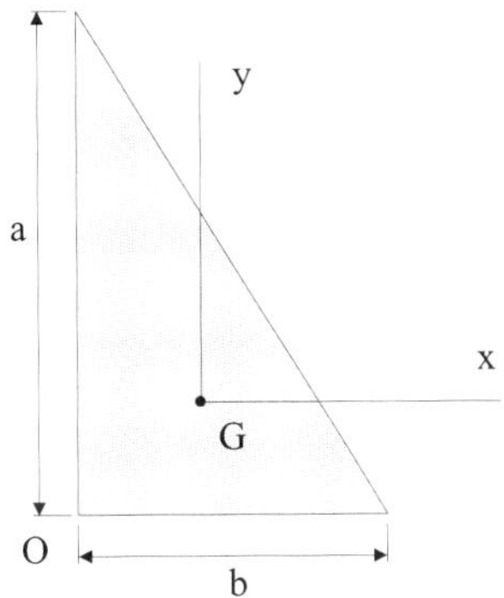

Figure 7.19: Example: A right-angled triangle with origin of coordinates (x, y) at the triangle centroid G.

7.7 Relationship Between Young's Modulus and the Shear Modulus

In the present section we derive the relationship $E = 2G(1 + \nu)$ between Young's modulus, E, and the shear modulus, G, first presented in §1.6. For an element $ABCD$ in a state of pure shear ($\sigma_{xx} = \sigma_{yy} = 0$, $\tau_{xy} = \tau$), Figure 7.20a), then the local stresses are, from (7.8) and (7.12)

$$\sigma_{x'x'y'y'} = \pm\tau \sin 2\theta; \quad \tau_{x'y'} = \tau \cos 2\theta \tag{7.47}$$

For $\theta = 0^\circ$ then $\sigma_{x'x'} = \sigma_{y'y'} = 0$ and $\tau_{x'y'} = \tau$ with the shear stress and shear strain related through the shear modulus, $G = \tau/\gamma$. For $\theta = 45^\circ$ then $\sigma_{x'x',y'y'} = \pm\tau$ and $\tau_{x'y'} = 0$ as shown in Figure 7.20b). With $\sigma_{x'x'} = \tau$ for $\theta = 45^\circ$ then the direct local strain is $\varepsilon_{x'x'} = \sigma_{x'x'}/E = \tau/E$. From Poisson's ratio ($\nu = \varepsilon_{y'y'}/\varepsilon_{x'x'}$) the associated direct local strain $\varepsilon_{y'y'}$ is $\varepsilon_{y'y'} = \nu\tau/E$. Thus, the resultant normal strain, ε, at $\theta = 45^\circ$ is

$$\varepsilon = \varepsilon_{x'x'} + \varepsilon_{y'y'} = \frac{\tau}{E} + \nu\frac{\tau}{E} = \frac{\tau}{E}(1 + \nu) \tag{7.48}$$

If the original element is square with edge lengths h then the distorted element forms a rhombus in which angles BAD and DCB reduce by γ, Figure 7.20c), with diagonal AC increasing in length from AC to $L_{AC} = AC + \delta_{AC}$ where $AC = \sqrt{2}h$. The extension of AC is equal to the normal strain multiplied by the length of AC; namely $\theta_{AC} = \sqrt{2}h\varepsilon$. With ε given by (7.48) then

$$\delta_{AC} = \frac{\sqrt{2}h\tau}{E}(1 + v) \tag{7.49}$$

In order to establish a relationship between E and G we require to express θ_{AC} in terms of the shear strain. From triangle ADE in Figure 7.20c)

$$\cos\left[\frac{1}{2}\left(\frac{\pi}{2}-\gamma\right)\right]=\frac{L_{AC}/2}{h} \tag{7.50}$$

Substituting $L_{AC}=\sqrt{2}h+\theta_{AC}$ and using the general trigonometric identity $\cos(\alpha-\beta)=\cos\alpha\cos\beta+\sin\alpha\sin\beta$ then (7.50) can be expressed as

$$\cos\left(\frac{\pi}{4}-\frac{\gamma}{2}\right)=\frac{1}{\sqrt{2}}\left(\cos\frac{\gamma}{2}+\sin\frac{\gamma}{2}\right)=\frac{\sqrt{2}h+\theta_{AC}}{2h} \tag{7.51}$$

For small shearing strains γ, $\cos(\gamma/2)\approx 1$ and $\sin(\gamma/2)\approx\gamma/2$ and (7.51) reduces to

$$\delta_{AC}=\frac{h\gamma}{\sqrt{2}} \tag{7.52}$$

With $G=\tau/\gamma$

$$\delta_{AC}=\frac{h\tau}{\sqrt{2}G} \tag{7.53}$$

Comparing (7.49) and (7.53) we arrive at the desired relationship between E and G

$$E=2G(1+\nu) \tag{7.54}$$

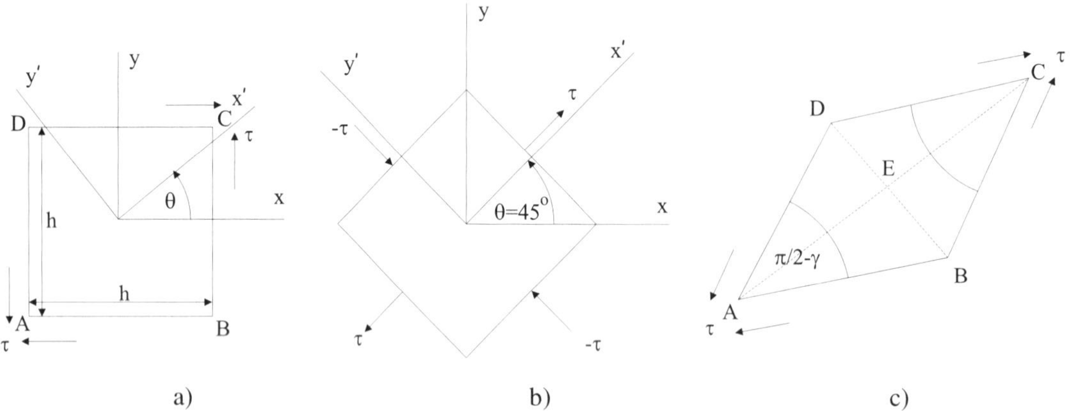

Figure 7.20: Element $ABCD$ subject to a state of pure shear. a) $\theta=0°$. b) $\theta=45°$. c) Distorted element.

7.8 Conclusion

This Chapter has introduced the stress transformation equations which enable the state of stress on a local set of axes to be determined from the state of stress with respect to the global set of axes. On consideration of the maximum and minimum local stresses we arrived at the principal stresses and maximum shear stress and their associated planes. An important property of the transformation equations is that they are derived from the equations of equilibrium and are therefore applicable to any material and not limited to linear elastic materials. The transformation equations can be illustrated conveniently and graphically by way of Mohr's circle from which the principal stresses, maximum shear stress and state of stress on an arbitrary plane can be determined. The Chapter also examined the determination of the second moments of area with respect to a set of local axes which have undergone a pure rotation with respect to the global set of axes. The Chapter concluded with an application of the transformation equations to determine a well-known relationship between Young's modulus and the shear modulus.

7.9 References and Further Reading

○ Benham, P. P., Crawford, R. J. and Armstrong, C. G. (1996) *Mechanics of Engineering Materials*, 2nd Ed., Longman, London.

○ Ugural, A. C. (1991) *Mechanics of Materials*, McGraw-Hill, Singapore.

○ Urry, S. A. and Turner, P. J. (1986) *Solving Problems in Solid Mechanics*, Volumes I and II, Pitman Press, Bath.

7.10 Exercises

7.1 A bar of cross-sectional area $A = 1,500\text{mm}^2$ is loaded in compression with a load of $P = 75\text{kN}$. If the bar is cut on a plane ab, as shown in Figure 7.21, determine the local state of stress for an element on the cut plane.

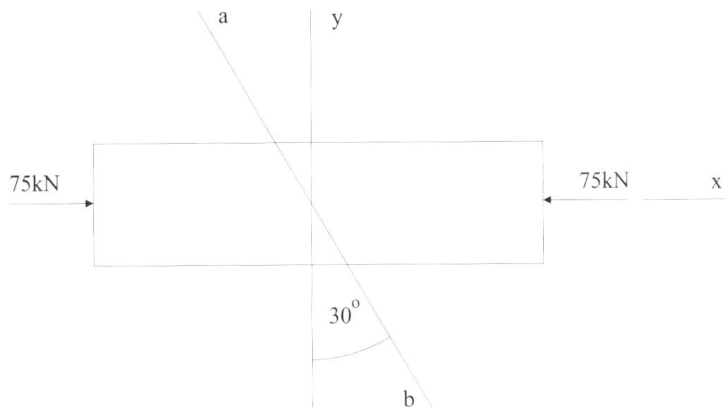

Figure 7.21: Exercise: A bar subject to a compressive load.

7.2 For the stress element shown in Figure 7.22 determine the stresses on an element that is rotated through a positive angle of $\theta = 45°$ with θ being assumed to be positive when measured in the anti-clockwise direction and measured from the x-axis. What do you note about the sum of the normal stresses in both the (x, y) and (x', y') coordinate systems?

7.3 Construct Mohr's circle for the plane element shown in Figure 7.23 in which the global stresses are $\sigma_{xx} = 50\text{MPa}$, $\sigma_{yy} = -100\text{MPa}$ and $\tau_{xy} = -50\text{MPa}$. Use Mohr's circle to determine the following:

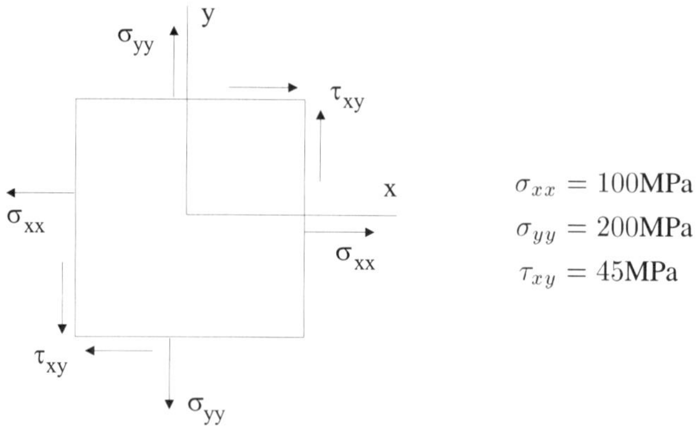

Figure 7.22: Exercise: A plane stress element.

- Principal stresses and associated planes.
- Maximum shear stress and associated plane.
- The stresses on an element which is rotated through an angle of $-45°$.

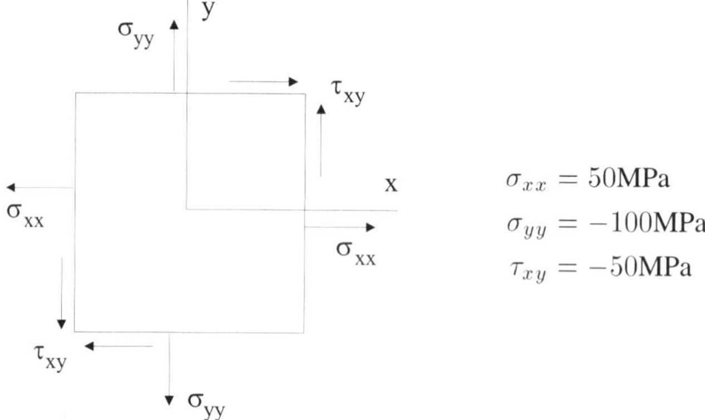

Figure 7.23: Exercise: Plane stress element with global stresses $\sigma_{xx} = 50$MPa, $\sigma_{yy} = -100$MPa and $\tau_{xy} = -50$MPa.

7.4 Schematically illustrate Mohr's circle for each of the three cases shown in Figure 7.24: a) uniaxial, b) equi-biaxial and c) pure shear loadings.

7.5 For a quadrant of a circle, Figure 7.25, with axes (x, y) centred at the centroid, G, then the second moments of area I_x, I_y and I_{xy} are given by

$$I_x = I_y = \left(\frac{\pi}{16} - \frac{4}{9\pi}\right) r^4; \quad I_{xy} = \left(\frac{1}{8} - \frac{4}{9\pi}\right) r^4$$

1. Show that the principal second moments of area, I_1 and I_2, and associated plane, θ_p, are given by

$$I_1 = \left(\frac{9\pi^2 + 18\pi - 128}{144\pi}\right) r^4; \quad I_2 = \left(\frac{\pi - 2}{16}\right) r^4; \quad \theta_p = 45°$$

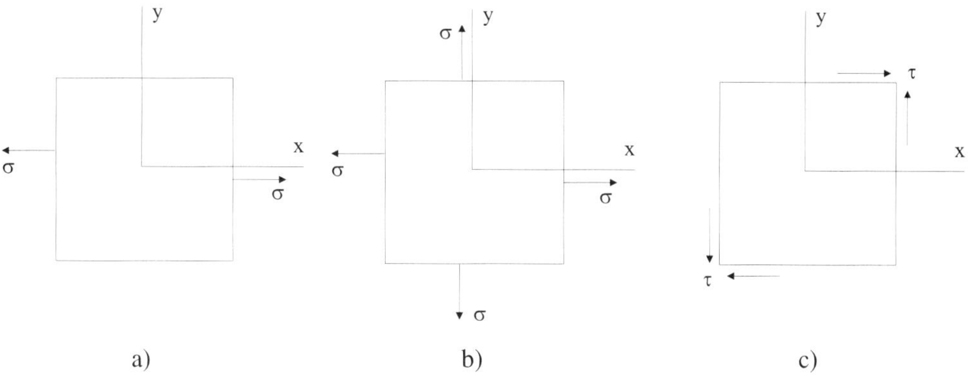

Figure 7.24: Exercise: Plane stress element loadings a) uniaxial, b) equi-biaxial and c) pure shear.

2. If $r = 10$mm then determine the second moments of area $I_{x'}$, I_y' and $I_{x'y'}$ for the set of axes (x', y') which are rotated in an anticlockwise direction through an angle of $\theta = 30°$ relative to the (x, y) axes.

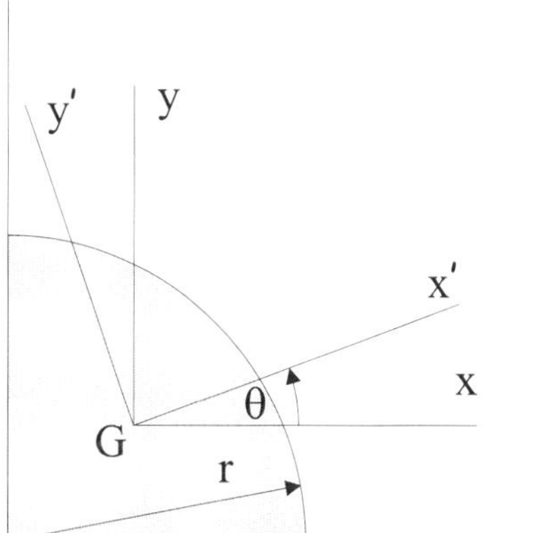

Figure 7.25: Exercise: Quadrant of a circle of radius r with axes (x, y) centred at the centroid G.

7.6 Figure 7.26 illustrates an equilateral triangle of edge length b with a vertex at the origin, O, of co-ordinates. Show that the values of $I_{x'}$, $I_{y'}$ and $I_{x'y'}$ for the local axes (x', y') with $\theta = 30°$ are given by

$$I_{x'} = I_y; \quad I_{y'} = I_x; \quad I_{x'y'} = I_{xy}$$

with I_x, I_y and I_{xy} given by

$$I_x = \frac{\sqrt{3}}{32}b^4; \quad I_y = \frac{7\sqrt{3}}{96}b^4; \quad I_{xy} = \frac{1}{16}b^4$$

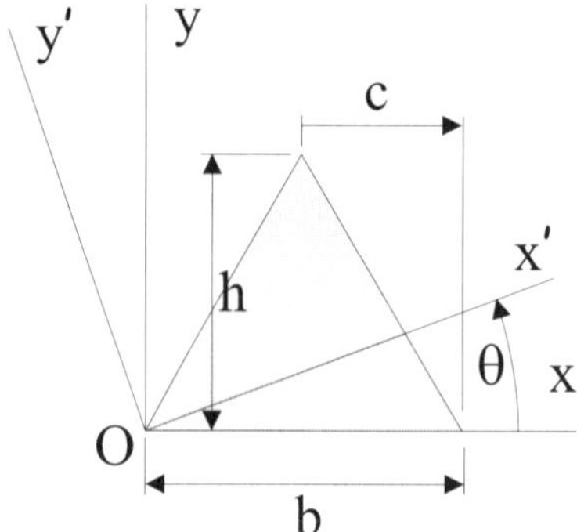

Figure 7.26: Exercise: An equilateral triangle of edge length b with the origin, O, of axes (x, y) centred at a vertex.

7.7 The Young's modulus, E, and Poisson's ratio, ν, of an aluminium alloy and steel are $(E, \nu) = (70\text{GPa}, 0.33)$ and $(210\text{GPa}, 0.3)$ respectively. Determine the shear modulus for both materials.

Chapter 8

Strain Analysis

8.1 Introduction

The analysis of strain at a point is similar to the analysis of stress at a point and as a result this Chapter relies heavily on a good understanding of the previous Chapter. The state of strain in an arbitrary three-dimensional body is generally represented by the three components of normal strain ε_{xx}, ε_{yy} and ε_{zz} and the three components of shear strain γ_{xy}, γ_{yz} and γ_{xz}. Consider the undeformed body shown in Figure 8.1a). Assume that the body consists of a number of small continuous hexahedral elements each of dimensions $\mathrm{d}x$, $\mathrm{d}y$ and $\mathrm{d}z$, Figure 8.1b). If the dimensions of the elements are small then the deformed shape of the elements will be a parallelepiped since straight line segments will remain straight after deformation, Figure 8.1c). The approximate lengths of the parallelepiped and the angles between the sides after deformation are shown in Figure 8.1c). Normal strains are produced by changes in length and shear strains are produced by relative rotations of two adjacent sides of an element.

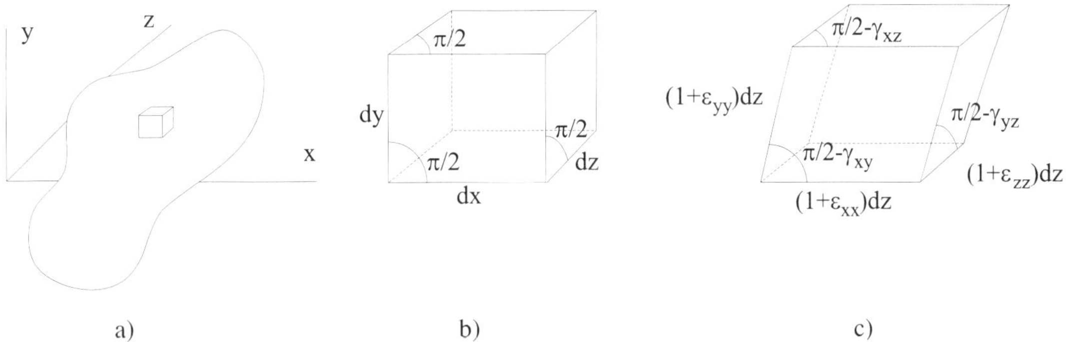

Figure 8.1: Strain in a body. a) A stress-strain element in a body. b) Initial undeformed element. c) Deformed element.

The following sections examine strain components in a plane and their associated sign convention. The strain transformation equations are then derived for a local set of axes. Thereafter these equations are used to determine the principal direct and maximum shear strains and their corresponding planes.

The construction of Mohr's circle is then discussed and the Chapter concludes with a discussion on the determination of the in-plane global strain components from a strain gauge rosette.

8.2 Strain in a Plane

Let us now restrict our attention to strain in a plane and hence neglect the effects of the components ε_{zz}, γ_{yz} and γ_{xz}. Consequently, an element in a plane is subject to normal strain components ε_{xx} and ε_{yy} and the shear strain γ_{xy}, Figure 8.2. It is worth noting that strain in a plane has the three components

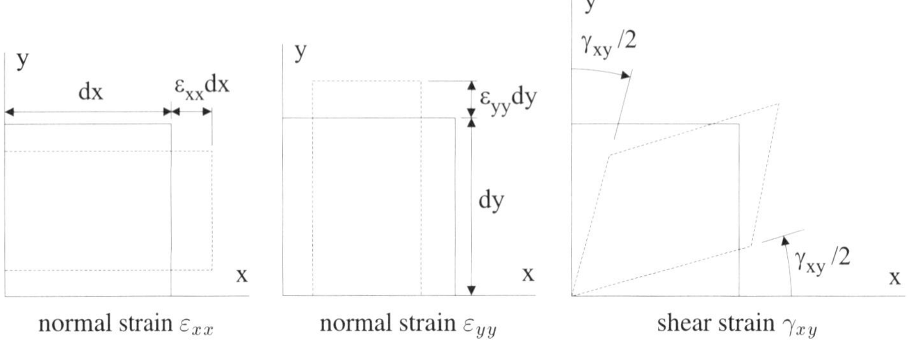

Figure 8.2: Normal and shear strain components.

ε_{xx}, ε_{yy} and τ_{xy} all lying in the same plane. Out of plane strains also occur as a result of the Poisson contraction. To illustrate this consider the element of Figure 8.3 subject to a uniaxial stress σ_{xx}. A normal

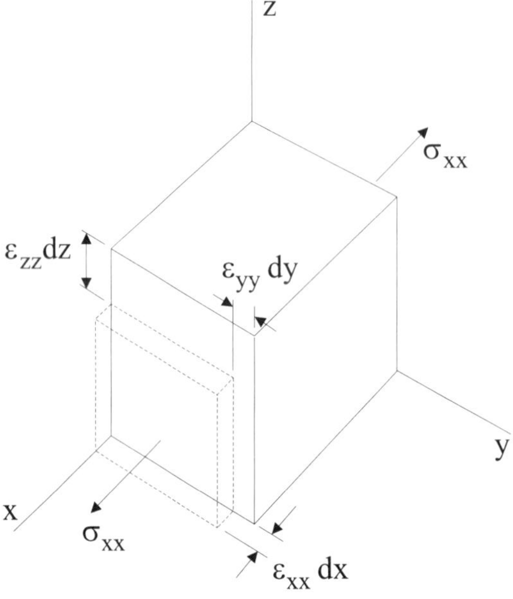

Figure 8.3: An element subject to the stress σ_{xx}.

stress σ_{xx} results not only in a normal strain ε_{xx} but also in the two associated strains $\varepsilon_{yy} = -\nu\varepsilon_{xx}$ and $\varepsilon_{zz} = -\nu\varepsilon_{xx}$. Thus, since $\varepsilon_{zz} \neq 0$ this is not simply a case of strain in the xy-plane. Furthermore, since shear stress and shear strain are not affected by Poisson's contraction a condition of $\tau_{yz} = \tau_{xz} = 0$ requires that $\gamma_{yz} = \gamma_{xz} = 0$.

As with stress at a point, the present Chapter addresses the problem of determining the (x', y') local components of normal and shear strain at a given point $(\varepsilon_{x'x'}, \varepsilon_{y'y'}, \gamma_{x'y'})$ given the (x, y) global components $(\varepsilon_{xx}, \varepsilon_{yy}, \gamma_{xy})$. However, before the transformation equations of strain are derived let us have a note on the sign convention used.

8.3 Sign Convention

Figure 8.4 illustrates the deformation of an element. Normal strains ε_{xx} and ε_{yy} are positive if they cause an elongation along the positive x and y axes respectively. The shear strain γ_{xy} is assumed to be positive if the interior angle AOB decreases as a result of deformation.

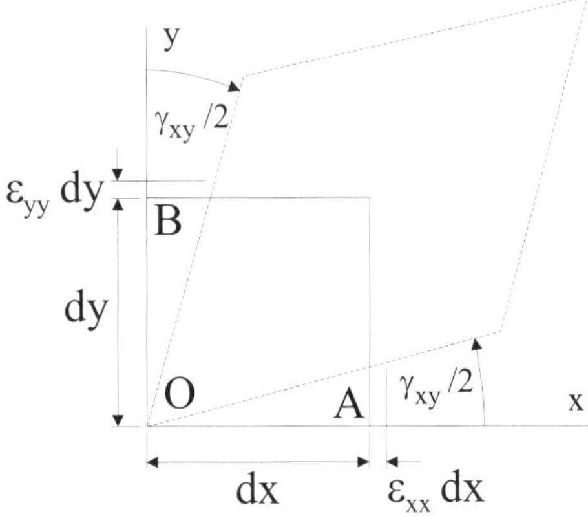

Figure 8.4: Strain sign convention.

8.4 Normal and Shear Strains

In this section we will develop the transformation equations of strain with respect to the global (x, y) axes and local (x', y') axes. For each of the three in-plane strains ε_{xx}, ε_{yy} and γ_{xy} we will examine their contribution to the extension of element $\mathrm{d}x'$ which lies on the x'-axis. Consider a plane subject to a positive normal strain ε_{xx}, Figurc 8.5a). The line $\mathrm{d}x$ is extended by $\varepsilon_{xx}\,\mathrm{d}x$ and the line $\mathrm{d}x'$ is extended by $\varepsilon_{xx}\,\mathrm{d}x\cos\theta$. Alternatively, if the strain ε_{yy} is now applied, Figure 8.5b), then the line $\mathrm{d}y$ is extended by $\varepsilon_{yy}\,\mathrm{d}y$ which results in the line $\mathrm{d}x'$ extending by $\varepsilon_{yy}\,\mathrm{d}y\sin\theta$. With $\mathrm{d}x$ remaining fixed the shear strain γ_{xy} causes the line $\mathrm{d}y$ to be displaced by $\gamma_{xy}\,\mathrm{d}y$, Figure 8.5c), from the general formula for arc length $s = r\theta$ with $r = \mathrm{d}y$ and $\theta = \gamma_{xy}$ for small θ and resulting in the line $\mathrm{d}x'$ extending by $\gamma_{xy}\,\mathrm{d}y\cos\theta$. Summing these three individual contributions to the extension, $\delta x'$, of line $\mathrm{d}x'$ we have

$$\delta x' = \varepsilon_{xx}\,\mathrm{d}x\cos\theta + \varepsilon_{yy}\,\mathrm{d}y\sin\theta + \gamma_{xy}\cos\theta \tag{8.1}$$

and with $\varepsilon_{x'x'} = \delta x' / \mathrm{d}x'$, $\mathrm{d}x = \mathrm{d}x'\cos\theta$ and $\mathrm{d}y = \mathrm{d}x'\sin\theta$ then

$$\varepsilon_{x'x'} = \varepsilon_{xx}\cos^2\theta + \varepsilon_{yy}\sin^2\theta + \gamma_{xy}\sin\theta\cos\theta \tag{8.2}$$

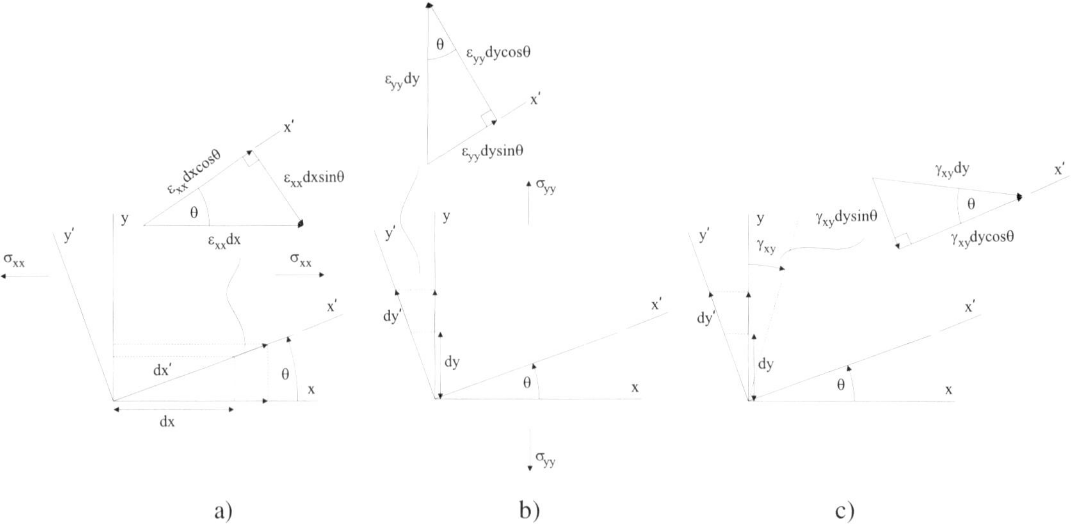

Figure 8.5: Strain components in the global and local axes. a) Normal strain ε_{xx}. b) Normal strain ε_{yy}. c) Shear strain γ_{xy}.

and is the *strain transformation equation* for determining the local strain $\varepsilon_{x'x'}$ from the global strains.

The local strain $\varepsilon_{y'y'}$ is determined from $\varepsilon_{x'x'}$ by substituting $(\theta + 90°)$ for θ in (8.2)

$$\varepsilon_{y'y'} = \varepsilon_{xx}\cos^2(\theta + 90°) + \varepsilon_{yy}\sin^2(\theta + 90°) + \gamma_{xy}\sin(\theta + 90°)\cos(\theta + 90°) \tag{8.3}$$

and with $\sin(\theta + 90°) = \cos\theta$ and $\cos(\theta + 90°) = -\sin\theta$ then

$$\varepsilon_{y'y'} = \varepsilon_{xx}\sin^2\theta + \varepsilon_{yy}\cos^2\theta - \gamma_{xy}\sin\theta\cos\theta \tag{8.4}$$

To determine the strain transformation equation for $\gamma_{x'y'}$ we will consider the rotation of the $\mathrm{d}x'$ and $\mathrm{d}y'$ line segments when subject to the ε_{xx}, ε_{yy} and γ_{xy} strain components, Figure 8.6. Following a similar procedure to that above, but now for the extension of the $\mathrm{d}y'$ segment then ε_{xx} extends $\mathrm{d}y'$ by $-\varepsilon_{xx}\,\mathrm{d}x\sin\theta$, ε_{yy} extends $\mathrm{d}y'$ by $\varepsilon_{yy}\,\mathrm{d}y\cos\theta$ and γ_{xy} extends $\mathrm{d}y'$ by $-\gamma_{xy}\,\mathrm{d}y\sin\theta$. Thus, the resultant elongation is

$$\delta y' = -\varepsilon_{xx}\,\mathrm{d}x\sin\theta + \varepsilon_{yy}\,\mathrm{d}y\cos\theta - \gamma_{xy}\,\mathrm{d}y\sin^2\theta \tag{8.5}$$

and again noting that $\mathrm{d}x = \mathrm{d}x'\cos\theta$ and $\mathrm{d}y = \mathrm{d}x'\sin\theta$ then

$$\delta y' = -\varepsilon_{xx}\,\mathrm{d}x'\sin\theta\cos\theta + \varepsilon_{yy}\,\mathrm{d}x'\sin\theta\cos\theta - \gamma_{xy}\,\mathrm{d}x'\sin^2\theta \tag{8.6}$$

From Figure 8.6 we have that $\alpha \approx \tan\alpha = \delta y'/\,\mathrm{d}x'$ for small angles and dividing (8.6) by $\mathrm{d}x'$

$$\alpha = -\varepsilon_{xx}\sin\theta\cos\theta + \varepsilon_{yy}\sin\theta\cos\theta - \gamma_{xy}\sin^2\theta \tag{8.7}$$

To find angle β we can make the substitution $(\theta + 90°)$ for θ in (8.7) and noting once more that $\sin(\theta + 90°) = \cos\theta$ and $\cos(\theta + 90°) = -\sin\theta$

$$\beta = \varepsilon_{xx}\sin\theta\cos\theta - \varepsilon_{yy}\sin\theta\cos\theta - \gamma_{xy}\cos^2\theta \tag{8.8}$$

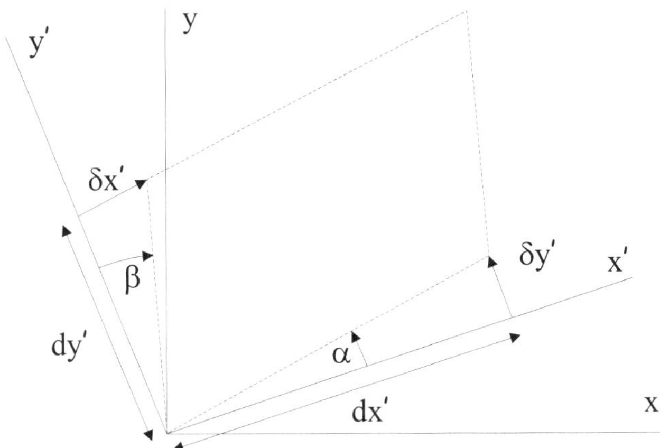

Figure 8.6: Rotation of line segments $\mathrm{d}x'$ and $\mathrm{d}y'$ through the angles α and β respectively.

With α and β acting in opposite directions then the total shear strain is

$$\gamma_{x'y'} = \alpha - \beta = -2\varepsilon_{xx}\sin\theta\cos\theta + 2\varepsilon_{yy}\sin\theta\cos\theta + \gamma_{xy}(\cos^2\theta - \sin^2\theta) \tag{8.9}$$

As with the stress transformation equations in the previous Chapter, the most convenient form of the above strain transformations is in terms of angle 2θ by making use of the trigonometric identities $\sin 2\theta = 2\sin\theta\cos\theta$, $\cos 2\theta = (1+\cos^2\theta)/2$ and $\sin^2\theta + \cos^2\theta = 1$

$$\begin{aligned} \varepsilon_{x'x',y'y'} &= \left(\frac{\varepsilon_{xx}+\varepsilon_{yy}}{2}\right) \pm \left(\frac{\varepsilon_{xx}-\varepsilon_{yy}}{2}\right)\cos 2\theta \mp \frac{\gamma_{xy}}{2}\sin 2\theta \\ \frac{\gamma_{x'y'}}{2} &= -\left(\frac{\varepsilon_{xx}-\varepsilon_{yy}}{2}\right)\sin 2\theta + \frac{\gamma_{xy}}{2}\cos 2\theta \end{aligned} \tag{8.10}$$

Comparing these transformations to the stress transformations of (7.8) we observe a direct equivalence between $(\varepsilon_{xx}, \varepsilon_{yy}, \varepsilon_{x'x'}, \varepsilon_{y'y'})$ and $(\sigma_{xx}, \sigma_{yy}, \sigma_{x'x'}, \sigma_{y'y'})$, and $(\tau_{xy}, \tau_{x'y'})$ and $(\gamma_{xy}/2, \gamma_{x'y'}/2)$ noting the factor of $\frac{1}{2}$ in the case of the shear strain.

8.5 Principal Strains and Maximum Shear Strain

As with stress, the orientation of an element can be such that deformation is composed solely of normal strains with no shear strain. In this case the normal strains are referred to as the *principal strains* and the planes along which they occur coincide with the planes of the principal stresses. Thus, with a direct analogy with the principal stresses and maximum shear stress of the previous Chapter we find that the principal strains, ε_1 and ε_2, and the principal planes, θ_p, are

$$\varepsilon_{1,2} = \left(\frac{\varepsilon_{xx}+\varepsilon_{yy}}{2}\right) \pm \sqrt{\left(\frac{\varepsilon_{xx}-\varepsilon_{yy}}{2}\right)^2 + \left(\frac{\gamma_{xy}}{2}\right)^2}; \quad \tan 2\theta_p = \frac{\gamma_{xy}}{\varepsilon_{xx}-\varepsilon_{yy}} \tag{8.11}$$

Similarly, the maximum shear strain, $(\gamma_{x'y'}/2)_{max}$, and associated plane, θ_s, are

$$\left(\frac{\gamma_{x'y'}}{2}\right)_{max} = \sqrt{\left(\frac{\varepsilon_{xx}-\varepsilon_{yy}}{2}\right)^2 + \left(\frac{\gamma_{xy}}{2}\right)^2}; \quad \tan 2\theta_p = -\frac{\varepsilon_{xx}-\varepsilon_{yy}}{\gamma_{xy}} \tag{8.12}$$

In addition, an average of the global normal strains, ε_{avg}, is occasionally used

$$\varepsilon_{avg} = \frac{\varepsilon_{xx} + \varepsilon_{yy}}{2} \tag{8.13}$$

8.6 Mohr's Circle for Strains in a Plane

Since there is a direct analogy between the stress and strain transformations equations we can similarly represent strain transformations graphically using Mohr's construction. Squaring both of (8.10), adding together and noting $\sin^2 2\theta + \cos^2 2\theta = 1$

$$\left[\varepsilon_{x'x'} - \left(\frac{\varepsilon_{xx} + \varepsilon_{yy}}{2}\right)\right]^2 + \left(\frac{\gamma_{x'y'}}{2}\right)^2 = \left[\left(\frac{\varepsilon_{xx} - \varepsilon_{yy}}{2}\right)^2 + \left(\frac{\gamma_{x'y'}}{2}\right)^2\right] \tag{8.14}$$

which can be expressed as

$$(\varepsilon_{x'x'} - \varepsilon_{avg})^2 + \left(\frac{\gamma_{x'y'}}{2}\right)^2 = R^2 \tag{8.15}$$

with ε_{avg} given by (8.13) and R as follows

$$R = \sqrt{\left(\frac{\varepsilon_{xx} - \varepsilon_{yy}}{2}\right)^2 + \left(\frac{\gamma_{xy}}{2}\right)^2} \tag{8.16}$$

Equation (8.15) is of the form $(x-a)^2 + (y-b)^2 = R^2$ which represents the equation of a plane circle with centre at $[a, b]$ and radius R.

Figure 8.7 illustrates Mohr's circle for the global strains ε_{xx}, ε_{yy} and γ_{xy}. As with Mohr's stress circle, the construction of Mohr's strain circle is best drawn by following the four key steps:

1. Locate the centre of the the circle $C[\varepsilon_{avg}, 0]$.
2. Locate a point A on the circle that represents the strain state on the x-face of the stress element $(\theta = 0°)$ in which case $\varepsilon_{x'x'} = \varepsilon_{xx}$ and $\gamma_{x'y'} = \gamma_{xy}$.
3. Locate a point B on the circle that is diametrically opposite A that represents the strain state on the y-face of the strain element $(\theta = 90°)$ in which case $\varepsilon_{x'x'} = \varepsilon_{yy}$ and $\gamma_{x'y'} = -\gamma_{xy}$.
4. Draw a circle through points A and B with centre C.

It follows that the principal strains are given by $\varepsilon_1 = OC + R$ and $\varepsilon_2 = OC - R$.

8.6.1 Calculation of Strains on an Arbitrary Plane

The components of strain $\varepsilon_{x'x'}$, $\varepsilon_{y'y'}$ and $\gamma_{x'y'}$ on a plane which is oriented at an angle of θ to the x-axis can be determined from the strain transformation equations but alternatively found from Mohr's circle in an identical manner to that outlined in §7.5. Figure 8.8 illustrates the state of strain at an angle θ in which $2\theta < 2\theta_{p1}$ with 2θ measured from the plane AB in an anti-clockwise direction.

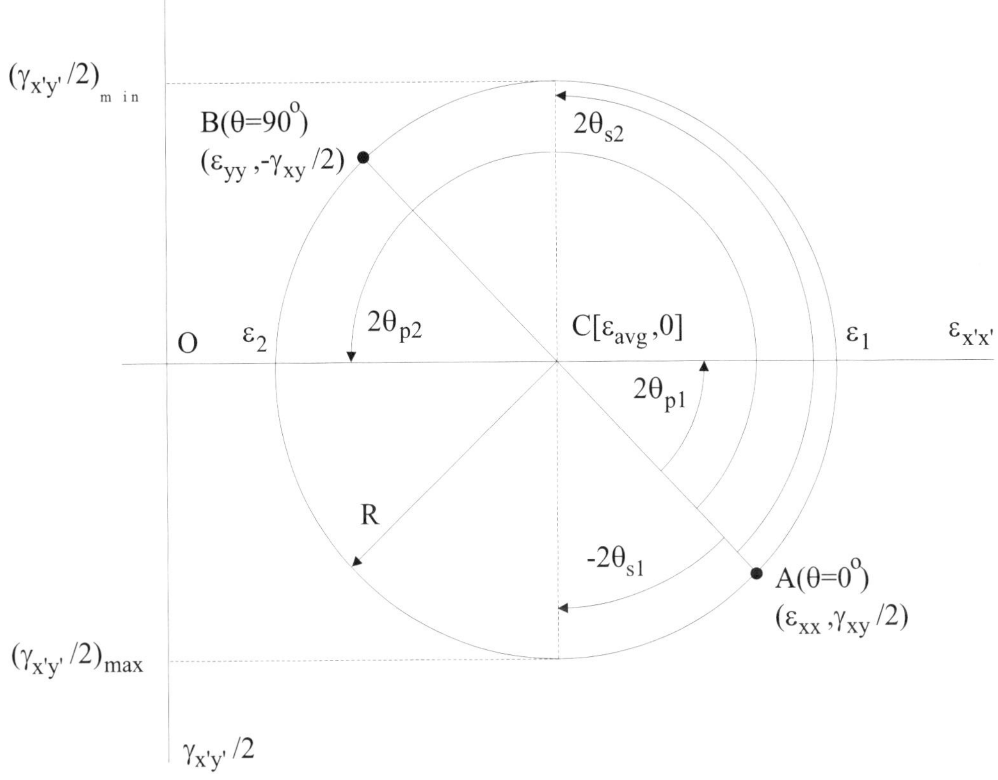

Figure 8.7: Mohr's strain circle for the global strains ε_{xx}, ε_{yy} and γ_{xy}.

Example 8.1 Construction of Mohr's circle

For the in-plane global strains $\varepsilon_{xx} = 75 \times 10^{-6}$, $\varepsilon_{yy} = 25 \times 10^{-6}$ and $\gamma_{xy} = 50 \times 10^{-6}$ determine, using Mohr's circle, the principal strains and their associated planes, the maximum and minimum shear strains and their associated planes and the strains on a plane inclined at $30°$ to the x-axis.

To construct Mohr's circle we first determine the circle centre C, radius R and points A and B

$$C = \left[\left(\frac{\varepsilon_{xx} + \varepsilon_{yy}}{2}\right), 0\right] = [50, 0]$$

$$A(\theta = 0°) = (\varepsilon_{xx}, \gamma_{xy}/2) = (75, 25)$$

$$B(\theta = 90°) = (\varepsilon_{yy}, -\gamma_{xy}/2) = (25, -25)$$

$$R = \sqrt{\left(\frac{\varepsilon_{xx} - \varepsilon_{yy}}{2}\right)^2 + \left(\frac{\gamma_{xy}}{2}\right)^2} = 35$$

with micro-strain assumed. Mohr's circle is now drawn in Figure 8.9. The principal strains and planes

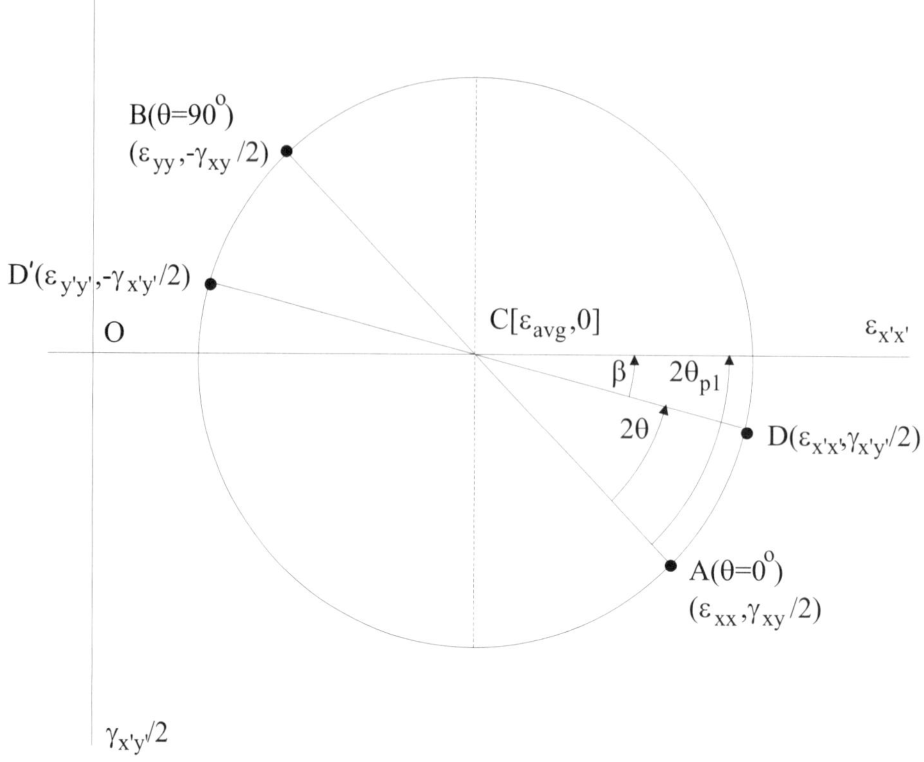

Figure 8.8: State of strain on an arbitrary plane θ.

are, from Mohr's circle

$$\varepsilon_1 = OC + R = 85; \quad \varepsilon_2 = OC - R = 15$$

$$2\theta_{p1} = \tan^{-1}\left(\frac{25}{25}\right); \quad \theta_{p1} = 22.5^\circ$$

$$2\theta_{p2} = 180^\circ + 2\theta_{p1}; \quad \theta_{p2} = 112.5^\circ$$

with the maximum and minimum shear strains and planes given by

$$\left(\frac{\gamma}{2}\right)_{max} = R = 35; \quad 2\theta_{s1} = -(90^\circ - 2\theta_{p1}); \quad \theta_{s1} = -22.5^\circ$$

$$\left(\frac{\gamma}{2}\right)_{min} = -R = -35; \quad 2\theta_2 = 90^\circ + 2\theta_{p1}; \quad \theta_{s2} = 67.5^\circ$$

To determine the strains at $\theta = 30^\circ$ we rotate line AB through the angle 2θ in an anticlockwise direction to line DD' as illustrated in Figure 8.9. With angle $\beta = 2\theta - 2\theta_{p1} = 15^\circ$ then from point D

$$\varepsilon_{x'x'} = OC + R\cos\beta = 83.8$$

$$\frac{\gamma_{x'y'}}{2} = -R\sin\beta = -9.1; \quad \gamma_{x'y'} = -18.3$$

and from point D'

$$\varepsilon_{y'y'} = OC - R\cos\beta = 16.2$$

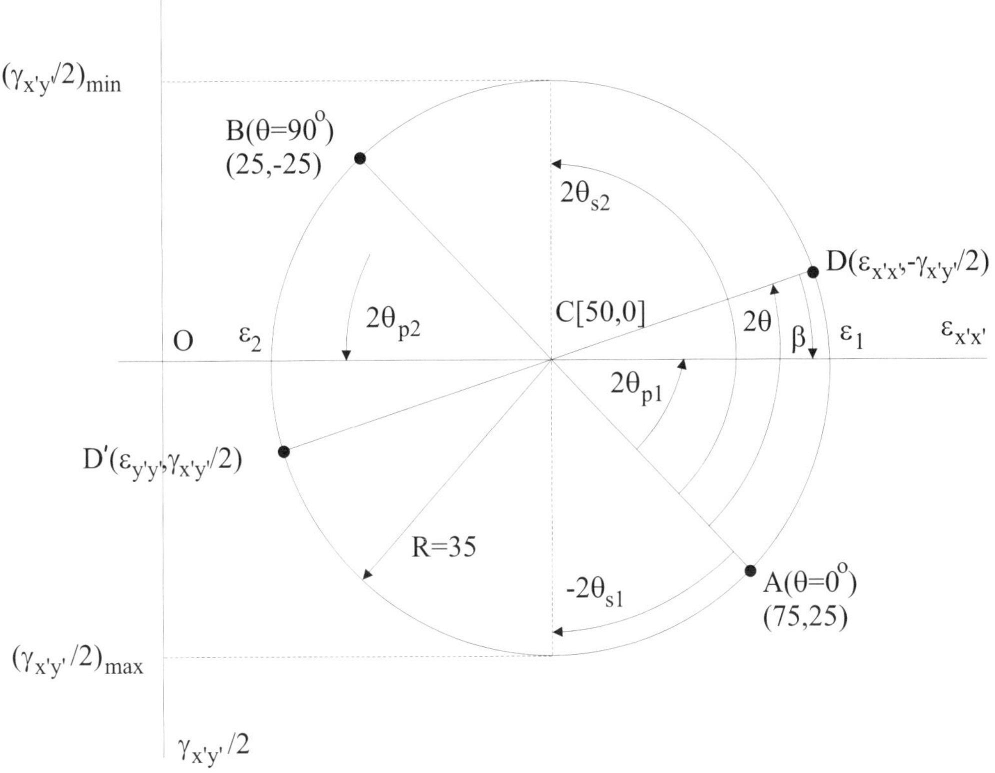

Figure 8.9: Example: Mohr's circle for the global strains $\varepsilon_{xx} = 75 \times 10^{-6}$, $\varepsilon_{yy} = 25 \times 10^{-6}$ and $\gamma_{xy} = 50 \times 10^{-6}$.

8.7 Strain Gauges

Strain components are frequently determined by the application of strain gauges to the surface of a body. A strain gauge measures the strain on the surface of a body in specified directions which can be transformed into strain components in other directions by the use of the strain transformation equations. Strain gauge measurements are generally made on a load-free surface and as a result only the strain on the surface is determined and not the strain perpendicular to the surface. Such measurements are a consequence of the surface of a body being in a state of plane stress.

Practically, strains on the surface of a body are measured using a set of three strain gauges arranged in a pattern referred to as a *strain gauge rosette*, an example of which is shown in Figure 8.10. In general, the axes of the three linear gauges a, b and c are arranged at the angles θ_a, θ_b and θ_c to the x-axis, Figure 8.11. If direct strain readings are measured for gauges a, b and c ($\varepsilon_a, \varepsilon_b$ and ε_c) then the in-plane strain components ε_{xx}, ε_{yy} and γ_{xy} can be determined from the strain transformation equation (8.2) applied to each of the three gauges. In terms of angle θ (8.2) can be written

$$\varepsilon_{x'x'} = \varepsilon_{xx}\cos^2\theta + \varepsilon_{yy}\sin^2\theta + \gamma_{xy}\sin\theta\cos\theta \tag{8.17}$$

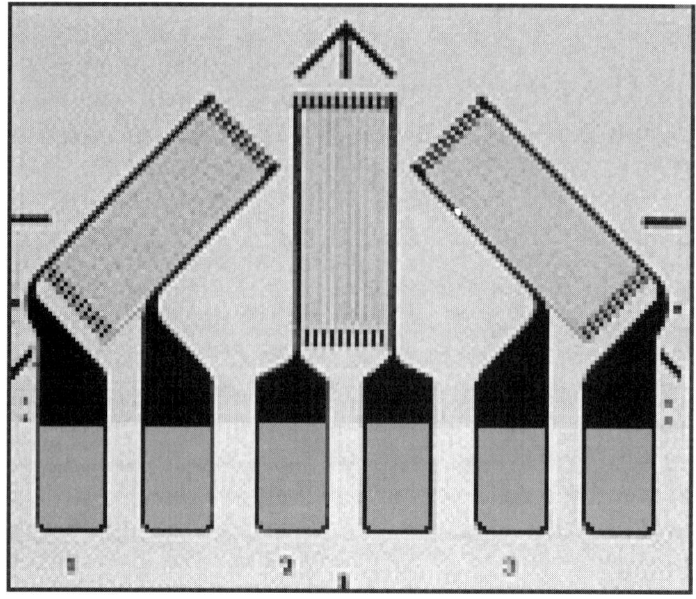

Figure 8.10: General purpose three element 45° rectangular rosette. Courtesy of Measurements Group (www.measurementsgroup.com).

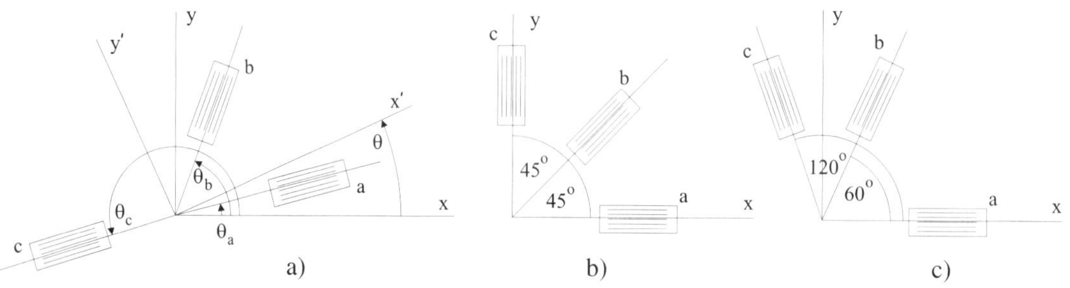

Figure 8.11: Strain gauge rosette. a) General strain gauge rosette with three linear gauges a, b and c at angles θ_a, θ_b and θ_c respectively. b) 45° strain gauge rosette configuration. c) 60° strain gauge rosette configuration.

and for each gauge we have

$$
\begin{aligned}
\varepsilon_a &= \varepsilon_{xx}\cos^2\theta_a + \varepsilon_{yy}\sin^2\theta_a + \gamma_{xy}\sin\theta_a\cos\theta_a \\
\varepsilon_b &= \varepsilon_{xx}\cos^2\theta_b + \varepsilon_{yy}\sin^2\theta_b + \gamma_{xy}\sin\theta_b\cos\theta_b \\
\varepsilon_c &= \varepsilon_{xx}\cos^2\theta_c + \varepsilon_{yy}\sin^2\theta_c + \gamma_{xy}\sin\theta_c\cos\theta_c
\end{aligned}
\tag{8.18}
$$

which can be expressed as the following linear system of equations for unknowns ε_{xx}, ε_{yy} and γ_{xy}

$$
\left\{ \begin{array}{c} \varepsilon_a \\ \varepsilon_b \\ \varepsilon_c \end{array} \right\} = \left[\begin{array}{ccc} \cos^2\theta_a & \sin^2\theta_a & \sin\theta_a\cos\theta_a \\ \cos^2\theta_b & \sin^2\theta_b & \sin\theta_b\cos\theta_b \\ \cos^2\theta_c & \sin^2\theta_c & \sin\theta_c\cos\theta_c \end{array} \right] \left\{ \begin{array}{c} \varepsilon_{xx} \\ \varepsilon_{yy} \\ \gamma_{xy} \end{array} \right\}
\tag{8.19}
$$

To assist in the solution of (8.19) strain gauge rosettes are frequently arranged in either $45°$ or $60°$ configurations, see Figure 8.11. For the $45°$ configuration with angles $\theta_a = 0°$, $\theta_b = 45°$ and $\theta_c = 90°$ then solving (8.19) gives

$$\varepsilon_{xx} = \varepsilon_a; \quad \varepsilon_{yy} = \varepsilon_c; \quad \gamma_{xy} = 2\varepsilon_b - (\varepsilon_a + \varepsilon_c) \tag{8.20}$$

whereas for the $60°$ configuration with angles $\theta_a = 0°$, $\theta_b = 60°$ and $\theta_c = 120°$ then solution of (8.19) gives

$$\varepsilon_{xx} = \varepsilon_a; \quad \varepsilon_{yy} = \frac{1}{3}(2\varepsilon_b + 2\varepsilon_c - \varepsilon_a); \quad \gamma_{xy} = \frac{2}{\sqrt{3}}(\varepsilon_b - \varepsilon_c) \tag{8.21}$$

Example 8.2 Strain gauge rosette

A bracket is subject to a given loading system which generates the strains $\varepsilon_a = 60 \times 10^{-6}$, $\varepsilon_b = 135 \times 10^{-6}$ and $\varepsilon_c = 264 \times 10^{-6}$ at gauges a, b and c respectively for a $60°$ strain gauge rosette configuration. Determine the in-plane principal strains and their associated directions using Mohr's circle.

Firstly, we need to determine the global strains components ε_{xx}, ε_{yy} and γ_{xy} and we will use (8.18) rather than the above formula (8.21). With $\theta_a = 0°$, $\theta_b = 60°$ and $\theta_c = 120°$ then from (8.18) with all subsequent strains in micro-strain

$$\begin{aligned} 60 &= \varepsilon_a = \varepsilon_{xx} \\ 135 &= \varepsilon_b = 0.25\varepsilon_{xx} + 0.75\varepsilon_{yy} + 0.433\gamma_{xy} \\ 264 &= \varepsilon_c = 0.25\varepsilon_{xx} + 0.75\varepsilon_{yy} - 0.433\gamma_{xy} \end{aligned}$$

Re-arranging the second equation for ε_{yy}

$$0.75\varepsilon_{yy} = 135 - 0.25\varepsilon_{xx} - 0.433\gamma_{xy}$$

which, upon substituting into the thrid equation leads to

$$\gamma_{xy} = -\frac{129}{0.866} = -149$$

Finally, with substitution of ε_{xx} and γ_{xy} we determine ε_{yy}

$$\varepsilon_{yy} = \frac{1}{0.75}\left(135 - 0.25\,(60) - 0.433\,(-149)\right) = 246$$

Summarising, the global strain components are $\varepsilon_{xx} = 60 \times 10^{-6}$, $\varepsilon_{yy} = 246 \times 10^{-6}$ and $\gamma_{xy} = -149 \times 10^{-6}$. To construct Mohr's circle we begin by determining the circle centre C, radius R and points A and B

$$\begin{aligned} C &= \left[\left(\frac{\varepsilon_{xx} + \varepsilon_{yy}}{2}\right), 0\right] = [153, 0] \\ A(\theta = 0°) &= (\varepsilon_{xx}, \gamma_{xy}/2) = (60, -74.5) \\ B(\theta = 90°) &= (\varepsilon_{yy}, -\gamma_{xy}/2) = (246, 74.5) \\ R &= \sqrt{\left(\frac{\varepsilon_{xx} - \varepsilon_{yy}}{2}\right)^2 + \left(\frac{\gamma_{xy}}{2}\right)^2} = 119 \end{aligned}$$

Mohr's circle is now drawn in Figure 8.12. The principal strains immediately follow

$$\varepsilon_1 = OC + R = 272; \quad \varepsilon_2 = OC - R = 34$$

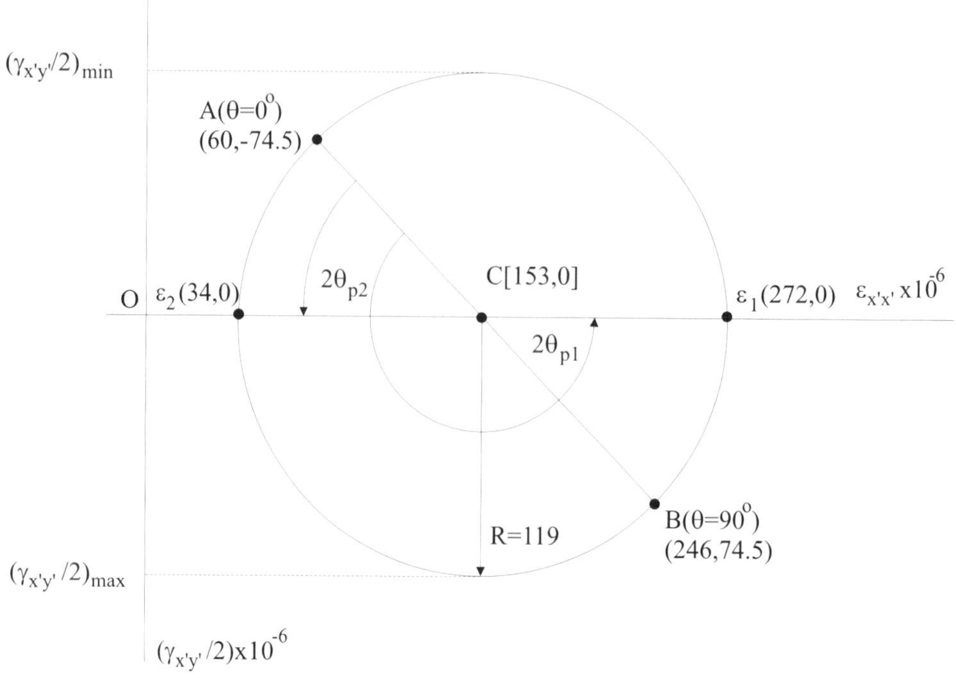

Figure 8.12: Example: Mohr's circle for the global strains $\varepsilon_{xx} = 60 \times 10^{-6}$, $\varepsilon_{yy} = 246 \times 10^{-6}$ and $\gamma_{xy} = -149 \times 10^{-6}$.

and with the principal angle θ_{p2} given by

$$\theta_{p2} = \frac{1}{2} \tan \left(\frac{74.5}{153 - 60} \right) = 19.3^\circ$$

and with $2\theta_{p1} = 180^\circ + 2\theta_{p2}$ then $\theta_{p1} = 109.3^\circ$.

8.8 Conclusion

This Chapter has examined strain components in a plane with particular emphasis on the strain transformation equations. The principal strains and maximum shear strain and their corresponding planes were then derived. By way of Mohr's circle a graphical illustration of strain transformation was then discussed. We concluded the Chapter with a discussion on strain gauge rosettes.

8.9 References and Further Reading

∘ For further information on strain gauges it is worth reading the *Interactive Guide to Strain Gauge Technology* at the Measurements Group web site, www.measurementsgroup.com.

8.10 Exercises

8.1 An element of material distorts to a state of strain given by $\varepsilon_{xx} = 500 \times 10^{-6}$, $\varepsilon_{yy} = -300 \times 10^{-6}$ and $\gamma_{xy} = 200 \times 10^{-6}$. Determine the in-plane strains $\varepsilon_{x'x'}$, $\varepsilon_{y'y'}$ and $\gamma_{x'y'}$ acting on an element whose local axes (x', y') are rotated clockwise by $30°$ from the global axes (x, y).

8.2 The state of strain on an in-plane element is given by $\varepsilon_{xx} = -350 \times 10^{-6}$, $\varepsilon_{yy} = -200 \times 10^{-6}$ and $\gamma_{xy} = 80 \times 10^{-6}$. Determine the principal strains and planes on which they act. Also, determine the maximum shear strain and the plane on which its acts. Illustrate your results graphically.

8.3 A body is subject to a given loading system which generates the strains $\varepsilon_a = 60 \times 10^{-6}$, $\varepsilon_b = 135 \times 10^{-6}$ and $\varepsilon_c = 264 \times 10^{-6}$ at the linear gauges a, b and c respectively for the strain gauge rosette shown in Figure 8.13. Determine the in-plane principal strains and the planes on which they occur using the transformation equations.

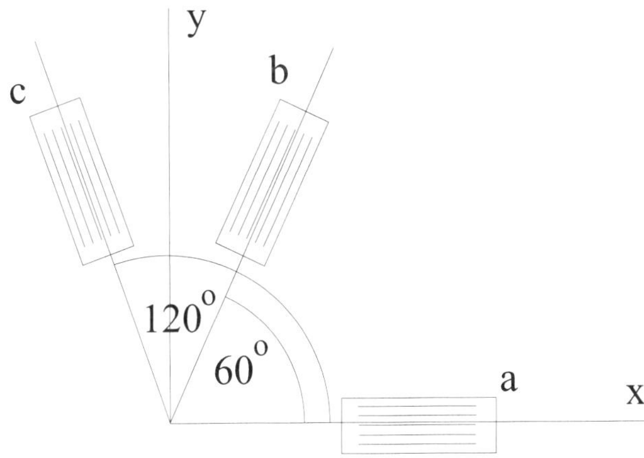

Figure 8.13: Exercise: A $0° - 60° - 120°$ strain gauge rosette for Exercise 8.3.

8.4 The state of in-plane strain in an element is given by $\varepsilon_{xx} = 250 \times 10^{-6}$, $\varepsilon_{yy} = -150 \times 10^{-6}$ and $\gamma_{xy} = 120 \times 10^{-6}$. Use Mohr's circle construction to determine the principal strains and maximum shear strain including their respective orientation angles.

8.5 The in-plane strain components at a given point are $\varepsilon_{xx} = -300 \times 10^{-6}$, $\varepsilon_{yy} = -100 \times 10^{-6}$ and $\gamma_{xy} = 100 \times 10^{-6}$. Using Mohr's circle determine the state of strain on an element rotated by $20°$ in a clockwise direction to the global axes.

8.6 The state of in-plane strain at a point is given by $\varepsilon_{xx} = 200 \times 10^{-6}$, $\varepsilon_{yy} = 400 \times 10^{-6}$ and $\gamma_{xy} = 100 \times 10^{-6}$. Determine the in-plane strains $\varepsilon_{x'x'}$, $\varepsilon_{y'y'}$ and $\gamma_{x'y'}$ acting on an element whose local axes (x', y') are rotated anticlockwise by $40°$ from the global axes (x, y). In addition, assuming that Young's modulus and Poisson's ratio are equal to 210GPa and 0.3 respectively, determine the in-plane stresses corresponding to $\varepsilon_{x'x'}$, $\varepsilon_{y'y'}$ and $\gamma_{x'y'}$.

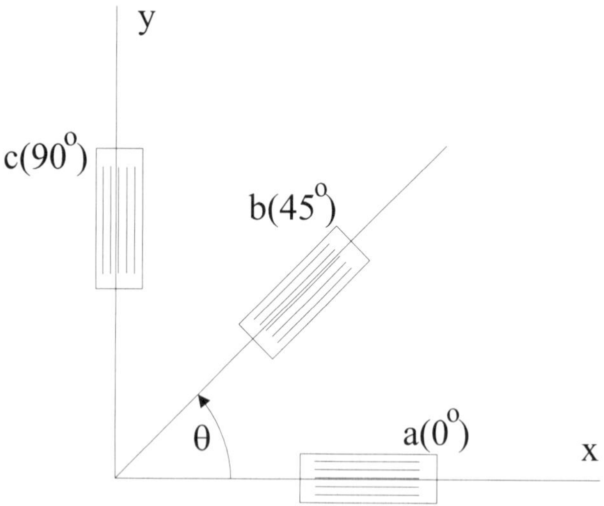

Figure 8.14: Exercise: Strain gauge rosette for Exercise 8.7.

8.7 The readings at gauges a, b and c for the $0° - 45° - 90°$ strain gauge rosette shown in Figure 8.14 are $\varepsilon_a = 65 \times 10^{-6}$, $\varepsilon_b = 95 \times 10^{-6}$ and $\varepsilon_c = 25 \times 10^{-6}$. Determine the in-plane strains ε_{xx}, ε_{yy} and γ_{xy} and the principal strains and their associated planes.

Chapter 9

Strain Energy

9.1 Introduction

This Chapter examines the strain energy of a deformable body and its application. The strain energy of a body is the internal energy corresponding to the deformation of the body and is a finite, bounded quantity. We begin by introducing the strain energy density which is the strain energy per unit volume. The strain energy of a body is then found by integrating the strain energy density throughout the entire body. We then discuss the complementary energy of a body. Although not as popular as strain energy, the complementary energy helps illustrate the components of the total energy. The Chapter then determines the strain energy for a bar subject to pure torsion and for a beam subject to pure bending. These results are then applied to the problem of helical springs. Next, Castigliano's first and second theorems are derived. Knowing the strain energy of a body then Castigliano's theorems provide an elegant method for quickly determining the displacement corresponding to point loading or the slope corresponding to a bending moment. Castigliano's theorems are applied to the problem of determining the displacements of the proving ring. The Chapter concludes with a brief discussion of the strain energy resulting from impact loading.

9.2 Strain Energy Density and Strain Energy

The strain energy of a deformable body is the energy stored during the process of deformation. To help understand what this actually represents it is useful to introduce the quantity of total energy or total *potential energy*, Π, which comprises both the energy as a result of externally applied loading, W, and internal energy, U, as a result of internal deformation

$$\Pi = U - W \tag{9.1}$$

The internal energy U is also termed the *strain energy*, that is the energy generated by the internal straining of the material. To calculate the strain energy of a body we integrate the *strain energy density* (strain energy per unit volume) over the entire volume of the body. Thus, let us begin by determining the strain energy density.

Consider a small differential element subject to the normal stress σ_{xx} only, Figure 9.1a). The stress σ_{xx} results in a force $\sigma_{xx}\,\mathrm{d}y\,\mathrm{d}z$ (force $=$ stress $\times$ area) that does work on an extension $\varepsilon_{xx}\,\mathrm{d}x$ as face $ABCD$ extends. The relation between these two quantities during loading is represented by the straight line OA in the force-extension diagram of Figure 9.1b) for a linear elastic material. Consequently, the work done during deformation is simply the force multiplied by the extension and is equal to the area of the triangle $OAB = \frac{1}{2}(\sigma_{xx}\,\mathrm{d}y\,\mathrm{d}z)(\varepsilon_{xx}\,\mathrm{d}x)$. Writing $\mathrm{d}U$ for this work we have

$$\mathrm{d}U = \frac{1}{2}\sigma_{xx}\varepsilon_{xx}\,\mathrm{d}x\,\mathrm{d}y\,\mathrm{d}z = \frac{1}{2}\sigma_{xx}\varepsilon_{xx}\,\mathrm{d}V \tag{9.2}$$

where $\mathrm{d}V = \mathrm{d}x\,\mathrm{d}y\,\mathrm{d}z$ is the elemental volume. The same amount of work is performed on all other similar elements if their volumes are equivalent.

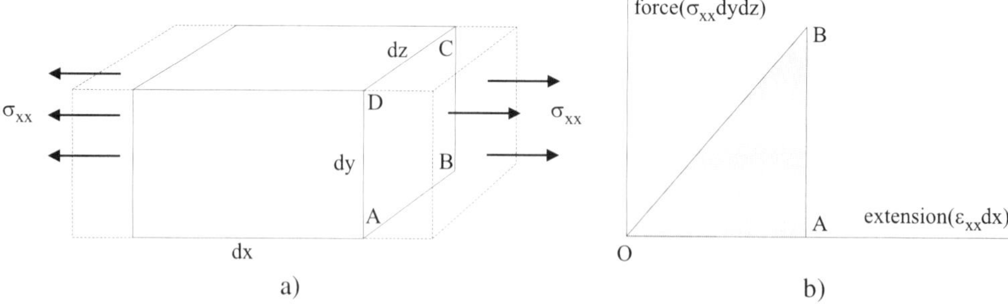

Figure 9.1: Strain energy. a) Stress element subject to the direct stress σ_{xx}. b) Force-extension diagram for a linear elastic material.

If now the body deforms from an initial strain of zero to a final strain of ε_f then (9.2) can be integrated to give the strain energy per unit volume in the final state of deformation

$$U_o = U\mid_{vol} = \frac{1}{2}\int_0^{\varepsilon_f} \sigma_{xx}\,\mathrm{d}\varepsilon_{xx} \tag{9.3}$$

where U_o denotes the strain energy density or strain energy per unit volume. Assuming that σ_{xx} and ε_{xx} now denote the final values of stress and strain then the subscript (f) can be removed and we therefore have for a linear elastic material

$$U_o = \frac{1}{2}\sigma_{xx}\varepsilon_{xx} \tag{9.4}$$

In general, for materials other than linear elastic we may have a non-linear force-extension diagram, Figure 9.2, for which it is necessary to model the exact relationship between stress and strain. However, to a first approximation the work done is $(\sigma_{xx}\,\mathrm{d}y\,\mathrm{d}z)(\varepsilon_{xx}\,\mathrm{d}x)$ since the element size is differentially small. Therefore, in this case the strain energy density is

$$U_o = \int_0^{\varepsilon_f} \sigma_{xx}\,\mathrm{d}\varepsilon_{xx} \tag{9.5}$$

Similarly, if the element is subject to shear, Figure 9.3, we have for U_o

$$U_o = \int_0^{\gamma_f} \tau_{xy}\,\mathrm{d}\gamma_{xy} \tag{9.6}$$

In the case of a linear elastic material the $(\tau_{xy} - \gamma_{xy})$ response is a straight line and we have

$$U_o = \frac{1}{2}\tau_{xy}\gamma_{xy} \tag{9.7}$$

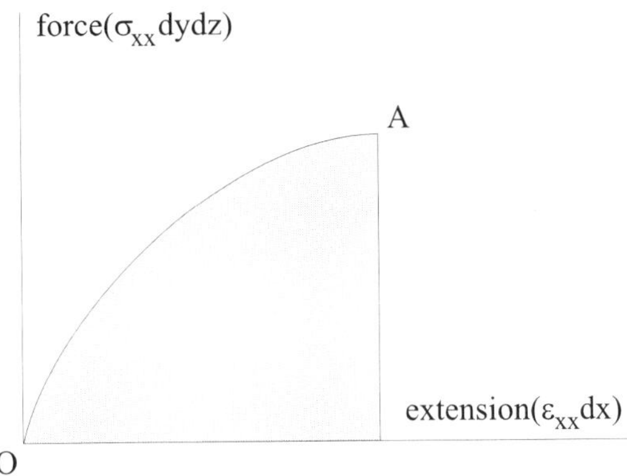

Figure 9.2: Non-linear force-extension response.

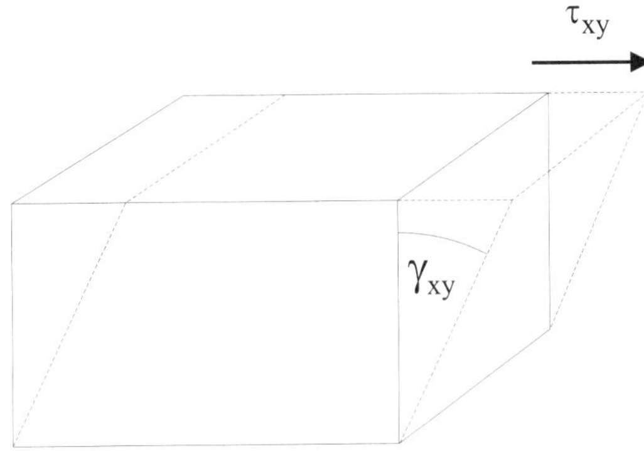

Figure 9.3: Stress element subject to the shear stress τ_{xy}.

We have now derived the strain energy per unit volume of an element subject to both normal and shear stresses. Generally, an element will be subject to the six components of stress σ_{xx}, σ_{yy}, σ_{zz}, τ_{xy}, τ_{yz}, τ_{xz} with corresponding strains ε_{xx}, ε_{yy}, ε_{zz}, γ_{xy}, γ_{yz}, γ_{xz}. Therefore, the elemental strain energy is

$$\mathrm{d}U = U_o \,\mathrm{d}x \,\mathrm{d}y \,\mathrm{d}z \tag{9.8}$$

where U_o is

$$U_o = \frac{1}{2}(\sigma_{xx}\varepsilon_{xx} + \sigma_{yy}\varepsilon_{yy} + \sigma_{zz}\varepsilon_{zz} + \tau_{xy}\gamma_{xy} + \tau_{yz}\gamma_{yz} + \tau_{xz}\gamma_{xz}) \tag{9.9}$$

for a linear elastic material. By means of the Hookian constitutive equations (1.16)

$$\begin{aligned}
\varepsilon_{xx} &= \frac{1}{E}\left[\sigma_{xx} - \nu(\sigma_{yy} + \sigma_{zz})\right]; \quad \gamma_{xy} = \frac{1}{G}\tau_{xy} \\
\varepsilon_{yy} &= \frac{1}{E}\left[\sigma_{yy} - \nu(\sigma_{xx} + \sigma_{zz})\right]; \quad \gamma_{yz} = \frac{1}{G}\tau_{yz} \\
\varepsilon_{zz} &= \frac{1}{E}\left[\sigma_{zz} - \nu(\sigma_{xx} + \sigma_{yy})\right]; \quad \gamma_{xz} = \frac{1}{G}\tau_{xz}
\end{aligned} \tag{9.10}$$

we can express U_o purely in terms of stress components by substituting (9.10) into (9.9)

$$U_o = \frac{1}{2E}(\sigma_{xx}^2 + \sigma_{yy}^2 + \sigma_{zz}^2) - \frac{\nu}{E}(\sigma_{xx}\sigma_{yy} + \sigma_{yy}\sigma_{zz} + \sigma_{xx}\sigma_{zz}) + \frac{1}{2G}(\tau_{xy}^2 + \tau_{yz}^2 + \tau_{xz}^2) \tag{9.11}$$

From (9.8) we find that the total strain energy, U, of a deformed body is obtained from the strain energy density per unit volume, U_o, or the strain energy density by an integration with respect to the element of volume V

$$U = \int_V U_0 \, \mathrm{d}V \tag{9.12}$$

Before we take a look at some examples of strain energy we will incorporate the effects of thermal strain into U_o and examine the complementary strain energy.

9.3 Two-Dimensional Thermal Strain

In our introduction to stress and strain in Chapter 1 we introduced the notion of one-dimensional thermal strain in §1.10. Let us now extend this definition of thermal strain in two-dimensions and see its effect on strain energy. For a two-dimensional state of stress that obeys Hooke's law with $\sigma_{zz} = \tau_{yz} = \tau_{xz} = 0$ then from (9.10)

$$\begin{aligned} \varepsilon_{xx} &= \frac{1}{E}(\sigma_{xx} - \nu\sigma_{yy}) + \alpha(\Delta T) \\ \varepsilon_{yy} &= \frac{1}{E}(\sigma_{yy} - \nu\sigma_{xx}) + \alpha(\Delta T) \\ \gamma_{xy} &= \frac{1}{G}\tau_{xy} = \frac{2(1+\nu)}{E}\tau_{xy} \end{aligned} \tag{9.13}$$

where α is the coefficient of thermal expansion. The above strains illustrate that the total strain is the superposition of the stress associated with the stress components and the strain due to the thermal effects. Note that the thermal expansion or contraction of a material has no effect on the shear deformation. Inverting (9.13) for σ we have

$$\begin{aligned} \sigma_{xx} &= \frac{E}{1-\nu^2}\left[(\varepsilon_{xx} - \alpha(\Delta T)) + \nu(\varepsilon_{yy} - \alpha(\Delta T))\right] \\ \sigma_{yy} &= \frac{E}{1-\nu^2}\left[(\varepsilon_{yy} - \alpha(\Delta T)) + \nu(\varepsilon_{xx} - \alpha(\Delta T))\right] \\ \tau_{xy} &= \frac{E}{2(1-\nu^2)}\gamma_{xy} \end{aligned} \tag{9.14}$$

From (9.9) the strain energy density for an elastic material is now given by

$$U_o = \frac{1}{2}\left[\sigma_{xx}(\varepsilon_{xx} - \varepsilon_{xx}^T) + \sigma_{yy}(\varepsilon_{yy} - \varepsilon_{yy}^T) + \sigma_{zz}(\varepsilon_{zz} - \varepsilon_{zz}^T)\right] + \frac{1}{2}[\tau_{xy}\gamma_{xy} + \tau_{yz}\gamma_{yz} + \tau_{xz}\gamma_{xz}] \tag{9.15}$$

where ε_{xx}^T, ε_{yy}^T and ε_{zz}^T are the strains due to thermal effects. In the above ΔT was assumed to be constant. Naturally, ΔT can be a function of, say, x and y in which case the thermal strain would be

$$\varepsilon^T = \alpha\Delta T(x, y) \tag{9.16}$$

9.4 Complementary Energy

The load-extension diagram of Figure 9.2 has been repeated in Figure 9.4. Strain energy is the work done during loading and is represented by the area below the stress-strain curve. The energy above the stress-strain curve (between the force-displacement curve and the force axis) is referred to as the *complementary energy* and is denoted by U^*. In an analogous manner to (9.12)

$$U^* = \int_V U_o^* \, \mathrm{d}V \tag{9.17}$$

where U_o^* is the complementary energy density (complementary energy per unit volume) following a similar derivation to that for U_o discussed earlier

$$U_o^* = \int_0^{\sigma_f} \varepsilon_{xx} \, \mathrm{d}\sigma_{xx} \tag{9.18}$$

noting that integration is now with respect to stress. In terms of a bar undergoing an axial extension of δ_1

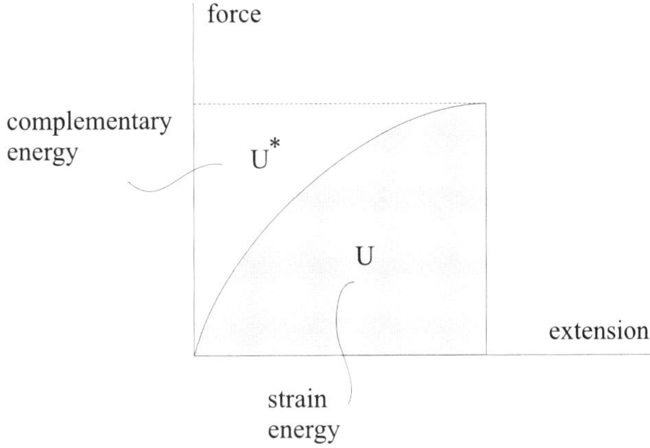

Figure 9.4: Non-linear force-extension response illustrating both strain and complementary energies.

subject to an axial force P_1 then the work done and strain and complementary energies are

$$W = \int_0^{\delta_1} P_1 \, \mathrm{d}\delta; \quad U = \int_0^{\delta_1} P_1 \, \mathrm{d}\delta; \quad U^* = \int_0^{P_1} \delta \, \mathrm{d}P; \tag{9.19}$$

Let us now have an example which determines both the strain and complementary energies for a bar in tension.

Example 9.1 Strain and complementary energy for a bar subject to an axial force

Consider the pin-jointed bar in Figure 9.5 which is of length L, cross-sectional area A, Young's modulus E and is fixed at one end. The bar is subject to an axial displacement of δ at the free end. With an axial strain of $\varepsilon_{xx} = \delta/L$ and stress of $\sigma_{xx} = E\varepsilon_{xx} = E\delta/L$ then the strain energy density is, (9.4)

$$U_o = \frac{1}{2}\sigma_{xx}\varepsilon_{xx} = \frac{1}{2}\left(\frac{E\delta}{L}\right)\left(\frac{\delta}{L}\right) = \frac{1}{2}E\left(\frac{\delta}{L}\right)^2$$

Figure 9.5: Example: A bar subject to an axial force P at the free end.

Integrating over the entire volume of the bar, AL, the strain energy is, (9.12)

$$U = \int_V U_o \, \mathrm{d}V = \int_0^L \frac{1}{2} E \left(\frac{\delta}{L}\right)^2 A \, \mathrm{d}x = \frac{1}{2} E \left(\frac{\delta}{L}\right)^2 AL = \frac{1}{2} k\delta^2; \quad k = \frac{EA}{L}$$

where k represents the bar stiffness. The complementary energy density is, (9.18)

$$U_o^* = \frac{1}{2}\varepsilon_{xx}\sigma_{xx} = \frac{1}{2}\left(\frac{\sigma_{xx}}{E}\right)\sigma_{xx} = \frac{\sigma_{xx}^2}{2E}$$

Integrating over the entire volume of the bar the complementary energy is, (9.17)

$$U^* = \int_V U_o^* \, \mathrm{d}V = \int_0^L \left(\frac{\sigma_{xx}^2}{2E}\right) A \, \mathrm{d}x = \frac{P^2}{2k}$$

Since $\delta = \varepsilon_{xx} L = \sigma_{xx} L/E = PL/AE$ then

$$\delta = \frac{PL}{AE}$$

and substituting into U above

$$U = \frac{1}{2} k\delta^2 = \frac{P^2}{2k}$$

noting that $U = U^*$. This result not only holds for the present example but is a general result which, when Hooke's law applies, the strain and complementary energies are equal. From U and U^* above it is also worth noting that the strain energy is generally expressed in terms of the displacement while the complementary energy is expressed in terms of the applied load.

9.5 Strain Energy due to Torsion

To determine the internal strain energy in a circular shaft or tube due to an applied torsional moment only, then from (9.7) the strain energy density is

$$U_o = \frac{\tau^2}{2G} \tag{9.20}$$

where τ is the shear stress acting on a cross-section of the bar, Figure 9.6. Integrating over the entire volume, V, the strain energy is

$$U = \int_V U_o \, \mathrm{d}V = \int_V \frac{\tau^2}{2G} \, \mathrm{d}V \tag{9.21}$$

With $dV = dA\,dx$ and $\tau = Tr/J$ from (3.11) then integrating throughout the total length of the bar

$$U = \int_V \frac{1}{2G}\left(\frac{Tr}{J}\right)^2 dA\,dx = \int_0^L \frac{T^2}{2GJ^2}\,dx \iint_A r^2\,dA \tag{9.22}$$

Since the area integral represents the polar moment of area, J, of the bar then (9.22) reduces to

$$U = \int_0^L \frac{T^2\,dx}{2GJ} \tag{9.23}$$

If the cross-sectional area is constant over the entire length of the bar then U is

$$U = \frac{T^2L}{2GJ} \tag{9.24}$$

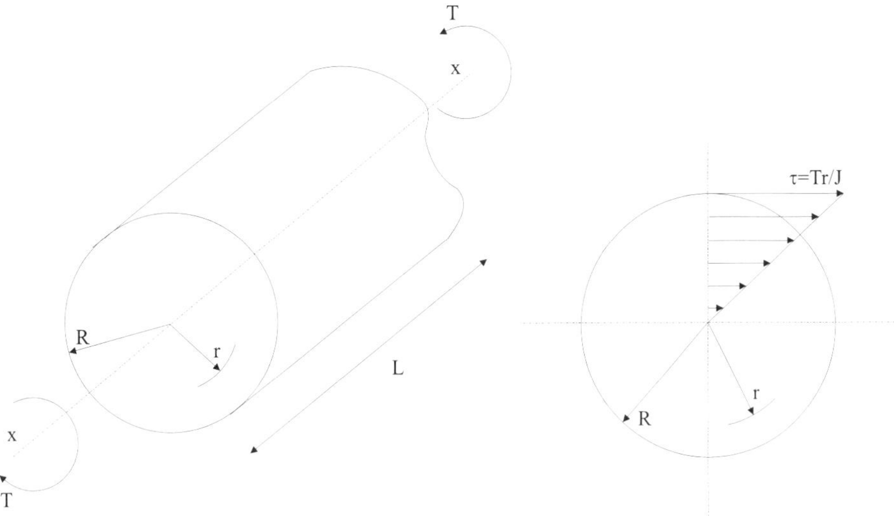

Figure 9.6: A solid shaft of circular cross-section of radius R, cross-sectional area A and length L subject to a torque T. Also shown is the linear variation of shear stress with radial position r on a given cross-section.

Example 9.2 Strain energy of a tubular bar subject to a torque

A tubular bar of length 1m, outside diameter 100mm and constant wall-thickness 10mm is subject to a torque of 50Nm. If the shear modulus is 75GPa then determine the strain energy of the bar.

The polar second moment of area is given by (2.12)

$$J = \frac{\pi}{2}(R_o^4 - R_i^4) = \frac{\pi}{2}(0.05^4 - 0.04^4) = 5.79 \times 10^{-6}\text{m}^4$$

The strain energy is, (9.24)

$$U = \frac{T^2L}{2GJ} = 2,879 \times 10^{-6}\text{Nm} = 2,879\mu\text{J}$$

9.6 Strain Energy due to Bending

In this section we will derive an expression for the strain energy in a beam which is subject to both axial and bending loads. Later, the contribution of axial loads to the strain energy will be neglected in which case the form of the strain energy will be seen to be analogous to that derived in the previous section under conditions of pure torsion.

If a beam is subject to both an axial load, P, and bending moment, M, then the axial strain ε_{xx} is the sum of the separate components (1.31) and (5.18)

$$\varepsilon_{xx} = \frac{\mathrm{d}u}{\mathrm{d}x} - \frac{y}{R} = \frac{\mathrm{d}u}{\mathrm{d}x} - y\left(\frac{\mathrm{d}^2 v}{\mathrm{d}x^2}\right) \tag{9.25}$$

and similarly the axial stress is, from (1.2) and (5.30)

$$\sigma_{xx} = \frac{P}{A} + \frac{My}{I} \tag{9.26}$$

For a linear elastic material the strain energy density is given by (9.4) after substituting ε_{xx} and σ_{xx}

$$U_o = \frac{1}{2}\sigma_{xx}\varepsilon_{xx} = \frac{1}{2}\left(\frac{P}{A} + \frac{My}{I}\right)\left(\frac{\mathrm{d}u}{\mathrm{d}x} - y\frac{\mathrm{d}^2 v}{\mathrm{d}x^2}\right) \tag{9.27}$$

Expanding as a quadratic function in y and integrating over the entire cross-sectional area A, we have the strain energy density per unit length

$$U_o(\text{length}) = \frac{1}{2}\left\{\frac{P}{A}\frac{\mathrm{d}u}{\mathrm{d}x}\iint_A \mathrm{d}A + \left[\frac{M}{I}\frac{\mathrm{d}u}{\mathrm{d}x} - \frac{P}{A}\frac{\mathrm{d}^2 v}{\mathrm{d}x}\right]\iint_A y\,\mathrm{d}A - \left[\frac{M}{I}\frac{\mathrm{d}^2 v}{\mathrm{d}x^2}\right]\iint_A y^2\,\mathrm{d}A\right\} \tag{9.28}$$

and noting that

$$\iint_A \mathrm{d}A = A; \quad \iint_A y^2\,\mathrm{d}A = I_x = I \tag{9.29}$$

and from (5.34)

$$\iint_A y\,\mathrm{d}A = 0 \tag{9.30}$$

then (9.28) reduces to

$$U_o(\text{length}) = \frac{1}{2}\left[P\frac{\mathrm{d}u}{\mathrm{d}x} + M\left(-\frac{\mathrm{d}^2 v}{\mathrm{d}x^2}\right)\right] \tag{9.31}$$

This expression can be written in a variety of forms since both P and M can be expressed in terms of u and v through the stress-strain and moment-curvature relationships. For instance, eliminating P and M gives

$$U_o(\text{length}) = \frac{1}{2}\left[EA\left(\frac{\mathrm{d}u}{\mathrm{d}x}\right)^2 + EI\left(\frac{\mathrm{d}^2 v}{\mathrm{d}x^2}\right)^2\right] \tag{9.32}$$

or conversely, eliminating $\mathrm{d}u/\,\mathrm{d}x$ and $d^2v/\,\mathrm{d}^2x$

$$U_o(\text{length}) = \frac{1}{2}\left[\frac{P}{A} + \frac{M^2}{EI}\right] \tag{9.33}$$

which is the most frequently used form of U_o.

Integrating over the entire length, L, of the beam we find the total strain energy

$$U = \frac{1}{2}\int_L \left[\frac{P^2}{EA} + \frac{M^2}{EI}\right] \mathrm{d}x \tag{9.34}$$

noting that both P and M are to the power two. In the case of pure bending we have

$$U = \int_L \frac{M^2\,\mathrm{d}x}{2EI} = \int_L \frac{EI}{2}\left(\frac{\mathrm{d}^2 v}{\mathrm{d}x^2}\right)^2 \mathrm{d}x \tag{9.35}$$

9.7 Strain Energy of Helical Springs

Let us now examine the deformation of both close-coil and open-coil helical springs. The analysis is based on the strain energy stored in an elastic spring and serves as an application of the above results for the strain energy due to both torsion and bending.

9.7.1 Close-Coil Helical Spring

Figure 9.7a) illustrates a helical spring subject to a tensile loading W. If the helix angle, α, is less than approximately 5° then the deformation of the spring wire is assumed to be due to torsion only and the spring is referred to as *close-coiled*. The relevant geometric parameters of the spring are the mean coil diameter D and radius R, wire diameter d, number of coils n, pitch h and the coil length, L.

The force W will twist the spring with a constant torque, T, along the entire length of the spring

$$T = WR = \frac{WD}{2} \tag{9.36}$$

If δ denotes the extension associated with W then the external work done is

$$\frac{1}{2}W\delta \tag{9.37}$$

The strain energy, U, stored in the spring wire is, from (9.23)

$$U = \frac{1}{2GJ}\int_L T^2\,\mathrm{d}x = \frac{T^2 L}{2GJ} \tag{9.38}$$

where x is chosen to pass through the wire. Equating the external work done to the internal strain energy

$$\frac{W\delta}{2} = \frac{T^2 L}{2GJ} \tag{9.39}$$

With $T = WD/2$ from (9.36) above, $L = \pi Dn$ and $J = \pi d^4/32$ then (9.39) reduces to

$$k = \frac{W}{\delta} = \frac{Gd^4}{8D^3 n} \tag{9.40}$$

where k is the *spring stiffness*.

From (3.11) we can determine the angular twist θ (in radians) between the ends of the wire

$$\theta = \frac{TL}{GJ} = \frac{16nWD^2}{Gd^4} \tag{9.41}$$

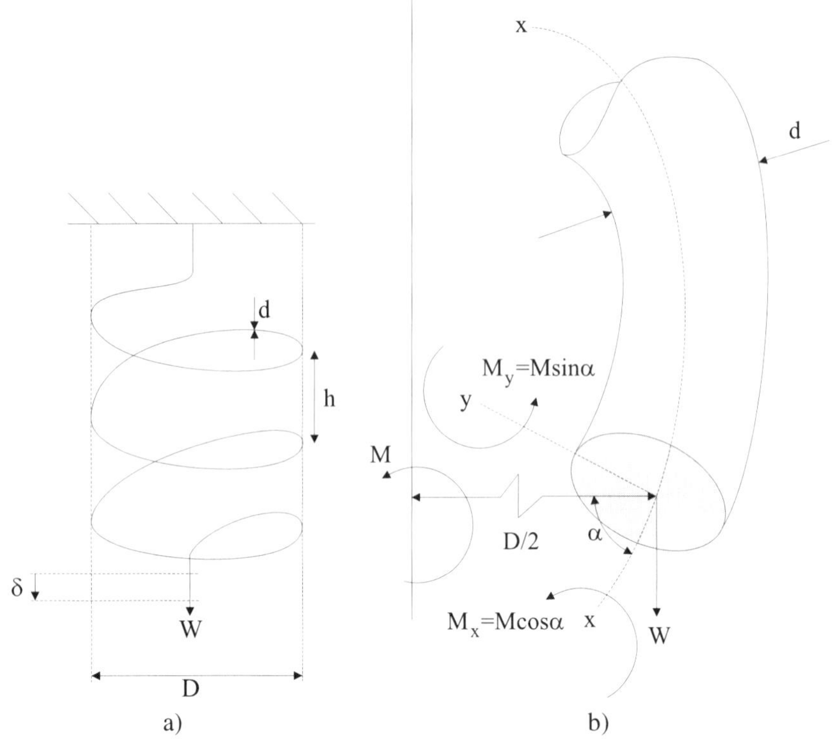

Figure 9.7: Strain energy of helical springs. a) A helical spring. b) A cross-section of the spring wire.

Also, from (3.11) the maximum torsional shear stress at the wire surface is

$$\tau = \frac{8WD}{\pi d^3} \tag{9.42}$$

In addition to this torsional shear stress there is also a shear stress induced by W

$$\tau = \frac{W}{A} = \frac{4W}{\pi d^2} \tag{9.43}$$

where A is the wire cross-sectional area. Summing these two shear stresses then the net shear stress is

$$\tau = \frac{8WD}{\pi d^3}\left[1 + \frac{d}{2D}\right] \tag{9.44}$$

9.7.2 Open-Coil Helical Spring

When the helix angle exceeds approximately $5°$ then both torsion and bending are induced at each section of the wire as illustrated in Figure 9.7b). When a plane cut through the wire lies in the vertical plane then the moment, M, due to the axial load W is

$$M = \frac{WD}{2} \tag{9.45}$$

The two resolved components M_x and M_y of M are:

- $M_x = M\cos\alpha$ and acts parallel to the helix axis at the cut plane (twists the wire)
- $M_y = M\sin\alpha$ and acts perpendicular to the helix axis at the cut plane (bends the wire)

The external work done is again given by (9.37) but the strain energy in the spring is now the sum of the torsion and bending components

$$U = \frac{1}{2GJ}\int_L T^2\,\mathrm{d}x + \frac{1}{2EI}\int_L M^2\,\mathrm{d}x \tag{9.46}$$

Substituting for $T(=M_x)$ and $M(=M_y)$ into (9.46) and equating with the external work done

$$\frac{W\delta}{2} = \frac{1}{2GJ}\int_L M_x^2\,\mathrm{d}x + \frac{1}{2EI}\int_L M_y^2\,\mathrm{d}x \tag{9.47}$$

Integrating over the entire length L and noting that both M_x and M_y are independent of L then

$$\delta = \frac{WD^2L}{4}\left[\frac{\cos^2\alpha}{GJ} + \frac{\sin^2\alpha}{EI}\right] \tag{9.48}$$

With $L = \pi Dn/\cos\alpha$, $J = \pi d^4/32$ and $I = \pi d^4/64$

$$\delta = \frac{8WD^3n}{d^4\cos\alpha}\left[\frac{\cos^2\alpha}{G} + \frac{2\sin^2\alpha}{E}\right] \tag{9.49}$$

Re-arranging in terms of stiffness, k, we have

$$k = \frac{W}{\delta} = \frac{d^4\cos\alpha}{8D^3n}\left[\frac{\cos^2\alpha}{G} + \frac{2\sin^2\alpha}{E}\right]^{-1} \tag{9.50}$$

Note that (9.50) tends to the close-coil value (9.40) as $\alpha \to 0$ since $\sin\alpha \to 0$ and $\cos\alpha \to 1$. Finally, note that the shear stress, τ, and angle of twist, θ, for an open-coil spring are

$$\tau = \frac{T(d/2)}{J} = \frac{M_x\,\mathrm{d}/2}{\pi d^4/32};\quad \theta = \frac{TL}{GJ} = \frac{M_x\pi Dn/\cos\alpha}{G\pi d^4/32} \tag{9.51}$$

while the bending stress and radius of curvature are

$$\sigma = \frac{M(d/2)}{I} = \frac{M_y d/2}{\pi d^4/64};\quad R = \frac{EI}{M_y} = \frac{E\pi d^4/64}{M_y} \tag{9.52}$$

9.7.3 Wahl's Correction Factor

We will conclude our discussion of springs by mentioning Wahl's correction factor, K. Letting $C = D/d$ denote the *spring index* (typically in the range [4:12]) then K is given by the empirical formula

$$K = \frac{4C-1}{4C-4} + \frac{0.615}{C} \tag{9.53}$$

and is illustrated in Figure 9.8. Wahl's correction factor accounts for the increasing contribution of shear force as C decreases. For example, for a close-coil spring then the shear stress (9.42) is modified accordingly

$$\tau = \left(\frac{8WD}{\pi d^3}\right)K \tag{9.54}$$

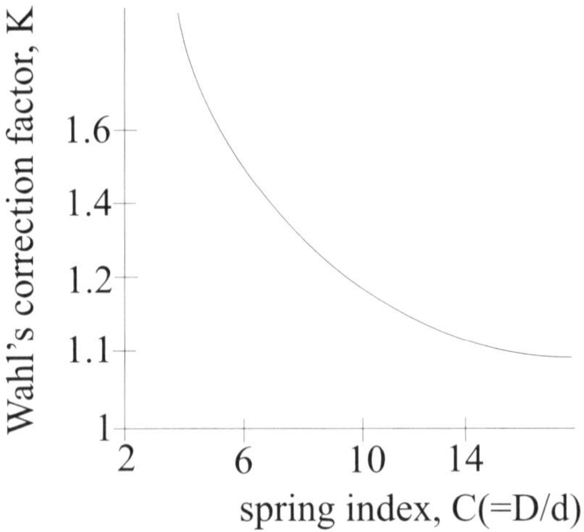

Figure 9.8: Wahl's correction factor.

Parameter	**Measured Value**
Young's modulus (E)	210GPa
Poisson's ratio (ν)	0.33
Wire diameter (d)	17.1mm
Mean radius (R)	58.56mm
Pitch (h)	52.5mm
Number of complete coils (n)	6
Helix angle (α)	11.1°

Table 9.1: Example: Measured spring properties.

Example 9.3 Comparison of the theoretical and experimental stiffnesses of a helical spring

A helical spring from the suspension system of a truck has the properties listed in Table 9.1 Since the helix angle is greater than 5° the spring is assumed to be open-coiled. From (9.50) the estimated or theoretical stiffness, k_t, for the spring is

$$k_t = \frac{W}{\delta} = 86,525\text{Nm}$$

The experimentally measured load-displacement response of the spring is listed in Table 9.2 with corresponding stiffness values. Performing a least squares straight line, $W = a + k_e\delta$, regression fit through the experimental load-displacement data then the normal equations for n data points are[1]

$$an + k_e \sum_{i=1}^{n} \delta_i = \sum_{i=1}^{n} W_i$$

$$a\sum_{i=1}^{n} \delta_i + k_e \sum_{i=1}^{n} \delta_i^2 = \sum_{i=1}^{n} \delta_i W_i$$

[1] Further details of the least-squares method can be found in a numerical methods text, such as O'Neil (1983)

Applied load, W (N)	Displacement, δ (mm)	Stiffness, k_e (Nm)
1,060	12	88,333
2,150	24	89,583
3,280	36	91,111
4,440	48	92,500

Table 9.2: Example: Experimental load-displacement response for the spring and corresponding stiffness values.

where, from Table 9.2, $\sum \delta_i = 120$, $\sum \delta_i^2 = 4,320$, $\sum W_i = 10,930$ and $\sum \delta_i W_i = 395,520$. Solving the above linear system of equations we arrive at $k_e = 91,559$Nm and $a = -14.25$. Normalising the experimental stiffness by the theoretical estimate for stiffness we have $k_e/k_t = 1.06$ which should be equal to unity if the experiment and theory are in perfect agreement.

9.8 Castigliano's First and Second Theorems

Consider the structure shown in Figure 9.9 which is subject to n point loads P_1, P_2, ..., P_n with corresponding displacements δ_1, δ_2, ..., δ_n. The structure may behave non-linearly and as a result the load-displacement response may be of any form. The strain energy of the structure is equal to the work

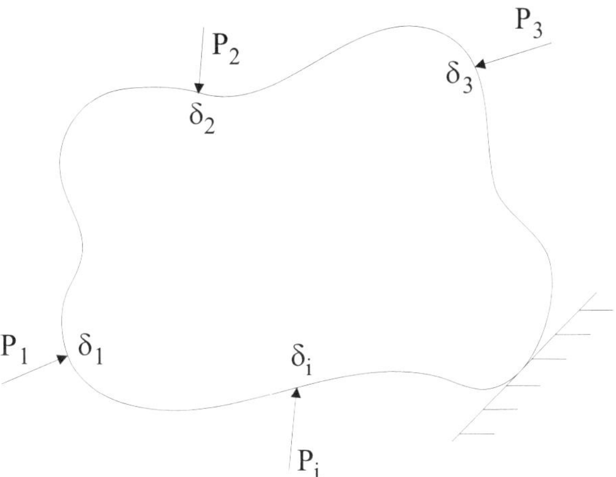

Figure 9.9: A structure subject to arbitrary point loading.

done, W, by the applied loads

$$W = \sum_{i=1}^{n} \int_0^{\delta_i} P \, \mathrm{d}\delta \tag{9.55}$$

with each force expressed as a function of its corresponding displacement through the load-displacement curve of the structure material. With the strain energy, U, in terms of displacement, δ, we can determine the change in strain energy when one displacement, δ_i, is increased by a small amount, $\mathrm{d}\delta_i$, while all other displacements are kept constant. This increase in strain energy, $\mathrm{d}U$, is

$$\mathrm{d}U = \frac{\partial U}{\partial \delta_i}\,\mathrm{d}\delta_i \tag{9.56}$$

where $\partial U/\partial \delta_i$ is the rate of change of U with respect to δ_i.

When δ_i is increased by $\mathrm{d}\delta_i$ then work is also done by the corresponding force P_i but only by this force since all other forces are kept constant. This work done by P_i is equal to $\mathrm{d}W = P_i\,\mathrm{d}\delta_i$ and is equivalent to the strain energy

$$\mathrm{d}U = P_i\,\mathrm{d}\delta_i \tag{9.57}$$

Equating (9.56) and (9.57) we arrive at the following

$$P_i = \frac{\partial U}{\partial \delta_i} \tag{9.58}$$

which is known as *Castigliano's first theorem* and illustrates that the first partial derivative of U with respect to any displacement, δ_i, is equal to the corresponding force P_i. Castigliano's second theorem can be derived by applying similar arguments as above and is of the form

$$\delta_i = \frac{\partial U}{\partial P_i} \tag{9.59}$$

illustrating that the first partial derivative of U with respect to any force P_i is equal to the corresponding displacement, δ_i.

In an analogous manner, instead of P and δ, we can derive Castigliano's second theorem in terms of bending moment, M, and slope, θ

$$\theta_i = \frac{\partial U}{\partial M_i} \tag{9.60}$$

Similarly, Castigliano's second theorem can be applied to the torsion of a bar in terms of torque T and angle of twist θ

$$\theta = \frac{\partial U}{\partial T_i} \tag{9.61}$$

In §9.5 we determined an expression for the strain energy of both solid or tubular bars of constant cross-sectional area subject to a torque, (9.24). Applying (9.61) to (9.24) we have

$$\theta = \frac{\partial}{\partial T}\left(\frac{T^2 L}{2GJ}\right) = \frac{TL}{GJ} \tag{9.62}$$

which is seen to agree with (3.11).

Example 9.4 Determination of the displacement and slope of a cantilever beam using Castigliano's theorems

In this example we will examine the determination of the displacement and slope at the free end of a cantilever beam by an application of Castigliano's second theorem. Figure 9.10 illustrates a cantilever beam of length L subject to a concentrated point load P at the free end. The bending moment at a distance x from the free end is $M = Px$. The strain energy of the beam is, from (9.35)

$$U = \int_L \frac{M^2\,\mathrm{d}x}{2EI} = \int_0^L \frac{(Px)^2\,\mathrm{d}x}{2EI} = \frac{P^2 L^3}{6EI}$$

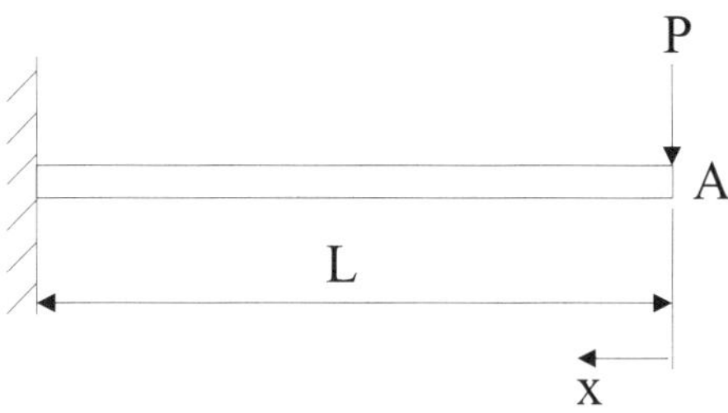

Figure 9.10: Example: A cantilever beam subject to a point load at the free end.

From Castigliano's second theorem (9.59) the displacement, δ_A, at the free end of the beam is

$$\delta_A = \frac{\partial U}{\partial P} = \frac{PL^3}{3EI}$$

and illustrates the simplicity and usefulness of Castigliano's theorems.

To determine the slope, θ_A, at the free end of the beam we have to apply a fictitious moment, say M_f, at the free end and then find an expression for θ_A in terms of P and M_f and then finally assign $M_f = 0$ since M_f is fictitious. The application of a fictitious moment is necessary because Castigliano's theorems require an applied moment in order to determine a slope. The bending moment of the beam is now given by

$$M = Px + M_f$$

The strain energy is now equal to

$$U = \int_L \frac{M^2\,\mathrm{d}x}{2EI} = \int_0^L \frac{(Px + M_f)^2\,\mathrm{d}x}{2EI} = \frac{P^2L^3}{6EI} + \frac{PM_fL^2}{2EI} + \frac{M_f^2}{2EI}$$

From Castigliano's second theorem, (9.60), the slope is

$$\theta_A = \frac{\partial U}{\partial M_f} = \frac{PL^2}{2EI} + \frac{M_fL}{EI}$$

and remembering that $M_f = 0$ then

$$\theta_A = \frac{PL^2}{2EI}$$

Thus, an expression for the slope, in terms of P, was determined when no moment was applied to the beam.

9.8.1 Castigliano's Second Theorem for Beams

Generalising the above example we observe that the displacement is

$$\delta = \frac{\partial U}{\partial P} = \frac{\partial}{\partial P}\int_0^L \frac{M^2\,\mathrm{d}x}{2EI} \tag{9.63}$$

Rather than squaring the expression for moment, M, performing the necessary integration and taking the partial derivative, alternatively we can perform the differentiation prior to the integration. Noting that

$$\frac{\partial U}{\partial P} = \frac{\partial U}{\partial M}\left(\frac{\partial M}{\partial P}\right) \tag{9.64}$$

then δ can be expressed as

$$\delta = \int_0^L \frac{M}{EI}\left(\frac{\partial M}{\partial P}\right) \mathrm{d}x \tag{9.65}$$

If the slope of the tangent θ at a point on the beam is required then the following partial derivative of M with respect to an externally applied couple moment M_0 acting at that point must be evaluated

$$\delta = \frac{\partial U}{\partial M_0} = \frac{\partial}{\partial M_0}\int_0^L \frac{M^2\,\mathrm{d}x}{2EI} \tag{9.66}$$

and noting that

$$\frac{\partial M}{\partial M_0} = \frac{\partial U}{\partial M}\left(\frac{\partial M}{\partial M_0}\right) \tag{9.67}$$

we have

$$\theta = \int_0^L \frac{M}{EI}\left(\frac{\partial M}{\partial M_0}\right) \mathrm{d}x \tag{9.68}$$

Example 9.5 Determination of the slope of a cantilever beam using Castigliano's theorems

Let us use the above expression (9.68) for the determination of θ for the cantilever beam of Example 9.4. With $M = Px + M_f$ the slope is

$$\theta_A = \int_0^L \frac{M}{EI}\left(\frac{\partial M}{\partial M_f}\right) \mathrm{d}x = \int_0^L \frac{(Px + M_f)}{EI}(1)\,\mathrm{d}x = \frac{1}{EI}\left[\frac{PL^2}{2} + M_f L\right]$$

and since $M_f = 0$ we arrive at the slope derived earlier

$$\theta_A = \frac{PL^2}{2EI}$$

Example 9.6 Determination of the displacement of a cantilever beam using Castigliano's theorems

As a further illustration of the above general formula, (9.65), consider the cantilever beam shown in Figure 9.11a) carrying a uniformly distributed load W. To determine the displacement at the free end point A we need to first apply a fictitious point load at A, Figure 9.11b). Taking moments at a distance x from the free end we have

$$M + Wx\left(\frac{x}{2}\right) + P_f x = 0$$

which leads to

$$M = -\frac{Wx^2}{2} - P_f x; \quad \frac{\partial M}{\partial P_f} = -x$$

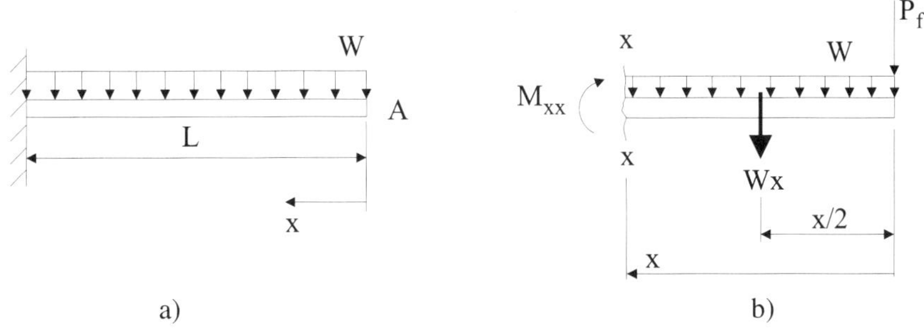

Figure 9.11: Example: a) A cantilever beam subject to a uniformly distributed load. b) Fictitious point load P_f at the free end.

and since $P_f = 0$

$$M = -\frac{Wx^2}{2}; \quad \frac{\partial M}{\partial P_f} = -x$$

From (9.65) the displacement at A is

$$\delta_A = \int_0^L \frac{M}{EI}\left(\frac{\partial M}{\partial P_f}\right) \mathrm{d}x = \int_0^L \frac{1}{EI}\left(\frac{-Wx^2}{2}\right)(-x)\,\mathrm{d}x = \frac{W}{2EI}\int_0^L x^3\,\mathrm{d}x = \frac{WL^4}{8EI}$$

9.9 Proving Ring

An interesting application of the strain energy in beams and Castigliano's theorems is the proving ring, Figure 9.12. It will be shown that a simple load-displacement relationship exists for the proving ring which is frequently used when calibrating tension-compression test machines. Due to the symmetry of the proving ring consider only the top-left quadrant assuming that it is built-in at point A. To ensure that this quadrant is equivalent to the original proving ring a restraining moment, M_0, is required which will be subsequently determined. Since the proving ring quadrant is in the form of a cantilever beam the strain energy is, from (9.35)

$$U = \int_0^L \frac{M^2\,\mathrm{d}x}{2EI} = \int_0^{\pi/2} \frac{M^2 R\,\mathrm{d}\theta}{2EI} \tag{9.69}$$

where the x-axis is measured along the mean radius, R, of the quadrant which is of length L. Thus, for the entire proving ring

$$U = 4\int_0^{\pi/2} \frac{M^2 R\,\mathrm{d}\theta}{2EI} \tag{9.70}$$

and from Castigliano's second theorem, (9.59), the vertical deflection, δ_v, at the points of application of point loads P is

$$\delta_v = \frac{\partial U}{\partial P} = 4\int_0^{\pi/2} \frac{M}{EI}\left(\frac{\partial M}{\partial P}\right) R\,\mathrm{d}\theta \tag{9.71}$$

Taking moments at a point along the quadrant results in

$$M + M_0 + \frac{P}{2}(R - R\cos\theta) = 0 \Rightarrow M = -\frac{P}{2}(R - R\cos\theta) - M_0 \tag{9.72}$$

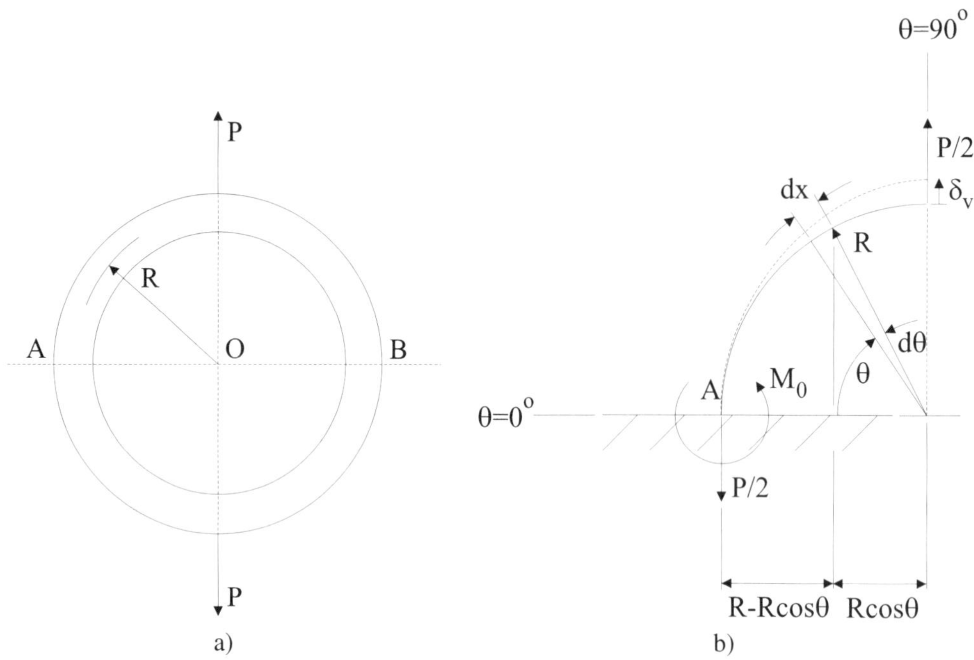

Figure 9.12: a) Proving ring. b) One quadrant of the proving ring.

To determine M_0 it is assumed that the angular rotation at point A is zero and from Castigliano's second theorem, (9.60)

$$\theta_A = 0 = \frac{\partial U}{\partial M_0} = 4\int_0^{\pi/2} \frac{M}{EI}\left(\frac{\partial M}{\partial M_0}\right) R\,\mathrm{d}\theta \tag{9.73}$$

Substitution of M from (9.72)

$$\frac{4}{EI}\int_0^{\pi/2}\left[-\frac{PR}{2}(1-\cos\theta) - M_0\right](-1)R\,\mathrm{d}\theta = 0 \tag{9.74}$$

Completing the integration we find

$$M_0 = -\frac{PR}{2}\left(1-\frac{2}{\pi}\right) \tag{9.75}$$

With $\partial M/\partial P$ given by, (9.72)

$$\frac{\partial M}{\partial P} = -\frac{R}{2}(1-\cos\theta) - \frac{\partial M_0}{\partial P} = \frac{R}{2}\left(\cos\theta - \frac{2}{\pi}\right) \tag{9.76}$$

then δ_v is given by, (9.71)

$$\delta_v = \frac{4}{EI}\int_0^{\pi/2} \frac{PR}{2}\left(\cos\theta-\frac{2}{\pi}\right)\frac{R}{2}\left(\cos\theta-\frac{2}{\pi}\right)R\,\mathrm{d}\theta = \frac{PR^3}{EI}\left[\frac{\pi}{4}-\frac{2}{\pi}\right] \tag{9.77}$$

noting that $(\pi/4 - 2/\pi) \approx 0.1488$ which results in a positive displacement, as expected.

To determine the horizontal displacement, δ_h, at points A and B using Castigliano's theorems we need to apply fictitious point loads, $2Q_f$, on either side of the proving ring, remembering to assign $Q_f = 0$. Considering once more the top left quadrant, Figure 9.13, but now with Q_f applied we find that the bending moment at a distance along the proving ring is

$$M = Q_f R \sin\theta - \frac{PR}{2}(1 - \cos\theta) - M_1 \tag{9.78}$$

with restraining moment M_1. With zero rotation at point A

$$\theta = 0 = \frac{\partial M}{\partial M_1} = \int_0^{\pi/2} \frac{M}{EI}\left(\frac{\partial M}{\partial M_0}\right) R\,\mathrm{d}\theta \tag{9.79}$$

and with $\partial M/\partial M_1 = -1$ from (9.78)

$$M_1 = \frac{2R}{\pi}\left(Q_f - \frac{\pi P}{4} + \frac{P}{2}\right) \tag{9.80}$$

With δ_h given by, for fictitious loads $2Q_f$

$$\delta_h = \frac{4}{2}\int_0^{\pi/2} \frac{M}{EI}\left(\frac{\partial M}{\partial Q_f}\right) R\,\mathrm{d}\theta \tag{9.81}$$

and noting from (9.78)

$$\frac{\partial M}{\partial Q_f} = R\sin\theta - \frac{2R}{\pi} \tag{9.82}$$

then δ_h is given by

$$\delta_h = \frac{2R}{EI}\int_0^{\pi/2} \left[Q_f R\sin\theta - \frac{PR}{2}(1-\cos\theta) - \frac{2R}{\pi}\left(Q_f - \frac{\pi P}{4} + \frac{P}{2}\right)\right] R\left(\sin\theta - \frac{2}{\pi}\right) \mathrm{d}\theta \tag{9.83}$$

and remembering that $Q_f = 0$

$$\delta_h = \frac{PR^3}{EI}\left[\frac{1}{2} - \frac{2}{\pi}\right] \tag{9.84}$$

noting that $(1/2 - 2/\pi) \approx -0.1366$ which results in a negative displacement.

Finally, the radial strains at points $\theta = 0^\circ$ and $\theta = 90^\circ$ are

$$\varepsilon_{rr}(\theta = 0^\circ) = \frac{\delta_h}{R};\quad \varepsilon_{rr}(\theta = 90^\circ) = \frac{\delta_v}{R} \tag{9.85}$$

Example 9.7 Determination of the displacement for a circular bar subject to a point force

In this example we will examine a circular rod which is subject to a point load, P, at the free end. The first case considered is that of Figure 9.14a) in which P acts normal to the free end. To determine the displacement at point A we will assume that the strain energy, U, of the rod is due to bending only. With the bending moment, M, at a point along the rod given by

$$M = (R - R\cos\theta)P$$

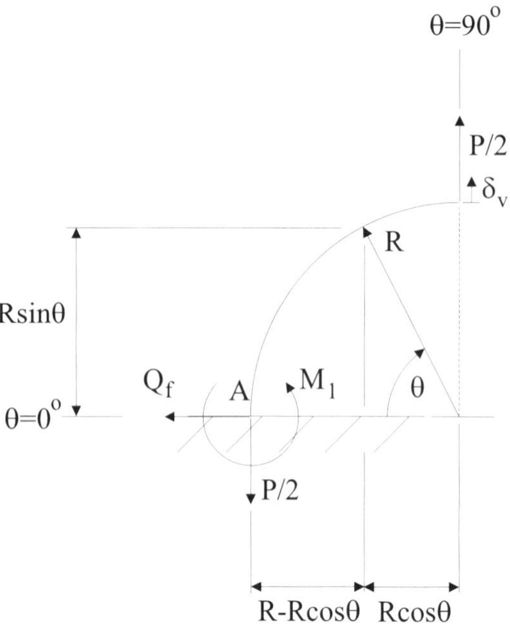

Figure 9.13: Top left quadrant of the proving ring with fictitious load Q_f applied.

then U is, (9.35)

$$U = \int_0^{\pi} \frac{M^2}{2EI} R\,\mathrm{d}\theta = \frac{1}{2EI}\int_0^{\pi} R^2P^2(1-\cos\theta)^2 R\,\mathrm{d}\theta = \frac{3\pi R^3P^2}{4EI}$$

where $\mathrm{d}x = R\,\mathrm{d}\theta$ and use is made of the trigonometric identity $\cos^2\theta = (1+\cos 2\theta)/2$. From Castigliano's second theorem, (9.59), the displacement, δ, at A is

$$\delta_A = \frac{\partial U}{\partial P} = \frac{\partial}{\partial P}\left(\frac{3\pi R^3P^2}{4EI}\right) = \frac{3\pi R^3P}{2EI}$$

When P acts parallel to the free end, Figure 9.14b), the bending moment is given by

$$M = (R\sin\theta)P$$

so that U is now

$$U = \int_0^{\pi} \frac{M^2}{2EI} R\,\mathrm{d}\theta = \frac{1}{2EI}\int_0^{\pi} R^2P^2\sin^2\theta R\,\mathrm{d}\theta = \frac{R^3P^2\pi}{4EI}$$

where use of the identity $\sin^2\theta = (1-\cos 2\theta)/2$ is made in performing the integration. From Castigliano's second theorem the displacement at A is

$$\theta_A = \frac{\partial U}{\partial P} = \frac{\pi R^3P}{2EI}$$

which is one-third of the displacement for case a) of Figure 9.14.

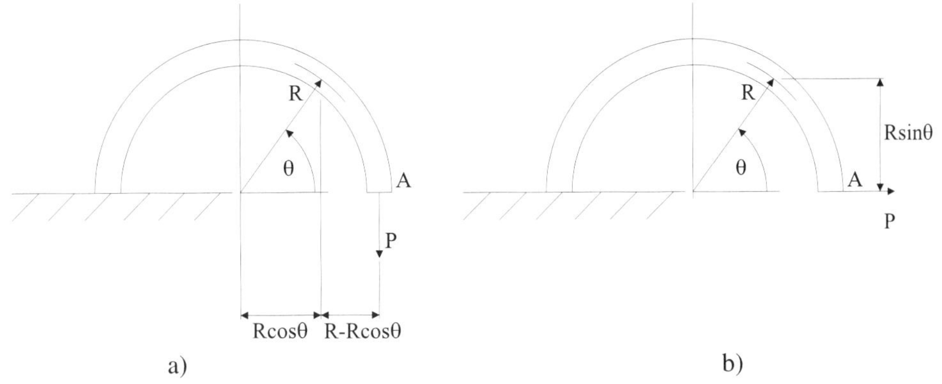

Figure 9.14: Example: Circular bar subject to a point load, P, which is a) perpendicular to the free end and b) parallel to the free end.

9.10 Impact Loading

When analysing impact loads there are two approaches that are generally taken: i) to estimate the loss of potential energy of a falling body and ii) to estimate the kinetic energy that is absorbed during impact. In the next two sub-sections we will consider these two approaches separately with application to the deflections of a beam induced by the impact of a body.

9.10.1 Potential Energy Approach

To develop a general result, consider a spring with stiffness k on which is dropped a body of mass m and weight W. The body is assumed to be at rest initially and falls from a height h, see Figure 9.15. Assume that all of the potential energy lost by the body is transformed into elastic strain energy stored in the spring. If the maximum dynamic displacement of the spring is denoted by δ_{max} then the potential energy lost by the body is

$$W(h + \delta_{max}) \tag{9.86}$$

There are clearly several assumptions behind (9.86) a few of which are: i) the spring remains elastic throughout impact, ii) the body remains in contact with the spring subsequent to impact, iii) the change in potential energy of the spring during impact can be neglected, iv) no external energy losses occur and v) the inertia of the spring during impact can be neglected.

The increase in strain energy of the spring is

$$\frac{1}{2}k\delta_{max}^2 \tag{9.87}$$

The static displacement, δ_{st}, of the spring due to a force of W is equal to W/k. Thus, equating (9.86) and (9.87) and substituting δ_{st} we have

$$k\delta_{st}(h + \delta_{max}) - \frac{1}{2}k\delta_{max}^2 = 0 \tag{9.88}$$

Solving the quadratic equation (9.88) for δ_{max} and choosing the largest value we find

$$\delta_{max} = \delta_{st} + \sqrt{\delta_{st}^2 + 2h\delta_{st}} = \left(1 + \sqrt{1 + \frac{2h}{\delta_{st}}}\right)\delta_{st} = F\delta_{st} \tag{9.89}$$

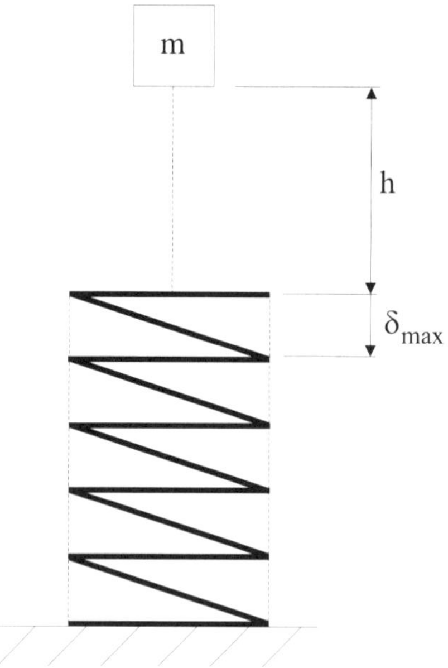

Figure 9.15: A mass m falling through a height h on to a spring and inducing the maximum spring displacement δ_{max}.

where F is referred to as the *impact factor*. Thus, δ_{max} can be expressed solely as a function of δ_{st} and h and (9.89) illustrates that $\delta_{max} > \delta_{st}$. If δ_{st} is small in comparison to h ($\delta_{st} \ll h$) then from (9.89)

$$\delta_{max} \approx \sqrt{2h\delta_{st}} \tag{9.90}$$

whereas if the weight of the body is suddenly applied with minimal free fall, that is $h = 0$, then

$$\delta_{max} \approx 2\delta_{st} \tag{9.91}$$

Finally, the maximum dynamic load, P_{max}, is W multiplied by the impact factor

$$P_{max} = FW \tag{9.92}$$

9.10.2 Kinetic Energy Approach

Let us now assume that a body is moving with a velocity v and comes into contact with another elastic body, after which the moving body is stationary. The kinetic energy of the moving body is $mv^2/2$ and equating this loss of energy with the increase in strain energy of the elastic body, as above, we have

$$\frac{1}{2}\left(\frac{W}{g}\right)v^2 - \frac{1}{2}k\delta_{max}^2 = 0 \tag{9.93}$$

With $\delta_{st} = W/k$ and solving for δ_{max} we have

$$\delta_{max} = \delta_{st}\sqrt{\frac{v^2}{g\delta_{st}}}; \quad P_{max} = m\sqrt{\frac{v^2 g}{\delta_{st}}} \tag{9.94}$$

where g is the gravitational constant.

Example 9.8 Impact loading on a simply supported beam

Figure 9.16 illustrates a simply supported beam of length $L = 1$m, square cross-section, $w = 25$mm, with Young's modulus $E = 200$GPa. A mass $m = 1$kg falls from a height $h = 0.5$m on to the mid-span of the beam. The static deflection, v, of the beam for a concentrated point load W acting at $x = L/2$ is, from Example 5.6

$$\nu = -\frac{Wx}{48EI}(3L^2 - 4x^2); \quad 0 \leq x \leq \frac{L}{2}$$

with the maximum static deflection, δ_{st}, occurring at $x = L/2$ with $I = w^4/12 = 3.253 \times 10^{-8}\text{m}^4$

$$\delta_{st} = \frac{WL^3}{48EI} = \frac{1 \times g \times 1^3}{48 \times 200 \times 10^9(3.253 \times 10^{-8})} = 0.0314\text{mm}$$

assuming the gravitational constant $g = 9.81\text{m/s}^2$. The impact factor F now follows from (9.89), $F = 179.5$ and the maximum displacement is $\delta_{max} = F\delta_{st} = 5.635$mm which is considerably greater than δ_{st}.

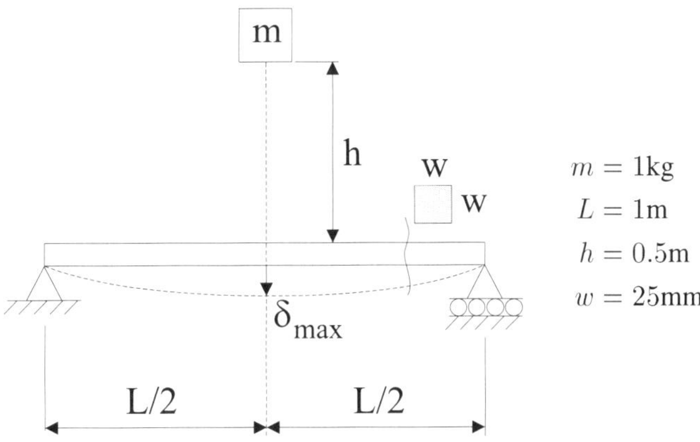

Figure 9.16: Example: A mass m falling through height h at the mid-span of a simply supported beam.

9.11 Conclusion

The present Chapter has examined strain energy and considered a few of its numerous applications. We also considered Castigiliano's first and second theorems which provide a convenient method of determining the displacement corresponding to point loading when the strain energy of the body is known. The applications considered were those of the strain energy of a bar subject to pure torsion, a beam subject to pure bending, the displacement of helical springs, the proving ring and impact loading.

9.12 References and Further Reading

- O'Neil, P. V. (1983) *Advanced Engineering Mathematics*, Wadsworth, Belmont, California.
- Chironis, N. P. (1961) *Spring Design and Application*, McGraw-Hill, London.

9.13 Exercises

9.1 Show that the strain energy, U, for a biaxially loaded linear elastic plate ($\sigma_{xx} = \sigma_{yy} = \sigma, \tau_{xy} = 0$) is given by

$$U = \frac{V}{2E}(\sigma_{xx}^2 - 2\nu\sigma_{xx}\sigma_{yy} + \sigma_{yy}^2)$$

where E, ν and V denote the Young's modulus, Poisson's ratio and volume of the plate respectively.

9.2 Show that the strain energy, U, for the linear elastic cylindrical tapered bar shown in Figure 9.17 subject to a compressive axial force of W is given by

$$U = \frac{2W^2L}{E\pi\, \mathrm{d}_1\, \mathrm{d}_2}$$

where L is the length of the bar. Using Castigliano's theorems determine the axial displacement of the bar. What do you observe about the displacement when $\mathrm{d}_1 = \mathrm{d}_2$?

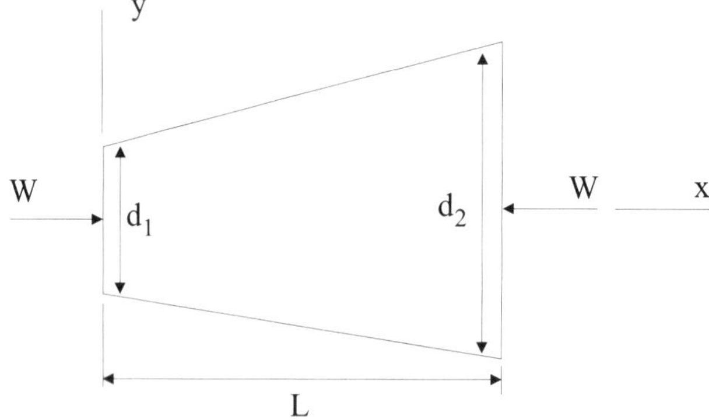

Figure 9.17: Exercise: A tapered bar subject to an axial compression.

9.3 Figure 9.18 illustrates a pin-jointed plane frame consisting of two members. Using Castigliano's second theorem determine the displacement at point B. Assume that the frame members have Young's modulus and cross-sectional area E and A respectively.

9.4 An open-coil helical spring has a wire diameter of 5mm, helix angle of 25°, 10 effective coils and a mean diameter of 50mm. Determine the axial load required to extend the spring by 10mm and the corresponding shear and bending stresses induced. Assume that $G = 85$GPa and $\nu = 0.33$.

9.5 A close-coil helical spring is required to have a stiffness of 2.5N/mm when subject to a maximum axial load of 35N without the maximum torsional shear stress exceeding 120MPa. If the solid length (number of coils times the wire diameter) is 50mm when unextended then determine a) the diameter of the spring wire, b) the mean coil diameter and c) the number of coils. Assume $G = 45$GPa.

9.6 A split steel piston ring of square cross-section (2.5mm×2.5mm), mean radius $R = 45$mm and Young's modulus 210GPa is initially closed as shown in Figure 9.19. Assuming that the strain energy of the ring is due to bending only, show that the total gap, δ, between the equal opposing forces P is given by

$$\delta = \frac{3\pi R^3 P}{EI}$$

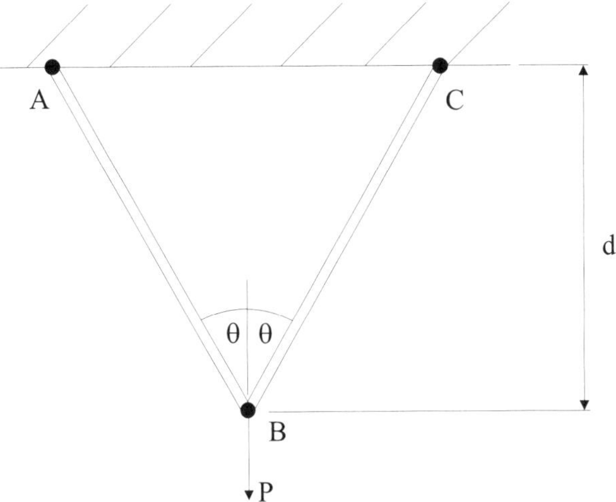

Figure 9.18: Exercise: A two member pin-jointed plane frame built-in at joints A and C and with a point force of P at joint B.

where I is the second moment of area. Determine the value of P to open the ring by a total gap of 10mm. Where does the maximum bending moment of the ring occur? At the location of maximum bending moment determine the maximum bending stress. Does this value exceed the tensile yield stress (σ_Y = 300MPa) of the material?

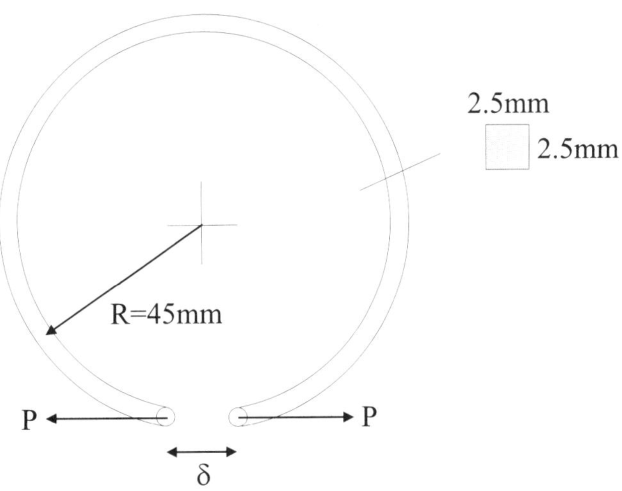

Figure 9.19: Exercise: A split piston ring.

9.7 Determine the maximum displacement at the free end of the cantilever beam shown in Figure 9.20 if a mass of 1kg falls through a height of 0.5m. The length of the beam is L = 1m, has a Young's modulus of E = 210GPa and a square cross-section of (30×30)mm. The static deflection, v, of the beam for a concentrated point load W acting at the free end is given by

$$v = -\frac{Wx^2}{6EI}(3L - x)$$

where I is the second moment of area. Assume the gravitational constant g to be equal to 9.81m/s^2.

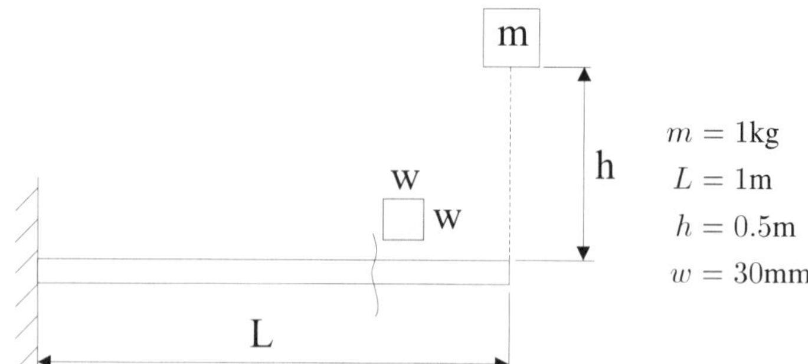

Figure 9.20: Exercise: A mass of !kg falling a height of 0.5m on to the free end of a cantilever beam.

Chapter 10

Virtual Displacements and the Principle of Virtual Work

10.1 Introduction

The present Chapter introduces the notion of a virtual displacement. The principles of virtual displacement, virtual work and stationary, total potential energy are examined. Use of the principle of stationary total potential energy is then made to determine the displacement in a structure when subject to general virtual loading. This result is subsequently tailored for the specific case of unit virtual loads for determining the displacement in beams via the *unit load method*.

A *virtual*[1] displacement is one that does not really exist but is arbitrarily imposed on a structure. The work done by real forces during a virtual displacement is referred to as virtual work. So why use an abstract idea such as applying a virtual displacement? In certain circumstances the principles of virtual work and virtual displacements can be used instead of the equations of equilibrium in the solution of problems. A virtual displacement must not exist although it must be admissible and it must be sufficiently small so that it does not generate any corresponding changes in the applied forces. Note in the following derivations that the material properties will never enter into the derivation. As a result, the following principles are applicable to all structures and materials.

There are generally two approaches that can be adopted when deriving the energy principles: i) a mathematical calculus of variations approach and ii) a derivation from physical considerations. The latter approach will be adopted in this Chapter. The variational approach will be used when deriving the element stiffness and force vector equations for the finite element method in §13.11.

10.2 Principle of Virtual Displacement for a Particle

Consider the particle in Figure 10.1 which is subject to a series of forces $P_1, P_2, \ldots, P_n$ with the particle remaining in a state of static equilibrium. If the particle is in a state of equilibrium then the vector sum of all the applied forces must equate to zero

$$P_1 + P_2 + P_3 + \cdots + P_n = 0 \tag{10.1}$$

[1] virtual means existing in effect but not in reality

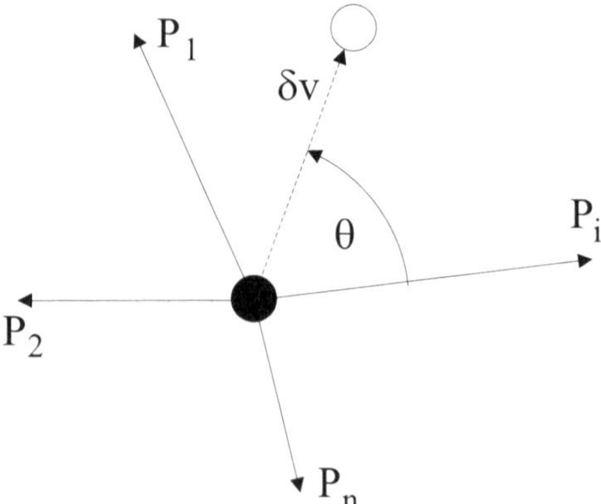

Figure 10.1: A particle subject to a series of forces P_1, P_2, ..., P_n.

If the particle is displaced through a virtual displacement of δv then the virtual work done, δW_i, by the real force P_i is

$$\delta W_i = \parallel P_i \parallel \bullet \parallel \delta v \parallel = P_i \delta v \cos\theta = P_{iv}\delta v \tag{10.2}$$

where θ is the angle between P_i and δv, Figure 10.1, and P_{iv} is the projection of P_i onto δv.

Assuming that δv is infinitesimally small then the forces P_i will remain unaltered by the virtual displacements. Thus, the total virtual work done by all of the forces is

$$\delta W = \sum_{i=1}^{n} \delta W_i = \left(\sum_{i=1}^{n} P_{iv} \right) \delta v = (P_{1v} + P_{2v} + \cdots + P_{nv})\delta v \tag{10.3}$$

However, from the equilibrium condition (10.1) we have

$$\delta W = 0 \tag{10.4}$$

and can be stated as the *principle of virtual displacements*:

A necessary and sufficient condition of the equilibrium of a particle is that the work done by the applied forces on a particle is zero for any displacement.

10.3 Principle of Virtual Work for a Continuous Body

Consider next the two-dimensional stress element ($\mathrm{d}x$, $\mathrm{d}y$, $\mathrm{d}z$) in Figure 10.2 acted on by the stress σ_{xx}. The real component of displacement is denoted by u whereas δu denotes the virtual component of displacement. The virtual work done corresponding to the virtual displacement is (force multiplied by displacement)

$$\delta W_{xx} = \sigma_{xx}\,\mathrm{d}y\,\mathrm{d}z \left(\delta u + \frac{\partial(\delta u)}{\partial x}\,\mathrm{d}x \right) - \sigma_{xx}\,\mathrm{d}y\,\mathrm{d}z\delta u = \sigma_{xx}\frac{\partial(\delta u)}{\partial x}\,\mathrm{d}x\,\mathrm{d}y\,\mathrm{d}z \tag{10.5}$$

The corresponding virtual strain arising from the virtual displacement is

$$\varepsilon_{xx} + \delta\varepsilon_{xx} = \frac{\partial u}{\partial x} + \frac{\left(\delta u + \frac{\partial(\delta u)}{\partial x}\right) - \delta u}{\mathrm{d}x} \tag{10.6}$$

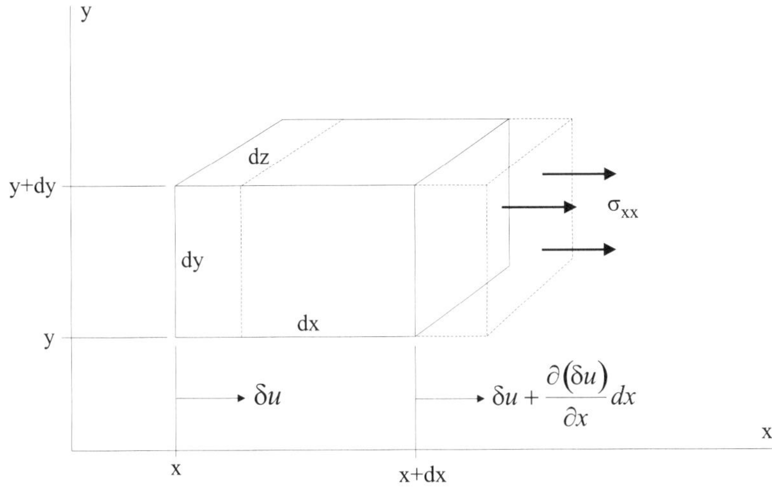

Figure 10.2: A stress element subject to stress σ_{xx} and undergoing a virtual displacement δu in the x-direction.

which reduces to, noting $\varepsilon_{xx} = \partial u/\partial x$

$$\delta\varepsilon_{xx} = \frac{\partial(\delta u)}{\partial x} \tag{10.7}$$

and substituting back into (10.5) the virtual work done is

$$\delta W_{xx} = \sigma_{xx}\delta\varepsilon_{xx}\,\mathrm{d}x\,\mathrm{d}y\,\mathrm{d}z \tag{10.8}$$

where σ_{xx} is the real applied stress and $\delta\varepsilon_{xx}$ is the virtual strain resulting from the virtual displacement.

From Chapter 9 it will be recalled that the internal work done or energy is referred to as the strain energy so that $\delta W_{xx} = \delta U_{xx}$ and that U_o represents the strain energy density. Following a similar procedure for σ_{yy} and τ_{xy} as outlined above for σ_{xx} we find that the virtual strain energy density is given by, in two-dimensions

$$\delta U_o = \sigma_{xx}\delta\varepsilon_{xx} + \sigma_{yy}\delta\varepsilon_{yy} + \tau_{xy}\delta\gamma_{xy} \tag{10.9}$$

The total strain energy resulting from a virtual displacement is, in an analogous manner to (9.12)

$$\delta U = \int_V \delta U_o\,\mathrm{d}V \tag{10.10}$$

where $\mathrm{d}V = \mathrm{d}x\,\mathrm{d}y\,\mathrm{d}z$. The work done by the internal forces, δW_I, is equal to minus the gain in strain energy

$$\delta W_I = -\int_V \delta U_o\,\mathrm{d}V \tag{10.11}$$

Alternatively, δU_o can be obtained from

$$\delta U_o = \frac{\partial U_o}{\partial\varepsilon_{xx}}\delta\varepsilon_{xx} + \frac{\partial U}{\partial\varepsilon_{yy}}\delta\varepsilon_{yy} + \cdots = \sigma_{xx}\delta\varepsilon_{xx} + \sigma_{yy}\delta\varepsilon_{yy} + \cdots \tag{10.12}$$

In terms of point forces $P_1, P_2, \ldots, P_n$ then δU can be expressed as a function of these forces

$$\delta U = \frac{\partial U}{\partial P_1}\delta P_1 + \frac{\partial U}{\partial P_2}\delta P_2 + \cdots \tag{10.13}$$

From Castigliano's second theorem, (9.59), we know that the corresponding virtual displacements are

$$\delta_1 = \frac{\partial U}{\partial P_1}; \quad \delta_2 = \frac{\partial U}{\partial P_2}; \quad \cdots \tag{10.14}$$

so that

$$\delta U = \delta_1 \delta P_1 + \delta_2 \delta P_2 + \cdots \tag{10.15}$$

Let us now consider the action of externally applied forces acting on the boundary of a structure, Figure 10.3. Denoting the externally applied boundary forces by X and Y with corresponding virtual displacements δu and δv then the virtual work done by the external forces is

$$\int_S (X\delta u + Y\delta v)\, \mathrm{d}s \tag{10.16}$$

over boundary S. Body forces (such as gravity) could also be accounted for but we will neglect such loading. For a state of static equilibrium to be ensured then the total work done during a virtual displacement

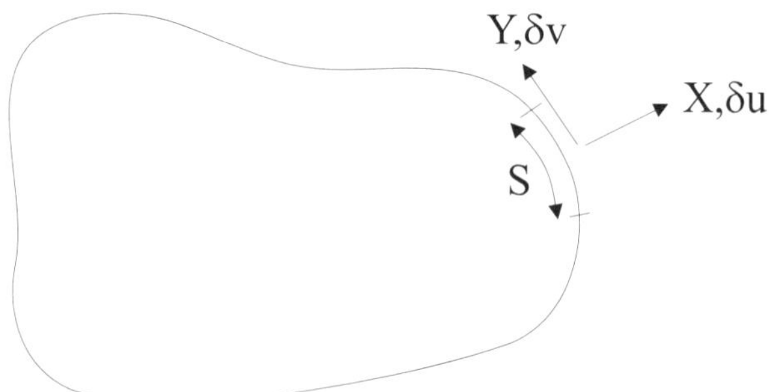

Figure 10.3: Externally applied forces X and Y, with corresponding virtual displacements δu and δv, acting on the boundary of a structure.

must vanish, from (9.12) and (10.16)

$$\int_S (X\delta u + Y\delta v)\, \mathrm{d}x - \int_V \delta U_o\, \mathrm{d}V = 0 \tag{10.17}$$

Since the externally applied forces and real stress components are taken as constant during a virtual displacement then the variation sign can be taken from under the integration

$$\delta \left[\int_V U_o\, \mathrm{d}V - \int_S (Xu + Yv) \right] = 0 \tag{10.18}$$

from which we have the *principle of virtual work*:

If a deformable continuum body is in a state of static equilibrium under the action of a system of applied forces and is given a virtual displacement then the virtual work done by the internal forces is equivalent to the virtual work done by the externally applied forces.

10.4 Principle of Stationary Total Potential Energy

Denoting the energy of deformation as

$$U = \int_V U_o \, \mathrm{d}V \tag{10.19}$$

and the energy of the applied forces as

$$W = \int_S (Xu + Yv) \, \mathrm{d}s \tag{10.20}$$

and the total potential energy as

$$\Pi = U - W \tag{10.21}$$

then

$$\delta\Pi = \delta(U - W) = 0 \tag{10.22}$$

which can be stated as the *principle of stationary total potential energy*:

> *For a body in a state of static equilibrium then the total potential energy must be stationary and a minimum as a result of a virtual displacement.*

This principle indicates that if the stationary value of the total potential energy is a minimum then the body is in a state of equilibrium. For this reason, this principle is often referred to as the principle of minimum total potential energy. We will see in §13.11 that the finite element method seeks such a minimum in the total potential energy of the system of elements. For instance, Figure 10.4 illustrates a mesh of elements for which the potential energy of an arbitrary element e is given by

$$\Pi^e = U^e - W \tag{10.23}$$

with the total potential energy for E elements given by

$$\Pi = \sum_{e=1}^{E} U^e - W \tag{10.24}$$

where W is the total work done by the applied forces $P_1, P_2, \ldots, P_n$

$$W = u_1 P_1 + u_2 P_2 + \cdots + u_n P_n = \{U\}^T\{P\} \tag{10.25}$$

The element strain energy is given by, (13.43)

$$U^e = \frac{1}{2}\{U\}^T[k^e]\{U\} \tag{10.26}$$

where k^e is the element stiffness matrix and U is the column vector of nodal displacements respectively. From (10.24), the minimisation of the total potential energy of the entire structure of elements gives

$$\frac{\partial\Pi}{\partial\{U\}} = \left[\sum_{e=1}^{E}[k^e]\right]\{U\} - \{P\} = 0 \tag{10.27}$$

Minimisation of Π with respect to U ensures that the structure obtains a state of equilibrium.

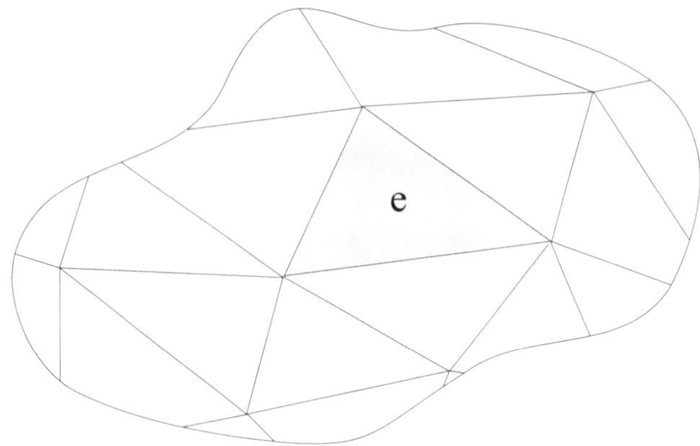

Figure 10.4: A structure of finite elements and an arbitrary element e.

10.5 Virtual Forces

Let a structure now be subject to a set of virtual forces $\delta P_1, \delta P_2, \ldots, \delta P_n$ with associated real displacements $\delta_1, \delta_2, \ldots, \delta_n$. The external work done, δW_E, by the virtual forces is therefore

$$\delta W_E = \int_0^{\delta P_n} \delta_n \, \mathrm{d}\delta P_n \tag{10.28}$$

For a state of static equilibrium then the virtual work done must vanish and from (10.11)

$$\delta W_E + \delta W_I = \int_0^{\delta P_n} \delta_n \, \mathrm{d}\delta P_n - \int_V \delta U_o \, \mathrm{d}V = 0 \tag{10.29}$$

The virtual forces $(\delta P_1, \delta P_2, \ldots, \delta P_n)$ produce a virtual stress system $(\delta\sigma_{xx}, \delta\sigma_{yy}, \ldots)$. Since $(\delta P_n, \ldots$ and $\delta\sigma_{xx}, \ldots)$ represent the virtual system and $(\delta_1, \ldots$ and $\varepsilon_{xx}, \ldots)$ represent the real system then there is no dependence between the two systems and the virtual external work in (10.29) integrates directly

$$\delta_n \delta P_n - \int_V \delta U_o \, \mathrm{d}V = 0 \tag{10.30}$$

or explicitly in terms of stress and strain

$$\delta_n \delta P_n - \int_V (\delta\sigma_{xx}\varepsilon_{xx} + \delta\sigma_{yy}\varepsilon_{yy} + \delta\tau_{xy}\gamma_{xy} + \cdots) \, \mathrm{d}V = 0 \tag{10.31}$$

Finally, since the virtual force-stress system is independent of the real system the real applied loading can be removed and replaced by a single virtual force, P^v, applied at the point where the actual displacement, δ, is required

$$P^v \delta - \int_V (\sigma^v_{xx}\varepsilon_{xx} + \sigma^v_{yy}\varepsilon_{yy} + \cdots) \, \mathrm{d}V = 0 \tag{10.32}$$

The following example will help demonstrate the use of (10.32).

Example 10.1 Determination of the displacement of a cantilever beam subject to linearly varying temperature distribution

Figure 10.5 illustrates a bar of constant cross-sectional area A and length L subject to a linearly varying temperature distribution of

$$\Delta T(x) = \Delta T_0 \frac{(L-x)}{L}$$

with the temperature equal to ΔT_0 at $x = 0$ and $\Delta T_0 = 0$ at $x = L$. Let us use (10.32) to determine the axial displacement at $x = L$ by applying a virtual force P^v at this point. The virtual axial stress and axial strain are

$$\sigma_{xx}^v = \frac{P^v}{A}; \quad \varepsilon_{xx}^v = \alpha \Delta T = \alpha \Delta T_0 \frac{(L-x)}{L}$$

Substituting these into (10.32) we have

$$P^v \delta - \int_0^L \left[\frac{P^v}{A} \alpha \Delta T_0 \frac{(L-x)}{L} \right] A \, \mathrm{d}x = 0$$

Performing the integration and re-arranging we arrive at

$$\delta = \frac{3}{2} \alpha \Delta T_0 L$$

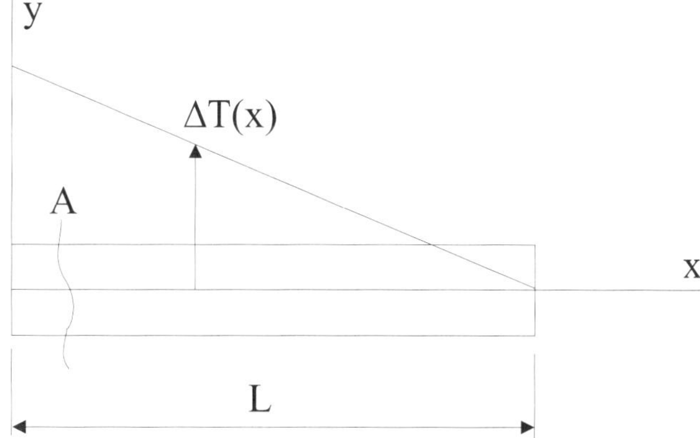

Figure 10.5: Example: A bar of constant cross-sectional subject to a linearly varying temperature distribution.

10.6 The Unit Load Method for the Determination of Deflections in Beams

In the previous example it was observed that the virtual force P^v cancels. This is no coincidence and it follows that P^v can be replaced by a unit virtual load such as force, moment and torque applied at the point where the displacement, rotation or twist is required. Hence, with a unit load for P^v we have from (10.32)

$$\delta - \int_V (\sigma^v_{xx}\varepsilon_{xx} + \sigma^v_{yy}\varepsilon_{yy} + \cdots)\,\mathrm{d}V = 0 \tag{10.33}$$

In the case of the pure bending of beams with $\sigma^v_{xx} = M^v y/I$, $\varepsilon_{xx} = \sigma_{xx}/E = My/IE$ we have from (10.33), neglecting other terms

$$\delta - \int_V \left(\frac{M^v y}{I}\right)\left(\frac{My}{IE}\right)\mathrm{d}V = 0 \tag{10.34}$$

With $\mathrm{d}V = \mathrm{d}A\,\mathrm{d}x$ then (10.34) is given by

$$\delta - \int_L \frac{M^v M}{EI}\frac{1}{I}\int_A y^2\,\mathrm{d}A\,\mathrm{d}x = 0 \tag{10.35}$$

Noting that the area integral is equivalent to the second moment of area I then we have for the displacement δ

$$\delta = \int_L \frac{M^v M}{EI}\,\mathrm{d}x \tag{10.36}$$

The unit load method is an extremely useful technique for determining displacements in beams as the next example will illustrate.

Example 10.2 Determination of the displacement of a cantilever beam using the unit load method

Let us show that the displacement, δ, at the free end of the cantilever beam in Figure 10.6a) is given by

$$\delta = \frac{Pa^2}{6EI}(3L - a)$$

where P is a point load acting at a distance b from the free end of the beam which is of length $L(= a + b)$. For the virtual unit load at the free end, Figure 10.6b), the virtual bending moment, M^v, is

$$M^v = x \quad 0 \le x \le L$$

with x measured from the free end. For the real applied loading the bending moment, M, is

$$M = 0 \quad 0 \le x \le b$$
$$M = P(x - b) \quad b \le x \le L$$

From (10.36) δ is

$$\delta = \frac{1}{EI}\int_b^L P(x-b)x\,\mathrm{d}x = \frac{P}{EI}\left[\frac{x^3}{3} - \frac{bx^2}{2}\right]_b^L = \frac{Pa^2}{6EI}(2a + 3b) = \frac{Pa^2}{6EI}(3L - a)$$

with no integral in the interval $0 \le x \le b$ because $M = 0$.

10.7 Potential Energy

Let us conclude this Chapter by presenting a general expression for the potential energy of a body. Equation (10.21) above noted that the potential energy, Π, is given by the strain energy due to deformation, U,

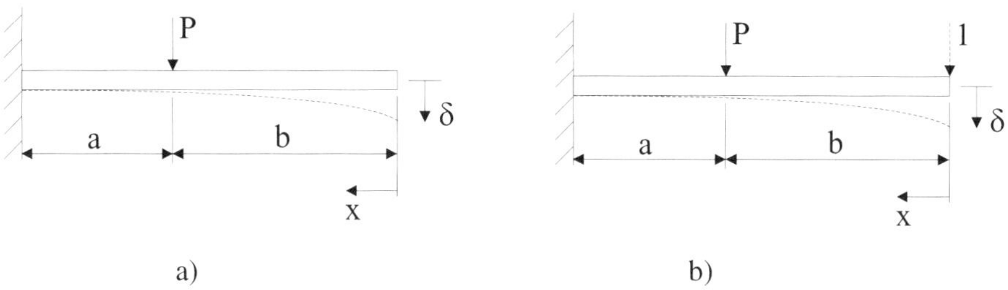

Figure 10.6: Example: The unit load method. a) A cantilever beam subject to a point load P at a distance b from the free end. b) The same cantilever beam as in case a) but with a unit virtual load applied at the free end.

minus the energy due to the applied forces, W. Accounting also for body forces then Π can be written in general form as follows

$$\Pi(u) = \int_V U(\varepsilon_{ij})\, \mathrm{d}V - \int_\Omega T_i u_i\, \mathrm{d}S - \int_V F_i u_i\, \mathrm{d}V \tag{10.37}$$

where Ω denotes the surface of the body on which the tractions are prescribed and V is the total volume of the three-dimensional body. The surface tractions are denoted by the traction vector $T_i = \sigma_{ij} n_j$ where n_j is the outward normal vector to the surface. The strain energy density, U, is given by, (9.5)

$$U(\varepsilon_{ij}) = \int_0^\varepsilon \sigma_{ij}\, \mathrm{d}\varepsilon_{ij} \tag{10.38}$$

for arbitrary materials and $U(\varepsilon_{ij}) = (1/2)\sigma_{ij}\varepsilon_{ij}$ for linear elastic materials. The third integral on the right hand side of (10.37) represents the contribution due to body forces with F_i denoting the body force in direction i. Collecting the volume integral terms then Π is

$$\Pi(u) = \int_V [U(\varepsilon_{ij}) - F_i u_i]\, \mathrm{d}V - \int_\Omega T_i u_i\, \mathrm{d}S \tag{10.39}$$

In the case of a two-dimensional body then (10.39) reduces to

$$\Pi(u) = \int_A [U(\varepsilon_{ij}) - F_i u_i]\, \mathrm{d}A - \int_\Gamma T_i u_i\, \mathrm{d}s \tag{10.40}$$

where Γ denotes the contour of the body on which the tractions are prescribed and A is the total area of the body.

We will make use of (10.40) in Chapter 17 when we derive the J-integral used in fracture mechanics and in Chapter 13 when we derive the finite element stiffness matrix and force vector equations using a variational approach. As an illustration of how the total potential energy of a body can be used as a variational function consider the simplest case of an axial bar, Figure 10.7. In one-dimension the stress and strain tensors, σ_{ij} and ε_{ij}, reduce to an axial stress and strain of σ_{xx} and ε_{xx}. The body force, F_i, and displacement, u_i, vectors similarly reduce to F_x and u_x. Assuming a linear elastic material then Hooke's constitutive law, $\sigma_{ij} = C_{ijkl}\varepsilon_{kl}$, reduces to $\sigma_{xx} = E\varepsilon_{xx}$ where E is Young's modulus. Thus, the volume integral of (10.39) reduces to

$$\int_V \left(\frac{1}{2}\sigma_{xx}\varepsilon_{xx} - F_x u_x\right) \mathrm{d}V \tag{10.41}$$

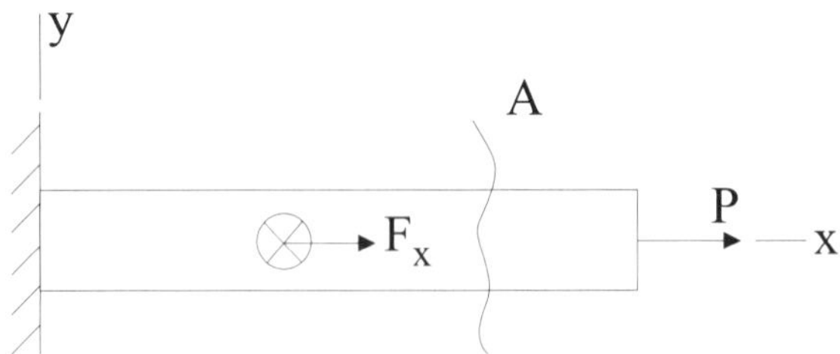

Figure 10.7: An axial bar of uniform cross-sectional area A subject to an axial load P and a body force F_x.

Letting $\mathrm{d}V = \mathrm{d}A\,\mathrm{d}x$ and assuming that the bar is of constant cross-sectional area A and of length L then (10.41) becomes

$$\int_0^L \left(\frac{1}{2}E\varepsilon_{xx}^2 - F_x u_x\right) A\,\mathrm{d}x \tag{10.42}$$

To express (10.42) solely in terms of displacement u_x then we note that $\varepsilon_{xx} = \mathrm{d}u/\,\mathrm{d}x$ from (1.31) and finally arrive at the variational function for unknown u_x

$$\int_0^L \left(\frac{1}{2}E\left(\frac{\mathrm{d}u_x}{\mathrm{d}x}\right)^2 - F_x u_x\right) A\,\mathrm{d}x \tag{10.43}$$

The surface integral of (10.39) is easily evaluated by noting that the surface traction vector $T_i = \sigma_{ij}n_j$ is equal to $\sigma_{xx} = P/A$ where P is the applied axial load. Thus, the surface integral reduces to

$$\int_A \frac{P}{A}u_x\,\mathrm{d}A = Pu_x \tag{10.44}$$

with Pu_x representing the external work done by P.

10.8 Conclusion

In this Chapter we have examined the notion of a virtual displacement and the principles of virtual displacement, virtual work and stationary and total potential energy. Use was made of the principle of stationary total potential energy to determine the displacement in a structure when subject to general virtual loading and tailored specifically for the case of unit virtual loads for determining the displacement in beams via the *unit load method*. The Chapter concluded with an examination of potential energy which will prove useful in later Chapters.

10.9 References and Further Reading

◦ Rees, D. W. A. (1990) *Mechanics of Solids and Structures*, McGraw-Hill, London.

10.10 Exercises

10.1 Describe the principle of virtual displacements.

10.2 Describe the principle of virtual work.

10.3 Describe the principle of stationary total potential energy.

10.4 For the cantilever beam of Figure 10.6a) determine the slope at the free end using the unit load method.

10.5 Using the unit load method to determine the horizontal deflection, δ, at the free end of the quarter circle beam shown in Figure 10.8.

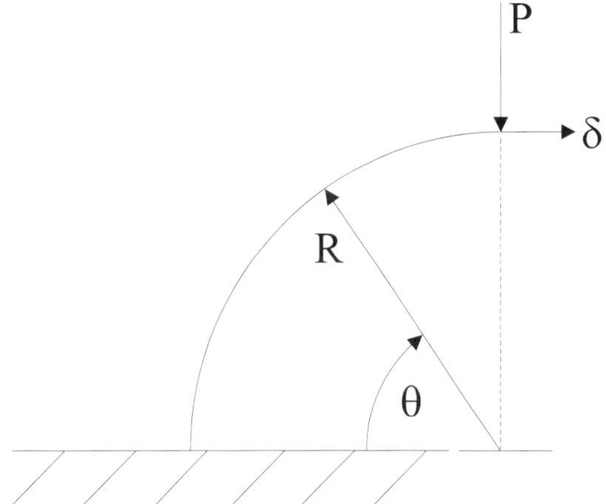

Figure 10.8: Exercise: A quarter circle beam subject to a concentrated force P at the free end.

10.6 Determine the deflection at the free end of the cantilever beam shown in Figure 10.9. Use the unit load method and assume that $I = 5 \times 10^9 \text{mm}^4$ and $E = 205\text{GPa}$.

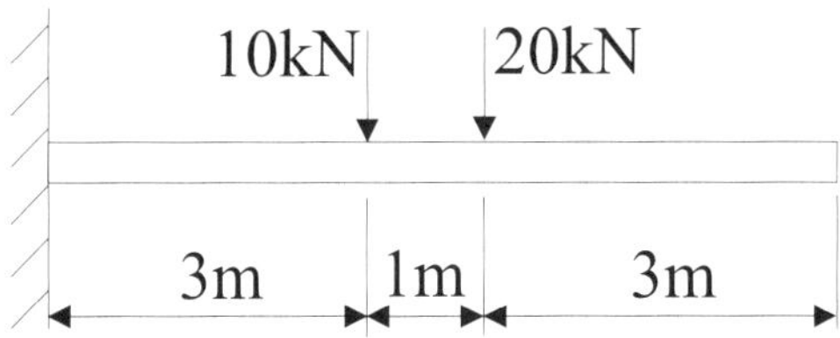

Figure 10.9: Exercise: A cantilever beam subject to concentrated loads.

10.7 Using the unit load method show that the deflection, δ, at the free end of the cantilever beam of Figure 10.10 is given by

$$\delta = \frac{6Wa^3}{EI}$$

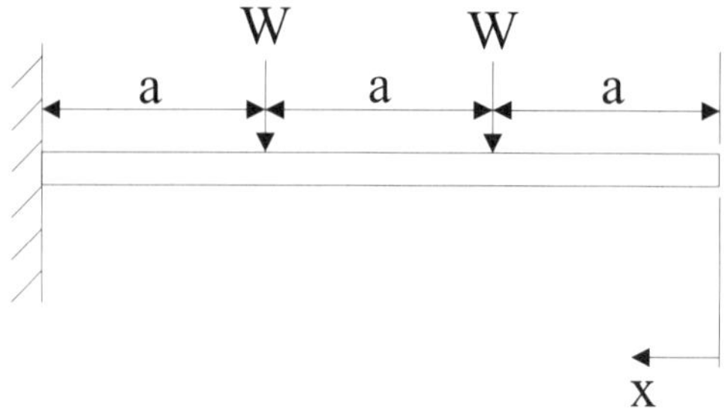

Figure 10.10: Exercise: A cantilever beam subject to concentrated loading.

where E and I are the Young's modulus and second moment of area respectively of the beam.

Chapter 11

Thick-Walled Pressure Vessels

11.1 Introduction

The present Chapter extends the thin-walled pressure vessel theory discussed in Chapter 3. Consider the thick-walled cylindrical vessel in Figure 11.1 with inner radius a and outer radius b. The internal and external pressures are denoted by p_i and p_o respectively. Figure 11.1 uses cylindrical coordinates (r, θ, z) with stress components σ_{rr}, $\sigma_{\theta\theta}$ and σ_{zz}. When we discussed thin-walled vessels we noted that a thin-walled vessel is one in which the ratio of the diameter $(d = 2b)$ to wall-thickness $(t = b - a)$ is much larger than unity, that is $d/t \gg 1$. A frequently used rule of thumb when categorising a pressure vessel is that it is thin-walled if $d/t > 10$ and thick-walled if $d/t < 10$. The subsequent sections will illustrate that the analysis of thick-walled vessels is significantly more involved than thin-walled vessels. As a consequence it is always advantageous to establish the type of vessel and not simply apply thick-walled vessel theory blindly.

If a cylindrical vessel is long in length with respect to the diameter then transverse plane sections will remain plane and unwarped when subject to pressure loading, Figure 11.2. There are radial (r, θ) variations but negligible axial changes. However, at the ends of the vessel the axial displacement ceases to be independent of the radius and as a result warping occurs. Thus, in the following sections it will be assumed that the only stresses present on a stress element that is sufficiently far from the ends are radial, σ_{rr}, circumferential, $\sigma_{\theta\theta}$, and axial, σ_{zz}.

The next section derives the equilibrium equation for a thick-walled vessel. This is followed by determining the strain components in a vessel and the compatibility equation. Lame's equations of stress are then derived in terms of general boundary conditions. The two boundary conditions of a vessel subject to both internal and external pressures, and internal pressure only are then examined separately. The effects of both open and closed end conditions on the state of stress and strain are then discussed. Lame's equations are then applied to the problems of shrink fitting a collar on to a shaft and compound cylindrical pressure vessels. Finally, the Chapter examines thick-walled spherical pressure vessels.

11.2 Equilibrium Equation

Figure 11.1 illustrates a differential stress element of angle $\mathrm{d}\theta$ and radial width $\mathrm{d}r$ at radial distance r. Throughout the element the radial stress varies from σ_{rr} to $\sigma_{rr} + (\partial\sigma_{rr}/\partial r)\,\mathrm{d}r$ and resolving forces in

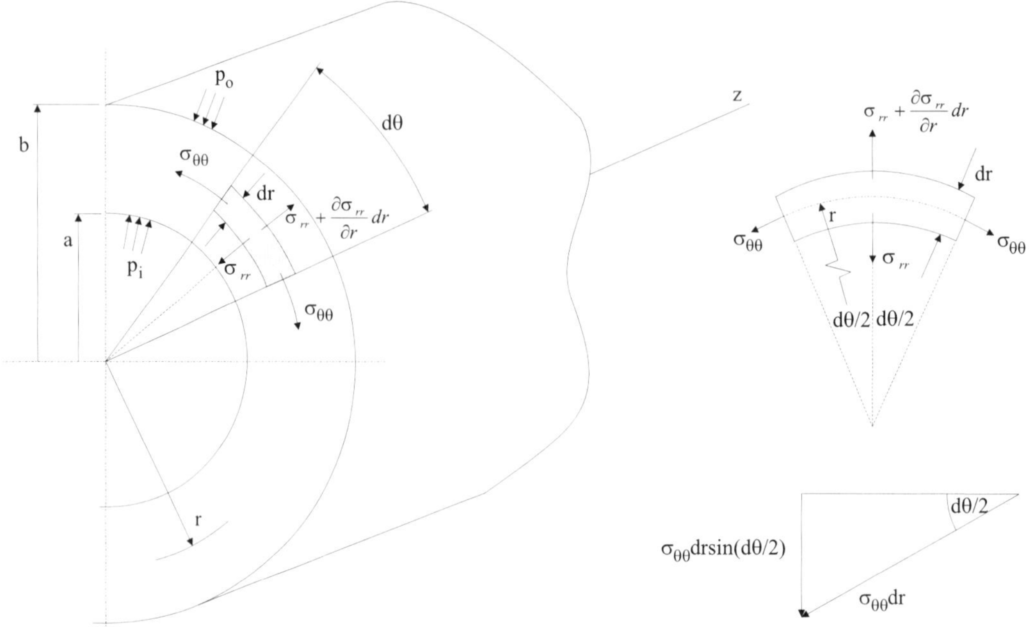

Figure 11.1: Thick-walled cylindrical vessel of inner radius a and outer radius b subject to internal and external pressures p_i and p_o respectively.

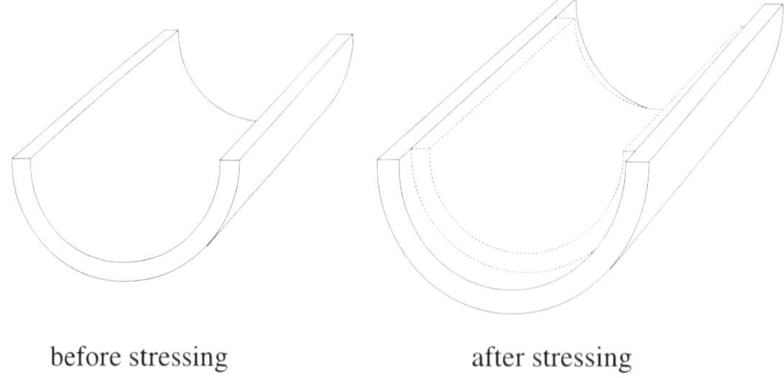

Figure 11.2: Section of a vessel before and after being pressurised.

the radial direction we have

$$\sigma_{rr} r \, \mathrm{d}\theta + 2\sigma_{\theta\theta} \, \mathrm{d}r \sin\frac{\mathrm{d}\theta}{2} = \left(\sigma_{rr} + \frac{\partial\sigma_{rr}}{\partial r} \, \mathrm{d}r\right)(r + \mathrm{d}r) \, \mathrm{d}\theta \tag{11.1}$$

For $\theta \approx 0^\circ$ then $\sin(\mathrm{d}\theta/2) \approx \mathrm{d}\theta/2$ and therefore (11.1) becomes

$$r\sigma_{rr} + \sigma_{\theta\theta} \, \mathrm{d}r = r\sigma_{rr} + \sigma_{rr} \, \mathrm{d}r + r\frac{\partial\sigma_{rr}}{\partial r} \, \mathrm{d}r + \frac{\partial\sigma_{rr}}{\partial r}(\mathrm{d}r)^2 \tag{11.2}$$

Cancelling terms and neglecting second order terms gives the following equilibrium equation

$$\frac{\partial \sigma_{rr}}{\partial r} = \frac{\sigma_{\theta\theta} - \sigma_{rr}}{r} \tag{11.3}$$

11.3 Strains and Compatibility Equation

The strains in a thick-walled vessel are easily found by considering the radial and circumferential displacements. Considering first the radial strain, Figure 11.3a) illustrates that a point at radial position r moves through a radial distance u_r whereas a point at position $r + \mathrm{d}r$ moves through a radial distance $u_r + \mathrm{d}u_r$. Thus, the radial strain ε_{rr} is

$$\varepsilon_{rr} = \frac{(u_r + \mathrm{d}u_r) - u_r}{u_r} = \frac{\mathrm{d}u_r}{u_r} \tag{11.4}$$

with $u_r \equiv \mathrm{d}r$.

The change in the radial direction correspondingly induces a circumferential strain $\varepsilon_{\theta\theta}$

$$\varepsilon_{\theta\theta} = \frac{2\pi(r + \mathrm{d}r) - 2\pi r}{2\pi r} = \frac{u_r}{r} \tag{11.5}$$

From (1.16) and in view of (11.4) and (11.5) we can express the Hookian equations for a cylindrical

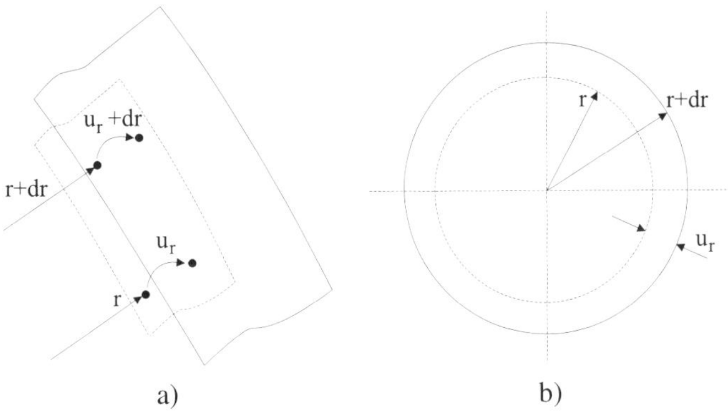

Figure 11.3: Displacement of a point in a thick-walled vessel. a) Radial displacement. b) Circumferential displacement.

coordinate system as follows

$$\begin{aligned}
\varepsilon_{rr} &= \frac{1}{E}\left[\sigma_{rr} - \nu\left(\sigma_{\theta\theta} + \sigma_{zz}\right)\right] = \frac{\mathrm{d}u_r}{\mathrm{d}r} \\
\varepsilon_{\theta\theta} &= \frac{1}{E}\left[\sigma_{\theta\theta} - \nu\left(\sigma_{rr} + \sigma_{zz}\right)\right] = \frac{u_r}{r} \\
\varepsilon_{zz} &= \frac{1}{E}\left[\sigma_{zz} - \nu\left(\sigma_{\theta\theta} + \sigma_{rr}\right)\right] = \text{const}
\end{aligned} \tag{11.6}$$

Re-arranging ε_{zz} for σ_{zz}

$$\sigma_{zz} = E\varepsilon_{zz} + \nu(\sigma_{\theta\theta} + \sigma_{rr}) \tag{11.7}$$

which, upon substitution into ε_{rr} and $\varepsilon_{\theta\theta}$ gives

$$\begin{aligned}\varepsilon_{rr} &= -\nu\varepsilon_{zz} + \left(\frac{1+\nu}{E}\right)[(1-\nu)\sigma_{rr} - \nu\sigma_{\theta\theta}] = \frac{\mathrm{d}u_r}{\mathrm{d}r} \\ \varepsilon_{\theta\theta} &= -\nu\varepsilon_{zz} + \left(\frac{1+\nu}{E}\right)[(1-\nu)\sigma_{\theta\theta} - \nu\sigma_{rr}] = \frac{u_r}{r}\end{aligned} \tag{11.8}$$

Since ε_{zz} is independent of r we can therefore eliminate u from (11.8) by observing that $\mathrm{d}\varepsilon_{\theta\theta}/\,\mathrm{d}r = \varepsilon_{rr}$ and upon substituting the equilibrium equation we have the following compatibility equation

$$\frac{\mathrm{d}}{\mathrm{d}r}(\sigma_{rr} + \sigma_{\theta\theta}) = 0 \tag{11.9}$$

which informs us that $(\sigma_{rr} + \sigma_{\theta\theta})$ must be a constant and independent of r.

11.4 Lame's Equations

To derive Lame's equations for the radial and circumferential stresses for a vessel let us begin by letting $(\sigma_{rr} + \sigma_{\theta\theta})$ equal $2A$ from the previously derived compatibility equation, (11.9)

$$\sigma_{rr} + \sigma_{\theta\theta} = 2A \tag{11.10}$$

and substituting into the equilibrium equation, (11.3)

$$2(A - \sigma_{rr}) = r\frac{\mathrm{d}\sigma_{rr}}{\mathrm{d}r} \tag{11.11}$$

Separating variables and integrating with respect to r

$$2\int \frac{\mathrm{d}r}{r} = \int \frac{\mathrm{d}\sigma_{rr}}{A - \sigma_{rr}} \tag{11.12}$$

which, after integration, leads to

$$2\ln r = -\ln(A - \sigma_{rr}) + \text{const} \tag{11.13}$$

Collecting logarithmic terms

$$\ln\left[r^2(A - \sigma_{rr})\right] = \text{const} \quad \text{or} \quad r^2(A - \sigma_{rr}) = \text{const} = B \tag{11.14}$$

and re-arranging for σ_{rr} we have

$$\sigma_{rr} = A - \frac{B}{r^2} \tag{11.15}$$

The circumferential stress $\sigma_{\theta\theta}$ follows from (11.10) so that we finally arrive at the well known *Lame's equations*

$$\sigma_{rr} = A - \frac{B}{r^2}; \quad \sigma_{\theta\theta} = A + \frac{B}{r^2} \tag{11.16}$$

where A and B are determined from the boundary conditions.

Before considering two different boundary conditions let us confirm that σ_{zz} is constant. From (11.6) and (11.10)

$$\varepsilon_{zz} = \text{const} = \frac{1}{E}\left[\sigma_{zz} - \nu(\sigma_{rr} + \sigma_{\theta\theta})\right] = \frac{1}{E}[\sigma_{zz} - 2\nu A] \tag{11.17}$$

illustrating that σ_{zz} is seen to be constant.

11.5 Internal and External Pressures

Various forms of boundary conditions can be prescribed to a cylindrical pressure vessel but we will consider just two cases. In this section we examine the case of both internal, p_i, and external, p_o, pressures, Figure 11.4. The constants A and B in Lame's equations (11.16) are determined by consideration of σ_{rr}

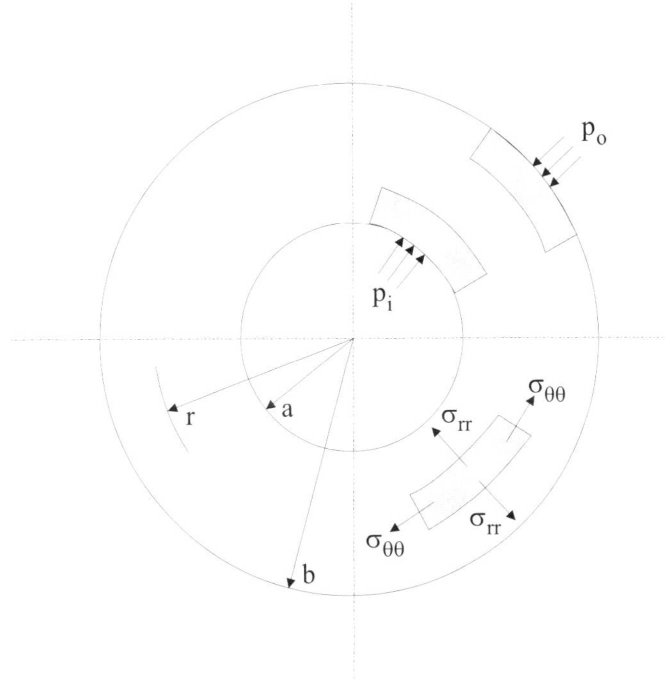

Figure 11.4: A thick-walled cylindrical vessel of inner radius a and outer radius b subject to internal and external pressures p_i and p_o respectively.

at both the inner and outer radii

$$\begin{aligned} \sigma_{rr} &= -p_i \quad \text{at} \quad r = a \\ \sigma_{rr} &= -p_o \quad \text{at} \quad r = b \end{aligned} \tag{11.18}$$

noting that σ_{rr} is negative at $r = a$ and $r = b$. The sign of σ_{rr} is established by letting a stress element approach the inner and outer vessel surfaces as shown in Figure 11.4. From σ_{rr} of (11.16)

$$-p_i = A - \frac{B}{a^2}; \quad -p_o = A - \frac{B}{b^2} \tag{11.19}$$

and solving for A and B

$$A = \frac{(b/a)^2 p_o - p_i}{1 - (b/a)^2}; \quad B = \frac{b^2(p_o - p_i)}{1 - (b/a)^2} \tag{11.20}$$

and substituting back into Lame's equations gives σ_{rr} at both the inner and outer radii

$$\begin{aligned}\sigma_{rr} &= A - \frac{B}{r^2} = \frac{(b/a)^2 p_o - p_i}{1-(b/a)^2} - \frac{(p_o - p_i)}{1-(b/a)^2}\left(\frac{b}{r}\right)^2 \\ \sigma_{\theta\theta} &= A + \frac{B}{r^2} = \frac{(b/a)^2 p_o - p_i}{1-(b/a)^2} + \frac{(p_o - p_i)}{1-(b/a)^2}\left(\frac{b}{r}\right)^2\end{aligned} \tag{11.21}$$

If the effects of the pressures p_i and p_o are resisted in the axial direction by the cylinder wall, Figure 11.5, then for equilibrium to be satisfied

$$(b^2 - a^2)\sigma_{zz} = (p_i a^2 - p_o b^2)\pi \tag{11.22}$$

or, re-arranging for σ_{zz}

$$\sigma_{zz} = \frac{p_o (b/a)^2 - p_i}{1-(b/a)^2} \tag{11.23}$$

and comparing with σ_{rr} and $\sigma_{\theta\theta}$ in (11.21) we find that

$$\sigma_{zz} = \frac{1}{2}(\sigma_{rr} + \sigma_{\theta\theta}) \tag{11.24}$$

which represents a hydrostatic component of stress.

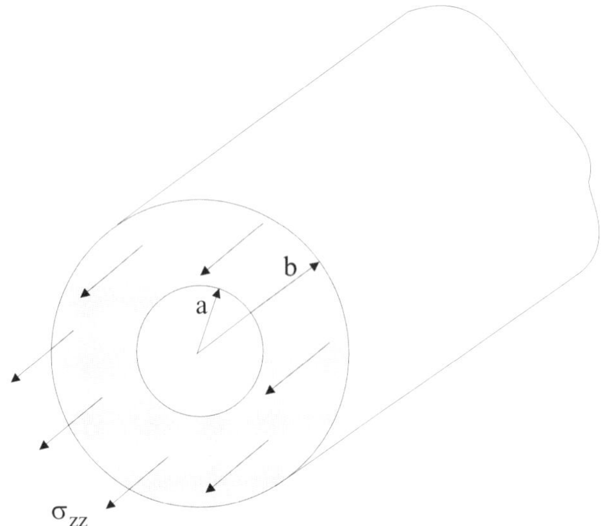

Figure 11.5: Axial stress σ_{zz} acting over the vessel cross-section.

11.6 Internal Pressure and Zero External Pressure

In this section we examine the case of an internal pressure, $p_i = p$, and atmospheric or zero external pressure, $p_o = 0$. As in the previous section, the constants A and B in Lame's equations (11.16) are determined by consideration of σ_{rr} at both the inner and outer radii

$$\begin{aligned}\sigma_{rr} &= -p \quad \text{at} \quad r = a \\ \sigma_{rr} &= 0 \quad \text{at} \quad r = b\end{aligned} \tag{11.25}$$

From σ_{rr} of (11.16)

$$A = \frac{p}{(b/a)^2 - 1}; \quad B = \frac{b^2 p}{(b/a)^2 - 1} \tag{11.26}$$

and substituting into Lame's equations gives

$$\sigma_{rr} = -p\left[\frac{(b/r)^2 - 1}{(b/a)^2 - 1}\right]; \quad \sigma_{\theta\theta} = p\left[\frac{(b/r)^2 + 1}{(b/a)^2 - 1}\right] \tag{11.27}$$

From (11.6) and (11.10)

$$\varepsilon_{zz} = \frac{1}{E}\left[\sigma_{zz} - \nu(\sigma_{rr} + \sigma_{\theta\theta})\right] = \frac{1}{E}\left[\sigma_{zz} - 2\nu A\right] \tag{11.28}$$

and re-arranging for σ_{zz}

$$\sigma_{zz} = E\varepsilon_{zz} + \frac{2\nu p}{(b/a)^2 - 1} \tag{11.29}$$

Let us conclude the present section by considering Figure 11.5 once more but on this occasion with an axial force P_{zz} acting on the vessel ends. From equilibrium considerations

$$P_{zz} = \pi(b^2 - a^2)\sigma_{zz} = \pi a^2\left[(b/a)^2 - 1\right]\left[E\varepsilon_{zz} + \frac{2\nu p}{(b/a)^2 - 1}\right] \tag{11.30}$$

Re-arranging for ε_{zz} we have the axial strain as a function of both internal pressure, p, and axial load, P_{zz}

$$\varepsilon_{zz} = \frac{\left(\frac{P_{zz}}{\pi a^2} - 2\nu p\right)}{E[(b/a)^2 - 1]} \tag{11.31}$$

Example 11.1 Stress components of a thick-walled pressure vessel subject to an internal pressure

A thick-walled vessel of inner radius $a = 25$mm and outer radius $b = 75$mm is subject to an internal pressure of $p = 75$bar and atmospheric external pressure which can be neglected. Tabulate the variation of σ_{rr} and $\sigma_{\theta\theta}$ across the wall of the vessel.

In this example we will start from Lame's equations and determine the constants A and B from the prescribed boundary conditions rather than blindly using the above expressions (11.27). Therefore, from σ_{rr} of (11.16) the two boundary conditions are

$$\sigma_{rr} = -p = A - \frac{B}{a^2} \quad \text{at} \quad r = a$$
$$\sigma_{rr} = 0 = A - \frac{B}{b^2} \quad \text{at} \quad r = b$$

Solving the first equation for A

$$A = \frac{B}{a^2} - p$$

and substituting into the second equation to determine B

$$B = \frac{(ab)^2}{b^2 - a^2}p$$

and after substituting back into A

$$A = \frac{a^2}{b^2 - a^2} p$$

Substituting the given values for a, b and p we find that $A = 9.375 \times 10^5$ and $B = 5,273$. With $\sigma_{rr,\theta\theta} = A \mp B/r^2$ then the variation of the radial and circumferential stresses across the wall thickness are tabulated below in Table 11.1.

r(mm)	σ_{rr}(MPa)	$\sigma_{\theta\theta}$(MPa)
25	-7.5	9.4
40	-2.4	4.2
50	-1.2	3.0
65	-0.3	2.2
75	0	1.9

Table 11.1: Variation of σ_{rr} and $\sigma_{\theta\theta}$ across the wall thickness of the pressure vessel in Example 11.1.

Example 11.2 Axial strain of a thick-walled pressure vessel

A thick-walled cylindrical pressure vessel of inner and outer radii $a = 0.2$m and $b = 0.5$m respectively is subject to an internal pressure, p, and zero external pressure and an axial force of $P_{zz} = 10$kN is applied. Determine the axial strain for the two internal pressures of zero and 0.5MPa. Assume that the vessel material has a Young's modulus and Poisson's ratio of $E = 350$GPa and $\nu = 0.33$ respectively.

From (11.31) for the case of zero internal pressure the axial strain is

$$\varepsilon_{zz} = \frac{P_{zz}}{\pi a^2 E\left[(b/a)^2 - 1\right]} = \frac{10 \times 10^3}{\pi 0.1^2 \times 350 \times 10^9 \left[(0.5/0.1)^2 - 1\right]} = 379 \times 10^{-6}$$

while, for the case of 0.5MPa internal pressure

$$\varepsilon_{zz} = \frac{\frac{P_{zz}}{\pi a^2} - 2\nu p}{E\left[(b/a)^2 - 1\right]} = \frac{\frac{10 \times 10^3}{\pi 0.1^2} - 2 \times 0.33 \times 0.25 \times 10^6}{350 \times 10^9 \left[(0.5/0.1)^2 - 1\right]} = 183 \times 10^{-6}$$

observing that a non-zero internal pressure reduces the axial strain when compared to the zero internal pressure case.

11.7 End Conditions

The state of axial strain and stress in a cylindrical vessel is dependent on the end conditions of the vessel. If the ends of a vessel are closed and open then $\sigma_{zz} \neq 0$ and $\sigma_{zz} = 0$ respectively.

In the case of closed ends the resultant axial force acting on the wall section is equal to the force exerted by the internal pressure p (assuming zero external pressure)

$$(b^2 - a^2)\pi\sigma_{zz} = p\pi a^2 \tag{11.32}$$

from which

$$\sigma_{zz} = \frac{p}{(b/a)^2 - 1} \tag{11.33}$$

Summing the radial and circumferential stresses, from (11.27)

$$\sigma_{rr} + \sigma_{\theta\theta} = \frac{2p}{(b/a)^2 - 1} \tag{11.34}$$

then, from (11.28), the axial strain is

$$\varepsilon_{zz} = \frac{p(1-2\nu)}{E\left[(b/a)^2 - 1\right]} \tag{11.35}$$

When the ends of a vessel are open then $\sigma_{zz} = 0$ and, from (11.6)

$$\varepsilon_{zz} = -\frac{\nu}{E}(\sigma_{rr} + \sigma_{\theta\theta}) = -\frac{2\nu p}{E\left[(b/a)^2 - 1\right]} \tag{11.36}$$

11.8 Variation of Stress Throughout the Wall Thickness

Let us conclude our discussion of single cylindrical pressure vessels by schematically illustrating the typical variation of stresses throughout the cross-section of a vessel which is subject to an internal pressure p, Figure 11.6. The figure illustrates that the circumferential stress is the largest of the three stresses and

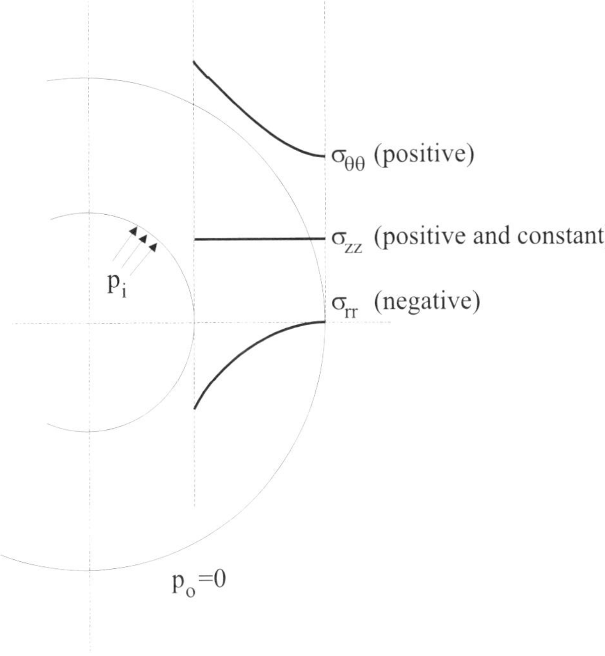

Figure 11.6: Variation of the radial, circumferential and axial stresses throughout the cross-section of a cylindrical vessel subject to an internal pressure.

hence the most detrimental from a design perspective. On the inner surface, $r = a$, $\sigma_{\theta\theta}$ achieves a maximum value of

$$\sigma_{\theta\theta,max}(r = a) = p\frac{(b/a)^2 + 1}{(b/a)^2 - 1} \tag{11.37}$$

with $\sigma_{\theta\theta,max} \to p$ as $(b/a) \to 0$. Equation (11.37) illustrates that $\sigma_{\theta\theta}$ is always greater than p.

11.9 Shrink Fit of a Collar on to a Shaft

Lame's equations can be applied to the problem of shrinking a collar of inner radius R_{ci} and outer radius R_{co} on to a shaft of radius $R_s(> R_{ci})$, Figure 11.7. Since the radius of the collar is less than the radius of the shaft there is an interference pressure, p, induced between the shaft and the collar. The interference

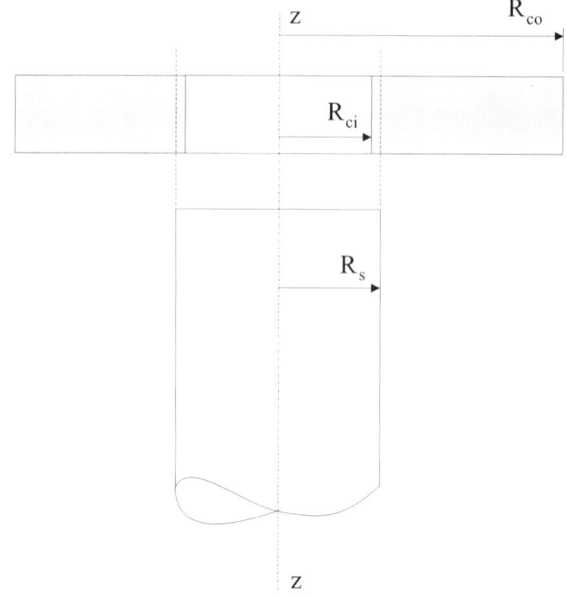

Figure 11.7: A shaft and collar.

fit results in a state of hydrostatic pressure in the shaft in which

$$\sigma_{rr,s} = \sigma_{\theta\theta,s} = -p; \quad \sigma_{zz,s} = 0 \tag{11.38}$$

From (11.5) the radial displacement, u_s, in the shaft is

$$\frac{u_s}{R_s} = \varepsilon_{\theta\theta} = \frac{1}{E}\left[\sigma_{\theta\theta,s} - \nu(\sigma_{rr,s} + \sigma_{zz,s})\right] = -\frac{p}{E}(1-\nu) \tag{11.39}$$

and re-arranging for u_s

$$u_s = -\frac{pR_s}{E}(1-\nu) \tag{11.40}$$

and is negative indicating a shrinking of the shaft, as expected.

Approximating the collar as a thick-walled pressure vessel with internal pressure p then the boundary conditions are, (11.16)

$$\begin{aligned} \sigma_{rr,c} &= -p = A - \frac{B}{R_{ci}^2} \quad \text{at} \quad r = R_{ci} \\ \sigma_{rr,c} &= 0 = A - \frac{B}{R_{co}^2} \quad \text{at} \quad r = R_{co} \end{aligned} \tag{11.41}$$

Solving for A and B

$$A = \frac{pR_{ci}^2}{R_{co}^2 - R_{ci}^2}; \quad B = \frac{pR_{ci}^2 R_{co}^2}{R_{co}^2 - R_{ci}^2} \tag{11.42}$$

As in the case of the shaft, the radial displacement in the collar is, from (11.5)

$$\frac{u_c}{R_{ci}} = \varepsilon_{\theta\theta,c} = \frac{1}{E}\left[\sigma_{\theta\theta,c} - \nu\sigma_{rr,c}\right] \tag{11.43}$$

With $\sigma_{rr,c}$ and $\sigma_{\theta\theta,c}$ given by (11.16) then

$$u_c = \frac{R_{ci}}{E}\left[\frac{pR_{ci}^2}{R_{co}^2 - R_{ci}^2}\right]\left[1 - \nu + (1+\nu)\left(\frac{R_{co}}{R_{ci}}\right)^2\right] \tag{11.44}$$

and is positive, indicating an expansion of the collar. The radial interference, δ, of the assembled shaft and collar is the total radial displacement of the shaft and collar

$$\delta = \| u_c \| + \| u_s \| = u_c - u_s \tag{11.45}$$

11.10 Compound Cylindrical Pressure Vessels

In this section we consider the problem of a thick-walled cylindrical pressure vessel formed by two thick-walled vessels, 1 and 2, as shown in Figure 11.8. As with the shaft and shrunk fit collar examined in the previous section, the assembly of the compound cylinders induces an interference pressure, p, at the common interface which is in addition to an applied internal pressure, p_i. Let us consider the stress distributions resulting from these two components separately.

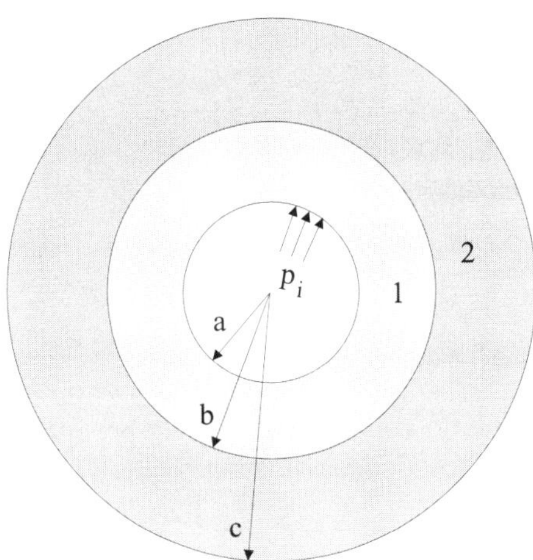

Figure 11.8: A compound cylindrical pressure vessel.

11.10.1 Assembly Pressure

Considering the inner cylinder of inner radius a and outer radius b then the boundary conditions are

$$\begin{aligned} \sigma_{rr} &= 0 \quad \text{at} \quad r = a \\ \sigma_{rr} &= -p \quad \text{at} \quad r = b \end{aligned} \tag{11.46}$$

Solving for A_1 and B_1 in Lame's equation (11.16) we have

$$A_1 = \frac{pb^2}{a^2 - b^2}; \quad B_1 = \frac{pa^2b^2}{a^2 - b^2} \tag{11.47}$$

with the circumferential stress given by $\sigma_{\theta\theta,1} = A_1 + B_1/r^2$ for $a \le r \le b$ and the radial displacement, u_1, at the interface

$$u_1 = \frac{b}{E}\left[\sigma_{\theta\theta,1} - \nu\sigma_{rr,1}\right] \tag{11.48}$$

For the outer cylinder the boundary conditions are

$$\begin{aligned} \sigma_{rr} &= -p \quad \text{at} \quad r = b \\ \sigma_{rr} &= 0 \quad \text{at} \quad r = c \end{aligned} \tag{11.49}$$

from which the Lame's constants are

$$A_2 = \frac{pb^2}{c^2 - b^2}; \quad B_1 = \frac{pb^2c^2}{c^2 - b^2} \tag{11.50}$$

with the circumferential stress given by $\sigma_{\theta\theta,2} = A_2 + B_2/r^2$ for $b \le r \le c$ and the radial displacement, u_2, at the interface

$$u_2 = \frac{b}{E}\left[\sigma_{\theta\theta,2} - \nu\sigma_{rr,2}\right] \tag{11.51}$$

Both A_1 and B_1 from (11.47) are negative since $a < b$ whereas both A_2 and B_2 from (11.50) are positive since $c > b$. It follows therefore from (11.48) and (11.51) that u_1 is negative and u_2 is positive. The total radial interference, δ, of the assembled compound cylinder is

$$\delta = \parallel u_1 \parallel + \parallel u_2 \parallel = u_2 - u_1 \tag{11.52}$$

11.10.2 Applied Internal Pressure

To determine the stress distribution of the compound cylindrical pressure vessel due to an internal pressure p_i we consider the vessel as a single cylinder of inner radius a and outer radius c with the following boundary conditions

$$\begin{aligned} \sigma_{rr} &= -p_i \quad \text{at} \quad r = a \\ \sigma_{rr} &= 0 \quad \text{at} \quad r = c \end{aligned} \tag{11.53}$$

and solving for Lame's constants, (11.16)

$$A_3 = \frac{p_i a^2}{c^2 - a^2}; \quad B_3 = \frac{p_i a^2 c^2}{c^2 - a^2} \tag{11.54}$$

with the circumferential stress given by $\sigma_{\theta\theta,3} = A_3 + B_3/r^2$ for $a \le r \le c$.

11.10.3 Total Stress Distribution

To determine the total stress distribution throughout a compound cylindrical pressure vessel we need to superimpose the stress distributions resulting from the assembly pressure and applied internal pressure derived in the previous sections. As an example, consider a compound vessel in which a = 50mm,

	Inner Cylinder, $\sigma_{\theta\theta,1}$ (MPa)		Outer Cylinder, $\sigma_{\theta\theta,2}$ (MPa)	
	$r = a$	$r = b$	$r = b$	$r = c$
Assembly pressure	-30	-21.7	41.7	30
Applied pressure	135	91.7	91.7	36
Total pressure	105	70	133.4	66

Table 11.2: Assembly, applied and total circumferential stresses for a compound cylindrical pressure vessel.

$b = 75$mm, $c = 100$mm, $p = 20$MPa and $p_i = 100$MPa. Table 11.2 lists the circumferential stresses for the assembly, applied and total pressures in both the inner and outer cylinders at the inner, interfacial and outer surfaces. The variation of circumferential stress through the compound vessel is shown in Figure 11.9, noting the discontinuity in the total stress at $r = b$. The total circumferential stress is observed to be less than the circumferential stress due to the applied internal pressure throughout the region $a \leq r < b$ whereas the total circumferential stress is greater than the circumferential stress due to the applied internal pressure throughout the region $b < r \leq c$.

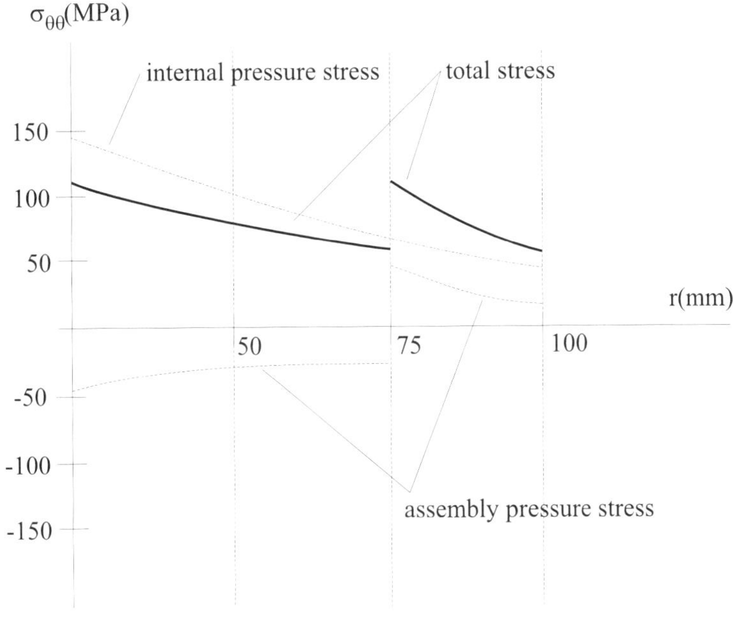

Figure 11.9: Assembly, applied and total circumferential stresses for a compound cylindrical pressure vessel.

11.11 Thick-Walled Spherical Pressure Vessels

Figure 11.10 illustrates a stress element of a thick-walled spherical pressure vessel of inner radius a and outer radius b. The main difference from the cylindrical case is that due to symmetry the circumferential stress $\sigma_{\theta\theta}$ now has a constant value in all tangential directions at a given radius r.

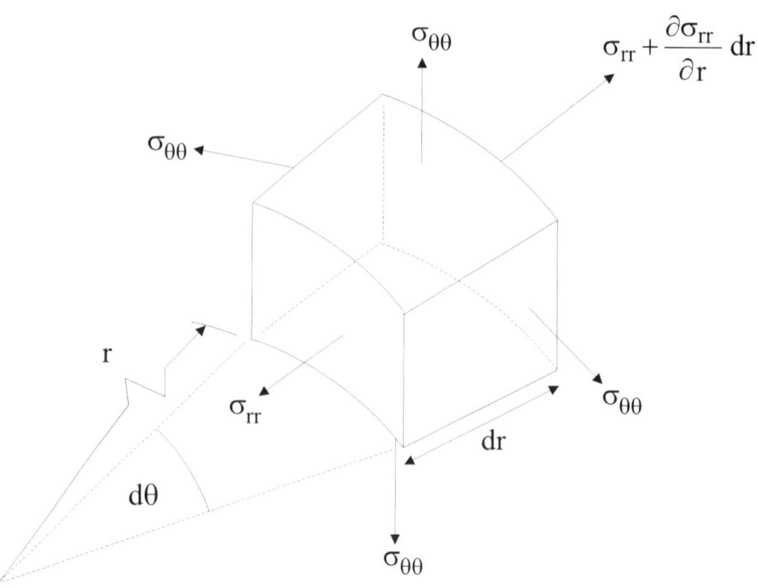

Figure 11.10: Stress element of a thick-walled spherical pressure vessel.

To determine the radial and circumferential stresses we will follow an equivalent procedure to that of §11.4 for the case of a thick-walled cylindrical pressure vessel. Considering the equilibrium of the stress element in the radial direction then we have

$$\left(\sigma_{rr} + \frac{\partial \sigma_{rr}}{\partial r}\,\mathrm{d}r\right)\pi(r + \mathrm{d}r)^2\,\mathrm{d}\theta - \sigma_{rr}\pi r^2\,\mathrm{d}\theta = 2\pi r\,\mathrm{d}r\,\mathrm{d}\theta \tag{11.55}$$

Cancelling terms and neglecting second order terms we arrive at the following equilibrium equation

$$\frac{\partial \sigma_{rr}}{\partial r} = \frac{2(\sigma_{\theta\theta} - \sigma_{rr})}{r} \tag{11.56}$$

and can be compared to (11.3).

From (11.6) the strains are

$$\begin{aligned}\varepsilon_{rr} &= \frac{\mathrm{d}u_r}{\mathrm{d}r} = \frac{1}{E}\left[\sigma_{rr} - \nu(\sigma_{\theta\theta} + \sigma_{zz})\right]\\ \varepsilon_{\theta\theta} &= \frac{u_r}{r} = \frac{1}{E}\left[\sigma_{\theta\theta} - \nu(\sigma_{rr} + \sigma_{\theta\theta})\right]\end{aligned} \tag{11.57}$$

Differentiating $\varepsilon_{\theta\theta}$ with respect to r and equating to ε_{rr} we have the compatibility equation

$$\frac{\mathrm{d}}{\mathrm{d}r}\left[r\sigma_{\theta\theta} - \nu r(\sigma_{rr} + \sigma_{\theta\theta})\right] = \sigma_{rr} - 2\nu\sigma_{\theta\theta} \tag{11.58}$$

Differentiating (11.58)

$$\sigma_{\theta\theta} + r\frac{\mathrm{d}\sigma_{\theta\theta}}{\mathrm{d}r} - \nu\sigma_{rr} - \nu r\frac{\mathrm{d}\sigma_{rr}}{\mathrm{d}r} - \nu\sigma_{\theta\theta} - \nu r\frac{\mathrm{d}\sigma_{\theta\theta}}{\mathrm{d}r} = \sigma_{rr} - 2\nu\sigma_{\theta\theta} \tag{11.59}$$

which reduces to

$$\nu r\frac{\mathrm{d}\sigma_{rr}}{\mathrm{d}r} - (1-\nu)r\frac{\mathrm{d}\sigma_{\theta\theta}}{\mathrm{d}r} = (1+\nu)(\sigma_{\theta\theta} - \sigma_{rr}) \tag{11.60}$$

Substituting (11.56) into (11.59)

$$\left[\frac{1}{2}\frac{\mathrm{d}\sigma_{rr}}{\mathrm{d}r} + \frac{\mathrm{d}\sigma_{\theta\theta}}{\mathrm{d}r}\right](1-\nu)r = 0 \tag{11.61}$$

such that

$$\frac{\mathrm{d}}{\mathrm{d}r}\left[\frac{\sigma_{rr}}{2} + \sigma_{\theta\theta}\right] = 0 \tag{11.62}$$

It follows that $(\sigma_{rr}/2 + \sigma_{\theta\theta})$ is a constant independent of r, say c_1

$$\frac{\sigma_{rr}}{2} + \sigma_{\theta\theta} = c_1 \tag{11.63}$$

Substituting into the equilibrium equation, (11.56)

$$2c_1 - 3\sigma_{rr} = r\frac{\mathrm{d}\sigma_{rr}}{\mathrm{d}r} \tag{11.64}$$

Separating variables and integrating with respect to r

$$\int \frac{\mathrm{d}r}{r} = \int \frac{\mathrm{d}\sigma_{rr}}{2c_1 - 3\sigma_{rr}} \tag{11.65}$$

Performing the integration

$$-3\ln r + \ln c_2 = \ln(2c_1 - 3\sigma_{rr}) \tag{11.66}$$

Collecting logarithmic terms

$$\frac{c_2}{r^3} = 2c_1 - 3\sigma_{rr} \quad \text{or} \quad \sigma_{rr} = \frac{1}{3}\left(2c_1 - \frac{c_2}{r^3}\right) \tag{11.67}$$

with $\sigma_{\theta\theta}$ following from (11.63)

$$\sigma_{\theta\theta} = c_1 - \frac{\sigma_{rr}}{2} = \frac{2c_1}{3} + \frac{c_2}{6r^3} \tag{11.68}$$

Finally, letting $A = 2c_1/3$ and $B = c_2/3$ then Lame's equations for a thick-walled spherical pressure vessel are

$$\sigma_{rr} = A - \frac{B}{r^3}; \qquad \sigma_{\theta\theta} = A + \frac{B}{2r^3} \tag{11.69}$$

and can be compared to (11.16). An alternative derivation of (11.69) can be found in Timoshenko and Goodier (1982).

The constants of integration A and B are determined from the prescribed boundary conditions in an identical manner to that outlined in §11.5. For example, in the case of an internal pressure, p, only then the boundary conditions are

$$\begin{aligned} \sigma_{rr} &= -p \quad \text{at} \quad r = a \\ \sigma_{rr} &= 0 \quad \text{at} \quad r = b \end{aligned} \tag{11.70}$$

From σ_{rr} of (11.69)

$$A = -\frac{a^3 p}{a^3 - b^3}; \qquad B = -\frac{a^3 b^3 p}{a^3 - b^3} \tag{11.71}$$

and substituting back into (11.69) gives

$$\sigma_{rr} = -p\left[\frac{(b/r)^3 - 1}{(b/a)^3 - 1}\right]; \qquad \sigma_{\theta\theta} = \frac{p}{2}\left[\frac{2 + (b/r)^3}{(b/a)^3 - 1}\right] \tag{11.72}$$

11.12 Conclusion

This Chapter has extended the discussion of Chapter 3 on thin-walled pressure vessels to that of thick-walled pressure vessels in which the wall-thickness is sufficiently thick so that the stress components vary across the wall of the vessel. The key equations derived in the Chapter are Lame's equations which are generally taken as the starting point when solving a thick-walled pressure vessel. The unknown constants, A and B, in Lame's equations are determined from the prescribed boundary conditions, which are generally an internal pressure with either zero or non-zero external pressure. Thick-walled pressure theory can be applied to the problem of shrink fitting a collar on to a shaft. The stress distribution throughout compound cylindrical pressure vessels was also considered, illustrating that the assembly pressure assists in reducing the circumferential stress throughout the inner cylinder. The Chapter concluded by deriving Lame's equations for thick-walled spherical pressure vessels.

11.13 Reference and Further Reading

- Blazynski, T. Z. (1983) *Applied Elasto-Plasticity of Solids*, Macmillan, London.
- Timoshenko, S. P. and Goodier, J. N. (1982) *Theory of Elasticity*, 3rd Ed., McGraw-Hill.

11.14 Exercises

11.1 Confirm that the radial and circumferential stresses (11.27) satisfy equilibrium.

11.2 A thick-walled cylindrical pressure vessel of internal radius 75mm and external radius 250mm is subject to an internal pressure of 75MPa with zero external pressure. Assuming the ends of the vessel to be closed then determine the radial, circumferential and axial stresses on the internal surface of the vessel.

11.3 A thick-walled cylindrical pressure vessel of inner radius 0.5m, outer radius 1m is subject to an internal pressure of 5MPa and an external pressure of 100kPa. Assuming the ends of the vessel to be closed then determine the radial, circumferential and axial stresses at a radius of 0.75m.

11.4 A thick-walled cylindrical pressure vessel of internal diameter 150mm and external diameter 200mm fails due to an internal pressure of 450bar and zero external pressure. Determine, for a safety factor of 2, the safe internal pressure for a second vessel made of the same material and internal diameter but with a wall thickness of 50mm.

11.5 A collar of inner radius 49.5mm and outer radius 100mm is shrunk on to a shaft of radius 50mm. Assuming that Lame's equations are applicable and that the interference pressure is not to exceed the material yield stress then determine the total radial interference of the assembled shaft and collar. Young's modulus, Poisson's ratio and yield stress of both the shaft and collar are 210GPa, 0.3 and 300MPa respectively.

11.6 A compound cylindrical pressure vessel consists of an inner cylinder (inner radius 50mm and outer radius 75mm) and outer cylinder (inner radius 74.9mm and outer radius 100mm). The resulting radial interference is therefore 0.1mm. Determine the interference pressure induced by the assembly of the two cylinders. Assume $E = 210$GPa and $\nu = 0.3$.

11.7 A thick-walled spherical pressure vessel of inner radius 100mm and outer radius 175mm is subject to an internal pressure of 100MPa with zero external pressure. Determine the circumferential stress on the inner radius of the vessel.

Chapter 12

Plasticity

12.1 Introduction

In this Chapter we provide an introduction to the elastic-plastic behaviour of materials. We begin by examining the features of the stress-strain diagram first discussed in §1.5. For mathematical computations it is necessary to represent the stress-strain curve by a suitable empirical equation. We present several of the more popular constitutive curves in §12.7. Due to the importance attached to the tensile test, we then determine the point of plastic instability for tensile test specimens. The Chapter then discusses five yield criteria, of which the Tresca and Huber-von Mises criteria are the most representative. The Chapter concludes by examining the spread of plasticity through a circular cross-section subject to a pure torque and through both rectangular and circular cross-sections subject to a pure bending moment.

12.2 Why Elasto-Plasticity?

Ever increasing industrial demand for more sophisticated machine components calls for a considerable degree of ingenuity on the part of the designer. This, in turn, calls for a higher degree of expertise in the analysis of a material in order to be able to push the boundaries of application of a material from one of purely elastic to one of elasto-plastic or purely plastic.

The standard elastic concept of design consists of estimating the level of the stress state within a given system and then, having chosen the material to be used, comparing the stress state of the system to the mechanical capabilities of the material. If a basic compatibility is found between the capabilities of the material and the ability of the material to respond to the applied stress state of the system then a solution is found. However, if no compatibility is found then maybe the dimensions of the designed component have to be altered (generally increased) or a different (possibly higher quality and consequently more expensive) material has to be chosen.

A solution to this common design dilemma (size against cost) is frequently found in the elasto-plastic analysis of a material. A material which has undergone a plastic deformation or 'permanent' deformation will have different material properties from that of the elastic material and, if skilfully manipulated, can be applied to produce improved material performance and hence improved design/component performance.

12.3 What is Plasticity?

The elasto-plastic condition is attained by increasing the loading (in a controlled manner!), either locally or globally, beyond the critical elastic value for the geometry and material under consideration. The theory which deals with plasticity in materials is one of calculating the stresses and strains of a ductile material which has been permanently deformed by the applied loading. The theory of plasticity tends to be based on experimental observations on the macro-scopic behaviour of materials, from which mathematical formulations have been formulated to describe this behaviour.

Unlike elastic materials, in which the state of strain depends purely on the final state of stress, the deformations which occur during plastic deformation depend on the complete history of the loading. Therefore, plasticity theory tends to be essentially incremental in nature, the final deformation of the material being the sum of the incremental deformations of the material along a prescribed strain path.

A polycrystalline metal can be regarded as macro-scopically homogeneous and isotropic in nature with the crystal grains of the material distributed throughout the aggregate with random orientations to one another. During plastic deformation, these random crystallographic directions gradually rotate toward a common axis, thus producing a preferred orientation. Hence an initially 'macroscopically' isotropic material becomes anisotropic and its mechanical properties vary with direction. This development of anisotropy, combined with the progressive cold work and resulting strain-hardening of the material are simply too complex to be successfully analysed. As a result, it is generally assumed that the material remains isotropic throughout the deformation process irrespective of the degree of hardening.

The strain-hardening characteristics of a material in a complicated state of stress are generally too complicated to characterise and are thus related to the uniaxial stress-strain hardening behaviour. Therefore, we will examine the uniaxial stress-strain behaviour of a material before considering in detail the general theory of plasticity.

12.4 The Macro Approach

Generally, engineers use tests such as tensile, compressive, torsion, impact and hardness to characterise material behaviour. Such tests can only provide information about the 'average' properties to be expected, but cannot provide us with the important explanation of why the material behaves the way that it does. To understand the mechanisms behind a material's behaviour requires a micro-scopic examination.

The macro approach is selected purely on the basis of being simpler. The estimation of basic properties (such as elastic constants) on an atomic or microscopic level is generally very difficult. For instance, the plastic deformation of a single crystal is generally produced by slip. Slip is the sliding of adjacent blocks of the crystal along definite crystallographic planes, known as slip planes. The boundary line between a slipped region and a neighbouring un-slipped region is called a *dislocation*.

Slip, caused by the movement of dislocations, is initiated by a line defect which causes a local concentration of stress, Figure 12.1. The magnitude and direction of the relative movement of the slip is specified by the Burgers vector. In a polycrystalline metal the crystallographic orientation changes from one grain to the next through a narrow transition zone (grain boundary) which acts as an effective barrier to slip and the movement of dislocations. During loading a polycrystalline material it is found that dislocations pile-up along the active slip planes at the grain boundaries. This pile-up of dislocations at grain boundaries effectively oppose the generation of new dislocations. If the applied loading is increased further then the shear stress at the head of the pile-up can become large enough to initiate further dislocations in a neighbouring grain and consequently cause dislocation movement across the grain boundary. It is this pile-up of dislocations at grain boundaries that is mainly responsible for the strain-hardening of a material in the early stages of plastic deformation.

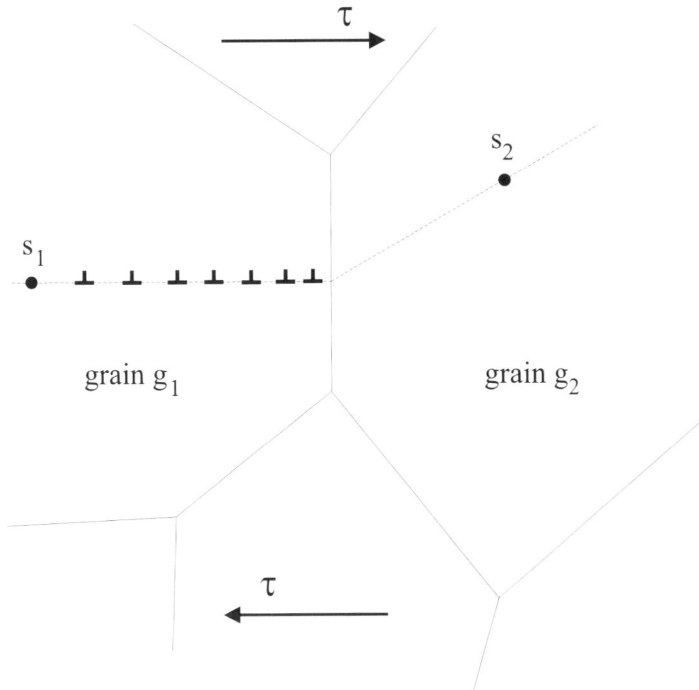

Figure 12.1: Two grains, g_1 and g_2, subject to a shear stress τ with a source of local stress concentration at s_1 in g_1. Dislocations emitted from s_1 pile-up at the boundary between the two grains with a further initiation of dislocations in the neighbouring grain g_2 from source s_2.

In the later stages of plastic deformation it is the interaction of dislocations that control the yield strength of a polycrystalline metal. The rate of hardening in polycrystalline metals is always higher than for single crystals due to the interaction of dislocations and the presence of foreign atoms (such as inclusions) acting as barriers to the movement of dislocations. For a further discussion on dislocations refer to Hull and Bacon (1984).

12.5 The Tensile Test and Behaviour of Metals in Static Loading Conditions

In Chapter 1 we introduced the stress-strain diagram for metallic engineering materials. Let us revisit this diagram with the emphasis now on elastic-plastic behaviour. Consider the characteristic features of a standard tensile test for a low carbon steel shown in Figure 12.2. The applied load is denoted by P and the original length and cross-sectional area of the tensile specimen are denoted by l_o and A_o respectively whereas the current length and cross-sectional area are denoted by l and A respectively. The true or natural strain and true stress are denoted by $\varepsilon(= \ln(l/l_o))$ and σ, the nominal stress by σ_o and the engineering strain by $e(= \Delta l/l_o, \Delta l = (l - l_o))$. As load is increased from zero, the specimen extends elastically in the range Oa. Point a is referred to as the *limit of proportionality*. Between a and a' linearity is no longer present but conditions of elasticity still prevail. At the point a' elasticity ends and we have the *onset of*

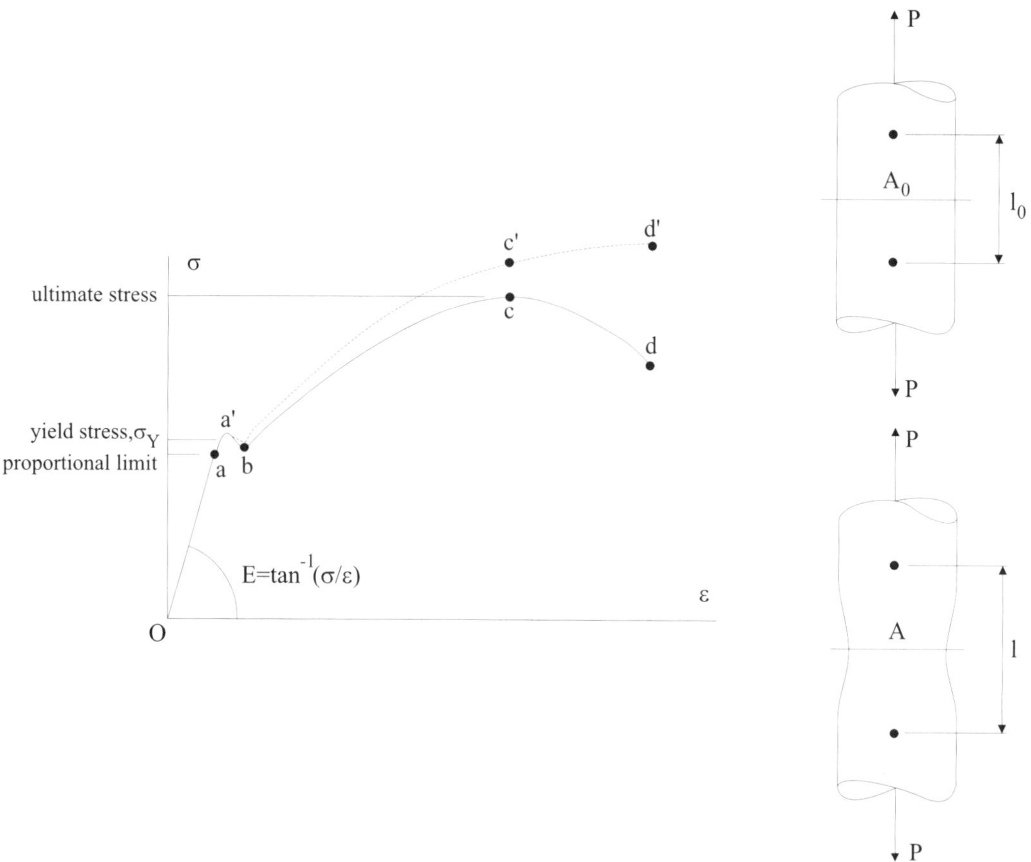

Figure 12.2: Typical stress-strain diagram for a steel.

yielding. Continued loading leads initially to a drop in the load. The stress corresponding to the lower load is known as the *lower yield stress*. As loading continues, the cross-sectional area reduces gradually, but because of strain-hardening, the load required to produce this change increases. The imbalance between these two effects occurs at a point c where the load is at a maximum and local instability sets in. Necking of the specimen occurs after point c until, eventually, it fractures at point d.

So far, we have been describing the stress-strain curve in terms of nominal stress-engineering strain. For a material in the plastic regime, this curve tends not to be truly representative of the strength potential of the material, because in this region the response of the material depends on incremental changes. For this reason it is more advantageous to plot the true stress against the natural strain ($\mathrm{d}\varepsilon = \mathrm{d}l/l$). This curve $(Oa'c'd')$ indicates that a much higher level of stress is permissible than that suggested by the nominal stress-engineering strain curve. The elastic regime of the two curves remains practically unchanged.

The actual location of the yield stress is largely a matter of convention. In conditions in which the incipient point of yielding remains undefinable, we introduce the concept of a *proof stress* which is generally defined as that stress for which a specified small amount of permanent deformation is observed, usually of the order of 0.2%.

If the tensile test specimen is stressed to some point C, in the plastic range, beyond the limit of proportionality with the load subsequently released, there is an elastic recovery following the path CD which is very nearly a straight line of slope equal to Young's modulus, E; see Figure 12.3. The permanent strain that remains on complete unloading is equal to OF. On re-application of the load, the specimen deforms elastically until a *new* yield point G is reached. Neglecting the small hysteresis loop formed during loading and unloading, then point G can be taken to be approximately coincident with point C. Upon further loading the stress-strain curve proceeds along GH which is approximately equivalent to curve BC. The

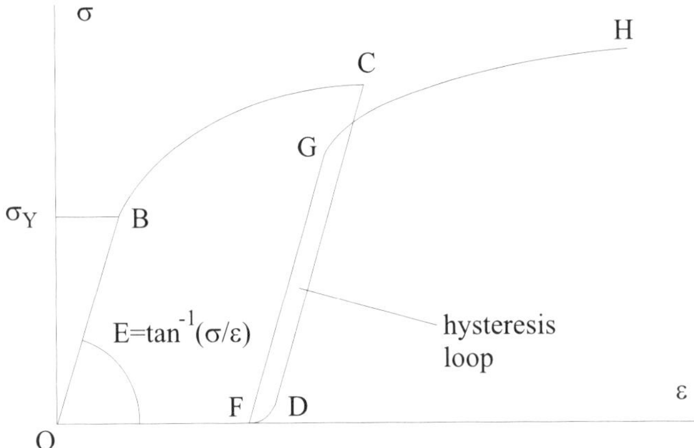

Figure 12.3: Unloading in the stress-strain diagram.

curve FGH may be regarded as the *new* stress-strain curve of the metal when pre-strained by an amount OF. The greater the degree of prestrain, the higher the new yield point and the flatter the strain-hardening curve. If the metal is heavily prestrained then the rate of strain-hardening is so small that the material can be regarded as approximately non-hardening or elastic-perfectly plastic.

12.6 Strain Rate Effects

An increase in the rate of straining of a material will increase the strain-hardening effect and therefore raise the yield stress. This property is irrespective of the method of testing. True stress-natural strain curves indicate that engineering alloys fall broadly into two types: i) stress-strain curves that show that an increase in strain rate initially has no noticeable effect on the strain to fracture, but a further increase reduces the strain to fracture and ii) stress-strain curves that indicate that the strain to fracture increases slightly with increasing strain rate and then decreases rapidly. These two cases are shown schematically in Figure 12.4. At higher strain rates, the material can support a higher value of stress but consequently fractures more easily.

12.7 Types of Stress-Strain Empirical Constitutive Curves

The following five cases illustrate the most frequently observed stress-strain responses of engineering materials.

1) Perfectly elastic

In a perfectly elastic material no non-linear behaviour is observed and is typically experienced by highly brittle ceramics and glass, Figure 12.5.

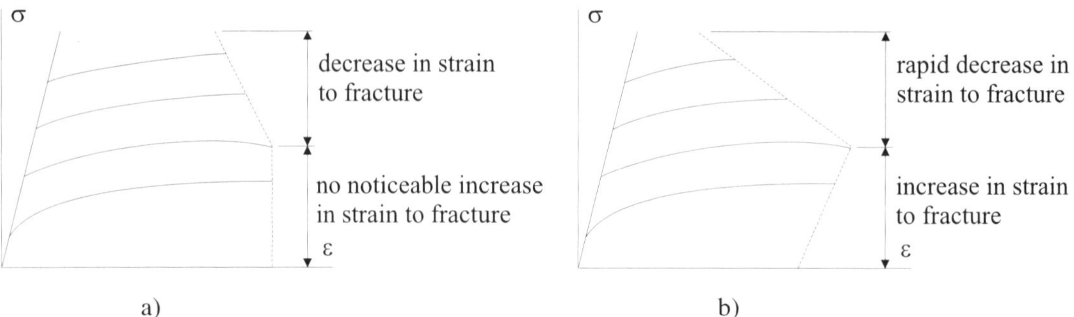

Figure 12.4: Strain rate effects. a) No initial effect followed by a decrease in strain to fracture. b) An initial increase in strain to fracture followed by a decrease in strain to fracture.

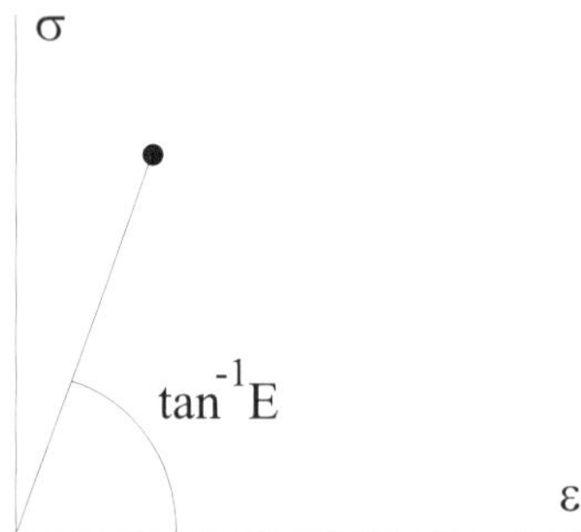

Figure 12.5: Perfectly elastic stress-strain response.

2) Rigid-perfectly plastic

A rigid-perfectly plastic material experiences no elastic behaviour with a flow stress that remains constant until failure and is generally exhibited by copper alloys, Figure 12.6.

Figure 12.6: Rigid-perfectly plastic stress-strain response.

3) Rigid-linear strain hardening

A rigid-linear strain hardening material experiences no elastic behaviour followed by a linear increase in stress with increasing strain, Figure 12.7.

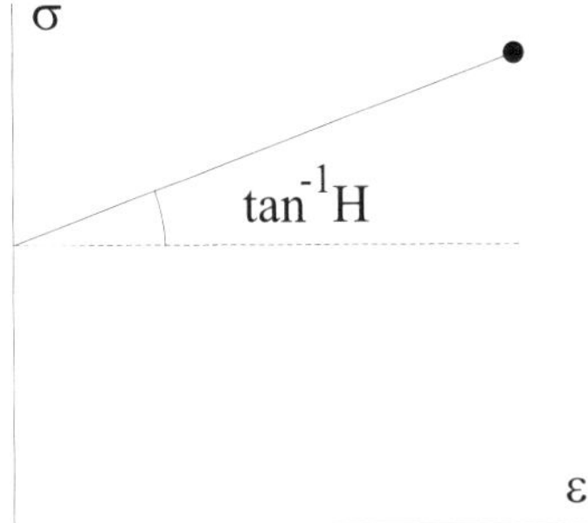

Figure 12.7: Rigid-linear strain hardening stress-strain response.

4) Elastic-perfectly plastic

An elastic-perfectly plastic material initially exhibits an elastic response followed by a constant flow stress, Figure 12.8.

Figure 12.8: Elastic-perfectly plastic stress-strain response.

5) Elastic-linear strain hardening

Finally, an elastic-linear strain hardening material initially responds in an elastic manner followed by linear stress-strain hardening, Figure 12.9.

Although idealised, these stress-strain relationships quite realistically reflect the behaviour of actual metallic materials. However, for practical reasons the constitutive relationships of the material have to be put into a suitable mathematical form for subsequent calculations and ease of use. There are several variations of empirical expressions available, of which Ludwik's, Swift's and linear approximations are the most commonly used.

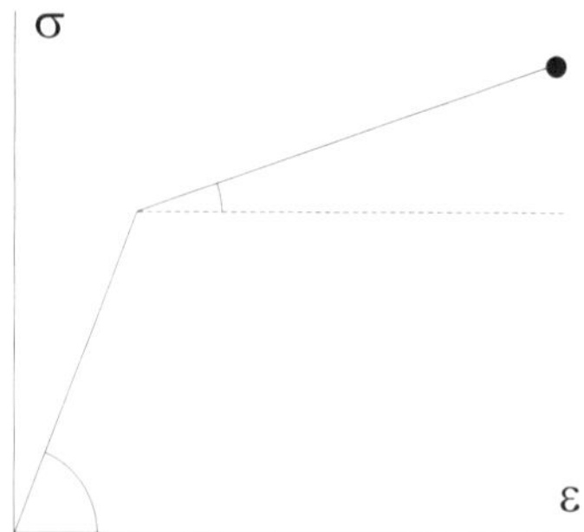

Figure 12.9: Elastic-linear strain hardening stress-strain response.

12.7.1 Ludwik's Power Law

The Ludwik power law is of the form

$$\sigma = C\varepsilon^n; \quad C > 0; \quad 0 \leq n \leq 1 \tag{12.1}$$

when a material is rigid-plastic. The term C is a constant stress and n is a strain-hardening exponent (usually between 0 and 0.5). Ludwik's power law is also known as the Holloman power law. The curve predicts a zero initial stress and an infinite initial slope (except for $n = 0$ which represents a non-hardening rigid-plastic material). The higher the value of n, the more pronounced is the strain hardening of the material, see Figure 12.10.

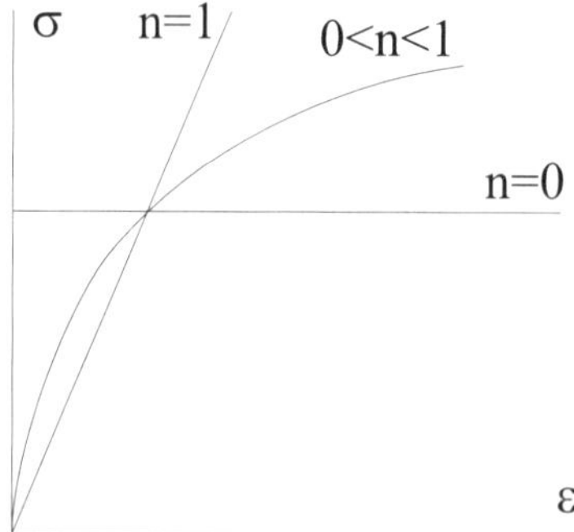

Figure 12.10: The Ludwik stress-strain response.

The Ludwik power law is easily modified by introducing a constant yield stress term σ_Y

$$\sigma = \sigma_Y + C\varepsilon^n \tag{12.2}$$

which is illustrated in Figure 12.11. Although this formula represents the strict rigid-plastic behaviour of metals, it does not give a better fit to actual stress-strain curve data over a wide range of strains. When $n = 1$ the curve is a close approximation for heavily prestrained metals.

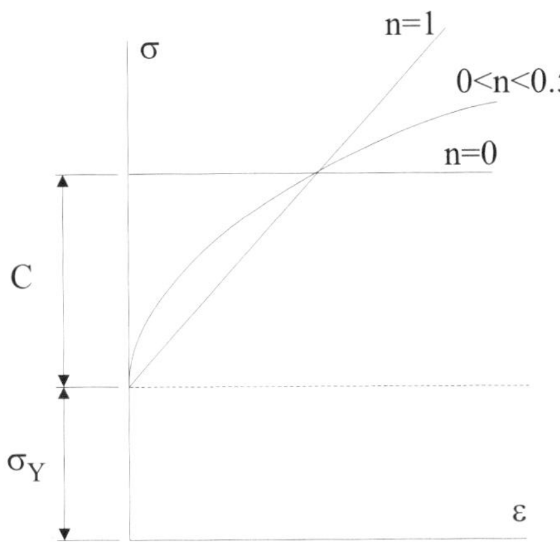

Figure 12.11: The Ludwik power law modified for an initial yield stress σ_Y.

12.7.2 Swift's Generalised Power Law

Swift's power law is of the form

$$\sigma = B(C_1 + \varepsilon)^n; \quad 0 \leq n \leq 1 \tag{12.3}$$

where B, C_1 and n are empirical constants. This expression can be obtained from Ludwik's expression if the stress axis is moved along the positive strain axis through a distance of C_1. Hence, C_1 can be regarded as the amount of pre-strain in a material whose stress-strain curve in the annealed state corresponds to $C_1 = 0$, the value of n, of course, remaining the same.

12.7.3 Other Stress-Strain Laws

Numerous other stress-strain laws exist, a few of which we will present in this section. The Voce law is as follows

$$\sigma = C(1 - me^{-ne}) \tag{12.4}$$

and provides an experimental fit which can generally be applied to a variety of materials. A law proposed by Prager is

$$\sigma = \sigma_Y \tanh(\frac{E\varepsilon}{\sigma_Y}) \tag{12.5}$$

and has an initial slope of gradient E and gradually bends over to approach the yield stress σ_Y in an asymptotic manner. Another expression that is very popular is that of Ramberg and Osgood

$$\varepsilon = \frac{\sigma}{E}\left[1 + \alpha\left(\frac{\sigma}{\sigma_Y}\right)^{m-1}\right] \tag{12.6}$$

where the elastic part of strain is equal to (σ/E) and the plastic part is assumed to vary, from Ludwik's law, as σ^m where $m = 1/n$. The term α is a dimensionless constant. The slope of the stress-strain curve continuously decreases from the value of E as shown in Figure 12.12. It is this property of a continuous curve that makes the Ramberg-Osgood equation so useful. For $m = \infty$ we have a non-hardening material and the equation degenerates into a pair of straight lines meeting at $\sigma = \sigma_Y$.

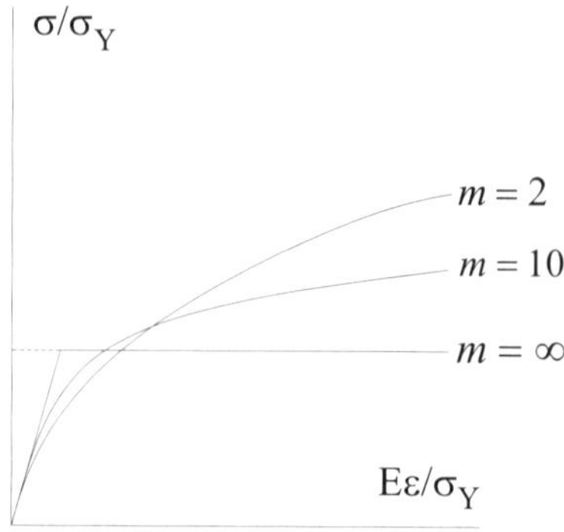

Figure 12.12: Ramberg-Osgood stress-strain response.

There are occasions in which it is simplest to express the stress-strain curve in the plastic regime by a simple power law and assuming linear elastic ($\sigma = E\varepsilon$) behaviour below yielding and with a definite yield point at $\sigma = \sigma_Y$

$$\begin{aligned} \sigma &= E\varepsilon \quad \text{for} \quad \varepsilon \le \sigma_Y/E \\ \sigma &= \sigma_Y \left(\frac{E\varepsilon}{\sigma_Y}\right)^n \quad \text{for} \quad \varepsilon \ge \sigma_Y/E \end{aligned} \tag{12.7}$$

Occasionally the following representation is used

$$\sigma = \sigma_Y + B\left(\varepsilon - \frac{\sigma_Y}{E}\right) = A + B\varepsilon; \quad \text{where} \quad A = \sigma_Y\left(1 - \frac{B}{E}\right) \tag{12.8}$$

and gives a fit between the two linear regions as shown in Figure 12.13.

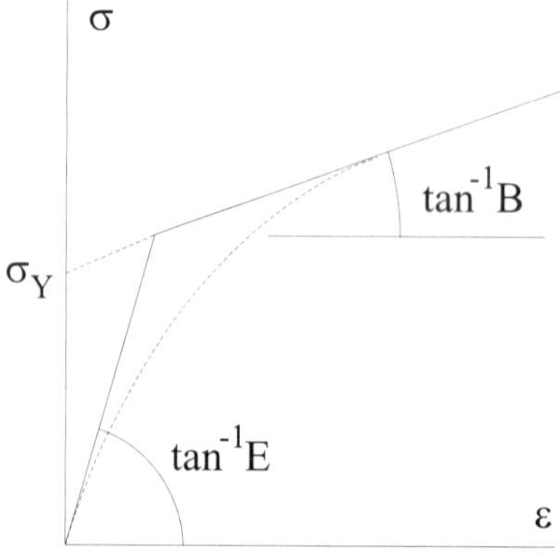

Figure 12.13: Elastic-linear strain hardening.

12.8 Mean Value of the Yield Stress

For a material that exhibits a continuous variation in stress with strain it is often desirable to obtain a *mean value* of the yield stress over the total range of strain imposed on a material and then to use this mean value in an appropriate criterion of yielding. If the total strain in a given operation is ε_T and the variable yield stress is σ_Y then the mean value of the yield stress, σ_m, is

$$\sigma_m = \frac{1}{\varepsilon_T} \int \sigma_Y \, \mathrm{d}\varepsilon \tag{12.9}$$

Example 12.1 Mean yield stress for a strain hardening material

In terms of strain ε and the first yield stress, σ_o, let us determine the mean yield stress of two materials whose empirical stress-strain curves are given by: 1) $\sigma = \sigma_o + H\varepsilon$ and 2) $\sigma = \sigma_o + H\varepsilon^n$ which are both illustrated in Figure 12.14.

For 1) $\sigma = \sigma_o + H\varepsilon$

$$\sigma_m = \frac{1}{\varepsilon} \int \sigma_Y \, \mathrm{d}\varepsilon = \frac{1}{\varepsilon} \int_0^{\varepsilon} (\sigma_o + H\varepsilon) \, \mathrm{d}\varepsilon = \sigma_o + \frac{1}{2} H\varepsilon$$

For 2) $\sigma = \sigma_o + H\varepsilon^n$

$$\sigma_m = \frac{1}{\varepsilon} \int_0^{\varepsilon} (\sigma_o + H\varepsilon^n) \, \mathrm{d}\varepsilon = \sigma_o + \frac{H\varepsilon^n}{(n+1)}$$

when $n = 1$ we revert to case 1).

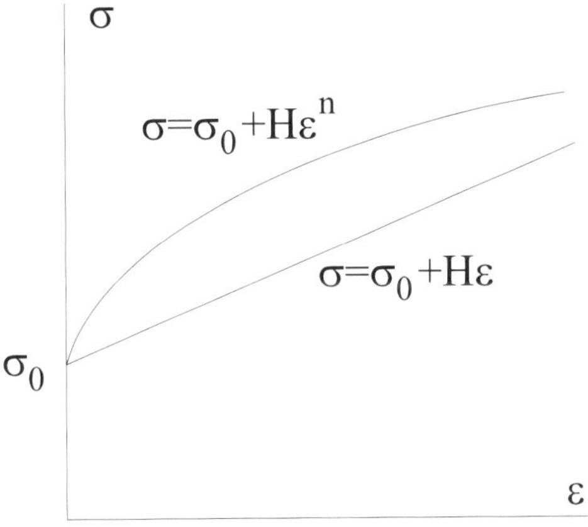

Figure 12.14: Example: Linear and power strain hardening.

12.9 Plastic Instability in Tension and the Point of Necking in a Tensile Test

The point at which necking occurs in the tensile test is the point at which local instability in the specimen first occurs. For such large values of strain, the tensile test stress-strain curve is unsuitable since the rate of hardening decreases to a critical value. At this stage, the increase in load due to strain-hardening is exactly balanced by the decrease in load caused by the local diminution of the cross-sectional area. If we take the initial cross sectional area and length to be denoted by A_o and L_o respectively and the current values by A and L, then using the constancy of volume condition of the specimen we have

$$A_o L_o = AL = \text{const} \tag{12.10}$$

Further, if σ is the current stress (true stress) and ε and e are the natural and engineering strains respectively then

$$\begin{aligned} \sigma &= \frac{P}{A} = \frac{PL}{A_o L_o} = \sigma_o(1+e_o); \quad \text{since} \quad e_o = \frac{(L-L_o)}{L_o} \\ \varepsilon &= \int \mathrm{d}L/L \end{aligned} \tag{12.11}$$

where σ_o and e_o denote the nominal engineering stress and strain respectively.

The instability is caused by the maximum tensile load applied, or $\mathrm{d}P = 0$. Therefore, differentiating the above

$$\begin{aligned} P = \sigma A; \quad \frac{\mathrm{d}P}{\mathrm{d}\sigma} = A; \quad \frac{\mathrm{d}P}{\mathrm{d}A} = \sigma \\ \text{hence} \quad \mathrm{d}P = A\,\mathrm{d}\sigma + \sigma\,\mathrm{d}A = 0; \quad \text{or} \quad A\,\mathrm{d}\sigma = -\sigma\,\mathrm{d}A \end{aligned} \tag{12.12}$$

Re-arranging and noting that $(L\,\mathrm{d}A + A\,\mathrm{d}L = 0$, or $\mathrm{d}A/A = -\,\mathrm{d}L/L)$ we find

$$\frac{\mathrm{d}\sigma}{\sigma} = -\frac{\mathrm{d}A}{A} = \frac{\mathrm{d}L}{L} \tag{12.13}$$

With $\mathrm{d}L/L = \mathrm{d}\varepsilon$

$$\varepsilon = \ln\left(\frac{L}{L_o}\right) = \ln(1+e_o) \tag{12.14}$$

then the condition of plastic instability in the tensile test specimen becomes

$$\frac{\mathrm{d}\sigma}{\mathrm{d}\varepsilon} = \sigma \quad \text{or} \quad \frac{\mathrm{d}\sigma}{\mathrm{d}e_o} = \frac{\sigma}{(1+e_o)} = \sigma_o \tag{12.15}$$

These results can be illustrated graphically using the fact that the slope at the point of instability is equal to the current stress, Figure 12.15. It follows that the maximum load corresponds to the point of contact of the tangent to the (σ, e) curve from the point (-1,0) on the negative strain axis, Figure 12.16. Additionally, we may compare the true and nominal stresses against the nominal engineering strain as illustrated in Figure 12.17.

The true-plastic component of strain ε_p is found by subtracting the elastic component, $\varepsilon_E(=\sigma/E)$, from the total strain, $\varepsilon = \ln(L/L_o) = \ln(1+e_o)$

$$\begin{aligned} \varepsilon_p &= \varepsilon - \varepsilon_E \\ \varepsilon_p &= \ln(L/L_o) - \sigma/E \\ \text{or} \quad \varepsilon_p &= \ln(1+e_o) - \frac{\sigma_o}{E}(1+e_o) \end{aligned} \tag{12.16}$$

If the plastic strain ε_p had been used instead of ε then the instability on true stress-true plastic strain is expressed as

$$\frac{\mathrm{d}\sigma}{\mathrm{d}\varepsilon_p} = \frac{\sigma}{(1-\sigma/E)} \tag{12.17}$$

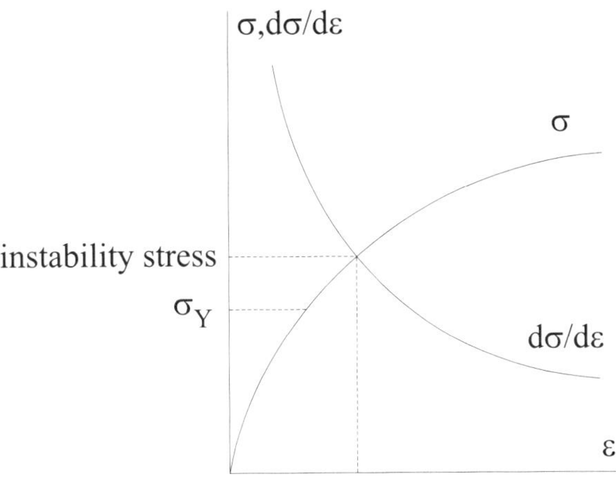

Figure 12.15: Plastic instability of the tensile test specimen in terms of true stress-true strain.

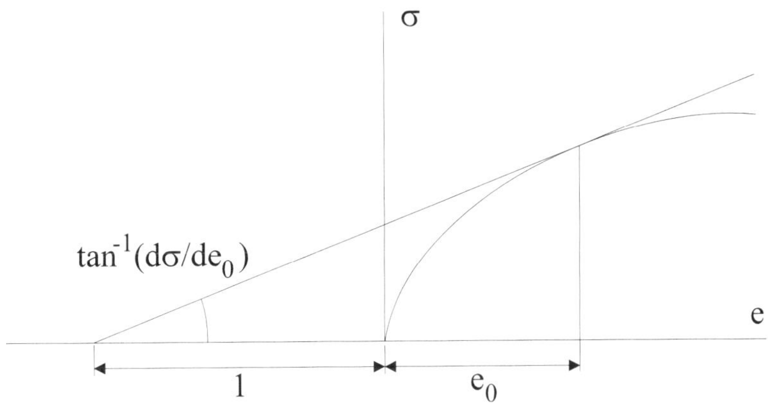

Figure 12.16: Plastic instability of the tensile test specimen in terms of true stress-engineering strain.

which is obtained from

$$\begin{aligned} \frac{\mathrm{d}\varepsilon_p}{\mathrm{d}\sigma} &= \frac{\mathrm{d}\varepsilon}{\mathrm{d}\sigma} - \frac{\mathrm{d}\varepsilon_p}{\mathrm{d}\sigma} \\ &= \frac{1}{\sigma} - \frac{1}{E} \\ &= \frac{E-\sigma}{\sigma E} = \frac{(1-\sigma/E)}{\sigma} \end{aligned} \tag{12.18}$$

and upon inverting, we have the above result (12.17).

The tensile test becomes unstable when the load reaches its maximum. The deformation is confined locally to the neck region, while the remainder of the specimen recovers elastically under decreasing load until fracture intervenes. The stress in the necked region assumes a triaxial state which varies through the cross-section of the neck and consequently the test no longer provides a direct measure of the stress-

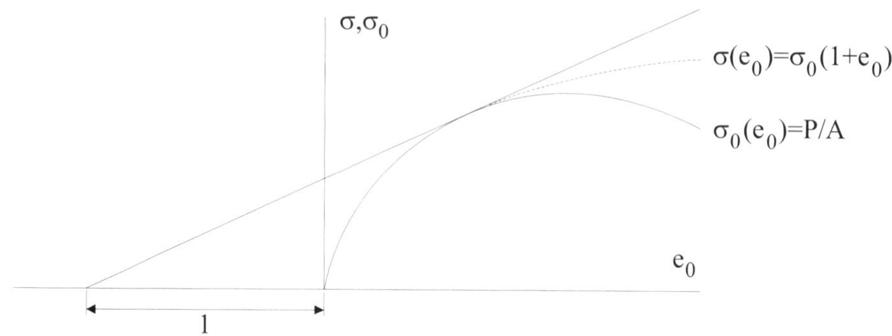

Figure 12.17: Stress-strain response in terms of true and nominal stresses.

strain behaviour. Although the stress-strain may be continued by introducing a correction factor that requires careful measurements of the geometry of the neck, the experimental difficulties rendered by this procedure make the method unsuitable for practical purposes. Briefly, Bridgman (1944) was the first to analyse the state of triaxial stress induced in the necking region of the tensile test specimen. He showed that radial (σ_{rr}) and tangential ($\sigma_{\theta\theta}$) stresses are induced under the applied axial stress (σ_o) as illustrated in Figure 12.18.

Example 12.2 Point of plastic instability for a material obeying Ludwik's power law

Determine the true stress and true strain at the point of tensile instability for the Ludwik power law $\sigma = C\varepsilon^n$.

We know from (12.15) that at the point of instability

$$\frac{\mathrm{d}\sigma}{\mathrm{d}\varepsilon} = Cn\varepsilon^{n-1} = \sigma = C\varepsilon^n$$

Cancelling terms we find $n\varepsilon^{-1} = 1$ or $\varepsilon = n$ and hence $\sigma = Cn^n$. Thus, at the point of tensile instability for a material which obeys the Ludwik power law the true stress and true strain are

$$\sigma = Cn^n; \quad \varepsilon = n$$

12.10 Hydrostatic and Deviatoric Stresses

When the cubic equation of stress is discussed in Chapter 14 the scalar stress invariants will be determined, which in terms of principal stresses are given by, (14.71)

$$\begin{aligned} I_1 &= \sigma_1 + \sigma_2 + \sigma_3 \\ I_2 &= -(\sigma_1\sigma_2 + \sigma_2\sigma_3 + \sigma_3\sigma_1) \\ I_3 &= \sigma_1\sigma_2\sigma_3 \end{aligned} \tag{12.19}$$

The invariants I_1 and I_2 acquire particular significance when considering criteria of yielding. Experimental evidence, backed up by analytical considerations, indicates that the application of a loading system to a solid body consists of two distinct component fields. The first of these is the effect of *hydrostatic*

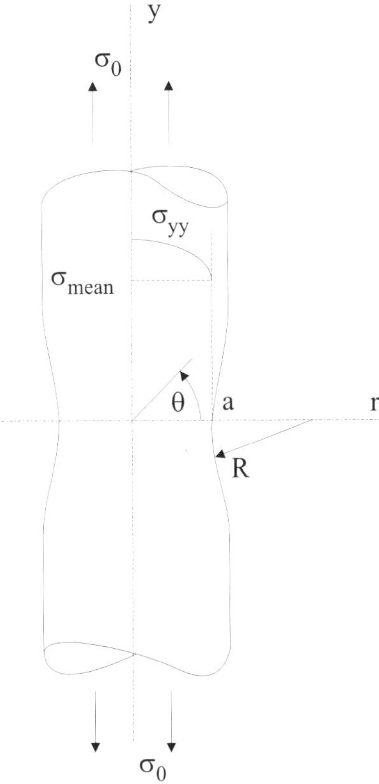

Figure 12.18: Stresses induced in a tensile test specimen which is undergoing extensive local necking.

pressure, Figure 12.19. Hydrostatic pressure can be identified, in terms of the principal stresses, by the action of a mean stress, σ_m

$$\sigma_m = \frac{\sigma_1 + \sigma_2 + \sigma_3}{3} \tag{12.20}$$

When $\sigma_1 = \sigma_2 = \sigma_3$ we have a hydrostatic state of stress in which any direction in space is a principal direction. The second component field is obtained by deducting the mean stress, σ_m, from the stress value in a given direction. This process leaves us with a component known as the *deviatoric stress*, σ'

$$\sigma_1' = \sigma_1 - \frac{I_1}{3}; \quad \sigma_2' = \sigma_2 - \frac{I_1}{3}; \quad \sigma_3' = \sigma_3 - \frac{I_1}{3} \tag{12.21}$$

The deviatoric stress component is responsible for dimensional changes by virtue of producing shearing stresses. It is the deviatoric component that is thus responsible for plastic flow. Bear in mind that the resultant of all the deviatoric stresses is always zero and therefore the deviatoric stress does not affect the change in volume of the stressed material.

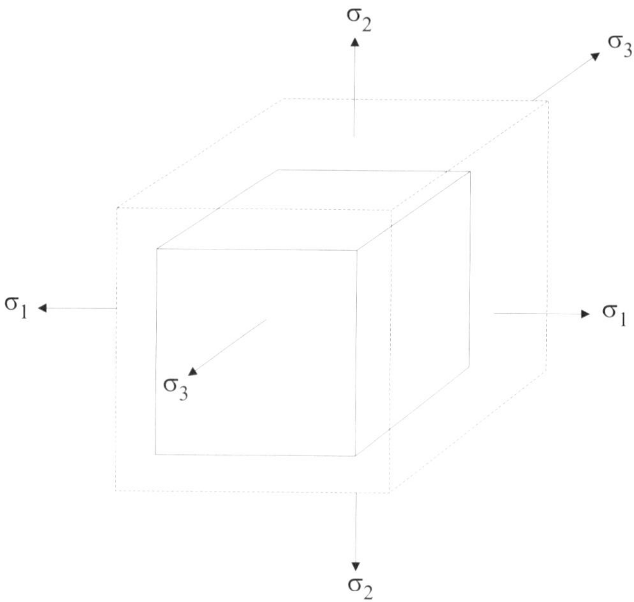

Figure 12.19: A three-dimensional stress element subject to a hydrostatic stress state.

12.11 Criteria for Yielding

It was shown in §12.5 and §12.9 that the elastic failure and the onset of plastic flow are relatively simple to account for in a tensile test. The effect of a loading system which introduces combined stresses (e.g. direct and shearing, biaxial or triaxial) is slightly more difficult. The link between these two states of stress, in terms of their combined effect, can be sought on the understanding that a hypothesis defining the critical conditions necessary for yielding must apply equally to the simple tests and to a complex loading system. It is required to find a suitable criterion based upon strain, stress or strain energy for the complex system that can be related to the corresponding quantity at the uniaxial yield point. It will be shown that the following criteria can all be related to the uniaxial yield stress which is most conveniently measured from a tensile test.

Experimental evidence indicates that it is shear that plays the dominant role in influencing the plastic behaviour of a material. This naturally implies that volumetric changes will not affect the onset of yielding and that plastic flow will be independent of the hydrostatic pressure. It follows that when the hydrostatic pressure is discounted, it will be the deviatoric quantities and their mutual relationship that will define the conditions of yielding. Although the value of I_1 is of no practical importance to yielding, the criterion of yielding remains a function of the deviatoric stress invariants (J_2 and J_3), particularly J_2, refer to §14.8.2.

The following sub-sections describe the main yield criteria that have been proposed, although it is now recognised that those attributed to Tresca and von Mises are the most representative of the initial yield behaviour of metallic materials. The yield criteria are normally expressed in terms of the principal stresses σ_1, σ_2 and σ_3 where $\sigma_1 > \sigma_2 > \sigma_3$.

12.11.1 Maximum Principal Stress Theory

The simplest criterion of yielding is due to Rankine and states that yielding commences either when the major principal stress σ_1 obtains the value of the tensile yield stress, σ_Y, or when the minor principal stress, σ_3, is compressive and of greater numerical magnitude than σ_1

$$\sigma_1 = \sigma_Y \quad \text{or} \quad \sigma_3 = -\sigma_Y \tag{12.22}$$

These equations ignore the intermediate principal stress (σ_2) and assume, along with all of the other yield criteria presented here, that the tensile and compressive yield stresses are equal which is not necessarily the case.

If the stress system is a general biaxial one (σ_{xx}, σ_{yy} and τ_{xy}) then the non-zero principal stresses are given by, (7.19)

$$\sigma_{1,2} = \frac{(\sigma_{xx} + \sigma_{yy})}{2} \pm \sqrt{\left(\frac{\sigma_{xx} - \sigma_{yy}}{2}\right)^2 + \tau_{xy}^2} \tag{12.23}$$

12.11.2 Maximum Principal Strain Theory

The maximum principal strain criterion postulates that yielding commences when the major principal strain attains the magnitude of the uniaxial strain at the elastic limit. When the limiting strain is equated to σ_Y/E it follows from the Hookian equations (1.16) that

$$\begin{aligned} \varepsilon_{xx} &= \frac{1}{E}\left[\sigma_{xx} - \nu(\sigma_{yy} + \sigma_{zz})\right] \\ \varepsilon_{zz} &= \frac{1}{E}\left[\sigma_{zz} - \nu(\sigma_{xx} + \sigma_{yy})\right] \end{aligned} \tag{12.24}$$

Assuming that x and z coincide with σ_1 and σ_3 respectively then

$$\begin{aligned} \varepsilon_1 &= \frac{\sigma_Y}{E} = \frac{1}{E}\left[\sigma_1 - \nu(\sigma_2 + \sigma_3)\right] \\ \varepsilon_3 &= -\frac{\sigma_Y}{E} = \frac{1}{E}\left[\sigma_3 - \nu(\sigma_1 + \sigma_2)\right] \end{aligned} \tag{12.25}$$

or

$$\begin{aligned} \left[\sigma_1 - \nu(\sigma_2 + \sigma_3)\right] &= \sigma_Y \\ \left[\sigma_3 - \nu(\sigma_1 + \sigma_2)\right] &= -\sigma_Y \end{aligned} \tag{12.26}$$

The maximum principal strain criterion was proposed by St. Venant and is seldom used in practice.

12.11.3 Maximum Shear Stress Theory

This popular theory simply states that yielding begins when the maximum shear stress reaches a critical value and is generally attributed to Tresca, Coulomb and Guest. The critical value is taken as the maximum shear stress, k, at the point of yielding under simple tension or compression. For a uniaxial stress system the stress transformation equations (7.8) give

$$\begin{aligned} \sigma_{x'x'} &= \sigma_{xx}\cos^2\theta; \quad & \sigma_{\theta=45^\circ} &= \frac{1}{2}\sigma_{xx} \\ \tau_{x'y'} &= \sigma_{xx}\sin\theta\cos\theta; \quad & \tau_{max} &= \frac{1}{2}\sigma_{xx} \end{aligned} \tag{12.27}$$

taking the absolute value of $\tau_{x'y'}$ and with τ_{max} determined from $\partial\tau_{x'y'}/\partial\theta = 0$. Therefore, it follows that $k = \sigma_Y/2$ and acts along planes which are at 45° to the uniaxial stress axis.

The maximum shear stress is given by $\tau_{max} = \tau_2 = (\sigma_1 - \sigma_3)/2$ (since $\sigma_1 > \sigma_2 > \sigma_3$) and acts along a plane inclined at 45° to the 1 and 3 directions. Equating the shear stresses for the uniaxial ($k = \sigma_Y/2$) and triaxial cases ($\tau_2 = (\sigma_1 - \sigma_3)/2 = k$) leads to the Tresca yield criterion

$$\sigma_1 - \sigma_3 = \sigma_Y \tag{12.28}$$

and can be simply stated as

(Greatest principal stress) – (Least principal stress) = (Tensile yield stress)

The criterion clearly neglects the effect of the intermediate principal stress σ_2 and consequently the criterion cannot define fully the actual physical situation. Numerical values must be substituted with their signs denoting tension or compression. For example, if $\sigma_1 = 20$MPa, $\sigma_2 = -10$MPa and $\sigma_3 = -12$MPa then $\sigma_1 - \sigma_3 = 20 - (-12) = 32$MPa, with the intermediate principal stress, σ_2, being irrelevant.

In a biaxial stress case ($\sigma_3 = 0$) then the Tresca criterion is given as

$$\sigma_1 - \sigma_2 = \sigma_Y \tag{12.29}$$

and in terms of stress components σ_{xx}, σ_{yy} and σ_{xy} we have from (7.19)

$$\sigma_{1,2} = \frac{(\sigma_{xx} + \sigma_{yy})}{2} \pm \sqrt{\left(\frac{\sigma_{xx} - \sigma_{yy}}{2}\right)^2 + \tau_{xy}^2} \tag{12.30}$$

so that

$$\sigma_1 - \sigma_2 = 2\sqrt{\left(\frac{\sigma_{xx} - \sigma_{yy}}{2}\right)^2 + \tau_{xy}^2} \equiv \sqrt{(\sigma_{xx} - \sigma_{yy})^2 + 4\tau_{xy}^2} \tag{12.31}$$

Thus, the Tresca criterion can be written

$$\sqrt{(\sigma_{xx} - \sigma_{yy})^2 + 4\tau_{xy}^2} = \sigma_Y \tag{12.32}$$

For the particular case of $\sigma_{yy} = 0$, squaring the above we have

$$\sigma_{xx}^2 + 4\tau_{xy}^2 = \sigma_Y^2 \tag{12.33}$$

12.11.4 Total Strain Energy Theory

This criterion, due to Beltrami and Haigh, assumes that yielding commences when the total strain energy stored attains the value of the strain energy for uniaxial yielding. For the principal system the total strain energy is, (9.4) and (9.12)

$$U = \int_\varepsilon \sigma_1 \,\mathrm{d}\varepsilon_1 + \sigma_2 \,\mathrm{d}\varepsilon_2 + \sigma_3 \,\mathrm{d}\varepsilon_3 = \frac{1}{2}(\sigma_1\varepsilon_1 + \sigma_2\varepsilon_2 + \sigma_3\varepsilon_3) \tag{12.34}$$

which is the sum of the areas enclosed by the linear stress-strain curve for each orthogonal direction. Substituting for the strain via the Hookian equations (1.16) we have

$$U = \frac{1}{2E}(\sigma_1^2 + \sigma_2^2 + \sigma_3^2) - \frac{\nu}{E}(\sigma_1\sigma_2 + \sigma_2\sigma_3 + \sigma_1\sigma_3) \tag{12.35}$$

Putting $\sigma_1 = \sigma_Y$ and $\sigma_2 = \sigma_3 = 0$ gives $U = \sigma_Y^2/2E$ at the point of yield in simple tension.

The criterion states that we equate the total strain energy to the strain energy for uniaxial yielding, which leads to the total strain energy criterion

$$(\sigma_1^2 + \sigma_2^2 + \sigma_3^2) - 2\nu(\sigma_1\sigma_2 + \sigma_2\sigma_3 + \sigma_1\sigma_3) = \sigma_Y^2 \tag{12.36}$$

We note that the intermediate principal stress σ_2 contributes to triaxial yielding. For the non-zero principal biaxial stresses σ_1 and σ_2, with $\sigma_3 = 0$ we have

$$\sigma_1^2 - 2\nu\sigma_1\sigma_2 + \sigma_2^2 = \sigma_Y^2 \tag{12.37}$$

12.11.5 Shear Strain Energy Theory

Clark Maxwell implied that hydrostatic stress plays no part in yielding when he was the first to propose that yielding was due only to the shear strain energy at the tensile yield point. Huber, later proposed that the total strain energy was composed of dilatational (volumetric) and distortional (shear) components. The former depends upon the mean or hydrostatic component of the applied stress system while the latter is due to the remaining deviatoric components of stress. This was later formulated by von Mises and Hencky. For ductile, initially isotropic metallic materials, there is now considerable experimental evidence to support this theory.

Let us begin by dividing the total strain energy, U, of the system into its volumetric and shear components U_v and U_s respectively

$$U = U_v + U_s \tag{12.38}$$

From (12.34) in the previous section the volumetric strain energy is given by

$$\begin{aligned} U_v &= \int_\varepsilon (\sigma_m \,\mathrm{d}\varepsilon + \sigma_m \,\mathrm{d}\varepsilon + \sigma_m \,\mathrm{d}\varepsilon) \\ &= \frac{1}{2}(\sigma_m\varepsilon + \sigma_m\varepsilon + \sigma_m\varepsilon) \\ &= \frac{3}{2}\sigma_m\varepsilon \end{aligned} \tag{12.39}$$

Since

$$\varepsilon_1 = \varepsilon_2 = \varepsilon_3 = \varepsilon = \frac{1}{E}\left[\sigma_m - \nu(\sigma_m + \sigma_m)\right] = \frac{\sigma_m}{E}(1 - 2\nu) \tag{12.40}$$

which, upon substitution into (12.39) gives

$$U_v = \frac{3}{2}\sigma_m\frac{\sigma_m}{E}(1 - 2\nu) = \frac{3(1-2\nu)}{2E}\sigma_m^2 = \frac{(1-2\nu)(\sigma_1 + \sigma_2 + \sigma_3)^2}{6E} \tag{12.41}$$

From the previous section, we know that the total strain energy is, (12.35)

$$U = \frac{1}{2E}(\sigma_1^2 + \sigma_2^2 + \sigma_3^2) - \frac{\nu}{E}(\sigma_1\sigma_2 + \sigma_2\sigma_3 + \sigma_1\sigma_3) \tag{12.42}$$

so that $U_s = U - U_v$

$$U_s = \frac{(1+\nu)}{6E}\left[(\sigma_1 - \sigma_2)^2 + (\sigma_2 - \sigma_3)^2 + (\sigma_1 - \sigma_3)^2\right] \tag{12.43}$$

The value of U_s at the uniaxial yield point is found by putting $\sigma_1 = \sigma_Y$ and $\sigma_2 = \sigma_3 = 0$ in U_s

$$U_s = \frac{(1+\nu)}{3E}\sigma_Y^2; \quad \sigma_1 = \sigma_Y, \sigma_2 = \sigma_3 = 0 \tag{12.44}$$

Equating (12.43) and (12.44) gives the usual form of the Huber-von Mises yield criterion

$$(\sigma_1 - \sigma_2)^2 + (\sigma_2 - \sigma_3)^2 + (\sigma_1 - \sigma_3)^2 = 2\sigma_Y^2 = 6k^2 \tag{12.45}$$

since $\sigma_Y = \sqrt{3}k$ in uniaxial loading for the Huber-von Mises criterion.

If $\sigma_3 = 0$, then the biaxial form becomes

$$\sigma_1^2 - \sigma_1\sigma_2 + \sigma_2^2 = \sigma_Y^2 \tag{12.46}$$

Substituting for $\sigma_{1,2}$; namely

$$\sigma_{1,2} = \left(\frac{\sigma_{xx} - \sigma_{yy}}{2}\right) \pm \sqrt{\left(\frac{\sigma_{xx} - \sigma_{yy}}{2}\right)^2 + \tau_{xy}^2} \tag{12.47}$$

into the biaxial expression (with $\sigma_{yy} = 0$) gives the following form of the Huber-von Mises criterion

$$\sigma_{xx}^2 + 3\tau_{xy}^2 = \sigma_Y^2 \tag{12.48}$$

Re-arranging the von Mises criterion into the form

$$(\sigma_1 - \sigma_2)^2 + (\sigma_2 - \sigma_3)^2 + (\sigma_1 - \sigma_3)^2 - 6k^2 = 0 \tag{12.49}$$

and noting that, (14.77)

$$J_2 = \frac{1}{6}\left[(\sigma_1 - \sigma_2)^2 + (\sigma_2 - \sigma_3)^2 + (\sigma_1 - \sigma_3)^2\right] \tag{12.50}$$

then the Huber-von Mises criterion can be expressed as

$$J_2 - k^2 = 0 \tag{12.51}$$

where J_2 is the second scalar invariant of the deviatoric stress and k is the yield stress in shear.

The introduction of the yield stress in shear, k, in addition to the yield stress in tension, σ_Y, is made to indicate the dependence of yielding on shear and also to emphasise the fact that in a number of operations the yield stress in shear is more representative of the behaviour of the material than is the tensile or compressive yield stress.

Comparing the Tresca and Huber-von Mises criteria for the case of uniaxial loading ($\sigma_1 = \sigma_Y, \sigma_2 = \sigma_3 = 0$) we have from the Tresca criterion

$$\sigma_1 - \sigma_3 = \sigma_Y = 2k \quad \text{or} \quad k = \frac{\sigma_Y}{2} \tag{12.52}$$

and from the Huber-von Mises criterion

$$\begin{aligned} 2\sigma_1^2 &= 2\sigma_Y^2 = 6k^2 \\ \text{or} \quad k &= \sqrt{\frac{1}{3}}\sigma_Y = (1.155)\left(\frac{\sigma_Y}{2}\right) \end{aligned} \tag{12.53}$$

indicating an average difference of 15% between the two criteria.

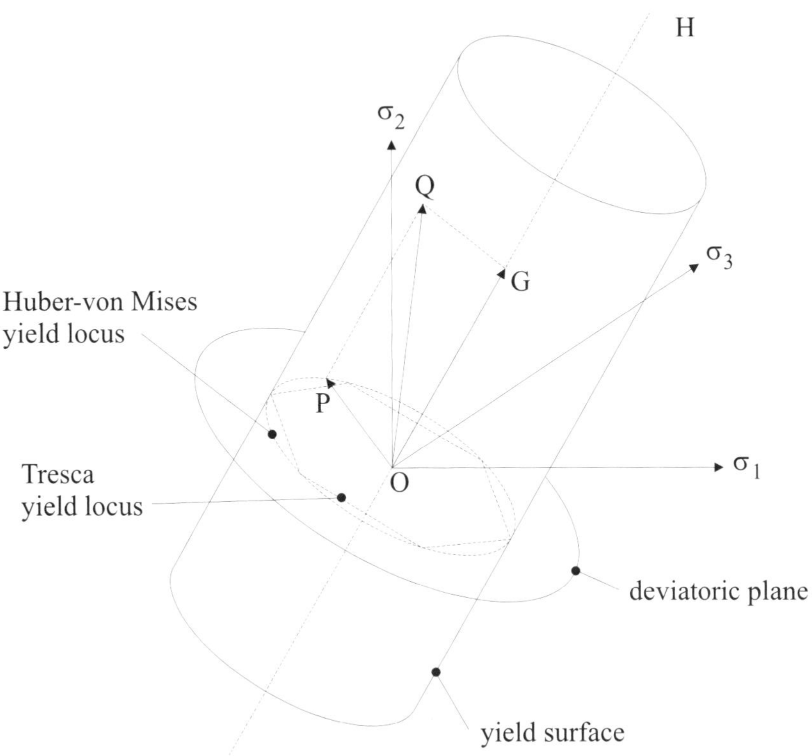

Figure 12.20: Principal stress space $(\sigma_1, \sigma_2, \sigma_3)$ illustrating the Huber-von Mises and Tresca loci.

12.12 A Geometrical Representation of the Tresca and Huber-von Mises Yield Criteria

Consideration of the analytical forms of the Tresca and Huber-von Mises criteria makes it clear that mathematically they both represent surfaces of revolution and that consequently we are in a position to produce a geometrical representation. Consider a system of three mutually perpendicular axes with the principal stresses taken as the rectangular coordinates as shown in Figure 12.20. The state of stress at any point may be represented by a vector emanating from the origin O. Consider the line OH equally inclined to the three axes so that its direction cosines are $(1/\sqrt{3}, 1/\sqrt{3}, 1/\sqrt{3})$. The stress vector OQ whose components are $(\sigma_1, \sigma_2, \sigma_3)$ may be resolved to a vector OG along OH and a vector OP perpendicular to OH. The vector OG represents the hydrostatic stress with components $(\sigma_o, \sigma_o, \sigma_o)$ and of magnitude $\sqrt{3}\sigma_o$. The vector OP is the deviatoric stress with components and magnitude $\sqrt{2J_2}$, since $\sigma_1' = \sigma_1 - I_1/3$. Therefore

$$OP^2 = \left(\sqrt{2J_2}\right)^2 = 2J_2 = \frac{1}{3}\left[(\sigma_1 - \sigma_2)^2 + (\sigma_2 - \sigma_3)^2 + (\sigma_1 - \sigma_3)^2\right] \tag{12.54}$$

Comparing this equation with the Huber-von Mises criterion we see that the criterion is represented by a cylindrical surface of radius $(\sqrt{2/3})\sigma_Y$, since $\left[(\sigma_1 - \sigma_2)^2 + (\sigma_2 - \sigma_3)^2 + (\sigma_1 - \sigma_3)^2\right] = 2\sigma_Y^2$.

Consider now the case where the third principal stress $\sigma_3 = 0$, which is the condition of *plane stress*; Figure 12.21. The Tresca criterion describes a regular hexagon inscribed within the cylinder. The set of

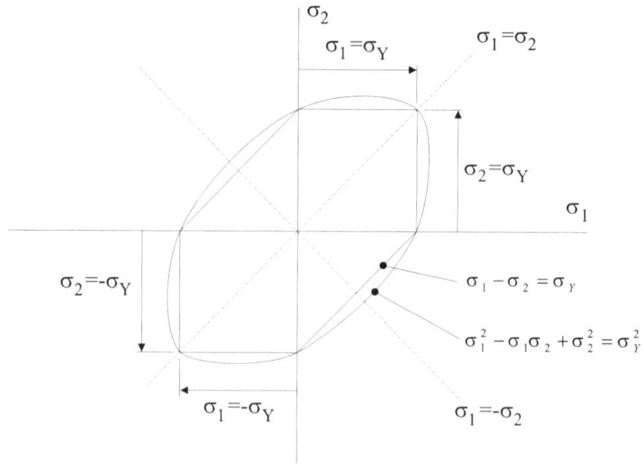

Figure 12.21: Principal stress space (σ_1, σ_2) illustrating the Huber-von Mises and Tresca loci.

magnitudes are given by

$$\sigma_1 = \sigma_Y, \sigma_2 = \sigma_Y, \sigma_1 - \sigma_2 = \sigma_Y \tag{12.55}$$

The Huber-von Mises criterion is described by an ellipse with a $45°$ inclination to the major principal axes. For this case the Huber-von Mises criterion simplifies to

$$\sigma_1^2 + \sigma_2^2 - \sigma_1\sigma_2 = \sigma_Y^2 \tag{12.56}$$

Consider now the upper right quadrant of the Huber-von Mises ellipse for $\sigma_3 = 0$ as shown in Figure 12.22. The following points are to be noted:

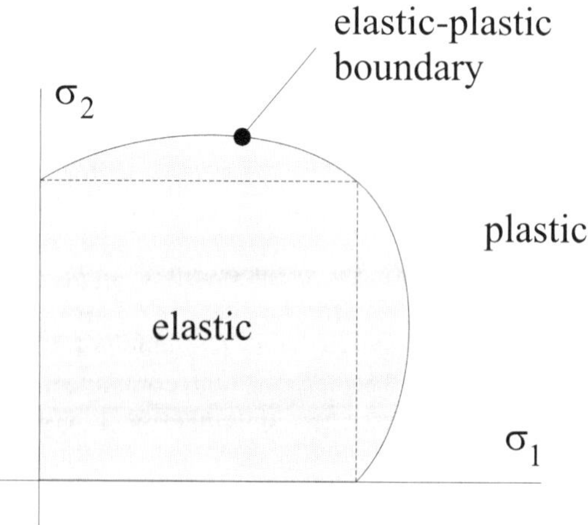

Figure 12.22: Huber-von Mises yield locus for $\sigma_3 = 0$.

- The Huber-von Mises ellipse describes a *yield locus* wherein elastic conditions prevail.

- A point on the surface of the ellipse represents the onset of yielding.
- Plastic behaviour occurs to the exterior of the yield locus.

12.13 Tresca and Huber-von Mises Criteria in Combined Tension and Torsion

By comparison of the Tresca and Huber-von Mises criteria we see that in principal stress space the simpler Tresca predictions are always the safest, but those from the Huber-von Mises criterion are generally the more realistic. We can compare the two criteria by considering a shaft subject to a state of combined torsion, τ, and tension, σ, as shown in Figure 12.23. From the Tresca criterion for biaxial stress ($\sigma_3 = 0$),

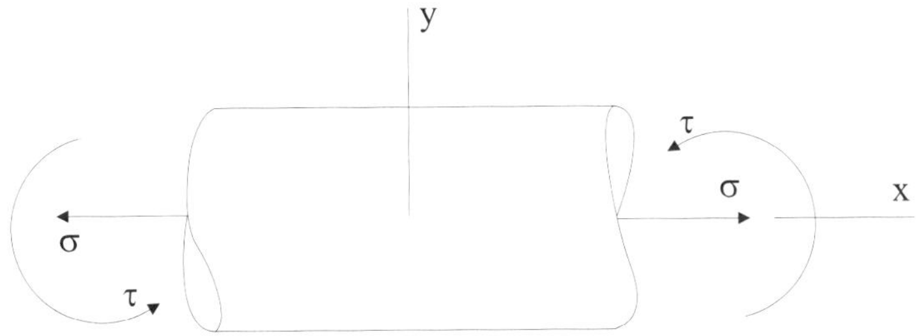

Figure 12.23: A shaft subjected to combined tension and torsion.

from (12.29)

$$\sigma_1 - \sigma_2 = \sigma_Y \tag{12.57}$$

and substituting for σ_1 and σ_2, (7.19), we find

$$\sqrt{(\sigma_{xx} - \sigma_{yy})^2 + 4\tau_{xy}^2} = \sigma_Y \tag{12.58}$$

or for $\sigma_{yy} = 0$ in the present case

$$\sigma_{xx}^2 + 4\tau_{xy}^2 = \sigma_Y^2; \quad \sigma^2 + 4\tau^2 = \sigma_Y^2 \tag{12.59}$$

using the notation above. Similarly, for the Huber-von Mises criterion in biaxial stress conditions ($\sigma_3 = 0$)

$$\sigma_1^2 - \sigma_1\sigma_2 + \sigma_2^2 = \sigma_Y^2 \tag{12.60}$$

and substituting for σ_1 and σ_2 with $\sigma_{yy} = 0$, we find

$$\sigma_{xx}^2 + 3\tau_{xy}^2 = \sigma_Y^2; \quad \sigma^2 + 3\tau^2 = \sigma_Y^2 \tag{12.61}$$

Thus, comparing the two expressions (12.59) and (12.61) we see that

$$\sigma^2 + n\tau^2 = \sigma_Y^2 \tag{12.62}$$

where $n = 4$ for Tresca's criterion and $n = 3$ for the Huber-von Mises criterion.

Figure 12.24 illustrates the early experimental results of Taylor and Quinney (1931) for combined torsion, τ, and tension, σ, tests on three different materials. From (12.62) the (τ, σ) curves are given by

$$\frac{\tau}{\sigma_Y} = \sqrt{\frac{1}{n}\left[1 - \left(\frac{\sigma}{\sigma_Y}\right)^2\right]} \tag{12.63}$$

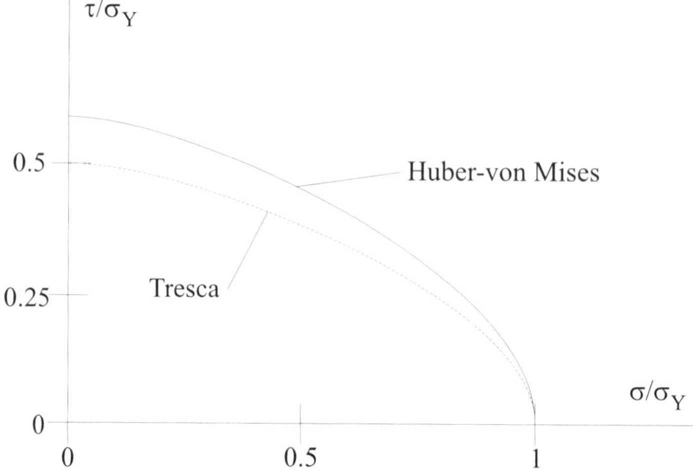

Figure 12.24: Combined torsion and tension tests of Taylor and Quinney (1931) with τ/σ_Y against σ/σ_Y.

12.14 Elastic-Plastic Torsion of Circular Components

In Chapter 3 we considered the torsion of an elastic solid bar. Consider now the solid bar of circular cross-sectional radius R in Figure 12.25 made of an elastic-perfectly plastic material and subject to a torque T. As the torque is increased, but with the material still responding elastically, the relationship between the basic parameters is $T/J = G\theta/L = \tau/r$, as derived in Chapter 3. On reaching a critical value the torque causes the onset of plasticity at the outer surface and the shear stress at this point reaches the value k for the material, where k is the yield stress in pure shear and is equal to $k = (1/2)\sigma_Y$ for the Tresca criterion and $k = (1/\sqrt{3})\sigma_Y$ for the Huber-von Mises criterion with σ_Y denoting the uniaxial yield stress. Throughout yielding we will assume that the material behaves in an elastic-perfectly plastic manner with no strain hardening, that is k = const. If then the plastic zone spreads inwards to a radius R_p, the total torque will be composed of the components sustained by the elastic core (T_E) and the plastic annulus (T_P).

At radius R_P, from $T = J\tau/r$,

$$T_E = T_Y = \frac{J_E k}{R_p} \tag{12.64}$$

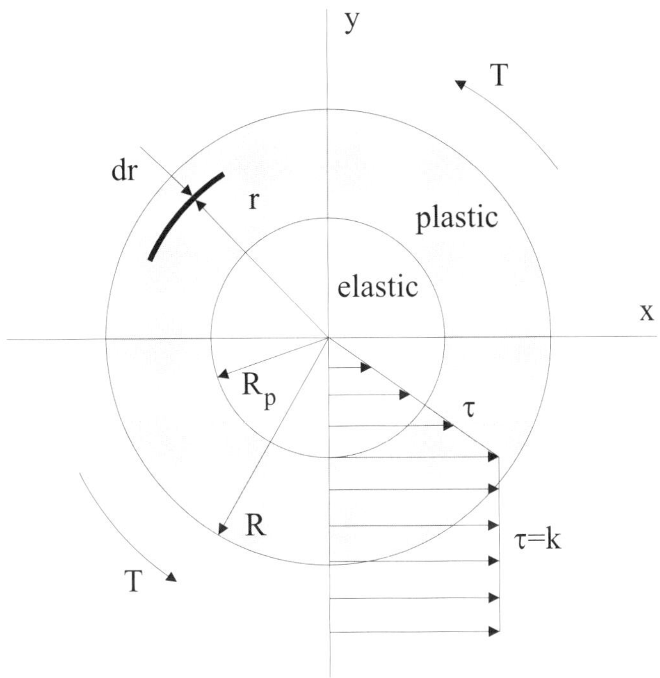

Figure 12.25: Circular cross-section subject to a torque T.

and the plastic component

$$T_P = \int_{R_P}^{R} 2\pi r^2 k \, \mathrm{d}r = \frac{2}{3}\pi k(R^3 - R_P^3) \tag{12.65}$$

T_p is obtained by a direct analogy with the elastic case, except that we integrate from the elastic-plastic boundary (R_P) to the outer radius. In the above integration it was assumed that k = const. Alternatively, if $k \neq$ const then its functional relationship must be introduced into the integration. The total torque of the shaft is obtained by summing the elastic and plastic components

$$\begin{aligned} T &= T_E + T_P \\ &= \frac{J_E k}{R_P} + \frac{2}{3}\pi k(R^3 - R_P^3) \\ &= \frac{\pi R_P^4}{2}\frac{k}{R_P} + \frac{2}{3}\pi k R^3 \left[1 - \left(\frac{R_P}{R}\right)^3\right] \\ T &= \frac{2}{3}\pi k R^3 \left[1 - \frac{1}{4}\left(\frac{R_P}{R}\right)^3\right] \end{aligned} \tag{12.66}$$

The total torque can alternatively be expressed in terms of the angle of twist. For a given value of R_P

at the elastic-plastic boundary

$$\frac{\theta_P}{L} = \frac{k}{GR_P}$$
$$T = \frac{2}{3}\pi k R^3 \left[1 - \frac{1}{4}\left(\frac{k}{R}\right)^3 \left(\frac{L}{G\theta_P}\right)^3\right] \tag{12.67}$$

since at the elastic-plastic boundary the elastic relation $\theta/L = \tau/rG$ still holds but with τ replaced by the yield stress in shear k. When $R_P = R$, at the onset of plasticity, then the angle of twist θ_Y is

$$\frac{\theta_Y}{L} = \frac{k}{GR} \tag{12.68}$$

and at $r = R_P$, eliminating G from the above two expressions

$$\frac{\theta_P}{L} = \frac{k}{R_P} \cdot \frac{\theta_Y R}{kL}; \quad \frac{\theta_P}{\theta_Y} = \frac{R}{R_P} \tag{12.69}$$

and therefore

$$T = \frac{2}{3}\pi k R^3 \left[1 - \frac{1}{4}\left(\frac{\theta_Y}{\theta_P}\right)^3\right] \tag{12.70}$$

For large values of applied torque the cylindrical shaft will tend to a situation of fully plastic, $R_P = 0$. In this case

$$T_{FP} = \frac{2}{3}\pi k R^3 \tag{12.71}$$

The ratio of *total* torque (T) to the *fully-plastic* torque (T_{FP}) is

$$\frac{T}{T_{FP}} = \left[1 - \frac{1}{4}\left(\frac{R_P}{R}\right)^3\right] \tag{12.72}$$

which is illustrated graphically in Figure 12.26.

Finally, consider the ratio of the torques for the fully plastic and the maximum allowable elastic torque (at the onset of plasticity)

$$\frac{\text{Fully plastic torque}}{\text{Maximum elastic torque}} = \frac{T_{FP}}{T_Y} = \frac{\frac{2}{3}\pi k R^3}{\frac{\pi}{2} k R^3} = \frac{2/3}{1/2} = \frac{4}{3} \tag{12.73}$$

Thus, the fully plastic torque is approximately 33% higher than the maximum allowable elastic torque. Note that the maximum allowable elastic torque can also be obtained by letting $R_P = R$ in (12.66)

$$T_Y = \frac{2}{3}\pi k R^3 \left[1 - \frac{1}{4}\left(\frac{1}{1}\right)^3\right] = \frac{2}{3}\pi k R^3 . \frac{3}{4} = \frac{\pi}{2} k R^3 \tag{12.74}$$

12.15 Elastic-Plastic Bending of Rectangular Cross-Section Beams

Concerning the bending of elastic-plastic beams we shall only consider beams of rectangular sections. Since a certain degree of uncertainty still exists about the exact nature of the elastic-plastic response to

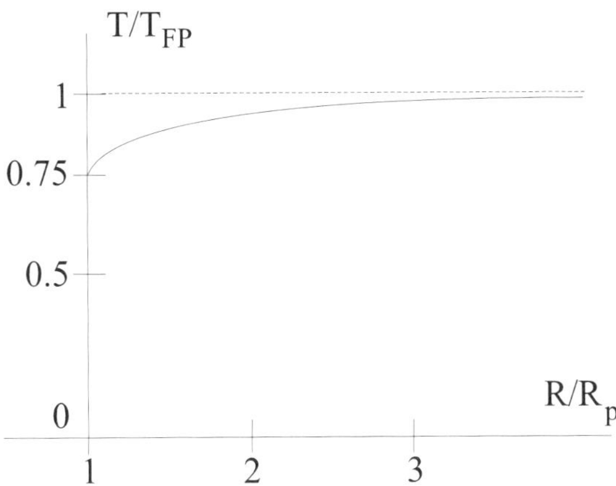

Figure 12.26: Variation of torque normalised with respect to the attained torque at a fully plastic cross-section against the outer radius normalised with respect to the radius of the elastic-plastic interface.

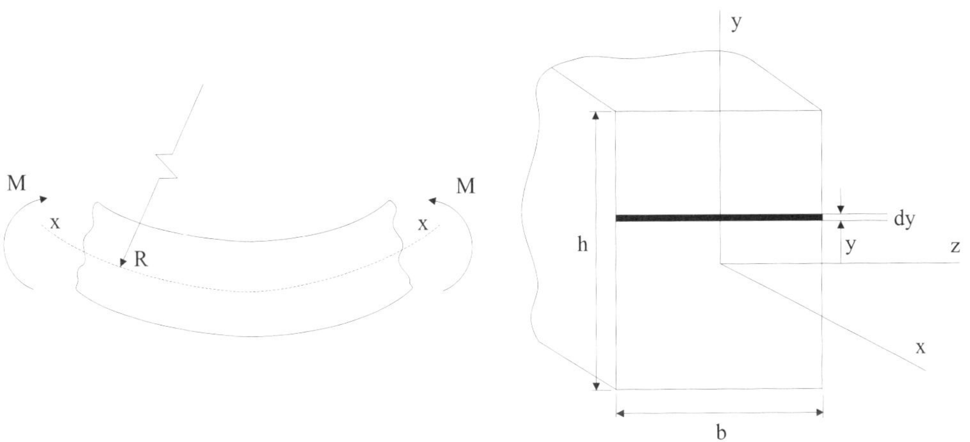

Figure 12.27: A beam of rectangular cross-section (b, h) subject to a pure bending moment M.

bending of a prismatic bar, the following discussion only offers an approximate solution to the problem. The approach is confined to pure bending and considers the two cases of a non-linear strain-hardening material and an idealised linear elastic-perfectly plastic material. Consider an initially straight beam of rectangular cross-section which is subject to a pure bending moment M, Figure 12.27.

Within the elastic range of deformations the bending stress, σ_B, is distributed across the section in accordance with the well-known relation, (5.30)

$$\sigma_B = \frac{My}{I} \tag{12.75}$$

where y is measured from the neutral axis xx that passes through the centroid of the section and I is the second moment of area. The maximum stress due to bending always occurs on the outer fibres of the beam, $y = h/2$

$$\sigma_{B_{max}} = \frac{Mh}{2I} \tag{12.76}$$

Thus the stress distribution remains linear as shown in case a) of the Figure 12.28. If the bending moment

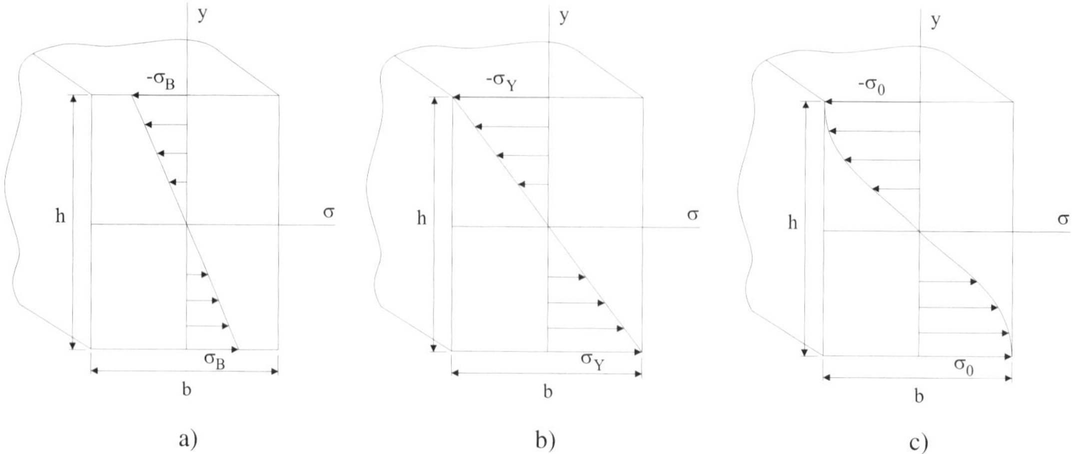

Figure 12.28: A beam of rectangular cross-section (b, h) subject to a pure bending moment. a) Elastic bending stress. b) Elastic bending stress at the incipient point of yielding. c) Bending stress subsequent to yielding.

is increased further until an increase in the bending moment brings $\sigma_{B_{max}}$ to its critical value (yield stress, σ_Y) then

$$\sigma_{B_{max}} = \sigma_Y \tag{12.77}$$

This will precipitate the onset of yielding in the outer fibres (see case b) of Figure 12.28) at which point termination of the linear stress distribution occurs. Further increases of the applied bending moment will cause the spread of the plastic zone inwards with σ_B being equal to the current value of the yield stress, σ_Y, within the plastic zone region and linearly distributed within the elastic zone, see case c) of Figure 12.28.

If throughout the loading the strain remains small then we have for the strain, (5.18)

$$\varepsilon = \frac{y}{R} \tag{12.78}$$

where R is a large radius of curvature of the deformed beam. Therefore, the strain on the outer fibres is

$$\varepsilon_o = \frac{h}{2R} \tag{12.79}$$

The corresponding stress on the outer fibres, σ_o, can then be obtained from the stress-strain curve of the material. For instance, consider the following stress-strain curve

$$\sigma = \varepsilon(A + B\varepsilon^{n-1}) \tag{12.80}$$

where A, B and n are material constants. This stress-strain relationship is illustrated in Figure 12.29. If an alternative idealised material is considered then the sequence of events described above will be the

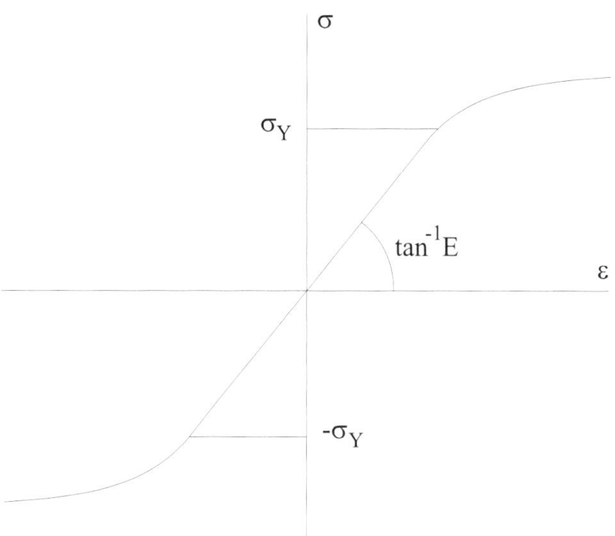

Figure 12.29: Strain-hardening stress-strain response.

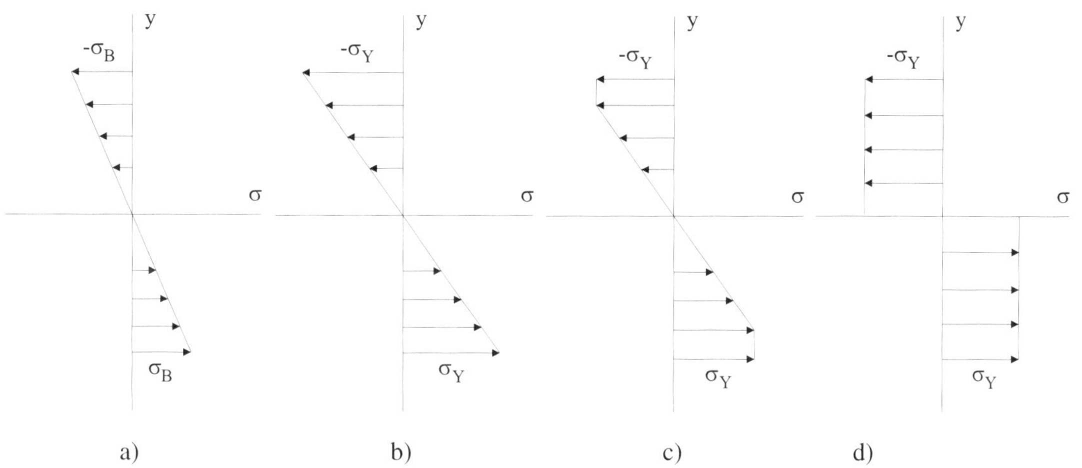

Figure 12.30: Spread of plasticity through the cross-section of a beam for an elastic-perfectly plastic material.

same but the stress distribution will be slightly different. For example, let us consider an elastic-perfectly plastic material shown in Figure 12.30. After the onset of yielding has occurred on the outer fibres, an increase in M causes a penetration of the plastic zone to a depth compatible with the value of M. As M is steadily increased the bands of plastic material will slowly approach the neutral axis until the entire section has become fully plastic. This model of deformation, however, does not agree with actual beams. Not only does the model indicate a stress discontinuity (from $+\sigma_Y$ to $-\sigma_Y$) at the neutral axis, it also postulates a total disappearance of the elastic core that is known to exist in beams which are subject to pure bending, even for very large values of M.

12.15.1 Determination of the Bending Moment

With reference to Figure 12.27 the bending moment acting on an element $\mathrm{d}y$ at a distance of y from the neutral axis is

$$\mathrm{d}M = (\sigma b \,\mathrm{d}y)y \tag{12.81}$$

Therefore, for the entire cross-section the total bending moment is

$$M = \int_{-h/2}^{+h/2} \sigma b y \,\mathrm{d}y \tag{12.82}$$

When $\sigma = \sigma_Y$ the moment associated with the onset of yielding, M_Y, is

$$M_Y = \int_{-h/2}^{+h/2} \sigma_Y b y \,\mathrm{d}y \tag{12.83}$$

In the following sub-sections we will consider the two cases of non-linear and linear stress-strain relationships.

12.15.1.1 Non-Linear Stress-Strain Relationship

If the stress-strain relation is given by (12.80) then from (12.82) and (12.78)

$$\begin{aligned} M &= \int_{-h/2}^{+h/2} \sigma b y \,\mathrm{d}y = \int_{-h/2}^{+h/2} \varepsilon(A + B\varepsilon^{n-1}) b y \,\mathrm{d}y \\ &= \frac{A}{R}\int_{-h/2}^{+h/2} b y^2 \,\mathrm{d}y + \frac{B}{R^n}\int_{-h/2}^{+h/2} b y^{n+1} \,\mathrm{d}y \end{aligned} \tag{12.84}$$

Noting that

$$I = \int b y^2 \,\mathrm{d}y \tag{12.85}$$

is the second moment of area of a rectangular section beam, and letting

$$I_0 = \int_{-h/2}^{+h/2} b y^{n+1} \,\mathrm{d}y \tag{12.86}$$

we have

$$M = \frac{A}{R} I + \frac{B}{R^n} I_0 \tag{12.87}$$

The integral I_0 reflects the non-linearity of the material properties. If the stress-strain curve is of the form of a Ludwik power law (namely $A = 0$ and $\sigma = B\varepsilon^n$) then we have

$$M = \frac{B I_0}{R^n} \tag{12.88}$$

where

$$\begin{aligned} I_0 &= b\int_{-h/2}^{+h/2} y^{n+1} \,\mathrm{d}y = \left(\frac{b}{n+2}\right)\left[y^{n+2}\right]_{-h/2}^{+h/2} \\ &= \left(\frac{b}{n+2}\right)\left(\frac{h}{2}\right)^{n+2}\left[1 - (-1)^{n+2}\right] = \left(\frac{b}{n+2}\right)\frac{h^{n+2}}{2^{n+1}} \end{aligned} \tag{12.89}$$

12.15.1.2 Linear Stress-Strain Relationship

In the case of a linear stress-strain curve then from (12.76) we have

$$M_Y = \frac{2I}{h}\sigma_{B_{max}} = \frac{2}{h}\left(\frac{bh^3}{12}\right)\sigma_Y = \frac{bh^2}{6}\sigma_Y \tag{12.90}$$

and constitutes the total resisting moment. When the applied moment M exceeds the value of M_Y then the beam section yields as described previously, say to a depth of $(h/2 - y)$ or y_o from the neutral axis, Figure 12.31. The total moment will consist of an elastic component, M_E, and a plastic component, M_P

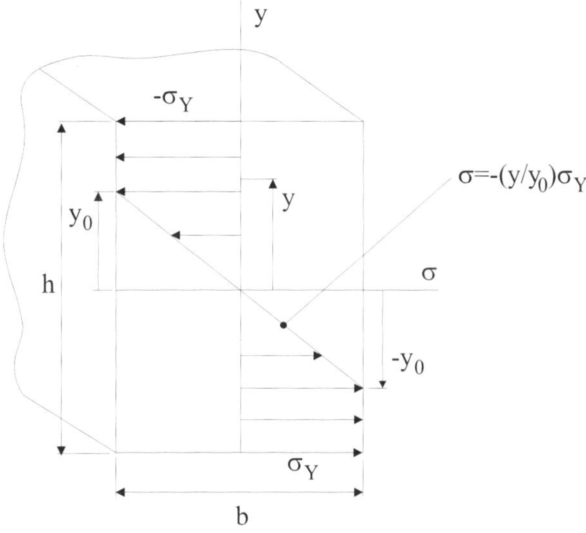

Figure 12.31: Elastic-perfectly plastic stress distribution acting over the cross-section of a beam. The beam material has yielded to a depth of y_o.

$$M = M_E + M_P \tag{12.91}$$

Considering the lower half of the beam in Figure 12.31 with an elemental strip of thickness dy at a distance y from the neutral axis then the bending moment is

$$\mathrm{d}M_{lower} = (\sigma b\,\mathrm{d}y)y \tag{12.92}$$

Integrating over the entire lower cross-section then the total bending moment is

$$M_{lower} = \frac{\sigma_Y b}{y_o}\int_0^{y_o} y^2\,\mathrm{d}y + \sigma_Y b\int_{y_o}^{h/2} y\,\mathrm{d}y \tag{12.93}$$

since $\sigma = (y/y_o)\sigma_Y$ for $y \le y_o$ and $\sigma = \sigma_Y$ for $y \ge y_o$ Performing the integration we have

$$M_{lower} = \sigma_Y b\left[\frac{h^2}{8} - \frac{y_o^2}{6}\right] \tag{12.94}$$

and since

$$M = M_{lower} + M_{upper} = 2M_{lower} = \sigma_Y b\left[\frac{h^2}{4} - \frac{y_o^2}{3}\right] = \frac{\sigma_Y bh^2}{4}\left[1 - \frac{4}{3}\left(\frac{y_o}{h}\right)^2\right] \tag{12.95}$$

or in terms of M_Y

$$M = \frac{2}{3}M_Y\left[1 - \frac{1}{3}\left(\frac{y_o}{h/2}\right)^2\right] \tag{12.96}$$

Let us conclude by considering the two limiting cases. At the onset of plasticity $y_o = h/2$

$$M = M_Y = \frac{1}{6}bh^2\sigma_Y \tag{12.97}$$

and for the fully plastic case $y_o = 0$

$$M_{FP} = \frac{3}{2}M_Y = \frac{1}{4}bh^2\sigma_Y \tag{12.98}$$

which shows that $M_{FP}/M_Y = 1.5$ for a rectangular section beam. Thus, the fully plastic bending moment is 50% higher than the maximum allowable elastic bending moment for a beam of rectangular cross-section.

12.16 Elastic-Plastic Bending of Circular Cross-Section Beams

Consider the shaft of circular cross-section and diameter D in Figure 12.32 subject to a pure bending moment M. From simple beam theory the axial stress is $\sigma_{xx} = My/I$ where $I(= \pi D^4/64)$ is the second moment of area, and noting that yielding will commence on the outermost fibres where $y = D/2$ then the bending moment at the point of first yielding is

$$M_Y = \frac{\pi D^3}{32}\sigma_Y \tag{12.99}$$

As illustrated in Figure 12.32, the elastic-plastic interface is assumed to be parallel to and at a depth c from the neutral axis. Considering the upper half of the beam with an elemental strip of thickness $\mathrm{d}y$ at a distance y from the neutral axis then the bending moment for the upper half is

$$M_{upper} = \int_0^R [\sigma b(y)\,\mathrm{d}y]\,y = \int_0^R \sigma b(y) y\,\mathrm{d}y \tag{12.100}$$

In the elastic region, $y \leq c$, σ varies linearly between $-\sigma_Y$ and σ_Y

$$\sigma = \left(\frac{y}{c}\right)\sigma_Y; \quad y \leq c \tag{12.101}$$

whereas in the plastic region, for an elastic-perfectly plastic material which neglects strain hardening

$$\sigma = \sigma_Y; \quad y \geq c \tag{12.102}$$

The integration in (12.100) is complicated by the fact that b is now a function of y whereas in the rectangular beam examined in the previous section b was constant. To find $b(y)$ consider the upper half of the beam once more in Figure 12.32. With $R^2 = z^2 + y^2$ then $z = \sqrt{R^2 - y^2}$. Thus, for both the left and right sides of the upper half $b(y) = 2\sqrt{R^2 - y^2}$. Hence, M_{upper} is now given by

$$M_{upper} = 2\int_0^R \sigma y\sqrt{R^2 - y^2}\,\mathrm{d}y \tag{12.103}$$

with the total bending moment given by

$$M = 4\int_0^R \sigma y\sqrt{R^2 - y^2}\,\mathrm{d}y \tag{12.104}$$

Substituting σ from (12.101) and (12.102) for the two intervals $[0, c]$ and $[c, R]$ we have

$$M = \frac{4\sigma_Y}{c}\int_o^c y^2\sqrt{R^2 - y^2}\,\mathrm{d}y + 4\sigma_Y\int_c^R y\sqrt{R^2 - y^2}\,\mathrm{d}y \tag{12.105}$$

The integration is assisted by making the substitution $y = R\sin\theta$ after which we arrive at

$$\frac{M}{M_Y} = \frac{2}{\pi}\left[\frac{1}{3}\left(5 - \frac{2c^2}{R^2}\right)\sqrt{1 - \frac{c^2}{R^2}} + \frac{R}{c}\sin^{-1}\left(\frac{c}{R}\right)\right] \tag{12.106}$$

where M_Y is given by (12.99). To determine the value of M/M_Y when the entire cross-section becomes fully plastic, M_{FP}/M_Y, then let $c \to 0$ and after taking limits we find

$$\frac{M_{FP}}{M_Y} = \frac{16}{3\pi} \approx 1.7 \quad \text{or} \quad M_{FP} = \frac{D^3}{6}\sigma_Y \tag{12.107}$$

which can be compared to the value of $M_{FP}/M_Y = 1.5$ for the rectangular section beam of the previous section.

Example 12.3 Elastic-plastic torsion of a solid shaft of circular cross-section

A solid shaft of circular cross-section is subject to a torque of $T = 10$kNm. The diameter of the shaft is $D = 50$mm and the yield stress of the material in pure shear is $k = 350$MPa. Determine the radius of the elastic-plastic interface.

From $\tau = Tr/J$, at the elastic-plastic boundary $r = R_p$ and $\tau = k$ then the elastic component of torque, T_E, is

$$T_E = T_Y = \frac{J_E k}{R_p}$$

Considering an elemental annular strip of thickness $\mathrm{d}r$ at radius r then the total plastic component of torque, T_P, is

$$T_P = \int_{R_p}^R 2\pi r^2 k\,\mathrm{d}r = \frac{2}{3}\pi k(R^3 - R_p^3)$$

The total torque is therefore

$$T = T_E + T_P = \frac{J_E k}{R_p} + \frac{2}{3}\pi k R^3\left[1 - \left(\frac{R_p}{R}\right)^3\right]$$

The elastic polar moment of area is

$$J_E = \frac{\pi R_p^4}{2}$$

and substituting into T we arrive at

$$T = \frac{2}{3}\pi k R^3\left[1 - \frac{1}{4}\left(\frac{R_p}{R}\right)^3\right]$$

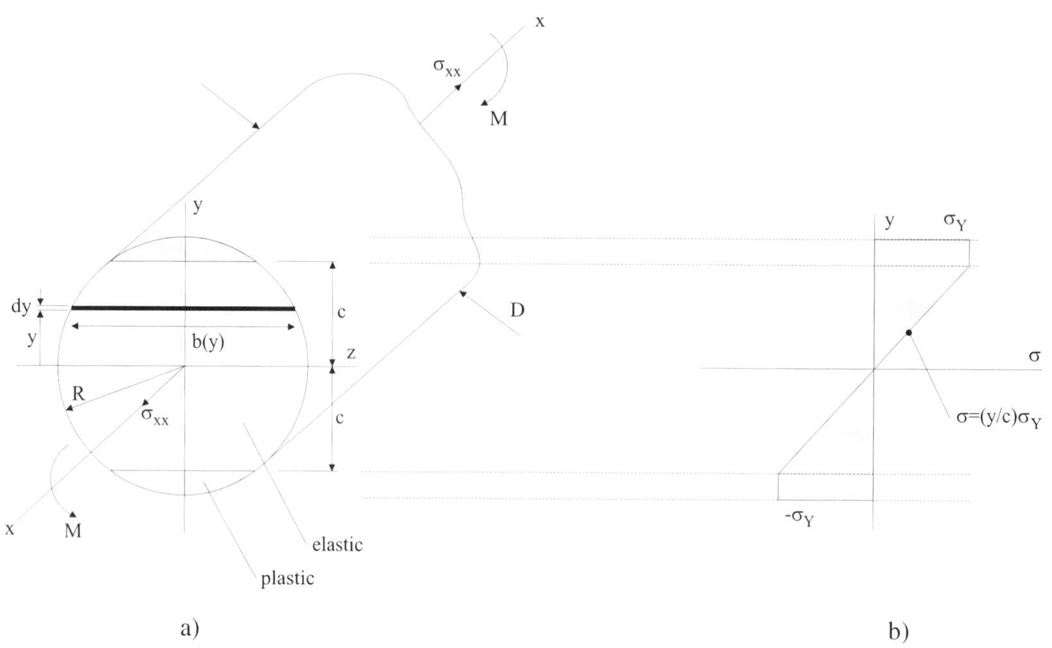

Figure 12.32: A circular cross-section bar subject to pure bending moment M. a) General configuration with an elastic-plastic interface at depth c. b) Stress distribution throughout the cross-section. Within the elastic region the stress distribution is linear whereas in the plastic region, for an elastic-perfectly plastic material, the stress is constant and equal to the yield stress, σ_Y.

Therefore, the torque at first yield occurs when $R_p = R$, at which

$$T = T_Y = \frac{2}{3}\pi k R^3 \left[1 - \frac{1}{4}\right] = \frac{\pi}{2} = 8.6\text{kNm}$$

The elastic-plastic boundary, R_p, is conveniently obtained in terms of T/T_Y by re-arranging the above expression

$$\left(\frac{R_p}{R}\right)^3 = 4 - 3\left(\frac{T}{T_Y}\right) \quad \text{or} \quad R_p = R^3\sqrt{4 - 3\left(\frac{T}{T_Y}\right)} = 20\text{mm}$$

The radius of the elastic-plastic boundary is therefore 20mm.

Example 12.4 Elastic-plastic bending of a beam of rectangular cross-section

A beam of rectangular cross-section ($b = 50$mm, $h = 100$mm) is subject to a bending moment of $M = 30$kNm. Determine the distance of the elastic-plastic boundary from the neutral axis. Assume that the material of the beam is elastic-perfectly plastic with a tensile yield stress of $\sigma_Y = 300$MPa.

Dividing the cross-section into lower and upper sections and considering the lower section then the bending moment is, (12.92)

$$M_{lower} = \int_0^{h/2} \sigma b y \, dy = \frac{\sigma_Y b}{y_o} \int_0^{y_o} y^2 \, dy - \sigma_Y b \int_{y_o}^{h/2} d \, dy = \sigma_Y b \left[\frac{h^2}{8} - \frac{y_o^2}{6} \right]$$

where y_o is the depth of the elastic-plastic boundary from the neutral axis. Summing the contributions from the lower and upper sections we have for the total bending moment

$$M = M_{lower} + M_{upper} = 2M_{lower} = \sigma_Y b \left[\frac{h^2}{4} - \frac{y_o^2}{3} \right] = \frac{\sigma_Y b h^2}{4} \left[1 - \frac{4}{3} \left(\frac{y_o}{h} \right)^2 \right]$$

Re-arranging in terms of y_o then

$$y_o = \sqrt{\frac{3h^2}{4} \left[1 - \frac{4M}{bh^2 \sigma_Y} \right]} = 38.7\text{mm}$$

12.17 Conclusion

This Chapter has provided an introduction to the extensive subject of the elastic-plastic behaviour of materials. The stress-strain diagram of typical engineering materials, first discussed in Chapter 1, was re-visited but with the emphasis now on the elastic-plastic behaviour. Frequently used empirical constitutive curves for the tensile stress-strain curve were presented as well as estimating a mean value for the yield stress. Five yield criteria were presented, with particular emphasis on the Tresca and Huber-von Mises criteria. Two applications of elastic-perfectly plastic behaviour under conditions of pure torsion and pure bending were examined.

12.18 References and Further Reading

- Blazynski, T. Z. (1983) *Applied Elasto-Plasticity of Solids*, Macmillan, London.
- Bridgman, P. W. (1944) *Trans., ASME*, 32, 553.
- Chakrabarty, J. (1987) *Theory of Plasticity*, McGraw-Hill, Singapore.
- Hill, R. (1950) *The Mathematical Theory of Plasticity*, Oxford University Press, Oxford.
- Honeycombe, R. W. K. (1984) *The Plastic Deformation of Metals*, 2nd Ed., Edward Arnold, London.
- Hull, D. and Bacon, D. J. (1984) *Introduction to Dislocations*, Pergamon Press, Oxford.
- Taylor, G. I. and Quinney, H. (1931) *Phil. Trans. Roy. Soc.*, A230, 323.

12.19 Exercises

12.1 For a given material the yield point is taken as that in which the permanent strain is equal to one-fifth of the recoverable elastic strain. The true stress-true strain constitutive curve in the plastic range is represented by the following

$$\sigma = \frac{E}{200} \varepsilon^{1/5}$$

where E is Young's modulus. Determine the stress, σ_Y, at the point of yield as a function of E.

12.2 Find the mean yield stress of a material which has the following empirical stress-strain curve

$$\sigma = \sigma_Y + B\left(\varepsilon - \frac{\sigma_Y}{E}\right)$$

where σ_Y is the stress at which yielding first occurs, E is Young's modulus and B is a constant.

12.3 For a material that obeys the Ludwik power law, $\sigma = C\varepsilon^n$, the point of tensile instability is found in terms of the engineering stress $\sigma_0 = 340$MPa and the engineering strain $e_0 = 30\%$. Determine the constants C and n for this material.

12.4 A solid bar of circular cross-section is subject to a bending moment M and a torque T with $M = cT$; where c is a positive real constant. Show that, according to the Huber-von Mises yield criterion, the bar will begin to yield when

$$\frac{\sigma_Y}{\tau_{max}} = \sqrt{3 + 4c^2}$$

where σ_Y is the tensile yield stress and τ_{max} is the maximum shear stress induced by the externally applied loading.

12.5 In a plane stress system the direct and shear stresses are given by $\sigma_{xx} = 50$MPa, $\sigma_{yy} = 100$MPa and $\tau_{xy} = 100$MPa. If the state of stress is just sufficient to cause the onset of yielding then determine the value of yield stress in simple tension that would accord with i) the Tresca yield criterion and ii) the Huber-von Mises yield criterion.

12.6 A solid circular bar of diameter 50mm is plastically deformed to a depth of 9mm by the application of a torque. Determine the values of the applied torque, the torque at first yield and the torque when the entire cross-section becomes plastic. Assume the yield stress in pure shear of the bar material to be 175MPa.

12.7 A beam of rectangular cross-section (b =2.5cm and h =4cm) is made of an elastic-perfectly plastic material with yield stress equal to 250MPa. The beam is loaded gradually such that plasticity spreads inwards until the entire cross-section has yielded. Determine the values of the applied bending moment that cause:

i) the initial yield of the beam;

ii) the spread of the plastic interface to a depth of 1cm;

iii) the beam to become fully plastic.

Chapter 13

Finite Elements

13.1 Introduction

The finite element method is now firmly established as one of the most powerful and frequently used tools available to engineers for performing a variety of analyses. The most popular analyses are those of time-independent continuum and structural mechanics and as a result these will be emphasised in the present Chapter although we will also address the finite element method applied to heat conduction and fluid flow problems. Furthermore, the present Chapter will focus on one-dimensional elements although we will repeatedly make reference to two-dimensional and higher order analyses to help demonstrate the generic nature of the finite element method.

The finite element method relies heavily on the use of a vector and matrix notation due to the computer being the driving force in its development. As a result, you may need to revise vector and matrix operations before reading the present Chapter.

The Chapter begins by presenting a brief history and overview of the finite element method. The process of discretisation is introduced through the example of a bar hanging under its own weight. The formulation and solution of a typical problem is then presented. Interpolation functions are at the core of field variable approximations throughout an element and as a consequence we will examine in detail the linear interpolation function for the one-dimensional bar element and also present an overview of general interpolation functions. The stress-strain relations for a linear elastic material are then presented in matrix form. This is followed by deriving the element stiffness matrix and force vector for a one-dimensional bar element and the assembly of elements into a structure. Pin-jointed plane frames are then examined with the Chapter concluding with an examination of the one-dimensional bar element for both heat transfer and fluid flow problems. These non-structural problems are analysed to illustrate the generic nature of the finite element method.

13.2 Background

The finite element method dates back to 1906 when a lattice analogy was proposed for stress analysis by Weighardt. The continuum was replaced by a regular pattern of elastic bars. Properties of the bars were chosen such that the displacement of the joints approximate the displacements of points in the continuum. Around the same period, Richardson in 1910 was using the finite difference technique to solve continuum problems, notably a stress analysis of the Aswan Dam, whereas Southwell (1946) developed the iterative relaxation method which was successfully used for solving larger systems of equations. In 1943 Courant was the first to propose the finite element method as we know it today via the principle of stationary potential energy and piecewise polynomial interpolations over triangular sub-regions. Courant's pioneering work remained unknown until engineers had independently developed the finite element method much later. Similarly, Rayleigh, Ritz and Galerkin demonstrated that continuum problems could be approached via the minimisation of potential energy or equivalent functionals.

The foregoing work was of limited practical use at the time due to the difficulties encountered in solving large systems of simultaneous equations. The development of the finite element method had to wait for the development of electronic computers and high-level programming languages. The 1950s marks the beginning of the widespread use of the finite element method, particularly by engineers in the aerospace industry. Unfortunately, much of this early work on finite element modelling went unrecognised because of aerospace company policies against publication, Robinson (1985). By 1953 Levy had written stiffness equations in matrix format which were solved using computers. Similarly, in the mid 1950s Turner *et al.* (1956) and Argyris (1960) were applying the finite element method to analyse aerospace structures.

The name *finite element* was first coined in 1960 by Clough (1960) with the term finite used to emphasise a clear distinction between differential elements used in calculus. In line with the increasing developments in computing power, the 1960s witnessed a rapid growth in the development of the finite element method. Key developments included new types of element and the analysis of shells and plate problems. The generic nature of the finite element method also became apparent with non-structural applications such as heat transfer. The 1960s and 1970s witnessed the emergence of several large commercial finite element programs such as ABAQUS and ANSYS. Today, such programs incorporate numerous element types for analysing a variety of problems such as static, dynamic, fluid flow, potential (such as heat transfer and electric and magnetic fields) and large derfomation.

For an overview of the history of the finite element method refer to Robinson (1985) and Zienkiewicz (1992).

13.3 Overview

Figure 13.1a) illustrates a tapered bar subject to an axial force and a bar element discretisation in Figure 13.1b) consisting of four uniform non-tapered elements. To determine the displacement throughout the bar the classical approach to this problem is to simultaneously satisfy the differential equations of equilibrium and compatibility subject to prescribed boundary conditions. The finite element approach *discretises* the bar of length L as a series of *finite elements*, each uniform but of a different cross-sectional area, A. In each element, the displacement is assumed to vary, say, linearly with x and as a result for $0 \leq x \leq L$ the displacement is a piecewise smooth function of x. The total elongation of the bar is simply the sum of the element elongations. The use of only four elements would lead to a crude approximation of the actual state of deformation in the bar with the approximation improving as the number of elements is increased.

In the foregoing example, and in general, the finite element method models a structure as an assemblage of small parts or elements. Each element is of simple geometry and therefore is much easier to analyse than the actual structure. In essence, we approximate a complicated solution by a model that consists of piecewise continuous simple solutions. In a heat transfer context, the above problem might represent a tapered cooling fin with a prescribed temperature at the left hand end, with the temperature distribution required as a function of x.

Figure 13.2 illustrates a slightly more complicated problem of a plate containing a circular hole subject to a uniformly applied tension with corresponding finite element model. The finite element model consists of plane areas, in the present case quadrilateral elements. The dots are referred to as *nodes* and indicate the connectivity of the elements. In this model each node has two degrees of freedom (dof), that is each node can displace in both the x and y directions. Therefore, if there are n nodes there are $2n$ degrees of freedom in the model minus the degrees of freedom due to nodes which have prescribed boundary

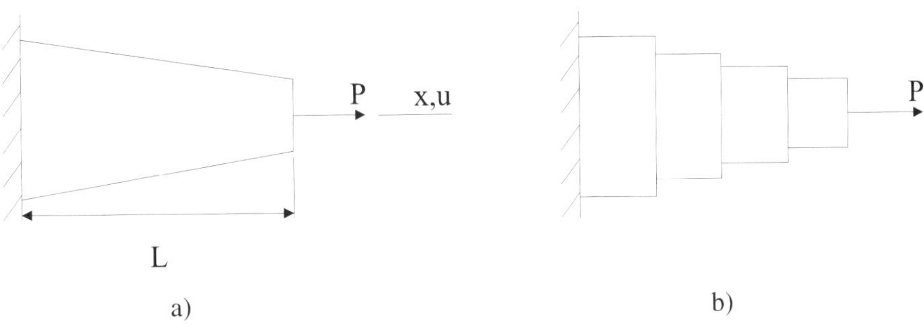

Figure 13.1: Finite element discretisation. a) A typical tapered bar subject to an axial force P. b) A model built of four uniform non-tapered elements of equal length.

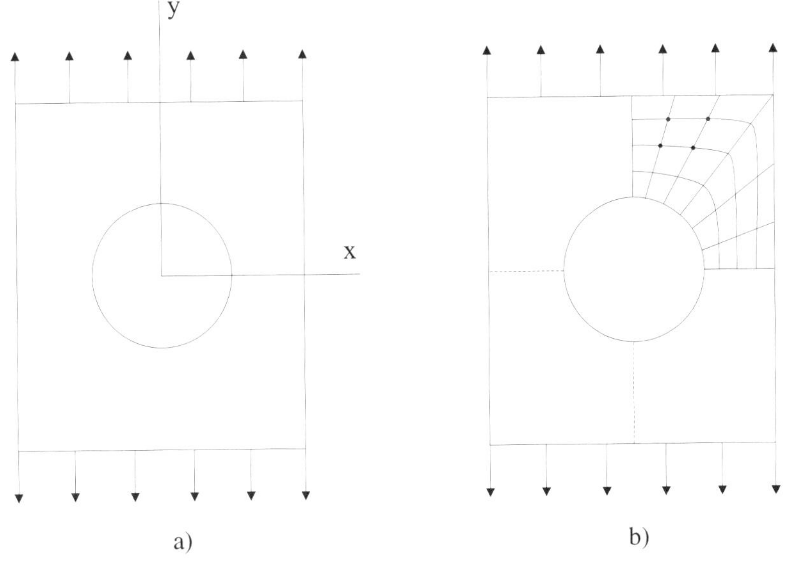

Figure 13.2: A finite element discretisation. a) A plate containing a circular hole subject to a uniform tension. b) A finite element model of the plate.

conditions. In real structures, however, there are infinitely many degrees of freedom. In a heat transfer problem each node has only one degree of freedom, namely the temperature of the node. Combining all of the elements in the model results in a system of algebraic equations that describe the finite element model. This system of equations is then solved for the nodal degrees of freedom. It is worth noting that the applied pressure in the model is converted into equivalent nodal forces at the corresponding nodes. The analysis procedure gives a prescription for making the conversion of pressure to nodal forces, as will be shown later.

It may appear that the process of discretisation is accomplished by simply cutting the continuum into pieces and then pinning the pieces together again at the nodes. However, such a model would not deform like that of a continuum. Under load, strain concentrations would appear at the nodes and the elements would tend to overlap or separate along the cuts. To prevent such behaviour the elements must be restricted in their deformation patterns. If the elements are restricted to deformation modes that keep edges straight then adjacent elements will neither overlap nor separate. In this way we satisfy the basic requirement that the deformations of a continuous medium must be compatible.

13.4 A Bar Under its Own Weight

As an introduction to the finite element method consider the intuitive example of a bar acting under its own weight. We will analyse the extension and strain energy of the bar by discretising the bar into a number of elements. First, let the bar be equally divided into four elements of constant cross sectional area A and density ρ, Figure 13.3. Element 1 has no forces acting on it as a result of adjacent elements

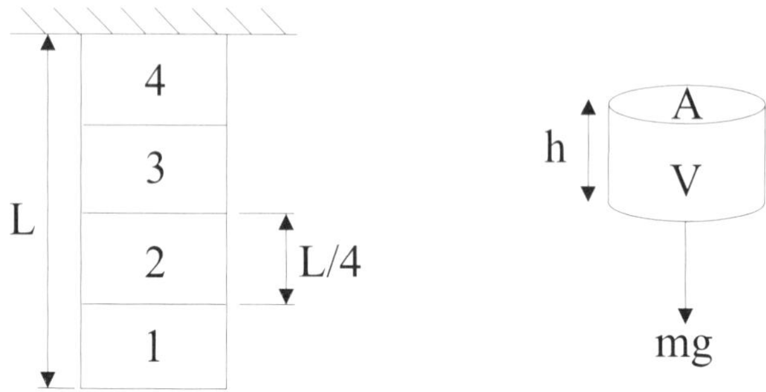

Figure 13.3: A bar hanging under its own weight and divided into four elements.

whereas element 2 has element 1 acting on it, and so on. If the loading on a given element is due to gravity alone then the axial force F is

$$F = mg = \rho V g = \rho g h A \tag{13.1}$$

where m is the mass, V the volume, ρ the density, h is the length and A the cross sectional area of an element, and where g is the gravitational constant. Thus, the force acting on the ith element is

$$F_i = (i-1)\rho g \left(\frac{L}{4}\right) A = (i-1)\frac{A\rho g L}{4} \tag{13.2}$$

From $\delta = FL/AE$ (see Example 9.1) the extension of the ith element is

$$\delta_i = (i-1)\frac{\rho g L^2}{16E} \tag{13.3}$$

The extension of the entire bar is therefore the sum of the four element extensions

$$\delta = \delta_1 + \delta_2 + \delta_3 + \delta_4 = (0+1+2+3)\frac{\rho g L^2}{16E} = \frac{3}{4}\left(\frac{\rho g L^2}{2E}\right) \tag{13.4}$$

Discretising the bar into eight elements of equal length then we find that the total extension is now given by

$$\delta = (0+1+2+3+4+5+6+7)\frac{\rho g L^2}{64} = \frac{7}{8}\left(\frac{\rho g L^2}{2E}\right) \tag{13.5}$$

Let us now compare these approximate estimates to the exact extension of the bar. Measuring x from the free-end of the bar then the force acting at position x is $Ax\rho g$ with corresponding stress and strain

$$\sigma_{xx} = x\rho g; \quad \varepsilon_{xx} = \frac{\sigma_{xx}}{E} = \frac{x\rho g}{E} \tag{13.6}$$

The extension of a differential element $\mathrm{d}x$ is $\varepsilon_{xx}\,\mathrm{d}x$ so that the total extension of the bar is

$$\delta = \int_0^L \frac{x\rho g}{E}\,\mathrm{d}x = \frac{\rho g L^2}{2E} \tag{13.7}$$

Comparing the approximate solutions with the exact extension we observe that the four and eight element approximations have associated errors of 25 and 12.5 per cent respectively, the error halving as the number of elements is doubled.

We will now examine both the approximate and exact strain energy of the bar. From (9.4) the strain energy density, U_o, of an axially loaded bar is known to be given by

$$U_o = \frac{\sigma_{xx}^2}{2E} \tag{13.8}$$

For element i of a four element model the stress σ_i is, (13.2)

$$\sigma_i = \frac{F_i}{A} = \frac{(i-1)A\rho g(L/4)}{A} = (i-1)\frac{\rho g L}{4} \tag{13.9}$$

The strain energy density of element i is therefore

$$(U_o)_i = \frac{1}{2E}(i-1)^2\frac{(\rho g L)^2}{16} \tag{13.10}$$

with the strain energy, U_i, of element i given by, (9.12)

$$U_i = \int_V (U_o)_i\,\mathrm{d}V = \int_0^{L/4} \frac{\sigma_i^2}{2E}A\,\mathrm{d}x = \frac{\sigma_i^2 AL}{8E} \tag{13.11}$$

and substituting for σ_i from (13.9)

$$U_i = \frac{(i-1)^2(\rho g L)^2 AL}{128E} \tag{13.12}$$

Thus, for all four elements the total strain energy is given by

$$U = (0^2+1^2+2^2+3^2)\frac{(\rho g L)^2 AL}{128} = \frac{14}{128}\left(\frac{A\rho^2 g^2 L^3}{E}\right) \approx 0.1094\left(\frac{A\rho^2 g^2 L^3}{E}\right) \tag{13.13}$$

and for eight elements U is

$$U \approx 0.1367\left(\frac{A\rho^2 g^2 L^3}{E}\right) \tag{13.14}$$

The exact strain energy is, with $\sigma_{xx} = \rho g x$ at distance x from the free end

$$\begin{aligned} U &= \int_V U_o\,\mathrm{d}V = \int_0^L \frac{\sigma_{xx}^2}{2E}A\,\mathrm{d}x = \int_0^L \frac{(\rho g x)^2}{2E}A\,\mathrm{d}x \\ &= \frac{1}{6}\left(\frac{A\rho^2 g^2 L^3}{E}\right) \approx 0.166\dot{6}\left(\frac{A\rho^2 g^2 L^3}{E}\right) \end{aligned} \tag{13.15}$$

Examining the above approximate and exact extensions and strain energies we note the following points:

- A simplifying assumption is made initially, but thereafter all of the arithmetic is precise.
- Even with the crude approximations of four and eight elements the approximate analyses are found to be reasonably accurate.
- The approximation is improved by increasing the number of elements.
- The approximate solutions of extension and strain energy both underestimate the exact solutions. This is because the finite element model is stiffer than the continuum.

13.5 Model, Analysis and Solution Breakdown

A finite element analysis typically includes the steps shown in Figure 13.4. Phases 1 to 4 in Figure 13.4

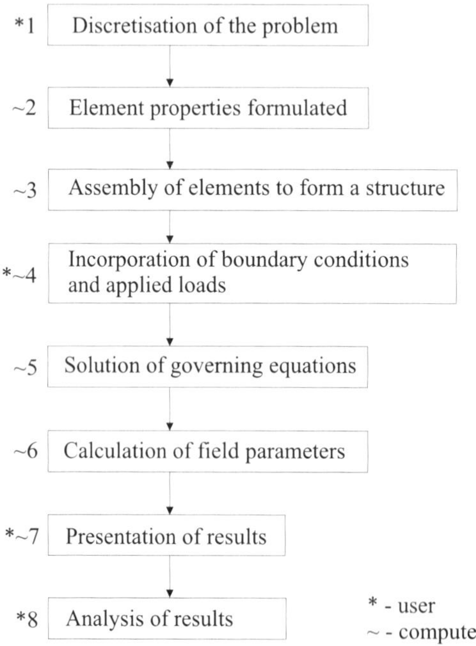

Figure 13.4: Typical finite element analysis flow chart.

occupy the majority of a user's time and effort in formulating the model, inputing the data and analysing the results. Phases 2, 3, 5 and 6 are performed automatically by the finite element program with undoubtedly the largest percentage of computing time spent in phase 5. Phases 1 to 8 are discussed in more detail below:

1. The structure or continuum is sub-divided into finite elements. Automatic mesh generators can assist the user in performing this task.
2. Formulation of element properties via element stiffness matrix and force vector.
3. Assemble all elements to obtain the entire structure model.
4. Incorporate prescribed boundary conditions.

5. Solution of a system of simultaneous algebraic equations to determine the nodal degrees of freedom such as displacement in stress analyses.

6. Calculation of field parameters. In stress analyses this generally requires the determination of element strains from the nodal displacements and stresses from strains. Element interpolation functions allow field parameters to de determined at points within an element other than the nodes.

7. Output of results in either non-graphical or graphical formats.

8. Analysis of the results. Clearly, the greater the number of degrees of freedom in a structure then the more data that requires analysis. In a large complex three-dimensional problem the solution data can be enormous.

13.6 Advantages and Disadvantages of the Finite Element Method

13.6.1 Advantages

The main advantage of the finite element method, compared to other numerical methods, is its versatility. The method can be applied to a variety of physical problems with a close resemblance between the actual structure and its finite element model. The structure analysed can be of arbitrary shape with arbitrary applied loading and restraints. The structure mesh can mix elements of different types, shapes and physical properties.

13.6.2 Disadvantages

A specific numerical result is obtained for a specific problem. In other words, a finite element analysis does not provide a closed form solution which permits an analytical solution. A computer and an extensive, reliable and general purpose program and an experienced user are all required. Experience and good engineering judgement are required by a user in order to define a representative model. Typical difficulties encountered by users are model simplification, geometric and loading symmetries, selection of element type, element sizes and number of elements, applied loading and prescribed boundary conditions.

13.7 Interpolation Function of the One-Dimensional Pin-Jointed Bar Element

Consider a one-dimensional bar element that is pin-jointed at both ends. The restriction that the element is pin-jointed requires that the element cannot transmit a moment at its ends or a lateral force but only tensile or compressive axial forces, see Figure 13.5a). The corresponding finite element model of the bar is shown in Figure 13.5b). The element e lies parallel to the x-axis with nodes i and j having coordinates x_i and x_j and corresponding nodal displacements u_i and u_j. Let the displacement, u, along the entire length of the element be assumed to have the form

$$u = a + bx \tag{13.16}$$

where a and b are constants. This is a linear interpolation function for u and consequently the variation of displacement along the element will be linear and the strain will be constant over the element since, (1.31), $\varepsilon_{xx} = \partial u/\partial x$. If the unknown displacements at the nodes are represented by u_i and u_j then

$$u_i = a + bx_i; \quad u_j = a + bx_j \tag{13.17}$$

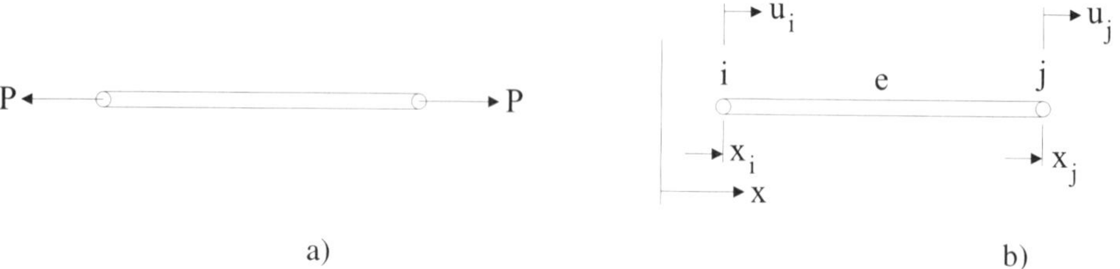

Figure 13.5: One-dimensional pin-jointed bar element. a) Element subject to tensile forces P at its ends. b) Finite element model of the bar element consisting of two end nodes i and j at positions x_i and x_j with corresponding axial displacements u_i and u_j respectively.

Solving for a and b we find

$$a = \frac{u_i x_j - u_j x_i}{L}; \quad b = \frac{u_j - u_i}{L} \tag{13.18}$$

which, upon substitution back into (13.16) gives

$$u = \left(\frac{x_j - x}{L}\right) u_i + \left(\frac{x - x_i}{L}\right) u_j = N_i u_i + N_j u_j \tag{13.19}$$

where N_i and N_j are referred to as the element *shape functions*.

13.8 Interpolation Functions

An essential ingredient of the finite element method is the behaviour of the individual elements. A few good higher order elements will generally produce better results than many lower order elements. Figure 13.6 illustrates approximating the function $\phi = \phi(x, y)$ using three different element types. The function ϕ will typically be displacement in a stress analysis and temperature in a heat transfer analysis and will generally vary smoothly in a structure. A finite element model yields a piecewise-smooth representation of ϕ. Within each element ϕ is a smooth function that is usually represented by a simple polynomial. Polynomials are mainly used due to their ease of differentiation, integration and computer implementation. Also, polynomial interpolation functions can easily be improved by increasing the order of the polynomial. For the one-dimensional bar element encountered in the previous section a linear interpolation function was used, generalised now for ϕ

$$\phi = \alpha_1 + \alpha_2 x \tag{13.20}$$

where α_1 and α_2 are determined by expressing them at terms of ϕ_i and ϕ_j at the nodes i and j as demonstrated previously. For a three-noded planar triangular element the following linear polynomial is used

$$\phi = \alpha_1 + \alpha_2 x + \alpha_3 y \tag{13.21}$$

with three unknowns α_1, α_2 and α_3. Figure 13.7 lists linear, quadratic and cubic interpolation functions for both one- and two-dimensional elements.

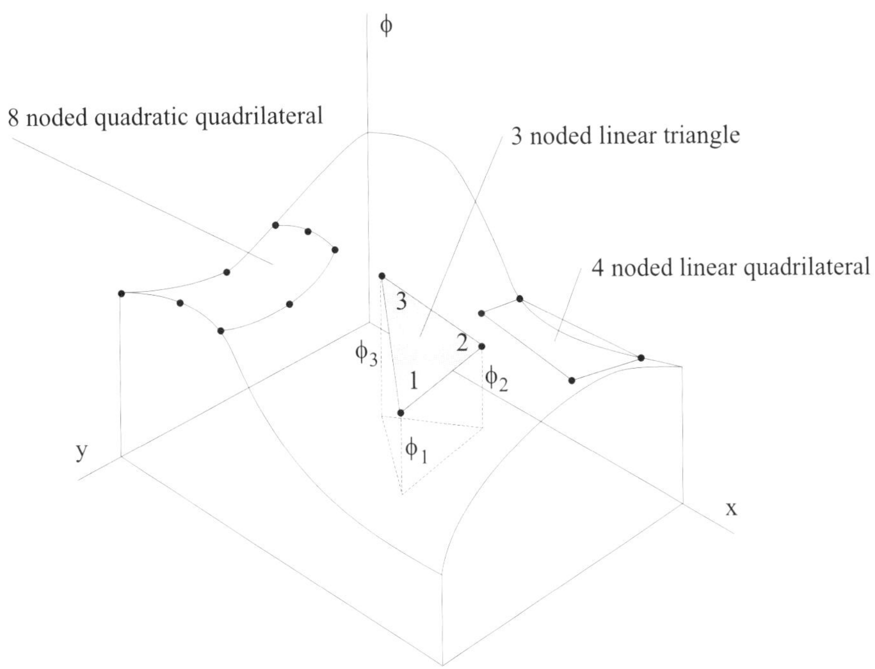

Figure 13.6: Variation of a function $\phi = \phi(x, y)$ and approximation by three different element types.

Complete interpolation functions of two-dimensional elements are conveniently described by Pascal's triangle, Figure 13.8a), with examples of the linear triangle and rectangle elements shown in Figure 13.8b). For example, a complete quadratic interpolation function has the following form

$$\phi = \alpha_1 + \alpha_2 x + \alpha_3 y + \alpha_4 x^2 + \alpha_5 xy + \alpha_6 y^2 \tag{13.22}$$

In general, from Pascal's triangle a complete polynomial of degree n in two-dimensions contains $(n + 1)(n + 2)/2$ terms.

When all of the α_i constants have been determined in terms of the nodal values ϕ_i then ϕ is represented within an element in terms of the nodal values ϕ_i with the aid of the shape functions N_i. If the mesh of elements is not too coarse and if the ϕ_i are exact then the non-nodal values of ϕ would produce a good approximation; see the one-dimensional approximation of ϕ with linear bar elements shown in Figure 13.9.

Elements are generally categorised by their interpolation functions into the following three groups: *simplex, complex* and *multiplex* and illustrated in Figure 13.10. Simplex elements have interpolation functions of constant and linear terms with nodes only located at the corners or vertices of the elements. Complex elements have quadratic, cubic and higher order interpolation functions whereas multiplex elements have higher order interpolation functions which do not use complete polynomial expressions as do complex elements. To ensure compatibility for rectangular and hexahedral elements the element edges are restricted to be parallel to the coordinate system. To overcome the restriction that this places on mesh generation, the multiplex elements are generally defined in terms of a local curvilinear coordinate system (ξ, η) which are then mapped onto the global coordinate system, Figure 13.11.

dimension	element	ϕ	order	no. of nodes
1		$\phi = \alpha_1 + \alpha_2 x$	linear	2
1		$\phi = \alpha_1 + \alpha_2 x + \alpha_3 x^2$	quadratic	3
1		$\phi = \alpha_1 + \alpha_2 x + \alpha_3 x^2 + \alpha_4 x^3$	cubic	4
2	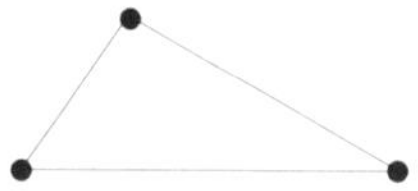	$\phi = \alpha_1 + \alpha_2 x + \alpha_3 y$	linear	3
2	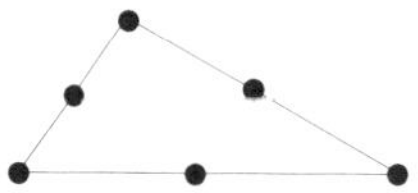	$\phi = \alpha_1 + \alpha_2 x + \alpha_3 y$ $+\alpha_4 xy + \alpha_5 x^2 + \alpha_6 y^2$	quadratic	6
2		$\phi = \alpha_1 + \alpha_2 x + \alpha_3 y + \alpha_4 xy$	linear	4

Figure 13.7: Interpolation functions for one-dimensional bar elements and two-dimensional triangle and rectangle elements.

13.9 One-Dimensional Simplex Bar Element

Let us now return to the one-dimensional bar element encountered in §13.7. Generalising the interpolation function (13.16) we have, Figure 13.12

$$\phi = \alpha_1 + \alpha_2 x \tag{13.23}$$

The element has two nodes i and j positioned at $x = x_i$ and $x = x_j$ from the origin O respectively. Substituting for ϕ at each node we determine α_1 and α_2

$$\alpha_1 = \frac{\phi_i x_j - \phi_j x_i}{L}; \quad \alpha_2 = \frac{\phi_j - \phi_i}{L} \tag{13.24}$$

and substituting α_1 and α_2 into (13.23) we have

$$\phi = \left(\frac{x_j - x}{L}\right)\phi_i + \left(\frac{x - x_i}{L}\right)\phi_j = N_i\phi_i + N_j\phi_j = [N]\{\phi\} \tag{13.25}$$

where N_i and N_j are known as the shape functions of the element and have the following properties:

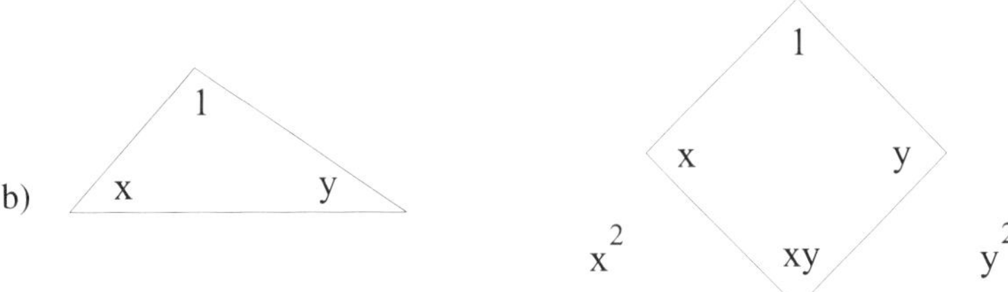

Figure 13.8: Interpolation functions. a) Pascal's triangle. b) Linear triangle and rectangle interpolation functions from Pascal's triangle.

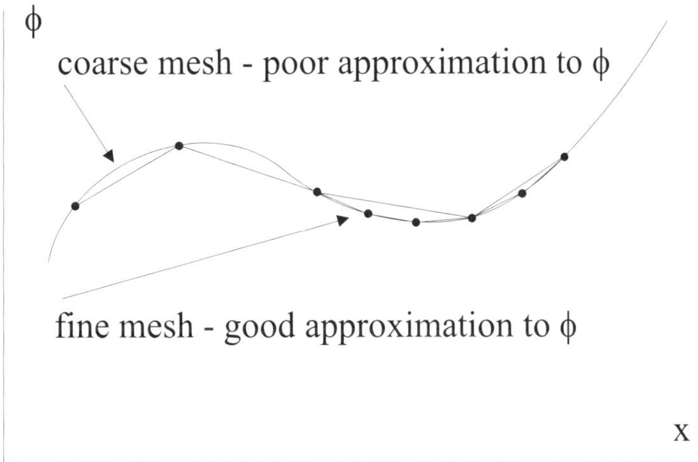

Figure 13.9: Coarse and fine approximations to ϕ.

- Each shape function is associated with a given node.
- The shape function is equal to unity at its associated node and zero at any other node, see Figure 13.13.
- The sum of all the element shape functions at any point within the element is equal to unity.
- The shape functions are positive and of the same order as the interpolation function.

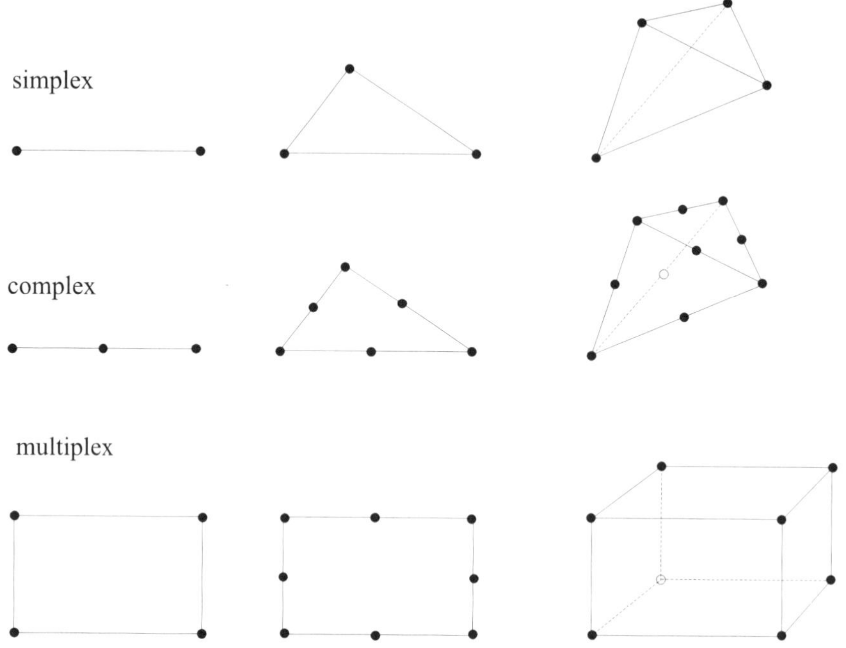

Figure 13.10: Simplex, complex and multiplex elements.

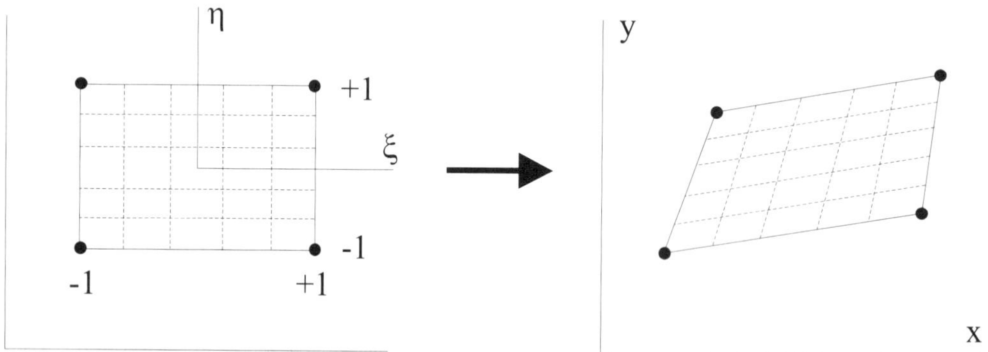

Figure 13.11: Transformation from local (ξ, η) to global (x, y) coordinates.

- The shape functions of an element are unique to that element. Figure 13.14 illustrates two adjacent triangle elements. Although elements 1 and 2 share the same node (2) the shape function for element 1, node 2 ($N_2^{(1)}$) is different from the shape function for element 2, node 2 ($N_2^{(2)}$).

13.10 Stress-Strain Relations

In the case of a stress analysis in which the solution vector is the structure displacements, $\{U\}$, then the strain, $\{\varepsilon\}$, and stress, $\{\sigma\}$, vectors need to be determined from $\{U\}$. In the case of an isotropic material

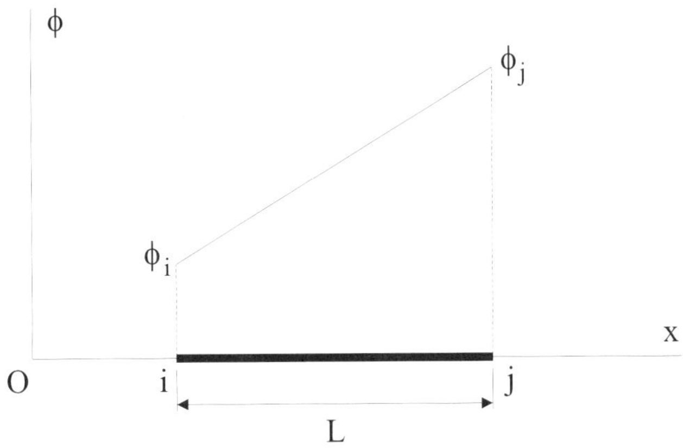

Figure 13.12: One-dimensional simplex bar element

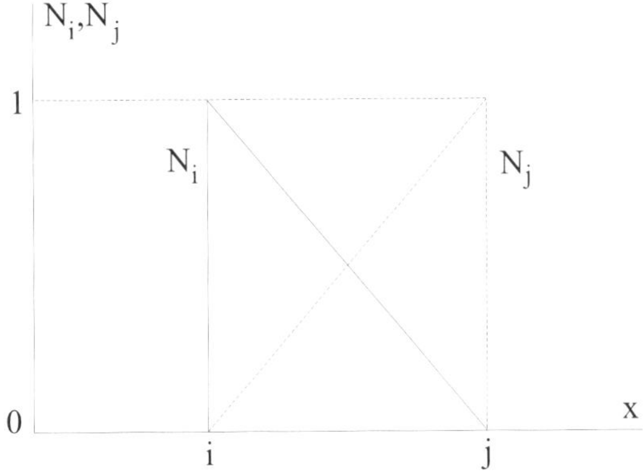

Figure 13.13: Variation of shape functions N_i and N_j with x for the one-dimensional simplex bar element.

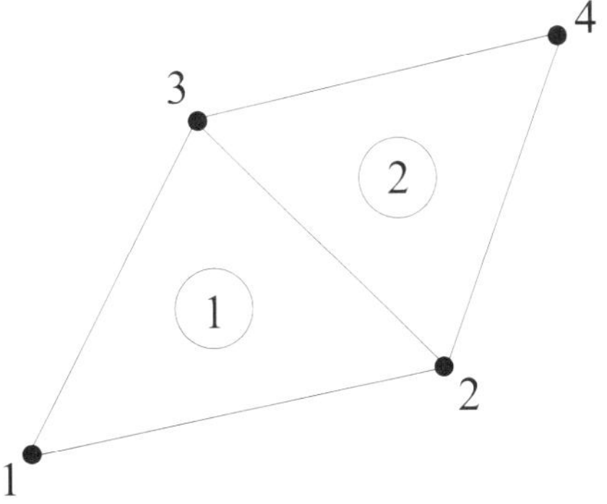

Figure 13.14: Two adjacent triangular elements sharing nodes 2 and 3.

the three-dimensional strain and stress vectors are defined as follows

$$\{\varepsilon\} = \begin{Bmatrix} \varepsilon_{xx} \\ \varepsilon_{yy} \\ \varepsilon_{zz} \\ \gamma_{xy} \\ \gamma_{yz} \\ \gamma_{xz} \end{Bmatrix}; \quad \{\sigma\} = \begin{Bmatrix} \sigma_{xx} \\ \sigma_{yy} \\ \sigma_{zz} \\ \tau_{xy} \\ \tau_{yz} \\ \tau_{xz} \end{Bmatrix} \tag{13.26}$$

noting that $\{\varepsilon\}$ is the total strain vector and therefore includes any initial strains. In the case of two-dimensional components then $\{\varepsilon\}$ and $\{\sigma\}$ reduce to

$$\{\varepsilon\} = \begin{bmatrix}\varepsilon_{xx} & \varepsilon_{yy} & \gamma_{xy}\end{bmatrix}^{\mathrm{T}}; \quad \{\sigma\} = \begin{bmatrix}\sigma_{xx} & \sigma_{yy} & \tau_{xy}\end{bmatrix}^{\mathrm{T}} \tag{13.27}$$

Let us also define an initial strain vector $\{\varepsilon_o\}$ which incorporates strains resulting from any initial deformation such as thermal strains

$$\{\varepsilon_o\} = \begin{bmatrix}\varepsilon_{xxo} & \varepsilon_{yyo} & \gamma_{xyo}\end{bmatrix}^{\mathrm{T}} \tag{13.28}$$

Once the displacement vector $\{U\}$ is known then the strains are found immediately from (1.31), (1.32) and (1.36)

$$\varepsilon_{xx} = \frac{\partial u}{\partial x}; \quad \varepsilon_{yy} = \frac{\partial v}{\partial y}; \quad \gamma_{xy} = \frac{\partial u}{\partial y} + \frac{\partial v}{\partial x} \tag{13.29}$$

or, in matrix form

$$\{\varepsilon\} = [B]\{U\} \tag{13.30}$$

The $\{\sigma\}$ vector for both plane stress and plane strain are considered separately in the following sub-sections.

13.10.1 Plane Stress

In the case of plane stress the Hookian equations are, (1.16)

$$\begin{aligned}
\varepsilon_{xx} &= \frac{1}{E}\left[\sigma_{xx} - \nu\sigma_{yy}\right] + \varepsilon_{xxo} \\
\varepsilon_{yy} &= \frac{1}{E}\left[\sigma_{yy} - \nu\sigma_{xx}\right] + \varepsilon_{yyo} \\
\gamma_{xy} &= \frac{\tau_{xy}}{G} = \frac{2(1+\nu)}{E}\tau_{xy}
\end{aligned} \tag{13.31}$$

In terms of matrices and inverting the above, the stress-strain relation is

$$\{\sigma\} = [D]\{\varepsilon\} - [D]\{\varepsilon_o\} = [D]\left(\{\varepsilon\} - \{\varepsilon_o\}\right) \tag{13.32}$$

where the D *matrix* is

$$[D] = \frac{E}{1-\nu^2}\begin{bmatrix} 1 & \nu & 0 \\ \nu & 1 & 0 \\ 0 & 0 & (1-\nu)/2 \end{bmatrix} \tag{13.33}$$

noting that $[D]$ is symmetric and contains the two material coefficients E and ν.

If an element is subject to a temperature increase of ΔT and the element material has a coefficient of thermal expansion of α then $\{\varepsilon_o\}$ is given by, (1.24)

$$\{\varepsilon_o\} = \alpha\Delta T \begin{Bmatrix} 1 \\ 1 \\ 0 \end{Bmatrix} \tag{13.34}$$

with $\gamma_{xy} = 0$ since there are no shear strains induced by a thermal dilatation.

13.10.2 Plane Strain

In the case of plane strain the normal stress σ_{zz} has also to be taken into account, (1.16)

$$\begin{aligned} \varepsilon_{xx} &= \frac{1}{E}\left[\sigma_{xx} - \nu(\sigma_{yy} + \sigma_{zz})\right] + \varepsilon_{xxo} \\ \varepsilon_{yy} &= \frac{1}{E}\left[\sigma_{yy} - \nu(\sigma_{xx} + \sigma_{zz})\right] + \varepsilon_{yyo} \\ \gamma_{xy} &= \frac{2(1+\nu)}{E}\tau_{xy} \end{aligned} \tag{13.35}$$

with the restraint $\varepsilon_{zz} = 0$. Eliminating σ_{zz} and solving for the in-plane stress components we find

$$[D] = \frac{E}{(1+\nu)(1-2\nu)} \begin{bmatrix} (1-\nu) & \nu & 0 \\ \nu & (1-\nu) & 0 \\ 0 & 0 & (1-2\nu)/2 \end{bmatrix}; \quad \{\varepsilon_o\} = (1+\nu)\alpha\Delta T \begin{Bmatrix} 1 \\ 1 \\ 0 \end{Bmatrix} \tag{13.36}$$

with the stress-strain relation (13.32) remaining the same.

13.11 Derivation of the Element Stiffness Matrix and Force Vector using a Variational Formulation

To derive the finite element equations of elasticity we begin by noting that the potential energy, Π, of an element is given by, (10.21)

$$\Pi = U - W \tag{13.37}$$

The total potential energy is equal to the strain energy, U, minus the external work done, W, by the externally applied forces. The following three sub-sections consider these three quantities separately.

13.11.1 Strain Energy

From §9.2 the strain energy of a linear elastic differential element is, (9.4)

$$\mathrm{d}U_e = \frac{1}{2}\left[\{\varepsilon\}^{\mathrm{T}}\{\sigma\} - \{\varepsilon_o\}^{\mathrm{T}}\{\sigma\}\right] \tag{13.38}$$

Let a one-dimensional bar element with cross-section A and length L have nodes i and j at coordinates x_i and x_j. Neglecting initial strain components and with $\{\varepsilon\} = \varepsilon_{xx}$ and $\{\sigma\} = \sigma_{xx}$ then the strain energy is given by

$$U_e = \frac{1}{2}\int_{x_i}^{x_j} \sigma_{xx}\varepsilon_{xx} A \,\mathrm{d}x \tag{13.39}$$

The strain vector $\{\varepsilon\}$ is given by (13.30) with $[B]$ equal to the first derivatives of $[N]$, (13.25)

$$[B] = \frac{\partial}{\partial x}[N] = \begin{bmatrix} \frac{\partial N_i}{\partial x} & \frac{\partial N_j}{\partial x} \end{bmatrix} = \begin{bmatrix} -\frac{1}{L} & \frac{1}{L} \end{bmatrix} \tag{13.40}$$

so that $\{\varepsilon\}$ is

$$\{\varepsilon\} = \varepsilon_{xx} = [B]\{U\} = \begin{bmatrix} -\frac{1}{L} & \frac{1}{L} \end{bmatrix} \begin{Bmatrix} u_i \\ u_j \end{Bmatrix} = \frac{-u_i + u_j}{L} \tag{13.41}$$

With $[D] = E$, (13.33), then $\{\sigma\} = \sigma_{xx} = E\sigma_{xx}$ and substituting ε_{xx} and σ_{xx} into (13.39)

$$U_e = \frac{1}{2}\int_{x_i}^{x_j} (E\varepsilon_{xx})\varepsilon_{xx} A \,\mathrm{d}x = \frac{EA}{2}\int_{x_i}^{x_j} \left(\frac{-u_i + u_j}{L}\right)^2 \mathrm{d}x = \frac{AE}{2L}(-u_i + u_j)^2 \tag{13.42}$$

or, in matrix form

$$U_e = \frac{AE}{2L}\begin{bmatrix} u_i & u_j \end{bmatrix} \begin{bmatrix} 1 & -1 \\ -1 & 1 \end{bmatrix} \begin{Bmatrix} u_i \\ u_j \end{Bmatrix} = \frac{1}{2}\{U\}^{\mathrm{T}}[K]\{U\} \tag{13.43}$$

where the stiffness matrix, $[K]$, is

$$[K] = \frac{AE}{L}\begin{bmatrix} 1 & -1 \\ -1 & 1 \end{bmatrix} \tag{13.44}$$

13.11.2 Force Vector

The only externally applied forces that can be applied to the pin-jointed element are the nodal forces P_i and P_j acting at nodes i and j respectively. The work done, W, by the external forces is therefore

$$W = u_i P_i + u_j P_j = \begin{bmatrix} u_i & u_j \end{bmatrix} \begin{Bmatrix} P_i \\ P_j \end{Bmatrix} = \{U\}^{\mathrm{T}}\{F\} \tag{13.45}$$

13.11.3 Total Potential Energy

Collecting the above results (13.37), (13.43) and (13.45) the total potential energy is

$$\Pi = U - W = \frac{1}{2}\{U\}^{\mathrm{T}}[K]\{U\} - \{U\}^{\mathrm{T}}\{F\} \tag{13.46}$$

For the element to be in a state of equilibrium we must have equivalently a state of minimum potential energy, §10.4. Thus, minimising the total potential energy with respect to the unknown displacements

$$\frac{\partial \Pi}{\partial u_i} = \frac{\partial \Pi}{\partial u_j} = 0 \quad \text{or} \quad \frac{\partial \Pi}{\partial \{U\}} = 0 \tag{13.47}$$

Differentiating (13.46) we have

$$\frac{\partial \Pi}{\partial \{U\}} = [K]\{U\} - \{F\} = 0 \quad \text{or} \quad [K]\{U\} = \{F\} \tag{13.48}$$

or, explicitly

$$\frac{AE}{L}\begin{bmatrix} 1 & -1 \\ -1 & 1 \end{bmatrix} \begin{Bmatrix} u_i \\ u_j \end{Bmatrix} = \begin{Bmatrix} P_i \\ P_j \end{Bmatrix} \tag{13.49}$$

For a structure consisting of E elements the total potential energy is

$$\Pi = \sum_{e=1}^{E} U_e - W \tag{13.50}$$

Denoting the structure displacement vector by $\{U\}_s$ and differentiating (13.50) with respect to $\{U\}_s$

$$\frac{\partial \Pi}{\partial \{U\}_s} = \left(\sum_{e=1}^{E} [K_e] \right) \{U\}_s - \{F\}_s = 0 \tag{13.51}$$

or

$$[K]_s \{U\}_s = \{F\}_s \tag{13.52}$$

and is the system of linear equations for the entire structure.

Let us conclude the present section with a note on the potential energy variational function. The variational functional for plane elasticity can be written in a general form using the potential energy of a body, (10.39)

$$\Pi(u) = \int_V \left(\frac{1}{2} \sigma_{ij} \varepsilon_{ij} - F_i u_i \right) \mathrm{d}V - \int_S T_i u_i \, \mathrm{d}S \tag{13.53}$$

where V is the volume of the body and S is the surface of the body over which the applied forces or tractions act. The body force and traction vectors are denoted by F_i and T_i respectively. Neglecting the body force term then (13.53) is seen to be a generalisation of (13.46). For one-dimensional bodies, neglecting body forces, $U = \sigma_{ij}\varepsilon_{ij}/2 = \sigma_{xx}\varepsilon_{xx}/2$ and the volume integral is seen to be equal to (13.42). Similarly, the surface integral is seen to be equal to (13.45) for the two nodal forces P_i and P_j.

13.12 Element Assembly

To date we have predominantly considered individual elements. By way of a simple example we will now illustrate how the structure system of equations is formed by the assembly of the element equations. An axially loaded bar structure is shown in Figure 13.15a) and a two-element model of the structure in Figure 13.15b). From (13.44) the stiffness coefficients associated with elements 1 and 2 are

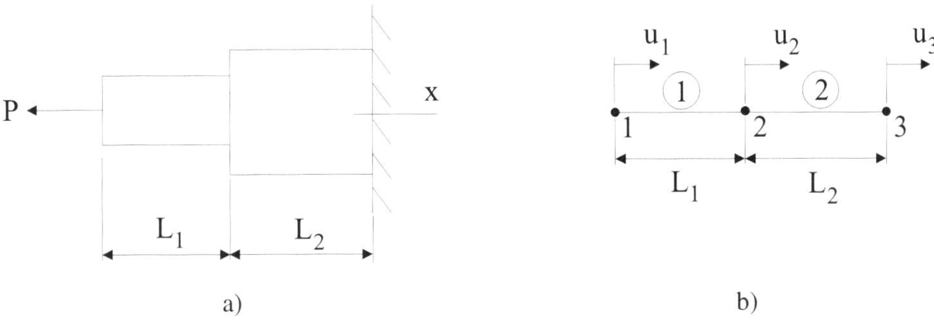

Figure 13.15: One-dimensional structure. a) Two bar structure. b) Finite element model.

$$K_1 = \frac{A_1 E_1}{L_1}; \quad K_2 = \frac{A_2 E_2}{L_2} \tag{13.54}$$

The degrees of freedom of the structure are the axial nodal displacements u_1, u_2 and u_3. The stiffness matrix for the structure will therefore be of size (3×3). Consider first element 1 and inserting into the structure stiffness matrix then we have

$$\begin{array}{c} \begin{matrix} u_1 & u_2 & u_3 \end{matrix} \\ \begin{bmatrix} K_1 & -K_1 & 0 \\ -K_1 & K_1 & 0 \\ 0 & 0 & 0 \end{bmatrix} \end{array} \begin{matrix} u_1 \\ u_2 \\ u_3 \end{matrix} \tag{13.55}$$

where the row and column headings indicate the degree of freedom associated with the matrix coefficients. Considering next element 2 independently

$$\begin{array}{c} \begin{matrix} u_1 & u_2 & u_3 \end{matrix} \\ \begin{bmatrix} 0 & 0 & 0 \\ 0 & K_2 & -K_2 \\ 0 & -K_2 & K_2 \end{bmatrix} \end{array} \begin{matrix} u_1 \\ u_2 \\ u_3 \end{matrix} \tag{13.56}$$

Combining the above element contributions by direct addition we generate the structure stiffness matrix

$$\begin{array}{c} \begin{matrix} u_1 & u_2 & u_3 \end{matrix} \\ \begin{bmatrix} K_1 & -K_1 & 0 \\ -K_1 & K_1 + K_2 & -K_2 \\ 0 & -K_2 & K_2 \end{bmatrix} \end{array} \begin{matrix} u_1 \\ u_2 \\ u_3 \end{matrix} \tag{13.57}$$

with no interaction between nodes 1 and 3. The structure stiffness matrix maintains all the properties of the element stiffness matrices in that the matrix is symmetric with the sum of row and column elements equal to zero.

At present $[K]_s$ is singular. Physically, this means that the structure is unsupported and can undergo a rigid body translation or rotation. As a result the structure must be restrained. To impose the displacement boundary condition we must enforce the constraint $u_3 = 0$. This is performed by discarding row 3 and column 3 from $[K]_s$ with the structure system of equations reducing to the following (2×2) system for the unknowns u_1 and u_2

$$\begin{bmatrix} -K_1 & -K_1 \\ -K_1 & K_1 + K_2 \end{bmatrix} \begin{Bmatrix} u_1 \\ u_2 \end{Bmatrix} = \begin{Bmatrix} -P \\ 0 \end{Bmatrix} \tag{13.58}$$

The right hand side force vector indicates that node 1 is subject to a point force of P acting in the negative x-direction, with node 2 carrying no applied forces. The structure stiffness matrix is now non-singular and can be solved for u_1 and u_2. Expanding (13.58)

$$\begin{aligned} K_1 u_1 - K_1 u_2 &= -P \\ -K_1 u_1 + (K_1 + K_2) u_2 &= 0 \end{aligned} \tag{13.59}$$

Solving (13.59) gives

$$u_1 = -\frac{P(K_1 + K_2)}{K_1 K_2}; \quad u_2 = -\frac{P}{K_2} \tag{13.60}$$

Clearly, for larger systems of simultaneous equations alternative automated solution techniques are required such as the Gaussian elimination direct method or the Gauss-Seidel iterative method. The element strains and stresses follow from (13.41) and (13.33)

$$\begin{aligned}\sigma_1 &= E_1\varepsilon_1 = E_1\left(\frac{u_2-u_1}{L_1}\right) = \frac{E_1}{L_1}\left(\frac{P}{A_1E_1/L_1}\right) = \frac{P}{A_1}\\ \sigma_2 &= E_2\varepsilon_2 = E_2\left(\frac{u_3-u_2}{L_2}\right) = \frac{E_2}{L_2}\left(\frac{P}{A_2E_2/L_2}\right) = \frac{P}{A_2}\end{aligned} \tag{13.61}$$

illustrating that the displacements (and consequently strains and stresses) are exact for this simple two-bar structure.

It is worth noting that prescribed boundary conditions are not always zero. When we discuss heat transfer and fluid flow in sections 13.14 and 13.15 the respective unknowns of temperature and potential are generally non-zero.

Example 13.1 Finite element analysis of a tapered bar

A tapered bar is built-in at one end and subject to an axial tensile force of 2kN at the free end, Figure 13.16. If the bar is modelled using two one-dimensional pin-jointed finite elements with linear interpolation functions then determine the axial displacements, strains and stresses at nodes 1, 2 and 3.

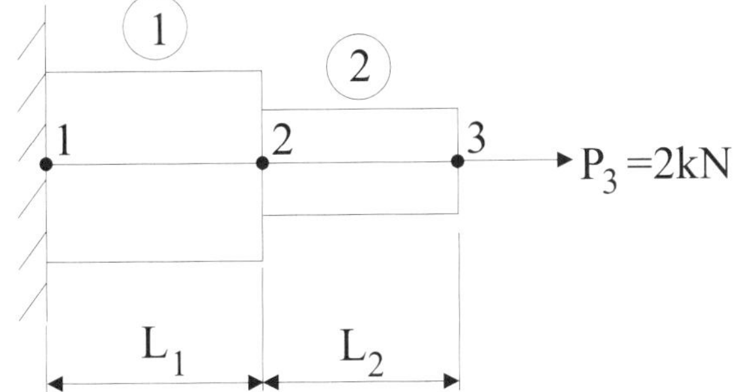

element 1	element 2
$L_1 = 50\text{mm}$	$L_2 = 50\text{mm}$
$A_1 = 80\text{mm}^2$	$A_2 = 40\text{mm}^2$
$E_1 = 175\text{GPa}$	$E_2 = 150\text{GPa}$

Figure 13.16: Example: A two bar approximation to a tapered bar built-in at one end and subject to a tensile force at the free end.

From (13.44) and (13.45) the element stiffness matrices and force vectors are

$$[K^1] = \frac{A_1E_1}{L_1}\begin{bmatrix}1 & -1\\ -1 & 1\end{bmatrix} = 280\times 10^3\begin{bmatrix}1 & -1\\ -1 & 1\end{bmatrix};\quad \{F^1\} = \begin{Bmatrix}0\\ 0\end{Bmatrix}$$

$$[K^2] = \frac{A_2E_2}{L_2}\begin{bmatrix}1 & -1\\ -1 & 1\end{bmatrix} = 120\times 10^3\begin{bmatrix}1 & -1\\ -1 & 1\end{bmatrix};\quad \{F^2\} = \begin{Bmatrix}0\\ 2000\end{Bmatrix}$$

The structure stiffness matrix and force vector are, using node ordering (1,2,3)

$$[K]_s = \begin{bmatrix}280 & -280 & 0\\ -280 & 280+120 & -120\\ 0 & -120 & 120\end{bmatrix}\times 10^3;\quad \{F\}_s = \begin{Bmatrix}0\\ 0\\ 2\end{Bmatrix}\times 10^3$$

Incorporating the boundary condition $u_1 = 0$ then the structure system of equations $\mathbf{K}_s\mathbf{U}_s = \mathbf{F}_s$ to be solved for $\mathbf{U}_s$ are

$$10^3\begin{bmatrix}280 & -280 & 0\\ -280 & 280+120 & -120\\ 0 & -120 & 120\end{bmatrix}\begin{Bmatrix}0\\ u_1\\ u_2\end{Bmatrix}\begin{Bmatrix}0+R_1\\ 0\\ 2\end{Bmatrix}\times 10^3$$

where R_1 is the reaction at node 1. Performing row multiplications we have

$$
\begin{aligned}
10^3(-280)u_2 &= R_1 \\
10^3(400u_2 - 120u_3) &= 0 \\
10^3(-120u_2 + 120u_3) &= 2 \times 10^3
\end{aligned}
$$

Solving these equations for $\mathbf{U}_s$ we find $u_1 = 0$, $u_2 = 1/140 \approx 7.1429 \times 10^{-3}$mm and $u_3 = 1/42 \approx 23.8095 \times 10^{-3}$mm. Both u_2 and u_3 are positive and $u_3 > u_2$ as expected. As a further check of the solution we can confirm that R_1 is equal and opposite to the applied force of 2kN

$$R_1 = -280 \times 10^3 u_2 = -2 \times 10^3\text{N}$$

From (13.41) the element strains are

$$
\begin{aligned}
\varepsilon^1_{xx} &= \frac{-u_1 + u_2}{L_1} = \frac{0 + 1/140}{50} = \frac{1}{7000} \approx 142.86 \times 10^{-6} \\
\varepsilon^2_{xx} &= \frac{-u_2 + u_3}{L_3} = \frac{-1/140 + 1/42}{50} = \frac{98}{294000} \approx 333.3\dot{3} \times 10^{-6}
\end{aligned}
$$

again, confirming that $\varepsilon^2_{xx} > \varepsilon^1_{xx}$. Finally, the stresses are, from (13.32)

$$
\begin{aligned}
\sigma^1_{xx} &= [D]\varepsilon^1_{xx} = E_1\varepsilon^1_{xx} = 175 \times 10^3 \frac{1}{7000} = 25\text{N/mm}^2 \\
\sigma^2_{xx} &= [D]\varepsilon^2_{xx} = E_2\varepsilon^2_{xx} = 175 \times 10^3 \frac{98}{294000} = 50\text{N/mm}^2
\end{aligned}
$$

and agree exactly with the known values $\sigma^1_{xx} = P_3/A_1 = 25\text{N/mm}^2$ and $\sigma^2_{xx} = P_3/A_2 = 50\text{N/mm}^2$.

13.13 Pin-Jointed Plane Frames

To date we have examined one-dimensional pin-jointed bar elements oriented so that the element is parallel to the global x-axis. We will now rotate the element by an angle θ in the anti-clockwise direction with respect to the x-axis, Figure 13.17. The axes (x, y) and (x', y') are referred to as the global and local axes respectively. To establish the global stiffness matrix and force vector from their local counterpart it is first necessary to examine coordinate transformations. Consider the vector $\mathbf{v}(v_x, v_y)$ with respect to the two coordinate systems (x, y) and (x', y') shown in Figure 13.18. The unit vectors in the positive (x, y) and (x', y') directions are denoted by $(\mathbf{i}, \mathbf{j})$ and $(\mathbf{i}', \mathbf{j}')$ respectively. Vector $\mathbf{v}$ is now given by, in both coordinate systems

$$\mathbf{v} = v_x\mathbf{i} + v_y\mathbf{j}; \quad \mathbf{v} = v'_x\mathbf{i}' + v'_y\mathbf{j}' \tag{13.62}$$

Multiplying both vectors of (13.62) by $\mathbf{i}'$ we have

$$\mathbf{i}' \cdot \mathbf{v} = v_x\mathbf{i}' \cdot \mathbf{i} + v_y\mathbf{i}' \cdot \mathbf{j}; \quad \mathbf{i}' \cdot \mathbf{v} = v'_x\mathbf{i}' \cdot \mathbf{i}' + v'_y\mathbf{i}' \cdot \mathbf{j}' \tag{13.63}$$

Since $\mathbf{i}' \cdot \mathbf{i}' = 1$ and $\mathbf{i}' \cdot \mathbf{j}' = 0$ then

$$\mathbf{i}' \cdot \mathbf{v} = v'_x \tag{13.64}$$

from which, and in view of (13.63)

$$v'_x = \mathbf{i}' \cdot \mathbf{i}\, v_x + \mathbf{i}' \cdot \mathbf{j}\, v_y \tag{13.65}$$

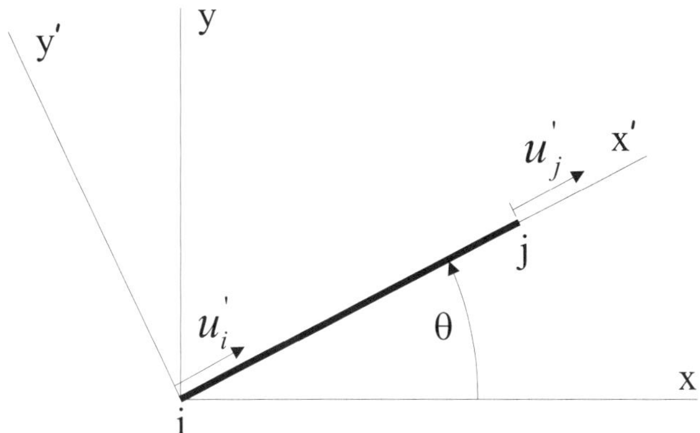

Figure 13.17: A two-dimensional bar element with local and global coordinate axes (x', y') and (x, y) respectively.

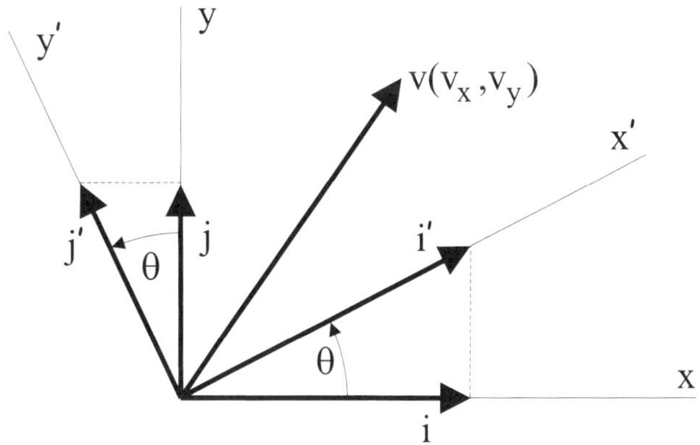

Figure 13.18: Vector **v** with respect to the coordinate systems (x, y) and (x', y').

and similarly for v'_y

$$v'_y = \mathbf{j}' \cdot \mathbf{i}\, v_x + \mathbf{j}' \cdot \mathbf{j}\, v_y \tag{13.66}$$

In matrix form and inspection of Figure 13.18 the transformation formula is

$$\begin{Bmatrix} v'_x \\ v'_y \end{Bmatrix} = \begin{bmatrix} \mathbf{i}' \cdot \mathbf{i} & \mathbf{i}' \cdot \mathbf{j} \\ \mathbf{j}' \cdot \mathbf{i} & \mathbf{j}' \cdot \mathbf{j} \end{bmatrix} \begin{Bmatrix} v_x \\ v_y \end{Bmatrix} = \begin{bmatrix} \cos\theta & \sin\theta \\ -\sin\theta & \cos\theta \end{bmatrix} \begin{Bmatrix} v_x \\ v_y \end{Bmatrix} \tag{13.67}$$

It is worth noting that for an element with nodes i and j at the global coordinates (x_i^0, y_i^0) and (x_j^0, y_j^0) then the length L and $\cos\theta$ and $\sin\theta$ are given by

$$L = \sqrt{(x_j^0 - y_i^0)^2 + (y_j^0 - y_i^0)^2}; \quad \cos\theta = \frac{x_j^0 - x_i^0}{L}; \quad \sin\theta = \frac{y_j^0 - y_i^0}{L} \tag{13.68}$$

From (13.49) the local stiffness matrix and force vector system of equations are

$$\frac{AE}{L} \begin{bmatrix} 1 & -1 \\ -1 & 1 \end{bmatrix} \begin{Bmatrix} u'_i \\ u'_j \end{Bmatrix} = \begin{Bmatrix} F'_i \\ F'_j \end{Bmatrix} \tag{13.69}$$

with displacement u' in the x'-axis. Similarly, let (u, v) denote the displacements in the global (x, y) axes, Figure 13.19. Thus, for the two nodes i and j there will be four components of nodal displacement

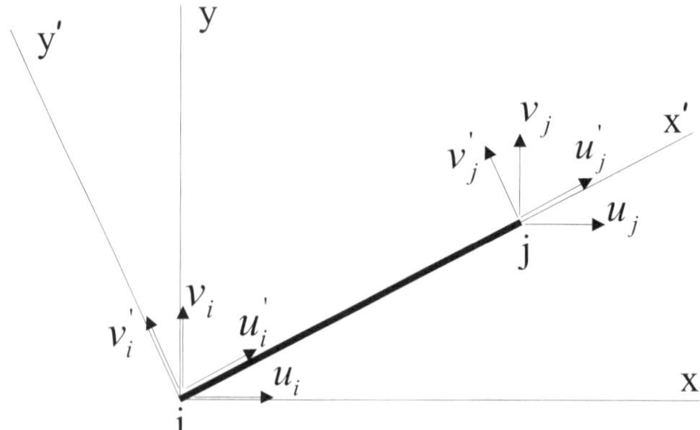

Figure 13.19: Local (u', v') and global (u, v) nodal displacements for a two-dimensional bar element.

and force for the global system. From (13.67) for nodes i and j

$$\begin{aligned} u_i' &= u_i \cos\theta + v_i \sin\theta \\ v_i' &= -u_i \sin\theta + v_i \cos\theta \\ u_j' &= u_j \cos\theta + v_j \sin\theta \\ v_j' &= -u_j \sin\theta + v_j \cos\theta \end{aligned} \tag{13.70}$$

or in matrix form, letting

$$\{U'\} = \begin{Bmatrix} u_i' \\ v_i' \\ u_j' \\ v_j' \end{Bmatrix}; \quad [T'] = \begin{bmatrix} \cos\theta & \sin\theta & 0 & 0 \\ -\sin\theta & \cos\theta & 0 & 0 \\ 0 & 0 & \cos\theta & \sin\theta \\ 0 & 0 & -\sin\theta & \cos\theta \end{bmatrix}; \quad \{U\} = \begin{Bmatrix} u_i \\ v_i \\ u_j \\ v_j \end{Bmatrix} \tag{13.71}$$

then (13.70) can be written as

$$\{U'\} = [T]\{U\} \tag{13.72}$$

Since forces are vectorial the global-local transformation of nodal forces is analogous to that of displacements

$$\{F'\} = \begin{Bmatrix} F_{ix}' \\ F_{iy}' \\ F_{jx}' \\ F_{jy}' \end{Bmatrix} = [T] \begin{Bmatrix} F_{ix} \\ F_{iy} \\ F_{jx} \\ F_{jy} \end{Bmatrix} = [T]\{F\} \tag{13.73}$$

Substituting (13.72) and (13.73) into (13.48) we have

$$[K'][T]\{U\} = [T]\{F\} \tag{13.74}$$

where $[K']$ is given by

$$\frac{AE}{L} \begin{bmatrix} 1 & 0 & -1 & 0 \\ 0 & 0 & 0 & 0 \\ -1 & 0 & 1 & 0 \\ 0 & 0 & 0 & 0 \end{bmatrix} \tag{13.75}$$

and is the local stiffness matrix of (13.44) expanded for both u' and v'. The global force vector $\{F\}$ is given by, (13.74)

$$\{F\} = [T]^{-1}[K'][T]\{U\} \tag{13.76}$$

where $[T]^{-1}$ is the inverse of $[T]$. Matrix $[T]$ is an orthogonal matrix for which its inverse is equal to its transpose, $[T]^{\mathrm{T}}$. Therefore

$$\{F\} = [T]^{\mathrm{T}}[K'][T]\{U\} \tag{13.77}$$

If the system is given by, (13.52)

$$\{F\} = [K]\{U\} \tag{13.78}$$

then equating to (13.77) we find that the global stiffness matrix is

$$[K] = [T]^{\mathrm{T}}[K'][T] \tag{13.79}$$

Performing the matrix multiplications, firstly $[K'][T]$

$$[K'][T] = \frac{AE}{L}\begin{bmatrix} 1 & 0 & -1 & 0 \\ 0 & 0 & 0 & 0 \\ -1 & 0 & 1 & 0 \\ 0 & 0 & 0 & 0 \end{bmatrix}\begin{bmatrix} C & S & 0 & 0 \\ -S & C & 0 & 0 \\ 0 & 0 & C & S \\ 0 & 0 & -S & C \end{bmatrix} = \frac{AE}{L}\begin{bmatrix} C & S & -C & -S \\ 0 & 0 & 0 & 0 \\ -C & -S & C & S \\ 0 & 0 & 0 & 0 \end{bmatrix} \tag{13.80}$$

where C and S denote $\cos\theta$ and $\sin\theta$ respectively. Finally, evaluating $[T]^{\mathrm{T}}([K'][T])$ we find

$$\begin{aligned} [T]^{\mathrm{T}}\left([K'][T]\right) &= \frac{AE}{L}\begin{bmatrix} C & -S & 0 & 0 \\ S & C & 0 & 0 \\ 0 & 0 & C & -S \\ 0 & 0 & S & C \end{bmatrix}\begin{bmatrix} C & S & -C & -S \\ 0 & 0 & 0 & 0 \\ -C & -S & C & S \\ 0 & 0 & 0 & 0 \end{bmatrix} \\ &= \frac{AE}{L}\begin{bmatrix} C^2 & CS & -C^2 & -CS \\ CS & S^2 & -CS & -S^2 \\ -C^2 & -CS & C^2 & CS \\ -CS & -S^2 & CS & S^2 \end{bmatrix} \end{aligned} \tag{13.81}$$

noting that $[K]$ remains symmetric.

13.13.1 Axial Strain and Stress

To determine the axial strain and stress in an element we require to use the local displacement vector $\{U'\}$. Extracting the axial displacements from (13.67))

$$\{U'\} = \left\{\begin{array}{c} u'_i \\ u'_j \end{array}\right\} = \begin{bmatrix} C & S & 0 & 0 \\ 0 & 0 & C & S \end{bmatrix}\left\{\begin{array}{c} u_i \\ v_i \\ u_j \\ v_j \end{array}\right\} \tag{13.82}$$

With $[B'] = [B] = [-1/L \; 1/L]$ from (13.40) the local strain vector is

$$\{\varepsilon'\} = [B']\{U'\} = \left[\; -\tfrac{1}{L} \quad \tfrac{1}{L} \;\right] = \begin{bmatrix} C & S & 0 & 0 \\ 0 & 0 & C & S \end{bmatrix}\left\{\begin{array}{c} u_i \\ v_i \\ u_j \\ v_j \end{array}\right\} = \frac{1}{L}\left[-C \quad -S \quad C \quad S\right]\{U\} \tag{13.83}$$

The axial stress, $\{\sigma\}$, is given by, from (13.32) and (13.83)

$$\{\sigma'\} = [D]\{\varepsilon'\} = \frac{E}{L}\begin{bmatrix}-C & -S & C & S\end{bmatrix}\{U\} \tag{13.84}$$

Example 13.2 Finite element analysis of a plane frame structure

Determine the displacement at point C and the member forces of the plane frame shown in Figure 13.20 which consists of two pin-jointed bar elements. The downward force acting at C is $P = 1$kN and the length of bar AC is 1m with $\angle ACB = 45°$. Both bars AC and BC are of solid circular cross-section with diameter D =25mm and have a Young's modulus of 210GPa.

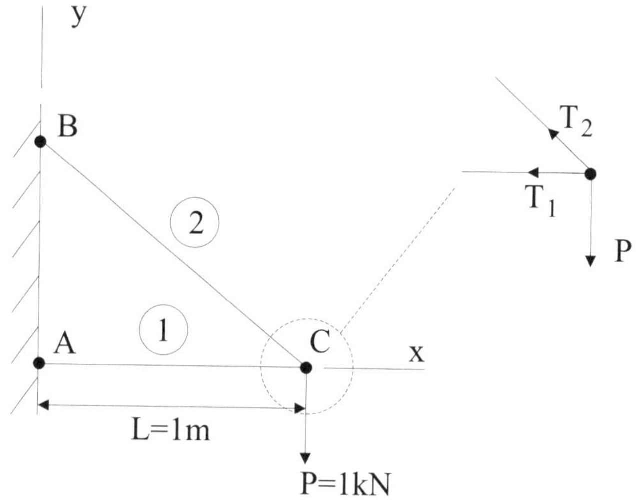

Figure 13.20: Example: A two bar plane frame with a force P acting at point C.

From (13.75) and (13.73) the stiffness matrix and force vector for element 1 are, with $A = \pi D^2/4 = 1.9635 \times 10^{-3}\text{m}^2$

$$[K^1] = 412.33 \times 10^6 \begin{bmatrix} 1 & 0 & -1 & 0 \\ 0 & 0 & 0 & 0 \\ -1 & 0 & 1 & 0 \\ 0 & 0 & 0 & 0 \end{bmatrix}; \quad \{F^1\} = \begin{Bmatrix} 0 \\ 0 \\ 0 \\ -1000 \end{Bmatrix}$$

whereas for element 2

$$[K^2] = 145.78 \times 10^6 \begin{bmatrix} 1 & -1 & -1 & 1 \\ -1 & 1 & 1 & -1 \\ -1 & 1 & 1 & -1 \\ 1 & -1 & -1 & 1 \end{bmatrix}; \quad \{F^2\} = \begin{Bmatrix} 0 \\ 0 \\ 0 \\ -1000 \end{Bmatrix}$$

Inserting the element stiffness matrices into the global matrix and inserting the boundary conditions

$u_1 = v_1 = u_2 = v_2 = 0$ we have

$$10^6 \begin{bmatrix} 412.33 & 0 & 0 & 0 & -412.33 & 0 \\ 0 & 0 & 0 & 0 & 0 & 0 \\ 0 & 0 & 145.78 & -145.78 & -145.78 & 145.78 \\ 0 & 0 & -145.78 & 145.78 & 145.78 & 145.78 \\ -412.33 & 0 & -145.78 & 145.78 & 558.11 & -145.78 \\ 0 & 0 & 145.78 & -145.78 & -145.78 & 145.78 \end{bmatrix} \begin{Bmatrix} 0 \\ 0 \\ 0 \\ 0 \\ u_3 \\ v_3 \end{Bmatrix} = \begin{Bmatrix} 0 \\ 0 \\ 0 \\ 0 \\ 0 \\ -1000 \end{Bmatrix}$$

with 1, 2 and 3 node ordering. Rows 5 and 6 of the system of equations yield

$$10^6[558.11u_3 - 145.78v_3] = 0$$
$$10^6[-145.78u_3 + 145.78v_3] = -1000$$

and solving for u_3 and v_3 gives $u_3 = -2.4253 \times 10^{-6}$m and $v_3 = -9.2850 \times 10^{-6}$m.

From (13.83) the axial strain for element 1 is

$$\{\varepsilon^1\} = \frac{1}{L_1}[-C \quad -S \quad C \quad S] \begin{Bmatrix} u_1 \\ v_1 \\ u_3 \\ v_3 \end{Bmatrix} = [-1\ 0\ 1\ 0] \begin{Bmatrix} 0 \\ 0 \\ -2.4253 \\ -9.2850 \end{Bmatrix} \times 10^{-6} = -2.4253 \times 10^{-6}$$

and for element 2

$$\{\varepsilon^2\} = \frac{1}{L_2}[-C \quad -S \quad C \quad S] \begin{Bmatrix} u_2 \\ v_2 \\ u_3 \\ v_3 \end{Bmatrix} = \frac{1}{\sqrt{2}}[-\frac{1}{\sqrt{2}} \ \frac{1}{\sqrt{2}} \ \frac{1}{\sqrt{2}} \ -\frac{1}{\sqrt{2}}] \begin{Bmatrix} 0 \\ 0 \\ -2.4253 \\ -9.2850 \end{Bmatrix} \times 10^{-6}$$
$$= 3.4298 \times 10^{-6}$$

From (13.84) the element axial stresses are

$$\{\sigma^1\} = E\{\varepsilon^1\} = 210 \times 10^9(-2.4253 \times 10^{-6}) = -0.5093\text{MPa}$$
$$\{\sigma^2\} = E\{\varepsilon^2\} = 210 \times 10^9(3.4298 \times 10^{-6}) = 0.7203\text{MPa}$$

Thus, the axial forces in the two elements are

$$F^1 = \sigma^1 A = -1\text{kN}; \quad F^2 = \sigma^2 A = 1.4142\text{kN}$$

These axial forces can be compared to the exact values by resolving forces, Figure 13.20. Resolving forces vertically

$$T_2 \cos 45° = P \quad \text{or} \quad T_2 = \sqrt{2}P$$

and resolving forces horizontally

$$T_1 + T_2 \cos° = 0 \quad \text{or} \quad T_1 = -P$$

Thus, the finite element solution is seen to be exact.

13.14 One-Dimensional Heat Transfer

In this section the finite element equations are derived for one-dimensional heat transfer. We will consider thermal effects due to conduction, convection and internal heat sources and sinks and begin by examining conduction.

13.14.1 Conduction

Figure 13.21 illustrates a plate of thickness $\mathrm{d}x$ and cross-sectional area A which lies perpendicular to the direction of flow. If the boundaries of the plate are fully insulated so that there is no loss through the boundaries then, by the conservation of energy

$$E_{in} + E_{generated} = \Delta U + E_{out} \tag{13.85}$$

where E denotes energy and ΔU is the change in internal stored energy. With q_x denoting the heat conducted (heat flux) into the plate and with Q denoting an internal heat source or sink then (13.85) gives

$$q_x A\,\mathrm{d}t + QA\,\mathrm{d}x\,\mathrm{d}t = \Delta U + q_{x+\mathrm{d}x} A\,\mathrm{d}t \tag{13.86}$$

for time $\mathrm{d}t$.

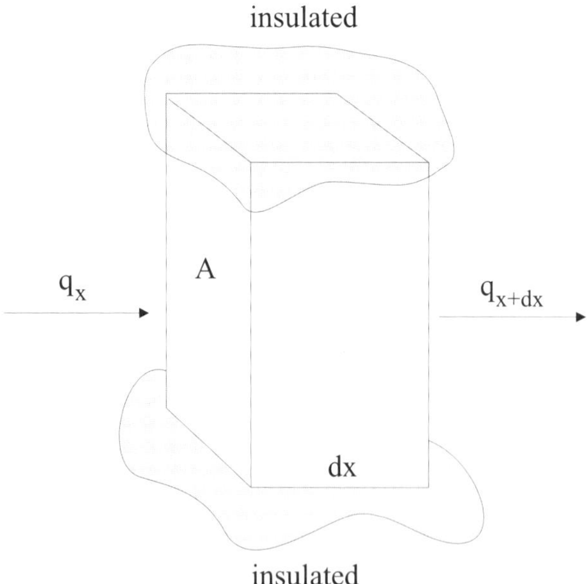

Figure 13.21: Heat conducted through an insulated plate.

Fourier's law of heat conduction states that the amount of heat flux is proportional to the area perpendicular to the direction of flow, x, time, t, and temperature, T, difference along the path of flow and inversely proportional to the length of the path

$$q_x = -k_{xx}\frac{\mathrm{d}T}{\mathrm{d}x} \tag{13.87}$$

where k_{xx} is the thermal conductivity of the material in the x-direction and has units of J/m or W/m per degree K or C. Table 13.1 lists typical values of thermal conductivities for both solids and fluids.

Material	k_{xx} **(W/m deg K)**
Gases at atmospheric pressure	0.007-0.17
Insulating materials	0.03-0.2
Non-metallic liquids	0.09-0.7
Non-metallic solids (e.g. concrete)	0.03-2.6
Liquid metals	9-80
Alloys	14-120
Pure metals	52-420

Table 13.1: Thermal conductivities for solids and fluids.

The negative sign in (13.87) indicates that heat transfer is positive in the direction of temperature drop. It is observed that Fourier's law is analogous to the one-dimensional stress-strain constitutive equation $\sigma_{xx} = E\varepsilon_{xx} = E(\,\mathrm{d}u/\,\mathrm{d}x)$. Similarly, on the plate face $x + \,\mathrm{d}x$

$$q_{x+\,\mathrm{d}x} = -k_{xx}\frac{\mathrm{d}T}{\mathrm{d}x} \tag{13.88}$$

Expanding $q_{x+\,\mathrm{d}x}$ using Taylor's series we have

$$q_{x+\,\mathrm{d}x} = -\left[k_{xx}\frac{\mathrm{d}T}{\mathrm{d}x} + \frac{\mathrm{d}}{\mathrm{d}x}\left(k_{xx}\frac{\mathrm{d}T}{\mathrm{d}x}\right)\,\mathrm{d}x + \cdots\right] \tag{13.89}$$

The drop in stored energy is

$$\Delta U = c(\rho A\,\mathrm{d}x)\,\mathrm{d}T \tag{13.90}$$

where c is the specific heat, $\rho A\,\mathrm{d}x$ is the mass and $\mathrm{d}T$ is the temperature change. Substituting (13.87), (13.89) and (13.90) into (13.86) we have

$$-k_{xx}\frac{\mathrm{d}T}{\mathrm{d}x}A\,\mathrm{d}t + QA\,\mathrm{d}x\,\mathrm{d}t = c\rho A\,\mathrm{d}x\,\mathrm{d}T - \left[k_{xx}\frac{\mathrm{d}T}{\mathrm{d}x} + \frac{\mathrm{d}}{\mathrm{d}x}\left(k_{xx}\frac{\mathrm{d}T}{\mathrm{d}x}\right)\,\mathrm{d}x\right]A\,\mathrm{d}t \tag{13.91}$$

which simplifies to the following governing second-order differential equation

$$\frac{\mathrm{d}}{\mathrm{d}x}\left(k_{xx}\frac{\mathrm{d}T}{\mathrm{d}x}\right) + Q = c\rho\frac{\mathrm{d}T}{\mathrm{d}t} \tag{13.92}$$

Under steady state conditions $(\,\mathrm{d}T/\,\mathrm{d}t = 0)$ equation (13.92) reduces to

$$\frac{\mathrm{d}}{\mathrm{d}x}\left(k_{xx}\frac{\mathrm{d}T}{\mathrm{d}x}\right) + Q = 0 \tag{13.93}$$

and if k_{xx} is a constant and not a function of x then

$$k_{xx}\frac{\mathrm{d}^2T}{\mathrm{d}x^2} + Q = 0 \tag{13.94}$$

Equation (13.94) can be compared to the one-dimensional governing second-order differential equation of elasticity

$$E\frac{\mathrm{d}^2u}{\mathrm{d}x^2} + F_x = 0 \tag{13.95}$$

where E is Young's modulus, u is the displacement and F_x is the x-component of the body force.

13.14.2 Heat Conduction Through Single and Composite Walls

Before developing the finite element expressions for heat conduction let us determine the heat conduction through both single and composite walls. These results will prove useful later when confirming the accuracy of the finite element approximation. For the single wall shown in Figure 13.22a) of thickness x and wall temperatures of T_{w1} and T_{w2} then from (13.87)

$$q_x = -kA\left(\frac{T_{w2} - T_{w1}}{x}\right) = kA\left(\frac{T_{w1} - T_{w2}}{x}\right) \tag{13.96}$$

noting that if q_x, k and A are all constant then $\mathrm{d}T/\mathrm{d}x$ is also constant, that is the temperature drops linearly with distance x.

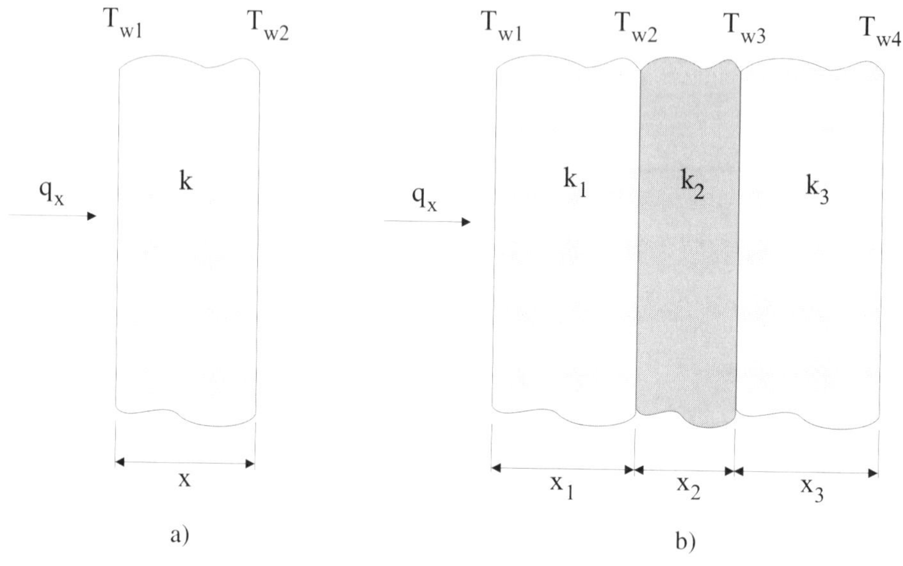

Figure 13.22: Conduction through a) a single wall and b) a composite wall.

For the composite wall of Figure 13.22b) consisting of three sections, q_x is the same for each section of the wall

$$q_x = \frac{k_1 A}{x_1}(T_{w1} - T_{w2}) = \frac{k_2 A}{x_2}(T_{w2} - T_{w3}) = \frac{k_3 A}{x_3}(T_{w3} - T_{w4}) \tag{13.97}$$

Re-arranging for the wall temperatures

$$T_{w1} - T_{w2} = \frac{q_x x_1}{k_1 A}; \quad T_{w2} - T_{w3} = \frac{q_x x_2}{k_2 A}; \quad T_{w3} - T_{w4} = \frac{q_x x_3}{k_3 A}; \tag{13.98}$$

Noting that the temperature difference between T_{w1} and T_{w4} is

$$T_{w1} - T_{w4} = (T_{w1} - T_{w2}) + (T_{w2} - T_{w3}) + (T_{w3} - T_{w4}) \tag{13.99}$$

and substituting (13.98)

$$T_{w1} - T_{w4} = q_x\left[\frac{x_1}{k_1 A} + \frac{x_2}{k_2 A} + \frac{x_3}{k_3 A}\right] = q_x \sum_{i=1}^{3} \frac{x_i}{k_i A} \tag{13.100}$$

Therefore, the heat flux through the composite wall is

$$q_x = \frac{T_{w1} - T_{w4}}{\sum_{i=1}^{3} \frac{x_i}{k_i A}} \tag{13.101}$$

which could be easily extended for n sections.

13.14.3 Convection

Heat transfer across a solid/fluid boundary occurs in the majority of heat transfer problems. Introducing the heat transfer or convection coefficient h (W/m^2 per degree K or C) then, by definition, the heat flow, q_h, by convection is

$$q_h = h(T - T_\infty) \tag{13.102}$$

where T_∞ is the temperature of the surrounding fluid. Denoting the perimeter of the area A by P then the conservation of energy expression (13.86) has now to be modified by the convection term $q_h P \,\mathrm{d}x\,\mathrm{d}t$

$$q_x A\,\mathrm{d}t + QA\,\mathrm{d}x\,\mathrm{d}t = \Delta U + q_{x+\mathrm{d}x} A\,\mathrm{d}t + q_h P\,\mathrm{d}x\,\mathrm{d}t \tag{13.103}$$

Following a similar procedure as in the derivation of (13.92) we arrive at the one-dimensional heat conduction and convection governing equation

$$\frac{\mathrm{d}}{\mathrm{d}x}\left(k_{xx}\frac{\mathrm{d}T}{\mathrm{d}x}\right) + Q = c\rho\frac{\mathrm{d}T}{\mathrm{d}t} + \frac{hP}{A}(T - T_\infty) \tag{13.104}$$

13.14.4 Derivation of the Element Stiffness Matrix and Force Vector using a Variational Formulation

The variational function, Π, that corresponds to the steady-state form of the equation (13.94) can be shown to be, Zienkiewicz and Taylor (1991)

$$\Pi(T) = \int_V \frac{1}{2}k_{xx}\left(\frac{\mathrm{d}T}{\mathrm{d}x}\right)^2 \mathrm{d}V - \int_V QT\,\mathrm{d}V + \int_{S_1} qT\,\mathrm{d}S + \int_{S_2} \frac{1}{2}h(T - T_\infty)^2\,\mathrm{d}S \tag{13.105}$$

where S_1 and S_2 are different surfaces over which conducted flow, q_c, and convective flow, q_h, occur. For the case of a one-dimensional two-noded element with a linear interpolation function the approximation temperature, T, at a point x in the element is, (13.25)

$$T = [N]\{T\} = [N_1\ N_2]\left\{\begin{array}{c} T_1 \\ T_2 \end{array}\right\} \tag{13.106}$$

and letting $[L] = \mathrm{d}/\mathrm{d}x$ and $[D] = k_{xx}$ then Π is

$$\begin{aligned}\Pi(T) = &\int_V \frac{1}{2}\{T\}^{\mathrm{T}}[N]^{\mathrm{T}}[L][D][L][N]\{T\}\,\mathrm{d}V - \int_V Q\{T\}^{\mathrm{T}}[N]^{\mathrm{T}}\,\mathrm{d}V + \\ &\int_{S_1} q\{T\}^{\mathrm{T}}[N]^{\mathrm{T}}\,\mathrm{d}S + \int_{S_2} \frac{1}{2}h\left[\{T\}^{\mathrm{T}}[N]^{\mathrm{T}}[N]\{T\} - \left(\{T\}^{\mathrm{T}}[N]^{\mathrm{T}} + [N]\{T\}\right)T_\infty + T_\infty^2\right]\mathrm{d}S\end{aligned} \tag{13.107}$$

Letting $[B] = [L][N]$ and minimising Π with respect to $\{T\}$

$$\begin{aligned}\frac{\partial \Pi(T)}{\partial \{T\}} = 0 = \int_V [B]^{\mathrm{T}}[D][B]\{T\}\,\mathrm{d}V - \int_V Q[N]^{\mathrm{T}}\,\mathrm{d}V \\ + \int_{S_1} q[N]^{\mathrm{T}}\,\mathrm{d}S + \int_{S_2} h\left[[N]^{\mathrm{T}}[N]\{T\} - [N]^{\mathrm{T}}T_\infty\right]\,\mathrm{d}S\end{aligned} \tag{13.108}$$

Denoting

$$\begin{aligned}[K] &= \int_V [B]^{\mathrm{T}}[D][B]\,\mathrm{d}V + \int_{S_2} h[N]^{\mathrm{T}}[N]\,\mathrm{d}S \\ \{F_Q\} &= \int_V Q[N]^{\mathrm{T}}\,\mathrm{d}V; \quad \{F_q\} = \int_{S_1} q[N]^{\mathrm{T}}\,\mathrm{d}S; \quad \{F_h\} = \int_{S_2} [N]^{\mathrm{T}}hT_\infty\,\mathrm{d}S\end{aligned} \tag{13.109}$$

then (13.108) can be written

$$[K]\{T\} = \{F\} = \{F_Q\} - \{F_q\} + \{F_h\} \tag{13.110}$$

and is analogous to the linear elastic bar element system of equations (13.48). The flux term is negative because heat is being removed from the system. To input heat into the system use a negative q.

With $N_1 = (x_2 - x)/L$ and $N_2 = (x - x_1)/L$ from (13.25) then (13.40) is

$$[B] = \left[\frac{\partial N_1}{\partial x}\ \frac{\partial N_2}{\partial x}\right] = \left[-\tfrac{1}{L}\quad \tfrac{1}{L}\right] \tag{13.111}$$

Since $[D] = k_{xx}$ is a matrix of one element in the one-dimensional element case and with $\mathrm{d}V = A\,\mathrm{d}x$ then the stiffness matrix is

$$\begin{aligned}[K] &= k_{xx}\int_L [B]^{\mathrm{T}}[B]A\,\mathrm{d}x + h\int_L [N]^{\mathrm{T}}[N]P\,\mathrm{d}x \\ &= k_{xx}A\int_L \left\{\begin{array}{c} -1/L \\ 1/L \end{array}\right\}[-1/L\ 1/L]\,\mathrm{d}x + hP\int_L \left\{\begin{array}{c} N_1 \\ N_2 \end{array}\right\}[N_1\ N_2]\,\mathrm{d}x \\ &= \frac{k_{xx}A}{L}\left[\begin{array}{cc} 1 & -1 \\ -1 & 1 \end{array}\right] + \frac{hPL}{6}\left[\begin{array}{cc} 2 & 1 \\ 1 & 2 \end{array}\right]\end{aligned} \tag{13.112}$$

with the first term equivalent to (13.44) with E replaced by k_{xx}.

Evaluating now the three terms of the force vector in (13.110), the first of which is the heat source/sink component

$$\{F_Q\} = \int_0^L [N]^{\mathrm{T}}QA\,\mathrm{d}x = \int_0^L \left\{\begin{array}{c} \frac{x_2 - x}{L} \\ \frac{x - x_1}{L} \end{array}\right\} QA\,\mathrm{d}x = \int_0^L \left\{\begin{array}{c} \frac{L-x}{L} \\ \frac{x}{L} \end{array}\right\} QA\,\mathrm{d}x = \frac{QAL}{2}\left\{\begin{array}{c} 1 \\ 1 \end{array}\right\} \tag{13.113}$$

The integration of $[N]^{\mathrm{T}}$ is identical for the remaining two components so that we have

$$\{F_q\} = \int_0^L [N]^{\mathrm{T}}qP\,\mathrm{d}x = \frac{qPL}{2}\left\{\begin{array}{c} 1 \\ 1 \end{array}\right\}; \quad \{F_h\} = \int_0^L [N]^{\mathrm{T}}hT_\infty P\,\mathrm{d}x = \frac{hT_\infty PL}{2}\left\{\begin{array}{c} 1 \\ 1 \end{array}\right\} \tag{13.114}$$

In addition to the above, we must also account for flux and convection at the free ends of the element. Assuming that heat flux occurs at node 2 of the element then the force vector is modified by, from $\{F_q\}$

$$-\int_A q[N]^{\mathrm{T}}\,\mathrm{d}A = -\int_A q\left\{\begin{array}{c} 0 \\ 1 \end{array}\right\}\,\mathrm{d}A = -qA\left\{\begin{array}{c} 0 \\ 1 \end{array}\right\} \tag{13.115}$$

since $N_1 = 0$ and $N_2 = 1$ at node 2. No modification to the stiffness matrix needs to be made for end heat flux. In the case of end convection the following contribution needs to be added to the stiffness matrix, from (13.112)

$$\int_A h[N]^{\mathrm{T}}[N]\,\mathrm{d}A = \int_A h\left\{\begin{array}{c}0\\1\end{array}\right\}[0\ 1]\,\mathrm{d}A = hA\begin{bmatrix}0 & 0\\0 & 1\end{bmatrix} \tag{13.116}$$

with the force vector modified by, from $\{F_h\}$

$$\int_A [N]^{\mathrm{T}} hT_\infty\,\mathrm{d}A = \int_A \left\{\begin{array}{c}0\\1\end{array}\right\} hT_\infty\,\mathrm{d}A = hT_\infty A\left\{\begin{array}{c}0\\1\end{array}\right\} \tag{13.117}$$

Summarising the above results, the stiffness matrix is given by, with end convection at node 2 only

$$\frac{k_{xx}A}{L}\left[\begin{array}{cc}1 & -1\\-1 & 1\end{array}\right] + \frac{hPL}{6}\left[\begin{array}{cc}2 & 1\\1 & 2\end{array}\right] + hA\left[\begin{array}{cc}0 & 0\\0 & 1\end{array}\right] \tag{13.118}$$

while the force vector is given by, with both heat flux and convection at node 2 only

$$\frac{QAL}{2}\left\{\begin{array}{c}1\\1\end{array}\right\} + \frac{qPL}{2}\left\{\begin{array}{c}1\\1\end{array}\right\} + \frac{hT_\infty PL}{2}\left\{\begin{array}{c}1\\1\end{array}\right\} - qA\left\{\begin{array}{c}0\\1\end{array}\right\} + hT_\infty A\left\{\begin{array}{c}0\\1\end{array}\right\} \tag{13.119}$$

Example 13.3 Finite element analysis of the temperature distribution throughout a bar

A one-dimensional bar of length $L = 75\text{mm}$ and circular cross-section of radius $R = 25\text{mm}$ is subject to a temperature of 250°C at one-end and free at the other end, Figure 13.23. Modelling the bar using three linear elements of equal length then determine the temperature distribution throughout the bar. Assume that the thermal conductivity, k_{xx}, is 35W/m°C, the convection coefficient, h, is 5W/m^2°C and the background temperature, T_∞, is 21°C.

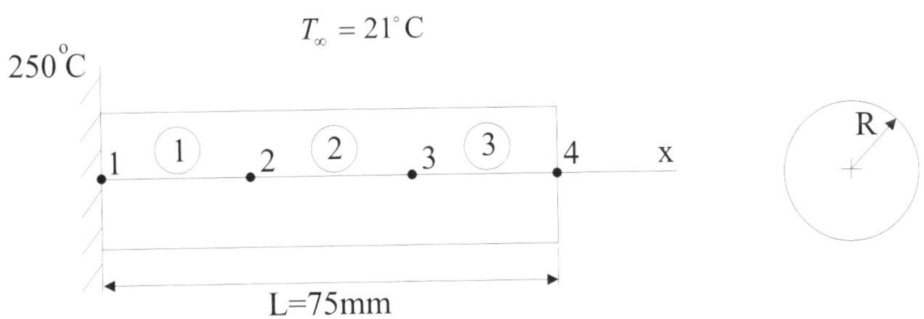

Figure 13.23: Example: A bar of circular cross-section subject to a temperature of 250°C at the built-in end. The bar is modelled using three linear bar elements of equal length.

The cross-sectional area, $A = \pi R^2$, and perimeter, $P = 2\pi R$, for all three elements are $1.96 \times 10^{-3}\text{m}^2$ and 0.1571m respectively. The stiffness matrix for element 1 has contributions due to conduction and perimeter convection, (13.118)

$$[K^1] = \frac{k_{xx}A}{L_1}\left[\begin{array}{cc}1 & -1\\-1 & 1\end{array}\right] + \frac{hPL_1}{6}\left[\begin{array}{cc}2 & 1\\1 & 2\end{array}\right] = \left[\begin{array}{cc}2.7554 & -2.7456\\-2.7456 & 2.7554\end{array}\right]$$

with the force vector due to $\{F_h\}$ only, (13.119)

$$\{F^1\} = \frac{hT_\infty PL_1}{2}\left\{\begin{array}{c}1\\1\end{array}\right\} = 0.2062\left\{\begin{array}{c}1\\1\end{array}\right\}$$

Elements 2 and 3 are equivalent to element 1 except that element 3 also experiences end-convection through node 4. Thus, from (13.118) the stiffness matrix of element 3 is

$$[K^3] = \left[\begin{array}{cc}2.7554 & -2.7456\\ -2.7456 & 2.7554\end{array}\right] + hA\left[\begin{array}{cc}0 & 0\\ 0 & 1\end{array}\right] = \left[\begin{array}{cc}2.7544 & -2.7456\\ -2.7456 & 2.7653\end{array}\right]$$

Similarly, adding the end-convection term to the force vector of element 3 we have

$$\{F^3\} = 0.2062\left\{\begin{array}{c}1\\1\end{array}\right\} + hT_\infty A\left\{\begin{array}{c}0\\1\end{array}\right\} = \left\{\begin{array}{c}0.2062\\0.412\end{array}\right\}$$

Assembling the element contributions into the structure stiffness matrix and force vector we find

$$\left[\begin{array}{cccc}2.7554 & -2.7456 & 0 & 0\\ -2.7456 & 5.5109 & -2.7456 & 0\\ 0 & -2.7456 & 5.5109 & -2.7456\\ 0 & 0 & -2.7456 & 2.7653\end{array}\right]\left\{\begin{array}{c}T_1\\T_2\\T_3\\T_4\end{array}\right\} = \left\{\begin{array}{c}0.2062\\0.4124\\0.4124\\0.4124\end{array}\right\}$$

with node ordering 1 to 4. We have a prescribed temperature of 250°C at node 1 which results in a non-homogeneous boundary condition. The stiffness matrix and force vector is modified by first setting all non-diagonal terms in the first row and column of the stiffness matrix to zero. Also, the term $(-2.7456) \times$ 250°C $= -3,186.4$ on the left hand side of the second equation is transposed to the right hand side as +3,186.4. The resulting system of equations is now given by

$$\left[\begin{array}{cccc}1 & 0 & 0 & 0\\ 0 & 5.5109 & -2.7456 & 0\\ 0 & -2.7456 & 5.5109 & -2.7456\\ 0 & 0 & -2.7456 & 2.7653\end{array}\right]\left\{\begin{array}{c}T_1\\T_2\\T_3\\T_4\end{array}\right\} = \left\{\begin{array}{c}250\\0.4124 + 3,186.4\\0.4124\\0.4124\end{array}\right\}$$

The second through to fourth equations are now solved in the usual manner, with the solution vector given by $\{T\} = \{250, 245.2453, 242.0943, 240.5244\}$.

The element heat flux, q_x, from Fourier's law is

$$q_x = -k_{xx}\frac{\mathrm{d}T}{\mathrm{d}x} = -k_{xx}[B]\{T\}$$

and is constant for an element since $[B]$ is constant for the linear element. Thus, the heat flux and flow (q_x, Q_x) for elements 1, 2 and 3 are (6656.5643,13.0701), (4411.4384,8.6618) and (2197.8603,4.3155) respectively.

Example 13.4 Finite element analysis of the temperature distribution throughout a composite wall

A composite wall consists of two layers of material with thermal conductivity equal to 9W/mm°C and thickness 100mm separated by a layer of insulation of thickness 100mm and with thermal conductivity of 4.5W/mm°C as shown in Figure 13.24. Determine the heat flow through the wall if the inside wall surface is at 21°C and the outside temperature is at 5°C with a convection coefficient of 50W/mm°C.

All three elements experience no perimeter convection ($P = 0$) with element 3 experiencing convection at node 4. With a unit cross-sectional area for all three elements then the stiffness matrix and force vector for element 1 are, (13.118) and (13.119)

$$[K^1] = \frac{k_{xx}A}{L_1}\left[\begin{array}{cc}1 & -1\\ -1 & 1\end{array}\right] = 0.09\left[\begin{array}{cc}1 & -1\\ -1 & 1\end{array}\right]; \quad \{F^1\} = \left\{\begin{array}{c}0\\0\end{array}\right\}$$

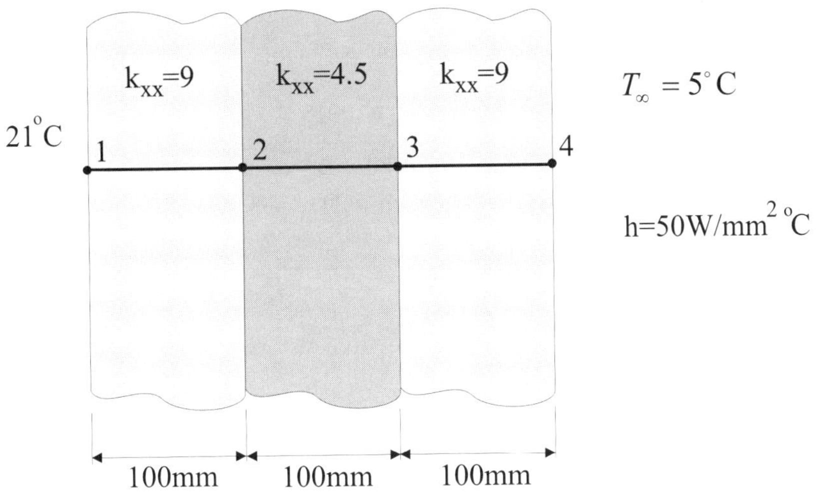

Figure 13.24: Example: A composite wall consisting of three layers of equal thickness subject to an internal temperature of 21°C.

and similarly for element 2

$$[K^2] = \frac{k_{xx}A}{L_2}\begin{bmatrix} 1 & -1 \\ -1 & 1 \end{bmatrix} = 0.045\begin{bmatrix} 1 & -1 \\ -1 & 1 \end{bmatrix}; \quad \{F^2\} = \begin{Bmatrix} 0 \\ 0 \end{Bmatrix}$$

whereas element 3 consists of the additional end-convection term

$$[K^3] = \frac{k_{xx}A}{L_3}\begin{bmatrix} 1 & -1 \\ -1 & 1 \end{bmatrix} + hA\begin{bmatrix} 0 & 0 \\ 0 & 1 \end{bmatrix} = \begin{bmatrix} 0.09 & -0.09 \\ -0.09 & 50.09 \end{bmatrix}; \quad \{F^3\} = hT_\infty A\begin{Bmatrix} 0 \\ 1 \end{Bmatrix} = \begin{Bmatrix} 0 \\ 250 \end{Bmatrix}$$

Assembly of the elements into the structure stiffness matrix and force vector gives

$$\begin{bmatrix} 0.09 & -0.09 & 0 & 0 \\ -0.09 & 0.135 & -0.045 & 0 \\ 0 & -0.045 & 0.135 & -0.09 \\ 0 & 0 & -0.09 & 50.09 \end{bmatrix}\begin{Bmatrix} T_1 \\ T_2 \\ T_3 \\ T_4 \end{Bmatrix} = \begin{Bmatrix} 0 \\ 0 \\ 0 \\ 250 \end{Bmatrix}$$

with node ordering 1 to 4. Incorporating the prescribed boundary condition T_1 =21°C then the system of equations is modified as follows

$$\begin{bmatrix} 1 & 0 & 0 & 0 \\ 0 & 0.135 & -0.045 & 0 \\ 0 & -0.045 & 0.135 & -0.09 \\ 0 & 0 & -0.09 & 50.09 \end{bmatrix}\begin{Bmatrix} T_1 \\ T_2 \\ T_3 \\ T_4 \end{Bmatrix} = \begin{Bmatrix} 21 \\ 1.89 \\ 0 \\ 250 \end{Bmatrix}$$

where the right hand side of the first equation is set to 21°C. The term $(-0.09) \times 21°\mathrm{C} = -1.89$ on the left hand side of the second equation is transposed to the right hand side as $+1.89$. Solution of the system of equations yields $\{T\} = \{21, 17.0018, 9.0054, 5.0072\}$. The heat flux, q_x, from Fourier's law (13.87) and (13.111) for an element of length L and nodes i and j is

$$q_x = -k_{xx}\frac{\mathrm{d}T}{\mathrm{d}x} = k_{xx}[B]\{T\} = -k_{xx}\left[-\tfrac{1}{L}\,\tfrac{1}{L}\right]\begin{Bmatrix} T_i \\ T_j \end{Bmatrix}$$

For example, for element 1 then q_x is

$$q_x = -9\left[-\tfrac{1}{100}\ \tfrac{1}{100}\right]\begin{Bmatrix} 21 \\ 17.0018 \end{Bmatrix} = 0.36$$

and with $Q_x = q_x A = q_x$. Evaluating q_x for elements 2 and 3 we find that the heat flow is constant for all three elements. This value of Q_x can be compared with the exact estimate (13.101)

$$Q_x = q_x A = q_x = \frac{T_{w1} - T_{w4}}{\sum_{i=1}^{3} \frac{x_i}{k_i}} = \frac{21 - 5}{\frac{100}{9} + \frac{100}{4.5} + \frac{100}{9}} = 0.36$$

The finite element solution is exact because the temperature distribution through the wall is linear with distance x.

13.15 One-Dimensional Fluid Flow

In this final section we examine one-dimensional fluid flow through porous media. As with heat transfer the finite element formulation of heat transfer will be seen to be analogous to that of the elasticity formulation derived in §13.11. Figure 13.25 illustrates a control volume of thickness $\mathrm{d}x$ and cross-sectional area A. If the boundaries of the volume are impermeable then by conservation of mass we have

$$M_{in} + M_{generated} = M_{out} \tag{13.120}$$

where M denotes mass. Letting v_x denote the velocity of the fluid flow in the x-direction, ρ the mass density of the fluid and Q denote an internal fluid source then (13.120) gives

$$\rho v_x A\,\mathrm{d}t + \rho Q\,\mathrm{d}t = \rho v_{x+\mathrm{d}x} A\,\mathrm{d}t \tag{13.121}$$

for time $\mathrm{d}t$.

Darcy's law of fluid flow relates the velocity to the fluid gradient in the fluid head, ϕ

$$v_x = -k_{xx}\frac{\mathrm{d}\phi}{\mathrm{d}x} \tag{13.122}$$

where k_{xx} is the permeability coefficient of the porous media in the x-direction and has units of m/s. Typical values of k_{xx} for granular materials are listed in Table 13.2. For $k_{xx} < 10^{-9}$m/s the material is essentially impermeable. When considering ideal flow through a pipe or over a solid then k_{xx} can be set arbitrarily and for convenience is generally set to unity. The negative sign in (13.122) indicates that fluid flow is positive in the direction of fluid head drop. Note the similarity between Darcy's law and the one-dimensional stress-strain constitutive equation $\sigma_{xx} = E\varepsilon_{xx} = E(\mathrm{d}u/\mathrm{d}x)$ and Fourier's law (13.87). On the face $x + \mathrm{d}x$ we have

$$v_{x+\mathrm{d}x} = -k_{xx}\frac{\mathrm{d}\phi}{\mathrm{d}x} \tag{13.123}$$

Expanding $v_{x+\mathrm{d}x}$ using Taylor's series

$$v_{x+\mathrm{d}x} = -\left[k_{xx}\frac{\mathrm{d}\phi}{\mathrm{d}x} + \frac{\mathrm{d}}{\mathrm{d}x}\left(k_{xx}\frac{\mathrm{d}\phi}{\mathrm{d}x}\right)\mathrm{d}x + \cdots\right] \tag{13.124}$$

Substituting (13.122) and (13.124) into (13.121) we have

$$\rho k_{xx}\frac{\mathrm{d}\phi}{\mathrm{d}x}A\,\mathrm{d}t + \rho Q\,\mathrm{d}t = -\rho\left[k_{xx}\frac{\mathrm{d}\phi}{\mathrm{d}x} + \frac{\mathrm{d}}{\mathrm{d}x}\left(k_{xx}\frac{\mathrm{d}\phi}{\mathrm{d}x}\right)\mathrm{d}x + \cdots\right]A\,\mathrm{d}t \tag{13.125}$$

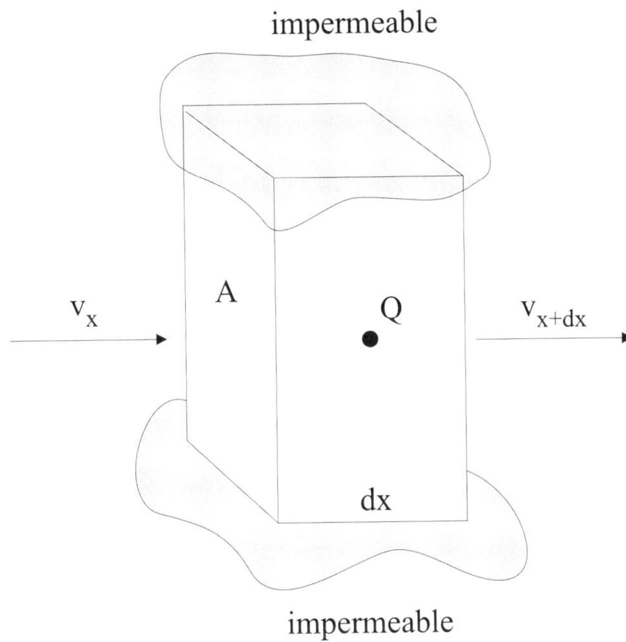

Figure 13.25: Fluid flow through a control volume with impermeable boundaries.

which simplifies to the following governing second-order differential equation

$$\frac{\mathrm{d}}{\mathrm{d}x}\left(k_{xx}\frac{\mathrm{d}\phi}{\mathrm{d}x}\right)+\frac{Q}{A\,\mathrm{d}x}=0 \tag{13.126}$$

If k_{xx} is a constant then

$$k_{xx}\frac{d^2\phi}{\mathrm{d}x^2}+\frac{Q}{A\,\mathrm{d}x}=0 \tag{13.127}$$

Material	k_{xx}**(m/s)**
clay	1×10^{-10}
sandy clay	1×10^{-5}
coarse gravel	1×10^{-2}

Table 13.2: Permeability coefficients for granular materials.

13.15.1 Derivation of the Element Stiffness Matrix and Force Vector

Let a one-dimensional bar element with cross-sectional area A and length L have nodes i and j at coordinates x_i and x_j. The nodal potentials or fluid heads at i and j are denoted by p_i and p_j. The potential function ϕ is now given by, (13.25)

$$\phi = N_i p_i + N_j p_j = [N_i \ N_j] \left\{ \begin{array}{c} p_i \\ p_j \end{array} \right\} \tag{13.128}$$

The fluid head gradient is

$$\{g\} = \left\{ \frac{\mathrm{d}\phi}{\mathrm{d}x} \right\} = \frac{\partial}{\partial x}[N]\{P\} = [B]\{P\} \tag{13.129}$$

with $[B]$ given by (13.40). From Darcy's law, (13.122), the velocity is given by

$$v_x = -[D]\{g\} \tag{13.130}$$

where $[D] = k_{xx}$.

To determine the stiffness matrix and force vector we note that the stiffness matrix for the elasticity problem of §13.11 relates the nodal forces to the nodal displacements, whereas for the heat transfer problem of §13.14 the stiffness matrix relates the nodal heat flux to the nodal temperatures. In the present fluid flow problem the stiffness matrix will relate the nodal fluid flow rates to the nodal potentials of fluid heads. The volumetric flow rate, f, of the element is equal to

$$f = v_x A \tag{13.131}$$

and substituting (13.122)

$$f = -k_{xx} g A; \quad g = \frac{p_j - p_i}{L} \tag{13.132}$$

Applying (13.132) to nodes i and j then

$$f_i = -Ak_{xx}\left(\frac{p_j - p_i}{L}\right); \quad f_j = -Ak_{xx}\left(\frac{p_j - p_i}{L}\right) \tag{13.133}$$

with f_i and f_j flowing in and out of the element respectively. Expressing (13.133) in matrix form

$$\frac{Ak_{xx}}{L}\left[\begin{array}{cc} 1 & -1 \\ -1 & 1 \end{array} \right] \left\{ \begin{array}{c} p_i \\ p_j \end{array} \right\} = \left\{ \begin{array}{c} f_i \\ f_j \end{array} \right\} \tag{13.134}$$

or, more concisely

$$[K]\{P\} = \{F\} \tag{13.135}$$

where the stiffness matrix $[K]$, potential vector $\{P\}$ and fluid head vector $\{F\}$ are

$$[K] = \frac{Ak_{xx}}{L}\left[\begin{array}{cc} 1 & -1 \\ -1 & 1 \end{array} \right]; \quad \{P\} = \left\{ \begin{array}{c} p_i \\ p_j \end{array} \right\}; \quad \{F\} = \left\{ \begin{array}{c} f_i \\ f_j \end{array} \right\} \tag{13.136}$$

The element may be subject to both an internal source/sink, Q, acting throughout the entire volume V of the element and surface flow rate, q, acting over the entire surface S of the element. In an analogous manner to the heat source/sink, $\{F_Q\}$, and conduction, $\{F_q\}$, components of heat transfer, (13.109), we have

$$\{F_Q\} = \int_V Q[N]^{\mathrm{T}}\,\mathrm{d}V; \quad \{F_q\} = \int_S q[N]^{\mathrm{T}}\,\mathrm{d}S \tag{13.137}$$

Evaluating the integrals in an analogous manner to (13.113) and (13.114) we have

$$\{F_Q\} = \frac{QAL}{2}\left\{ \begin{array}{c} 1 \\ 1 \end{array} \right\}; \quad \{F_q\} = \frac{qLt}{2}\left\{ \begin{array}{c} 1 \\ 1 \end{array} \right\} \tag{13.138}$$

Example 13.5 Finite element analysis of the fluid flow through a pipe

Figure 13.26 illustrates a smooth pipe of diameter 50mm. The velocity of the fluid entering the pipe at the left hand end is 5m/s. By discretising the pipe using two one-dimensional finite elements then determine the velocities at the centre and right hand end of the pipe.

The cross-sectional area, A, of the pipe is

$$A = \frac{\pi(50 \times 10^{-3})^2}{4} = 1.96 \times 10^{-3}\text{m}^2$$

Setting $k_{xx} = 1$ then the stiffness matrices and flow rate vectors for both elements are, (13.136)

$$[K^1] = [K^2] = \frac{1.96 \times 10^{-3}(1)}{0.75}\begin{bmatrix} 1 & -1 \\ -1 & 1 \end{bmatrix}; \quad \{F^1\} = \begin{Bmatrix} f_1 \\ f_2 \end{Bmatrix}; \quad \{F^2\} = \begin{Bmatrix} f_2 \\ f_3 \end{Bmatrix}$$

Assembling the element components then the structure matrix and force vector are, using node ordering (1,2,3)

$$2.613 \times 10^{-3}\begin{bmatrix} 1 & -1 & 0 \\ -1 & 2 & -1 \\ 0 & -1 & 1 \end{bmatrix}\begin{Bmatrix} p_1 \\ p_2 \\ p_3 \end{Bmatrix} = \begin{Bmatrix} f_1 \\ f_2 \\ f_3 \end{Bmatrix}$$

The prescribed boundary condition is $v_x^1 = 5\text{m/s}$ at node 1 so that the flow rate f_1 is, (13.131)

$$f_1 = v_x^1 A = 5(1.96 \times 10^{-3}) = 9.8 \times 10^{-3}\text{m}^3/\text{s}$$

In order to solve the above system of equations for unknowns $\{P\}$ it is first necessary to incorporate the prescribed boundary conditions. However, the boundary condition is in terms of velocity and not potential. It is necessary therefore to specify a value for p_3 leaving the two unknowns p_1 and p_2. From (13.122) we observe that the velocities are proportional to the derivatives or differences in potential so that it suffices to assign a value of zero to p_3. The solution of the system of equations will then result in p_1 and p_2 relative to p_3. Setting $p_3 = 0$ we have

$$2.613 \times 10^{-3}\begin{bmatrix} 1 & -1 \\ -1 & 2 \end{bmatrix}\begin{Bmatrix} p_1 \\ p_2 \end{Bmatrix} = \begin{Bmatrix} 9.8 \times 10^{-3} \\ 0 \end{Bmatrix}$$

Solving for p_1 and p_2 we find

$$\begin{aligned} 2.613 \times 10^{-3}(p_1 - p_2) &= 9.8 \times 10^{-3} \\ 2.613 \times 10^{-3}(-p_1 + p_2) &= 0 \end{aligned}$$

from which $p_1 = 7.5\text{m}$ and $p_2 = 3.75\text{m}$ relative to $p_3 = 0$. From (13.130) the element velocities are

$$v_x^1 = -[D]\{g\} = -k_{xx}\left[-\tfrac{1}{L_1} \quad \tfrac{1}{L_1}\right]\begin{Bmatrix} p_1 \\ p_2 \end{Bmatrix} = -\frac{(-p_1 + p_2)}{L_1} = 5\text{m/s}$$

$$v_x^2 = -[D]\{g\} = -k_{xx}\left[-\tfrac{1}{L_1} \quad \tfrac{1}{L_1}\right]\begin{Bmatrix} p_2 \\ 0 \end{Bmatrix} = -\frac{p_2}{L_2} = 5\text{m/s}$$

Although the potentials are relative to $p_3 = 0$ the velocities are absolute since from (13.122) they depend on the differences in potential. This can be verified by alternatively selecting $p_3 = 10\text{m}$ and repeating the above solution procedure.

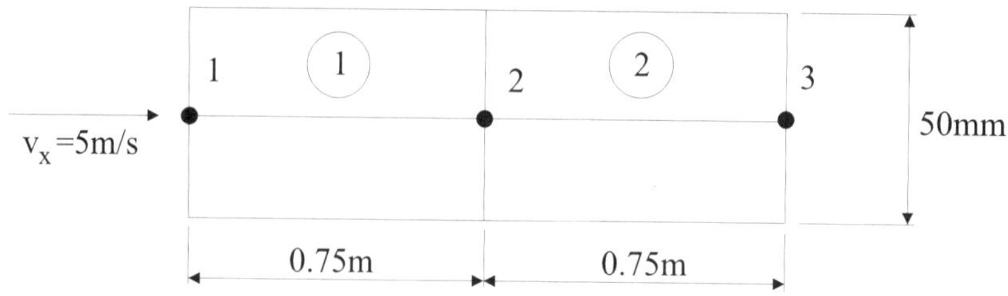

Figure 13.26: Example: A pipe of circular cross-section has a fluid entering at the left hand end at 5m/s. The pipe is modelled using two linear elements of equal length.

13.16 Conclusion

This Chapter has presented an introduction to the finite element method. Essentially, the method discretises a complex structure into a series of small simpler finite elements. The field variable, such as displacement or temperature, is approximated throughout each element by way of an interpolation function which leads to the element shape functions. By using a variational formulation we derived the stiffness matrix and force vector for the one-dimensional simplex bar element which has been the emphasis of the present Chapter. Finally, to solve the original structure then all the element stiffness and force contributions are assembled together to form a structure stiffness/force system of linear equations which is solved for the unknown field variable.

The analysis of forces presented in §13.13 is easily extended to the general case of an arbitrarily oriented pin-jointed bar element in three dimensions, although we have refrained from examining three-dimensional frames due to the cumbersome arithmetic involved. The interested reader should refer to Logan (1993) for an introductory examination of three-dimensional frames using the finite element method.

The final two sections of the present Chapter derived the finite element equations for both heat transfer and fluid flow for one-dimensional systems. Both derivations used a control volume approach with respect to conservation of energy and mass through the volume. In the case of heat transfer Hooke's law is replaced by Fourier's law whereas Darcy's law is used for fluid flow. The structural system of equations were derived for heat transfer using a variational function whereas the system of equations for fluid flow were derived by making a direct comparison with the system of equations for elasticity. Comparing equations (13.44), (13.118) and (13.136) we observe the similarity in problem formulation for the one-dimensional finite elements of elasticity, heat transfer and fluid flow.

13.17 References and Further Reading

∘ Argyris, J. H. (1960) *Energy Theorems and Structural Analysis*, Butterworth.

∘ Clough, R. W. (1960) *The Finite Element in Plane Stress Analysis,* Proc. Second ASCE Conf. on Electronic Computation, Pittsburgh, PA.

∘ Courant, R. (1943) *Variational Methods for the Solution of Problems of Equilibrium and Vibration,* Bull. Am. Math. Soc., 49, 1-23.

∘ Fagan, M. J. (1992) *Finite Element Analysis: Theory and Practice*, Longman Scientific and Technical.

∘ Galerkin, B. G. (1915) *Series Solution of Some Problems of Elastic Equilibrium of Rods and Plates,* Vestn. Inzh. Tech., 19, 897-908.

o Levy, S. (1953) *Structural Analysis and Influence Coefficients for Delta Wings,* J. Aero. Sci., 20, 449-454.

o Logan, D. L. (1993) *A First Course in the Finite Element Method,* PWS Publishing, Boston.

o Lord Rayleigh (1870) *On the Theory of Resonance,* Trans. Roy. Soc. (London), A161, 77-118.

o Richardson, L. F. (1910) *The Approximate Arithmetical Solution by Finite Differences of Physical Problems,* Trans. Roy. Soc., London, A210, 305-357.

o Ritz, W. (1909) *Über eine neue Methode zur Lösung gewissen Variations-Probleme der mathematischen Physik,* J. Rene Angew. Math., 135, 1-61.

o Robinson, J. (1985) *Early FEM Pioneers,* Robinsion & Associates, Dorset, England.

o Southwell, R. V. (1946) *Relaxation Methods in Theoretical Physics*, Clarendon Press.

o Turner, M. J., Clough, R. W., Martin, H. C. and Topp, L. J. (1956) *Stiffness and Defletion Analysis of Complex Structures*, J. Areo. Sci., 23, 805-823.

o Weighardt, K. (1906) *Über einen Grenzübergang der Elastizitätslehre und seine Anwendung auf die Statik hochgradig statisch ungestimmter Fachwerke,* Verhandlungen des Vereins z. Beförderung des Gewerbefleisses, Abhandlungen, 85, 139-176.

o Zienkiewcz, O. C. (1992) *The Finite Element Method: Its Genesis and Future,* The Structural Engineer, 70(20), 355-360.

o Zienkiewicz, O. C. and Taylor, R. L. (1991) *The Finite Element Method,* 4th Ed., 2 Volumes, McGraw-Hill.

13.18 Exercises

13.1 If a one-dimensional finite element, consisting of two nodes at either end ($x = x_i = 2$ and $x = x_j = 6$), is to be modelled by the following interpolation function

$$\phi = \alpha_1 + \alpha_2 x$$

then derive the shape functions of the element. Note the salient properties of the shape functions. If the value of ϕ at the nodes is $\phi_i = 10$ and $\phi_j = 20$ then determine the value of ϕ at the point $x = 3$.

13.2 The state of strain at a point is given by $\varepsilon_{xx} = 60 \times 10^{-6}$, $\varepsilon_{yy} = 80 \times 10^{-6}$ and $\gamma_{xy} = 55 \times 10^{-6}$. Determine the corresponding stress components for both plane stress and plane strain states assuming that Young's modulus and Poisson's ratio are 70GPa and 0.28 respectively.

13.3 The composite steel/copper bar shown in Figure 13.27 is fully built-in at one end and free at the other end. The free end of the bar is subject to a compressive load of 185kN. The length, diameter and Young's modulus of the two components of the composite bar are denoted by L_i, D_i and E_i $(i = 1, 2)$ respectively with the assigned values shown in the Table in Figure 13.27. By representing the composite bar as two one-dimensional pin-jointed simplex finite elements determine the axial displacements of points 1, 2 and 3 and the strains and stresses of elements 1 and 2.

13.4 Determine the vertical displacement at joint 2 for the pin-jointed plane frame shown in Figure 13.28. Joints 1 and 3 are built-in and joint 2 has a point force of $P = 1$kN acting in the downward vertical direction. The frame is symmetric about the y-axis with angle θ equal to 30° and $d = 1$m. The cross-sectional area and Young's modulus of both members are $A = 2 \times 10^{-3}$m^2 and $E = 200$GPa respectively.

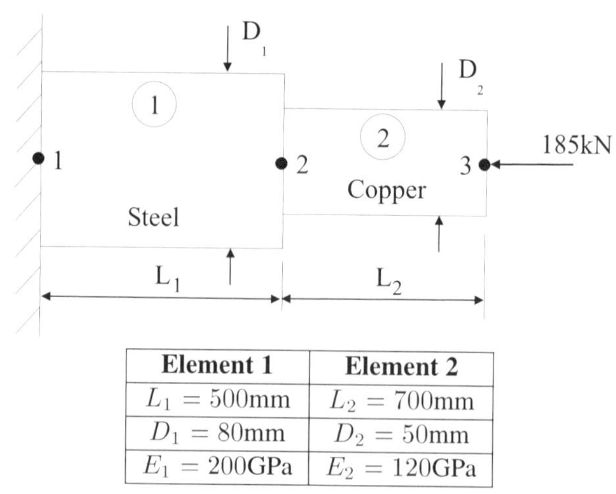

Element 1	Element 2
$L_1 = 500\text{mm}$	$L_2 = 700\text{mm}$
$D_1 = 80\text{mm}$	$D_2 = 50\text{mm}$
$E_1 = 200\text{GPa}$	$E_2 = 120\text{GPa}$

Figure 13.27: Exercise: A composite steel/copper bar in compression.

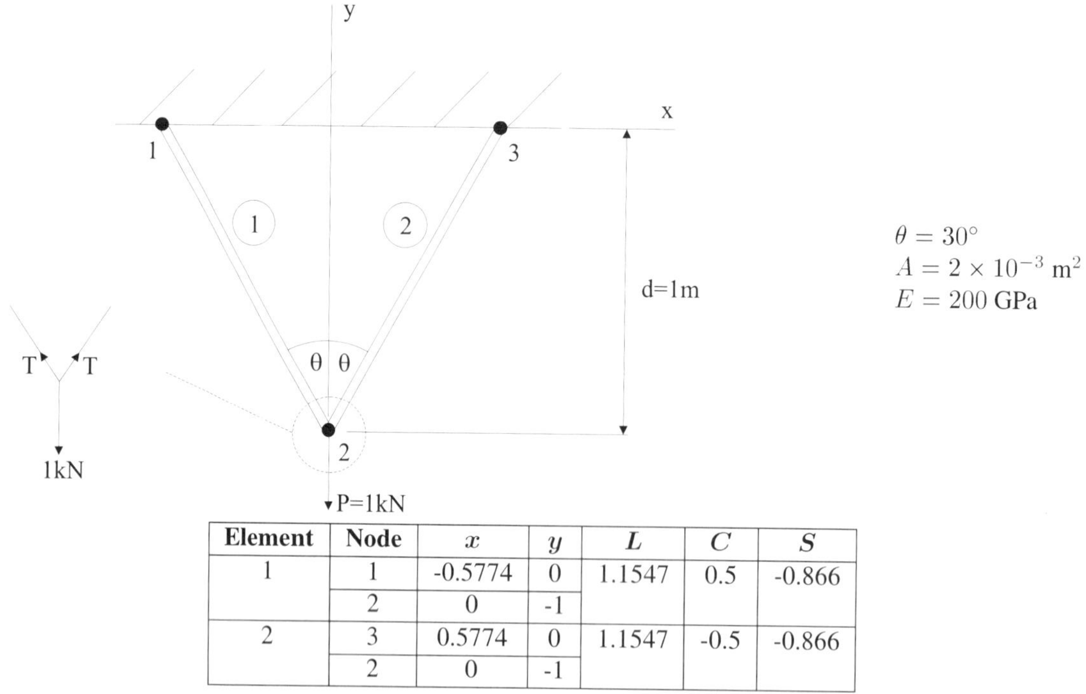

Element	Node	x	y	L	C	S
1	1	-0.5774	0	1.1547	0.5	-0.866
	2	0	-1			
2	3	0.5774	0	1.1547	-0.5	-0.866
	2	0	-1			

Figure 13.28: Exercise: Pin-jointed plane frame with point force P acting at joint 2.

13.5 A one-dimensional bar of length L =50mm and square cross-section (5mm×5mm) is subject to a temperature of 100°C at the left hand end and free at the right hand end. Modelling the bar using two linear elements of equal length then determine the temperature distribution throughout the bar. Assume that the thermal conductivity, k_{xx}, is 5W/m°C, the convection coefficient, h, is 1W/m^2°C and the background temperature, T_∞, is 0°C.

13.6 A composite wall consists of two layers of material with thermal conductivity equal to 10W/mm°C and thickness 100mm separated by a layer of insulation of thickness 50mm and with thermal conductivity of 5W/mm°C as shown in Figure 13.29. Determine the heat flow through the wall if the inside wall surface is at 25°C and the outside temperature is at -5°C with a convection coefficient of 50W/mm°C.

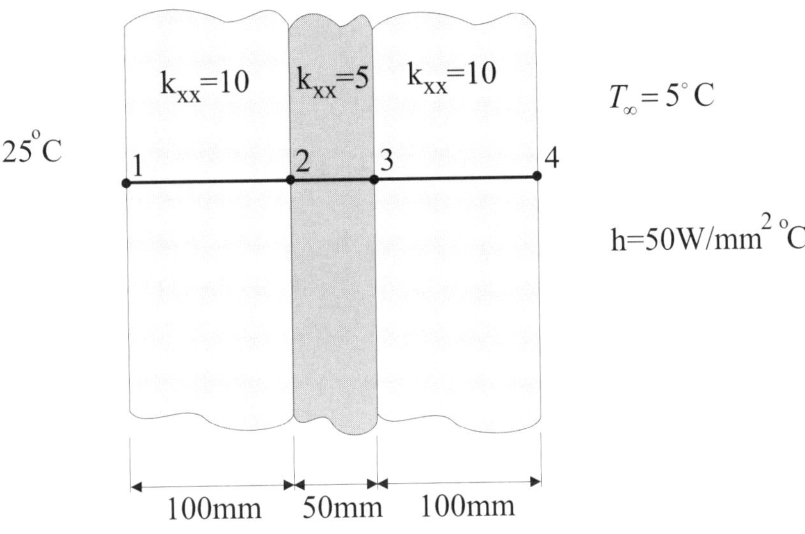

Figure 13.29: Exercise: A composite wall consisting of three layers subject to an internal temperature of 25°C.

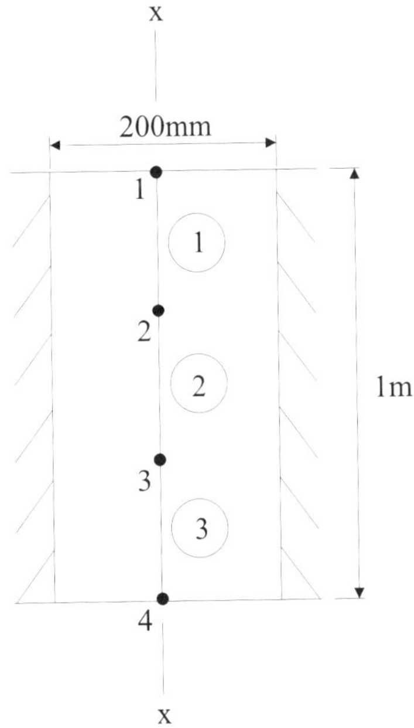

Figure 13.30: Exercise: Fluid flow in a porous medium.

13.7 Figure 13.30 illustrates a 1m high sample of coarse gravel (permeability coefficient equal to 1×10^{-2}m/s) of circular cross-section and diameter 200mm. The fluid head at the top and bottom of the sample is 0.2m and 0.1m respectively. The permeability coefficient of the sample is 1×10^{-2}m/s and the surrounding boundary is impermeable. By representing the sample as three simplex one-dimensional finite elements of equal length then determine both the fluid heads and velocities throughout the sample.

Part III

Advanced Topics

In this part, we examine more advanced topics of the strength of materials. The first three Chapters continue our examination of elasticity, plasticity and the finite element method presented in previous Chapters whereas the last seven Chapters provide introductions to selected topics in the discipline of *Strength of Materials*. We begin with Chapter 14 by providing a more in-depth analysis of elasticity than presented in previous Chapters. Throughout the Chapter a tensorial representation of the stress-strain components is emphasised. A more general form of the transformation equations of Chapters 7 and 8 suitable for three-dimensional systems is presented and the elegant complex-variable representation of elasticity is introduced. The latter part of Chapter 14 examines the problems of the bending of plates, the torsion of prismatic bars, a hole in a plate subject to uniform loading and a point-loaded wedge. Chapter 15 extends the examination of plasticity presented in Chapter 12 and derives the Levy-von Mises and Prandtl-Reuss flow rules. Similarly, Chapter 16 continues the discussion of the finite element method of Chapter 13 with the focus now on two-dimensional and higher-order elements. A thorough introduction is given in Chapter 17 to the fracture and fatigue of engineering materials. All of the key fracture parameters are introduced with their application to several well-known configurations. The latter half of the Chapter addresses both long and short fatigue crack growth. Chapter 18 examines the creep of engineering materials while Chapter 19 introduces the key models of the viscoelastic behaviour of materials. The subject of continuum damage mechanics is introduced in Chapter 20. An overview of the important subject area of composite materials is presented in Chapter 21 with a focus on unidirectional laminae. In Chapter 22 we provide an introduction to the subject of contact mechanics. The key topics covered are concentrated and distributed normal and tangential forces acting on the surface of a half-plane, the indentation of a flat rigid indenter, Hertzian contact and indentation fracture. In Chapter 23, finally, we look at the statistical behaviour of brittle materials.

Chapter 14

Further Elasticity

14.1 Introduction

The present Chapter consists of two key parts. The first part presents an overview of the theory of elasticity. The second part is more applied and presents solutions for the bending of rectangular plates, the torsion of prismatic bars, circular and elliptical holes in a plate and concludes with a discussion of the so-called Boussinesq wedge problem. These latter problems will lead our discussion towards the stress analysis of idealised line cracks which are discussed in Chapter 17. A key objective throughout this Chapter is to familiarise the reader with the use of a more convenient tensor notation.

The Chapter begins by introducing the stress element in three-dimensions. The general equations of equilibrium and compatibility equations are then derived. The stress transformations equations of Chapter 7 are then revisited for three-dimensional systems. Hooke's law is also re-examined with the stress components now conveniently expressed in tensorial form. The Airy stress function and Airy stresses are then introduced for both Cartesian and cylindrical coordinate systems. The cubic equation of stress is then derived for three-dimensional stress states and shown to reduce to the principal stresses σ_1 and σ_2 previously encountered in §7.3 for two dimensions. The cubic equation of stress then leads to a discussion on the scalar stress invariants. The previous Chapter illustrated the importance of a matrix representation of stress and strain for the finite element method. In §14.9 we present the constitutive equations in matrix form for a variety of different materials. The reciprocal displacement and work theorems are then presented and their usefulness demonstrated by considering the deflection of a simply supported beam. Next, an overview of the complex variable representation of the equations of elasticity is presented and the method's conciseness illustrated by way of the two problems of a point-loaded wedge and a centre cracked plate. Sections 14.12 and 14.13 present the theory of bending of rectangular plates and the torsion of prismatic bars. The final two sections take a more applied approach by examining the problems of a hole in a plate and a point-loaded wedge.

14.2 Stress Element

A three-dimensional hexahedral stress element will, in general, have both direct and shear stresses acting on each of its six faces. It was illustrated in Chapter 1 that to form a coherent reference system it is necessary to resolve the stress acting on each face in accordance with the coordinate system using the *face-direction* rule as shown in Figure 14.1. The stress components for a Cartesian stress element have

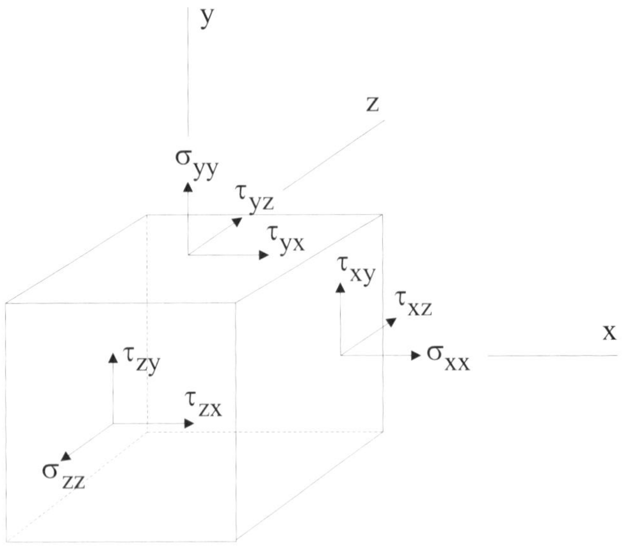

Figure 14.1: Cartesian stress element.

nine components and can be represented by the following (3×3) stress tensor σ_{ij}

$$\sigma_{ij} = \begin{bmatrix} \sigma_{xx} & \tau_{xy} & \tau_{xz} \\ \tau_{yx} & \sigma_{yy} & \tau_{yz} \\ \tau_{zx} & \tau_{zy} & \sigma_{zz} \end{bmatrix} \tag{14.1}$$

To preserve moment equilibrium in the stress element we have

$$\tau_{yx} = \tau_{xy}; \quad \tau_{yz} = \tau_{zy}; \quad \tau_{zx} = \tau_{xz} \tag{14.2}$$

In certain circumstances it may be more advantageous to use an alternative coordinate system such as cylindrical coordinates, Figure 14.2. Moment equilibrium is ensured if $\tau_{r\theta} = \tau_{\theta r}$, $\tau_{\theta z} = \tau_{z\theta}$ and $\tau_{zr} = \tau_{rz}$.

Clearly, the stress components in different coordinate systems are interchangeable. For example, consider the two-dimensional plate shown in Figure 14.3 subject to uniform tension. The Cartesian stress coordinates are $\sigma_{xx} = 0$, $\sigma_{yy} = \sigma_P$, $\tau_{xy} = 0$ and $\sigma_{zz} = 0$ for plane stress and $\sigma_{zz} = \nu\left(\sigma_{xx} + \sigma_{yy}\right) = \nu\sigma_P$ for plane strain. To determine the stress components in terms of the polar coordinate system use can be made of the transformation equations, (7.8), with a direct analogy between the local Cartesian coordinates (x', y') and polar coordinates (r, θ)

$$\begin{aligned} \sigma_{rr} &= \sigma_{xx}\cos^2\theta + \sigma_{yy}\sin^2\theta + 2\tau_{xy}\sin\theta\cos\theta = \sigma_P\sin^2\theta = \frac{1}{2}\sigma_P(1-\cos 2\theta) \\ \sigma_{\theta\theta} &= \sigma_{xx}\sin^2\theta + \sigma_{yy}\cos^2\theta - 2\tau_{xy}\sin\theta\cos\theta = \sigma_P\cos^2\theta = \frac{1}{2}\sigma_P(1+\cos 2\theta) \\ \tau_{r\theta} &= \left(\sigma_{yy} - \sigma_{xx}\right)\sin\theta\cos\theta + \tau_{xy}\left(\cos^2\theta - \sin^2\theta\right) = \sigma_P\sin\theta\cos\theta = \frac{1}{2}\sigma_P\sin 2\theta \end{aligned} \tag{14.3}$$

In §1.11 we established that the in-plane strains $(\varepsilon_{xx}, \varepsilon_{yy}, \gamma_{xy})$ are given by

$$\varepsilon_{xx} = \frac{\partial u}{\partial x}; \quad \varepsilon_{yy} = \frac{\partial v}{\partial y}; \quad \gamma_{xy} = \frac{1}{2}\left(\frac{\partial u}{\partial y} + \frac{\partial v}{\partial x}\right) \tag{14.4}$$

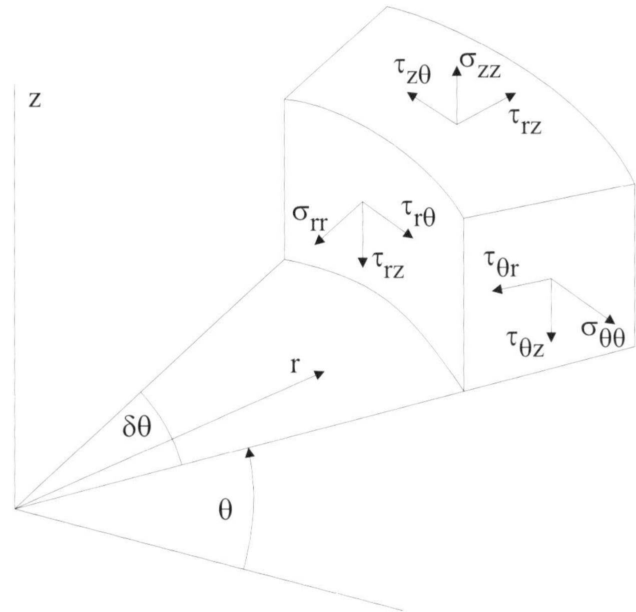

Figure 14.2: Cylindrical stress element.

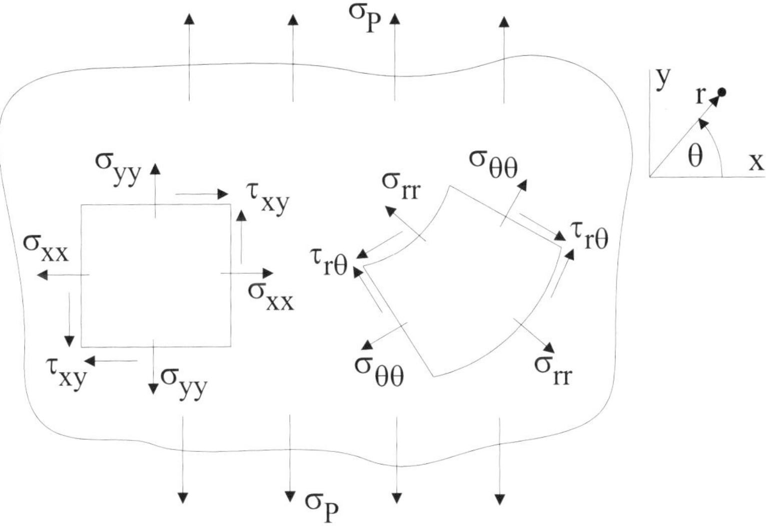

Figure 14.3: A plate subject to a uniaxial tension illustrating Cartesian and polar stress elements.

where u and v are the displacements in the x and y directions respectively. In tensor form these can be written

$$\varepsilon_{ij} = \frac{1}{2}\left(\frac{\partial u_i}{\partial x_j} + \frac{\partial u_j}{\partial x_i}\right) \tag{14.5}$$

To determine the in-plane polar strain components $(\varepsilon_{rr}, \varepsilon_{\theta\theta}, \gamma_{r\theta})$ in terms of the radial, u_r, and circumferential, u_θ, displacements we will refer to Figure 14.4. The radial strain is the extension of segment AB

$$\varepsilon_{rr} = \frac{[\mathrm{d}r + (\partial u_r/\partial r)\,\mathrm{d}r] - \mathrm{d}r}{\mathrm{d}r} = \frac{\partial u_r}{\partial r} \tag{14.6}$$

The circumferential strain is the extension of segment AD (original length $r\,\mathrm{d}\theta$) due to a change in radius from r to $(r + u_r)$ and the change of u_θ with respect to θ

$$\varepsilon_{\theta\theta} = \frac{[(r + u_r)\,\mathrm{d}\theta + (\partial u_\theta/\partial\theta)\,\mathrm{d}\theta] - r\,\mathrm{d}\theta}{r\,\mathrm{d}\theta} = \frac{u_r}{r} + \frac{1}{r}\frac{\partial u_\theta}{\partial\theta} \tag{14.7}$$

The shear strain is the change in angle BAD minus the rigid body rotation u_θ/r

$$\gamma_{r\theta} = \frac{\frac{\partial u_r}{\partial\theta}\,\mathrm{d}\theta}{r\,\mathrm{d}\theta} + \frac{\frac{\partial u_\theta}{\partial r}\,\mathrm{d}r}{\mathrm{d}r} - \frac{u_\theta}{r} = \frac{1}{r}\frac{\partial u_r}{\partial\theta} + \frac{\partial u_\theta}{\partial r} - \frac{u_\theta}{r} \tag{14.8}$$

for small $\mathrm{d}\theta$ in which $\tan\,\mathrm{d}\theta \approx \mathrm{d}\theta$.

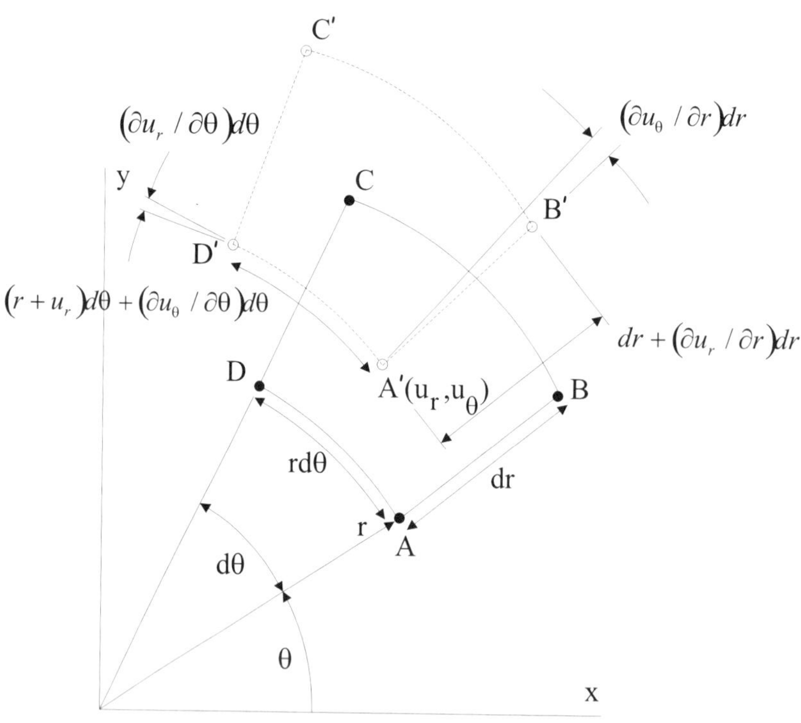

Figure 14.4: A stress element in polar coordinates.

14.3 Equations of Equilibrium

Consider a differential stress element of sides δx, δy and δz which is subject to forces in the x-direction only, Figure 14.5. Neglecting body forces, equilibrium in the x-direction is satisfied by

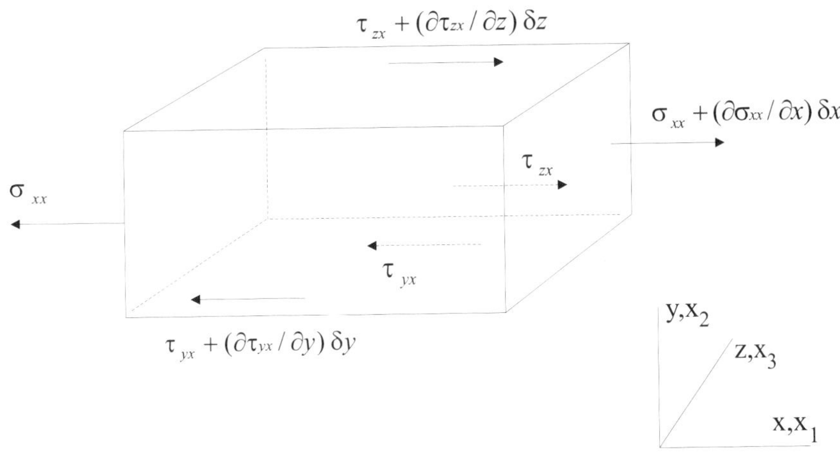

Figure 14.5: A stress element subject to forces in the x-direction only.

$$\frac{\partial \sigma_{xx}}{\partial x} + \frac{\partial \tau_{yx}}{\partial y} + \frac{\partial \tau_{zx}}{\partial z} = 0 \tag{14.9}$$

and states that the rate of change of stress in the x-direction is zero. Similarly, for the y and z directions

$$\begin{aligned} \frac{\partial \sigma_{yy}}{\partial y} + \frac{\partial \tau_{zy}}{\partial z} + \frac{\partial \tau_{xy}}{\partial x} &= 0 \\ \frac{\partial \sigma_{zz}}{\partial z} + \frac{\partial \tau_{xz}}{\partial x} + \frac{\partial \tau_{yz}}{\partial y} &= 0 \end{aligned} \tag{14.10}$$

Alternatively, in terms of coordinates $x_1(=x)$, $x_2(=y)$ and $x_3(=z)$

$$\begin{aligned} \frac{\partial \sigma_{11}}{\partial x_1} + \frac{\partial \sigma_{21}}{\partial x_2} + \frac{\partial \sigma_{31}}{\partial x_3} &= 0 \\ \frac{\partial \sigma_{22}}{\partial x_2} + \frac{\partial \sigma_{32}}{\partial x_3} + \frac{\partial \sigma_{12}}{\partial x_1} &= 0 \\ \frac{\partial \sigma_{33}}{\partial x_3} + \frac{\partial \sigma_{13}}{\partial x_1} + \frac{\partial \sigma_{23}}{\partial x_2} &= 0 \end{aligned} \tag{14.11}$$

Adopting this notation reduces the use of the three coordinate variables (x, y, z) to a single variable x_i and the stress variables σ_{ij} and τ_{ij} to the single variable σ_{ij} for $i = 1, 2, 3$.

Equations (14.11) are referred to as the *differential equations of equilibrium* and can be more concisely written with the help of a tensor notation as follows

$$\sum_{i=1}^{3} \frac{\partial \sigma_{ij}}{\partial x_j} = 0 \tag{14.12}$$

where j is equal to any of the three coordinate directions. Summation is generally assumed and (14.12) is written as

$$\frac{\partial \sigma_{ij}}{\partial x_j} = 0 \tag{14.13}$$

Being even more concise (14.13) can be written as $\sigma_{ij,j} = 0$ with partial differentiation with respect to a coordinate variable indicated by using the subscript of that variable, preceded by a comma.

If a stress element is not in a state of equilibrium then there will be an acceleration of the element $\mathrm{d}^2 u_j / \mathrm{d}t^2$. A body force may also be considered, where the j component of the force per unit volume is denoted by F_j. The resulting equation is a form of the momentum equation since it gives the rate of change of momentum of a particular element

$$\frac{\partial \sigma_{ij}}{\partial x_j} + F_j = \rho \frac{d^2 u_j}{dt^2} \tag{14.14}$$

where ρ is the density of the stress element.

Let us conclude this section by deriving the differential equations of equilibrium in terms of polar coordinates. These are derived in an analogous manner to the derivation of the equilibrium equation in §11.2 but now taking account for $\tau_{r\theta}$. Figure 14.6 illustrates a differential stress element of angle $\mathrm{d}\theta$ and radial width $\mathrm{d}r$ at radial distance r. Over the element the radial stress varies from σ_{rr} to $\sigma_{rr} + (\partial \sigma_{rr}/\partial r)\,\mathrm{d}r$ and resolving forces in the radial direction

$$\sigma_{rr} r\,\mathrm{d}\theta + 2\sigma_{\theta\theta}\,\mathrm{d}r \sin\frac{\mathrm{d}\theta}{2} = \left(\sigma_{rr} + \frac{\partial \sigma_{rr}}{\partial r}\,\mathrm{d}r\right)(r + \mathrm{d}r)\,\mathrm{d}\theta + \partial \tau_{r\theta}\,\mathrm{d}r \tag{14.15}$$

For $\theta \approx 0^\circ$ then $\sin(\mathrm{d}\theta/2) \approx \mathrm{d}\theta/2$ and therefore (14.15) becomes

$$r\sigma_{rr} + \sigma_{\theta\theta}\,\mathrm{d}r - \partial \tau_{r\theta}\,\mathrm{d}r = r\sigma_{rr} + \sigma_{rr}\,\mathrm{d}r + r\frac{\partial \sigma_{rr}}{\partial r}\,\mathrm{d}r + \frac{\partial \sigma_{rr}}{\partial r}(\mathrm{d}r)^2 \tag{14.16}$$

Cancelling terms and neglecting second order terms we arrive at the following equilibrium equation

$$\frac{\partial \sigma_{rr}}{\partial r} + \frac{1}{r}\frac{\partial \tau_{r\theta}}{\partial \theta} + \frac{\sigma_{rr} - \sigma_{\theta\theta}}{r} = 0 \tag{14.17}$$

Performing a similar resolving of forces in the tangential direction we find

$$\frac{1}{r}\frac{\partial \sigma_{\theta\theta}}{\partial \theta} + \frac{\partial \tau_{r\theta}}{\partial r} + \frac{2\tau_{r\theta}}{r} = 0 \tag{14.18}$$

Finally, in terms of cylindrical coordinates and using a tensor notation the equations of equilibrium are, Timoshenko and Goodier (1982)

$$\begin{aligned} \sigma_{rr,r} + \frac{1}{r}\sigma_{r\theta,\theta} + \sigma_{zr,z} + \frac{\sigma_{rr} - \sigma_{\theta\theta}}{r} &= 0 \\ \sigma_{r\theta,r} + \frac{1}{r}\sigma_{\theta\theta,\theta} + \sigma_{z\theta,z} + \frac{2}{r}\sigma_{r\theta} &= 0 \\ \sigma_{zr,r} + \frac{1}{r}\sigma_{z\theta,\theta} + \sigma_{zz,z} + \frac{\sigma_{zr}}{r} &= 0 \end{aligned} \tag{14.19}$$

14.4 Compatibility Equations

In the solution of a problem it is necessary to solve the differential equations of equilibrium while simultaneously satisfying the boundary conditions. These equations are derived by an application of the equations of statics and are not sufficient for the unique determination of the stress components. The problem is a statically indeterminate one and in order to obtain a unique solution then the elastic deformation of the body must also be considered. For a two-dimensional system the strain components are

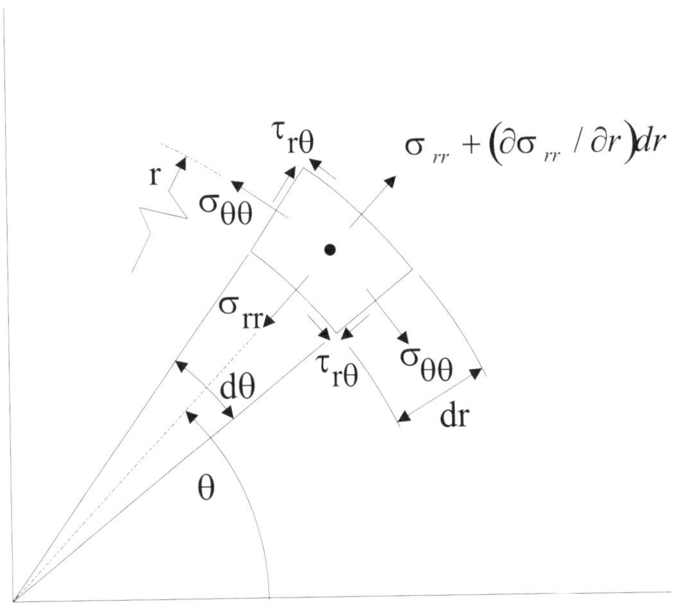

Figure 14.6: A polar stress element.

expressed by the two displacements u and v ($\varepsilon_{ij} = (u_{i,j} + u_{j,i})/2$) and cannot therefore be taken arbitrarily. A relationship between the strain components can be obtained by differentiating ε_{xx} twice with respect to y, ε_{yy} twice with respect to x and γ_{xy} with respect to x and y

$$\frac{\partial^2 \varepsilon_{xx}}{\partial y^2} = \frac{\partial^3 u}{\partial x \partial y^2}; \quad \frac{\partial^2 \varepsilon_{yy}}{\partial x^2} = \frac{\partial^3 v}{\partial x^2 \partial y}; \quad 2\frac{\partial^2 \gamma_{xy}}{\partial x \partial y} = \frac{\partial^3 u}{\partial x \partial y^2} + \frac{\partial^3 v}{\partial y \partial x^2} \tag{14.20}$$

from which we have

$$\frac{\partial^2 \varepsilon_{xx}}{\partial y^2} + \frac{\partial^2 \varepsilon_{yy}}{\partial x^2} - 2\frac{\partial^2 \gamma_{xy}}{\partial x \partial y} = 0 \tag{14.21}$$

and is known as the *condition of compatibility* and must be satisfied by the strain components.

By using Hooke's law, (1.16), the above relation can be expressed in terms of stress

$$\varepsilon_{xx} = \frac{1}{E}\left[\sigma_{xx} - \nu\sigma_{yy}\right]; \quad \varepsilon_{yy} = \frac{1}{E}\left[\sigma_{yy} - \nu\sigma_{xx}\right]; \quad \gamma_{xy} = \frac{\tau_{xy}}{G} = \frac{2(1+\nu)}{E}\tau_{xy} \tag{14.22}$$

which, upon substituting into (14.21), gives

$$\frac{\partial^2\left(\sigma_{xx} - \nu\sigma_{yy}\right)}{\partial y^2} + \frac{\partial^2\left(\sigma_{yy} - \nu\sigma_{xx}\right)}{\partial x^2} = 2(1+\nu)\frac{\partial^2 \tau_{xy}}{\partial x \partial y} \tag{14.23}$$

With the help of the differential equations of equilibrium, (14.11)

$$\frac{\partial \sigma_{xx}}{\partial x} + \frac{\partial \tau_{xy}}{\partial y} = 0; \quad \frac{\partial \sigma_{yy}}{\partial y} + \frac{\partial \tau_{xy}}{\partial x} = 0 \tag{14.24}$$

then (14.21) can be alternatively expressed. Differentiate the first of (14.24) with respect to x and the second of (14.24) with respect to y

$$\frac{\partial^2 \sigma_{xx}}{\partial x^2} + \frac{\partial^2 \tau_{xy}}{\partial x \partial y} = 0; \quad \frac{\partial^2 \sigma_{yy}}{\partial y^2} + \frac{\partial^2 \tau_{xy}}{\partial x \partial y} = 0 \tag{14.25}$$

Adding these two equations together we have

$$2\frac{\partial^2\tau_{xy}}{\partial x\partial y} = -\frac{\partial^2\sigma_{xx}}{\partial x^2} - \frac{\partial^2\sigma_{yy}}{\partial y^2} \tag{14.26}$$

and substituting into (14.23) we arrive at the compatibility equation in terms of stress

$$\frac{\partial^2\left(\sigma_{xx} - \nu\sigma_{yy}\right)}{\partial y^2} + \frac{\partial^2\left(\sigma_{yy} - \nu\sigma_{xx}\right)}{\partial x^2} = -(1+\nu)\left(\frac{\partial^2\sigma_{xx}}{\partial x^2} + \frac{\partial^2\sigma_{yy}}{\partial y^2}\right) \tag{14.27}$$

which reduces to

$$\left(\frac{\partial^2}{\partial x} + \frac{\partial^2}{\partial y}\right)\left(\sigma_{xx} + \sigma_{yy}\right) = 0 \quad \text{or} \quad \nabla^2\left(\sigma_{xx} + \sigma_{yy}\right) = 0 \tag{14.28}$$

where ∇^2 is the Laplacian operator.

14.5 Stress and Strain Transformations Revisited

In Chapter 7 we examined the transformation equations of stress in a two-dimensional system and arrived at the following set of equations which transform a state of stress $(\sigma_{xx}, \sigma_{yy}, \tau_{xy})$ with respect to the global axes (x, y) to the local axes (x', y'), (7.6)

$$\begin{aligned}
\sigma_{x'x'} &= \sigma_{xx}\cos^2\theta + \sigma_{yy}\sin^2\theta + 2\tau_{xy}\sin\theta\cos\theta \\
\sigma_{y'y'} &= \sigma_{xx}\sin^2\theta + \sigma_{yy}\cos^2\theta - 2\tau_{xy}\sin\theta\cos\theta \\
\tau_{x'y'} &= -\left(\sigma_{xx} - \sigma_{yy}\right)\sin\theta\cos\theta + \tau_{xy}\left(\cos^2\theta - \sin^2\theta\right)
\end{aligned} \tag{14.29}$$

expressed in terms of θ rather than 2θ. The fact that the stress transformation equations are more complicated than those for a vector should not be surprising since the stress components involve two directions (normal and parallel forces to a section) whereas the components of a vector are related to only one direction. Equations (14.29) can be greatly simplified by consideration of direction cosines. Consider the two sets of coordinates axes (x, y, z) and (x', y', z') in Figure 14.7. The sets of cosines between various coordinate axes are shown in Table 14.1. From Figure 14.7 we observe that $l_{x'x} = l_{xx'}$, $l_{x'y} = l_{yx'}$ and

	x	y	z
x'	$l_{x'x} = \cos\theta_{x'x}$	$l_{x'y} = \cos\theta_{x'y}$	$l_{x'z} = \cos\theta_{x'z}$
y'	$l_{y'x} = \cos\theta_{y'x}$	$l_{y'y} = \cos\theta_{y'y}$	$l_{y'z} = \cos\theta_{y'z}$
z'	$l_{z'x} = \cos\theta_{z'x}$	$l_{z'y} = \cos\theta_{z'y}$	$l_{z'z} = \cos\theta_{z'z}$

Table 14.1: Direction cosines between axes (x, y, z) and (x', y', z').

so on, but it is not necessarily true that $l_{x'y} = l_{xy'}$.

If a vector **a** has components (a_x, a_y, a_z) relative to the coordinates (x, y, z) then the components (a'_x, a'_y, a'_z) of a vector **a**$'$ relative to the coordinates (x', y', z') are

$$\begin{aligned}
a'_x &= l_{x'x}a_x + l_{x'y}a_y + l_{x'z}a_z \\
a'_y &= l_{y'x}a_x + l_{y'y}a_y + l_{y'z}a_z \\
a'_z &= l_{z'x}a_x + l_{z'y}a_y + l_{z'z}a_z
\end{aligned} \tag{14.30}$$

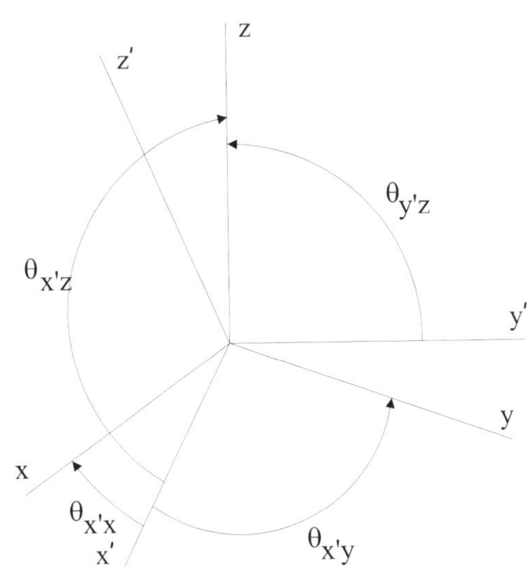

Figure 14.7: Angles between two sets of coordinate axes (x, y, z) and (x', y', z').

or, more concisely

$$a'_i = l_{i'j}a_j; \quad i' = x', y', z' \tag{14.31}$$

noting that $l_{i'j}l_{i'j} = 1$ and $l_{i'j}l_{k'j} = 0$ for $i' \neq k'$.

It is now possible to re-write the stress transformation equations, (14.29), as

$$\sigma_{i'j} = \sigma_{kl}l_{i'k}l_{j'l} \quad \text{for} \quad k, l = x, y, z \tag{14.32}$$

As an illustration of (14.32) let us consider a two-dimensional system. For $i = x$ and $j = x$

$$\sigma_{x'x'} = \sum \sigma_{kl}l_{x'k}l_{x'l} \tag{14.33}$$

For subscript $l = x$ and $l = y$

$$\sigma_{x'x'} = \sigma_{kx}l_{x'k}l_{x'x} + \sigma_{ky}l_{x'k}l_{x'y} \tag{14.34}$$

For $k = x$ and $k = y$

$$\sigma_{x'x'} = \sigma_{xx}l_{x'x}l_{x'x} + \sigma_{xy}l_{x'x}l_{x'y} + \sigma_{yx}l_{x'y}l_{x'x} + \sigma_{yy}l_{x'y}l_{x'y} \tag{14.35}$$

With $l_{x'x}l_{x'x} = \cos^2\theta_{x'x}$, $l_{x'x}l_{x'y} = \cos\theta_{x'x}\cos\theta_{x'y}$, $l_{x'y}l_{x'x} = \cos\theta_{x'y}\cos\theta_{x'x}$ and $l_{x'y}l_{x'y} = \cos^2\theta_{x'y}$, and letting $\theta = \theta_{x'x}$ and $\theta - 90° = \theta_{x'y}$ then we arrive at the local stress component $\sigma_{x'x'}$ given in (14.29).

The displacements, being vectors, transform according to the vector rule

$$u'_i = l_{i'j}u_j \tag{14.36}$$

where u and u' are with respect to axes (x, y, z) and (x', y', z') respectively. For example, Figure 14.8 illustrates a point p with respect to two sets of axes with $(p_{x'}, p_{y'})$ given by, (14.30)

$$\begin{aligned} p_{x'} &= l_{x'x}p_x + l_{x'y}p_y = p_x\cos\theta + p_y\sin\theta \\ p_{y'} &= l_{y'x}p_x + l_{y'y}p_y = -p_x\sin\theta + p_y\cos\theta \end{aligned} \tag{14.37}$$

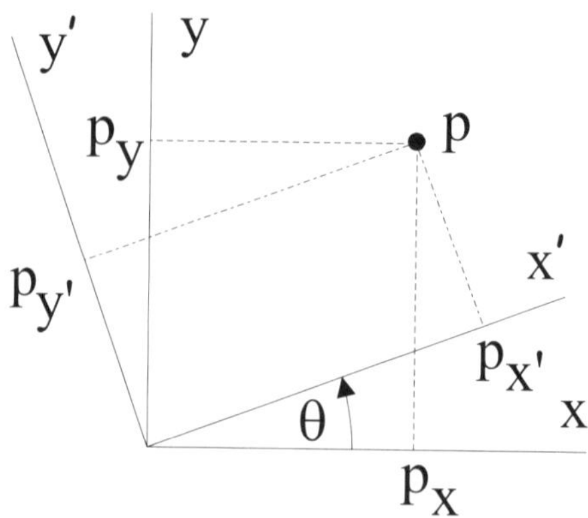

Figure 14.8: A point p with respect to the two sets of coordinate axes (x, y) and (x', y').

Since

$$\frac{\partial}{\partial x'_j} = \left(\frac{\partial}{\partial x}\right)\frac{\partial x}{\partial x'_j} + \left(\frac{\partial}{\partial y}\right)\frac{\partial y}{\partial x'_j} + \left(\frac{\partial}{\partial z}\right)\frac{\partial z}{\partial x'_j} \tag{14.38}$$

and

$$\frac{\partial x}{\partial x'_j} = l_{j'x}; \quad \cdots \tag{14.39}$$

then the strain transformation tensor is, $(\varepsilon_{ij} = (u_{i,j} + u_{j,i})/2)$

$$\varepsilon_{i',j'} = \varepsilon_{kl} l_{i'k} l_{j'l} \quad \text{for} \quad k, l = x, y, z \tag{14.40}$$

and is directly analogous to the stress transformation tensor, (14.32).

Let us conclude our discussion of stress-strain transformations by expressing the polar stresses $(\sigma_{rr}, \sigma_{\theta\theta}, \tau_{r\theta})$ in terms of the Cartesian stresses $(\sigma_{xx}, \sigma_{yy}, \tau_{xy})$. By a direct analogy between the local coordinates (x', y') and (r, θ) we have from (14.29)

$$\begin{aligned}
\sigma_{rr} &= \sigma_{xx}\cos^2\theta + \sigma_{yy}\sin^2\theta + 2\tau_{xy}\sin\theta\cos\theta \\
\sigma_{\theta\theta} &= \sigma_{xx}\sin^2\theta + \sigma_{yy}\cos^2\theta - 2\tau_{xy}\sin\theta\cos\theta \\
\tau_{r\theta} &= -\left(\sigma_{xx} - \sigma_{yy}\right)\sin\theta\cos\theta + \tau_{xy}\left(\cos^2\theta - \sin^2\theta\right)
\end{aligned} \tag{14.41}$$

14.6 Hooke's Law Revisited

Hooke's law has been examined in detail previously in §1.7. In the present section we are interested in obtaining a more concise tensor representation for the stress components. Recall that the Hookian equations for a three-dimensional system are, (1.16)

$$\begin{aligned}
\varepsilon_{xx} &= \frac{1}{E}\left[\sigma_{xx} - \nu\left(\sigma_{yy} + \sigma_{zz}\right)\right] \\
\varepsilon_{yy} &= \frac{1}{E}\left[\sigma_{yy} - \nu\left(\sigma_{xx} + \sigma_{zz}\right)\right] \\
\varepsilon_{zz} &= \frac{1}{E}\left[\sigma_{zz} - \nu\left(\sigma_{xx} + \sigma_{yy}\right)\right] \\
\gamma_{xy} &= \frac{\tau_{xy}}{G}; \quad \gamma_{yz} = \frac{\tau_{yz}}{G}; \quad \gamma_{zx} = \frac{\tau_{zx}}{G}
\end{aligned} \tag{14.42}$$

where $G = E/(2(1+\nu))$ denotes the modulus of rigidity. Solving the normal strains of (14.42) for σ_{xx}, σ_{yy} and σ_{zz} we find

$$\begin{aligned}\sigma_{xx} &= 2G\left[\varepsilon_{xx} + \frac{\nu}{(1-2\nu)}\varepsilon\right]\\ \sigma_{yy} &= 2G\left[\varepsilon_{yy} + \frac{\nu}{(1-2\nu)}\varepsilon\right]\\ \sigma_{zz} &= 2G\left[\varepsilon_{zz} + \frac{\nu}{(1-2\nu)}\varepsilon\right]\end{aligned} \tag{14.43}$$

where $\varepsilon = \varepsilon_{xx} + \varepsilon_{yy} + \varepsilon_{zz}$. Alternatively, by introducing Lame's constant λ

$$\lambda = \frac{\nu E}{(1+\nu)(1-2\nu)} = \frac{2G\nu}{1-2\nu} \tag{14.44}$$

then we can express the above equations, (14.43), as follows

$$\sigma_{ii} = \lambda\varepsilon + 2G\varepsilon_{ii}; \quad i = x, y, z \tag{14.45}$$

Note that $\lambda \to \infty$ as $\nu \to 1/2$ and we reach the upper limit of linear elasticity and achieve a state of incompressibility, §15.3.1.

14.7 Airy Stress Function

In the preceding sections we have seen that, in addition to the prescribed boundary conditions, the solution of two-dimensional problems reduces to the integration of the differential equations of equilibrium, (14.13), combined with satisfying the compatibility equation, (14.28). Both of these functions can be solved simultaneously by the introduction of a new function ϕ generally referred to as the *Airy stress function* after G. B. Airy. It is easily confirmed that the differential equations of equilibrium are satisfied by the following stress components

$$\sigma_{xx} = \frac{\partial^2\phi}{\partial y^2}; \quad \sigma_{yy} = \frac{\partial^2\phi}{\partial x^2}; \quad \tau_{xy} = \frac{\partial^2\phi}{\partial x\partial y} \tag{14.46}$$

and are generally referred to as the Airy stresses. The unique solution due to ϕ must satisfy simultaneously the compatibility equation. Therefore, substituting the Airy stresses into the compatibility equation, (14.28), we have

$$\frac{\partial^4\phi}{\partial x^4} + 2\frac{\partial^4\phi}{\partial x^2\partial y^2} + \frac{\partial^4\phi}{\partial y^4} = 0 \tag{14.47}$$

or

$$\left(\frac{\partial^2}{\partial x^2} + \frac{\partial^2}{\partial y^2}\right)\left(\frac{\partial^2}{\partial x^2} + \frac{\partial^2}{\partial y^2}\right)\phi = 0 \quad \text{or} \quad \nabla^4\phi = 0 \tag{14.48}$$

where ∇^2 is the Laplacian operator.

The Airy stresses can be expressed in terms of polar coordinates. With $x = r\cos\theta$ and $y = r\sin\theta$ then

$$r^2 = x^2 + y^2; \quad \theta = \tan^{-1}\frac{y}{x} \tag{14.49}$$

with partial derivatives

$$
\begin{aligned}
\frac{\partial r}{\partial x} &= \frac{x}{\sqrt{x^2+y^2}} = \frac{x}{r} = \cos\theta; \quad \frac{\partial r}{\partial y} = \frac{y}{\sqrt{x^2+y^2}} = \frac{y}{r} = \sin\theta \\
\frac{\partial \theta}{\partial x} &= \frac{1}{1+(y/x)^2}\frac{\mathrm{d}}{\mathrm{d}x}\left(\frac{y}{x}\right) = -\frac{y}{x^2+y^2} = -\frac{y}{r^2} = -\frac{\sin\theta}{r} \\
\frac{\partial \theta}{\partial y} &= \frac{x}{x^2+y^2} = \frac{x}{r^2} = \frac{\cos\theta}{r}
\end{aligned}
\tag{14.50}
$$

The partial derivative of the Airy function $\phi(r,\theta)$ in terms of polar coordinates with respect to x is

$$
\frac{\partial \phi}{\partial x} = \frac{\partial \phi}{\partial r}\frac{\partial r}{\partial x} + \frac{\partial \phi}{\partial \theta}\frac{\partial \theta}{\partial x} = \cos\theta\frac{\partial \phi}{\partial r} - \frac{\sin\theta}{r}\frac{\partial \phi}{\partial \theta}
\tag{14.51}
$$

To determine $\partial^2\phi/\partial x^2$ we repeat the above partial derivative but now with $\partial\phi/\partial x$ as the operator

$$
\frac{\partial^2 \phi}{\partial x^2} = \left(\cos\theta\frac{\partial}{\partial r} - \frac{\sin\theta}{r}\frac{\partial}{\partial \theta}\right)\left(\cos\theta\frac{\partial \phi}{\partial r} - \frac{\sin\theta}{r}\frac{\partial \phi}{\partial \theta}\right)
\tag{14.52}
$$

and is found to reduce to

$$
\frac{\partial^2 \phi}{\partial x^2} = \cos^2\theta\frac{\partial^2 \phi}{\partial r^2} + \sin^2\theta\left(\frac{1}{r}\frac{\partial \phi}{\partial r} + \frac{1}{r^2}\frac{\partial^2 \phi}{\partial \theta^2}\right) - 2\sin\theta\cos\theta\frac{\partial}{\partial r}\left(\frac{1}{r}\frac{\partial \phi}{\partial \theta}\right)
\tag{14.53}
$$

Following a similar procedure for $\partial^2\phi/\partial y^2$ and $\partial^2\phi/\partial y\partial x$ we find

$$
\begin{aligned}
\frac{\partial^2 \phi}{\partial y^2} &= \sin^2\theta\frac{\partial^2 \phi}{\partial r^2} + \cos^2\theta\left(\frac{1}{r}\frac{\partial \phi}{\partial r} + \frac{1}{r^2}\frac{\partial^2 \phi}{\partial \theta^2}\right) + 2\sin\theta\cos\theta\frac{\partial}{\partial r}\left(\frac{1}{r}\frac{\partial \phi}{\partial \theta}\right) \\
-\frac{\partial^2 \phi}{\partial y\partial x} &= \sin\theta\cos\theta\left(\frac{1}{r}\frac{\partial \phi}{\partial r} + \frac{1}{r^2}\frac{\partial^2 \phi}{\partial \theta^2} - \frac{\partial^2 \phi}{\partial r^2}\right) - \left(\cos^2\theta - \sin^2\theta\right)\frac{\partial}{\partial r}\left(\frac{1}{r}\frac{\partial \phi}{\partial \theta}\right)
\end{aligned}
\tag{14.54}
$$

Comparing (14.53) and (14.54) to (14.46) we see that these are the Airy stresses in terms of Cartesian coordinates. Substituting (14.53) and (14.54) into (14.41) we arrive at the Airy stresses in terms of polar stresses

$$
\begin{aligned}
\sigma_{rr} &= \frac{1}{r}\frac{\partial \phi}{\partial r} + \frac{1}{r^2}\frac{\partial^2 \phi}{\partial \theta^2}; \quad \sigma_{\theta\theta} = \frac{\partial^2 \phi}{\partial r^2} \\
\tau_{r\theta} &= \frac{1}{r^2}\frac{\partial \phi}{\partial \theta} - \frac{1}{r}\frac{\partial^2 \phi}{\partial r\partial \phi} = -\frac{\partial}{\partial r}\left(\frac{1}{r}\frac{\partial \phi}{\partial \theta}\right)
\end{aligned}
\tag{14.55}
$$

Finally, from (14.53) and (14.54) above we find that

$$
\left(\frac{\partial^2}{\partial x^2} + \frac{\partial^2}{\partial y^2}\right)\phi = \left(\frac{\partial^2}{\partial r^2} + \frac{1}{r}\frac{\partial}{\partial r} + \frac{1}{r^2}\frac{\partial^2}{\partial \theta^2}\right)\phi
\tag{14.56}
$$

Noting that, (7.13)

$$
\sigma_{xx} + \sigma_{yy} = \sigma_{rr} + \sigma_{\theta\theta}
\tag{14.57}
$$

then we arrive at the equivalent form of (14.48) in polar coordinates

$$
\left(\frac{\partial^2}{\partial r^2} + \frac{1}{r}\frac{\partial}{\partial r} + \frac{1}{r^2}\frac{\partial^2}{\partial \theta^2}\right)\left(\frac{\partial^2 \phi}{\partial r^2} + \frac{1}{r}\frac{\partial \phi}{\partial r} + \frac{1}{r^2}\frac{\partial^2 \phi}{\partial \theta^2}\right) = 0
\tag{14.58}
$$

Example 14.1 Airy stress functions

Discuss the following algebraic Airy stress functions and what type of stress they describe, determining the stresses where possible:

i) $\phi =$ constant, $\phi = Ax$ and $\phi = By$

ii) $\phi = Ax^2$

iii) $\phi = Ax^3$

i) The Airy stresses are given by (14.46) from which it follows that $\phi =$ constant, $\phi = Ax$ and $\phi = By$ all correspond to zero stresses because they each describe a plane without curvature.

ii) Substitution of ϕ into $\nabla^4\phi = 0$, (14.48), shows that $\phi = Ax^2$ is an admissible Airy stress function with stresses given by

$$\sigma_{xx} = 0; \quad \sigma_{yy} = 2A; \quad \tau_{xy} = 0$$

and is therefore a uniform stress of $2A$ in the y-direction. Similarly, $\phi = By^2$ gives a constant stress of $2B$ in the x-direction while $\phi = Cxy$ gives the stress system $(\sigma_{xx} = 0, \sigma_{yy} = 0, \tau_{xy} = C)$ of a constant shear parallel to the x and y axes.

iii) The stress function $\phi = Ax^3$ also satisfies $\nabla^4\phi = 0$ with the following stresses

$$\sigma_{xx} = 0; \quad \sigma_{yy} = 6Ax; \quad \tau_{xy} = 0$$

which describe a plate subject to pure bending.

14.8 Three Dimensional Stress and the Cubic Equation of Stress

The state of stress in three dimensions can be obtained by considering the equilibrium of a tetrahedron $OABC$ with the origin of coordinates at O. If the tetrahedron is isolated then equilibrium will be restored by the application of a resultant stress, S, acting on plane ABC which can be resolved into both normal, S_N, and shear, S_S, components on the face. To determine the state of stress on face ABC we first let face ABC have unit area and then require to find the areas of the other three faces, see Figure 14.9. The normal, **n**, to plane ABC is defined in terms of the three direction cosines l, m and n

$$l = \cos\alpha; \quad m = \cos\beta; \quad n = \cos\gamma \tag{14.59}$$

The area of $ABC = \frac{1}{2}AB \times CD$ and the area of $OAB = \frac{1}{2}AB \times OD$ so that Area$(OAB)/$Area$(ABC) = OD/CD = cos\gamma = n$. Therefore, Area$(OAB) = n$ since Area$(ABC) = 1$ and similarly Area$(OBC) = l$ and Area$(OAC) = m$.

Before continuing with our derivation of the cubic equation of stress let us examine normal vectors for a moment. A vector **n** (n_x, n_y, n_z) is given by $\mathbf{n} = n_x\mathbf{i} + n_y\mathbf{j} + n_z\mathbf{k}$ where **i** $(1,0,0)$, **j** $(0,1,0)$ and **k** $(0,0,1)$ are the unit vectors. The unit vector is obtained by normalising the components of **n**:

$$\hat{\mathbf{n}} = \frac{n_x}{||\mathbf{n}||}\mathbf{i} + \frac{n_y}{||\mathbf{n}||}\mathbf{j} + \frac{n_z}{||\mathbf{n}||}\mathbf{k} \tag{14.60}$$

where $||\mathbf{n}||$ denotes the norm of **n**. However, $l = \cos\alpha = n_x/||\mathbf{n}||$, $m = \cos\beta = n_y/||\mathbf{n}||$ and $n = \cos\gamma = n_z/||\mathbf{n}||$ so that

$$\hat{\mathbf{n}} = l\,\mathbf{i} + m\,\mathbf{j} + n\mathbf{k} \tag{14.61}$$

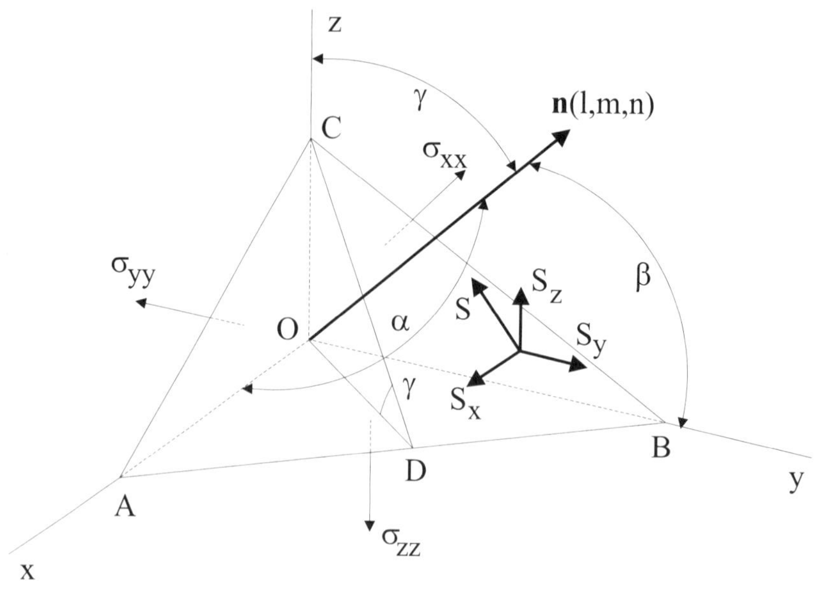

Figure 14.9: Tetrahedron $OABC$ subject to normal stresses σ_{xx}, σ_{yy} and σ_{zz}. The resultant stress S (with corresponding components S_x, S_y and S_z) acts on face ABC which has outward normal **n**.

Furthermore,

$$n_x^2 + n_y^2 + n_z^2 = ||\mathbf{n}||^2 \quad \text{or} \quad \left(\frac{n_x}{||\mathbf{n}||}\right)^2 + \left(\frac{n_y}{||\mathbf{n}||}\right)^2 + \left(\frac{n_z}{||\mathbf{n}||}\right)^2 = 1 \quad \text{or} \quad l^2 + m^2 + n^2 = 1 \quad (14.62)$$

Returning now to tetrahedron $OABC$, if S_x, S_y and S_z are the coordinate components of S then

$$\begin{aligned} S_x &= l\sigma_{xx} + m\tau_{xy} + n\tau_{xz} \\ S_y &= m\sigma_{yy} + n\tau_{yz} + l\tau_{yx} \\ S_z &= n\sigma_{zz} + l\tau_{zx} + m\tau_{zy} \end{aligned} \quad (14.63)$$

Since the area of ABC is unity then the normal stress to ABC, S_N, is

$$S_N = S_x \cos\alpha + S_y \cos\beta + S_z \cos\gamma = S_x l + S_y m + S_z n \quad (14.64)$$

substituting S_x, S_y and S_z

$$S_N = \sigma_{xx} l^2 + \sigma_{yy} m^2 + \sigma_{zz} n^2 + 2\left(lm\tau_{xy} + mn\tau_{yz} + ln\tau_{zx}\right) \quad (14.65)$$

with $S_S = S^2 - S_N^2$ where $S^2 = S_x^2 + S_y^2 + S_z^2$.

When the shear stress S_S on face ABC is zero then the normal stress S_N becomes the principal stress. With $S_S = 0$, $S = S_N$ and substituting S for S_x, S_y and S_z in (14.63) where $S_x = lS$, $S_y = mS$ and $S_z = nS$

$$\begin{aligned} S_x = lS &= l\sigma_{xx} + m\tau_{xy} + n\tau_{xz} \\ S_y = mS &= m\sigma_{yy} + n\tau_{yz} + l\tau_{yx} \\ S_z = nS &= n\sigma_{zz} + l\tau_{zx} + m\tau_{zy} \end{aligned} \quad (14.66)$$

or

$$
\begin{aligned}
l\left(\sigma_{xx}-S\right)+m\tau_{xy}+n\tau_{xz}&=0\\
l\tau_{yx}+m\left(\sigma_{yy}-S\right)+n\tau_{yz}&=0\\
l\tau_{zx}+m\tau_{zy}+n\left(\sigma_{zz}-S\right)&=0
\end{aligned}
\tag{14.67}
$$

Solving for S, from Cramer's rule for example, the following determinant is determined

$$
\begin{vmatrix}
(\sigma_{xx}-S) & \tau_{xy} & \tau_{xz}\\
\tau_{yx} & (\sigma_{yy}-S) & \tau_{yz}\\
\tau_{zx} & \tau_{zy} & (\sigma_{zz}-S)
\end{vmatrix}=0
\tag{14.68}
$$

which upon expanding gives the *cubic equation of stress*

$$
S^3-I_1S^2+I_2S-I_3=0 \tag{14.69}
$$

where I_1, I_2 and I_3 are the *scalar stress invariants*

$$
\begin{aligned}
I_1&=\sigma_{xx}+\sigma_{yy}+\sigma_{zz}=\sigma_{ii}\\
I_2&=\sigma_{xx}\sigma_{yy}+\sigma_{yy}\sigma_{zz}+\sigma_{zz}\sigma_{xx}-\tau_{xy}^2-\tau_{yz}^2-\tau_{zx}^2=\frac{1}{2}\left(\sigma_{ii}\sigma_{jj}-\sigma_{ij}\sigma_{ji}\right)\\
I_3&=\sigma_{xx}\sigma_{yy}\sigma_{zz}+2\tau_{xy}\tau_{yz}\tau_{zx}-\sigma_{xx}\tau_{yz}^2-\sigma_{yy}\tau_{zx}^2-\sigma_{zz}\tau_{xy}^2=\det\left[\sigma_{ij}\right]
\end{aligned}
\tag{14.70}
$$

or in terms of principal stresses

$$
\begin{aligned}
I_1&=\sigma_1+\sigma_2+\sigma_3\\
I_2&=\sigma_1\sigma_2+\sigma_2\sigma_3+\sigma_1\sigma_3\\
I_3&=\sigma_1\sigma_2\sigma_3
\end{aligned}
\tag{14.71}
$$

The values of the principal stresses induced are independent of the orientation of the original stress element and hence independent of the coordinate axes. Thus, I_1, I_2 and I_3 are constants for any given coordinate system and hence the reason for being referred to as invariant.

Let us conclude this section by noting that the cubic equation of stress was derived purely from equilibrium considerations and is therefore applicable to all types of materials and not restricted to linear elastic materials.

14.8.1 Hydrostatic and Deviatoric Stresses

When we discussed plasticity in Chapter 12 it was noted that the dominant component in the plastic deformation of materials is the shear component. The stress at a point can be separated into purely hydrostatic and purely shear components by subtracting a hydrostatic or mean stress component from the total stress. In terms of principal stresses the mean stress, σ_m, is given by

$$
\sigma_m=\frac{\sigma_1+\sigma_2+\sigma_3}{3}=\frac{I_1}{3} \tag{14.72}
$$

Subtracting this mean stress from the total stress components leads to the shear or *deviatoric* stress components

$$
\sigma_1'=\sigma_1-\frac{I_1}{3};\quad \sigma_2'=\sigma_2-\frac{I_1}{3};\quad \sigma_3'=\sigma_3-\frac{I_1}{3} \tag{14.73}
$$

14.8.2 Scalar Invariants of the Deviatoric Stress

The general cubic equation of stress can now be re-arranged by replacing the actual stress components by the corresponding deviatoric stress components. Thus, substituting for $S\ (=\sigma' + I_1/3)$ into the cubic equation (14.69) we obtain

$$\left(\sigma' + \frac{I_1}{3}\right)^3 - I_1\left(\sigma' + \frac{I_1}{3}\right)^2 + I_2\left(\sigma' + \frac{I_1}{3}\right) - I_3 = 0 \tag{14.74}$$

Expanding and re-arranging this relation reduces to

$$(\sigma')^3 - \frac{(I_1^2 + 3I_2)}{3}\sigma' - \frac{(2I_1^3 + 9I_1I_2 + 27I_3)}{27} = 0 \tag{14.75}$$

with the term involving $(\sigma')^2$ vanishing. Alternatively this can be written as

$$(\sigma')^3 - J_2\sigma' - J_3 = 0 \tag{14.76}$$

where the coefficients J of σ' are known as the *scalar invariants of the deviatoric stress*.

Comparing the above two equations we find that J_2 is

$$J_2 = \frac{I_1^3 + 3I_2}{3} = \frac{1}{6}\left[(\sigma_1 - \sigma_2)^2 + (\sigma_2 - \sigma_3)^2 + (\sigma_1 - \sigma_3)^2\right] \tag{14.77}$$

Comparing the coefficient J_2 with the Huber-von Mises yield criterion, (12.45), we observe

$$(\sigma_1 - \sigma_2)^2 + (\sigma_2 - \sigma_3)^2 + (\sigma_1 - \sigma_3)^2 = 2\sigma_Y^2 = 6k^2 \tag{14.78}$$

where σ_Y and k denote the yield stress in pure tension and pure shear respectively. Comparing (14.77) and (14.78) we observe that J_2 has particular physical significance when analysing the yielding of materials and the Huber-von Mises criterion can be conveniently expressed as

$$J_2 - k^2 = 0 \tag{14.79}$$

14.8.3 Maximum Shear Stresses

When the principal stresses and their directions are known, it is convenient to take the principal axes as the axes of reference. If Ox, Oy and Oz denote the coordinate axes associated with the principal stresses σ_1, σ_2 and σ_3, respectively, the corresponding components of the stress vector across a plane whose normal is in the direction (l, m, n) are

$$S_x = l\sigma_1; \quad S_y = m\sigma_2; \quad S_z = n\sigma_3 \tag{14.80}$$

The normal stress $(\sigma = S_N)$ across the oblique plane therefore becomes

$$\sigma = S_N = lS_x + mS_y + nS_z = l^2\sigma_1 + m^2\sigma_2 + n^2\sigma_3 \tag{14.81}$$

If the magnitude of the shear stress across the plane is denoted by $(\tau = S_S)$, then

$$\begin{aligned}\tau^2 = S_S^2 &= S^2 - S_N^2 = \left(S_x^2 + S_y^2 + S_z^2\right) - S_N^2 \\ &= (\sigma_1 - \sigma_2)^2 l^2m^2 + (\sigma_2 - \sigma_3)^2 m^2n^2 + (\sigma_3 - \sigma_1)^2 n^2l^2\end{aligned} \tag{14.82}$$

in view of the relation $(l^2 + m^2 + n^2 = 1)$. To find the stationary values of the shear stress, for varying orientations of the oblique plane, we put $n^2 = l^2 - m^2$ in the above

$$\tau^2 = l^2\left(\sigma_1^2 - \sigma_3^2\right) + m^2\left(\sigma_2^2 - \sigma_3^2\right) + \sigma_3^2 - \left[l^2(\sigma_1 - \sigma_3) + m^2(\sigma_2 - \sigma_3) + \sigma_3\right]^2 \tag{14.83}$$

where l and m are treated as the independent variables. Equating to zero the derivatives of τ^2 with respect to l and m, we obtain

$$\begin{aligned} l\left(\sigma_1-\sigma_3\right)\left[\left(1-2l^2\right)\left(\sigma_1-\sigma_3\right)-2m^2\left(\sigma_2-\sigma_3\right)\right]=0 \\ m\left(\sigma_2-\sigma_3\right)\left[\left(1-2m^2\right)\left(\sigma_2-\sigma_3\right)-2l^2\left(\sigma_1-\sigma_3\right)\right]=0 \end{aligned} \tag{14.84}$$

These equations are obviously satisfied for $l = m = 0$ and hence $n = 1$ which corresponds to a principal stress direction for which the shear stress has a minimum value of zero (since the shear stresses are zero on principal planes).

To obtain a maximum value of the shear stress, we set $l = 0$ satisfying the first equation of (14.84) and use this value in the second equation to obtain $1 - 2m^2 = 0$. This gives $l = 0$, $m^2 = n^2 = 1/2$ corresponding to a maximum shear stress equal to $(\sigma_2 - \sigma_3)/2$. Following a similar procedure for $(m = 0,\ n^2 = l^2 = 1/2)$ and $(n = 0,\ l^2 + m^2 = 1)$ we find the other maximum values of the shear stress. These are

$$\tau_1=\frac{1}{2}\left(\sigma_2-\sigma_3\right); \quad \tau_2=\frac{1}{2}\left(\sigma_1-\sigma_3\right); \quad \tau_3=\frac{1}{2}\left(\sigma_1-\sigma_2\right) \tag{14.85}$$

Following the convention $(\sigma_1 > \sigma_2 > \sigma_3)$ then the greatest shear stress is of magnitude $(\sigma_1 - \sigma_3)/2$ and acts across a plane whose normal bisects the angle between the directions of σ_1 and σ_3.

14.8.4 Plane Stress States

For a plane system $\sigma_{zz} = \tau_{zx} = \tau_{zy} = 0$ the scalar stress invariants reduce to $I_1 = \sigma_{xx} + \sigma_{yy}$, $I_2 = \sigma_{xx}\sigma_{yy} - \tau_{xy}^2$ and $I_3 = 0$ and the cubic equation of stress reduces to the quadratic $S^2 - I_1 S + I_2 = 0$ with solution

$$S=\frac{I_1 \pm \sqrt{I_1^2-4I_2}}{2}=\frac{\sqrt{\left(\sigma_{xx}+\sigma_{yy}\right)^2-4\left(\sigma_{xx}\sigma_{yy}-\tau_{xy}^2\right)}}{2} \tag{14.86}$$

which is found to reduce to

$$\begin{aligned} S &= \left(\frac{\sigma_{xx}+\sigma_{yy}}{2}\right) \pm \frac{1}{2}\sqrt{\left(\sigma_{xx}-\sigma_{yy}\right)^2+4\tau_{xy}^2} = \\ &= \left(\frac{\sigma_{xx}+\sigma_{yy}}{2}\right) \pm \sqrt{\left(\frac{\sigma_{xx}-\sigma_{yy}}{2}\right)^2+\tau_{xy}^2} \end{aligned} \tag{14.87}$$

and are seen to agree with σ_1 and σ_2 of (7.19). The maximum shear stress in the present case is, τ_3

$$\tau_{max}=\tau_3=\frac{\sigma_1-\sigma_2}{2}=\sqrt{\left(\frac{\sigma_{xx}-\sigma_{yy}}{2}\right)^2+\tau_{xy}^2} \tag{14.88}$$

where the positive root has been taken as the maximum value. Similarly, this result agrees with the previously found expression for τ_{max} for a two-dimensional analysis, (7.25).

14.9 Matrix Representation of Stress and Strain

For a three-dimensional body of an isotropic material the Hookian equations are given by (14.42). Alternatively, these stress-strain constitutive equations can be written in matrix form as

$$\begin{Bmatrix} \varepsilon_{xx} \\ \varepsilon_{yy} \\ \varepsilon_{zz} \\ \gamma_{xy} \\ \gamma_{yz} \\ \gamma_{xz} \end{Bmatrix} = \begin{bmatrix} 1/E & -\nu/E & -\nu/E & 0 & 0 & 0 \\ -\nu/E & 1/E & -\nu/E & 0 & 0 & 0 \\ -\nu/E & -\nu/E & 1/E & 0 & 0 & 0 \\ 0 & 0 & 0 & 1/2G & 0 & 0 \\ 0 & 0 & 0 & 0 & 1/2G & 0 \\ 0 & 0 & 0 & 0 & 0 & 1/2G \end{bmatrix} \begin{Bmatrix} \sigma_{xx} \\ \sigma_{yy} \\ \sigma_{zz} \\ \tau_{xy} \\ \tau_{yz} \\ \tau_{xz} \end{Bmatrix} \tag{14.89}$$

or, more concisely

$$\{\varepsilon_i\} = [G_{ij}]\{\sigma_j\}; \quad i,j = 1,\cdots,6 \tag{14.90}$$

where $\varepsilon_1 = \varepsilon_{xx}$, $\varepsilon_2 = \varepsilon_{yy}$, $\cdots$, $\varepsilon_4 = \gamma_{xy}$, $\cdots$, $\sigma_1 = \sigma_{xx}$, $\cdots$ The **G** matrix is referred to as the *compliance matrix*. Inverting (14.90) expresses the stresses in terms of the strains

$$\{\sigma_i\} = [G_{ij}]^{-1}\{\varepsilon_j\} = [C_{ij}]\{\varepsilon_j\} \tag{14.91}$$

where **C** is referred to as the *stiffness matrix*; refer to §13.11 for a discussion on the use of the stiffness matrix in the finite element method.

In the most general case a stress or strain component with subscript ij differs from a stress or strain component with subscript ji; for example $\tau_{xy} \neq \tau_{yx}$. In this case **C** and **G** would both be of size (9×9) with (14.91) written in its most general tensorial form

$$\sigma_{ij} = C_{ijkl}\varepsilon_{kl} \tag{14.92}$$

When $\tau_{xy} = \tau_{yx}$ and $\varepsilon_{xy} = \varepsilon_{yx}$ then both **C** and **G** reduce to (6×6) matrices and in the most general case will consist of 36 different elements. Compatibility requires that both **C** and **G** are symmetric. The most general case is for an *anisotropic* or *triclinic* material which has no planes of symmetry and requires a total of 21 material constants to fully describe the stress-strain behaviour

$$\begin{Bmatrix} \sigma_{xx} \\ \sigma_{yy} \\ \sigma_{zz} \\ \tau_{xy} \\ \tau_{yz} \\ \tau_{xz} \end{Bmatrix} = \begin{bmatrix} C_{11} & C_{12} & C_{13} & C_{14} & C_{15} & C_{16} \\ & C_{22} & C_{23} & C_{24} & C_{25} & C_{26} \\ & & C_{33} & C_{34} & C_{35} & C_{36} \\ & & & C_{44} & C_{45} & C_{46} \\ & \text{sym} & & & C_{55} & C_{56} \\ & & & & & C_{66} \end{bmatrix} \begin{Bmatrix} \varepsilon_{xx} \\ \varepsilon_{yy} \\ \varepsilon_{zz} \\ \gamma_{xy} \\ \gamma_{yz} \\ \gamma_{xz} \end{Bmatrix} \tag{14.93}$$

If the material has 1 plane of symmetry the material is referred to as *monoclinic* with the material described by 13 unique constants

$$\begin{Bmatrix} \sigma_{xx} \\ \sigma_{yy} \\ \sigma_{zz} \\ \tau_{xy} \\ \tau_{yz} \\ \tau_{xz} \end{Bmatrix} = \begin{bmatrix} C_{11} & C_{12} & C_{13} & 0 & 0 & C_{16} \\ & C_{22} & C_{23} & 0 & 0 & C_{26} \\ & & C_{33} & 0 & 0 & C_{36} \\ & & & C_{44} & C_{45} & 0 \\ & \text{sym} & & & C_{55} & 0 \\ & & & & & C_{66} \end{bmatrix} \begin{Bmatrix} \varepsilon_{xx} \\ \varepsilon_{yy} \\ \varepsilon_{zz} \\ \gamma_{xy} \\ \gamma_{yz} \\ \gamma_{xz} \end{Bmatrix} \tag{14.94}$$

When the material has 3 planes of symmetry the material is referred to as *orthotropic* (in an analogous manner to orthogonal Cartesian coordinate axes) and requires 9 unique material constants

$$\begin{Bmatrix} \sigma_{xx} \\ \sigma_{yy} \\ \sigma_{zz} \\ \tau_{xy} \\ \tau_{yz} \\ \tau_{xz} \end{Bmatrix} = \begin{bmatrix} C_{11} & C_{12} & C_{13} & 0 & 0 & 0 \\ & C_{22} & C_{23} & 0 & 0 & 0 \\ & & C_{33} & 0 & 0 & 0 \\ & & & C_{44} & 0 & 0 \\ & \text{sym} & & & C_{55} & 0 \\ & & & & & C_{66} \end{bmatrix} \begin{Bmatrix} \varepsilon_{xx} \\ \varepsilon_{yy} \\ \varepsilon_{zz} \\ \gamma_{xy} \\ \gamma_{yz} \\ \gamma_{xz} \end{Bmatrix} \quad (14.95)$$

When the material has 1 plane of isotropy the material is referred to as *transversely isotropic* with 5 material constants

$$\begin{Bmatrix} \sigma_{xx} \\ \sigma_{yy} \\ \sigma_{zz} \\ \tau_{xy} \\ \tau_{yz} \\ \tau_{xz} \end{Bmatrix} = \begin{bmatrix} C_{11} & C_{12} & C_{13} & 0 & 0 & 0 \\ & C_{22} & C_{23} & 0 & 0 & 0 \\ & & C_{33} & 0 & 0 & 0 \\ & & & C_{44} & 0 & 0 \\ & \text{sym} & & & C_{55} & 0 \\ & & & & & (C_{11} - C_{12}) \end{bmatrix} \begin{Bmatrix} \varepsilon_{xx} \\ \varepsilon_{yy} \\ \varepsilon_{zz} \\ \gamma_{xy} \\ \gamma_{yz} \\ \gamma_{xz} \end{Bmatrix} \quad (14.96)$$

We will examine transversely isotropic materials when we discuss composite materials in Chapter 21.

Finally, let us consider the case of an *isotropic* material which has an infinite number of planes of symmetry. The number of unique material constants reduces to 2

$$\begin{Bmatrix} \sigma_{xx} \\ \sigma_{yy} \\ \sigma_{zz} \\ \tau_{xy} \\ \tau_{yz} \\ \tau_{xz} \end{Bmatrix} = \begin{bmatrix} C_{11} & C_{12} & C_{12} & 0 & 0 & 0 \\ & C_{11} & C_{12} & 0 & 0 & 0 \\ & & C_{11} & 0 & 0 & 0 \\ & & & (C_{11} - C_{12}) & 0 & 0 \\ & \text{sym} & & & (C_{11} - C_{12}) & 0 \\ & & & & & (C_{11} - C_{12}) \end{bmatrix} \begin{Bmatrix} \varepsilon_{xx} \\ \varepsilon_{yy} \\ \varepsilon_{zz} \\ \gamma_{xy} \\ \gamma_{yz} \\ \gamma_{xz} \end{Bmatrix} \quad (14.97)$$

or in terms of the more familiar constants, Young's modulus E, and Poisson's ratio ν, we have

$$\begin{Bmatrix} \sigma_{xx} \\ \sigma_{yy} \\ \sigma_{zz} \\ \tau_{xy} \\ \tau_{yz} \\ \tau_{xz} \end{Bmatrix} = \frac{E}{(1+\nu)(1-2\nu)} \begin{bmatrix} 1-\nu & \nu & \nu & 0 & 0 & 0 \\ & 1-\nu & \nu & 0 & 0 & 0 \\ & & 1-\nu & 0 & 0 & 0 \\ & & & (1-2\nu)/2 & 0 & 0 \\ & \text{sym} & & & (1-2\nu)/2 & 0 \\ & & & & & (1-2\nu)/2 \end{bmatrix} \begin{Bmatrix} \varepsilon_{xx} \\ \varepsilon_{yy} \\ \varepsilon_{zz} \\ \gamma_{xy} \\ \gamma_{yz} \\ \gamma_{xz} \end{Bmatrix} \quad (14.98)$$

In the case of plane stress there exists no through-thickness stress components and $\sigma_{zz} = \tau_{yz} = \tau_{xz} = 0$ with (14.98) reducing to

$$\begin{Bmatrix} \sigma_{xx} \\ \sigma_{yy} \\ \tau_{xy} \end{Bmatrix} = \frac{E}{(1-\nu^2)} \begin{bmatrix} 1 & \nu & 0 \\ \nu & 1 & 0 \\ 0 & 0 & (1-\nu)/2 \end{bmatrix} \begin{Bmatrix} \varepsilon_{xx} \\ \varepsilon_{yy} \\ \gamma_{xy} \end{Bmatrix} \quad (14.99)$$

with the **C** matrix equal to the **D** matrix used in the finite element method. In the case of plane strain the through-thickness strain $\varepsilon_{zz} = 0$ and $\sigma_{zz} = \nu\left(\sigma_{xx} + \sigma_{yy}\right)$ with (14.98) reducing to

$$\begin{Bmatrix} \sigma_{xx} \\ \sigma_{yy} \\ \tau_{xy} \end{Bmatrix} = \frac{E}{(1+\nu)(1-2\nu)} \begin{bmatrix} 1-\nu & \nu & 0 \\ \nu & 1-\nu & 0 \\ 0 & 0 & (1-2\nu)/2 \end{bmatrix} \begin{Bmatrix} \varepsilon_{xx} \\ \varepsilon_{yy} \\ \gamma_{xy} \end{Bmatrix} \quad (14.100)$$

14.10 Reciprocal Theorems

The following two sub-sections examine the reciprocal displacement and work theorems. Both theorems are applicable to linear elastic materials in which small-displacement theory applies.

14.10.1 Reciprocal-Displacement Theorem

The reciprocal displacement theorem is best illustrated by way of an example. Consider the simply supported beam of Figure 14.10. In case a) of Figure 14.10 the concentrated load P acts at point 2 and the displacement, δ_{12}, is measured at point 1. For the particular case of $a = 3L/4, b = L/4$ and $x = L/2$ with $0 \leq x \leq a$ then from (5.56) δ_{12} is

$$\delta_{12} = \frac{11PL^3}{768EI} \tag{14.101}$$

The subscript notation used with δ indicates that δ is measured at point 1 and P is applied at point 2. In case b) of Figure 14.10 P now acts at point 1 and the displacement, δ_{21}, is measured at point 2. From (5.56) with $a = b = L/2, x = 3L/4$ and x now in the range $a \leq x \leq L$ then δ_{21} is given by

$$\delta_{21} = \frac{11PL^3}{768EI} \tag{14.102}$$

Thus, comparing (14.101) and (14.102) we observe that the deflection at point 1 due to P acting at point 2 is equivalent to the deflection at point 2 due to P acting at point 1; namely $\delta_{12} = \delta_{21}$.

To develop a more general result for the reciprocal displacement theorem consider once again the simply supported beam of Figure 14.10 but with load P simultaneously applied at points 1 and 2. Assuming the principle of superposition applies for small displacements then the deflection at point 1 is now $\delta_{22} + \delta_{21}$. It follows that the work or strain energy, U, for a linear elastic material due to the two loads is

$$U = \frac{1}{2}P\left(\delta_{11} + \delta_{12}\right) + \frac{1}{2}P\left(\delta_{22} + \delta_{21}\right) \tag{14.103}$$

Assuming that the beam responds in a linear manner the total strain energy due to the two loads is independent of the order in which the loads are applied. Therefore, let us first assume that the first load P is applied at point 1 with corresponding strain energy

$$\frac{1}{2}P\delta_{11} \tag{14.104}$$

Maintaining the load at point 1 then the strain energy due to the second load applied at point 2 is

$$\frac{1}{2}P\delta_{22} \tag{14.105}$$

In addition there is the contribution to the strain energy due to the additional deflection δ_{12} at point 1 when the load at point 2 is being applied

$$P\delta_{12} \tag{14.106}$$

and does not contain a factor of $\frac{1}{2}$ since P remains constant throughout the additional deflection. Summing (14.104), (14.105) and (14.106) then the strain energy due to sequential loading is

$$U = \frac{1}{2}P\delta_{11} + \frac{1}{2}P\delta_{22} + P\delta_{12} \tag{14.107}$$

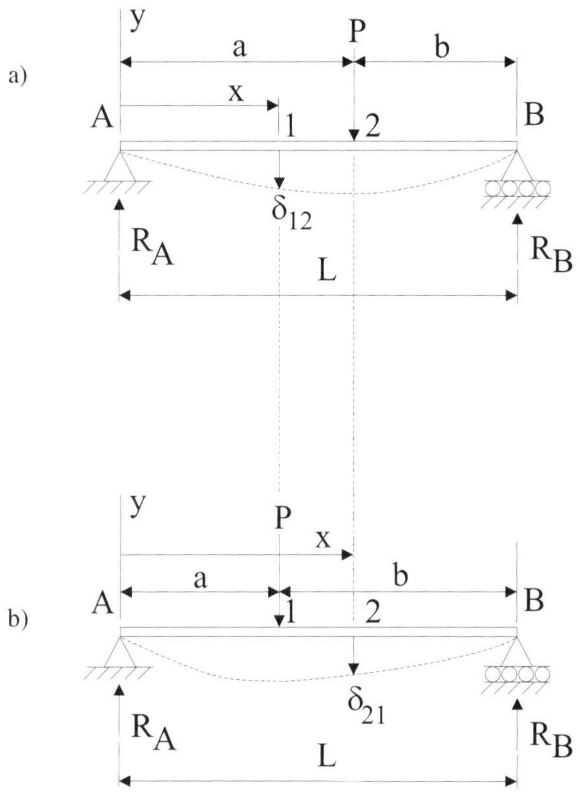

Figure 14.10: Reciprocal displacement theorem. a) Simply supported beam with concentrated load P at point 2 and displacement δ_{12} measured at point 1. b) Simply supported beam with concentrated load P at point 1 and displacement δ_{21} measured at point 2.

Finally, equating (14.103) and (14.107) we arrive at the *reciprocal displacement theorem*

$$\delta_{12} = \delta_{21} \tag{14.108}$$

The reciprocal displacement theorem is also referred to as Maxwell's reciprocal theorem. Although we considered the deflection of a beam in deriving (14.108) it is emphasised that the reciprocal displacement theorem is applicable to all linear elastic structures in which the principle of superposition applies.

14.10.2 Reciprocal-Work Theorem

We will derive the reciprocal work theorem for the two geometrically equivalent elastic states of equilibrium shown in Figure 14.11. The first state consists of m loads $P_1^{(1)}, \cdots, P_m^{(1)}$ and corresponding displacements $\delta_{p1}^{(1)}, \cdots, \delta_{pm}^{(1)}$ and the second state consists of n loads $Q_1^{(2)}, \cdots, Q_n^{(2)}$ and corresponding displacements $\delta_{q1}^{(2)}, \cdots, \delta_{qn}^{(2)}$. The loads in the second state cause displacements in the first state and vice versa. Adopting a similar approach to that used when deriving the reciprocal displacement theorem, we

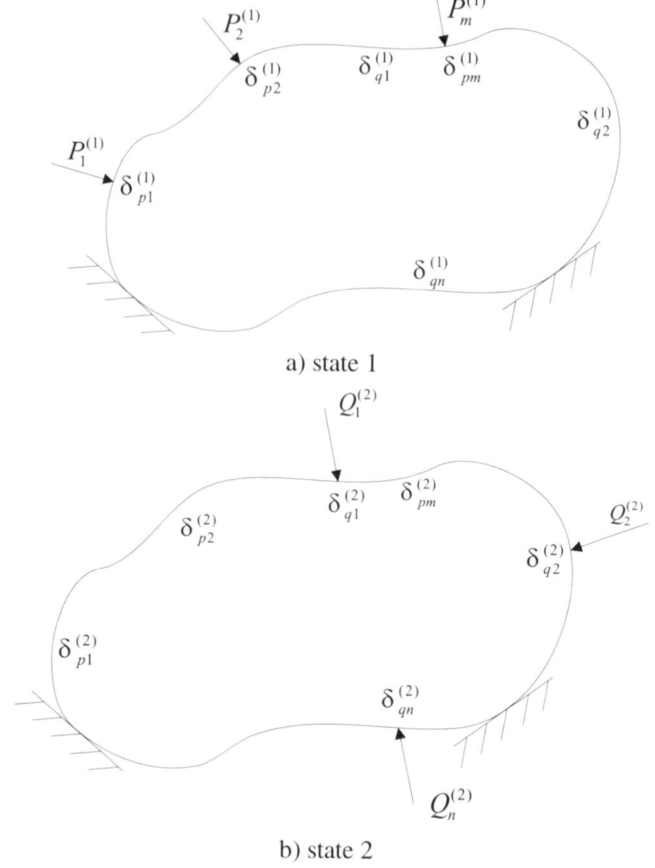

Figure 14.11: Reciprocal work theorem. a) State 1 with m loads $P_1, \cdots, P_m$. b) State 2 with n loads $Q_1, \cdots, Q_n$.

will consider the strain energy due to simultaneous and sequential loading of the structure. Firstly, when the P and Q loads are applied simultaneously the strain energy is

$$\begin{aligned} U = &\frac{1}{2}P_1^{(1)}\left(\delta_{p1}^{(1)} + \delta_{p1}^{(2)}\right) + \cdots + \frac{1}{2}P_m^{(1)}\left(\delta_{pm}^{(1)} + \delta_{pm}^{(2)}\right) \\ &+ \frac{1}{2}Q_1^{(2)}\left(\delta_{q1}^{(1)} + \delta_{q1}^{(2)}\right) + \cdots + \frac{1}{2}Q_n^{(2)}\left(\delta_{qn}^{(1)} + \delta_{qn}^{(2)}\right) \end{aligned} \tag{14.109}$$

When the P loads are applied alone the strain energy is

$$\frac{1}{2}P_1^{(1)}\delta_{p1}^{(1)} + \cdots + \frac{1}{2}P_m^{(1)}\delta_{pm}^{(1)} \tag{14.110}$$

followed by the subsequent application of the Q loads then the strain energy due to the Q loads alone is

$$\frac{1}{2}Q_1^{(2)}\delta_{q1}^{(2)} + \cdots + \frac{1}{2}Q_n^{(2)}\delta_{qn}^{(2)} \tag{14.111}$$

with the additional strain energy due to the work done by the P loads

$$P_1^{(1)}\delta_{p1}^{(2)} + \cdots + P_m^{(1)}\delta_{pm}^{(2)} \tag{14.112}$$

Summing the strain energies due to (14.110), (14.111) and (14.112) and equating to (14.109) we arrive at the *reciprocal work theorem*

$$\sum_{i=1}^{m} P_i^{(1)} \delta_{pi}^{(2)} = \sum_{j=1}^{n} Q_j^{(2)} \delta_{qj}^{(1)} \tag{14.113}$$

which is also referred to as the Betti-Rayleigh reciprocal theorem.

In conclusion, the reciprocal work theorem can be expressed more generally as

$$\oint_{\Gamma} \left(u_i^{(1)} \sigma_{ij}^{(2)} - u_i^{(2)} \sigma_{ij}^{(1)} \right) n_j \, \mathrm{d}s = 0 \tag{14.114}$$

where $u_i = u_i(x, y)$ now denotes the two-dimensional displacement vector and $\sigma_{ij} = \sigma_{ij}(x, y)$ denotes the two-dimensional stress tensor. The contour Γ is assumed to be continuous, closed and encompassing no singularities or body forces. The unit outward normal to Γ is n_j and $\mathrm{d}s$ is an element of arc length along Γ. Equation (14.114) is verified with the help of Gauss' divergence theorem to transform the line integral to a surface integral

$$\begin{aligned} \oint_{\Gamma} u_i^{(1)} \sigma_{ij}^{(2)} n_j \, \mathrm{d}s &= \oint_{\Omega} u_i^{(1)} \frac{\partial \sigma_{ij}^{(2)}}{\partial x_j} \, \mathrm{d}S + \oint_{\Omega} \sigma_{ij}^{(2)} \frac{\partial u_i^{(1)}}{\partial x_j} \, \mathrm{d}S \\ \oint_{\Gamma} u_i^{(2)} \sigma_{ij}^{(1)} n_j \, \mathrm{d}s &= \oint_{\Omega} u_i^{(2)} \frac{\partial \sigma_{ij}^{(1)}}{\partial x_j} \, \mathrm{d}S + \oint_{\Omega} \sigma_{ij}^{(1)} \frac{\partial u_i^{(2)}}{\partial x_j} \, \mathrm{d}S \end{aligned} \tag{14.115}$$

where Ω represents a continuous surface with elemental area $\mathrm{d}S$. From the differential equations of equilibrium (14.13) then $\sigma_{ij,j}^{(1)} = 0$ and $\sigma_{ij,j}^{(2)} = 0$ and noting that $u_{i,j}^{(1)} \sigma_{ij}^{(2)} = u_{i,j}^{(2)} \sigma_{ij}^{(1)}$ in virtue of the symmetry of the elastic moduli tensor. Upon substitution of these results into (14.115) we observe that (14.114) is valid.

14.11 Complex Variable Representation of the Equations of Plane Elasticity

This section is concerned with representing the general solution of the equations of plane elasticity in terms of complex stress functions. This complex representation follows that of Muskhelishvili (1953) and is found useful for the solution of several important boundary value problems. In the next sub-section we derive expressions for the in-plane displacements in terms of the Airy stress function which will prove necessary in the subsequent sub-section.

14.11.1 Displacements from the Airy Stress Function

The Airy stresses, (14.46), express the in-plane stresses in terms of the Airy stress function, U[1]. We will now derive formulae which express the in-plane displacements in terms of the Airy stress function. Substituting the stress relations (14.45) into (14.46) we have

$$\lambda \varepsilon + 2\mu \frac{\partial u}{\partial x} = \frac{\partial^2 U}{\partial y^2}; \quad \lambda \varepsilon + 2\mu \frac{\partial v}{\partial y} = \frac{\partial^2 U}{\partial x^2}; \quad \mu \left(\frac{\partial u}{\partial y} + \frac{\partial v}{\partial x} \right) = -\frac{\partial^2 U}{\partial x \partial y} \tag{14.116}$$

[1]The Airy stress function will be denoted by U rather than ϕ to eliminate a clash of notation in later sections

where μ denotes the shear modulus. Substituting $\varepsilon = \varepsilon_{xx} + \varepsilon_{yy} = \partial u/\partial x + \partial v/\partial y$ into the first two equations we have

$$\lambda\left(\frac{\partial u}{\partial x} + \frac{\partial v}{\partial y}\right) + 2\mu\frac{\partial u}{\partial x} = \frac{\partial^2 U}{\partial y^2}; \quad \lambda\left(\frac{\partial u}{\partial x} + \frac{\partial v}{\partial y}\right) + 2\mu\frac{\partial v}{\partial y} = \frac{\partial^2 U}{\partial x^2} \tag{14.117}$$

and solving for $\partial u/\partial x$ and $\partial v/\partial y$ gives

$$2\mu\frac{\partial u}{\partial x} = \frac{\partial^2 U}{\partial y^2} - \frac{\lambda}{2(\lambda+\mu)}\nabla^2 U; \quad 2\mu\frac{\partial v}{\partial y} = \frac{\partial^2 U}{\partial x^2} - \frac{\lambda}{2(\lambda+\mu)}\nabla^2 U \tag{14.118}$$

Letting $\nabla^2 U = P$ and substituting into (14.118) with $\partial^2 U/\partial y^2 = P - \partial^2 U/\partial x^2$ and $\partial^2 U/\partial x^2 = P - \partial^2 U/\partial y^2$ then

$$2\mu\frac{\partial u}{\partial x} = -\frac{\partial^2 U}{\partial x^2} + \frac{\lambda+2\mu}{2(\lambda+\mu)}P; \quad 2\mu\frac{\partial v}{\partial y} = -\frac{\partial^2 U}{\partial y^2} + \frac{\lambda+2\mu}{2(\lambda+\mu)}P \tag{14.119}$$

Since $\nabla^2 P = \nabla^4 U = 0$ then P is harmonic and letting Q also be a harmonic function conjugate to P such that P and Q satisfy the Cauchy-Riemann equations

$$\frac{\partial P}{\partial x} = \frac{\partial Q}{\partial y}; \quad \frac{\partial P}{\partial y} = -\frac{\partial Q}{\partial x} \tag{14.120}$$

It follows therefore that P and Q satisfy the analytic complex function

$$f(z) = P(x,y) + iQ(x,y) \tag{14.121}$$

where $i = \sqrt{-1}$ is the imaginary unit. For a discussion of the Cauchy-Riemann equations refer to Kreyszig (1983). In addition, let

$$\phi(z) = p + qi = \frac{1}{4}\int f(z)\,\mathrm{d}z \tag{14.122}$$

with

$$\phi'(z) = \frac{\partial p}{\partial x} + \frac{\partial q}{\partial y}i = \frac{1}{4}(P + iQ) \tag{14.123}$$

A further application of the Cauchy-Riemann equations for $\phi(z)$ gives

$$\frac{\partial p}{\partial x} = \frac{\partial q}{\partial y} = \frac{1}{4}P; \quad \frac{\partial p}{\partial y} = -\frac{\partial q}{\partial x} = -\frac{1}{4}Q \tag{14.124}$$

so that

$$P = 4\frac{\partial p}{\partial x} = 4\frac{\partial q}{\partial y} \tag{14.125}$$

Substituting into (14.119) we have

$$2\mu\frac{\partial u}{\partial x} = -\frac{\partial^2 U}{\partial x^2} + \frac{2(\lambda+2\mu)}{\lambda+\mu}\frac{\partial p}{\partial x}; \quad 2\mu\frac{\partial v}{\partial y} = -\frac{\partial^2 U}{\partial y^2} + \frac{2(\lambda+2\mu)}{\lambda+\mu}\frac{\partial q}{\partial y} \tag{14.126}$$

and integrating

$$2\mu u = -\frac{\partial U}{\partial x} + \frac{2(\lambda+2\mu)}{\lambda+\mu}p + f_1(y); \quad 2\mu v = -\frac{\partial U}{\partial y} + \frac{2(\lambda+2\mu)}{\lambda+\mu}q + f_2(x) \tag{14.127}$$

Substituting these into the third equation of (14.116) and noting from the Cauchy-Riemann equations $\partial p/\partial y + \partial q/\partial x = 0$ we find that

$$f_1'(y) + f_2'(x) = 0 \tag{14.128}$$

The functions f_1 and f_2 are responsible for rigid body displacement only so that the u and v in-plane displacements in terms of the Airy stress function U are finally given by

$$2\mu u = -\frac{\partial U}{\partial x} + \frac{2(\lambda + 2\mu)}{\lambda + \mu}p; \quad 2\mu v = -\frac{\partial U}{\partial y} + \frac{2(\lambda + 2\mu)}{\lambda + \mu}q \tag{14.129}$$

14.11.2 Complex Representation of the Airy Stress Function, Displacements and Stresses

From (14.129) we observe that the function $U - px - qy$ is harmonic; namely

$$\nabla^2(U - px - qy) = 0 \tag{14.130}$$

so that $U = px + qy + p_1$ where p_1 is a known function. Letting $\chi(z)$ denote the real part of p_1 then U is given by

$$U = \Re\left[\bar{z}\phi(z) + \chi(z)\right] \tag{14.131}$$

where $\Re[f]$ denotes the real part of f and $\bar{z}(= x - iy)$ denotes the complex conjugate of $z(= x + iy)$. Furthermore, $\Re[f] = (f + \bar{f})/2$ and $\Im[f] = (f - \bar{f})/2i$ where $\Im[f]$ denotes the imaginary part of f. In (14.131) $\bar{z}\phi(z) = px + qy - i(py + qx)$ the real part of which is $px + qy$. In view of the following

$$\bar{z}\phi(z) + z\overline{\phi(z)} = 2\Re\left[\bar{z}\phi(z)\right]; \quad \chi(z) + \overline{\chi(z)} = 2\Re\left[\chi(z)\right] \tag{14.132}$$

then U can be written

$$2U = \bar{z}\phi(z) + z\overline{\phi(z)} + \chi(z) + \overline{\chi(z)} \tag{14.133}$$

To determine the displacements we require $\partial U/\partial x$ and $\partial U/\partial y$ which can be obtained by noting the following for an arbitrary function f

$$\frac{\partial f}{\partial z} = \frac{1}{2}\left(\frac{\partial f}{\partial x} - i\frac{\partial f}{\partial y}\right); \quad \frac{\partial f}{\partial \bar{z}} = \frac{1}{2}\left(\frac{\partial f}{\partial x} + i\frac{\partial f}{\partial y}\right) \tag{14.134}$$

Re-arranging for $\partial f/\partial x$ and $\partial f/\partial y$ we have

$$\frac{\partial f}{\partial x} = \frac{\partial f}{\partial z} + \frac{\partial f}{\partial \bar{z}}; \quad \frac{\partial f}{\partial y} = i\left(\frac{\partial f}{\partial z} - \frac{\partial f}{\partial \bar{z}}\right) \tag{14.135}$$

Therefore, with $\partial U/\partial z$ and $\partial U/\partial \bar{z}$ are given by

$$2\frac{\partial U}{\partial z} = \bar{z}\phi'(z) + \overline{\phi(z)} + \chi'(z); \quad 2\frac{\partial U}{\partial \bar{z}} = \phi(z) + z\overline{\phi'(z)} + \overline{\chi'(z)} \tag{14.136}$$

it follows that $\partial U/\partial x$ and $\partial U/\partial y$ are

$$\begin{aligned} 2\frac{\partial U}{\partial x} &= \bar{z}\phi'(z) + \overline{\phi(z)} + \chi'(z) + \phi(z) + z\overline{\phi'(z)} + \overline{\chi'(z)} \\ 2\frac{\partial U}{\partial y} &= i\left[\bar{z}\phi'(z) + \overline{\phi(z)} + \chi'(z) - \phi(z) - z\overline{\phi'(z)} - \overline{\chi'(z)}\right] \end{aligned} \tag{14.137}$$

Combining $\partial U/\partial x$ and $\partial U/\partial y$ we arrive at

$$\frac{\partial U}{\partial x}+i\frac{\partial U}{\partial y}=\phi(z)+z\overline{\phi'(z)}+\overline{\psi(z)} \tag{14.138}$$

where $\psi(z)=\mathrm{d}\chi/\mathrm{d}z$.

Returning now to the displacement expressions (14.129) and multiplying the second equation by i and adding it to the first we have

$$2\mu(u+iv)=-\left(\frac{\partial U}{\partial x}+i\frac{\partial U}{\partial y}\right)+\frac{2(\lambda+2\mu)}{\lambda+\mu}\phi(z) \tag{14.139}$$

and substituting (14.138)

$$2\mu(u+iv)=\left(\frac{\lambda+3\mu}{\lambda+\mu}\right)\phi(z)+z\overline{\phi'(z)}-\overline{\psi(z)}=\kappa\phi(z)-z\overline{\phi'(z)}-\overline{\psi(z)} \tag{14.140}$$

where κ denotes Muskhelishvili's constant, $\kappa=(\lambda+3\mu)/(\lambda+\mu)$.

To determine the stresses in terms of complex stress functions consider the coordinate components S_x and S_y of the resultant stress S acting on a curve C which is the two-dimensional analogue of the three-dimensional stress analysis discussed in §14.8

$$\begin{aligned} S_x&=l\sigma_{xx}+m\tau_{xy}=l\frac{\partial^2 U}{\partial y^2}-m\frac{\partial^2 U}{\partial x\partial y}\\ S_x&=l\tau_{xy}+m\sigma_{yy}=-l\frac{\partial^2 U}{\partial x\partial y}+m\frac{\partial^2 U}{\partial x^2}\end{aligned} \tag{14.141}$$

where $l(=\mathrm{d}y/\mathrm{d}s)$ and $m(=-\mathrm{d}x/\mathrm{d}s)$ are the direction cosines and $\mathrm{d}s$ is the elemental arc length of C. Substituting l and m

$$S_x=\frac{\mathrm{d}}{\mathrm{d}s}\left(\frac{\partial U}{\partial y}\right);\quad S_y=-\frac{\mathrm{d}}{\mathrm{d}s}\left(\frac{\partial U}{\partial x}\right) \tag{14.142}$$

or

$$S_x+iS_y=\frac{\mathrm{d}}{\mathrm{d}s}\left(\frac{\partial U}{\partial y}-i\frac{\partial U}{\partial x}\right)=-i\frac{\mathrm{d}}{\mathrm{d}s}\left(\frac{\partial U}{\partial x}+i\frac{\partial U}{\partial y}\right) \tag{14.143}$$

Substituting (14.138)

$$(S_x+iS_y)\,\mathrm{d}s=-i\,\mathrm{d}\left[\phi(z)+z\overline{\phi'(z)}+\overline{\psi(z)}\right] \tag{14.144}$$

Firstly, letting $\mathrm{d}s$ lie in the direction of the y-axis then $\mathrm{d}s=\mathrm{d}y$, $\mathrm{d}z=-i\,\mathrm{d}y$, $\mathrm{d}\bar{z}=-i\,\mathrm{d}y$, $S_x=\sigma_{xx}$ and $S_y=\tau_{xy}$ so that

$$\sigma_{xx}+i\tau_{xy}=\phi'(z)+\overline{\phi'(z)}-z\overline{\phi''(z)}-\overline{\psi'(z)} \tag{14.145}$$

Next, let $\mathrm{d}s$ lie in the direction of the x-axis then $\mathrm{d}s=\mathrm{d}x$, $\mathrm{d}z=\mathrm{d}\bar{z}$, $S_x=-\tau_{xy}$ and $S_y=-\sigma_{yy}$ so that

$$\sigma_{yy}-i\tau_{xy}=\phi'(z)+\overline{\phi'(z)}+z\overline{\phi''(z)}+\overline{\psi'(z)} \tag{14.146}$$

Letting $\Phi(z)=\phi'(z)$ and $\Psi(z)=\psi'(z)$ then (14.145) and (14.146) can be written

$$\begin{aligned}\sigma_{xx}+i\tau_{xy}&=\Phi(z)+\overline{\Phi(z)}-z\overline{\Phi'(z)}-\overline{\Psi(z)}\\ \sigma_{yy}-i\tau_{xy}&=\Phi(z)+\overline{\Phi(z)}+z\overline{\Phi'(z)}+\overline{\Psi(z)}\end{aligned} \tag{14.147}$$

Adding these two formulae together we have

$$\sigma_{xx} + \sigma_{yy} = 2\left[\Phi(z) + \overline{\Phi(z)}\right] = 4\Re\left[\Phi(z)\right] \tag{14.148}$$

To determine the individual stress components we take the complex conjugate of (14.147); for example from the first of (14.147)

$$\sigma_{xx} - i\tau_{xy} = \overline{\Phi(z)} + \Phi(z) - \bar{z}\Phi(z) - \Psi(z) \tag{14.149}$$

and adding to (14.147) gives σ_{xx}. Following a similar procedure for σ_{yy} we arrive at

$$\begin{aligned}\sigma_{xx} &= \frac{1}{2}\left[\left(\Phi(z) + \overline{\Phi(z)}\right) - z\overline{\Phi'(z)} - \bar{z}\Phi'(z) - \Psi(z) - \overline{\Psi(z)}\right]\\ \sigma_{yy} &= \frac{1}{2}\left[\left(\Phi(z) + \overline{\Phi(z)}\right) + z\overline{\Phi'(z)} + \bar{z}\Phi'(z) + \Psi(z) + \overline{\Psi(z)}\right]\end{aligned} \tag{14.150}$$

and substituting either of these into (14.149) we find τ_{xy}

$$\tau_{xy} = \frac{1}{2i}\left[\bar{z}\Phi'(z) - z\overline{\Phi'(z)} + \Psi(z) - \overline{\Psi(z)}\right] \tag{14.151}$$

The three stresses σ_{xx}, σ_{yy} and τ_{xy} can be coupled to give the expression

$$\sigma_{yy} - \sigma_{xx} + 2i\tau_{xy} = 2\left[\bar{z}\Phi'(z) + \Psi(z)\right] \tag{14.152}$$

Let us conclude by determining expressions for the polar displacements (u_r, u_θ) and polar stresses $(\sigma_{rr}, \sigma_{\theta\theta}, \tau_{r\theta})$ with polar coordinates (r, θ) centred at the origin

$$z = x + iy = re^{i\theta} \tag{14.153}$$

If u_r and u_θ denote the radial and circumferential displacements then the Cartesian and polar displacements are related by

$$\begin{Bmatrix} u \\ v \end{Bmatrix} = \begin{bmatrix} \cos\theta & -\sin\theta \\ \sin\theta & \cos\theta \end{bmatrix} \begin{Bmatrix} u_r \\ u_\theta \end{Bmatrix} \tag{14.154}$$

such that

$$(u + iv) = (u_r + iu_\theta)e^{i\theta}; \quad (u_r + iu_\theta) = (u + iv)e^{-i\theta} \tag{14.155}$$

where, from De Moivre's formula, $e^{i\theta} = \cos\theta + i\sin\theta$. Combining (14.140) and (14.155) we have

$$2\mu\left(u_r + iu_\theta\right) = \left[\kappa\phi(z) - z\overline{\phi'(z)} - \overline{\psi(z)}\right]e^{-i\theta} \tag{14.156}$$

The stress transformation equations (14.29) which map the global Cartesian stresses $(\sigma_{xx}, \sigma_{yy}, \tau_{xy})$ to the local Cartesian stresses $(\sigma_{x'x'}, \sigma_{y'y'}, \tau_{x'y'})$ can be expressed alternatively in terms of complex variables as

$$\begin{aligned}\sigma_{x'x'} + \sigma_{y'y'} &= \sigma_{xx} + \sigma_{yy}\\ \sigma_{y'y'} - \sigma_{x'x'} + 2i\tau_{x'y'} &= \left[\sigma_{yy} - \sigma_{xx} + 2i\tau_{xy}\right]e^{2i\theta}\end{aligned} \tag{14.157}$$

where the local axes (x', y') are rotated through an angle of θ to the (x, y) global axes. Subtracting the second equation from the first equation of (14.157) we obtain

$$2\left(\sigma_{x'x'} - i\tau_{x'y'}\right) = \sigma_{xx} + \sigma_{yy} - \left(\sigma_{yy} - \sigma_{xx} + 2i\tau_{xy}\right)e^{2i\theta} \tag{14.158}$$

By making a direct comparison between the polar axes (r, θ) and the local Cartesian axes (x', y') then from (14.152), (14.157) and (14.158) the polar stresses in terms of the complex stress functions $\Phi(z)$ and $\Psi(z)$ are

$$\begin{aligned} \sigma_{rr} + \sigma_{\theta\theta} &= 2\left[\Phi(z) + \overline{\Phi(z)}\right] = 4\Re\left[\Phi(z)\right] \\ \sigma_{\theta\theta} - \sigma_{rr} + 2i\tau_{r\theta} &= 2\left[\bar{z}\Phi'(z) + \Psi(z)\right]e^{2i\theta} \\ \sigma_{rr} - i\tau_{r\theta} &= \Phi(z) + \overline{\Phi(z)} - \left[\bar{z}\Phi'(z) + \Psi(z)\right]e^{2i\theta} \end{aligned} \tag{14.159}$$

As an illustration of the complex-variable approach to the theory of elasticity we will consider the two problems of concentrated forces acting at the apex of a wedge and a centre cracked plate. We will not attempt to derive $\Phi(z)$ and $\Psi(z)$ but simply quote and refer the reader to Muskhelishvili (1953) for derivations and further details.

14.11.3 Concentrated Forces acting at the Apex of a Symmetrical Wedge

Let us examine the Boussinesq wedge which will be discussed later in §14.15. Figure 14.12 illustrates a wedge of included angle 2α subject to concentrated forces P and Q at the wedge apex. Forces P and Q act parallel and perpendicular to the axis of symmetry of the wedge respectively. In terms of the stress

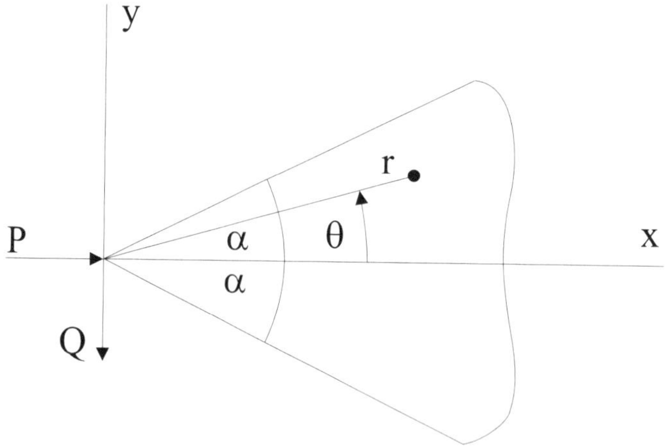

Figure 14.12: Wedge geometry of included angle 2α subject to concentrated forces P and Q.

functions $\Phi(z)$ and $\Psi(z)$ the solution is given by

$$\Phi(z) = -\frac{P}{2Ez} + i\frac{Q}{2Fz}; \quad \Psi(z) = \frac{P}{2Ez} + i\frac{Q}{2Fz} \tag{14.160}$$

where $E = 2\alpha + \sin 2\alpha$ and $F = 2\alpha - sin2\alpha$. For point force P only with $z = re^{i\theta}$ then from (14.159)

$$\sigma_{rr} + \sigma_{\theta\theta} = 4\Re\left[\Phi(z)\right] = 4\Re\left[-\frac{P}{2Ere^{i\theta}}\right] = 2\Re\left[-\frac{Pe^{i\theta}}{Er}\right] = -\frac{2P\cos\theta}{Er} \tag{14.161}$$

From (14.159)

$$\begin{aligned}\sigma_{\theta\theta} - \sigma_{rr} + 2i\tau_{r\theta} &= 2e^{2i\theta}\left[z\Phi'(z) + \Psi(z)\right] \\ &= 2e^{2i\theta}\left[re^{-i\theta}\frac{P}{2Er^2e^{2i\theta}} + \frac{P}{2Ere^{i\theta}}\right]\frac{P}{Er}\left[e^{-i\theta} + e^{i\theta}\right] \\ &= \frac{2P\cos\theta}{Er}\end{aligned} \tag{14.162}$$

and with no imaginary part then $\tau_{r\theta} = 0$. Solving for σ_{rr} and $\sigma_{\theta\theta}$ we find that $\sigma_{\theta\theta} = 0$ and σ_{rr} is given by

$$\sigma_{rr} = -\frac{2P\cos\theta}{Er} \tag{14.163}$$

and is seen to be equivalent to the purely radial stress distribution derived in §14.15. Following a similar procedure for Q then the polar stresses are

$$\sigma_{rr} = -\frac{2P\cos\theta}{Er} + \frac{2Q\sin\theta}{Fr}; \quad \sigma_{\theta\theta} = \tau_{r\theta} = 0 \tag{14.164}$$

and the Cartesian displacements for force P are, from (14.140)

$$u = \frac{P}{4\mu E}\left[-(\kappa+1)\ln r + \cos 2\theta\right]; \quad v = \frac{P}{4\mu E}\left[(\kappa-1)\theta + \sin 2\theta\right] \tag{14.165}$$

and for force Q

$$u = \frac{Q}{4\mu F}\left[(1-\kappa)\theta - \sin 2\theta\right]; \quad v = \frac{Q}{4\mu F}\left[(\kappa-1)\ln r + \cos 2\theta\right] \tag{14.166}$$

14.11.4 A Centre Cracked plate

Chapter 17 examines the subject of fracture mechanics in more detail but as an illustration of the usefulness of adopting a complex-variable approach, consider the centre cracked plate shown in Figure 14.13. The plate contains a traction-free crack of length $2a$ lying along the x-axis with centre at the origin O. The plate is subject to far-field biaxial loadings of σ and $\lambda\sigma$ and pure shear loading of τ. The local crack tip coordinate, centred at $x = a$ is $z_R \left(= re^{i\theta}\right)$ where $r = |x - a|$. The solution to the problem is given by

$$\Phi(z) = \frac{K_I - iK_{II}}{2\sqrt{2\pi z_R}} - \frac{(1-\lambda)\sigma}{4} + \frac{i\tau}{2}; \quad \Psi(z) = \frac{K_I + 3iK_{II}}{2\sqrt{2\pi z_R}} + \frac{(1-\lambda)}{2} \tag{14.167}$$

assuming that $r/a \ll 1$ and zero rotation at infinity. For this infinite plate $K_I = \sigma\sqrt{\pi a}$ and $K_{II} = \tau\sqrt{\pi a}$ are the mode I and mode II stress intensity factors. With the aid of (14.148), (14.149) and (14.152) then the Cartesian crack tip stresses can be expressed in the familiar form (see §17.2)

$$\sigma_{ij} = \frac{K_I}{\sqrt{2\pi r}}f_{ij}(\theta) + \frac{K_{II}}{\sqrt{2\pi r}}g_{ij}(\theta) + T\delta_{ix}\delta_{jx} \tag{14.168}$$

where f_{ij} and g_{ij} are universal functions of θ. The term $T(= -(1-\lambda)\sigma)$ is referred to as the T-stress and δ_{ij} as the Kronecker delta tensor.

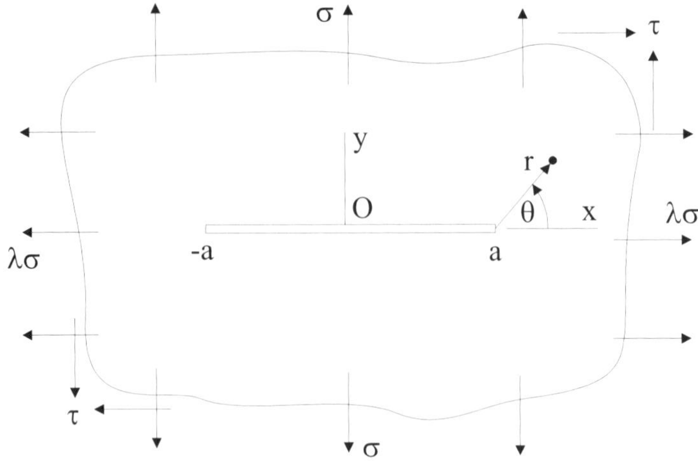

Figure 14.13: Infinite plate containing a crack of length $2a$ subject to biaxial and pure shear loading.

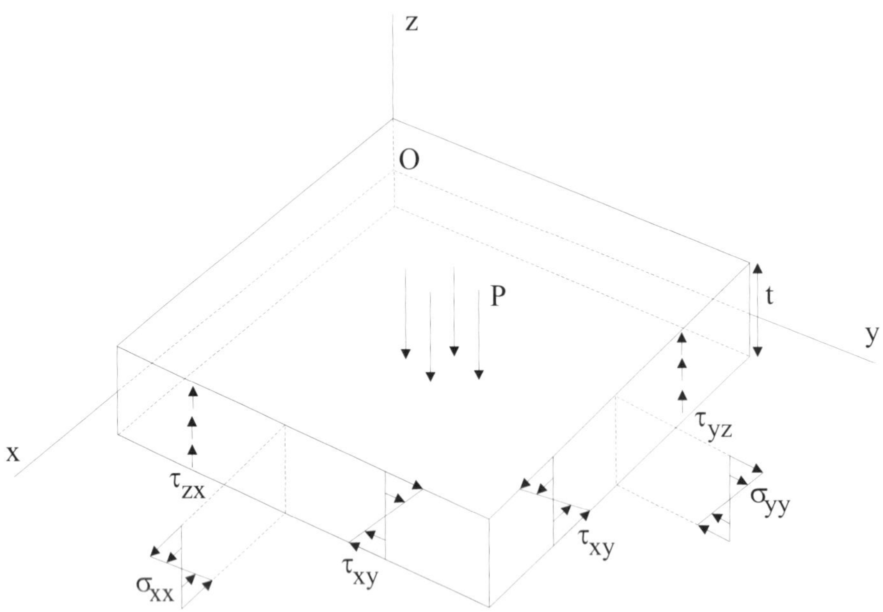

Figure 14.14: Stresses in a rectangular plate.

14.12 Bending of Rectangular Plates

The stresses induced in a rectangular element subject to bending moments and transverse loading are shown in Figure 14.14. As in simple beam theory the direct stresses σ_{xx} and σ_{yy} are assumed to vary linearly throughout the cross-section whereas the shear stresses τ_{xz}, τ_{yz} and τ_{xy} are assumed to vary parabolically. The bending of a plate is thus seen to be a generalisation of the bending of a beam. These stresses result in the following shear forces, Q, and bending moments, M

$$
\begin{aligned}
Q_x &= \int_{-t/2}^{t/2} \tau_{xz}\,\mathrm{d}z; \quad Q_y = \int_{-t/2}^{t/2} \tau_{yz}\,\mathrm{d}z \\
M_x &= \int_{-t/2}^{t/2} \sigma_{xx} z\,\mathrm{d}z; \quad M_y = \int_{-t/2}^{t/2} \sigma_{yy} z\,\mathrm{d}z; \quad M_{xy} = \int_{-t/2}^{t/2} \tau_{xy} z\,\mathrm{d}z
\end{aligned}
\tag{14.169}
$$

and are illustrated in Figure 14.15. From equilibrium we arrive at the following differential equations

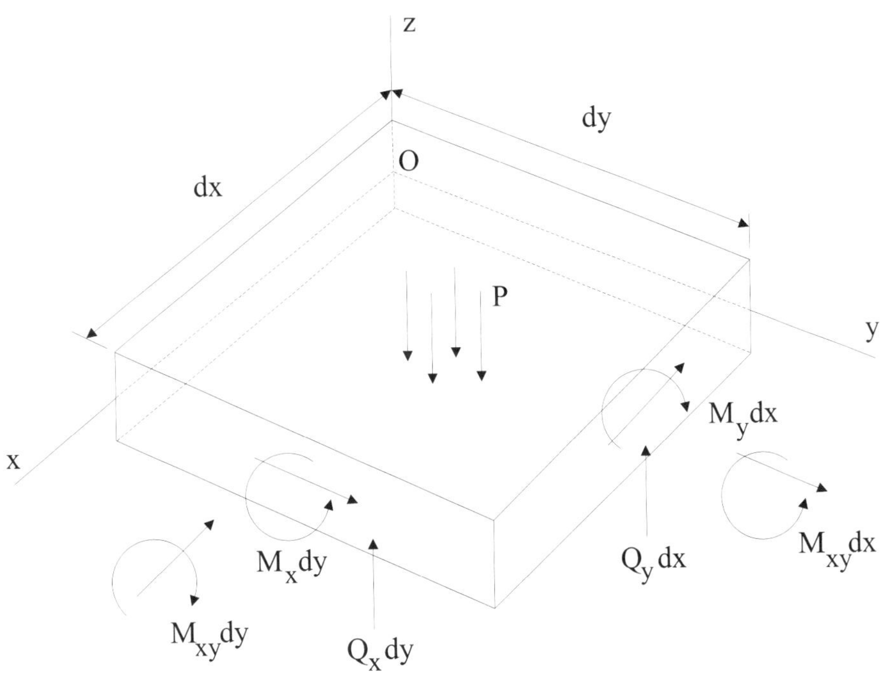

Figure 14.15: Shear forces and bending moments in a rectangular plate element ($\mathrm{d}x \times \mathrm{d}y$).

$$
\begin{aligned}
\frac{\partial Q_x}{\partial x} + \frac{\partial Q_y}{\partial y} + p = 0 \\
\frac{\partial M_x}{\partial x} + \frac{\partial M_{xy}}{\partial y} = Q_x \\
\frac{\partial M_{xy}}{\partial x} + \frac{\partial M_y}{\partial y} = Q_y
\end{aligned}
\tag{14.170}
$$

where p is the distributed surface load.

Now consider the strain-displacement and stress relations assuming that the element is thin (Kirchoff plate theory) and that the displacements are small. A point within the element has the following in-plane displacements, u and v

$$
u = -z\frac{\partial w}{\partial x}; \quad v = -z\frac{\partial w}{\partial y} \tag{14.171}
$$

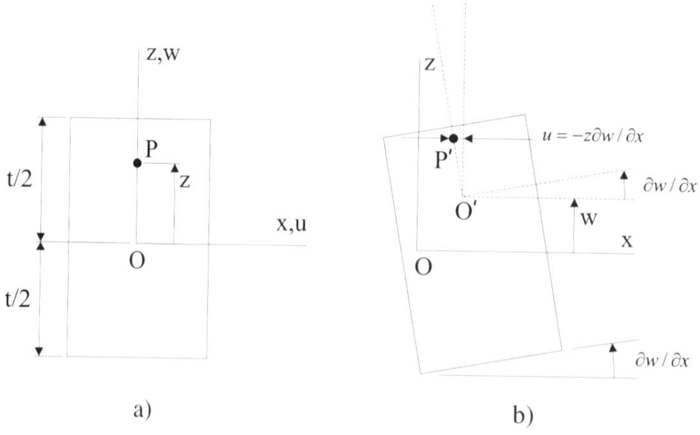

Figure 14.16: Plate deformation. a) Undeformed plate. b) Deformed plate.

and assume that points on the mid-surface cannot displace in the xy-plane. However, at distance z from the mid-surface u and v develop as the plate bends, as shown in Figure14.16. Therefore, the in-plane strains are, §1.11

$$\begin{aligned} \varepsilon_{xx} &= \frac{\partial u}{\partial x} = -z\frac{\partial^2 w}{\partial x^2} \\ \varepsilon_{yy} &= \frac{\partial v}{\partial y} = -z\frac{\partial^2 w}{\partial y^2} \\ \gamma_{xy} &= \left(\frac{\partial u}{\partial y} + \frac{\partial v}{\partial x}\right) = -2z\frac{\partial^2 w}{\partial x \partial y} \end{aligned} \tag{14.172}$$

with no transverse shear deformation. The strains illustrate that the transverse displacement, w, completely describes the state of deformation.

The stress-strain relations are of the usual form, (13.32)

$$\{\sigma\} = \begin{Bmatrix} \sigma_{xx} \\ \sigma_{yy} \\ \tau_{xy} \end{Bmatrix} = [D]\left(\{\varepsilon\} - \{\varepsilon_0\} + \{\sigma_0\}\right) \tag{14.173}$$

where, assuming a state of plane stress, $[D]$ is given by, (13.33)

$$[D] = \frac{E}{1-\nu^2}\begin{bmatrix} 1 & \nu & 0 \\ \nu & 1 & 0 \\ 0 & 0 & (1-\nu)/2 \end{bmatrix} \tag{14.174}$$

The moment-displacement relations can be obtained by substituting (14.172) into (14.173) followed by substitution into (14.169)

$$\begin{aligned} M_x &= \int_{-t/2}^{t/2} \sigma_{xx} z \, \mathrm{d}z \\ \sigma_{xx} &= \frac{E}{1-\nu^2}\left(\varepsilon_{xx} + \nu\varepsilon_{yy}\right) = -\left(\frac{E}{1-\nu^2}\right)\left[z\frac{\partial^2 w}{\partial x^2} + z\nu\frac{\partial^2 w}{\partial y^2}\right] \end{aligned} \tag{14.175}$$

from which we arrive at

$$M_x = \int_{-t/2}^{t/2} -\left(\frac{E}{1-\nu^2}\right)\left(\frac{\partial^2 w}{\partial x^2} + \nu\frac{\partial^2 w}{\partial y^2}\right) z^2\,\mathrm{d}z = -D_t\left(\frac{\partial^2 w}{\partial x^2} + \nu\frac{\partial^2 w}{\partial y^2}\right) \tag{14.176}$$

where D_t is

$$D_t = \frac{Et^3}{12(1-\nu^2)} \tag{14.177}$$

Similarly, M_y is found

$$M_y = -D_t\left(\frac{\partial^2 w}{\partial y^2} + \nu\frac{\partial^2 w}{\partial x^2}\right) \tag{14.178}$$

and M_{xy}

$$M_{xy} = \int_{-t/2}^{t/2} \tau_{xy} z\,\mathrm{d}z = \int_{-t/2}^{t/2}\left(\frac{E}{1-\nu^2}\right)\left(\frac{1-\nu}{2}\right)\left(-2z\frac{\partial^2 w}{\partial x \partial y}\right) z\,\mathrm{d}z = -D_t(1-\nu)\frac{\partial^2 w}{\partial x \partial y} \tag{14.179}$$

From $(14.170)_2$ of the equilibrium equations we now have, upon substitution of M_x and M_{xy}

$$\frac{\partial}{\partial x}\left[-D_t\left(\frac{\partial^2 w}{\partial x^2} + \nu\frac{\partial^2 w}{\partial y^2}\right)\right] + \frac{\partial}{\partial y}\left[-D_t(1-\nu)\frac{\partial^2 w}{\partial x \partial y}\right] = Q_x \tag{14.180}$$

which is found to reduce to

$$-D_t\frac{\partial}{\partial x}\left[\frac{\partial^2 w}{\partial x^2} + \frac{\partial^2 w}{\partial y^2}\right] = Q_x \tag{14.181}$$

Performing a similar procedure for $(14.170)_3$ we have

$$-D_t\frac{\partial}{\partial y}\left[\frac{\partial^2 w}{\partial x^2} + \frac{\partial^2 w}{\partial y^2}\right] = Q_y \tag{14.182}$$

Substituting (14.181) and (14.182) into $(14.170)_1$

$$\frac{\partial}{\partial x}\left[-D_t\frac{\partial}{\partial x}\left(\frac{\partial^2 w}{\partial x^2} + \frac{\partial^2 w}{\partial y^2}\right)\right] + \frac{\partial}{\partial y}\left[-D_t\frac{\partial}{\partial y}\left(\frac{\partial^2 w}{\partial x^2} + \frac{\partial^2 w}{\partial y^2}\right)\right] + p = 0 \tag{14.183}$$

which reduces to

$$\begin{aligned} &\frac{\partial^4 w}{\partial x^4} + 2\frac{\partial^4 w}{\partial x^2 \partial y^2} + \frac{\partial^4 w}{\partial y^4} = \frac{p}{D_t} \\ \Rightarrow\quad &\left(\frac{\partial^2}{\partial x^2} + \frac{\partial^2}{\partial y^2}\right)\left(\frac{\partial^2 w}{\partial x^2} + \frac{\partial^2 w}{\partial y^2}\right) = \frac{p}{D_t} \\ \Rightarrow\quad &\nabla^4 w = \frac{p}{D_t} \end{aligned} \tag{14.184}$$

Thus, the problem of a thin plate reduces to the solution of a fourth-order partial differential equation, i.e. a biharmonic equation, in conjunction with the appropriate boundary conditions. In general, the following three boundary conditions are most commonly encountered:

1) Simply supported along an edge

1.1) simply supported along y =constant, $w = 0$ and $M_y = 0$

1.2) simply supported along x =constant, $w = 0$ and $M_x = 0$

2) Clamped edge

2.1) clamped edge, y =constant, $w = 0$ and $dw/dy = 0$

2.2) clamped edge, x =constant, $w = 0$ and $dw/dx = 0$

3) Free edge

3.1) free edge, y =constant, $M_y = 0$

3.2) free edge, x =constant, $M_x = 0$

Example 14.2 Bending of a rectangular plate

In this example we will investigate the solution of (14.184) for the quadratic functions $w = Cx^2$, $w = C\left(x^2 + y^2\right)$ and $w = Cxy$; where C is a constant.

Considering the first function $w = Cx^2$ then we find that $\nabla^2 w = 2C$, a constant, so that $\nabla^2(\nabla^2 w) = 0$ which therefore satisfies (14.184) for zero pressure. From (14.176), (14.178) and (14.179) the bending moments M_x, M_y and M_{xy} are

$$M_x = -2CD_t; \quad M_y = -2\nu CD_t; \quad M_{xy} = 0$$

From (14.170) the shear forces, Q_x and Q_y, are zero everywhere. Furthermore, $\partial^2 w/\partial x^2$ and $\partial^2 w/\partial y^2$ are the curvatures of the w surface with vertical planes parallel to the x and y axes respectively. Therefore, the reciprocals of the curvatures are the radii of curvature

$$\frac{1}{R_x} = \frac{\partial^2 w}{\partial x^2} = 2C; \quad \frac{1}{R_y} = \frac{\partial^2 w}{\partial y^2} = 0$$

and therefore $R_x = 1/(2C)$ and $R_y = \infty$. The mixed second derivative is called the *twist* of the surface and is denoted by $1/T_{xy}$

$$\frac{1}{T_{xy}} = \frac{\partial^2 w}{\partial x \partial y} = 0$$

Figure 14.17a) illustrates the bending of a plate described by $w = Cx^2$.

The function $w = Cy^2$ is analogous to $w = Cx^2$ so we next investigate the function $w = C\left(x^2 + y^2\right) = Cr^2$. In this case we have

$$M_x = M_y = -2CD_t(1+\nu); \quad M_{xy} = 0; \quad \frac{1}{R_x} = \frac{1}{R_y} = 2C; \quad \frac{1}{T_{xy}} = 0$$

and is referred to as *spherical bending* in which a plate is subject to bending moments on the outside edge only, Figure 14.17b).

For the function $w = Cxy$ we have

$$M_x = M_y = 0; \quad M_{xy} = -CD_t(1-\nu); \quad \frac{1}{R_x} = \frac{1}{R_y} = 0; \quad \frac{1}{T_{xy}} = C$$

and represents a state of pure uniform twist, Figure 14.17c).

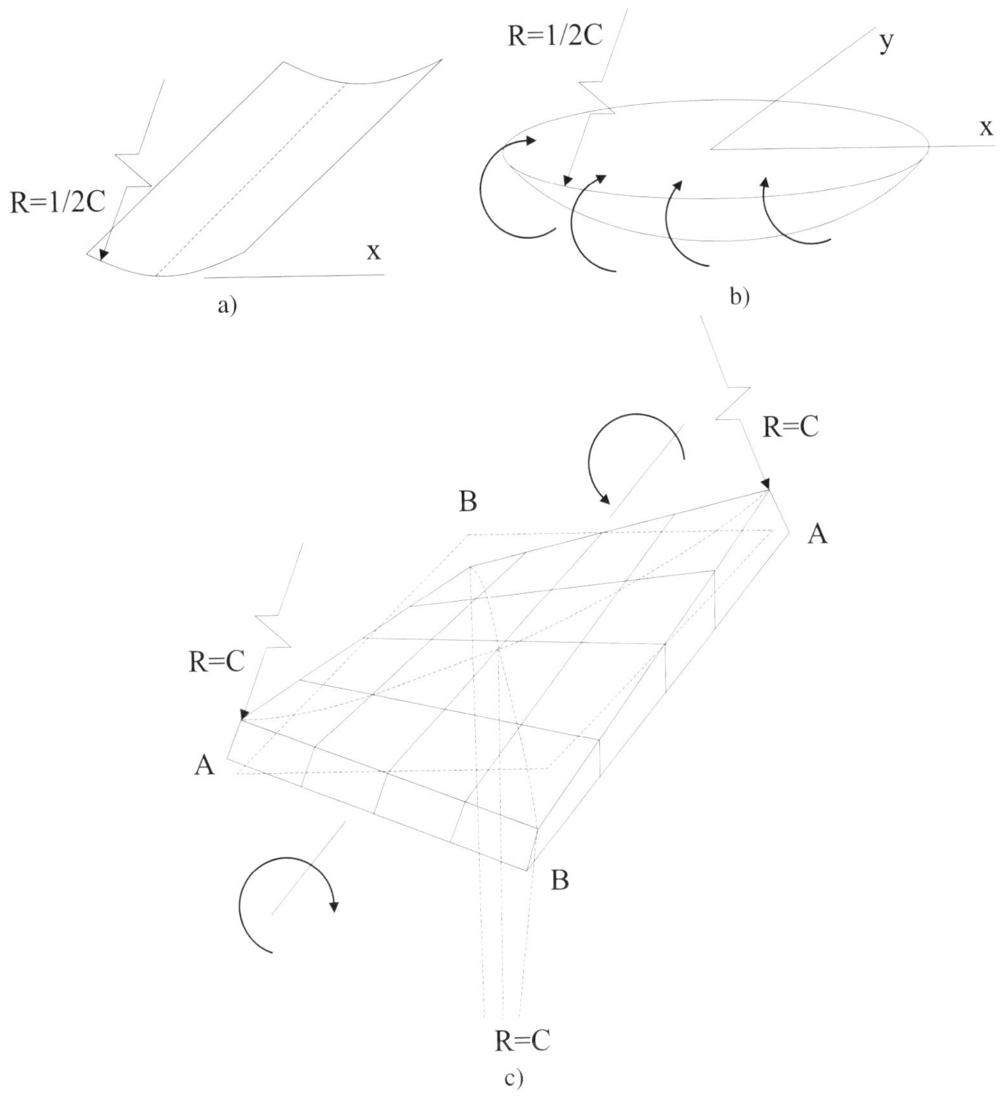

Figure 14.17: Example: Bending of rectangular plates. a) The surface is described by $w = Cx^2$. b) The surface is described by $w = Cr^2$. c) The surface is described by $w = Cxy$ noting that corners A (B) are higher (lower) than the undeformed plate.

14.13 Torsion of Prismatic Bars

Consider the solid prismatic bar of arbitrary cross-section subject to a torque T about the z-axis shown in Figure 14.18. Point p is displaced to point q with displacements u and v in the x and y directions

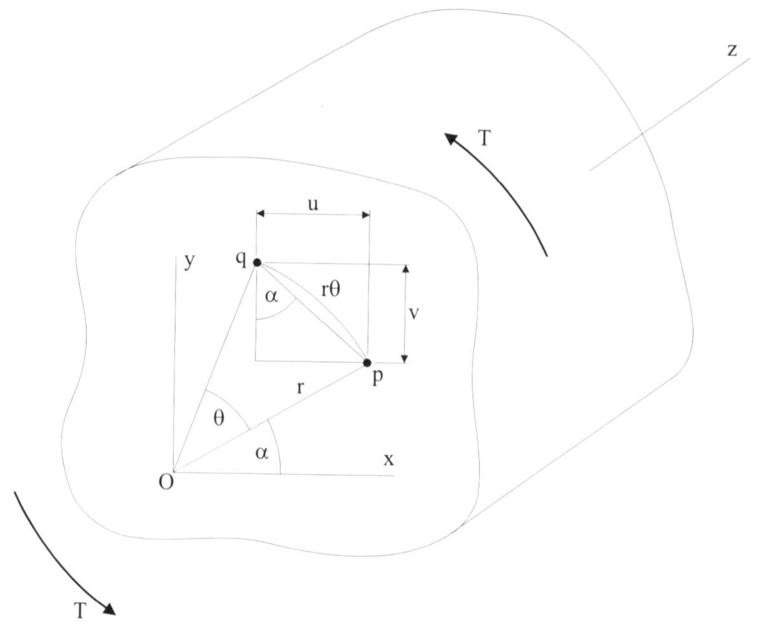

Figure 14.18: A prismatic bar subject to a torque T.

respectively

$$
\begin{aligned}
u &= -r\theta_z \sin\alpha = -r\theta_z\left(\frac{y}{r}\right) = -y\theta_z \\
v &= r\theta_z \cos\alpha = r\theta_z\left(\frac{x}{r}\right) = x\theta_z
\end{aligned}
\tag{14.185}
$$

where θ denotes the angle of twist per unit length and θ_z is the angle of twist at a distance z from the origin O. The displacement w in the z-direction is assumed to be proportional to the angle of twist and defined by a function $\psi(x, y)$, which is to be determined

$$
w = \psi(x, y)\theta \tag{14.186}
$$

The components of strain are given by, §1.11

$$
\begin{aligned}
\varepsilon_{xx} &= \frac{\partial u}{\partial x} = 0; \quad \varepsilon_{yy} = \frac{\partial v}{\partial y} = 0; \quad \varepsilon_{zz} = \frac{\partial w}{\partial z} = 0 \\
\gamma_{xy} &= \frac{1}{2}\left(\frac{\partial u}{\partial y} + \frac{\partial v}{\partial x}\right) = \frac{1}{2}(-\theta + \theta) = 0 \\
\gamma_{xz} &= \frac{1}{2}\left(\frac{\partial u}{\partial z} + \frac{\partial w}{\partial x}\right) = \frac{\theta}{2}\left(\frac{\partial \psi}{\partial x} - y\right) \\
\gamma_{yz} &= \frac{1}{2}\left(\frac{\partial v}{\partial z} + \frac{\partial w}{\partial y}\right) = \frac{\theta}{2}\left(\frac{\partial \psi}{\partial x} + x\right)
\end{aligned}
\tag{14.187}
$$

which illustrate that all of the axial strains are zero and $\gamma_{xy} = 0$ which indicates that there is no distortion of the cross-section in the xy-plane. The only non-zero strains are γ_{xz} and γ_{yz} which have associated shear stresses τ_{xz} and τ_{yz}

$$\begin{aligned}
\tau_{xz} &= G\gamma_{xz} = G\frac{\theta}{2}\left(\frac{\partial\psi}{\partial x} - y\right) \\
\tau_{yz} &= G\gamma_{yz} = G\frac{\theta}{2}\left(\frac{\partial\psi}{\partial x} + x\right) \\
\sigma_{xx} &= \sigma_{yy} = \sigma_{zz} = \tau_{xy} = 0
\end{aligned} \tag{14.188}$$

The differential equations of equilibrium, in the absence of body forces, reduce to; see (14.11)

$$\frac{\partial\tau_{yz}}{\partial z} = 0; \quad \frac{\partial\tau_{yz}}{\partial z} = 0; \quad \frac{\partial\tau_{xz}}{\partial x} + \frac{\partial\tau_{yz}}{\partial y} = 0 \tag{14.189}$$

The first two equations are satisfied since both τ_{xz} and τ_{yz} are independent of z and the third equation is satisfied by the introduction of the Prandtl stress function ϕ

$$\tau_{xz} = \frac{\partial\phi}{\partial y}; \quad \tau_{yz} = -\frac{\partial\phi}{\partial x} \tag{14.190}$$

with the first two equations of equilibrium (14.190) still satisfied. From (14.190) and (14.188) we observe that

$$\frac{\partial\phi}{\partial y} = \frac{G\theta}{2}\left(\frac{\partial\psi}{\partial x} - y\right); \quad -\frac{\partial\phi}{\partial x} = \frac{G\theta}{2}\left(\frac{\partial\psi}{\partial y} + x\right) \tag{14.191}$$

Differentiating the first of (14.191) with respect to y and the second of (14.191) with respect to x and adding together we arrive at

$$\nabla^2\phi = \frac{\partial^2\phi}{\partial x^2} + \frac{\partial^2\phi}{\partial y^2} = -G\theta \tag{14.192}$$

The shear stress normal to the boundary of the bar must be zero so that the bar is free from external forces. As a result equations (14.190) illustrate that ϕ must be constant on the boundary and since this constant can be arbitrarily fixed then it will be assumed to equal zero.

The total torque acting on a section is given by

$$T = \iint_A (x\tau_{yz} - y\tau_{xz})\, \mathrm{d}x\, \mathrm{d}y \tag{14.193}$$

where A is the cross-sectional area of the bar. In terms of the stress function ϕ then T is

$$T = \iint_A \left(x\frac{\partial\phi}{\partial x} + y\frac{\partial\phi}{\partial y}\right) \mathrm{d}x\, \mathrm{d}y = -\int \mathrm{d}y \int \frac{\partial\phi}{\partial x} x\, \mathrm{d}x - \int \mathrm{d}x \int \frac{\partial\phi}{\partial y} y\, \mathrm{d}y \tag{14.194}$$

Integrating by parts we have

$$-\int \mathrm{d}y\left(x\phi \int \phi\, \mathrm{d}x\right) - \int \mathrm{d}x\left(y\phi - \int \phi\, \mathrm{d}y\right) = -\int \phi\, \mathrm{d}(xy) + 2\iint_A \phi\, \mathrm{d}x\, \mathrm{d}y \tag{14.195}$$

and with $\phi = 0$ on the boundary of the bar

$$T = 2\iint_A \phi\, \mathrm{d}x\, \mathrm{d}y \tag{14.196}$$

Once the angle of twist has been determined then the u and v components of displacement follow from (14.185). The w displacement is determined from (14.186) and (14.191)

$$\begin{aligned}\tau_{xz} &= \frac{\partial \phi}{\partial y} = G\theta\left(\frac{\partial \psi}{\partial x} - y\right) = G\theta\left(\frac{1}{\theta}\frac{\partial w}{\partial x} - y\right)\\ \tau_{yz} &= -\frac{\partial \phi}{\partial x} = G\theta\left(\frac{\partial \psi}{\partial y} + x\right) = G\theta\left(\frac{1}{\theta}\frac{\partial w}{\partial y} + x\right)\end{aligned} \tag{14.197}$$

which, upon re-arranging, give

$$\frac{\partial w}{\partial x} = \frac{1}{G}\frac{\partial \phi}{\partial y} + \theta y; \quad \frac{\partial w}{\partial y} = -\frac{1}{G}\frac{\partial \phi}{\partial x} - \theta x \tag{14.198}$$

Example 14.3 Torsion of an elliptical cross-section

In this example we will examine the torsion of an elliptical cross-section whose boundary in the xy-plane is given by

$$\frac{x^2}{a^2} + \frac{y^2}{b^2} - 1 = 0$$

with half major and minor axes denoted by a and b respectively. Equation (14.192) is given by

$$\nabla^2\phi = \frac{\partial^2 \phi}{\partial x^2} + \frac{\partial^2 \phi}{\partial y^2} = -2G\theta = c$$

and is satisfied by a Prandtl stress function of the form

$$\phi = m\left(\frac{x^2}{a^2} + \frac{y^2}{b^2} - 1\right)$$

where m is a constant. Substituting ϕ into $\nabla^2\phi = c$ we find that m is given by

$$m = \frac{a^2b^2}{2\left(a^2+b^2\right)}c; \quad c = -2G\theta$$

If the angle of twist is defined then $c = -2G\theta$ and ϕ is known. However, if T is specified then we need to express c in terms of T. From (14.196) the torque is

$$T = 2\iint_A \phi \, \mathrm{d}x \, \mathrm{d}y = \frac{a^2b^2}{a^2+b^2}c\left[\frac{1}{a^2}\iint_A x^2 \, \mathrm{d}x \, \mathrm{d}y + \frac{1}{b^2}\iint_A y^2 \, \mathrm{d}x \, \mathrm{d}y - \iint_A \mathrm{d}x \, \mathrm{d}y\right]$$

The first, second and third integrals on the right hand side are seen to be equal to $I_y(= \pi ba^3/4)$, $I_x(= \pi ab^3/4)$ and $A(= \pi ab)$ respectively; refer to Exercise 2.3. Substituting I_x, I_y and A into T we have

$$T = \frac{\pi a^3 b^3 c}{2\left(a^2+b^2\right)}$$

Re-arranging for c and noting that $c = -2G\theta$ then

$$c = -\frac{2T\left(a^2+b^2\right)}{\pi a^3 b^3}; \quad \theta = \frac{T\left(a^2+b^2\right)}{\pi a^3 b^3 G}$$

Substituting both m and c into ϕ

$$\phi = -\frac{T}{\pi ab}\left(\frac{x^2}{a^2} + \frac{y^2}{a^2} - 1\right)$$

The stress components now follow from (14.190)

$$\tau_{xz} = \frac{\partial \phi}{\partial y} = -\frac{2Ty}{\pi ab^3}; \quad \tau_{yz} = -\frac{\partial \phi}{\partial x} = \frac{2Tx}{\pi a^3 b}$$

The maximum shear stress occurs at the ends of the minor axis ($y = \pm b$) and is therefore

$$\tau_{max} = \frac{2T}{\pi ab^2}$$

When $a = b$ and the ellipse reduces to a circle of radius a then $\tau_{max} = 2T/\pi a^3$ which agrees with the formula (3.11), $\tau_{max} = Ta/J$ with $J = \pi a^3/2$.

14.14 Linear Elastic Continuum Plate with a Hole

In this section we will present an overview of the popular problem of a hole in a plate which is subject to both uniaxial and biaxial loadings. The introduction of a hole into an otherwise uniform plate creates a more complicated three-dimensional stress-strain field than originally may be appreciated. The implications and dominant features of such a configuration have to be understood fully before analysing the addition of a crack to such a feature. Furthermore, a hole is a popular engineering configuration which generates a localised stress concentration from which cracks will initiate and propagate. An elliptical hole will also be examined and an elliptical hole (major axis a and minor axis b) with aspect ratio (a/b) degenerates to a line crack of total length $2a$ as b tends to zero.

14.14.1 Circular Hole in a Plate

The following presents an overview of a solution due to Kirsch (1898). Consider a plate containing a circular hole of radius a subject to a uniform tension in the y-direction, Figure 14.19. By Saint Venant's principle the change in stress distribution resulting from the circular hole will be negligible at distances which are large compared to the radius of the hole.

> **Saint Venant's principle**
> *If a set of forces acting on a given boundary are redistributed without changing the resulting force or moment then the effect of the change will be negligible at a distance equal to a few times the dimension of the boundary.*

Thus, the stresses at a radius $R(\gg a)$ are effectively those of the plate without the hole present

$$\begin{aligned} \sigma_{xx} &= 0; \quad \sigma_{rr} = \frac{\sigma_P}{2}(1 - \cos 2\theta) \\ \sigma_{yy} &= \sigma_P; \quad \sigma_{\theta\theta} = \frac{\sigma_P}{2}(1 + \cos 2\theta) \\ \tau_{xy} &= 0; \quad \tau_{r\theta} = \frac{\sigma_P}{2}\sin 2\theta \end{aligned} \tag{14.199}$$

The stresses in the ring $(R - a)$ may be considered as consisting of two parts: i) the normal component $\sigma_P/2$ and ii) the normal and shear components $(\sigma_P/2)\cos 2\theta$ and $(\sigma_P/2)\sin 2\theta$ respectively. The second part of the stress components can be derived from an Airy stress function of the form

$$\phi = f(r)\cos 2\theta \tag{14.200}$$

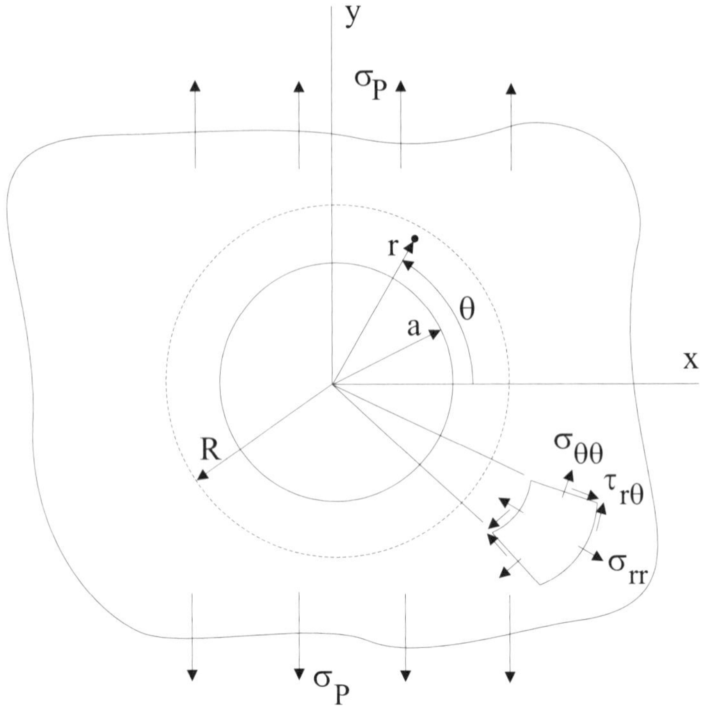

Figure 14.19: A hole of radius a in a plate.

Substitution of ϕ into the differential equations of compatibility, $\nabla^4\phi = 0$, in terms of polar coordinates, (14.58), we have

$$\left(\frac{\partial^2}{\partial r^2} + \frac{1}{r}\frac{\partial}{\partial r} - \frac{4}{r^2}\right)\left(\frac{\partial^2 f}{\partial r^2} + \frac{1}{r}\frac{\partial f}{\partial r} - \frac{4f}{r^2}\right) = 0 \tag{14.201}$$

with general solution

$$f(r) = Ar^2 + Br^4 + \frac{C}{r^2} + D; \quad \phi = \left(Ar^2 + Br^4 + \frac{C}{r^2} + D\right)\cos 2\theta \tag{14.202}$$

The stress components are determined from the Airy stresses, (14.55), in terms of polar coordinates

$$\begin{aligned}
\sigma_{rr} &= \frac{1}{r}\frac{\partial\phi}{\partial r} + \frac{1}{r^2}\frac{\partial^2\phi}{\partial\theta^2} = -\left(2A + \frac{6C}{r^4} + \frac{4D}{r^2}\right)\cos 2\theta \\
\sigma_{\theta\theta} &= \frac{\partial^2\phi}{\partial r^2} = \left(2A + 12Br^2 + \frac{6C}{r^4}\right)\cos 2\theta \\
\tau_{r\theta} &= -\frac{\partial}{\partial r}\left(\frac{1}{r}\frac{\partial\phi}{\partial r}\right) = \left(2A + 6Br^2 - \frac{6C}{r^4} - \frac{2D}{r^2}\right)\sin 2\theta
\end{aligned} \tag{14.203}$$

The constants A, B, C and D are determined from the boundary conditions $\sigma_{rr} = \tau_{r\theta} = 0$ at $r = a$ and the conditions (14.199) at $r = R$

$$\begin{aligned} 2A + \frac{6C}{a^4} + \frac{4D}{a^2} &= 0 \\ 2A + 6Ba^2 - \frac{6C}{a^4} - \frac{2D}{a^2} &= 0 \\ 2A + \frac{6C}{R^4} + \frac{4D}{R^2} &= -\frac{\sigma_P}{2} \\ 2A + 6BR^2 - \frac{6C}{R^4} - \frac{2D}{R^2} &= -\frac{\sigma_P}{2} \end{aligned} \tag{14.204}$$

Solving for A, B, C and D and letting $a/R = 0$, that is $R \to \infty$ we find

$$A = -\frac{\sigma_P}{4}; \quad B = 0; \quad C = -\frac{a^4\sigma_P}{4}; \quad D = -\frac{a^2\sigma_P}{2} \tag{14.205}$$

Substituting these back into (14.203) and adding the stresses produced by the uniform tension, (14.199), we arrive at

$$\begin{aligned} \sigma_{rr} &= \frac{\sigma_P}{2}\left(1 - \frac{a^2}{r^2}\right) + \frac{\sigma_P}{2}\left(1 - \frac{4a^2}{r^2} + \frac{3a^4}{r^4}\right)\cos 2\theta \\ \sigma_{\theta\theta} &= \frac{\sigma_P}{2}\left(1 + \frac{a^2}{r^2}\right) + \frac{\sigma_P}{2}\left(1 + \frac{3a^4}{r^4}\right)\cos 2\theta \\ \tau_{r\theta} &= \frac{\sigma_P}{2}\left(1 + \frac{2a^2}{r^2} - \frac{3a^4}{r^4}\right)\sin 2\theta \end{aligned} \tag{14.206}$$

where, for a state of plane strain

$$\sigma = \nu\left(\sigma_{rr} + \sigma_{\theta\theta}\right) = \nu\left(1 + 2\left(\frac{a^2}{r^2}\right)\cos 2\theta\right) \tag{14.207}$$

Figure 14.20 illustrates the variation of $\sigma_{\theta\theta}$ with θ on the hole surface, $r = a$. From (14.206) we have the following well-known results

$$\frac{(\sigma_{\theta\theta})_{\theta=0^\circ, r=a}}{\sigma_P} = 3; \quad \frac{(\sigma_{\theta\theta})_{\theta=90^\circ, r=a}}{\sigma_P} = -1 \tag{14.208}$$

which are the stress concentration on the hole surface at $\theta = 0^\circ$ and $\theta = 90^\circ$. Note that $\sigma_{\theta\theta}$ changes sign at $\theta = (1/2)cos^{-1}(1/2) = 60^\circ$.

14.14.2 Elliptical Hole in a Plate

Determining the stress distribution for an elliptical hole, which in the limit becomes a crack, was first due to Inglis (1913). The analysis is more complicated than that for a circular hole but is generally performed in terms of curvilinear coordinates which help simplify the boundary conditions. The solution method is too detailed to describe here and the interested reader is referred to Muskhelishvili (1953) for the exact details of the solution. Briefly, the technique of conformal mapping is used to transform between a unit circle in an auxiliary logical ξ-plane to an ellipse in the physical z-plane, Figure 14.21. The mapping function is given by

$$z = \omega(\xi) = c\left(\xi + \frac{m}{\xi}\right); \quad c = \frac{a+b}{2}; \quad m = \frac{a-b}{a+b}; \quad m < 1 \tag{14.209}$$

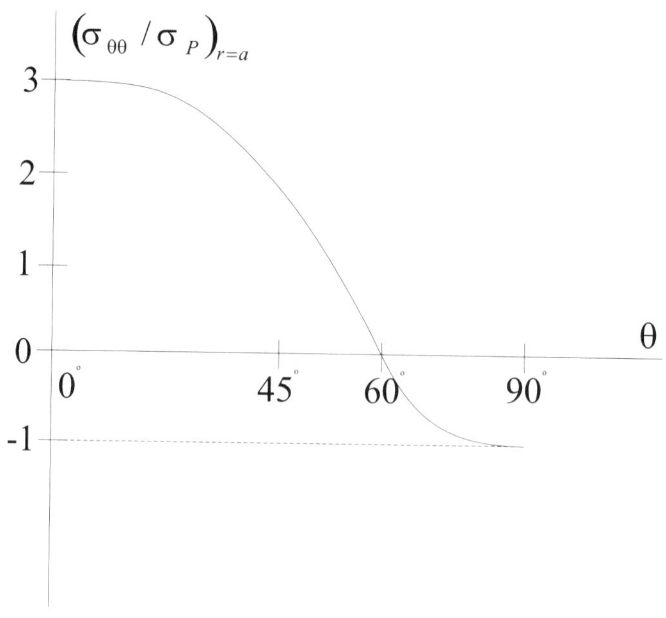

Figure 14.20: Variation of $\sigma_{\theta\theta}$ with θ on the surface of a circular hole subject to a far-field uniform tension of σ_P.

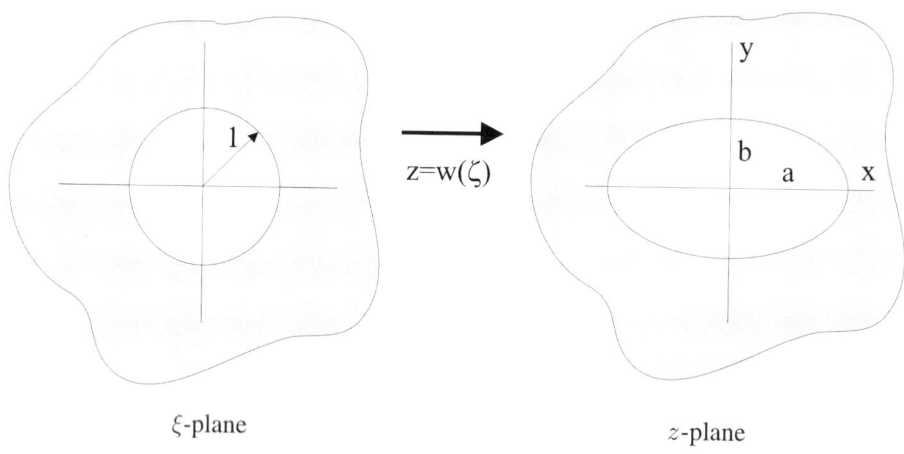

Figure 14.21: Conformal mapping of a unit circle in a logical ξ-plane to an ellipse in the physical z-plane.

Because the actual stress components are so complicated for an elliptical hole several approximate solutions have been proposed of which the following due to Neuber (1946) is one of the more popular

$$\sigma_{yy} = \sigma_P K_t \sqrt{\frac{\rho}{\rho + 4x}} \tag{14.210}$$

where $\rho = b^2/a$ is the *notch root radius* and x is measured from the hole/notch root as shown in Figure 14.22. The *stress concentration factor* (after Neuber) is denoted by K_t and simply defines the ratio

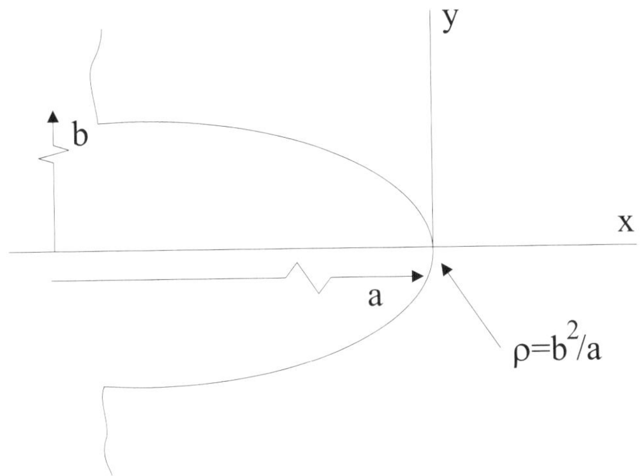

Figure 14.22: The root of an elliptical hole/notch with notch root radius ρ.

of stress at a given point on the surface of the elliptical hole to the stress at the same point without a hole, that is a magnification factor. The stress concentration factor for an elliptical hole is given by

$$K_t = \frac{1 - 2m - m^2 + 2\cos 2\theta}{1 + m^2 - 2m\cos 2\theta}; \quad m = \frac{a-b}{a+b} \tag{14.211}$$

Figure 14.23 illustrates the variation of K_t given by (14.211) for a circular hole $(a/b = 1)$ and two ellipses (a *sharp* ellipse $a/b = 2$ and a *blunt* ellipse $a/b = 0.5$) for the range $0° \leq \theta \leq 90°$. Figure 14.23 illustrates that all ellipses experience the same compressive stress at $\theta = 90°$ and that the largest stress concentration occurs along the major axis at $\theta = 0°$, from (14.211)

$$K_t = 1 + 2\left(\frac{a}{b}\right) \tag{14.212}$$

or alternatively, in terms of $\rho = b^2/a$

$$K_t = 1 + 2\sqrt{\frac{a}{\rho}} \tag{14.213}$$

illustrating that $K_t \to \infty$ as $b \to 0$ and the ellipse reduces to a line crack.

Let us now introduce the *biaxiality parameter* $\lambda(= \sigma_Q/\sigma_P)$ which is the ratio of parallel to normal applied stresses, Figure 14.24. By inspection we observe that K_t is linearly related to λ as follows

$$K_t = 1 + 2\left(\frac{a}{b}\right) - \lambda \tag{14.214}$$

Table 14.2 illustrates the effect of λ on K_t for a circular hole subject to uniaxial, equibiaxial and shear loadings and demonstrates that shear loading is the most detrimental. It is worth noting that when $\lambda = -1$ then for both $\theta = 0°$ and $\theta = 90°$ we find

$$K_t = \pm\sigma_P \frac{(a+b)^2}{ab} \tag{14.215}$$

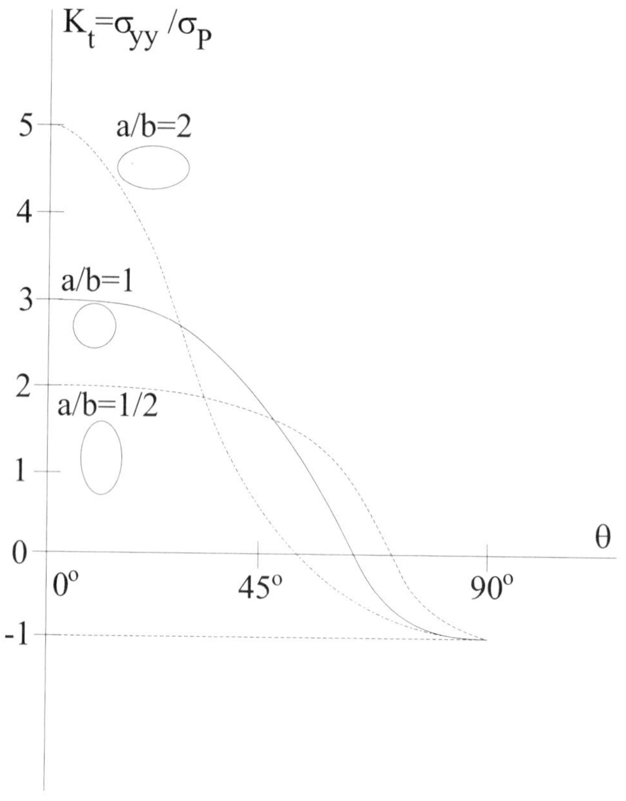

Figure 14.23: Variation of K_t for various elliptical holes.

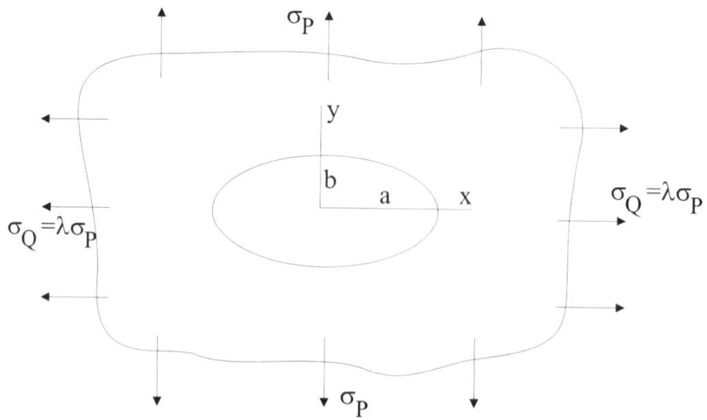

Figure 14.24: A plate with an elliptical hole subject to biaxial loading.

illustrating that cracks will nucleate equally along the major and minor axes.

Figure 14.25 illustrates the variation of σ_{yy} with $(r_{\theta=0^\circ}/a)$ for several circular and elliptical holes subject to a far-field uniform tension. As an ellipse becomes sharper then the more localised its effect on the stress field. In the case of a circular hole the variation of σ_{xx} with $(r_{\theta=0^\circ}/a)$ is illustrated in

Loading	λ	$K_t(a/b = 1)$
uniaxial	0	3
equibiaxial	1	2
shear	-1	4

Table 14.2: Variation of K_t with λ for uniaxial, equibiaxial and shear loadings.

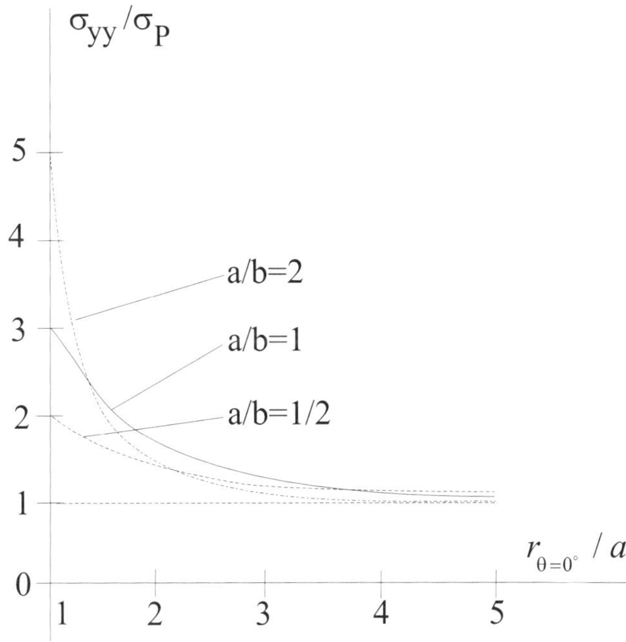

Figure 14.25: Variation in K_t for circular and elliptical holes subject to a far-field uniform tension.

Figure 14.26. The σ_{xx} stress is zero at the hole surface and tends to zero at large distances from the hole, although we observe that σ_{xx} attains a maximum within the proximity of the hole. This maximum is obtained by differentiating σ_{rr} in (14.206) with respect to r and letting $\theta = 0°$

$$\left(\frac{\sigma_{xx}}{\sigma_P}\right)_{max} = \frac{3}{8} \approx 0.375 \quad \text{at} \quad \left(\frac{r_{\theta=0°}}{a}\right) = \sqrt{2} \approx 1.414 \tag{14.216}$$

This variation in σ_{xx} illustrates that although the hole in a plate geometry is subject to uniaxial loading only, a complex biaxial stress field is induced.

Finally, we will conclude our discussion of circular and elliptical holes with a brief discussion on the effects of plate geometry on stress concentration. Figure 14.27 illustrates the effect of plate geometry (H/W) on the σ_{yy} stress distribution of a circular hole; where W and H are the half width and height of the plate respectively. Observe that K_t increases with decreasing H/W. Also, a similar increase in K_t is observed by increasing a/W as the plate width approaches the hole diameter. As a result, in practice K_t

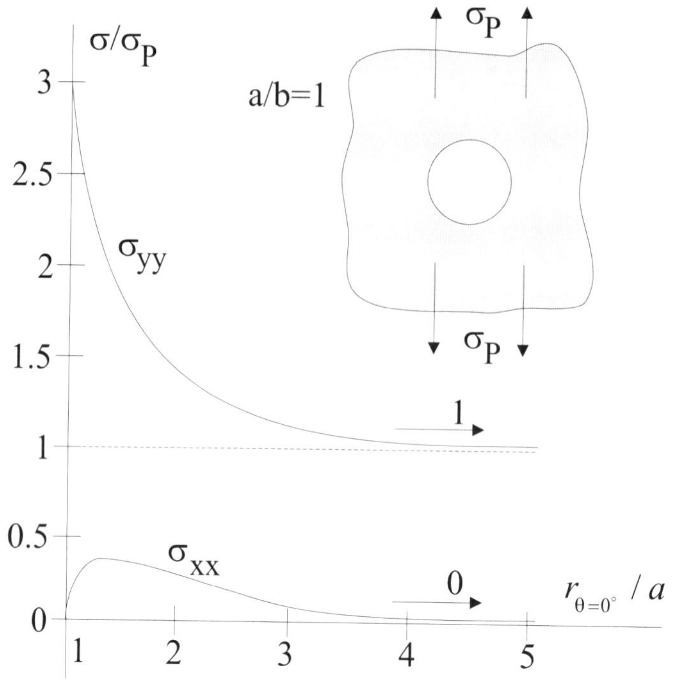

Figure 14.26: Variation of σ_{xx} and σ_{yy} with $(r_{\theta=0^\circ}/a)$ for a circular hole subject to a far-field tension.

can be a complex combination of several geometric and loading parameters

$$K_t\left(\frac{a}{b},\frac{a}{W},\frac{H}{W},\frac{a}{t},\lambda\right) \tag{14.217}$$

where t is the plate thickness.

14.15 The Boussinesq Wedge

In 1885 J. Boussinesq obtained a closed form solution for the problem of a concentrated point force, f, applied to the apex of a symmetrical wedge in a direction parallel to the axis of symmetry of the wedge, Figure 14.28. Boussinesq showed that this problem is uniquely solved by an Airy stress function of the form $\phi = -(k/2)fr\theta\sin\theta$ with polar coordinates (r,θ) centred at the wedge apex; where k is a constant to be determined. From the Airy stresses, (14.55), the traction-free conditions along the wedge faces $(\theta = \pm\alpha)$ are satisfied by the following purely radial stress distribution

$$\begin{aligned}
\sigma_{rr} &= \frac{1}{r}\frac{\partial\phi}{\partial r} + \frac{1}{r^2}\frac{\partial^2\phi}{\partial\theta^2} = -\frac{kf\cos\theta}{r}\\
\sigma_{\theta\theta} &= \frac{\partial^2\phi}{\partial r^2} = 0\\
\tau_{r\theta} &= -\frac{\partial}{\partial r}\left(\frac{1}{r}\frac{\partial\phi}{\partial r}\right) = 0
\end{aligned} \tag{14.218}$$

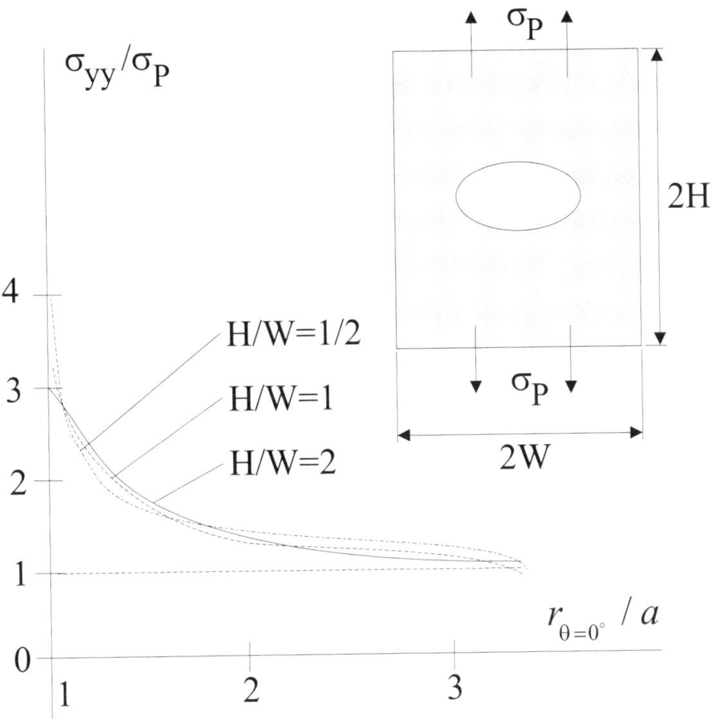

Figure 14.27: Effect of plate geometry on the σ_{yy} stress distribution for a circular hole subject to a far-field uniform tension with $a/W = 0.2$.

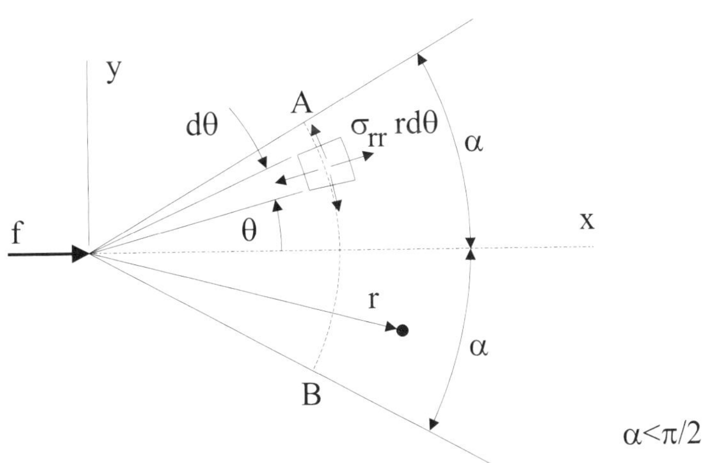

Figure 14.28: A wedge of included angle 2α subject to a point force f acting at the apex of the wedge.

For equilibrium to be satisfied then the resultant of all resolved forces acting on any radius r of arc AB of width $\mathrm{d}\theta$ (i.e., $(\sigma_{rr} r\,\mathrm{d}\theta)\cos\theta$) must equal the total force transmitted $(-f)$

$$\int_{-\alpha}^{\alpha} \sigma_{rr} r \cos\theta \,\mathrm{d}\theta = -f \tag{14.219}$$

Substituting $(14.218)_1$ into (14.219)

$$-\int_{-\alpha}^{\alpha} \frac{kf\cos^2\theta}{r} = -f \tag{14.220}$$

from which we find the following expression for k

$$k = \frac{1}{\alpha + \frac{1}{2}\sin 2\alpha} \tag{14.221}$$

When $\alpha = \pi/2$ then the configuration equates to a semi-infinite half-plane acted upon by a concentrated point force acting on the surface, from (14.218) with $k = 2/\pi$

$$\sigma_{rr} = -\frac{2f\cos\theta}{\pi r}; \quad \sigma_{\theta\theta} = \tau_{r\theta} = 0 \tag{14.222}$$

with σ_{rr} illustrated graphically in Figure 14.29. The Cartesian stress components are obtained from the stress transformation equations, (14.29)

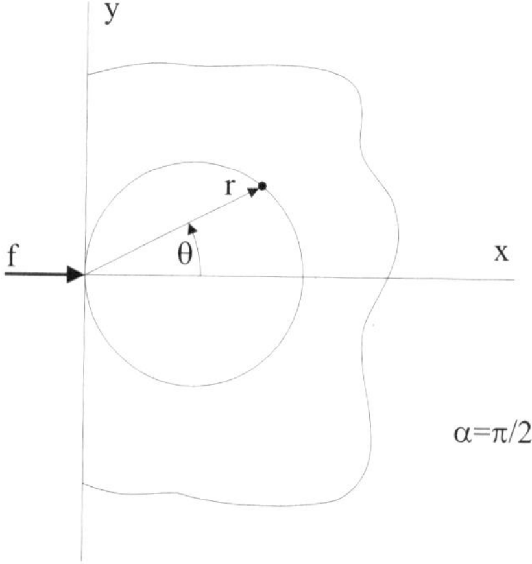

Figure 14.29: Compressive radial distribution of σ_{rr} when $\alpha = \pi/2$.

$$\begin{aligned}
\sigma_{xx} &= \sigma_{rr}\cos^2\theta + \sigma_{\theta\theta}\sin^2\theta + 2\tau_{r\theta}\sin\theta\cos\theta = \sigma_{rr}\cos^2\theta = -\frac{2f\cos^3\theta}{\pi r} = -\frac{2fx^3}{\pi r^4} \\
\sigma_{yy} &= \sigma_{rr}\sin^2\theta + \sigma_{\theta\theta}\cos^2\theta + 2\tau_{r\theta}\sin\theta\cos\theta = \sigma_{rr}\sin^2\theta = -\frac{2f\cos\theta\sin^2\theta}{\pi r} = -\frac{2fxy^2}{\pi r^4} \\
\tau_{xy} &= (\sigma_{rr} - \sigma_{\theta\theta})\sin\theta\cos\theta + \tau_{r\theta}\left(\cos^2\theta\sin^2\theta\right) = \sigma_{rr}\sin\theta\cos\theta = -\frac{2f\cos^2\theta\sin\theta}{\pi r} = -\frac{2fx^2y}{\pi r^4}
\end{aligned} \tag{14.223}$$

Furthermore, when $\alpha = \pi$ (see Figure 14.30) then the configuration equates to a semi-infinite cracked geometry with a point force acting at the crack tip, from (14.218) with $k = 1/\pi$

$$\sigma_{rr} = -\frac{f\cos\theta}{\pi r}; \quad \sigma_{\theta\theta} = \tau_{r\theta} = 0 \tag{14.224}$$

with the Cartesian stress components

$$\begin{aligned}
\sigma_{xx} &= \sigma_{rr}\cos^2\theta = -\frac{f\cos^3\theta}{\pi r} = -\frac{fx^3}{\pi r} \\
\sigma_{yy} &= \sigma_{rr}\sin^2\theta = -\frac{f\cos\theta\sin^2\theta}{\pi r} = -\frac{fxy^2}{\pi r} \\
\tau_{xy} &= \sigma_{rr}\sin\theta\cos\theta = -\frac{f\cos^2\theta\sin\theta}{\pi r} = -\frac{fx^2y}{\pi r}
\end{aligned} \tag{14.225}$$

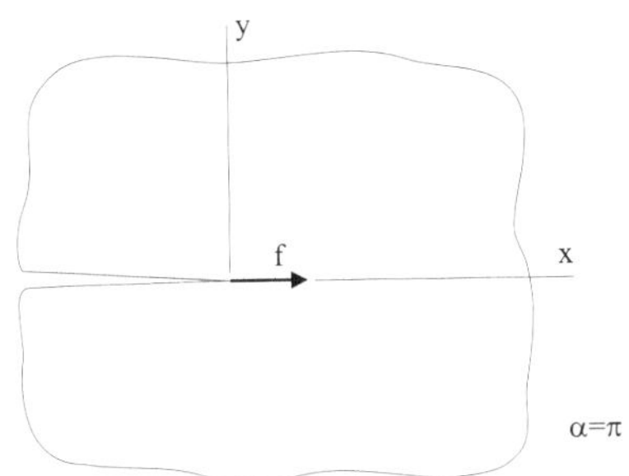

Figure 14.30: Wedge geometry when $\alpha = \pi$.

14.16 Conclusion

The aim of this Chapter was to present a comprehensive overview of the theory of elasticity. We began by examining a three-dimensional stress element, the equations of equilibrium and compatibility equations. The two-dimensional stress transformation equations of Chapters 7 and 8 were then generalised for three-dimensional states of stress and strain. The Hookian equations, first introduced in Chapter 1, were then revisited and conveniently expressed in terms of stress. The Airy stresses and Airy stress function were then introduced. The cubic equation of stress was derived for three-dimensional stress states and shown to reduce to the principal stresses previously encountered in Chapter 7 for two-dimensional stress states. The next key section presented the constitutive equations for a variety of different materials in matrix form which is ideally suited for computational modelling. This was followed by discussions on the reciprocal displacement and work theorems. The complex variable method of elasticity was discussed and shown to provide an elegant representation and solution to several important problems in the theory of elasticity. Next, we examined the bending of rectangular plates and the torsion of prismatic bars. The Chapter concluded by examining the problems of a hole in a plate and a point-loaded wedge. We will make use of both of these applications when we examine fracture mechanics in Chapter 17.

14.17 References and Further Reading

- Hearn, E. J. (1985) *Mechanics of Materials*, Pergamon Press, Oxford.
- Inglis, C. E. (1913) *Stresses in a Plate due to the Presence of Cracks and Sharp Corners*, Trans. Instituion Naval Architects, London, England, 60, 219-241.
- Kirsch, G. (1898) Zeitschrift des Ver. D. Ing.

∘ Kreyszig, E. (1983) *Advanced Engineering Mathematics*, 5th Ed., Wiley & Sons, Canada.

∘ Muskhelishvili, N. I. (1953) *Some Basic Problems of the Mathematical Theory of Elasticity*, 3rd Ed., (Translation by J. R. M. Radok), Noordhoff, Groningen, Holland.

∘ Neuber, H. (1946) *Theory of Notch Stresses*, (Translation by F. A. Raven), Edwards, Ann Arbor, Michigan.

∘ Timoshenko, S. P. and Goodier, J. N. (1982) *Theory of Elasticity*, 3rd Ed., McGraw-Hill.

14.18 Exercises

14.1 Show that the plane displacements $u = ax^2y^2$ and $v = byx^3$ are compatible.

14.2 A plane Airy stress function is given by

$$\phi(x, y) = Ax^2 + Bxy + Cy^2$$

Confirm that ϕ satisfies $\nabla^4\phi = 0$ and determine the associated stresses, strains and displacements. Assume plane stress conditions.

14.3 The state of stress at a point is given by the following stress tensor

$$\sigma_{ij} = \begin{bmatrix} 10 & 5 & 15 \\ 5 & 20 & 25 \\ 15 & 25 & 30 \end{bmatrix}$$

Determine both the magnitude and direction of the normal and shear stresses on a plane ABC whose unit normal is $\mathbf{N}_u = 1/\sqrt{3}(\mathbf{i} + \mathbf{j} + \mathbf{k})$.

14.4 Determine the stiffness matrix **C** for an isotropic material in which plane strain conditions prevail. If the state of strain at a point is given by $\varepsilon_{xx} = -19 \times 10^{-6}$, $\varepsilon_{yy} = 64 \times 10^{-6}$ and $\gamma_{xy} = 3 \times 10^{-6}$ then determine the stress vector. Assume Young's modulus and Poisson's ratio to be 210GPa and 0.3 respectively.

14.5 The following Prandtl stress function ϕ provides a solution to the torsion of a solid equilateral triangular cross-section of height a, edge length $2a/\sqrt{3}$ and origin of coordinates at the triangle centroid

$$\phi = -G\theta \left[\frac{1}{2}\left(x^2 + y^2\right) - \frac{1}{2a}\left(x^3 - 3xy^2\right) - \frac{2}{27}a^2 \right]$$

Determine expressions for the shearing stresses τ_{xz} and τ_{yz} and what do you note about the value of τ_{yz} at the centroid and corners of the triangle?

14.6 A circular hole of radius a(=12.5mm) is present in a plate which is subject to a uniaxial tension, σ_P. If the yield stress, σ_Y=300MPa, of the plate material is not to be exceeded then what is the maximum value of σ_P that can be applied to the plate? Use Neuber's expression to determine the σ_{yy} stress at a distance $a/2$ from the root of the hole when the maximum value of σ_P is applied.

14.7 Show that when $\alpha = \pi/2$ and the Boussinesq wedge is equivalent to a semi-infinite half-plane then the principal stresses are given by

$$\sigma_1 = 0; \quad \sigma_2 = -\frac{2fx}{\pi r^2}$$

Chapter 15

Further Plasticity

15.1 Introduction

This Chapter extends the discussion of Chapter 12 on the plastic deformation of materials. We begin by examining the notion of equivalent stress and strain for yielding beyond the initial yield stress. Elastic-plastic constitutive relationships and rules of flow are then introduced. The two flow rules examined are that of Levy-von Mises and Prandtl-Reuss. The Chapter concludes with the two applications of an elastic-plastic thick-walled pressure vessel and the plastic instability of both cylindrical and spherical pressure vessels.

15.2 Equivalent Stress and Strain

In accordance with the Huber-von Mises criterion with σ_1, σ_2 and σ_3 denoting the principal stresses then yielding commences in proportion to the root-mean-square of these stresses. For a hardening material this root-mean-square equivalent or effective stress continues beyond the initial yield stress σ_Y within the plastic regime. The value of the equivalent stress is found be replacing σ_Y by $\overline{\sigma}$ in the Huber-von Mises yield criterion, (12.45)

$$\overline{\sigma} = \frac{1}{\sqrt{2}}\sqrt{(\sigma_1 - \sigma_2)^2 + (\sigma_2 - \sigma_3)^2 + (\sigma_1 - \sigma_3)^2} \tag{15.1}$$

and implies that the initial yield surface expands uniformly to contain the current stress point (isotropic hardening).

A similar expression holds for the equivalent plastic strains

$$\mathrm{d}\overline{\varepsilon}^p = \frac{\sqrt{2}}{3}\sqrt{(\mathrm{d}\varepsilon_1^p - \mathrm{d}\varepsilon_2^p)^2 + (\mathrm{d}\varepsilon_2^p - \mathrm{d}\varepsilon_3^p)^2 + (\mathrm{d}\varepsilon_1^p - \mathrm{d}\varepsilon_3^p)^2} \tag{15.2}$$

where $\mathrm{d}\varepsilon_1^p$, $\mathrm{d}\varepsilon_2^p$ and, $\mathrm{d}\varepsilon_3^p$ are the incremental plastic strains following a corresponding change in stress $\mathrm{d}\sigma_1$, $\mathrm{d}\sigma_2$ and $\mathrm{d}\sigma_3$. The constant $\sqrt{2}/3$ ensures that $\mathrm{d}\overline{\varepsilon}^p$ correctly defines the equivalent plastic strain under uniaxial tension. For instance, if a unit cube is expanded by $\mathrm{d}\varepsilon_1^p$, $\mathrm{d}\varepsilon_2^p$ and, $\mathrm{d}\varepsilon_3^p$ without changing volume, Figure 15.1, then

$$(1 + \mathrm{d}\varepsilon_1^p)(1 + \mathrm{d}\varepsilon_2^p)(1 + \mathrm{d}\varepsilon_3^p) - 1 = 0 \tag{15.3}$$

which upon expanding gives

$$\mathrm{d}\varepsilon_1^p + \mathrm{d}\varepsilon_2^p + \mathrm{d}\varepsilon_3^p + \mathrm{d}\varepsilon_1^p\,\mathrm{d}\varepsilon_2^p + \mathrm{d}\varepsilon_1^p\,\mathrm{d}\varepsilon_3^p + \mathrm{d}\varepsilon_2^p\,\mathrm{d}\varepsilon_3^p + \mathrm{d}\varepsilon_1^p\,\mathrm{d}\varepsilon_2^p\,\mathrm{d}\varepsilon_3^p = 0 \tag{15.4}$$

Neglecting infinitesimal products, the *constancy of volume condition* reduces to

$$\mathrm{d}\varepsilon_1^p + \mathrm{d}\varepsilon_2^p + \mathrm{d}\varepsilon_3^p = 0 \tag{15.5}$$

which alternatively could have been arrived at by assuming a zero volumetric strain, (1.20).

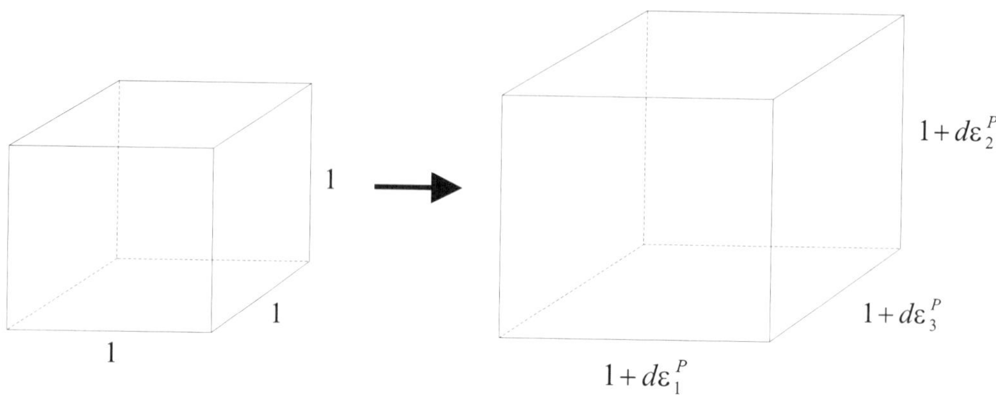

Figure 15.1: A cube of unit volume subjected to plastic strains $\mathrm{d}\varepsilon_1^p$, $\mathrm{d}\varepsilon_2^p$ and $\mathrm{d}\varepsilon_3^p$.

In the case of uniaxial tension $\mathrm{d}\varepsilon_1^p$ then from the constancy of volume condition $\mathrm{d}\varepsilon_2^p + \mathrm{d}\varepsilon_3^p = -\,\mathrm{d}\varepsilon_1^p$ and assuming that $\mathrm{d}\varepsilon_2^p = \mathrm{d}\varepsilon_3^p$ for an isotropic material then

$$\mathrm{d}\varepsilon_2^p = \mathrm{d}\varepsilon_3^p = -\frac{\mathrm{d}\varepsilon_1^p}{2} \tag{15.6}$$

Finally, substitution of $\mathrm{d}\varepsilon_1^p$, $\mathrm{d}\varepsilon_2^p$ and $\mathrm{d}\varepsilon_3^p$ into (15.2) verifies that $\mathrm{d}\bar{\varepsilon}^p = \mathrm{d}\varepsilon_1^p$ in uniaxial tension, as required.

15.3 Elastic-Plastic Stress-Strain Contitutive Relationships

To establish general stress-strain relationships for a material that experiences hardening in the plastic regime then a hypothesis is required that takes into account the degree of hardening as a function of the total plastic work done. Assume that the yield stress or equivalent stress is, at any moment, a function of the plastic work work, W^P, only. Thus, let us now examine the total, elastic and plastic components of work done during loading.

The increment of total work, $\mathrm{d}W$, per unit volume is the product of the stress and strain increment components

$$\mathrm{d}W = \sigma_{xx}\,\mathrm{d}\varepsilon_{xx} + \sigma_{yy}\,\mathrm{d}\varepsilon_{yy} + \sigma_{zz}\,\mathrm{d}\varepsilon_{zz} + 2\sigma_{xy}\,\mathrm{d}\varepsilon_{xy} + 2\sigma_{yz}\,\mathrm{d}\varepsilon_{yz} + 2\sigma_{zx}\,\mathrm{d}\varepsilon_{zx} = \sigma_{ij}\,\mathrm{d}\varepsilon_{ij} \tag{15.7}$$

The elastic work done consists of the elastic strain increments

$$\mathrm{d}W^E = \sigma_{ij}\,\mathrm{d}\varepsilon_{ij}^E \tag{15.8}$$

whereas the plastic work done can be determined from subtracting the elastic work from the total work

$$\mathrm{d}W^P = \mathrm{d}W - \mathrm{d}W^E = \sigma_{ij}\,\mathrm{d}\varepsilon_{ij} - \sigma_{ij}\,\mathrm{d}\varepsilon_{ij}^E = \sigma_{ij}(\,\mathrm{d}\varepsilon_{ij} - \,\mathrm{d}\varepsilon_{ij}^E) = \sigma_{ij}\,\mathrm{d}\varepsilon_{ij}^P \tag{15.9}$$

Therefore, the total plastic work is

$$W^P = \int_\varepsilon \sigma_{ij}\,\mathrm{d}\varepsilon_{ij}^P \tag{15.10}$$

or, in terms of equivalent stress and strain

$$W^P = \int_\varepsilon \sigma\,\mathrm{d}\bar{\varepsilon}^P \tag{15.11}$$

15.3.1 Rules of Flow

The relationship between stress and strain for linear elastic materials is governed by the Hookian equations, (1.16), which in terms of principal stresses and strains are

$$\begin{aligned} \varepsilon_1^E &= \frac{1}{E}[\sigma_1 - \nu(\sigma_2 + \sigma_3)] \\ \varepsilon_2^E &= \frac{1}{E}[\sigma_2 - \nu(\sigma_1 + \sigma_3)] \\ \varepsilon_3^E &= \frac{1}{E}[\sigma_3 - \nu(\sigma_1 + \sigma_2)] \end{aligned} \tag{15.12}$$

If the magnitude of the plastic deformation is so large that the elastic component of strain is numerically insignificant then the relationship between stress and strain can be expressed in a similar manner to the Hookian equations. However, instead of the multiplicative factor $(1/E)$ we use the coefficient λ which has the properties: i) instantaneous constant and varies with σ, ii) always positive and iii) dependent on yield stress, equilibrium and boundary conditions.

Before discussing two flow rules in the following two sub-sections let us note that the value of Poisson's ratio for an incompressible plastic material is equal to $\frac{1}{2}$. From the constancy of volume condition (15.5)

$$\begin{aligned} &\varepsilon_1 + \varepsilon_2 + \varepsilon_3 = 0 \\ \Rightarrow\; &\frac{1}{E}[\sigma_1 - \nu(\sigma_2 + \sigma_3)] + \frac{1}{E}[\sigma_2 - \nu(\sigma_1 + \sigma_3)] + \frac{1}{E}[\sigma_3 - \nu(\sigma_1 + \sigma_2)] = 0 \end{aligned} \tag{15.13}$$

which reduces to

$$\frac{1}{E}(1 - 2\nu)(\sigma_1 + \sigma_2 + \sigma_3) = 0 \tag{15.14}$$

implying that $\nu = \frac{1}{2}$ for incompressible plasticity.

15.3.1.1 Levy-von Mises Flow Rule

Assume that the form of the incremental plastic strains is similar to their elastic counterparts, (15.12)

$$\begin{aligned} \mathrm{d}\varepsilon_1^P &= \lambda[\sigma_1 - \nu(\sigma_2 + \sigma_3)] \\ \mathrm{d}\varepsilon_2^P &= \lambda[\sigma_2 - \nu(\sigma_1 + \sigma_3)] \\ \mathrm{d}\varepsilon_3^P &= \lambda[\sigma_3 - \nu(\sigma_1 + \sigma_2)] \end{aligned} \tag{15.15}$$

Eliminating λ and substituting $\nu = \frac{1}{2}$

$$\frac{d\varepsilon_1^P}{\sigma_1 - \frac{1}{2}(\sigma_2 + \sigma_3)} = \frac{d\varepsilon_2^P}{\sigma_2 - \frac{1}{2}(\sigma_1 + \sigma_3)} = \frac{d\varepsilon_3^P}{\sigma_3 - \frac{1}{2}(\sigma_1 + \sigma_2)} = \lambda \tag{15.16}$$

Remembering that plasticity is a function of the deviatoric stresses σ'_{ij}, §12.10

$$\sigma'_{ij} = \sigma_{ij} - \frac{I_1}{3}; \quad I_1 = \sigma_{xx} + \sigma_{yy} + \sigma_{zz} \tag{15.17}$$

where I_1 is the first scalar stress invariant. For example, $\sigma'_1 = \sigma_1 - (\sigma_1 + \sigma_2 + \sigma_3)/3 = (2/3)[\sigma_1 - (\sigma_2 + \sigma_3)/2]$. Substituting σ'_{ij} into (15.16)

$$\frac{d\varepsilon_1^P}{\sigma'_1} = \frac{d\varepsilon_2^P}{\sigma'_2} = \frac{d\varepsilon_3^P}{\sigma'_3} = \frac{3}{2}\lambda = \lambda' \tag{15.18}$$

or in terms of applied external stresses

$$\frac{d\varepsilon_{xx}^P}{\sigma'_{xx}} = \frac{d\varepsilon_{yy}^P}{\sigma'_{yy}} = \frac{d\varepsilon_{zz}^P}{\sigma'_{zz}} = \frac{d\varepsilon_{xy}^P}{\sigma'_{xy}} = \cdots = \lambda' \tag{15.19}$$

or

$$d\varepsilon_{ij}^P = \lambda' \sigma'_{ij} \tag{15.20}$$

which is the Levy-von Mises flow rule.

Alternatively, the Levy-von Mises flow rule can be expressed in terms of equivalent stress and strain. For simple tension $(\overline{\sigma}, \overline{\varepsilon}^P)$ then $d\varepsilon_1^P = d\overline{\varepsilon}^P$ and from (15.17) $\sigma'_1 = (2/3)\overline{\sigma}, \sigma'_2 = \sigma'_3 = -\overline{\sigma}/3$. From (15.18)

$$\lambda = \frac{2}{3}\frac{d\varepsilon_1^P}{\sigma'_1} = \frac{2}{3}\frac{d\varepsilon^P}{(2/3)\overline{\sigma}} = \frac{d\overline{\varepsilon}^P}{\overline{\sigma}} \tag{15.21}$$

and substituting into (15.15)

$$\begin{aligned} d\varepsilon_1^P &= \frac{d\overline{\varepsilon}^P}{\overline{\sigma}}[\sigma_1 - \nu(\sigma_2 + \sigma_3)] \\ d\varepsilon_2^P &= \frac{d\overline{\varepsilon}^P}{\overline{\sigma}}[\sigma_2 - \nu(\sigma_3 + \sigma_1)] \\ d\varepsilon_3^P &= \frac{d\overline{\varepsilon}^P}{\overline{\sigma}}[\sigma_3 - \nu(\sigma_1 + \sigma_2)] \end{aligned} \tag{15.22}$$

15.3.1.2 Prandtl-Reuss Flow Rule

When both the elastic and plastic strains are of the same order then an alternative flow rule is required that accounts for the elastic strain increments. For the principle direction 1 the total deformation is

$$d\varepsilon_1 = d\varepsilon_1^E + d\varepsilon_1^P \tag{15.23}$$

However, the elastic component of strain has an associated change in volume whereas the plastic component does not. Thus, eliminating the volumetric contribution from the elastic component gives the deviatoric elastic strain increment

$$\left(d\varepsilon_1^E\right)' = d\varepsilon_1^E - d\varepsilon_m^E \tag{15.24}$$

where the mean elastic strain increment is

$$\mathrm{d}\varepsilon_m^E = \frac{\mathrm{d}\varepsilon_1^E + \mathrm{d}\varepsilon_2^E + \mathrm{d}\varepsilon_3^E}{3} \tag{15.25}$$

with $\mathrm{d}\varepsilon_i^E$ given by the Hookian relations. For example, $\mathrm{d}\varepsilon_1^E$ is given by

$$\mathrm{d}\varepsilon_1^E = \frac{1}{E}[\mathrm{d}\sigma_1 - \nu(\mathrm{d}\sigma_2 + \mathrm{d}\sigma_3)] \tag{15.26}$$

and similarly for $\mathrm{d}\varepsilon_2^E$ and $\mathrm{d}\varepsilon_3^E$. Upon substitution into (15.25)

$$\mathrm{d}\varepsilon_m^E = \frac{1}{E}(1-2\nu)\,\mathrm{d}\sigma_m; \qquad \mathrm{d}\sigma_m = \frac{\mathrm{d}\sigma_1 + \mathrm{d}\sigma_2 + \mathrm{d}\sigma_3}{3} \tag{15.27}$$

and now substituting for $\mathrm{d}\varepsilon_m^E$ and $\mathrm{d}\varepsilon_1^E$ into (15.24)

$$\left(\mathrm{d}\varepsilon_1^E\right)' = \frac{1}{E}[\mathrm{d}\sigma_1 - \nu(\mathrm{d}\sigma_2 + \mathrm{d}\sigma_3)] - \frac{1}{E}(1-2\nu)\,\mathrm{d}\sigma_m \tag{15.28}$$

Upon reducing and use of (15.27) and noting $E = 2G(1+\nu)$ we have

$$\left(\mathrm{d}\varepsilon_1^E\right)' = \frac{(1+\nu)}{E}\,\mathrm{d}\sigma_1' = \frac{\mathrm{d}\sigma_1'}{2G} \tag{15.29}$$

Returning to (15.23), modified for the deviatoric stresses and use of the Levy-von Mises flow rule gives

$$\mathrm{d}\varepsilon_1' = \mathrm{d}\varepsilon_1^E + \mathrm{d}\varepsilon_1^P = \frac{\mathrm{d}\sigma_1'}{2G} + \lambda\sigma_1' \tag{15.30}$$

or in general, we have the Prandtl-Reuss flow rule

$$\mathrm{d}\varepsilon_{ij}' = \frac{\mathrm{d}\sigma_{ij}'}{2G} + \lambda\sigma_{ij}' \tag{15.31}$$

In terms of the absolute stresses rather than the deviatoric stresses, the strains are

$$\begin{aligned} \mathrm{d}\varepsilon_{xx} &= \frac{1}{E}\left[\mathrm{d}\sigma_{xx} - \nu(\mathrm{d}\sigma_{yy} + \mathrm{d}\sigma_{zz})\right] + \frac{2}{3}\lambda\left[\sigma_{xx} - \frac{1}{2}(\sigma_{yy} + \sigma_{zz})\right] \\ \mathrm{d}\gamma_{xy} &= \frac{\mathrm{d}\tau_{xy}}{2G} + \lambda\tau_{xy} \end{aligned} \tag{15.32}$$

and similarly for $\mathrm{d}\varepsilon_{yy}$, $\mathrm{d}\varepsilon_{zz}$, $\mathrm{d}\gamma_{xz}$ and $\mathrm{d}\gamma_{yz}$.

The Prandtl-Reuss flow rule constitutes the most general statement of the stress-strain constitutive relationships and applies equally to elastic, elastic-plastic and purely-plastic modes of deformation.

15.4 Elastic-Plastic Thick-Walled Pressure Vessel

Consider a thick-walled pressure vessel of inner radius a, outer radius b subject to an internal pressure p and zero external pressure with an elastic-plastic interface at radius c, Figure 15.2

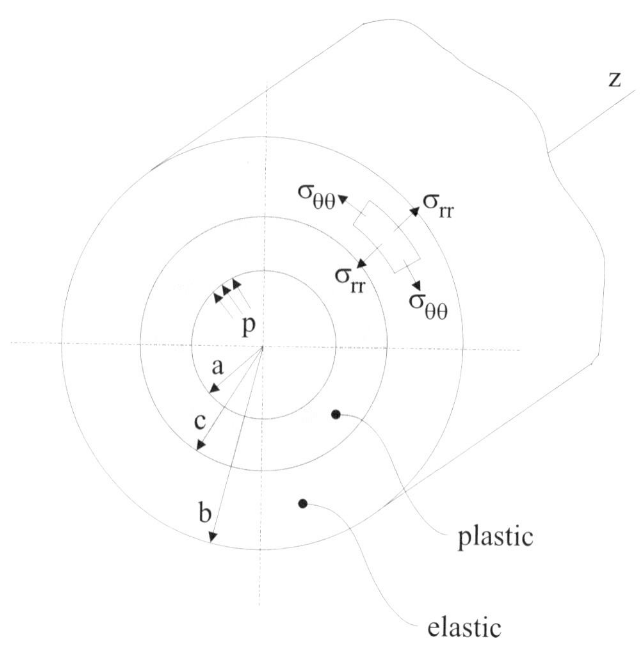

Figure 15.2: Thick-walled pressure vessel of inner radius a and outer radius b subject to an internal pressure p with an elastic-plastic interface at radius c.

15.4.1 Elastic Range

Of the three stress components $(\sigma_{\theta\theta}, \sigma_{rr}, \sigma_{zz})$ in the vessel, $\sigma_{\theta\theta}$ is the maximum and σ_{rr} is the minimum. Assuming that the vessel material is elastic-plastic and yields in accordance with Tresca's criterion, §12.11.3

$$\sigma_{\theta\theta} - \sigma_{rr} = \sigma_Y \tag{15.33}$$

where $\sigma_{\theta\theta}$ and σ_{rr} at $r = a$ are given by Lame's equation, §11.4

$$\sigma_{rr} = -p; \quad \sigma_{\theta\theta} = p\left(\frac{k^2+1}{k^2-1}\right); \quad k = \frac{b}{a} \tag{15.34}$$

Substituting $\sigma_{\theta\theta}$ and σ_{rr} into (15.33) the pressure at the incipient point of yielding is

$$p_Y = \frac{\sigma_Y}{2}\left(\frac{k^2-1}{k^2}\right) = \frac{\sigma_Y}{2}\left(1 - \frac{1}{k^2}\right) \tag{15.35}$$

15.4.2 Plastic Range

From Lame's analysis of thick-walled vessels the equation of equilibrium is, (11.3), and in view of (15.33)

$$r\frac{\mathrm{d}\sigma_{rr}}{\mathrm{d}r} = \sigma_{\theta\theta} - \sigma_{rr} = \sigma_Y \tag{15.36}$$

which, upon integrating

$$\sigma_{rr} = \sigma_Y \int \frac{\mathrm{d}r}{r} = \sigma_Y[\ln r + A] \tag{15.37}$$

noting that the constant of integration is multiplied by σ_Y in order to assist cancellation of σ_Y later. When $r = c$ on the elastic-plastic interface then the outer elastic annulus is on the point of yielding under pressure, p_c, and from (15.35)

$$\sigma_{rr} = -p_c = -\frac{\sigma_Y}{2}\left(1 - \frac{1}{k_1^2}\right); \quad k_1 = \frac{b}{c} \tag{15.38}$$

Equating (15.37) and (15.38) leads to the value of A

$$A = -\frac{1}{2}\left[1 - \left(\frac{c}{b}\right)^2\right] - \ln c \tag{15.39}$$

noting that the last term is $\ln c$ since $r = c$.

Finally, substituting A into (15.37) and from Tresca's criterion (15.33)

$$\begin{aligned}\sigma_{rr} &= \sigma_Y\left[\ln r - \frac{1}{2}\left[1 - \left(\frac{c}{b}\right)^2\right] - \ln c\right] = -\sigma_Y\left[\ln\left(\frac{c}{r}\right) + \frac{1}{2}\left[1 - \left(\frac{c}{b}\right)^2\right]\right]\\ \sigma_{\theta\theta} &= \sigma_Y + \sigma_{rr} = \sigma_Y\left[-\ln\left(\frac{c}{r}\right) + \frac{1}{2}\left[1 + \left(\frac{c}{b}\right)^2\right]\right]\end{aligned} \tag{15.40}$$

15.5 Plastic Instability of Thin-Walled Pressure Vessels

The following two sub-sections examine the plastic instability of thin-walled cylindrical and spherical pressure vessels and serve as an illustraiion of the Levy-von Mises flow rule.

15.5.1 Cylindrical Pressure Vessels

Figure 15.3 illustrates a thin-walled cylindrical pressure vessel of mean radius r and wall-thickness t subject to an internal pressure p. The stresses are given by (4.2) and (4.4)

$$\sigma_{\theta\theta} = \frac{pr}{r}; \quad \sigma_{zz} = \frac{pr}{2t} = \frac{\sigma_{\theta\theta}}{2}; \quad \sigma_{rr} = 0 \tag{15.41}$$

From (15.1) and (15.2) the equivalent stress and strain are

$$\begin{aligned}\overline{\sigma} &= \frac{1}{\sqrt{2}}\left[(\sigma_{\theta\theta} - \sigma_{zz})^2 + (\sigma_{zz} - \sigma_{\theta\theta})^2 + (\sigma_{\theta\theta} - \sigma_{rr})^2\right]^{\frac{1}{2}} = \frac{\sqrt{3}\sigma_{\theta\theta}}{2}\\ \mathrm{d}\overline{\varepsilon}^P &= \frac{\sqrt{2}}{3}\left[(\mathrm{d}\varepsilon_{\theta\theta}^P - \mathrm{d}\varepsilon_{zz}^P)^2 + (\mathrm{d}\varepsilon_{zz}^P - \mathrm{d}\varepsilon_{rr}^P)^2 + (\mathrm{d}\varepsilon_{\theta\theta}^P - \mathrm{d}\varepsilon_{rr}^P)^2\right]^{\frac{1}{2}}\end{aligned} \tag{15.42}$$

By definition, the circumferential and radial strain components are

$$\mathrm{d}\varepsilon_{\theta\theta} = \frac{\mathrm{d}r}{r} \Rightarrow \varepsilon_{\theta\theta} = \ln\left(\frac{r}{r_0}\right); \quad \mathrm{d}\varepsilon_{rr} = \frac{\mathrm{d}t}{t} \Rightarrow \varepsilon_{rr} = \ln\left(\frac{t}{t_0}\right) \tag{15.43}$$

where r_0 and t_0 denote the initial mean radius and wall thickness respectively. From (15.41) and (15.43) the pressure p can be expressed as

$$p = \frac{t\sigma_{\theta\theta}}{r} = \frac{t_0 e^{\varepsilon_{rr}^P}\sigma_{\theta\theta}}{r_0 e^{\varepsilon_{\theta\theta}^P}\sigma_{\theta\theta}} = \frac{t_0\sigma_{\theta\theta}e^{(\varepsilon_{rr}^P - \varepsilon_{\theta\theta}^P)}}{r_0} \tag{15.44}$$

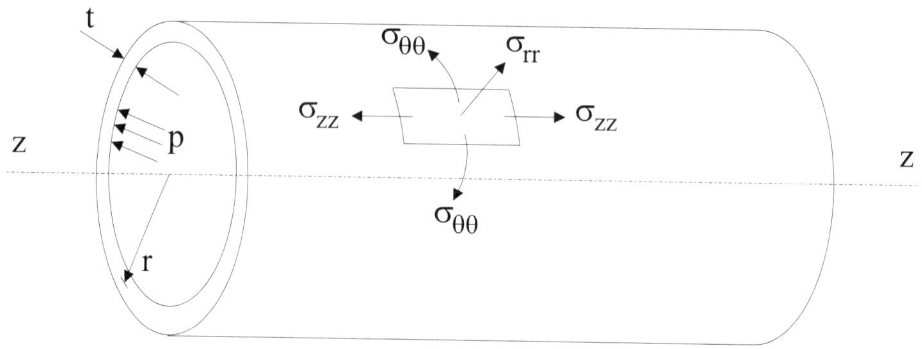

Figure 15.3: Thin-walled cylindrical pressure vessel of mean radius r, wall thickness t subject to an internal pressure p.

For a bar under a tensile load F instability is a result of the maximum tensile load, that is, when $\mathrm{d}F/\mathrm{d}\varepsilon = 0$; refer to §12.9. At the point of plastic instability for a cylindrical pressure vessel then $\mathrm{d}p/\mathrm{d}\overline{\varepsilon}^P = 0$. Therefore, from (15.44) and in view of $\sigma_{\theta\theta} = 2\overline{\sigma}/\sqrt{3}$ from (15.42) we have

$$\begin{aligned}\frac{\mathrm{d}p}{\mathrm{d}\overline{\varepsilon}^P} = 0 \Rightarrow \frac{\mathrm{d}}{\mathrm{d}\overline{\varepsilon}^P}\left[\frac{t_0\sigma_{\theta\theta}e^{\left(\varepsilon_{rr}^P-\varepsilon_{\theta\theta}^P\right)}}{r_0}\right] &= \frac{2t_0}{r_0}\frac{\mathrm{d}}{\mathrm{d}\overline{\varepsilon}^P}\left[\overline{\sigma}e^{\left(\varepsilon_{rr}^P-\varepsilon_{\theta\theta}^P\right)} = 0\right] \\ \Rightarrow \frac{2t_0}{r_0}\left[\frac{\mathrm{d}\sigma}{\mathrm{d}\overline{\varepsilon}^P}e^{\left(\varepsilon_{rr}^P-\varepsilon_{\theta\theta}^P\right)} + \sigma e^{\left(\varepsilon_{rr}^P-\varepsilon_{\theta\theta}^P\right)}\left(\frac{\mathrm{d}\varepsilon_{rr}^P}{\mathrm{d}\overline{\varepsilon}^P} - \frac{\mathrm{d}\varepsilon_{\theta\theta}^P}{\mathrm{d}\overline{\varepsilon}^P}\right)\right] &= 0\end{aligned} \tag{15.45}$$

from which we have

$$\frac{\mathrm{d}\overline{\sigma}}{\mathrm{d}\overline{\varepsilon}^P} = -\sigma\left(\frac{\mathrm{d}\varepsilon_{rr}^P}{\mathrm{d}\overline{\varepsilon}^P} - \frac{\mathrm{d}\varepsilon_{\theta\theta}^P}{\mathrm{d}\overline{\varepsilon}^P}\right) \tag{15.46}$$

From the Levy-von Mises flow rule, (15.20), we have

$$\begin{aligned}\frac{\mathrm{d}\varepsilon_{rr}^P}{\mathrm{d}\overline{\varepsilon}^P} &= \frac{1}{\overline{\sigma}}\left[\sigma_{rr} - \frac{1}{2}(\sigma_{\theta\theta}+\sigma_{zz})\right] = -\frac{3\sigma_{\theta\theta}}{4\overline{\sigma}} = -\frac{\sqrt{3}}{2} \\ \frac{\mathrm{d}\varepsilon_{\theta\theta}^P}{\mathrm{d}\overline{\varepsilon}^P} &= \frac{1}{\overline{\sigma}}\left[\sigma_{\theta\theta} - \frac{1}{2}(\sigma_{rr}+\sigma_{zz})\right] = \frac{3\sigma_{\theta\theta}}{4\overline{\sigma}} = \frac{\sqrt{3}}{2}\end{aligned} \tag{15.47}$$

Substituting (15.47) into (15.46) we arrive at the following condition of plastic instability

$$\frac{\mathrm{d}\overline{\sigma}}{\mathrm{d}\overline{\varepsilon}^P} = \sqrt{3}\overline{\sigma} \tag{15.48}$$

and should be compared to the condition of plastic instability in tension (12.15).

Finally, from (15.47) and (15.43) then t and r can be expressed in terms of

$$t = t_0 e^{-\sqrt{3}\overline{\varepsilon}^P/2}; \quad r = r_0 e^{\sqrt{3}\overline{\varepsilon}^P/2} \tag{15.49}$$

and provide geometric measures of the plastic instability condition.

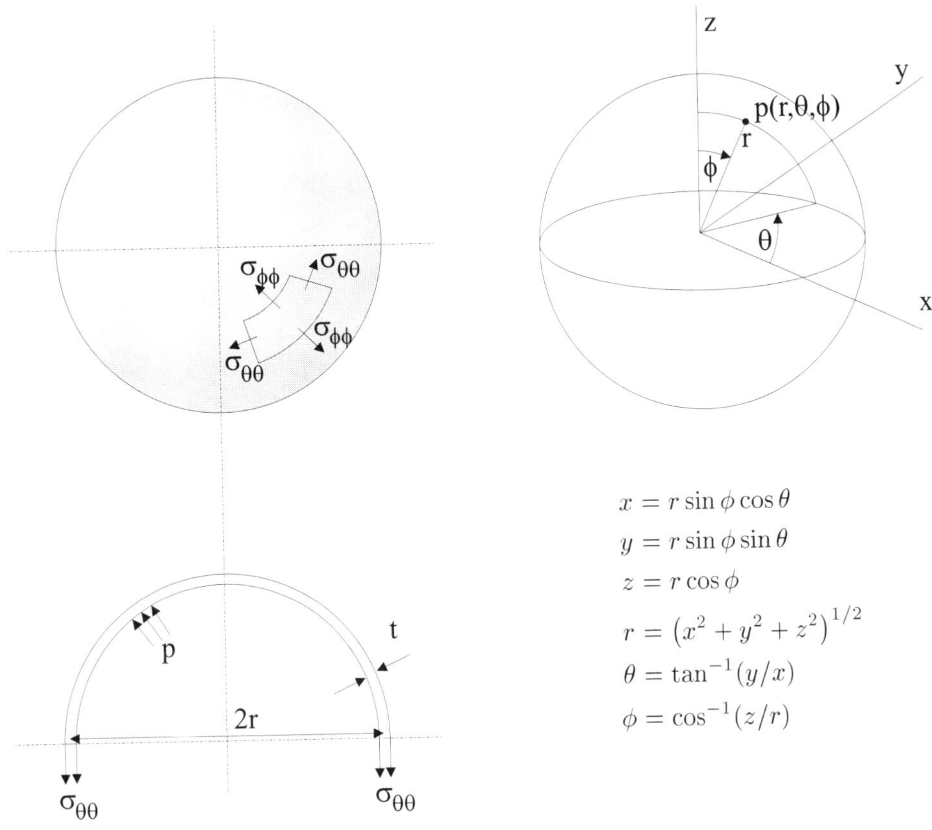

Figure 15.4: Thin-walled spherical pressure vessel of mean radius r and wall thickness t subject to an internal pressure p. Also shown is the spherical coordinate system (r, θ, ϕ).

15.5.2 Spherical Pressure Vessels

With reference to Figure 15.4 the stresses for a thin-walled spherical pressure vessel of mean radius r and wall thickness t subject to an internal pressure p are, §4.3

$$\sigma_{\theta\theta} = \sigma_{\phi\phi} = \frac{pr}{2t}; \quad \sigma_{rr} = 0 \tag{15.50}$$

Since we are accounting for plasticity and yielding beyond the yield stress, σ_Y, we will express plastic instability in terms of the equivalent stress and strain; which in the present case are

$$\begin{aligned} \overline{\sigma} &= \frac{1}{\sqrt{2}}\left[(\sigma_{\theta\theta} - \sigma_{\phi\phi})^2 + (\sigma_{\phi\phi} - \sigma_{rr})^2 + (\sigma_{\theta\theta} - \sigma_{rr})^2\right]^{\frac{1}{2}} = \sigma_{\theta\theta} \\ \mathrm{d}\overline{\varepsilon}^P &= \frac{\sqrt{2}}{3}\left[(\,\mathrm{d}\varepsilon^P_{\theta\theta} - \mathrm{d}\varepsilon^P_{\phi\phi})^2 + (\,\mathrm{d}\varepsilon^P_{\phi\phi} - \mathrm{d}\varepsilon^P_{rr})^2 + (\,\mathrm{d}\varepsilon^P_{\theta\theta} - \mathrm{d}\varepsilon^P_{rr})^2\right]^{\frac{1}{2}} \end{aligned} \tag{15.51}$$

From the Levy-von Mises flow rule, (15.20)

$$\mathrm{d}\varepsilon^P_{ij} = \lambda\sigma'_{ij}; \quad \lambda = \frac{\mathrm{d}\overline{\varepsilon}^P}{\overline{\sigma}}; \quad \mathrm{d}\varepsilon^P_1 = \frac{\mathrm{d}\overline{\varepsilon}^P}{\overline{\sigma}}[\sigma_1 - \nu(\sigma_2 + \sigma_3)] \tag{15.52}$$

The plastic strain increment is therefore, in view of $\sigma_{\theta\theta} = \sigma_{\phi\phi}$ and $\sigma_{rr} = 0$

$$\mathrm{d}\varepsilon_{rr}^P = \frac{\mathrm{d}\overline{\varepsilon}^P}{\overline{\sigma}}\left[\sigma_{rr} - \frac{1}{2}(\sigma_{\theta\theta} + \sigma_{\phi\phi})\right] = -\frac{\sigma_{\theta\theta}}{\overline{\sigma}}\,\mathrm{d}\overline{\varepsilon}^P = -\,\mathrm{d}\overline{\varepsilon}^P \tag{15.53}$$

since $\overline{\sigma} = \sigma_{\theta\theta}$ from (15.51). Since the radial strain is the change in t

$$\mathrm{d}\varepsilon_{rr} = -\,\mathrm{d}\overline{\varepsilon}^P = \frac{\mathrm{d}t}{t} \tag{15.54}$$

which upon integrating gives

$$\varepsilon_{rr} = -\overline{\varepsilon}^P = \ln\left(\frac{t}{t_0}\right) \tag{15.55}$$

where t_0 is the initial wall thickness. From the constancy of volume condition $(\varepsilon_{rr} + \varepsilon_{\theta\theta} + \varepsilon_{\phi\phi} = 0)$ with $\varepsilon_{\theta\theta} = \varepsilon_{\phi\phi}$

$$\varepsilon_{\theta\theta} = -\frac{1}{2}\varepsilon_{rr} = \frac{1}{2}\overline{\varepsilon}^P = \frac{1}{2}\ln\left(\frac{r}{r_0}\right) \tag{15.56}$$

Thus, t and r can now be expressed in terms of $\overline{\varepsilon}^P$

$$t = t_0 e^{-\overline{\varepsilon}^P}; \quad r = r_0 e^{-\overline{\varepsilon}^P/2}; \tag{15.57}$$

The instability condition for a spherical pressure vessel with $p = 2t\sigma_{\theta\theta}/r = 2t\overline{\sigma}/r$ from (15.50) is given by

$$\frac{\mathrm{d}p}{\mathrm{d}\overline{\varepsilon}^P} = 0$$

$$\Rightarrow \frac{\mathrm{d}p}{\mathrm{d}\overline{\varepsilon}^P}\left[\frac{2t_0 e^{-\overline{\varepsilon}^P}}{r_0 e^{\overline{\varepsilon}^P/2}}\overline{\sigma}\right] = 2\frac{t_0}{r_0}\frac{\mathrm{d}}{\mathrm{d}\overline{\varepsilon}^P}\left[\overline{\sigma}e^{-3\overline{\varepsilon}^P/2}\right] = 2\frac{t_0}{r_0}\left[\frac{\mathrm{d}\overline{\sigma}}{\mathrm{d}\overline{\varepsilon}^P}e^{-3\overline{\varepsilon}^P/2} - \frac{3}{2}e^{-3\overline{\varepsilon}^P/2}\right] = 0 \tag{15.58}$$

Cancelling terms, we find that the condition of plastic of instability is

$$\frac{\mathrm{d}\overline{\sigma}}{\mathrm{d}\overline{\varepsilon}^P} = \frac{3}{2}\overline{\sigma} \tag{15.59}$$

which can be compared to the condition of plastic instability for the cylindrical pressure vessel, (15.48).

Example 15.1 Plastic instability for a thin-walled spherical pressure vessel

Assuming a material response of $\overline{\sigma} = C(\overline{\varepsilon}^P)^n$ with $C = 725$ and $n = 0.15$ then determine the condition of plastic instability for a thin-walled spherical pressure vessel.

From the condition of plastic instability (15.59), with $\mathrm{d}\overline{\sigma}/\,\mathrm{d}\overline{\varepsilon}^P = nC(\overline{\varepsilon}^P)^{n-1}$, then

$$\frac{\mathrm{d}\overline{\sigma}}{\mathrm{d}\overline{\varepsilon}^P} = \frac{3}{2}\overline{\sigma} \Rightarrow nC(\overline{\varepsilon}^P)^{n-1} = \frac{3}{2}C(\overline{\varepsilon}^P)^n$$

Solving for equivalent plastic strain we have

$$\overline{\varepsilon}^P = 2n/3 = 0.1$$

The wall thickness and radius at instability are equal to $t = t_0 e^{-\overline{\varepsilon}^P} = 0.905t_0$ and $r = r_0 e^{\overline{\varepsilon}^P/2} = 1.051r_0$ from (15.57).

15.6 Conclusion

This Chapter has extended our discussion of the elastic-plastic behaviour of materials presented in Chapter 12 to cover the Levy-von Mises and Prandtl-Reuss flow rules. These flow rules are the elastic-plastic equivalents of the Hookian constitutive equations for linear elastic materials. The flow rules are expressed in terms of deviatoric components and equivalent stress and strain. The deviatoric components account for the fact that plasticity is governed by the shear components of deformation whereas the equivalent stress and plastic strain components provides measures of elastic-plastic deformation beyond the initial yield stress. The Chapter concluded by examining the two applications of an elastic-plastic thick-walled pressure vessel and the plastic instability of both cylindrical and spherical pressure vessels.

15.7 References and Further Reading

∘ Alexander, J. M. (1981) *Strength of Materials, Volume I: Fundamentals*, Ellis Horwood Series in Engineering Science.

∘ Blazynski, T. Z. (1983) *Applied Elasto-Plasticity of Solids*, Macmillan, London.

∘ Chakrabarty, J. (1987) *Theory of Plasticity*, McGraw-Hill, Singapore.

15.8 Exercises

15.1 Describe what is meant by equivalent stress and equivalent strain.

15.2 What is the constancy of volume condition? Use the constancy of volume condition to show that for incompressible plasticity Poisson's ratio is equal to $\frac{1}{2}$.

15.3 Determine expressions for the equivalent stress and strain of a thin-walled cylindrical pressure vessel.

15.4 A thick-walled cylinder with open ends is subject to an internal pressure $p(=100\text{MPa})$ and zero external pressure. The inner radius of the cylinder is denoted by a and the outer radius by ka; where $k(>1)$ is a constant. Assume that the yield stress of the vessel material is $\sigma_Y(=250\text{MPa})$ in pure tension and that the material does not strain-harden. Determine the critical value of k for which the inner radius will be at the point of first yielding according to both the Tresca and Huber-von Mises yield criteria.

15.5 A thick-walled cylinder of inner radius 50mm and outer radius 100mm is to be pressurised by an internal pressure only such that an elastic-plastic boundary is developed to a radius of 70mm. Determine the required internal pressure if the tensile yield stress of the vessel material is 300MPa. Also, determine the required internal pressure which will make the cylinder fully plastic.

15.6 A thin-walled cylindrical pressure vessel is of initial mean radius 0.45m and initial wall thickness 1mm. If the vessel material obeys the following Ludwik power law

$$\overline{\sigma} = C(\overline{\varepsilon}^P)^n; \quad C = 825; \quad n = 0.45$$

then determine the mean radius and wall thickness at the point of plastic instability.

15.7 Determine the mean radius and wall thickness at the point of plastic instability for a thin-walled spherical pressure vessel having the same mean radius and wall thickness and obeying the same material power law as in Exercise 15.6.

Chapter 16

Further Finite Elements

16.1 Introduction

In this Chapter we extend our examination of the finite element method presented in Chapter 13 by considering both two-dimensional and higher order elements. The two-dimensional element considered is the three noded constant strain triangle while the higher order elements considered are the quadratic and cubic bar and triangle elements. The discussion of higher order elements will illustrate the use of higher order interpolation functions and their derivation and the use of numerical integration.

16.2 Three Noded Constant Strain Triangle

16.2.1 Element Interpolation Function

Figure 16.1 illustrates a straight-sided triangular element with three nodes (i, j, k) at the triangle vertices. The interpolation function, ϕ, for this element is

$$\phi = \alpha_1 + \alpha_2 x + \alpha_3 y \tag{16.1}$$

Substituting the (x, y) coordinates of the three nodes (x_i, y_i), (x_j, y_j) and (x_k, y_k) into (16.1) we find

$$\begin{aligned} \phi_i &= \alpha_1 + \alpha_2 x_i + \alpha_3 y_i \\ \phi_j &= \alpha_1 + \alpha_2 x_j + \alpha_3 y_j \\ \phi_k &= \alpha_1 + \alpha_2 x_k + \alpha_3 y_k \end{aligned} \tag{16.2}$$

Eliminating α_1

$$\begin{aligned} \phi_j - \phi_i &= \alpha_2(x_j - x_i) + \alpha_3(y_j - y_i) \\ \phi_k - \phi_i &= \alpha_2(x_k - x_i) + \alpha_3(y_k - y_i) \end{aligned} \tag{16.3}$$

and then eliminating α_2 we find α_3 and performing back substitution to obtain α_2 and α_1 we arrive at

$$\begin{aligned} \alpha_1 &= \frac{1}{2A}(a_i\phi_i + a_j\phi_j + a_k\phi_k) \\ \alpha_2 &= \frac{1}{2A}(b_i\phi_i + b_j\phi_j + b_k\phi_k) \\ \alpha_3 &= \frac{1}{2A}(c_i\phi_i + c_j\phi_j + c_k\phi_k) \end{aligned} \tag{16.4}$$

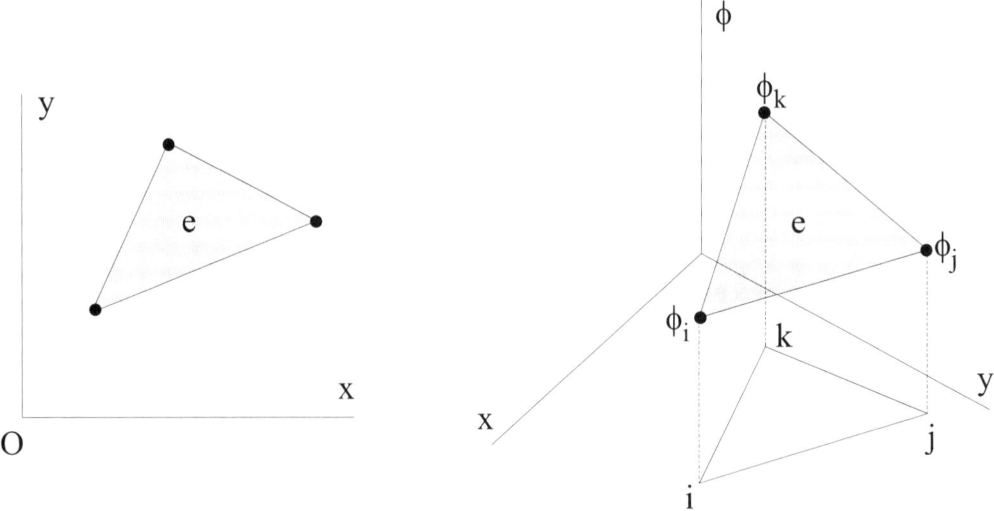

Figure 16.1: Three noded triangular element, e, and nodal values of ϕ.

where

$$\begin{aligned} a_i &= x_j y_k - x_k y_j; \quad b_i = y_j - y_k; \quad c_i = x_k - x_j \\ a_j &= x_k y_i - x_i y_k; \quad b_j = y_k - y_i; \quad c_j = x_i - x_k \\ a_k &= x_i y_j - x_j y_i; \quad b_k = y_i - y_j; \quad c_k = x_j - x_i \end{aligned} \tag{16.5}$$

and where A is the area of the triangle and is given by

$$A = \frac{1}{2} \begin{vmatrix} 1 & x_i & y_i \\ 1 & x_j & y_j \\ 1 & x_k & y_k \end{vmatrix} = \frac{1}{2}(x_i y_j + x_j y_k + x_k y_i - x_i y_k - x_j y_i - x_k y_j) \tag{16.6}$$

The interpolation function (16.1) can now be re-written in terms of the element shape functions and nodal values of ϕ

$$\phi = [N]\{\phi\} = \begin{bmatrix} N_i & N_j & N_k \end{bmatrix} \left\{ \begin{array}{c} \phi_i \\ \phi_j \\ \phi_k \end{array} \right\} = N_i \phi_i + N_j \phi_j + N_k \phi_k \tag{16.7}$$

where

$$\begin{aligned} N_i &= \frac{1}{2A}(a_i + b_i x + c_i y) \\ N_j &= \frac{1}{2A}(a_j + b_j x + c_j y) \\ N_k &= \frac{1}{2A}(a_k + b_k x + c_k y) \end{aligned} \tag{16.8}$$

or in general

$$N_\beta = \frac{1}{2A}(a_\beta + b_\beta + c_\beta y); \quad \beta = i, j, k \tag{16.9}$$

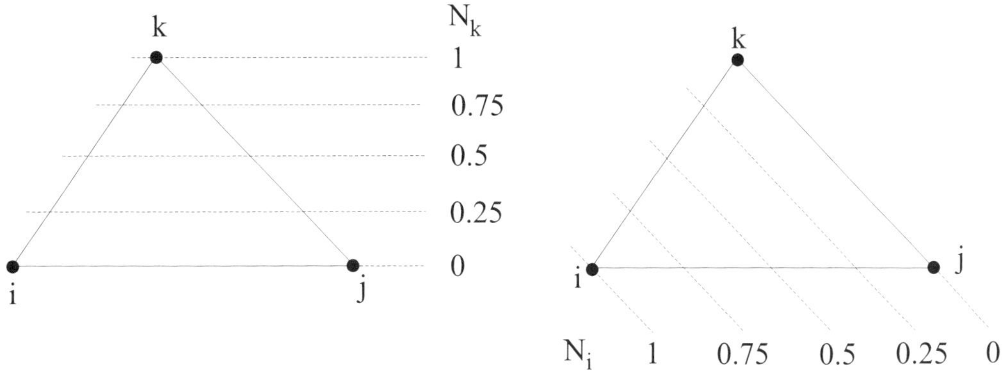

Figure 16.2: Variation of shape functions N_k and N_i for planes parallel to edges (i, j) and (j, k) respectively.

Equation (16.7) allows a value for ϕ to be determined at any point within an element based on the three nodal values ϕ_i, ϕ_j and ϕ_k. As for the one-dimensional bar element considered in Chapter 13, each nodal shape function is equal to unity at its respective node and zero at any other node, Figure 16.2.

So far, we have only considered the variation of a scalar quantity ϕ. However, for stress analyses in two-dimensions we are interested in the displacements u and v in both the x and y directions respectively, Figure 16.3. The element therefore requires two interpolation functions, one for each direction, (16.7)

$$\begin{aligned} u &= N_i u_i + N_j u_j + N_k u_k \\ v &= N_i v_i + N_j v_j + N_k v_k \end{aligned} \tag{16.10}$$

noting that the same shape functions (16.10) are used for both u and v since the shape functions are dependent on the geometry and independent of the displacements of the element. In matrix-vector format (16.10) is expressed as

$$\left\{ \begin{matrix} u \\ v \end{matrix} \right\} = [N]\{U\} \begin{bmatrix} N_i & 0 & N_j & 0 & N_k & 0 \\ 0 & N_i & 0 & N_j & 0 & N_k \end{bmatrix} \left\{ \begin{matrix} u_i \\ v_i \\ u_j \\ v_j \\ u_k \\ v_k \end{matrix} \right\} \tag{16.11}$$

16.2.2 Element Strains and Stresses

Section 13.10 showed that the element strain vector, $\{\varepsilon\}$, is given by, (13.30)

$$\{\varepsilon\} = [B]\{U\} \tag{16.12}$$

where $[B]$ is a matrix of first partial derivatives. For the present triangular element the $[B]$ matrix is obtained by considering the nodal displacements (16.10) and components of $\{\varepsilon\} = \{\varepsilon_{xx}, \varepsilon_{yy}, \gamma_{xy}\}$ separately. Considering ε_{xx} then

$$\varepsilon_{xx} = \frac{\partial u}{\partial x} = \frac{\partial N_i}{\partial x} u_i + \frac{\partial N_j}{\partial x} u_j + \frac{\partial N_k}{\partial x} u_k \tag{16.13}$$

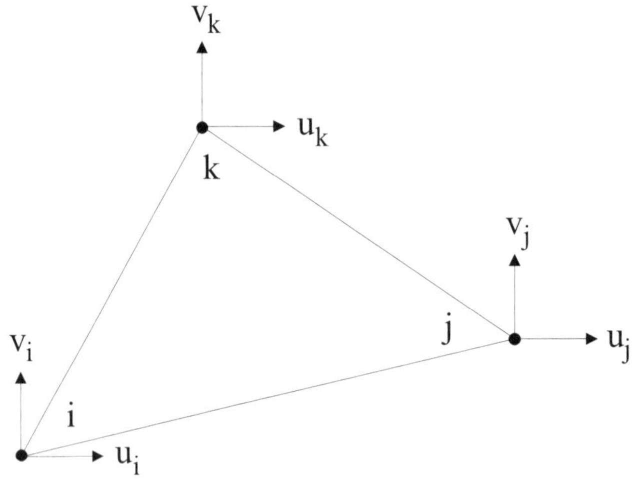

Figure 16.3: Nodal displacements (u_i, v_i), (u_j, v_j) and (u_k, v_k).

or in matrix format

$$\varepsilon_{xx} = \begin{bmatrix} \frac{\partial N_i}{\partial x} & 0 & \frac{\partial N_j}{\partial x} & 0 & \frac{\partial N_k}{\partial x} & 0 \end{bmatrix} \{U\} \tag{16.14}$$

The derivatives of N_β with respect to x are determined from (16.8)

$$\frac{\partial N_i}{\partial x} = \frac{b_i}{2A}; \quad \frac{\partial N_j}{\partial x} = \frac{b_j}{2A}; \quad \frac{\partial N_k}{\partial x} = \frac{b_k}{2A}; \tag{16.15}$$

Therefore, substituting (16.15) into (16.14) we find that ε_{xx} is given by

$$\varepsilon_{xx} = \frac{1}{2A} \begin{bmatrix} b_i & 0 & b_j & 0 & b_k & 0 \end{bmatrix} \{U\} \tag{16.16}$$

Performing a similar procedure for the other two strain components ε_{yy} and γ_{xy} then $\{\varepsilon\}$ is given by

$$\{\varepsilon\} = \left\{ \begin{array}{c} \varepsilon_{xx} \\ \varepsilon_{yy} \\ \gamma_{xy} \end{array} \right\} = \frac{1}{2A} \begin{bmatrix} b_i & 0 & b_j & 0 & b_k & 0 \\ 0 & c_i & 0 & c_j & 0 & c_k \\ c_i & b_i & c_j & b_j & c_k & b_k \end{bmatrix} \left\{ \begin{array}{c} u_i \\ v_i \\ u_j \\ v_j \\ u_k \\ v_k \end{array} \right\} = [B]\{U\} \tag{16.17}$$

Note that $[B]$ is independent of (x, y) and as a result the strains are constant throughout the entire element. Thus, a triangular element with linear interpolation function (16.1) is frequently referred to as a *constant strain triangle*.

The stress vector $\{\sigma\}$ follows immediately from $\{\varepsilon\}$ via the constitutive law (13.32)

$$\{\sigma\} = \left\{ \begin{array}{c} \sigma_{xx} \\ \sigma_{yy} \\ \tau_{xy} \end{array} \right\} = [D]\left(\{\varepsilon\} - \{\varepsilon_o\}\right) + \{\sigma_o\} \tag{16.18}$$

where $\{\varepsilon_o\}$ and $\{\sigma_o\}$ are initial strain and stress vectors.

16.2.3 Derivation of the Element Stiffness and Force Vector for Elasticity Problems using a Variational Formulation

The previous sub-section illustrated that once the displacement vector is known then the element strains and stresses immediately follow. In the present section we illustrate how to determine the displacement vector using a variational formulation similar to that used in §13.11 in the case of one-dimensional bar elements. In order to derive the finite element equations of elasticity and ultimately the entire structure equations we begin by considering the potential energy, Π_e, of an element, (9.1)

$$\Pi_e = U_e - W_e \tag{16.19}$$

and is equal to the strain energy, U_e, minus the work done, W_e, by the externally applied forces. From (9.4) the strain energy of a linear elastic differential element is

$$\mathrm{d}U_e = \frac{1}{2}\left[\{\varepsilon\}^{\mathrm{T}}\{\sigma\} - \{\varepsilon_o\}^{\mathrm{T}}\{\sigma\}\right] \tag{16.20}$$

Integrating over the entire elemental volume, $\mathrm{d}V_e$

$$U_e = \frac{1}{2}\int_{V_e}\left[\{\varepsilon\}^{\mathrm{T}}\{\sigma\} - \{\varepsilon_o\}^{\mathrm{T}}\{\sigma\}\right]\,\mathrm{d}V_e \tag{16.21}$$

Substituting the stress-strain relation (16.18) into (16.21) we have, neglecting the initial stress vector $\{\sigma_0\}$

$$U_e = \frac{1}{2}\int_{V_e}\left[\{\varepsilon\}^{\mathrm{T}}[D]\{\varepsilon\} - 2\{\varepsilon\}^{\mathrm{T}}[D]\{\varepsilon\} + \{\varepsilon_o\}^{\mathrm{T}}[D]\{\varepsilon_o\}\right]\,\mathrm{d}V_e \tag{16.22}$$

and substituting in the strain vector $\{\varepsilon\}$ from (16.17)

$$U_e = \frac{1}{2}\int_{V_e}\left[\{U\}^{\mathrm{T}}[B]^{\mathrm{T}}[D][B]\{U\} - 2\{U\}^{\mathrm{T}}[B][D]\{\varepsilon_o\} + \{\varepsilon_o\}^{\mathrm{T}}[D]\{\varepsilon_o\}\right]\,\mathrm{d}V_e \tag{16.23}$$

and is the required expression for the strain energy of the element. The strain energy of the structure is found by summing all element strain energies.

The work done by the applied loading is separated into its respective components. Although several different forms of applied loadings can be considered we will restrict the present discussion to the most popular of body forces, concentrated nodal forces and distributed edge pressures; see Figure 16.4. If the body forces are denoted by X and Y in the x and y directions respectively then the work done, W_e^B, by the body forces is

$$W_e^B = \int_{V_e}(uX + vY)\,\mathrm{d}V_e = \int_{V_e}\{U\}^{\mathrm{T}}[N]\left\{\begin{array}{c} X \\ Y \end{array}\right\}\mathrm{d}V_e \tag{16.24}$$

The work done, W_e^N, by the concentrated nodal forces is

$$W_e^N = \{U\}^{\mathrm{T}}\{P\} \tag{16.25}$$

where $\{P\} = \{P_x^i, P_y^i, P_x^j, P_y^j, P_x^k, P_y^k\}$ is the element nodal force vector. The work done, W_e^P, by the pressures acting over a triangle edge, S_e, is

$$W_e^P = \int_{S_e}(up_x + vp_y)\,\mathrm{d}S_e = \int_{S_e}\{U\}^{\mathrm{T}}[N]^{\mathrm{T}}\left\{\begin{array}{c} p_x \\ p_y \end{array}\right\}\mathrm{d}S_e \tag{16.26}$$

If pressures act on more than one edge the contributions have to be added. The potential energy (16.19) can now be written as

$$\Pi_e = U_e - W_e = U_e - (W_e^B + W_e^N + W_e^P) \tag{16.27}$$

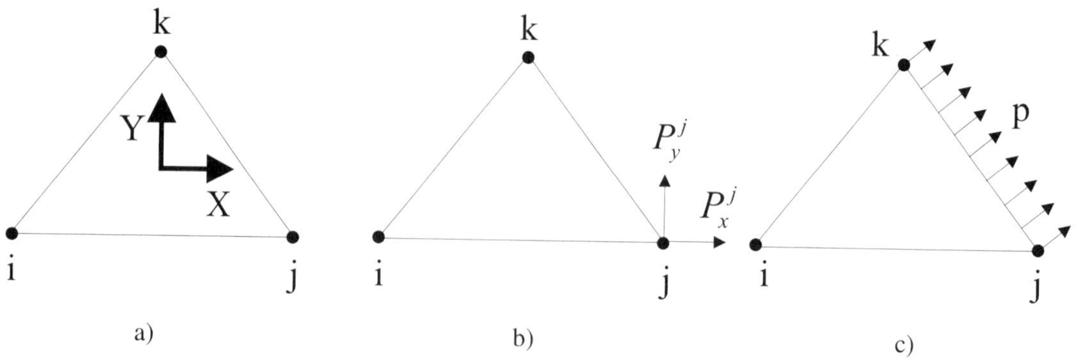

Figure 16.4: Applied loadings on a triangular element. a) Body forces X and Y. b) Nodal forces P_x^j and P_y^j acting at node j. c) Normal pressure p acting on edge (j, k).

Substituting the above results (16.23), (16.24), (16.25) and (16.26) into (16.27) the potential energy of a structure consisting of E elements is

$$\Pi = \sum_{e=1}^{E} \left\{ \frac{1}{2} \int_{V_e} \left[[\{U\}^{\mathrm{T}}[B]^{\mathrm{T}}[D][B]\{U\} - 2\{U\}^{\mathrm{T}}[B]^{\mathrm{T}}[D]\{\varepsilon_o\} + \{\varepsilon_o\}^{\mathrm{T}}[D]\{\varepsilon_o\}] \right] \mathrm{d}V_e \right\} - \sum_{e=1}^{E} \left\{ \int_{V_e} \{U\}^{\mathrm{T}}[N] \left\{ \begin{array}{c} X \\ Y \end{array} \right\} \mathrm{d}V_e + \{U\}^{\mathrm{T}}\{P\} + \int_{S_e} \{U\}^{\mathrm{T}}[N] \left\{ \begin{array}{c} p_x \\ p_y \end{array} \right\} \mathrm{d}S_e \right\} \tag{16.28}$$

To ensure that the structure is in a state of equilibrium the total potential energy must be a minimum. Since the primary unknowns in a stress analysis are the nodal displacements then we obtain equilibrium by differentiating Π with respect to the structure displacement vector $\{U\}_s$ which contains all the degrees of freedom in the structure

$$\frac{\partial \Pi}{\partial \{U\}_s} = 0 \tag{16.29}$$

Therefore, differentiating Π in (16.28) gives

$$\frac{\partial \Pi}{\partial \{U\}_s} = 0 = \sum_{e=1}^{E} \left[\int_{V_e} [B]^{\mathrm{T}}[D][B]\,\mathrm{d}V_e \right] \{U\}_s - \sum_{e=1}^{E} \left[\int_{V_e} [B]^{\mathrm{T}}[D]\{\varepsilon_o\}\,\mathrm{d}V_e \right] - \sum_{e=1}^{E} \left[\int_{V_e} [N]^{\mathrm{T}} \left\{ \begin{array}{c} X \\ Y \end{array} \right\} \mathrm{d}V_e \right] - \{P\}_s - \sum_{e=1}^{E} \left[\int_{S_e} [N]^{\mathrm{T}} \left\{ \begin{array}{c} p_x \\ p_y \end{array} \right\} \mathrm{d}S_e \right] \tag{16.30}$$

noting that the element nodal force vector has been taken outside of the summation sign and replaced by the structure nodal force vector $\{P\}_s$. The above equation (16.30) can be written as

$$[K]_s\{U\}_s = \{F\}_s \tag{16.31}$$

where $[K]_s$ denotes the structure stiffness matrix

$$[K]_s = \sum_{e=1}^{E} \left[\int_{V_e} [B]^{\mathrm{T}}[D][B]\,\mathrm{d}V_e \right] \tag{16.32}$$

and $\{F\}_s$ denotes the structure force vector

$$\{F\} = \sum_{e=1}^{E} \left[\int_{V_e} [B]^{\mathrm{T}}[D]\{\varepsilon_o\}\,\mathrm{d}V_e + \int_{V_e} [N]^{\mathrm{T}} \left\{ \begin{array}{c} X \\ Y \end{array} \right\} \mathrm{d}V + \int_{S_e} [N]^{\mathrm{T}} \left\{ \begin{array}{c} p_x \\ p_y \end{array} \right\} \mathrm{d}S_e \right] + \{P\}_s \quad (16.33)$$

The above variational formulation is still in a general format and we will now tailor the formulation specifically for the two-dimensional triangular element with the linear interpolation function (16.7). Considering first the stiffness matrix we observe from (16.17), (13.33) and (13.36) that the $[B]$ and $[D]$ matrices can be taken outside the volume integration because they contain only constant terms. Furthermore, for an element of uniform thickness t and cross-sectional area A then the volume integral is equal to tA

$$[K]_e = \int_{V_e} [B]^{\mathrm{T}}[D][B]\,\mathrm{d}V_e = [B]^{\mathrm{T}}[D][B] \int_{V_e} \mathrm{d}V_e = [B]^{\mathrm{T}}[D][B]tA \quad (16.34)$$

With $[B]$ and $[D]$ being of sizes (3×6) and (3×3) respectively then $[K]_e$ is seen to be of size (6×6).

The element nodal force vector simply consists of applied nodal forces

$$\{P\}_e = \left[P_x^i \quad P_y^i \quad P_x^j \quad P_y^j \quad P_x^k \quad P_y^k \right]^{\mathrm{T}} \quad (16.35)$$

The body force term is, with $[N]$ given by (16.11)

$$\int_{V_e} [N]^{\mathrm{T}} \left\{ \begin{array}{c} X \\ Y \end{array} \right\} \mathrm{d}V_e = \int_{A_e} \left[\begin{array}{cc} N_i & 0 \\ 0 & N_i \\ N_j & 0 \\ 0 & N_j \\ N_k & 0 \\ 0 & N_k \end{array} \right] \left\{ \begin{array}{c} X \\ Y \end{array} \right\} t\,\mathrm{d}A = \frac{At}{3} \left\{ \begin{array}{c} X \\ Y \\ X \\ Y \\ X \\ Y \end{array} \right\} \quad (16.36)$$

and is seen to equally distribute the body forces between the three nodes. The integration of the shape functions in (16.36) will be covered in §16.3.

The initial strain vector term can accommodate various forms of initial strain but we will only consider initial strains due to thermal expansion. As in the case of evaluating the volume integral for the stiffness matrix, the initial strain force vector term is similarly found to reduce to

$$\int_{V_e} [B]^{\mathrm{T}}[D]\{\varepsilon_o\}\,\mathrm{d}V_e = [B]^{\mathrm{T}}[D]\{\varepsilon_o\}tA \quad (16.37)$$

In the case of plane stress the $[D]$ matrix and thermal strain vector $\{\varepsilon_o\}$ are given by (13.33) and (13.34) respectively, which upon multiplication gives

$$[D]\{\varepsilon_o\} = \frac{E}{1-\nu^2} \left[\begin{array}{ccc} 1 & \nu & 0 \\ \nu & 1 & 0 \\ 0 & 0 & (1-\nu)/2 \end{array} \right] \alpha\Delta T \left\{ \begin{array}{c} 1 \\ 1 \\ 0 \end{array} \right\} = \frac{E\alpha\Delta T}{1-\nu^2} \left\{ \begin{array}{c} 1 \\ 1 \\ 0 \end{array} \right\} \quad (16.38)$$

and multiplying by $[B]^{\mathrm{T}}tA$ we have, with $[B]$ given by (16.17)

$$[B]^{\mathrm{T}}\left([D]\{\varepsilon_o\}\right)tA = \frac{1}{2A} \left[\begin{array}{ccc} b_i & 0 & c_i \\ 0 & c_i & b_i \\ b_j & 0 & c_j \\ 0 & c_j & b_j \\ b_k & 0 & c_k \\ 0 & c_k & b_k \end{array} \right] \frac{E\alpha\Delta T}{1-\nu^2} \left\{ \begin{array}{c} 1 \\ 1 \\ 0 \end{array} \right\} tA = \frac{E\alpha\Delta T}{1-\nu^2} \left\{ \begin{array}{c} b_i \\ c_i \\ b_j \\ c_j \\ b_k \\ c_k \end{array} \right\} \quad (16.39)$$

In the case of plane strain, $[D]$ and $\{\varepsilon_o\}$ are alternatively given by (13.36) so that

$$[B]^{\mathrm{T}}[D]\{\varepsilon_o\}tA = \frac{E\alpha\Delta Tt}{2(1-2\nu)}\left\{\begin{array}{c} b_i \\ c_i \\ b_j \\ c_j \\ b_k \\ c_k \end{array}\right\} \tag{16.40}$$

If normal pressure loading of magnitude p acts on edge (i, j), which is of length L_{ij}, and the element is of constant thickness t then the area of the face S_{ij} associated with edge (i, j) is $S_{ij} = L_{ij}t$, Figure 16.5. Also, on edge (i, j) the shape function N_k is zero and we have from (16.33)

$$\begin{aligned} \int_{S_{ij}} [N]^{\mathrm{T}} \left\{\begin{array}{c} p_x \\ p_y \end{array}\right\} \mathrm{d}S &= \int_{S_{ij}} \left[\begin{array}{cc} N_i & 0 \\ 0 & N_i \\ N_j & 0 \\ 0 & N_j \\ N_k & 0 \\ 0 & N_k \end{array}\right] \left\{\begin{array}{c} p_x \\ p_y \end{array}\right\} \mathrm{d}S \\ &= \int_{L_{ij}} \left[\begin{array}{cc} N_i & 0 \\ 0 & N_i \\ N_j & 0 \\ 0 & N_j \\ 0 & 0 \\ 0 & 0 \end{array}\right] \left\{\begin{array}{c} p_x \\ p_y \end{array}\right\} \mathrm{d}L = \frac{L_{ij}t}{2}\left\{\begin{array}{c} p_x \\ p_y \\ p_x \\ p_y \\ 0 \\ 0 \end{array}\right\} \end{aligned} \tag{16.41}$$

illustrating that a pressure loading is equally distributed between the two edge nodes. Combining the

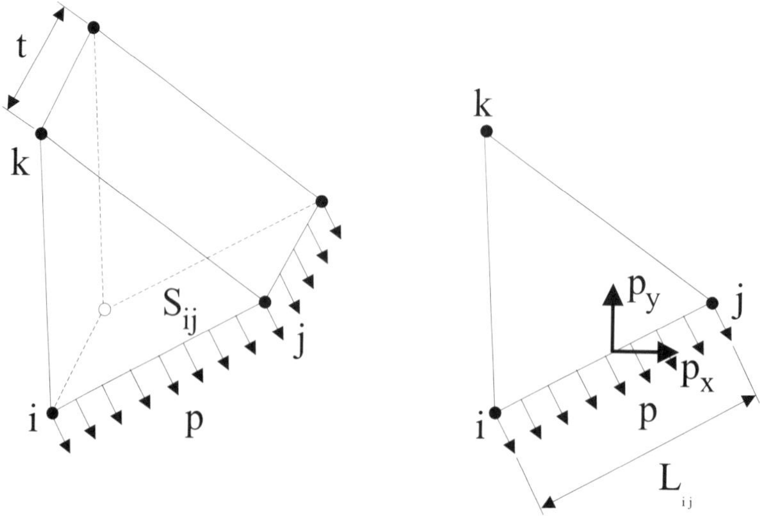

Figure 16.5: Normal pressure loading p acting on edge (i, j).

above results then the element equations are

$$[K]\begin{Bmatrix} u_i \\ v_i \\ u_j \\ v_j \\ u_k \\ v_k \end{Bmatrix} = \begin{Bmatrix} P_x^i \\ P_y^i \\ P_x^j \\ P_y^j \\ P_x^k \\ P_y^k \end{Bmatrix} + \frac{At}{3}\begin{Bmatrix} X \\ Y \\ X \\ Y \\ X \\ Y \end{Bmatrix} + \frac{E\alpha\Delta Tt}{2(1-\nu^2)}\begin{Bmatrix} b_i \\ c_i \\ b_j \\ c_j \\ b_k \\ c_k \end{Bmatrix} + \frac{L_{ij}t}{2}\begin{Bmatrix} p_x \\ p_y \\ p_x \\ p_y \\ 0 \\ 0 \end{Bmatrix} \tag{16.42}$$

for plane-stress thermal strain and pressure loading on edge (i, j).

16.3 Area Coordinates and Integration of Shape Functions

Figure 16.6 illustrates a triangle (i, j, k) with an internal point p which sub-divides the triangle into the three sub-triangles (i, j, p), (j, k, p) and (k, i, p). Point p can be located in terms of its (x, y) coordinates but alternatively in terms of the *barycentric* or *area* coordinates ξ_i, ξ_j and ξ_k. The area coordinates are defined by the ratios of the sub-triangle areas A_i, A_j and A_k opposite each node relative to the total triangle area A

$$\xi_i = \frac{A_i}{A}; \quad \xi_j = \frac{A_j}{A}; \quad \xi_k = \frac{A_k}{A} \tag{16.43}$$

where, clearly $A_i + A_j + A_k = A$ and hence $\xi_3 = 1 - \xi_1 - \xi_2$ which illustrates that ξ_1, ξ_2 and ξ_3 are not independent. If p is restricted to the inside or boundary of triangle $(1, 2, 3)$ then the area coordinates are within the range $[0 : 1]$ whereas if p lies outside the triangle then ξ_i will be negative. Each area coordinate achieves its maximum value of unity at its associated node; for example $\xi_i = 1$ when $p = i$. Thus, a direct analogy is observed between the variation of the shape functions N_i and the area coordinates

$$\xi_i = N_i; \quad \xi_j = N_j; \quad \xi_k = N_k \tag{16.44}$$

It can be shown that the integration of the area coordinates between two nodes on an edge of a triangle of length L and over the area A of the triangle are

$$\int_L \xi_i^\alpha \xi_j^\beta \, \mathrm{d}L = \frac{\alpha!\beta!}{(\alpha+\beta+1)!}L; \quad \int_A \xi_i^\alpha \xi_j^\beta \xi_k^\gamma \, \mathrm{d}A = \frac{\alpha!\beta!\gamma!}{(\alpha+\beta+\gamma+2)!}2A \tag{16.45}$$

For example, the body force term in (16.36) required the integration of N_i

$$\int_A N_i \, \mathrm{d}A = \int_A \xi_i \, \mathrm{d}A = \int_A \xi_i^1 \xi_j^0 \xi_k^0 \, \mathrm{d}A = \frac{1!0!0!}{(1+0+0+2)!}2A = \frac{A}{3} \tag{16.46}$$

Example 16.1 Stiffness matrix and force vector for a constant strain triangle

Consider the triangular element shown in Figure 16.7 with nodes i, j and k at $(2, 1)$, $(4, 1)$ and $(3, 3)$ respectively, with all coordinates in mm. The loading on the element consists of a point force of 100N at node i acting in the y-direction, a normal pressure of 10MPa on edge (j, k) and an increase in temperature of 50°C throughout the element. The thickness of the element is 2mm, Young's modulus is 200GPa, Poisson's ratio is 0.3, the coefficient of thermal expansion is 2×10^{-6}°C^{-1} and plane stress conditions are to be assumed.

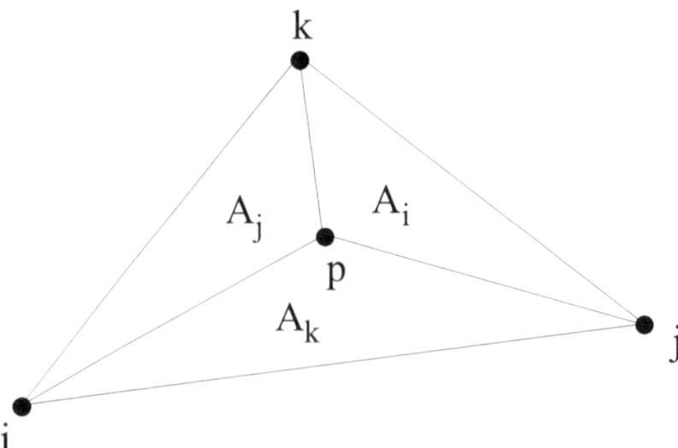

Figure 16.6: Triangle (i, j, k) with a point p inside sub-dividing the triangle into three sub-triangles (i, j, p), (j, k, p) and (k, i, p).

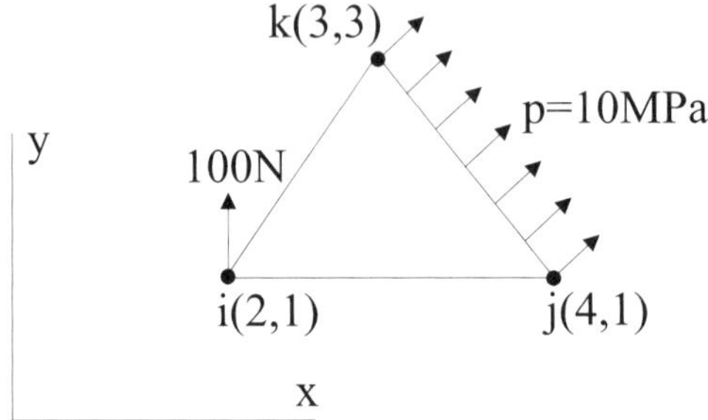

Figure 16.7: Example: Triangular element (i, j, k) subject to a point force of 100N at node i and a normal pressure of 10MPa on edge (j, k).

With $x_i = 2$, $y_i = 1$, $x_j = 4$, $y_j = 1$, $x_k = 3$ and $y_k = 3$ let us first determine the constants a, b and c in N_β (16.8) and the area, A, of the element

$$
\begin{aligned}
&a_i = 9 \times 10^{-6}\text{m}^2; \quad &&b_i = -2 \times 10^{-3}\text{m}; \quad &&c_i = -1 \times 10^{-3}\text{m} \\
&a_j = -3 \times 10^{-6}\text{m}^2; \quad &&b_j = 2 \times 10^{-3}\text{m}; \quad &&c_j = -1 \times 10^{-3}\text{m} \\
&a_k = -2 \times 10^{-6}\text{m}^2; \quad &&b_k = 0\text{m}; \quad &&c_k = 2 \times 10^{-3}\text{m} \\
& && && A = 2 \times 10^{-6}\text{m}^2
\end{aligned}
$$

Therefore, the shape functions are

$$
\begin{aligned}
N_i &= \frac{1}{2A}(a_i + b_i x + c_i y) = \frac{1}{4 \times 10^{-6}}[9 \times 10^{-6} - 2 \times 10^{-3}x - 1 \times 10^{-3}y] \\
N_j &= \frac{1}{2A}(a_j + b_j x + c_j y) = \frac{1}{4 \times 10^{-6}}[-3 \times 10^{-6} + 2 \times 10^{-3}x - 1 \times 10^{-3}y] \\
N_k &= \frac{1}{2A}(a_k + b_k x + c_k y) = \frac{1}{4 \times 10^{-6}}[-2 \times 10^{-6} + 0x + 2 \times 10^{-3}y]
\end{aligned}
$$

As a check we can confirm that the sum of the shape functions is equal to unity at the arbitrary point x =3mm and y =2mm. From the shape functions above N_i =0.25, N_j =0.25 and N_k =0.5 and hence $N_i + N_j + N_k = 1$.

From (16.17) the $[B]$ matrix is

$$[B] = \frac{1}{2A}\begin{bmatrix} b_i & 0 & b_j & 0 & b_k & 0 \\ 0 & c_i & 0 & c_j & 0 & c_k \\ c_i & b_i & c_j & b_j & c_k & b_k \end{bmatrix} = 250\begin{bmatrix} -2 & 0 & 2 & 0 & 0 & 0 \\ 0 & -1 & 0 & -1 & 0 & 2 \\ -1 & -2 & -1 & 2 & 2 & 0 \end{bmatrix}$$

and from (16.38) the $[D]$ matrix is

$$[D] = \frac{E}{1-\nu^2}\begin{bmatrix} 1 & \nu & 0 \\ \nu & 1 & 0 \\ 0 & 0 & (1-\nu)/2 \end{bmatrix} = 2.1978\times 10^{11}\begin{bmatrix} 1 & 0.3 & 0 \\ 0.3 & 1 & 0 \\ 0 & 0 & 0.35 \end{bmatrix}$$

The stiffness matrix is, (16.34)

$$[K] = [B]^{\mathrm{T}}[D][B]tA =$$

$$=250\begin{bmatrix} -2 & 0 & -1 \\ 0 & -1 & -2 \\ 2 & 0 & -1 \\ 0 & -1 & 2 \\ 0 & 0 & 2 \\ 0 & 2 & 0 \end{bmatrix}(2.1978\times 10^{11})\begin{bmatrix} 1 & 0.3 & 0 \\ 0.3 & 1 & 0 \\ 0 & 0 & 0.35 \end{bmatrix}\times$$

$$250\begin{bmatrix} -2 & 0 & 2 & 0 & 0 & 2 \\ 0 & -1 & 0 & -1 & 0 & 2 \\ -1 & -2 & -1 & 2 & 2 & 0 \end{bmatrix}(2\times 10^{-3})(2\times 10^{-6})$$

Performing the multiplications we find

$$[K] = 5.4945\times 10^7\begin{bmatrix} 4.35 & & & \text{sym} & & \\ 1.3 & 2.4 & & & & \\ -3.65 & 0.1 & 4.35 & & & \\ -0.1 & -0.4 & -1.3 & 2.4 & & \\ -0.7 & -1.4 & -0.7 & 1.4 & 1.4 & \\ -1.2 & -2 & 1.2 & -2 & 0 & 4 \end{bmatrix}$$

noting that the matrix is symmetric $(K_{ij} = K_{ji})$, the principal terms are all positive and non-zero $(K_{ii} > 0)$ and the sum of all terms in either a row or column are zero $\left(\sum_{j=1}^{6} K_{ij} = 0\right)$.

To determine the force vector we require the thermal, nodal and edge pressure components. From (16.42) the nodal forces term is

$$\{F\}_{nodal} = \begin{Bmatrix} P_x^i \\ P_y^i \\ P_x^j \\ P_y^j \\ P_x^k \\ P_y^k \end{Bmatrix} = \begin{Bmatrix} 0 \\ 100 \\ 0 \\ 0 \\ 0 \\ 0 \end{Bmatrix}$$

The thermal force vector term is, (16.42)

$$\{F\}_{thermal} = \frac{E\alpha\Delta Tt}{2(1-\nu^2)}\begin{Bmatrix} b_i \\ c_i \\ b_j \\ c_j \\ b_k \\ c_k \end{Bmatrix} = 21.9780\begin{Bmatrix} -2 \\ 2 \\ 2 \\ 4 \\ 0 \\ 3 \end{Bmatrix}$$

and finally the normal pressure term for edge (j,k) is, (16.42)

$$\{F\}_{pressure} = \frac{L_{jk}t}{2}\begin{Bmatrix} 0 \\ 0 \\ p_x \\ p_y \\ p_x \\ p_y \end{Bmatrix} = \sqrt{5}\begin{Bmatrix} 0 \\ 0 \\ 8.9433 \\ 4.4721 \\ 8.9433 \\ 4.4721 \end{Bmatrix}$$

Summing the above terms the total force vector is

$$\{F\} = \{F\}_{nodal} + \{F\}_{thermal} + \{F\}_{pressure} = \begin{Bmatrix} -43.9560 \\ 143.9560 \\ 63.9561 \\ 97.9120 \\ 20 \\ 75.9340 \end{Bmatrix}$$

Example 16.2 Assembly of three constant strain triangular elements

In this example we will examine assembly of three triangular elements. Assembly of two-dimensional elements is similar to the assembly of one-dimensional elements discussed in §13.12. The structure with edge pressure loadings is shown in Figure 16.8 with $\Delta T = 50°$C and $\alpha = 2 \times 10^{-3}$°C^{-1} and plane stress conditions assumed.

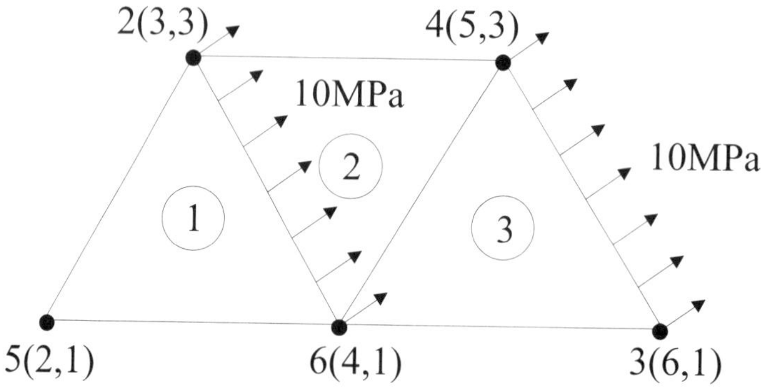

Figure 16.8: Example: A structure consisting of three triangular elements with normal pressures on edges (6,2) and (3,4).

Following a similar derivation to the previous example the stiffness matrix of element 1 is, with node ordering (5,6,2)

$$[K^1] = \begin{bmatrix} 2.39 & & & \text{sym} & & \\ 71.4 & 1.32 & & & & \\ -2 & 549 & 2.39 & & & \\ 549 & -21.9 & -71.4 & 1.32 & & \\ -38.5 & -76.9 & -38.5 & 76.9 & 76.9 & \\ -65.9 & -1.09 & 65.9 & -1.09 & 0 & 2.19 \end{bmatrix} \times 10^8$$

The stiffness matrix of element 3, with node ordering (6,3,4), is identical to that of element 1 because element 3 is translated by 2mm in the x-direction. The stiffness matrix of element 2 is found from (16.34) as in the previous example

$$[K^2] = \begin{bmatrix} 1.91 & & & & & \\ 57.1 & 1.05 & & \text{sym} & & \\ -1.5 & 19.7 & 2.1 & & & \\ 879 & 439 & -85.7 & 1.6 & & \\ -38.4 & -76.9 & -57.6 & 76.9 & 96.1 & \\ -65.9 & -1.09 & 65.9 & -1.64 & 0 & 2.74 \end{bmatrix} \times 10^8$$

with node ordering (6,4,2). Assembling the element stiffness matrices we arrive at the structure stiffness matrix with node ordering (5,6,2,4,3)

$$[K]_s = \begin{bmatrix} 2.39 & & & & & & & & & \\ 71.4 & 1.31 & & & & & & & & \\ -2 & 549 & 6.69 & & & & & & & \\ -549 & -21.9 & 57.1 & 3.69 & & \text{sym} & & & & \\ -38.4 & -76.9 & -76.9 & 0 & 1.73 & & & & & \\ -65.9 & -1.09 & 0 & -2.19 & 0 & 4.94 & & & & \\ 0 & 0 & -1.91 & -57.1 & -57.6 & 65.9 & 2.87 & & & \\ 0 & 0 & -57.1 & -1.05 & 76.9 & -1.64 & -85.7 & 3.8 & & \\ 0 & 0 & -2 & 549 & 0 & 0 & -38.4 & 65.9 & 2.39 & \\ 0 & 0 & -549 & -21.9 & 0 & 0 & 76.9 & -1.09 & -71.4 & 1.31 \end{bmatrix} \times 10^8$$

The structure stiffness matrix contains a large number of zeros where there is no connection between element nodes (for example, nodes 2 and 3 and nodes 5 and 3). The force vectors for the three elements are, (16.42)

$$\{F\}_1 = \begin{Bmatrix} -43.936 \\ -21.978 \\ 65.976 \\ -5.978 \\ 20.02 \\ 53.956 \end{Bmatrix}; \quad \{F\}_2 = \begin{Bmatrix} -37.3155 \\ -13.658 \\ 39.316 \\ -29.316 \\ 0 \\ 49.144 \end{Bmatrix}; \quad \{F\}_3 = \begin{Bmatrix} -41.956 \\ -15.978 \\ 63.956 \\ -11.978 \\ 20 \\ 53.956 \end{Bmatrix}$$

Summing the element force vectors we arrive at the structure force vector

$$\{F\}_s = \begin{Bmatrix} -43.936 \\ -21.978 \\ -13.296 \\ -35.614 \\ 20.02 \\ 103.1 \\ 59.316 \\ 24.469 \\ 63.956 \\ -11.978 \end{Bmatrix}$$

Hopefully this example has helped demonstrate the need for a computer!

16.4 Higher Order Elements

Our discussion to date has focused on elements with linear interpolation functions. As we have seen in the case of stress analyses, these simplex elements lead to constant strain and hence constant stress throughout the element. In order to analyse a structure with greater accuracy and have varying strain and stress throughout an element we require the use of higher order elements. Although higher order elements can in theory be constructed to accommodate any order of interpolation function, the order is generally restricted to quadratic and cubic.

As a result we will consider only quadratic and cubic interpolation functions and as in Chapter 13 and in previous sections restrict ourselves to bar and triangular elements. The quadratic interpolation order is generally chosen in preference to cubic because it offers a good balance between improved accuracy and increased analysis complexity. As we will see in subsequent sections, higher order elements require significantly more analysis and consequently computing time than simplex elements. However, the disadvantage of increased analysis and computing requirements is compensated by the property that in the majority of cases a few higher order elements will be more accurate than many simplex elements. Furthermore, higher order elements have the advantage of allowing the field variable interpolation functions to also be used for modelling the element geometry. For example, Figure 16.9 illustrates the curved surface of a hole in a plate modelled using both simplex and quadratic order elements. Triangular simplex elements have three nodes with straight edges whereas triangular quadratic elements have six nodes (three corner and three mid-side) and allow the modelling of curved boundaries. Elements which use the same order for both field variable and geometric interpolation functions are referred to as *isoparametric*. If the order of the geometric functions is less than and greater than the field variable functions then an element is referred to as *subparametric* and *superparametric* respectively. In the following sections we will consider only isoparametric elements since they are the most frequently used in practice.

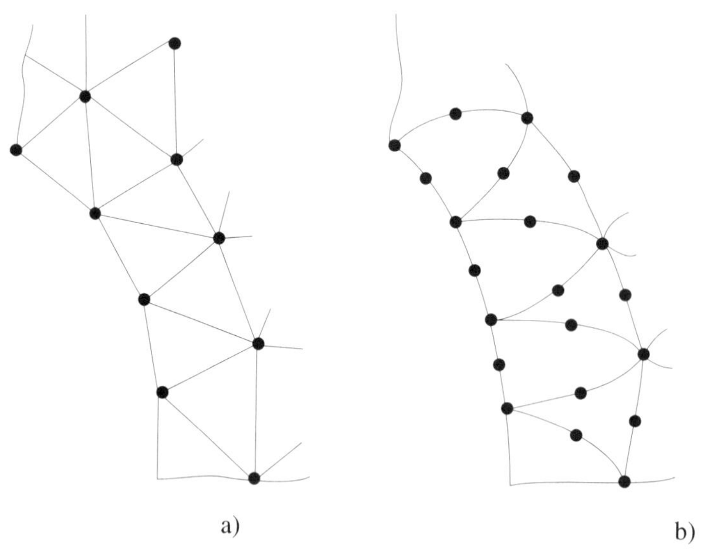

Figure 16.9: Curved surface of a hole modelled by a) simplex elements and b) quadratic elements.

When evaluating the integration of shape functions for higher order elements we frequently have to resort to numerical techniques due to the increased complexity of the shape functions. Thus, before examining the bar and triangular higher order element equations we will examine numerical integration in the next section.

16.4.1 Numerical Integration

Various methods exist for numerically integrating a function over a region such as the Trapezoidal, Simpson's and Gaussian methods. In general, Gaussian integration is the most accurate of these three methods and consists of a weighted sum of the function evaluated at prescribed sampling points. Consider the function $f(\xi)$ of one-variable ξ evaluated over the normalised interval of $-1 \leq \xi \leq +1$

$$\int_{-1}^{+1} f(\xi)\,\mathrm{d}\xi = \sum_{i=1}^{n} w_i f(\xi_i) \tag{16.47}$$

where n is the number of sampling points ξ_i with associated weights w_i. Table 16.1 lists both ξ_i and w_i for n up to and including 3 for the two most frequently used intervals of integration $[-1:+1]$ and $[0:1]$. As an example, consider the following integral

Order	n	ξ_i	w_i
$[-1:+1]$			
1	1	0	2
≤ 3	2	$\pm$ 0.577350	1
≤ 5	3	0	0.888889
		$\pm$0.774597	0.555556
$[0:1]$			
1	1	0.5	1
≤ 3	2	0.211325	0.5
		0.788675	0.5
≤ 5	3	0.112702	0.277778
		0.5	0.444444
		0.887298	0.277778

Table 16.1: Integration points and weights for Gaussian integration for functions of one variable.

$$I = \int_{3}^{6} x^2\,\mathrm{d}x \tag{16.48}$$

Before performing the integration we need to convert the interval of integration to either $[-1:+1]$ or $[0:1]$. For an integral over the interval $a \leq x \leq b$ the change of argument

$$x = a + \frac{b-a}{2}(\xi+1) \tag{16.49}$$

will produce the standard interval $-1 \leq \xi \leq +1$. Thus

$$\int_{a}^{b} f(x)\,\mathrm{d}x = \int_{-1}^{+1} \frac{b-a}{2} f\left[a + \frac{b-a}{2}(\xi+1)\right]\mathrm{d}\xi \tag{16.50}$$

In (16.50) $f(x) = x^2$, $a = 3$ and $b = 6$ so that

$$I = \int_3^6 x^2 \, \mathrm{d}x = \int_{-1}^{+1} \frac{3}{2} f\left[3 + \frac{3}{2}(\xi + 1)\right] \mathrm{d}\xi \tag{16.51}$$

Choosing two integration points ($n = 2$) we have from Table 16.1 $\xi_1 = 0.577350$, $w_1 = 1$, $\xi_2 = -0.577350$ and $w_2 = 1$. From (16.47) the numerical integration of I is

$$I = \frac{3}{2}\left\{1 \times f\left[3 + \frac{3}{2}(0.577350 + 1)\right] + 1 \times f\left[3 + \frac{3}{2}(-0.577350 + 1)\right]\right\} = 63 \tag{16.52}$$

and is exact.

If the region of integration is triangular then the Gaussian integration formula is now of the form

$$\int_0^1 \int_0^1 f(\xi_1, \xi_2) \, \mathrm{d}\xi_1 \, \mathrm{d}\xi_2 = 2A \sum_{i=1}^{n} w_i f(\xi_{1i}, \xi_{2i}) \tag{16.53}$$

where ξ_1 and ξ_2 are the triangle area coordinates and A is the area of the triangle. Integration points and weights are listed in Table 16.2 for linear, quadratic and cubic orders.

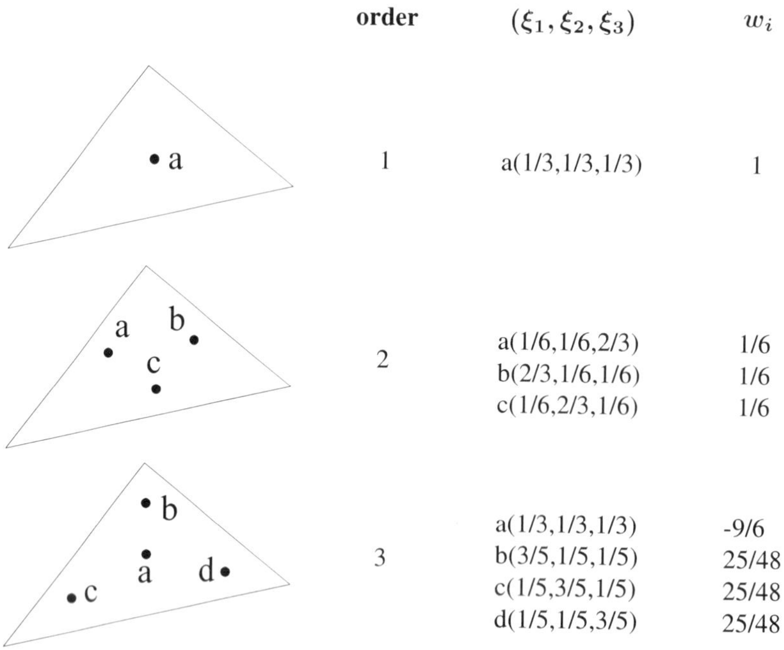

order	(ξ_1, ξ_2, ξ_3)	w_i
1	a(1/3,1/3,1/3)	1
2	a(1/6,1/6,2/3)	1/6
	b(2/3,1/6,1/6)	1/6
	c(1/6,2/3,1/6)	1/6
3	a(1/3,1/3,1/3)	-9/6
	b(3/5,1/5,1/5)	25/48
	c(1/5,3/5,1/5)	25/48
	d(1/5,1/5,3/5)	25/48

Table 16.2: Integration points and weights for Gaussian integration over triangular regions.

16.4.2 Bar Element

The following two sub-sections derive the interpolation functions, stiffness matrix and force vector for a one-dimensional bar element.

16.4.2.1 Quadratic and Cubic Interpolation Functions

Figure (16.10) illustrates a quadratic bar element with three nodes: two end nodes and one mid-point node. In an analogous manner to the triangle area coordinates discussed in §16.3 the length coordinates

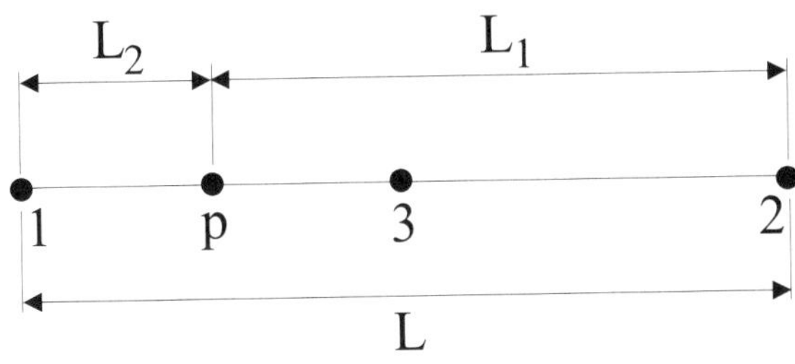

Figure 16.10: Quadratic bar element.

of a point p on the bar are ξ_1 and ξ_2

$$\xi_1 = \frac{L_1}{L}; \quad \xi_2 = \frac{L_2}{L} \tag{16.54}$$

Since $L_1 + L_2 = L$ then ξ_1 and ξ_2 are not independent

$$\xi_1 + \xi_2 = 1 \tag{16.55}$$

Thus, with $\xi_1 = 1 - \xi_2$ point p is uniquely described in terms of the single parametric variable $\xi = \xi_2$ which increases from zero at node 1 to unity at node 2.

The interpolation function ϕ for a quadratic bar element can now be written as

$$\phi = \alpha_1 + \alpha_2\xi + \alpha_3\xi^2 \tag{16.56}$$

where α_1, α_2 and α_3 are constants. Letting the nodal values of ϕ be denoted by ϕ_1, ϕ_2 and ϕ_3 at $\xi = 0$, $\xi = 0.5$ and $\xi = 1$ respectively, then

$$\begin{aligned} \phi_1 &= \alpha_1 \\ \phi_2 &= \alpha_1 + \frac{\alpha_2}{2}\xi + \frac{\alpha_3}{4}\xi^2 \\ \phi_3 &= \alpha_1 + \alpha_2 + \alpha_3 \end{aligned} \tag{16.57}$$

Solving for α_1, α_2 and α_3

$$\begin{aligned} \alpha_1 &= \phi_1 \\ \alpha_2 &= (4\phi_2 - 3\phi_1 - \phi_3)/\xi \\ \alpha_3 &= (2\phi_1 + 2\phi_3 - 4\phi_2)/\xi \end{aligned} \tag{16.58}$$

Substituting (16.58) back into (16.56) we have

$$\phi = N_1\phi_1 + N_2\phi_2 + N_3\phi_3 \tag{16.59}$$

where the shape functions are given by

$$\begin{aligned} N_1 &= 2\xi^2 - 3\xi + 1 \\ N_2 &= 2\xi^2 - \xi \\ N_3 &= 4\xi - 4\xi^2 \end{aligned} \tag{16.60}$$

A cubic interpolation function is given by

$$\phi = \alpha_1 + \alpha_2\xi + \alpha_3\xi^2 + \alpha_4\xi^3 \tag{16.61}$$

for the four node element shown in Figure 16.11. Following a similar procedure to that outlined above we find that ϕ, in terms of shape functions, is

$$\phi = N_1\phi_1 + N_2\phi_2 + N_3\phi_3 + N_4\phi_4 \tag{16.62}$$

where

$$\begin{aligned} N_1 &= 1 - 5.5\xi + 9\xi^2 - 4.5\xi^3 \\ N_2 &= \xi - 4.5\xi^2 + 4.5\xi^3 \\ N_3 &= 9\xi - 22.5\xi^2 + 13.5\xi^3 \\ N_4 &= -4.5\xi + 18\xi^2 - 13.5\xi^3 \end{aligned} \tag{16.63}$$

It is instructive to confirm that the shape functions (16.60) and (16.63) are equal to unity at their associated node and zero at all other nodes.

Figure 16.11: Cubic bar element.

16.4.2.2 Stiffness Matrix and Force Vector

The variational formulation for stress analysis in §16.2.3 arrived at the following stiffness matrix, $[K]$, and force vector, $\{F\}$

$$\begin{aligned} [K] &= \int_V [B]^{\mathrm{T}}[D][B]\,\mathrm{d}V \\ \{F\} &= \{F\}_B + \{F\}_N + \{F\}_P + \{F\}_T \end{aligned} \tag{16.64}$$

where $\{F\}_B$, $\{F\}_N$, $\{F\}_P$ and $\{F\}_T$ are the force vector terms due to body forces, nodal forces, edge pressures and thermal strain, and are of the form

$$\begin{aligned} \{F\}_B &= \int_V [N]^{\mathrm{T}} X\,\mathrm{d}V \\ \{F\}_N &= \left\{ \begin{array}{c} P_x^1 \\ P_x^2 \\ P_x^3 \end{array} \right\} \\ \{F\}_P &= \int_A [N]^{\mathrm{T}} p_x\,\mathrm{d}A \\ \{F\}_T &= \int_V [B]^{\mathrm{T}}[D]\{\varepsilon_0\}\,\mathrm{d}V = \alpha\Delta T \int_V [B]^{\mathrm{T}}[D]\,\mathrm{d}V \end{aligned} \tag{16.65}$$

Thus, we require the evaluation of integrals involving both the $[B]$ matrix and $[N]$ vector. The $[B]$ matrix for a one-dimensional quadratic bar element is, (13.40)

$$[B] = \left[\frac{\partial N_1}{\partial x} \quad \frac{\partial N_2}{\partial x} \quad \frac{\partial N_3}{\partial x} \right] \tag{16.66}$$

Unlike the simplex bar element in §13.11 the shape functions are now in terms of the length coordinate ξ so that a transformation of coordinates is required. By the chain rule of differentiation

$$\frac{\partial N_i}{\partial \xi} = \frac{\partial N_i}{\partial x}\frac{\partial x}{\partial \xi} \tag{16.67}$$

and therefore

$$\frac{\partial N_i}{\partial x} = \left(\frac{1}{\partial x/\partial \xi}\right)\frac{\partial N_i}{\partial \xi} = [J]^{-1}\frac{\partial N_i}{\partial \xi} \tag{16.68}$$

where $[J]$ is referred to as the Jacobian matrix. For the bar element $[J]$ is of size (1×1) and for the quadratic bar element with variable x then a point on the element is given by

$$x = \sum_{i=1}^{3} N_i x_i = N_1 x_1 + N_2 x_2 + N_3 x_3 \tag{16.69}$$

The Jacobian matrix for this element is therefore

$$[J] = \frac{\partial x}{\partial \xi} = \frac{\partial N_1}{\partial \xi}x_1 + \frac{\partial N_2}{\partial \xi}x_2 + \frac{\partial N_3}{\partial \xi}x_3 \tag{16.70}$$

For example, if nodes 1, 2 and 3 are at $x_1 = 3$, $x_2 = 6$ and $x_3 = 4.5$ then from (16.60)

$$\frac{\partial N_1}{\partial \xi} = 4\xi - 3; \quad \frac{\partial N_2}{\partial \xi} = 4\xi - 1; \quad \frac{\partial N_3}{\partial \xi} = 4 - 8\xi \tag{16.71}$$

with $[J]$ equal to, (16.70)

$$[J] = (4\xi - 3)3 + (4\xi - 1)6 + (4 - 8\xi)4.5 = 3 \tag{16.72}$$

and, in the present case, is observed to be independent of ξ and equal to the length of the element. The $[B]$ matrix is, (16.66) and (16.68), with $[J]^{-1} = 1/3$

$$[B] = [J]^{-1}\begin{bmatrix}\partial N_1/\partial \xi & \partial N_2/\partial \xi & \partial N_3/\partial \xi\end{bmatrix} = \frac{1}{3}\begin{bmatrix}4\xi - 3 & 4\xi - 1 & 4 - 8\xi\end{bmatrix} \tag{16.73}$$

Earlier in §16.4.1 we examined numerical integration and the transformation of the interval of integration. We can now make use of $[J]$ for evaluating a function $f(x)$ over the entire length of a bar element, (16.47)

$$\int_L f(x)\,\mathrm{d}x = \int_0^1 f(\xi) \mid J \mid \mathrm{d}\xi \tag{16.74}$$

Revisiting the example of $f(x) = x^2$ over the entire length of the quadratic bar element with nodes, as above, at $x_1 = 3$, $x_2 = 6$ and $x_3 = 4.5$ then

$$\begin{aligned}\int_L f(x)\,\mathrm{d}x &= \int_L x^2\,\mathrm{d}x = \int_0^1 f(\xi) \mid J \mid \mathrm{d}\xi \\ &= 3\int_0^1 f(\xi)\,\mathrm{d}\xi = 3\left\{\frac{1}{2}(3.633995)^2 + \frac{1}{2}(5.366025)^2\right\} = 63\end{aligned} \tag{16.75}$$

where the integration points for $n = 2$ are $\xi_1 = 0.211325$ and $\xi_2 = 0.788675$ from Table 16.1 and noting that $x = 3.633975$ at $\xi = 0.211325$ and $x = 5.366025$ at $\xi = 0.788675$.

Let us now return to the evaluation of the stiffness matrix and force vector terms for a quadratic interpolation function. If the bar is of constant cross-sectional area A then $\mathrm{d}V = A\,\mathrm{d}x$ and $[D] = E$ so that (16.64) reduces to, (16.74)

$$[K] = \int_V [B]^{\mathrm{T}}[D][B]\,\mathrm{d}V = AE\int_L [B]^{\mathrm{T}}[B]\,\mathrm{d}x = AE\int_0^1 [B]^{\mathrm{T}}[B] \mid J \mid \mathrm{d}\xi \tag{16.76}$$

Substituting $[B]$ from (16.73), with $\mid J \mid = L$

$$[K] = \frac{AE}{L}\int_0^1 \left\{ \begin{array}{c} 4\xi - 3 \\ 4\xi - 1 \\ 4 - 8\xi \end{array} \right\} [4\xi - 3 \quad 4\xi - 1 \quad 4 - 8\xi]\,\mathrm{d}\xi \tag{16.77}$$

Because $[J]$ is constant then the integration can be performed exactly with the help of (16.45)

$$\begin{aligned}[K] &= \frac{AE}{L}\int_0^1 \begin{bmatrix} (4\xi-3)^2 & (4\xi-3)(4\xi-1) & (4\xi-3)(4-8\xi) \\ (4\xi-3)(4\xi-1) & (4\xi-1)^2 & (4-8\xi)(4\xi-1) \\ (4\xi-3)(4-8\xi) & (4-8\xi)(4\xi-1) & (4-8\xi) \end{bmatrix} \mathrm{d}\xi \\ &= \frac{AE}{L}\begin{bmatrix} 7/3 & 1/3 & -8/3 \\ 1/3 & 7/3 & -8/3 \\ -8/3 & -8/3 & 16/3 \end{bmatrix}\end{aligned} \tag{16.78}$$

noting that $[K]$ is symmetric. To illustrate integrating $[K]$ numerically the two integration points from Table 16.1 are $\xi_1 = 0.211325$ and $\xi_2 = 0.788675$ each with weight $w_1 = w_2 = 1/2$, (16.47)

$$[K] = \frac{AE}{L}\left\{ \begin{array}{l} \frac{1}{2}\left\{ \begin{array}{c} -2.1547 \\ 2.3094 \\ -0.1547 \end{array} \right\} [-2.1547 \quad 2.3094 \quad -0.1547] + \\ \frac{1}{2}\left\{ \begin{array}{c} 0.1547 \\ -2.3094 \\ 2.1547 \end{array} \right\} [0.1547 \quad -2.3094 \quad 2.1547] \end{array} \right\} \tag{16.79}$$

and is found to be equivalent to (16.78).

The body force and edge pressure force vectors do not require a transformation of coordinates and can be evaluated using (16.36) and (16.41)

$$\begin{aligned}\{F\}_B &= \int_V [N]^{\mathrm{T}} X\,\mathrm{d}V = \int_L \left\{ \begin{array}{c} 3\xi - 3\xi + 1 \\ 2\xi^2 - \xi \\ 4\xi - 4\xi^2 \end{array} \right\} XA\,\mathrm{d}x = XAL\left\{ \begin{array}{c} 1/6 \\ 1/6 \\ 2/3 \end{array} \right\} \\ \{F\}_P &= \int_A [N]^{\mathrm{T}} p_x\,\mathrm{d}A = Ap_x\left\{ \begin{array}{c} 1 \\ 0 \\ 0 \end{array} \right\}\end{aligned} \tag{16.80}$$

assuming a pressure p_x acting at node 1 only. Observe that two-thirds of the body force is distributed to the mid-point node. The thermal strain does require a coordinate transformation due to the $[B]$ matrix

$$\begin{aligned}&\{F\}_T = \alpha\Delta T\int_V [B]^{\mathrm{T}}[D]\,\mathrm{d}V = \\ &E\alpha\Delta TA\int_0^1 [B]^{\mathrm{T}} \mid J \mid \mathrm{d}\xi = E\alpha\Delta TAL\int_0^1 \left\{ \begin{array}{c} 4\xi - 3 \\ 4\xi - 1 \\ 4 - 8\xi \end{array} \right\} \mathrm{d}\xi = E\alpha\Delta TA\left\{ \begin{array}{c} -1 \\ 1 \\ 0 \end{array} \right\}\end{aligned} \tag{16.81}$$

again noting that $\mid J \mid = L$.

Example 16.3 A tapered bar with quadratic bar elements

This example re-examines the problem of Example 13.1 but now using quadratic bar elements. Figure 16.12 illustrates a two bar approximation of a tapered bar that is built-in at one end and subject to a tensile force of 2kN at the free end. From (16.78) and (16.65) the element stiffness matrices and force vectors are

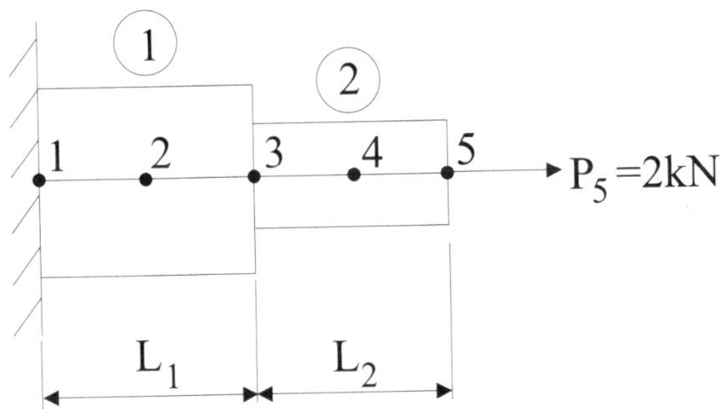

element 1	element 2
$L_1 = 50$mm	$L_2 = 50$mm
$A_1 = 80$mm^2	$A_2 = 40$mm^2
$E_1 = 175$GPa	$E_2 = 150$GPa

Figure 16.12: Example: A two bar approximation of a tapered bar built-in at one end and subject to a tensile force at the free end.

$$[K^1] = 280 \times 10^3 \begin{bmatrix} 7/3 & & \text{sym} \\ 1/3 & 7/3 & \\ -8/3 & -8/3 & 16/3 \end{bmatrix}; \quad \{F^1\} = \begin{Bmatrix} 0 \\ 0 \\ 0 \end{Bmatrix}$$

$$[K^2] = 120 \times 10^3 \begin{bmatrix} 7/3 & & \text{sym} \\ 1/3 & 7/3 & \\ -8/3 & -8/3 & 16/3 \end{bmatrix}; \quad \{F^2\} = \begin{Bmatrix} 0 \\ 2000 \\ 0 \end{Bmatrix}$$

The structure stiffness matrix and force vector are therefore

$$[K_s] = \begin{bmatrix} 653.33 & & & & \\ -746.66 & 1493.33 & & \text{sym} & \\ 93.33 & -746.66 & 653.33+280 & & \\ 0 & 0 & -320 & 640 & \\ 0 & 0 & 40 & -320 & 280 \end{bmatrix} \times 10^3; \quad \{F_s\} = \begin{Bmatrix} 0 \\ 0 \\ 0 \\ 0 \\ 2 \end{Bmatrix} \times 10^3$$

Incorporating the boundary condition $u_1 = 0$ and associated reaction R_1 then the structure system of equations are

$$\times 10^3 \begin{bmatrix} 653.33 & & & & \\ -746.66 & 1493.33 & & \text{sym} & \\ 93.33 & -746.66 & 933.33 & & \\ 0 & 0 & -320 & 640 & \\ 0 & 0 & 40 & -320 & 280 \end{bmatrix} \begin{Bmatrix} 0 \\ u_2 \\ u_3 \\ u_4 \\ u_5 \end{Bmatrix} = \begin{Bmatrix} 0 + R_1 \\ 0 \\ 0 \\ 0 \\ 2 \end{Bmatrix} \times 10^3$$

with node ordering (1,2,3,4,5). Multiplication of the rows and solving reveals that $u_2 = 0.5u_3$ and $u_4 = 0.65u_5$ with the structure nodal displacements (in mm) given by

$$\{U\} = \begin{bmatrix} 0 & \frac{1}{280} & \frac{1}{140} & \frac{13}{840} & \frac{1}{42} \end{bmatrix}^T$$

From the first row of the above structure system of linear equations we can confirm that R_1 is equal and opposite to the applied force

$$R_1 = 10^3(-746.66u_2 + 93.33u_3) = -2 \times 10^3$$

The Jacobian matrices of both elements are given by (16.70)

$$[J^1] = \frac{\partial N_1}{\partial \xi}x_1 + \frac{\partial N_3}{\partial \xi}x_3 + \frac{\partial N_2}{\partial \xi}x_2 = (4\xi - 3)0 + (4\xi - 1)50 + (4 - 8\xi)25 = 50$$
$$[J^2] = \frac{\partial N_3}{\partial \xi}x_3 + \frac{\partial N_5}{\partial \xi}x_5 + \frac{\partial N_4}{\partial \xi}x_4 = (4\xi - 3)50 + (4\xi - 1)100 + (4 - 8\xi)75 = 50$$

with $| J^1 |= L_1$ and $| J^2 |= L_2$. The $[B]$ matrix for both elements is therefore, (16.73)

$$[B^1] = [B^2] = \frac{1}{50}\begin{bmatrix}4\xi - 3 & 4\xi - 1 & 4 - 8\xi\end{bmatrix}$$

The strains of both elements at their mid-point, $\xi = 0.5$, are, (16.17)

$$\varepsilon^1_{xx} = [B^1]\begin{Bmatrix}u_1\\u_3\\u_2\end{Bmatrix} = \begin{bmatrix}-0.02 & 0.02 & 0\end{bmatrix}\begin{Bmatrix}0\\1/140\\1/280\end{Bmatrix} = 142.86 \times 10^{-6}$$
$$\varepsilon^2_{xx} = [B^2]\begin{Bmatrix}u_3\\u_5\\u_4\end{Bmatrix} = \begin{bmatrix}-0.02 & 0.02 & 0\end{bmatrix}\begin{Bmatrix}1/40\\1/42\\13/840\end{Bmatrix} = 333.33 \times 10^{-6}$$

with the element stresses given by (16.18)

$$\sigma^1_{xx} = [D^1]\varepsilon^1_{xx} = E_1\varepsilon^1_{xx} = 25\text{N/mm}^2$$
$$\sigma^2_{xx} = [D^2]\varepsilon^1_{xx} = E_2\varepsilon^2_{xx} = 50\text{N/mm}^2$$

and are exact.

16.4.3 Triangular Element

The following two sub-sections derive the interpolation functions, stiffness matrix and force vector for a two-dimensional triangular element.

16.4.3.1 Quadratic and Cubic Interpolation Functions

Figure 16.13 illustrates a two-dimensional quadratic triangular element with three vertex nodes (1, 2 and 3) and three mid-side nodes (4, 5 and 6). The interpolation function is

$$\phi = \alpha_1 + \alpha_2 x + \alpha_3 y + \alpha_4 xy + \alpha_5 x^2 + \alpha_6 y^2 \tag{16.82}$$

Following a similar procedure to that of the bar element then ϕ can alternatively be expressed in terms of the element shape functions

$$\phi = \sum_{i=1}^{6} N_i \phi_i \tag{16.83}$$

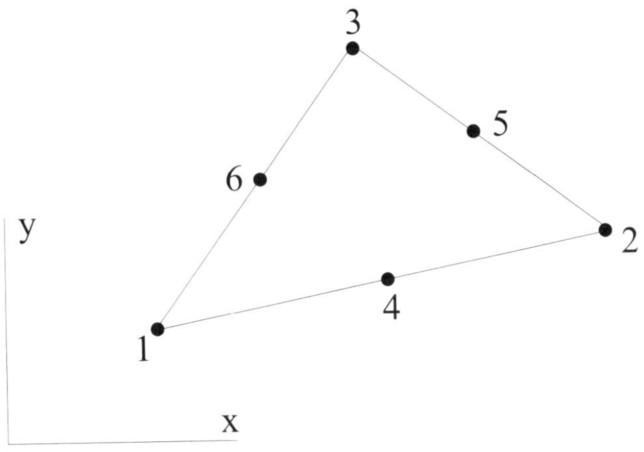

Figure 16.13: Quadratic triangular element.

where ϕ_i are the nodal values of ϕ and N_i are given by

$$\begin{aligned} N_1 &= (2\xi - 1)\xi_1 \\ N_2 &= (2\xi - 1)\xi_2 \\ N_3 &= (2\xi - 1)\xi_3 = 1 - 3(\xi_1 + \xi_2) + 2(\xi_1 + \xi_2)^2 \\ N_4 &= 4\xi_1\xi_2 \\ N_5 &= 4\xi_2\xi_3 = 4\xi_2(1 - \xi_1 - \xi_2) \\ N_6 &= 4\xi_1\xi_3 = 4\xi_1(1 - \xi_1 - \xi_2) \end{aligned} \tag{16.84}$$

Using a cubic interpolation function ϕ is given by

$$\phi = \sum_{i=1}^{10} N_i \phi_i \tag{16.85}$$

where N_i are

$$\begin{aligned} N_1 &= (3\xi_1 - 1)(3\xi_1 - 2)\xi_1/2 \\ N_2 &= (3\xi_2 - 1)(3\xi_2 - 2)\xi_2/2 \\ N_3 &= (3\xi_3 - 1)(3\xi_3 - 2)\xi_3/2 \\ N_4 &= 9(3\xi_1 - 1)\xi_1\xi_2/2 \\ N_5 &= 9(3\xi_2 - 1)\xi_1\xi_2/2 \\ N_6 &= 9(3\xi_2 - 1)\xi_2\xi_3/2 \\ N_7 &= 9(3\xi_3 - 1)\xi_2\xi_3/2 \\ N_8 &= 9(3\xi_3 - 1)\xi_1\xi_3/2 \\ N_9 &= 9(3\xi_1 - 1)\xi_1\xi_3/2 \\ N_{10} &= 27\xi_1\xi_2\xi_3 \end{aligned} \tag{16.86}$$

The cubic triangular element is shown in Figure 16.14 and consists of three vertex nodes (1, 2 and 3), six edge nodes (4, 5, 6, 7, 8 and 9) and a centroidal node (10).

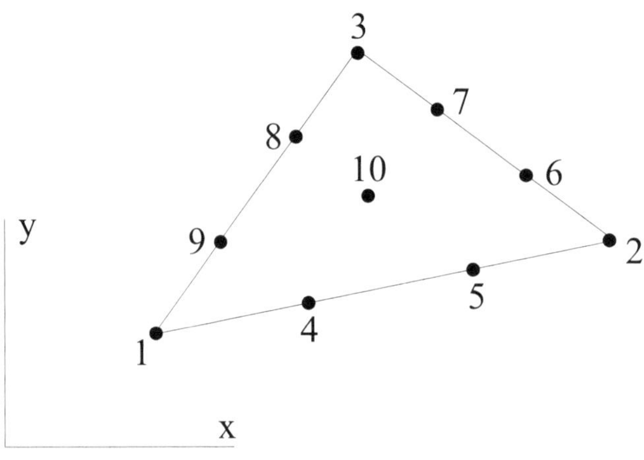

Figure 16.14: Cubic triangular element.

16.4.3.2 Stiffness Matrix and Force Vector

Before we can evaluate the stiffness matrix and force vector we need to determine the $[B]$ matrix which, in the case of two-dimensional stress analysis, is the first derivatives of the shape functions

$$\{\varepsilon\} = \left\{ \begin{array}{c} \varepsilon_{xx} \\ \varepsilon_{yy} \\ \gamma_{xy} \end{array} \right\} = [B]\{U\} \tag{16.87}$$

For a quadratic element consisting of six nodes, each with (x, y) degrees of freedom, then the nodal displacement vector, $\{U\}$, has twelve elements

$$\{U\} = \left\{ \begin{array}{c} u_1 \\ v_1 \\ u_2 \\ \vdots \\ u_6 \\ v_6 \end{array} \right\} \tag{16.88}$$

Thus, $[B]$ must be of size (3×12) and is of the form

$$[B] = \begin{bmatrix} \frac{\partial N_1}{\partial x} & 0 & \frac{\partial N_2}{\partial x} & 0 & \cdots & \frac{\partial N_6}{\partial x} & 0 \\ 0 & \frac{\partial N_1}{\partial y} & 0 & \frac{\partial N_2}{\partial y} & \cdots & 0 & \frac{\partial N_6}{\partial y} \\ \frac{\partial N_1}{\partial y} & \frac{\partial N_1}{\partial x} & \frac{\partial N_2}{\partial y} & \frac{\partial N_2}{\partial x} & \cdots & \frac{\partial N_6}{\partial y} & \frac{\partial N_6}{\partial x} \end{bmatrix} \tag{16.89}$$

As with the quadratic bar element the triangular element shape functions are in terms of the area coordinates ξ_1, ξ_2 and $\xi_3 (= 1 - \xi_1 - \xi_2)$ so that a transformation of coordinates is required. Considering shape function N_i, then from the chain rule of differentiation

$$\begin{aligned} \frac{\partial N_i}{\partial \xi_1} &= \frac{\partial N_i}{\partial x}\frac{\partial x}{\partial \xi_1} + \frac{\partial N_i}{\partial y}\frac{\partial y}{\partial \xi_1} \\ \frac{\partial N_i}{\partial \xi_2} &= \frac{\partial N_i}{\partial x}\frac{\partial x}{\partial \xi_2} + \frac{\partial N_i}{\partial y}\frac{\partial y}{\partial \xi_2} \end{aligned} \tag{16.90}$$

and re-arranging for $\partial N_i/\partial x$ and $\partial N_i/\partial y$

$$\left\{ \begin{array}{c} \frac{\partial N_i}{\partial x} \\ \frac{\partial N_i}{\partial x} \end{array} \right\} = [J]^{-1} \left\{ \begin{array}{c} \frac{\partial N_i}{\partial \xi_1} \\ \frac{\partial N_i}{\partial \xi_2} \end{array} \right\} \tag{16.91}$$

where the Jacobian matrix $[J]$ is

$$[J] = \left[\begin{array}{cc} \frac{\partial x}{\partial \xi_1} & \frac{\partial y}{\partial \xi_1} \\ \frac{\partial x}{\partial \xi_2} & \frac{\partial y}{\partial \xi_2} \end{array} \right] \tag{16.92}$$

Since the x and y coordinates of a point inside the element are given by

$$x = \sum_{i=1}^{6} N_i x_i; \quad y = \sum_{i=1}^{6} N_i y_i \tag{16.93}$$

then $[J]$ is

$$[J] = \left[\begin{array}{cc} \frac{\partial x}{\partial \xi_1} & \frac{\partial y}{\partial \xi_1} \\ \frac{\partial x}{\partial \xi_2} & \frac{\partial y}{\partial \xi_2} \end{array} \right] = \left[\begin{array}{cccc} \frac{\partial N_1}{\partial \xi_1} & \frac{\partial N_2}{\partial \xi_1} & \cdots & \frac{\partial N_6}{\partial \xi_1} \\ \frac{\partial N_1}{\partial \xi_2} & \frac{\partial N_2}{\partial \xi_2} & \cdots & \frac{\partial N_6}{\partial \xi_2} \end{array} \right] \left[\begin{array}{cc} x_1 & y_1 \\ x_2 & y_2 \\ \vdots & \vdots \\ x_6 & y_6 \end{array} \right] \tag{16.94}$$

As in the case of the quadratic bar element (16.76) the stiffness matrix integral requires transformation from global to local coordinates because the $[B]$ matrix is in terms of area coordinates

$$[K] = \int_V [B]^{\mathrm{T}}[D][B]\,\mathrm{d}V = t\int_V [B]^{\mathrm{T}}[D][B]\,\mathrm{d}x\,\mathrm{d}y = t\int_0^1\int_0^1 [B]^{\mathrm{T}}[D][B] \mid J \mid \,\mathrm{d}\xi_1\,\mathrm{d}\xi_2 \tag{16.95}$$

where t is the constant thickness of the element. The integral can be evaluated numerically using Gaussian integration (16.53) and noting that $\mid J \mid = 2A$ then

$$[K] = 2At\sum_{i=1}^{n} w_i [B]^{\mathrm{T}}[D][B] \tag{16.96}$$

From (16.65) the thermal force vector term, $\{F\}_T$, is

$$\{F\}_T = \alpha\Delta T\int_V [B]^{\mathrm{T}}[D]\,\mathrm{d}V = \alpha\Delta T t\int_0^1\int_0^1 [B]^{\mathrm{T}}[D] \mid J \mid \,\mathrm{d}\xi_1\,\mathrm{d}\xi_2 = 2A\alpha\Delta T t\sum_{i=1}^{n} w_i [B]^{\mathrm{T}}[D] \tag{16.97}$$

The body force, $\{F\}_B$, and edge pressure, $\{F\}_P$, force vectors are more straightforward because they can be expressed solely in terms of local coordinates and hence do not require numerical integration. From (16.65) $\{F\}_B$ is

$$\{F\}_B = \int_V [N]^{\mathrm{T}} \left\{ \begin{array}{c} X \\ Y \end{array} \right\} \mathrm{d}V \tag{16.98}$$

where X and Y are the body force components acting in the x and y directions respectively. Substituting

$[N]$ and with $\mathrm{d}V = t\,\mathrm{d}A$

$$\{F\}_B = t\int_A [N]^{\mathrm{T}}\left\{\begin{array}{c} X \\ Y \end{array}\right\}\mathrm{d}A = t\int_A \begin{bmatrix} N_1 & 0 \\ 0 & N_1 \\ N_2 & 0 \\ 0 & N_2 \\ \vdots & \vdots \\ N_6 & 0 \\ 0 & N_6 \end{bmatrix}\left\{\begin{array}{c} X \\ Y \end{array}\right\}\mathrm{d}A = \frac{At}{3}\left\{\begin{array}{c} 0 \\ 0 \\ 0 \\ 0 \\ 0 \\ 0 \\ 0 \\ X \\ Y \\ X \\ Y \\ X \\ Y \end{array}\right\} \quad (16.99)$$

and illustrates that the body force is equally divided between the mid-side nodes only. The integral in (16.99) can be evaluated using (16.46). The edge pressure force vector term, $\{F\}_P$, is of the form, (16.65)

$$\{F\}_P = \int_S [N]^{\mathrm{T}}\left\{\begin{array}{c} p_x \\ p_y \end{array}\right\} \quad (16.100)$$

where p_x and p_y are the x and y components of the applied pressure. Consider the case of the normal pressure $p(p_x, p_y)$ acting on edge (1,4,2) only as shown in Figure 16.15. On edge (1,4,2) $\xi_3 = 0$ and by inspection of (16.84) $N_3 = N_5 = N_6 = 0$ and (16.100) becomes

$$\{F\}_P = \int_{S_{142}} \begin{bmatrix} N_1 & 0 \\ 0 & N_1 \\ N_2 & 0 \\ 0 & N_2 \\ 0 & 0 \\ 0 & 0 \\ N_4 & 0 \\ 0 & N_4 \\ 0 & 0 \\ 0 & 0 \\ 0 & 0 \\ 0 & 0 \end{bmatrix}\left\{\begin{array}{c} p_x \\ p_y \end{array}\right\}\mathrm{d}S = \int_{L_{142}} \left\{\begin{array}{c} N_1 p_x \\ N_1 p_y \\ N_2 p_x \\ N_2 p_y \\ 0 \\ 0 \\ N_4 p_x \\ N_4 p_y \\ 0 \\ 0 \\ 0 \\ 0 \end{array}\right\} t\,\mathrm{d}L = \frac{L_{142}t}{6}\left\{\begin{array}{c} p_x \\ p_y \\ p_x \\ p_y \\ 0 \\ 0 \\ p_x \\ p_y \\ 0 \\ 0 \\ 0 \\ 0 \end{array}\right\} \quad (16.101)$$

where the integral in (16.101) can be evaluated using (16.45).

Example 16.4 Stiffness matrix for a triangular element with quadratic interpolation function

In this example we will evaluate the stiffness matrix $[K]$ for the quadratic triangular element shown in Figure 16.16 with an element thickness of t =1mm and assuming plane stress conditions with Young's modulus E =200GPa and Poisson's ratio ν =0.3.

From (16.96) $[K]$ was shown to be equal to

$$[K] = 2At\sum_{i=1}^{n} w_i[B]^{\mathrm{T}}[D][B]$$

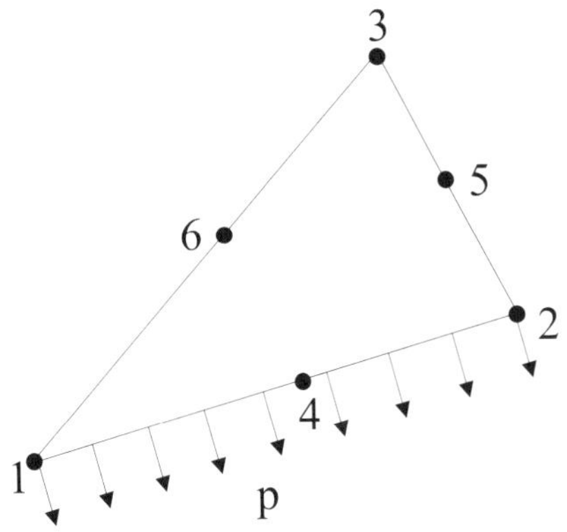

Figure 16.15: Normal edge pressure p acting on edge (1,4,2) only.

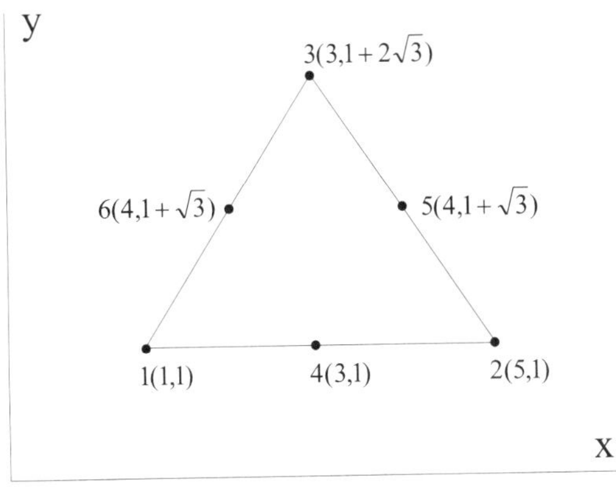

Figure 16.16: Example: Quadratic triangular element.

for n integration points. From Table 16.2 the $n = 3$ integration points (ξ_1, ξ_2) are $(1/6, 1/6)$, $(2/3, 1/6)$ and $(1/6, 2/3)$ with corresponding weights all equal to 1/6. From (16.84) the derivatives of the shape functions are

$$\frac{\partial N_1}{\partial \xi_1} = 4\xi_1 - 1; \quad \frac{\partial N_2}{\partial \xi_1} = 0; \quad \frac{\partial N_3}{\partial \xi_1} = -3 + 4(\xi_1 + \xi_2)$$
$$\frac{\partial N_4}{\partial \xi_1} = 4\xi_2; \quad \frac{\partial N_5}{\partial \xi_1} = -4\xi_1; \quad \frac{\partial N_6}{\partial \xi_1} = 4 - 8\xi_1 - 4\xi_2$$
$$\frac{\partial N_1}{\partial \xi_2} = 0; \quad \frac{\partial N_2}{\partial \xi_2} = 4\xi_2 - 1; \quad \frac{\partial N_3}{\partial \xi_2} = -3 + 4(\xi_1 + \xi_2)$$
$$\frac{\partial N_4}{\partial \xi_2} = 4\xi_1; \quad \frac{\partial N_5}{\partial \xi_2} = 4 - 8\xi_2 - 4\xi_1; \quad \frac{\partial N_6}{\partial \xi_2} = -4\xi_1$$

The Jacobian matrix $[J]$ at $(\xi_1, \xi_2) = (1/6, 1/6)$ is given by (16.94)

$$[J] = \begin{bmatrix} \frac{\partial N_1}{\partial \xi_1} & \frac{\partial N_2}{\partial \xi_1} & \cdots & \frac{\partial N_6}{\partial \xi_1} \\ \frac{\partial N_1}{\partial \xi_2} & \frac{\partial N_2}{\partial \xi_2} & \cdots & \frac{\partial N_6}{\partial \xi_2} \end{bmatrix} \begin{bmatrix} x_1 & y_1 \\ x_2 & y_2 \\ \vdots & \vdots \\ x_6 & y_6 \end{bmatrix} =$$

$$\begin{bmatrix} -1/3 & 0 & -5/3 & 2/3 & -2/3 & 2 \\ 0 & -1/3 & -5/3 & 2/3 & 2 & -2/3 \end{bmatrix} \begin{bmatrix} 1 & 1 \\ 5 & 1 \\ 3 & 1+2\sqrt{3} \\ 3 & 1 \\ 4 & 1+\sqrt{3} \\ 2 & 1+\sqrt{3} \end{bmatrix} = \begin{bmatrix} -2 & -2\sqrt{3} \\ 2 & -2\sqrt{3} \end{bmatrix}$$

The determinant of $[J]$ is $\mid J \mid = 8\sqrt{3}$ illustrating that $\mid J \mid = 2A$, where $A = 4\sqrt{3}$ is the area of the triangle. The inverse of $[J]$, when $[J]$ is of size (2×2), is

$$[J]^{-1} = \frac{\text{adjoint } J}{\det J} = \frac{1}{\mid J \mid} \begin{bmatrix} j_{22} & -j_{12} \\ -j_{21} & j_{11} \end{bmatrix}; \quad [J] = \begin{bmatrix} j_{11} & j_{12} \\ j_{21} & j_{22} \end{bmatrix}$$

so that in the present case

$$[J]^{-1} = \frac{1}{8\sqrt{3}} \begin{bmatrix} -2\sqrt{3} & 2\sqrt{3} \\ -2 & -2 \end{bmatrix} = \begin{bmatrix} -1/4 & 1/4 \\ -1/4\sqrt{3} & -1/4\sqrt{3} \end{bmatrix}$$

The $[B]$ matrix is given by (16.89) with the terms of $[B]$ from (16.91). Considering shape function N_1

$$\left\{ \begin{array}{c} \frac{\partial N_1}{\partial x} \\ \frac{\partial N_1}{\partial x} \end{array} \right\} = [J]^{-1} \left\{ \begin{array}{c} \frac{\partial N_1}{\partial \xi_1} \\ \frac{\partial N_1}{\partial \xi_2} \end{array} \right\} = \begin{bmatrix} -1/4 & 1/4 \\ -1/4\sqrt{3} & -1/4\sqrt{3} \end{bmatrix} \left\{ \begin{array}{c} -1/3 \\ 0 \end{array} \right\} = \left\{ \begin{array}{c} 1/12 \\ 1/12\sqrt{3} \end{array} \right\}$$

Determining the remaining terms of $[B]$ in an analogous manner we arrive at

$$[B] = \begin{bmatrix} \frac{1}{6} & 0 & -\frac{1}{12} & 0 & 0 & 0 & 0 & 0 & \frac{2}{3} & 0 & -\frac{2}{3} & 0 \\ 0 & \frac{1}{6\sqrt{3}} & 0 & \frac{1}{12\sqrt{3}} & 0 & \frac{5}{6\sqrt{3}} & 0 & -\frac{1}{3\sqrt{3}} & 0 & -\frac{1}{3\sqrt{3}} & 0 & \frac{1}{3\sqrt{3}} \\ \frac{1}{6\sqrt{3}} & \frac{1}{6} & \frac{1}{12\sqrt{3}} & -\frac{1}{12} & \frac{5}{6\sqrt{3}} & 0 & -\frac{1}{3\sqrt{3}} & 0 & -\frac{1}{3\sqrt{3}} & \frac{2}{3} & \frac{1}{3\sqrt{3}} & -\frac{2}{3} \end{bmatrix}$$

The $[D]$ matrix in the case of plane stress is, (16.38)

$$[D] = \frac{E}{1-\nu^2} \begin{bmatrix} 1 & \nu & 0 \\ \nu & 1 & 0 \\ 0 & 0 & \frac{1-\nu}{2} \end{bmatrix} = 2.1978 \times 10^{11} \begin{bmatrix} 1 & 0.3 & 0 \\ 0.3 & 1 & 0 \\ 0 & 0 & 0.35 \end{bmatrix}$$

The stiffness matrix $[K]$ at the integration point $w_1 = 1/2$ can now be determined from (16.96)

$$[K_{w1}] = 2Atw_1[B]^{\mathrm{T}}[D][B] = 1.5227 \times 10^7$$

$$\times \begin{bmatrix} 3.10 & & & & & & & & & & & \\ 1.04 & 1.90 & & & & & & & & & & \\ -1.23 & 0.04 & 0.78 & & & & & & & & & \\ -0.04 & -0.02 & -0.26 & 0.47 & & & & & & & & \\ 1.62 & 2.81 & 0.81 & -1.40 & 8.10 & & & & \text{sym} & & & \\ 2.41 & 4.63 & -1.20 & 2.31 & 0 & 23.15 & & & & & & \\ -0.65 & -1.12 & -0.32 & 0.56 & -3.24 & 0 & 1.30 & & & & & \\ -0.96 & -1.85 & 0.48 & -0.93 & 0 & -9.26 & 0 & 3.71 & & & & \\ 10.46 & 0.80 & -5.88 & 1.52 & -3.24 & 9.62 & 1.30 & -3.85 & 45.74 & & & \\ 1.28 & 2.04 & 1.60 & -2.87 & 11.23 & -9.26 & -4.49 & 3.71 & -8.34 & 19.26 & & \\ -10.46 & -0.80 & 5.88 & -1.52 & 3.24 & -9.62 & -1.30 & 3.85 & -45.74 & 8.34 & 45.74 & \\ -1.28 & -2.04 & -1.60 & 2.87 & -11.23 & 9.26 & 4.49 & -3.71 & 8.34 & -19.26 & -8.34 & 19.26 \end{bmatrix}$$

Similar stiffness matrices, $[K_{w2}]$ and $[K_{w3}]$, are further required at the integration points $2(w_2 = 1/6)$ and $3(w_3 = 1/6)$. The matrices $[K_{w2}]$ and $[K_{w3}]$ are equivalent to $[K_{w1}]$ except that the multiplicative term is now equal to 5.0756×10^6. Summing the stiffness contributions at the three integration points we finally arrive at the stiffness matrix for the element

$$[K_{w1}] + [K_{w2}] + [K_{w3}] = 1 \times 10^7$$

$$\times \begin{bmatrix} 7.87 & & & & & & & & & & & \\ 2.64 & 4.82 & & & & & & & & & & \\ -3.11 & 0.10 & 1.97 & & & & & & & & & \\ -0.10 & 0.06 & -0.66 & 1.20 & & & & & & & & \\ 4.11 & 7.12 & 2.06 & -3.56 & 20.56 & & & & \text{sym} & & & \\ 6.10 & 11.75 & -3.05 & 5.87 & 0 & 58.74 & & & & & & \\ -1.64 & -2.85 & -0.82 & 1.42 & -8.23 & 0 & 3.29 & & & & & \\ -2.44 & -4.70 & 1.22 & -2.35 & 0 & -23.50 & 0 & 9.40 & & & & \\ 26.55 & 2.03 & -14.92 & 3.87 & -8.23 & 24.42 & 3.29 & -9.77 & 116.08 & & & \\ 3.25 & 5.17 & 4.07 & -7.28 & 28.49 & -23.50 & -11.40 & 9.40 & -21.17 & 48.88 & & \\ -26.55 & -2.03 & 14.92 & -3.87 & 8.23 & -24.42 & -3.29 & 9.77 & -116.08 & 21.17 & 116.08 & \\ -3.25 & -5.17 & -4.07 & 7.28 & -28.49 & 23.50 & 11.40 & -9.40 & 21.17 & -48.88 & -21.17 & 48.88 \end{bmatrix}$$

16.5 Conclusion

This Chapter has extended the introduction of Chapter 13 on the finite element method by considering higher order elements in both one and two dimensions. The two-dimensional element considered is the popular three noded constant strain triangle. The higher order elements considered were the quadratic and cubic bar and triangle elements.

The finite element method is an active area of research in engineering and an up to date overview of the subject is best found in scientific journals. Several key journals are the *International Journal of Numerical Methods in Engineering, Computer Methods in Applied Mechanics and Engineering, Computers and Structures* and *Advances in Engineering Software*. Four current active areas of research are shape optimisation using the mathematical theory of optimisation, Vanderplaats (1984), adaptive mesh refinement, Zienkiewicz and Taylor (1991), the application of artificial neural networks, Topping and Bahreininejad (1997), and the use of parallel computing to solve large problems, Topping and Khan (1996).

16.6 References and Further Reading

∘ Cook, R. D., Malkus, D. S. and Plesha, M. E. (1989) *Concepts and Applications of Finite Element Analysis*, 3rd Ed., Wiley & Sons, Singapore.

∘ Fagan, M. J. (1992) *Finite Element Analysis: Theory and Practice*, Longman Scientific and Technical.

∘ Rao, S. S. (1999) *The Finite Element Method in Engineering*, 3rd Ed., Butterworth, Woburn, MA.

∘ Topping, B. H. V. and Khan, A. I. (1996) *Parallel Finite Element Computations*, Saxe-Coburg Publications, Edinburgh.

∘ Topping, B. H. V. and Bahreininejad, A. (1997) *Neural Computing for Structural Mechanics*, Saxe-Coburg Publications, Edinburgh.

∘ Vanderplaats, G. N. (1984) *Numerical Optimisation Techniques for Engineering Design: with applications*, McGraw-Hill, New York.

∘ Zienkiewicz, O. C. and Taylor, R. L. (1991) *The Finite Element Method*, 4th Ed., 2 Volumes, McGraw-Hill.

16.7 Exercises

16.1 For the constant strain triangular element shown in Figure 16.17 evaluate the element shape functions at point $p(2,3)$ using a linear interpolation function. If the same interpolation function is used to represent

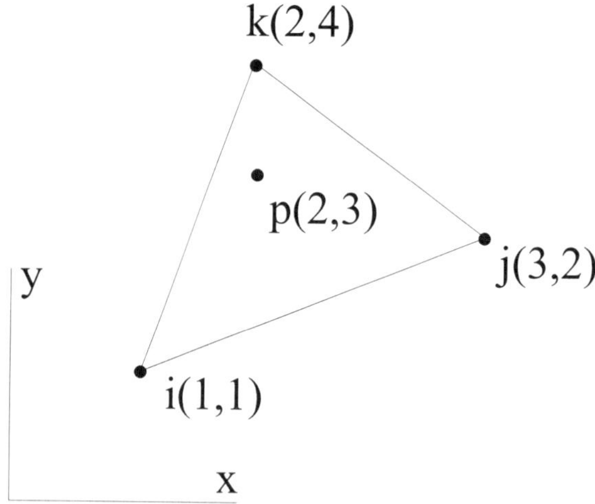

Figure 16.17: Exercise: A constant strain triangular element (i, j, k) with nodes at (x, y) coordinates (1,1), (3,2) and (2,4).

the u and v displacements within the element then calculate the displacements at the point (2,3) given the following nodal displacements

$$\left\{\begin{array}{c} u_i \\ v_i \\ u_j \\ v_j \\ u_k \\ v_k \end{array}\right\} = \left\{\begin{array}{c} 1 \\ 1 \\ 3 \\ 4 \\ 2 \\ 2 \end{array}\right\}$$

16.2 For the two-dimensional plane stress constant strain triangular element shown in Figure 16.18 calculate the shape functions matrix $[N]$, the shape functions derivatives matrix $[B]$ and the material matrix $[D]$. If the displacement of nodes 1, 2 and 3 are those given in Table 16.3 then calculate the total strain

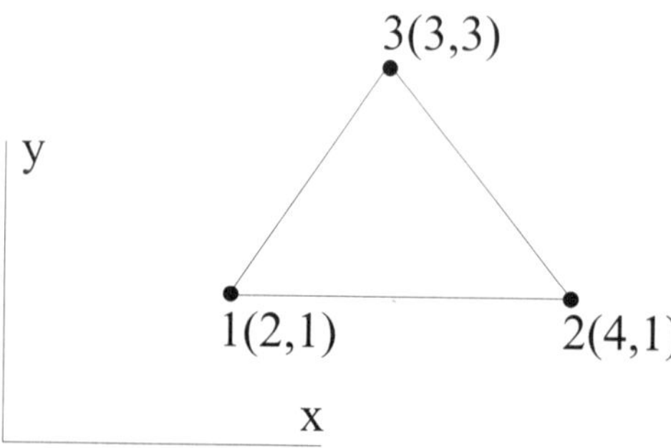

Figure 16.18: Exercise: A constant strain triangular element (1,2,3) with nodes at (x, y) coordinates (2,1), (4,1) and (3,3).

and stress vectors assuming that no initial strains or stresses are present. Assume that E =200GPa and $\nu = 0.3$.

Node	$u(1 \times 10^{-3})$mm	$v(1 \times 10^{-3})$mm
1	1	1
2	2	3
3	4	5

Table 16.3: Exercise: Nodal displacements for the triangular element of Figure 16.18.

16.3 For the two-dimensional plane stress constant strain triangular element shown in Figure 16.19 form the stiffness matrix and force vector. What properties do you observe about the stiffness matrix? Assume the thickness of the element to be 2mm, Young's modulus 200GPa and Poisson's ratio 0.3. The nodal coordinates of the element are in mm.

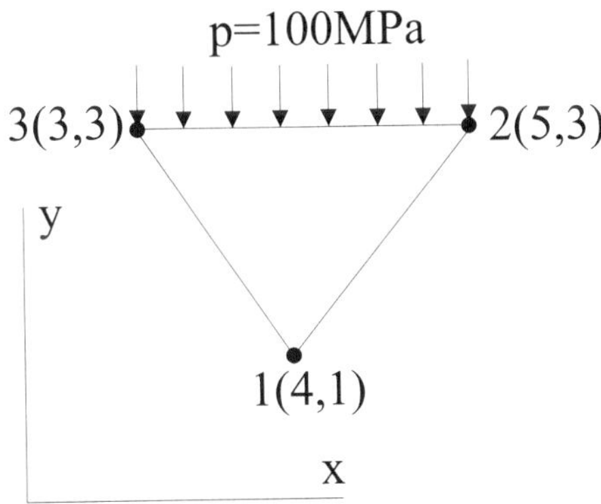

Figure 16.19: Exercise: A constant strain triangular element (1,2,3) with nodes at (x, y) coordinates (2,1), (4,1) and (3,3) with all coordinates in mm.

16.4 Figure 16.20 illustrates a quadratic bar element with nodes 1(1,1), 2(5,1) and 3(3,1). Determine the displacement at a point $p(4,1)$ if the displacements at nodes 1, 2 and 3 are $u_1 = 2$, $u_2 = 2.25$ and $u_3 = 2.55$.

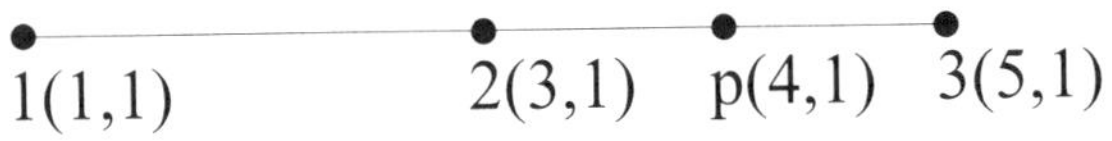

1(1,1) 2(3,1) p(4,1) 3(5,1)

Figure 16.20: Exercise: A quadratic bar element (1,2,3) with nodes at (x, y) coordinates (1,1), (3,1) and (5,1).

16.5 A quadratic triangular element is shown in Figure 16.21. Determine the displacement at a point $p(2,3)$ if the (u, v) displacements at nodes $1, \ldots, 6$ are those given in Table 16.4.

16.6 Evaluate the Jacobian matrix for the quadratic bar element of Exercise 16.4.

16.7 Evaluate the following integral I_y for the right-angled triangle shown in Figure 16.22 using quadratic

Node	u	v
1	1	1
2	2	3
3	4	5
4	1.4	0.8
5	3.1	4.2
6	2.75	3.2

Table 16.4: Exercise: Nodal displacements for the triangular element of Figure 16.21.

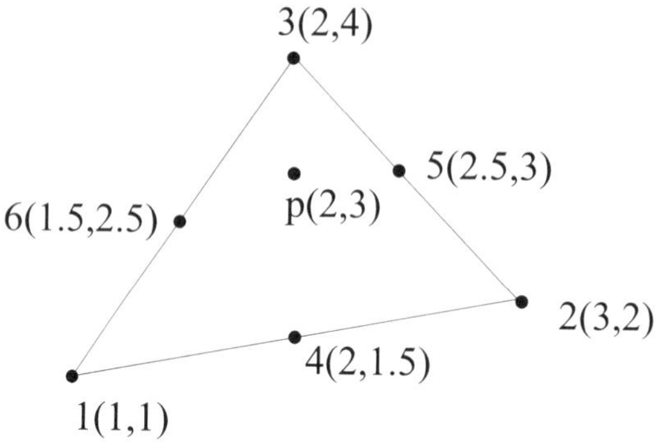

Figure 16.21: Exercise: A quadratic triangular element (1,2,3,4,5,6) with nodes at (x, y) coordinates (1,1), (3,2), (2,4), (2,1.5), (2.5,3) and (1.5,2.5).

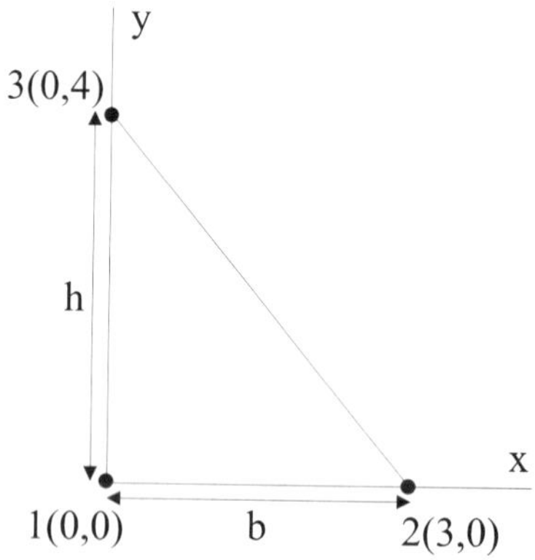

Figure 16.22: Exercise: A right-angled triangle with the origin of coordinates centred at vertex 1.

Gaussian integration

$$I_y = \iint_A x^2 \, \mathrm{d}A$$

Compare your estimate of I_y with the exact integration of I_y.

Chapter 17

Fracture and Fatigue

17.1 Introduction

The presence of crack-like flaws and defects cannot be precluded in any engineering component or structure. Increasing demands for energy and material conservation are requiring that engineers design structures with ever tighter safety margins. Thus, there is an increasing need for accurate quantitative estimates of the defect tolerance of structures. Previously, design procedures avoided performing detailed defect assessments by over-design and avoiding large stress concentrations and the immediate repair or removal from service of components that exhibited cracks and defects. However, recent improvements in non-destructive testing procedures have enabled defects to be found which previously would have gone unnoticed. Also, the presence of a defect in a structure does not necessarily mean that the structure should be removed from service. The concept of damage tolerance has arisen to provide engineers with quantitative measures for assessing the structural integrity of structures. The cost of repair or replacement of a flawed structure is assessed relative to the possibility that continued service may lead to failure. Damage tolerance is largely based on the topic of fracture mechanics.

For excellent discussions on the history of the strength of materials and the fracture of materials refer to the works of Gordon (1976), Timoshenko (1953) and Irwin (1964). Griffith (1921) was the first to make a quantitative connection between crack size and the strength of a structure via the concept of energy-release rate. However, linear elastic fracture mechanics became largely an engineering discipline following the work of Irwin and the introduction of the stress intensity factor. The next major development in fracture mechanics was the treatment of non-linear problems. Key works which addressed the non-linearity of fracture where Rice (1968a,b) and Hutchinson (1968). Current fracture mechanics analyses are used to predict the rate at which a crack can approach a critical size in fatigue or by environmental influences such as stress corrosion cracking. Extensive elastic-plastic and time-dependent analyses are now possible and can include the treatments of the cracking of welds and other areas where residual stresses are present, anisotropic and heterogeneous materials such as fibre-reinforced composite materials, adhesives and viscoelastic materials.

The Chapter begins by introducing the stress intensity factor which has become the cornerstone of fracture mechanics. The examination of local crack tip stress and displacement fields is accompanied by a discussion of the non-singular T-stress. We then examine crack extension based on strain energy, namely Griffith's energy release rate. Section 17.5 presents stress intensity expressions for some well-known configurations. The three methods of Green's functions, Bueckner weight function and compounding method are then introduced as methods for determining stress intensity factors. This is followed by an overview of crack tip plasticity. Section 17.8 discusses various fracture criteria and the plane strain fracture toughness. Due to the importance attached to performing damage analysis assessments of pressure vessels section 17.10 examines the application of fracture mechanics to pressure vessels. Section 17.11 is a major section of this Chapter and presents an overview of both long and short fatigue cracks. The Chapter concludes by examining the J, L and M conservation integrals and their application to well-known fracture mechanics problems.

17.2 The Stress Intensity Factor

Consider the remotely loaded infinite plate containing a central crack shown in Figure 17.1. This configuration is frequently referred to as a *centre cracked plate* or the *Inglis-Kolosov configuration* after the first investigators. The normal stress, σ_{yy}, ahead of the crack of length $2a$ and the displacement on the crack

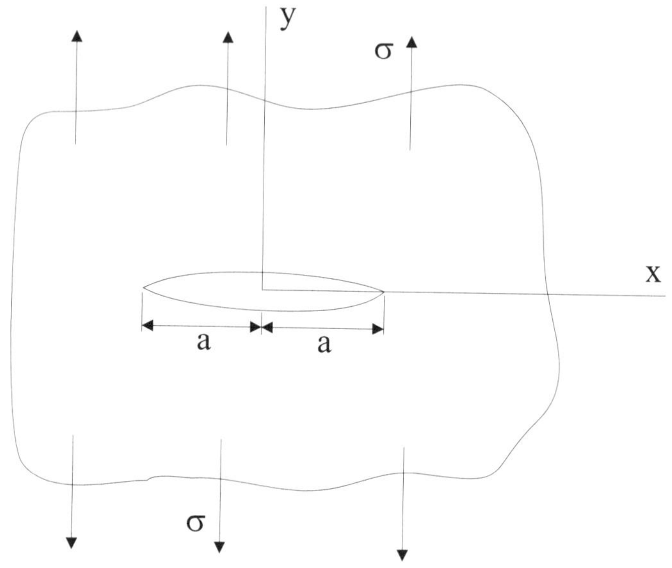

Figure 17.1: A centre cracked plate.

faces, v, are obtained from the solution of an elliptical hole in a plate by letting the minor axis $b \to 0$ and are given by

$$\begin{aligned} \sigma_{yy} &= \frac{x\sigma}{\sqrt{x^2 - a^2}}; \quad x > a \\ v &= \frac{(1+\nu)(\kappa+1)}{E}\frac{\sigma}{2}\left(a^2 - x^2\right); \quad x < a \end{aligned} \tag{17.1}$$

where x is taken from the centre of the crack. The material constant $\kappa = 3 - 4\nu$ for plane strain and $\kappa = (3 - \nu)/(1 + \nu)$ for plane stress offers a convenient notation since it incorporates both plane stress and plane strain into a single parameter. If we restrict ourselves to regions near the crack tip ($|x - a| \ll a$) then (17.1) can be approximately written as

$$\begin{aligned} \sigma_{yy} &= \frac{K_I}{\sqrt{2\pi(x-a)}}; \quad x > a \\ v &= \frac{(1+\nu)(\kappa+1)}{E} K_I \sqrt{\frac{a-x}{2\pi}}; \quad x < a \end{aligned} \tag{17.2}$$

where K_I is referred to as the *stress intensity factor* and is a geometry and loading dependent quantity that has the value $\sigma\sqrt{\pi a}$ with units of $\mathrm{Nm}^{-3/2}$. The term stress intensity factor was coined by Irwin (1957) and named after one of his co-investigators, Kies. Numerous other works (notably Sneddon (1946) and Williams (1957)) have determined more exact series expansions for the stresses and displacements in the vicinity of a crack tip and are generally expressed as follows

$$\begin{aligned}
\sigma_{xx} &= \frac{K_I}{\sqrt{2\pi r}}\cos\frac{\theta}{2}\left[1-\sin\frac{\theta}{2}\sin\frac{3\theta}{2}\right]+T+O\left(r^{1/2}\right)\\
\sigma_{yy} &= \frac{K_I}{\sqrt{2\pi r}}\cos\frac{\theta}{2}\left[1+\sin\frac{\theta}{2}\sin\frac{3\theta}{2}\right]+O\left(r^{1/2}\right)\\
\tau_{xy} &= \frac{K_I}{\sqrt{2\pi r}}\cos\frac{\theta}{2}\sin\frac{\theta}{2}\sin\frac{3\theta}{2}+O\left(r^{1/2}\right)\\
u &= \frac{K_I}{2\mu}\sqrt{\frac{r}{2\pi}}\cos\frac{\theta}{2}\left[\kappa-1+2\sin^2\frac{\theta}{2}\right]+O\left(r^{3/2}\right)\\
v &= \frac{K_I}{2\mu}\sqrt{\frac{2}{2\pi}}\sin\frac{\theta}{2}\left[\kappa+1-2\cos^2\frac{\theta}{2}\right]+O\left(r^{3/2}\right)
\end{aligned} \tag{17.3}$$

where the coordinate system shown in Figure 17.2 is used and where $O\left(r^{1/2}\right)$ indicates higher order terms of power $1/2$ or above.

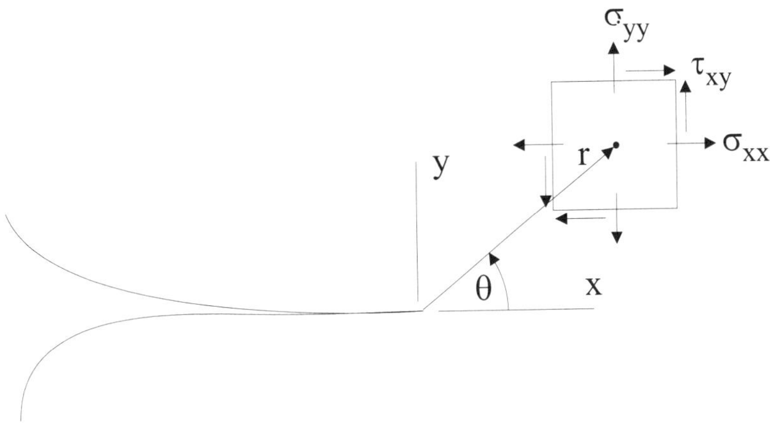

Figure 17.2: Crack tip coordinate system.

The stress intensity factor for this centre cracked plate is given by

$$K_I = \sigma\sqrt{\pi a} \tag{17.4}$$

and defines the magnitude of the crack tip stress field singularity

$$K_I = \lim_{r\to 0}\sqrt{2\pi r}\sigma_{yy}\Big|_{\theta=0^\circ} \tag{17.5}$$

The stress distribution at the tip of a crack for an arbitrary mode of loading and geometry of both body and crack would be difficult to analyse. As a consequence, the following three key modes of deformation are typically considered: i) crack faces pulled apart, ii) crack faces sheared parallel to the leading edge of the crack and iii) cracks faces sheared perpendicular to the leading edge of the crack. These three cases correspond to the three stress components indicated in Figure 17.3 below and are illustrated schematically below:

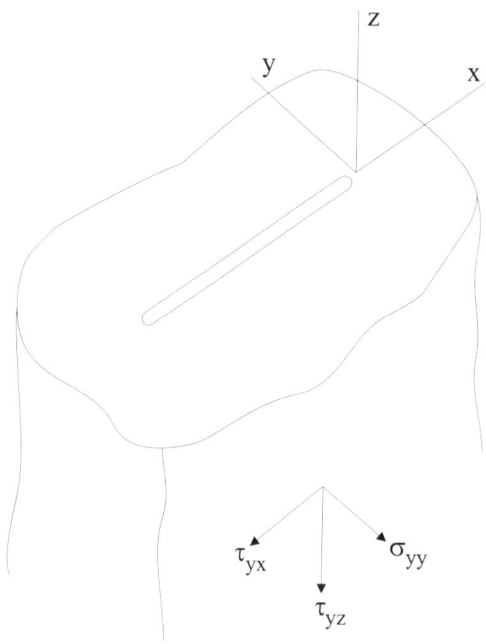

Figure 17.3: A crack in a finite solid.

i) Opening mode (mode I): symmetry about the (x, y) and (x, z) planes

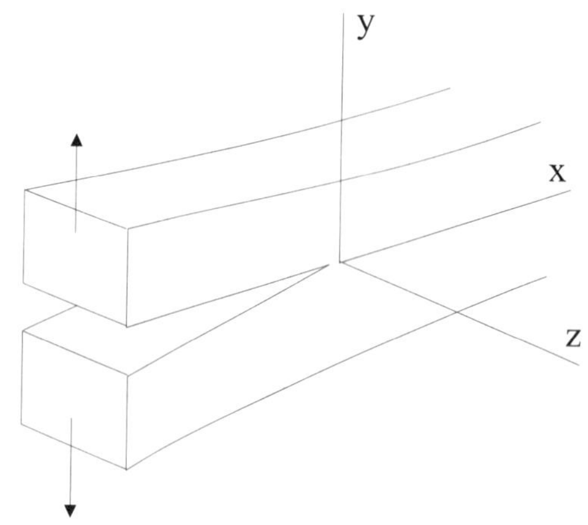

ii) Shear/sliding mode (mode II): anti-symmetry about the (x, z) plane and symmetry about the the (x, y) plane

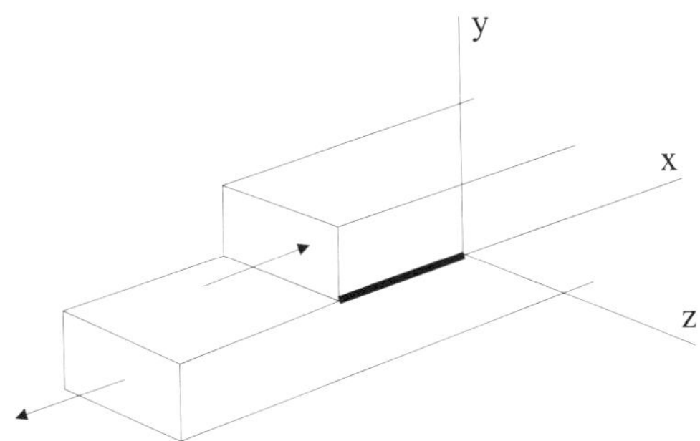

iii) Tearing mode (mode III): anti-symmetry about the (x, y) and (x, z) planes

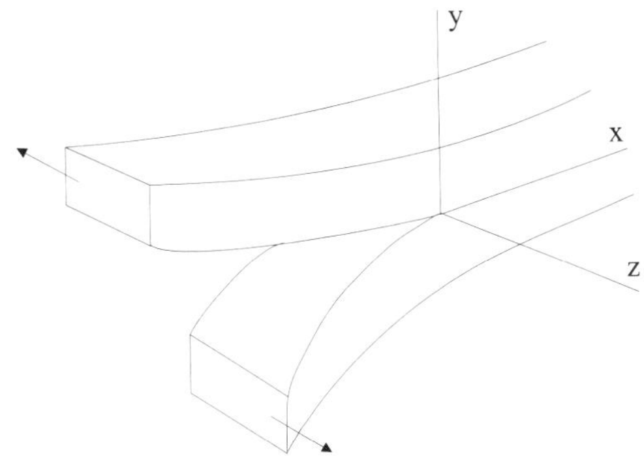

By analysing these three modes independently we can then superimpose them to determine the state of stress for more complex loadings. Note that the stress concentrations under tension and shear parallel to the leading edge produce maxima at the crack tip while shear normal to the leading edge gives a tension maxima on one side of the crack and compression maxima on the other side of the crack plane.

It can be shown that crack tip stresses and displacements for mode II loading are of an analogous form to those for mode I loading

$$\begin{Bmatrix} \sigma_{xx} \\ \sigma_{yy} \\ \tau_{xy} \end{Bmatrix} = \frac{K_{II}}{\sqrt{2\pi r}} \begin{Bmatrix} -\sin(\theta/2)\left[2+\cos(\theta/2)\cos(3\theta/2)\right] \\ \sin(\theta/2)\left[\cos(\theta/2)\cos(3\theta/2)\right] \\ \cos(\theta/2)\left[1-\sin(\theta/2)\sin(3\theta/2)\right] \end{Bmatrix} + \dots$$
$$\begin{Bmatrix} u \\ v \end{Bmatrix} = \frac{K_{II}}{2\mu}\sqrt{\frac{r}{2\pi}} \begin{Bmatrix} \sin(\theta/2)\left[\kappa+1+2\cos^2(\theta/2)\right] \\ -\cos(\theta/2)\left[\kappa-1-2\sin^2(\theta/2)\right] \end{Bmatrix} + \dots \tag{17.6}$$

with K_{II} defined as

$$K_{II} = \lim_{r\to 0} \sqrt{2\pi r}\, \tau_{xy}\Big|_{\theta=0^\circ} \tag{17.7}$$

Similarly, for mode III

$$\begin{Bmatrix} \tau_{xz} \\ \tau_{yz} \end{Bmatrix} = \frac{K_{III}}{\sqrt{2\pi r}} \begin{Bmatrix} -\sin(\theta/2) \\ \cos(\theta/2) \end{Bmatrix} + \dots$$
$$w = \frac{2K_{III}}{\mu}\sqrt{\frac{r}{2\pi}}\sin\frac{\theta}{2}; \quad u = v = 0 \tag{17.8}$$

with K_{III} defined as

$$K_{III} = \lim_{r\to 0} \sqrt{2\pi r}\, \tau_{zy}\Big|_{\theta=0^\circ} \tag{17.9}$$

There are occasions in which the stresses and displacements are required with respect to a polar coordinate system

$$\left\{\begin{matrix}\sigma_{rr}\\ \sigma_{\theta\theta}\\ \tau_{r\theta}\end{matrix}\right\} = \frac{K_I}{\sqrt{2\pi r}}\begin{bmatrix}\cos(\theta/2)\,(1+\sin^2(\theta/2))\\ \cos^3(\theta/2)\\ \sin(\theta/2)\cos^2(\theta/2)\end{bmatrix} + \frac{K_{II}}{\sqrt{2\pi r}}\begin{bmatrix}\sin(\theta/2)\,(1-3\sin^2(\theta/2))\\ -3\sin(\theta/2)\cos^2(\theta/2)\\ \cos(\theta/2)\,(1-3\sin^2(\theta/2))\end{bmatrix} + \ldots \tag{17.10}$$

with $\sigma_{zz} = 0$ for plane stress and $\sigma_{zz} = \nu(\sigma_{rr} + \sigma_{\theta\theta})$ for plane strain and the through-thickness shear stresses given by

$$\left\{\begin{matrix}\tau_{rz}\\ \tau_{\theta z}\end{matrix}\right\} = \frac{K_{III}}{\sqrt{2\pi r}}\begin{bmatrix}\sin(\theta/2)\\ \cos(\theta/2)\end{bmatrix} + \ldots \tag{17.11}$$

The displacements are given by

$$\begin{aligned}\left\{\begin{matrix}u_r\\ u_\theta\end{matrix}\right\} &= \frac{(1+\nu)}{2E}K_I\sqrt{\frac{r}{2\pi}}\begin{bmatrix}(2\kappa-1)\cos(\theta/2) - \cos(3\theta/2)\\ -(2\kappa+1)\cos(\theta/2) + \sin(3\theta/2)\end{bmatrix}\\ &+ \frac{(1+\nu)}{2E}K_{II}\sqrt{\frac{r}{2\pi}}\begin{bmatrix}-(2\kappa-1)\sin(\theta/2) + 3\sin(3\theta/2)\\ -(2\kappa+1)\cos(\theta/2) + 3\cos(3\theta/2)\end{bmatrix} + \ldots\end{aligned} \tag{17.12}$$

with u_z for plane stress given by

$$u_z = -\frac{\nu z}{E}(\sigma_{rr} + \sigma_{\theta\theta}) + \frac{K_{III}}{2E}\sqrt{\frac{r}{2\pi}}\,[2(1+\nu)\sin(\theta/2)] + \ldots \tag{17.13}$$

and for plane strain

$$u_z = \frac{K_{III}}{2E}\sqrt{\frac{r}{2\pi}}\,[2(1+\nu)\sin(\theta/2)] + \ldots \tag{17.14}$$

17.2.1 The Configuration Correction Factor

In the previous section we focused on the centre cracked plate configuration in which the mode I stress intensity factor is given by $K_I = \sigma\sqrt{\pi a}$. The stress intensity factor, however, is generally written $K_I = Y\sigma\sqrt{\pi a}$ where Y denotes a non-dimensional *configuration correction factor* which is a function of both specimen geometry and applied stress biaxiality. The use of Y helps generalise K_I since the term $\sigma\sqrt{\pi a}$ is characteristic for a variety of different configurations, as we will see shortly. For example, consider the finite plate containing an elliptical hole subject to biaxial loading shown in Figure 17.4. The stress intensity factor for this configuration is of the form

$$K_I = \sigma_P\sqrt{\pi a}\,Y_a\left\{\frac{a}{\alpha}, \frac{\alpha}{\beta}, \frac{\alpha}{W}, \frac{H}{W}, \lambda\right\} \tag{17.15}$$

noting that with $c = \alpha + a$ then $Y_a = Y_c\sqrt{c/a}$.

17.3 The T-stress

Let us recap on the previous discussion on the stress intensity factor and T-stress for an infinite centre cracked plate with remotely applied stresses σ_P and σ_Q acting normal and parallel to the crack plane

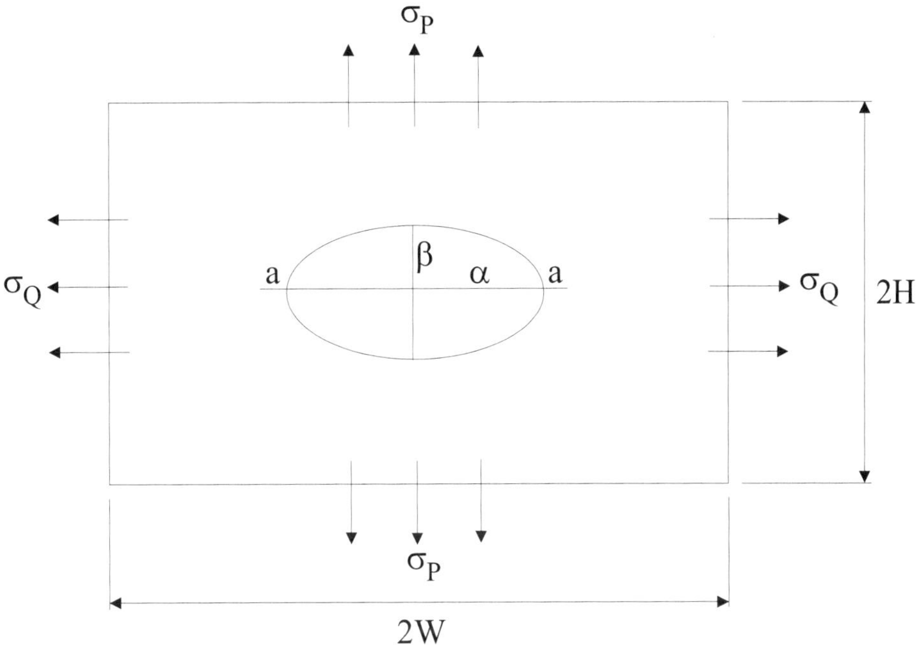

Figure 17.4: An elliptical hole in a finite plate of width $2W$ and height $2H$ subject to biaxial loading.

respectively

$$
\begin{aligned}
\sigma_{xx} &= \frac{K_I}{\sqrt{2\pi r}} f_{xx}(\theta) + T + O\left(r^{1/2}\right) \\
\sigma_{yy} &= \frac{K_I}{\sqrt{2\pi r}} f_{yy}(\theta) + O\left(r^{1/2}\right) \\
\tau_{xy} &= \frac{K_I}{\sqrt{2\pi r}} f_{xy}(\theta) + O\left(r^{1/2}\right) \\
u &= \frac{K_I}{\mu}\sqrt{\frac{r}{2\pi}} \cos\frac{\theta}{2}\left[\frac{1}{2}(\kappa-1) + \sin^2\frac{\theta}{2}\right] + \frac{T}{8\mu}(\kappa+1)(a + r\cos\theta) + O\left(r^{3/2}\right) \\
v &= \frac{K_I}{\mu}\sqrt{\frac{r}{2\pi}} \sin\frac{\theta}{2}\left[\frac{1}{2}(\kappa+1) - \cos^2\frac{\theta}{2}\right] - \frac{T}{8\mu}(3-\kappa) r\sin\theta + O\left(r^{3/2}\right)
\end{aligned}
\tag{17.16}
$$

For plane strain $\kappa = 3 - 4\nu$ and

$$
\begin{aligned}
\tau_{xz} &= \tau_{yz} = 0 \\
\sigma_{zz} &= \nu\left(\sigma_{xx} + \sigma_{yy}\right) = \frac{\nu K_I}{\sqrt{2\pi r}}\left[f_{xx}(\theta) + f_{yy}(\theta)\right] + S + O\left(r^{1/2}\right)
\end{aligned}
\tag{17.17}
$$

where

$$
K_I = Y\sigma\sqrt{\pi a}; \quad T = (\sigma_Q - \sigma_P) = (\lambda - 1)\sigma_P; \quad S = \nu T; \quad \lambda = \frac{\sigma_Q}{\sigma_P} \tag{17.18}
$$

The T-stress, T, is the first non-singular term in the series expansion of σ_{xx} and has the following properties:

- T is a non-singular stress and acts parallel to the crack plane.
- T is independent of r and is therefore constant over the crack tip region.
- T does not affect the crack tip singularity.
- T is directly proportional to the applied stresses.

With $K_I = Y\sigma_P\sqrt{\pi a}$ we observe that K_I is independent of biaxiality since Y=1 for all biaxial loads whereas the value of T for uniaxial, equibiaxial and shear loadings is different and summarised in Table 17.1. Table 17.1 illustrates that biaxiality effects can be modelled by considering both K_I and T but not

Loading	$\lambda(=\sigma_Q/\sigma_P)$	Y	T
uniaxial	0	1	$-\sigma_P$
equibiaxial	1	1	0
shear	-1	1	$-2\sigma_P$

Table 17.1: Comparison of Y and T for a centre cracked plate subject to uniaxial, equibiaxial and shear loadings.

with consideration of K_I alone.

Generally, the T-stress is expressed in terms of the B parameter (inherent stress biaxiality ratio) and is simply T normalised with respect to the nominal stress $\sigma = K_I/\sqrt{\pi a}$ associated with the centre cracked plate geometry

$$B = \frac{T}{\sigma} = \frac{T\sqrt{\pi a}}{K_I} \tag{17.19}$$

An approximate expression for T can be found by re-arranging the expression for u in terms of T given by (17.16), ignoring higher order terms

$$T = \frac{8\mu}{(\kappa+1)(a + r\cos\theta)}\left\{u - \frac{K_I}{\mu}\sqrt{\frac{r}{2\pi}}\cos\frac{\theta}{2}\left[\frac{1}{2}(\kappa-1) + \sin^2\frac{\theta}{2}\right]\right\} \tag{17.20}$$

In the particular case of $r = \theta = 0$ then (17.20) reduces to

$$T = \frac{8\mu}{(\kappa+1)}\left(\frac{u}{a}\right) \tag{17.21}$$

If the displacement u of the crack tip is determined via a method such as the finite element method then an approximate estimate for T follows. Note that for uniaxial loading, $\lambda = 0$, u is negative which leads to a negative T-stress as expected.

17.4 Energy Considerations of Cracks

Previous sections have demonstrated that the stresses at the tip of a crack are infinite. However, such infinitely sharp line cracks can be present within a body without resulting in component failure. Griffith (1921) addressed this problem by considering the change in energy as a crack extends. For a crack to extend then the following two conditions must be satisfied:

i) The stresses ahead of the crack must reach a critical value.

ii) The total energy of the system must be reduced by an incremental extension of the crack.

To determine the energy released per unit volume let us consider a single crack in an infinite plate subject to a far-field uniaxial tension as shown in Figure 17.5c). Stress free crack faces can be generated by introducing stresses on the faces which cancel out those along the crack line in the unflawed plate, see cases a) and b) in Figure 17.5. The difference between the strain energy of the uncracked and cracked

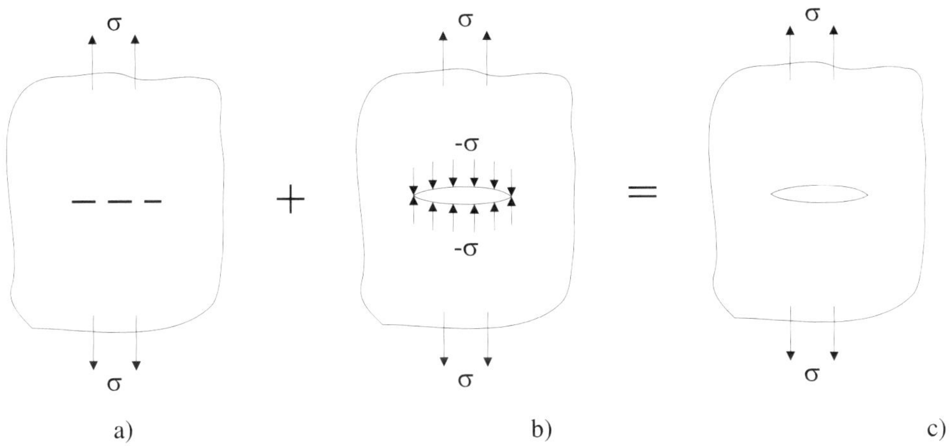

Figure 17.5: Superposition of a) an uncracked plate and b) a pressurised crack to give c) a cracked plate subject to a far-field stress.

plates is the strain energy of the pressurised crack which can be found from, (9.5)

$$U = \frac{1}{2}\int_a p(x)v(x,a)\,\mathrm{d}x \tag{17.22}$$

where $p(x)$ is the stress distribution along the crack line and $v(x,a)$ is the vertical displacement of the crack faces. Since we will be using an elastic displacement then the factor of $1/2$ is introduced with integration taken over both the lower and upper crack faces. From the displacement solution of an elliptical hole subject to a far-field uniaxial loading of σ and letting the hole degenerate into a line crack then the displacement is given by

$$v = \frac{\sigma}{2}\left(\frac{\kappa+1}{4\mu}\right)\sqrt{a^2-x^2} \tag{17.23}$$

Substituting v into (17.22) we have

$$U = \frac{1}{2}\int_{-a}^{a}(-\sigma)2\frac{\sigma}{2}\left(\frac{\kappa+1}{4\mu}\right)\sqrt{a^2-x^2}\,\mathrm{d}x = -\frac{\sigma^2}{2}\left(\frac{\kappa+1}{4\mu}\right)\int_{-a}^{a}\sqrt{a^2-x^2}\,\mathrm{d}x \tag{17.24}$$

where the additional factor of 2 has been introduced to take account of both the lower and upper crack faces. To evaluate the integral in (17.24) we make use of the standard integral

$$\int_{-a}^{a}\sqrt{a^2-x^2}\,\mathrm{d}x = \left[\frac{x}{2}\sqrt{a^2-x^2}+\frac{a^2}{2}\sin^{-1}\frac{x}{a}\right]_{-a}^{a} = \frac{\pi a^2}{2} \tag{17.25}$$

Thus, evaluating U in (17.24) we arrive at, dropping the negative sign

$$U = \left(\frac{\kappa + 1}{8\mu}\right) \frac{\sigma^2 \pi a^2}{2} \tag{17.26}$$

With $\mu = E/(2(1+\nu))$ and $\kappa = 3 - 4\nu$ for plane strain and $\kappa = (3-\nu)/(1+\nu)$ for plane stress then U is given by

$$U = \begin{cases} \dfrac{\sigma^2 \pi a^2}{E} & \text{for plane stress} \\ \left(\dfrac{1-\nu^2}{E}\right) \sigma^2 \pi a^2 & \text{for plane strain} \end{cases} \tag{17.27}$$

The variation of energy with crack length can be considered as consisting of three components: i) a surface energy, ii) an energy released and iii) the total energy; refer to Figure 17.6. Although energy is released as a result of the relaxation of the material due to the presence of the crack an energy input, W, is also required to extend the crack. Assuming that the energy required for each equal increment of crack length is constant then we have a linear increase in W with crack length. The surface energy W is designated as positive because it indicates an energy input and is employed in the creation of new crack surface whereas U is associated with the energy released due to crack extension. For a crack length within the region $0 \leq a \leq a_c$ an input of energy is required to the system. For $a > a_c$ then energy is released as a result of crack extension and the crack will propagate.

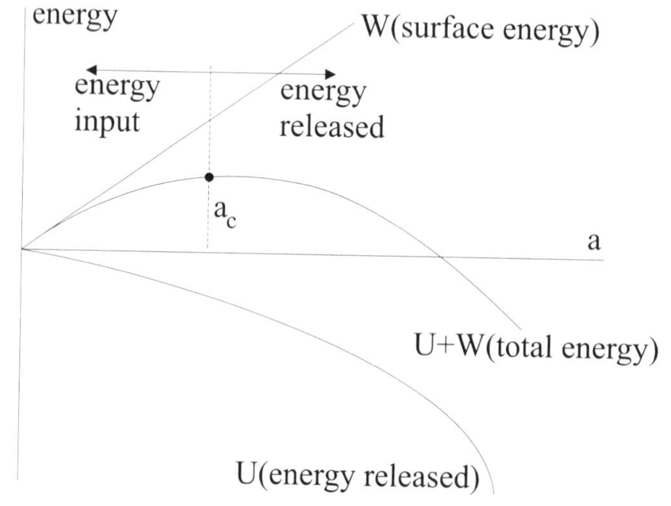

Figure 17.6: Surface energy, energy released and total energy as a function of crack length.

17.4.1 Griffith's Energy Release Rate

Crack stability-instability is associated with the stationary value a_c of the total energy, beyond which the energy released during an incremental extension of the crack exceeds the energy required to create new crack surface. The strain energy release rate is referred to as *Griffith's energy release rate* and is denoted by G and equal to, (17.27)

$$G = \frac{\partial U}{\partial a} = \begin{cases} \dfrac{\sigma^2 \pi a}{E} & \text{for plane stress} \\ \left(\dfrac{1-\nu^2}{E}\right) \sigma^2 \pi a & \text{for plane strain} \end{cases} \tag{17.28}$$

Let $G_{critical}$ be the value of G at which unstable crack propagation will occur $(a > a_c)$. Parameter $G_{critical}$ is a material property since it depends on the energy to create new surfaces of the material. From (17.28) we have

$$G_{critical} = \begin{cases} G_C = \dfrac{\sigma_c^2 \pi a}{E} & \text{for plane stress} \\ G_{IC} = \left(\dfrac{1-\nu^2}{E}\right) \sigma_c^2 \pi a & \text{for plane strain} \end{cases} \tag{17.29}$$

If the stress at extension, σ_c, is noted for a given geometry with a crack of length $2a$ then either G_C or G_{IC} can be determined for a given material.

Noting the similarity between G and K we have the following results

$$\begin{aligned} &\text{mode I:} \quad G_I = \frac{K_I^2}{E} \quad \text{for plane stress} \quad , \quad G_I = \left(\frac{1-\nu^2}{E}\right) K_I^2 \quad \text{for plane strain} \\ &\text{mode II:} \quad G_{II} = \frac{K_{II}^2}{E} \quad \text{for plane stress} \quad , \quad G_{II} = \left(\frac{1-\nu^2}{E}\right) K_{II}^2 \quad \text{for plane strain} \\ &\text{mode III:} \quad G_{III} = \left(\frac{1+\nu}{E}\right) K_{III}^2 = \frac{1}{2\mu} K_{III}^2 \end{aligned} \tag{17.30}$$

If the crack extends under combined loading then

$$G = G_I + G_{II} + G_{III} \tag{17.31}$$

When K_I, for example, attains its critical value then G_I must also reach its critical value and we have

$$G_{IC} = \left(\frac{1-\nu^2}{E}\right) K_{IC}^2 \tag{17.32}$$

From the above discussion we have seen that for linear elastic systems the stress intensity factor and the energy balance approaches to fracture mechanics are equivalent.

17.5 Some Well-Known Values for K and T

The following sections present both stress intensity factor and T-stress expressions (where available) and values for several well-known geometric and loading conditions.

17.5.1 Biaxially Loaded Centre Cracked Plate

Figure 17.7 illustrates a centre cracked plate geometry subject to biaxial loading for which we have

$$K_I = \sigma_P \sqrt{\pi a}; \quad K_{II} = 0; \quad T = (\lambda - 1)\sigma_P \tag{17.33}$$

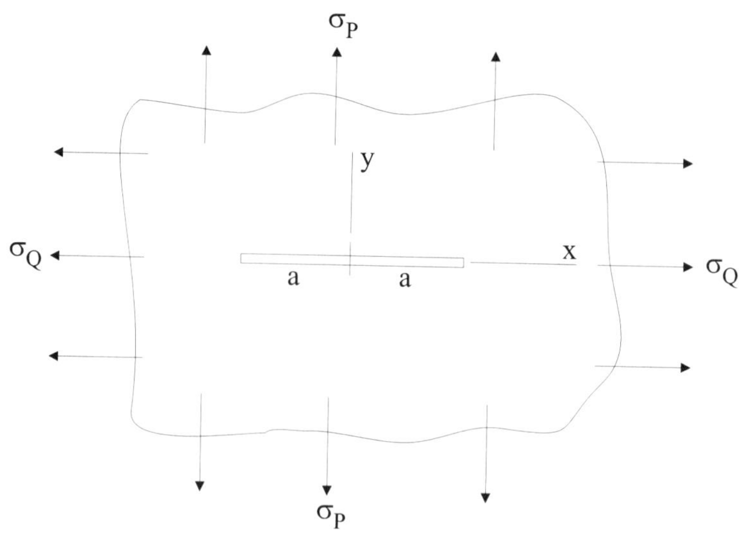

Figure 17.7: A centre cracked plate subject to biaxial loading.

17.5.2 Biaxially Loaded Centre Cracked Plate Containing an Inclined Crack

The biaxially-loaded centre cracked plate of Figure 17.8 contains an inclined crack with local coordinate system (x', y'). From the previous case we know that

$$K_I^{(x,y)} = \sigma_P\sqrt{\pi a}; \quad K_{II}^{(x,y)} = 0; \quad T^{(x,y)} = (\lambda - 1)\sigma_P \tag{17.34}$$

with respect to the global axes (x, y). To determine the stress intensity factors $K_I^{(x',y')}$ and $K_{II}^{(x',y')}$, and T-stress, $T^{(x',y')}$, with respect to the local set of axes (x', y') we make use of the stress transformation equations, (7.8), with $\sigma_{xx} = \sigma_Q$, $\sigma_{yy} = \sigma_P$ and $\tau_{xy} = 0$

$$\begin{aligned} \sigma_{x'x',y'y'} &= \left(\frac{\sigma_P + \sigma_Q}{2}\right) \pm \left(\frac{\sigma_P - \sigma_Q}{2}\right)\cos 2\theta = \frac{(\lambda+1)}{2}\sigma_P \pm \frac{\lambda - 1}{2}\sigma_P\cos 2\theta \\ \tau_{x'y'} &= -\left(\frac{\sigma_Q - \sigma_P}{2}\right)\sin 2\theta = -\frac{(\lambda-1)}{2}\sigma_P\sin 2\theta \end{aligned} \tag{17.35}$$

and substituting into the local stress intensity factors and T-stress

$$\begin{aligned} K_I^{(x',y')} &= \sigma_{y'y'}\sqrt{\pi a}\left[\frac{(\lambda+1)}{2}\sigma_P - \frac{(\lambda-1)}{2}\sigma_P\cos 2\theta\right]\sqrt{\pi a} \\ K_{II}^{(x',y')} &= \tau_{x'y'}\sqrt{\pi a} = -\frac{(\lambda-1)}{2}\sigma_P\sin 2\theta\sqrt{\pi a} \\ T^{(x',y')} &= \sigma_{x'x'} - \sigma_{y'y'} = (\lambda-1)\sigma_P\cos 2\theta \end{aligned} \tag{17.36}$$

Examining the specific case of $\lambda = 0$ these equations reduce to

$$K_I^{(x',y')} = \sigma_P\cos^2\theta\sqrt{\pi a}; \quad K_{II}^{(x',y')} = \sigma_P\sin\theta\cos\theta\sqrt{\pi a}; \quad T^{(x',y')} = -\sigma_P\cos 2\theta \tag{17.37}$$

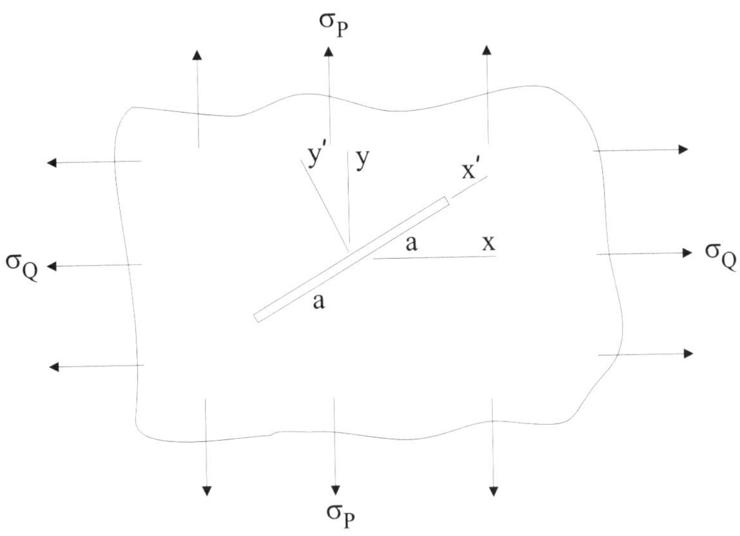

Figure 17.8: A centre cracked plate subject to biaxial loading and containing an inclined crack.

17.5.3 Point Loaded Crack

Figure 17.9 illustrates a centre cracked plate with point loaded crack faces. It can be shown that the mode I stress intensity factor for this configuration is

$$K_I = \frac{P}{\sqrt{\pi a}} \tag{17.38}$$

noting that K_I is proportional to $1/\sqrt{\pi a}$ and not $\sqrt{\pi a}$ as is the case for remotely loaded cracks. It may appear at first that K_I given by (17.38) is dimensionally incorrect but note that P is assumed to be an applied load per unit metre as illustrated by the inset in Figure 17.9. The more general case of point loads

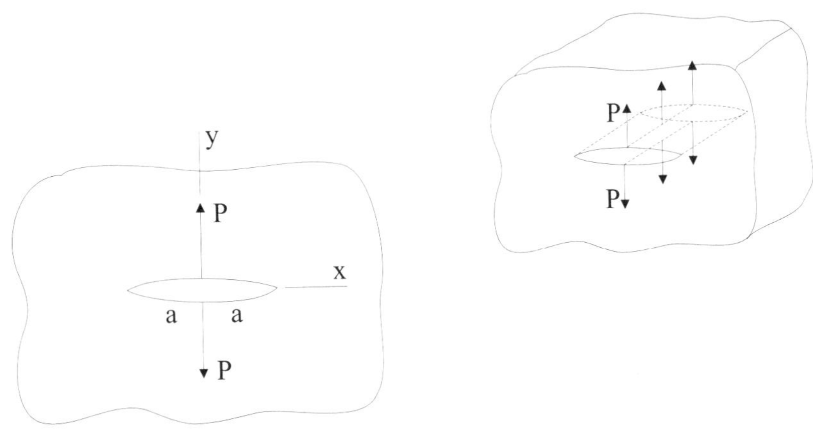

Figure 17.9: A point loaded crack.

P and Q acting at an arbitrary position, b, on one of the crack faces is also known and given by, Figure

17.10

$$K_I = \frac{P}{2\sqrt{\pi a}}\left(\frac{a+b}{a-b}\right)^{1/2} + \frac{Q}{2\sqrt{\pi a}}\left(\frac{\kappa-1}{\kappa+1}\right)^{1/2}$$
$$K_{II} = -\frac{P}{2\sqrt{\pi a}}\left(\frac{\kappa-1}{\kappa+1}\right)^{1/2} + \frac{Q}{2\sqrt{\pi a}}\left(\frac{a+b}{a-b}\right)^{1/2} \tag{17.39}$$

noting that the solution is obtained by conformally mapping from a unit circle in the logical ξ-plane to the physical z-plane. At first these expressions may appear to be of little practical value but they are in fact used frequently when prescribing a distributed pressure over the faces of a crack as a series of point loads.

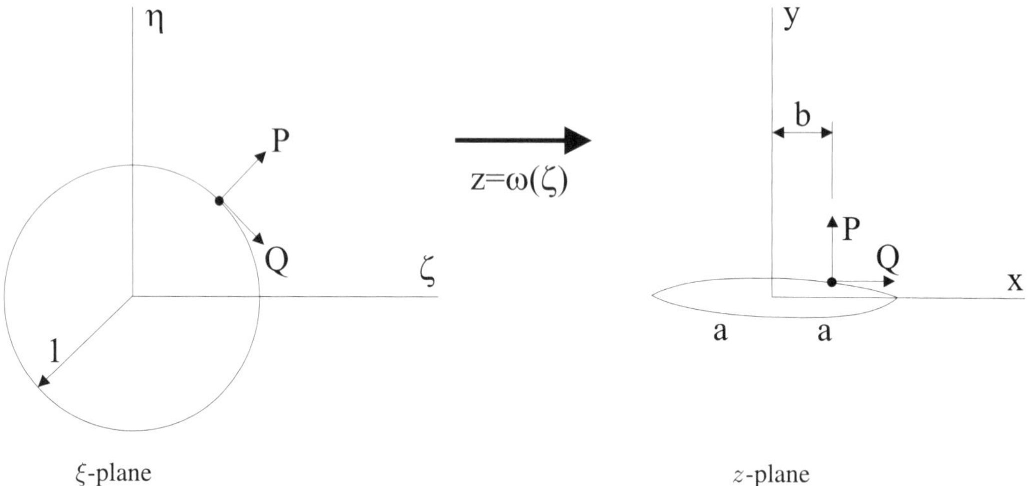

Figure 17.10: A point loaded crack.

17.5.4 Centre Cracked Plate with Opposing Point Forces

Figure 17.11 illustrates a centre cracked plate with opposing point loading at a distance y_0 from the crack line with a mode I stress intensity factor given by

$$K_I = P\sqrt{\frac{a}{\pi}}\left\{\frac{1}{\sqrt{a^2+y_0^2}} + \frac{y_0^2}{2(1-\nu)\left(a^2+y_0^2\right)^{3/2}}\right\}; \quad K_{II} = 0 \tag{17.40}$$

When $y_0 = 0$ then $K_I = P/\sqrt{\pi a}$ as expected.

17.5.5 Centre Cracked Plate of Finite Width

All of the configurations that we have seen so far have been idealised cases in which the boundary of the specimen geometry does not influence the crack tip stress field. Figure 17.12 illustrates a centre cracked plate of finite width $2W$ subject to a remote loading. The mode I stress intensity factor is given by

$$K_I = \sqrt{\sec\left(\frac{\pi a}{2W}\right)}\,\sigma\sqrt{\pi a} \approx \left[1 + 0.256\left(\frac{a}{W}\right) - 1.152\left(\frac{a}{W}\right)^2 + 12.2\left(\frac{a}{W}\right)^3 - \ldots\right]\sigma\sqrt{\pi a} \tag{17.41}$$

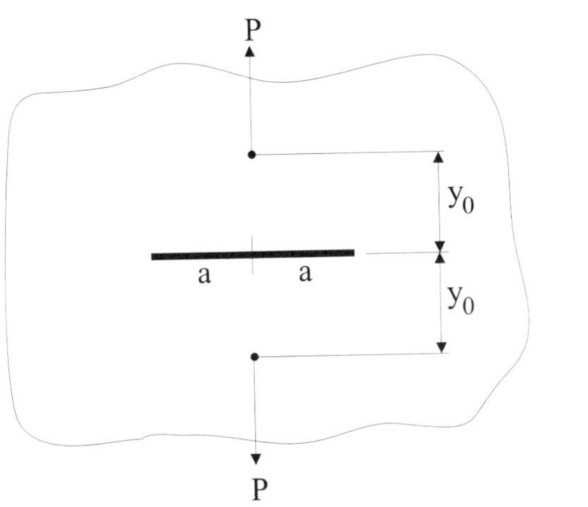

Figure 17.11: A centre cracked plate with opposing point loads.

If either $a \to 0$ or $W \to \infty$ then $Y = \sqrt{\sec(\pi a/2W)} \to 1$ and the solution returns to that of an infinite plate.

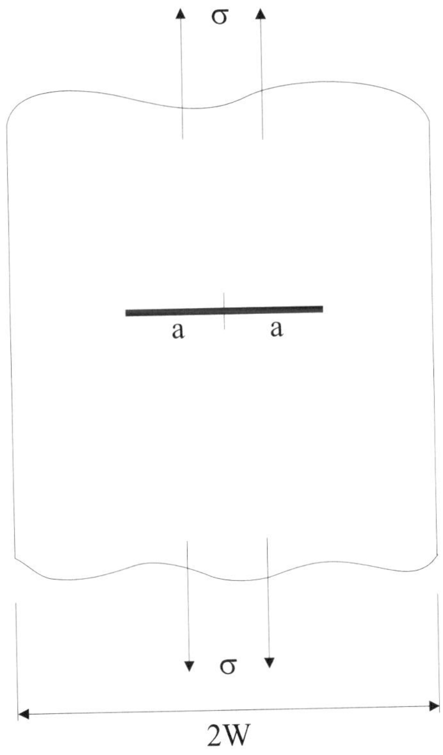

Figure 17.12: A centre cracked plate of finite width subject to remote loading.

17.5.6 Periodic Array of Cracks Subject to Equibiaxial Loading

For the periodic array of cracks subject to equibiaxial loading shown in Figure 17.13 the mode I stress intensity factor is given by

$$K_I = \sqrt{\frac{2b}{\pi a}\tan\left(\frac{\pi a}{2b}\right)}\,\sigma\sqrt{\pi a} \tag{17.42}$$

with the single crack solution obtained by letting $b \to \infty$ or $b/a \to \infty$

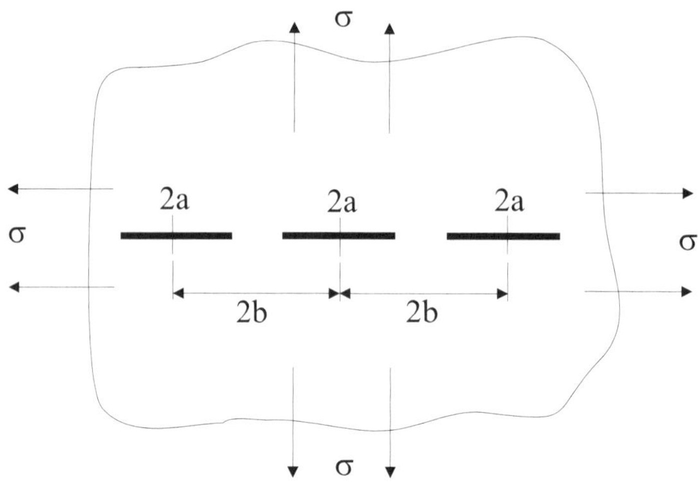

Figure 17.13: A periodic array of cracks each of length $2a$ separated by $2b$ subject to remote equibiaxial loading.

17.5.7 Single Edge Crack

Figure 17.14 illustrates an edge crack of length a in a semi-infinite plate subject to a far-field uniaxial tension σ. The mode I stress intensity factor and T-stress are approximately given by

$$K_I = 1.1215\sigma\sqrt{\pi a}; \quad T = -0.5258\sigma \tag{17.43}$$

observing that a 12% increase in K_I is experienced for the edge crack when compared to the infinite centre cracked plate whereas T is approximately half the value for the centre cracked plate.

If the edge crack is within a plate of finite width, W, and height, H, then account must be taken of the boundary effects a/W and H/W, see Figure 17.15a). For example, for $H/W = 2$ the configuration correction factor is given by the interpolating polynomial

$$Y = 1.1215 - 0.23\left(\frac{a}{W}\right) + 10.156\left(\frac{a}{W}\right)^2 - 21.74\left(\frac{a}{W}\right)^3 + 30.42\left(\frac{a}{W}\right)^4 - \dots \tag{17.44}$$

The variation of Y against a/W is shown is Figure 17.15b) noting that Y is greater for the finite plate than its semi-infinite counterpart for $a/W > 0$ and noting that as $a/W \to 0$ then $Y \to 1.1215$. Also shown in Figure 17.15b) is the semi-infinite case of $H/W = \infty$ which has been approximated as an edge correction of the finite width centre cracked plate case, (17.41)

$$K_I \approx 1.1215\sqrt{\sec\left(\frac{\pi a}{2W}\right)}\,\sigma\sqrt{\pi a} \tag{17.45}$$

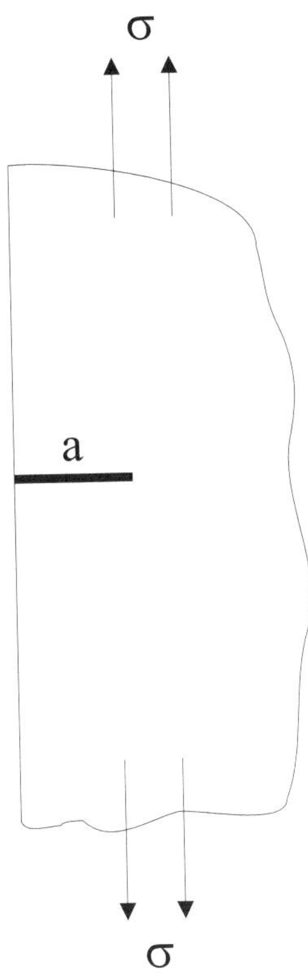

Figure 17.14: A semi-infinite edge cracked plate subject to uniaxial tension.

17.5.8 Double Edge Crack

Although not so popular, Figure 17.16 illustrates a double edge cracked plate of finite width for which the Y factor is

$$Y = 1.1215 + 0.43\left(\frac{a}{W}\right) - 4.79\left(\frac{a}{W}\right)^2 + 15.46\left(\frac{a}{W}\right)^3 - \dots \tag{17.46}$$

17.5.9 Cracked Pressure Vessels

Numerous stress intensity factor solutions have been developed for cracks in pressure vessels but we will examine two cases for which approximate expressions are known. The first case is that of a through-wall axial crack of length $2a$ in a vessel having diameter D and wall-thickness t subject to an internal pressure p, Figure 17.17a)

$$K \approx \frac{pD}{2t}\sqrt{\pi a}\sqrt{1 + 3.22\frac{a^2}{Dt}} \tag{17.47}$$

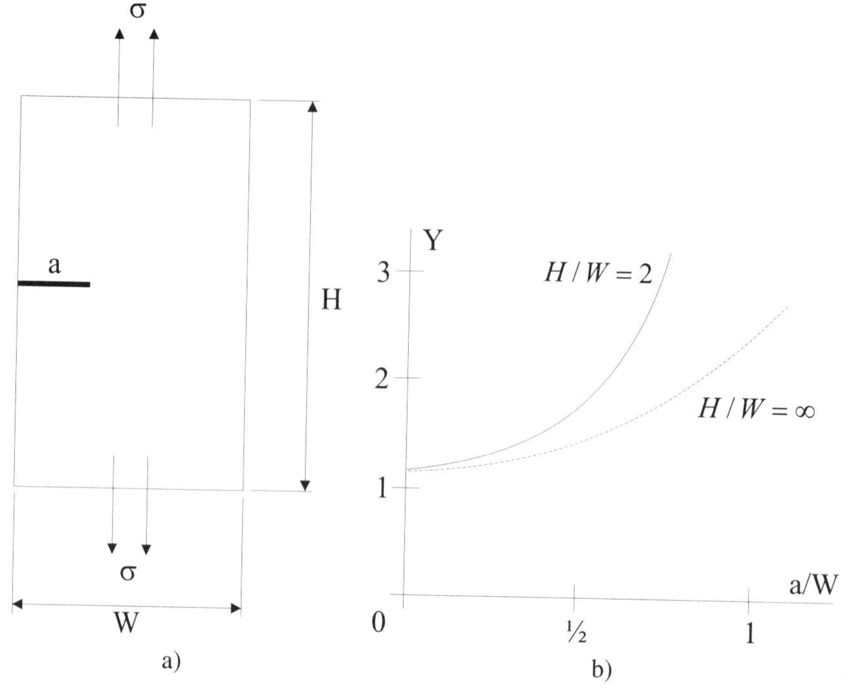

Figure 17.15: A finite edge cracked plate subject to uniaxial tension. a) Plate configuration. b) Variation of Y against a/W.

The second case is a through-wall circumferential crack of length $2a$ for which the mode I stress intensity factor is, Figure 17.17b)

$$K_I \approx \frac{pD}{4t}\sqrt{\pi a}\left\{1 + 1.1215\frac{a}{\sqrt{Dt}}\left[1 - \exp\left(-1.54\frac{a}{\sqrt{Dt}}\right)\right]\right\} \tag{17.48}$$

17.5.10 Symmetrically Cracked Elliptical and Circular Holes Subject to Remote Uniaxial Loading

Figure 17.18 is reproduced from Newman (1971) who performed an extensive study of elliptical holes containing cracks. The Figure illustrates the variation of Y_c against c/α for a crack of actual length a and effective crack length $c(= \alpha + a)$ where α and β are the half major and minor axes of the elliptical hole respectively. As $\beta \to 0$ the elliptical hole reduces to a crack of total length c and $Y_c \to 1$. As $\beta \to \infty$ the cracked elliptical hole reduces an edge crack in a semi-infinite plate with $Y_a \to 1.1215$. For large values of c/α then $Y_c \to 1$ for all α/β as the effect of the hole on the crack tip becomes negligible. For short cracks $(c/\alpha \to 1)$ then the cracks behave as if they are growing from a edge cracked geometry but with the far-field stress magnified by the stress concentration factor, K_t, of the hole. Thus, accounting for both the effects of K_t and edge-crack correction an approximate expression for K_I for short cracks growing from elliptical holes is

$$K_I \approx 1.1215 K_t \sigma_P \sqrt{\pi a} \tag{17.49}$$

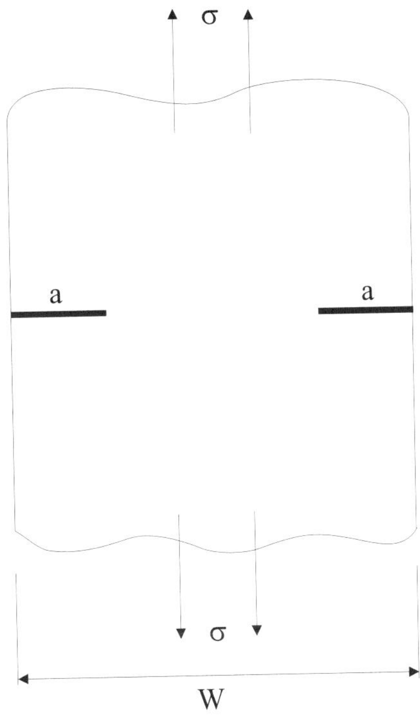

Figure 17.16: A double edge cracked plate of finite width subject to uniaxial tension.

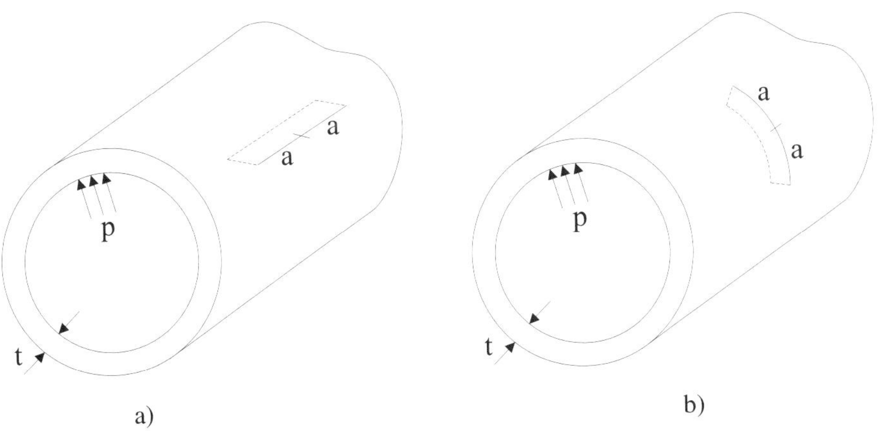

Figure 17.17: Cracks in pressure vessels. a) An axial crack. b) A circumferential crack.

with $K_t(= 1 + 2\alpha/\beta)$ from (14.212). Similarly, an expression for T follows

$$T \approx -0.5258 K_t \sigma_P \tag{17.50}$$

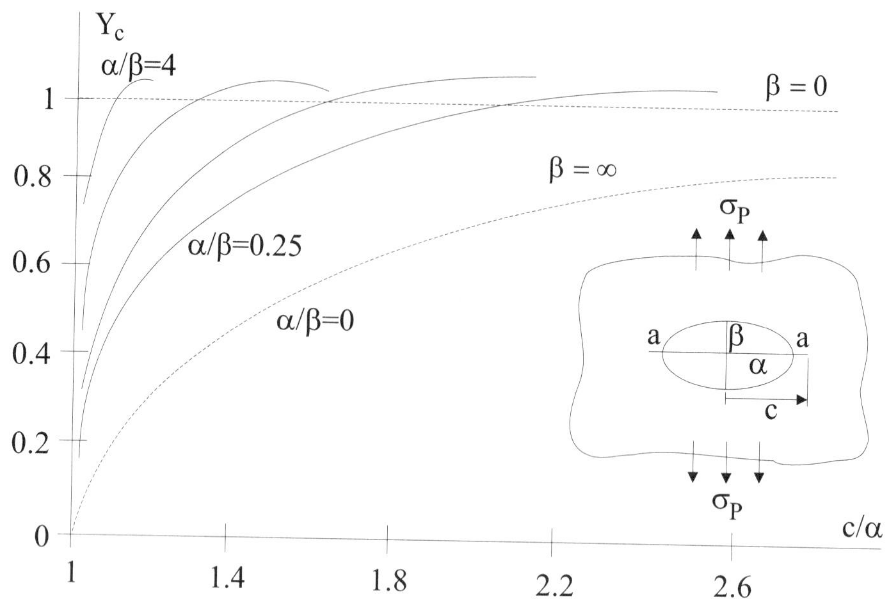

Figure 17.18: Stress intensity configuration correction factors for cracks emanating from an elliptical hole in an infinite plate subject to uniaxial loading.

In the case of a circular hole ($\alpha/\beta = 1$) then an approximate expression for K_I is known

$$K_I \approx \left[1 + 2.365 \left(\frac{\alpha}{\alpha + a} \right)^{2.4} \right] \sigma_P \sqrt{\pi a} \tag{17.51}$$

Comparing (17.49) and (17.51) for short cracks we find that (17.49) is equal to

$$K_I = 3.3645 \sigma_P \sqrt{\pi a} \tag{17.52}$$

whereas (17.51) is, with $((\alpha + a)/\alpha) \to 1$

$$K_I = 3.365 \sigma_P \sqrt{\pi a} \tag{17.53}$$

with the two approximations observed to be in close agreement.

17.6 Determination of Stress Intensity Factors

The previous section summarised several stress intensity factors for various geometries. In the present section we will examine three different methods for determining stress intensity factors.

17.6.1 Green's Functions

Section 17.5.3 examined a crack subject to point forces on one of its faces. Such general solutions may provide so-called *Green's functions* to solve the same geometric problem subject to arbitrary distributed loading. For instance, the stress intensity factors of (17.39) provide expressions for K_I and K_{II} for a crack subject to distributed tractions

$$K_I = \frac{1}{\sqrt{\pi a}} \int_{-a}^{a} \sigma_{yy}(x,0) \left(\frac{a+x}{a-x} \right)^{1/2} \mathrm{d}x$$
$$K_{II} = \frac{1}{\sqrt{\pi a}} \int_{-a}^{a} \tau_{xy}(x,0) \left(\frac{a+x}{a-x} \right)^{1/2} \mathrm{d}x \tag{17.54}$$

where σ_{yy} and τ_{xy} are the direct and shear stresses acting along the crack line in the absence of the crack. In general, the expression for, say, K_I in terms of the Green's function $G(x)$ is

$$K_I = \frac{1}{\sqrt{\pi a}} \int_a p(x) G(x)\, \mathrm{d}x \tag{17.55}$$

where $p(x)$ is the pressure acting normal to the crack faces and can found by a variety of methods such as the finite element method.

17.6.2 Bueckner Weight Function

H. F. Bueckner demonstrated that a function (termed the *Bueckner weight function*) is a property of a cracked geometry and is independent of applied loading. Therefore, a given weight function can be used in the derivation of additional stress intensity factors. Consider the cracked and uncracked bodies shown in Figure 17.19. The difference in elastic strain energy between the cracked and uncracked structures is

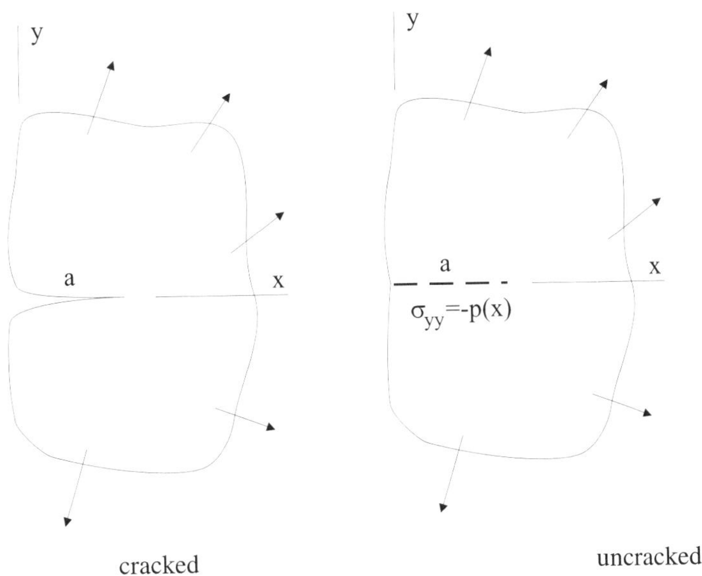

Figure 17.19: A cracked and uncracked body.

given by, (17.22)

$$U = \frac{1}{2}\int_a p(x)v(x,a)\,\mathrm{d}x \tag{17.56}$$

where $p(x)$ is the stress distribution along the x-axis in the uncracked structure and $v(x,a)$ is the displacement in the y-direction of the crack faces induced by the prescribed loading. Griffith's energy release rate is therefore, (17.28)

$$G = \frac{\partial U}{\partial a} = \frac{1}{2}\int_a p(x)\frac{\partial v(x,a)}{\partial a}\,\mathrm{d}x \tag{17.57}$$

in which it is assumed that $p(x)$ is not a function of a. Letting $E' = E$ for plane stress and $E' = E/(1-\nu^2)$ for plane strain then the relationship between K_I and G is given by

$$G = \frac{K_I^2}{E'} \tag{17.58}$$

Substituting (17.58) into (17.57) we have

$$K_I^2 = \int_a p(x)\frac{E'}{2}\frac{\partial v(x,a)}{\partial a}\,\mathrm{d}x \tag{17.59}$$

Alternatively, this can be written

$$K_I = \int_a p(x)m(x,a)\,\mathrm{d}x; \quad m(x,a) = \frac{E'}{2K_I}\frac{\partial v(x,a)}{\partial a} \tag{17.60}$$

where m is the Bueckner weight function. It can be shown that the Bueckner weight function is independent of loading and as a result additional stress intensity factor solutions, K_I^*, subject to crack-line loading $p^*(x)$ can be obtained

$$K_I^* = \int_a p^*(x)m(x,a)\,\mathrm{d}x \tag{17.61}$$

Let us conclude by noting that the crack face displacement can be determined by re-arranging m

$$\frac{\partial v(x,a)}{\partial a} = 2\frac{K_I}{E'}m(x,a) \tag{17.62}$$

17.6.3 The Compounding Method

The compounding method is a quick approximate technique for obtaining stress intensity factors in complex configurations from available solutions of simpler sub-systems. Cartwright and Rooke (1974) have shown that separate solutions can be *compounded* to produce a required solution, K_r, in accordance with

$$K_r = K_0 + \sum_{i=1}^{n}(K_i - K_0) + K_e \tag{17.63}$$

where K_0 is the stress intensity factor in the absence of all boundaries; for example $K_0 = \sigma\sqrt{\pi a}$ for remote loading and $K_0 = P/\sqrt{\pi a}$ for point loading. The ith. auxiliary configuration is denoted by K_i whereas K_e accounts for interaction effects between boundaries. Normalising (17.63) with respect to K_0 we have

$$\frac{K_r}{K_0} = 1 + \sum\left(\frac{K_i}{K_0} - 1\right) + \frac{K_e}{K_0} \tag{17.64}$$

The only unknown on the right hand side of (17.64) is K_e but for the majority of configurations in which auxiliary boundaries are sufficiently separated from each other relative to the crack size then neglecting K_e underestimates the compounded solution K_r by approximately 10%.

Example 17.1 Determination of the Bueckner weight function for a cracked body

Determine the Beuckner weight function for the configuration shown in Figure 17.20a) given that the stress intensity factor, K_I, and crack face displacement, v, are given by

$$K_I = \sigma\sqrt{\pi a}; \quad v = \left(\frac{\kappa+1}{4\mu}\right)\sigma\sqrt{a^2-x^2}; \quad -a \leq x \leq +a$$

Differentiating v with respect to a we have

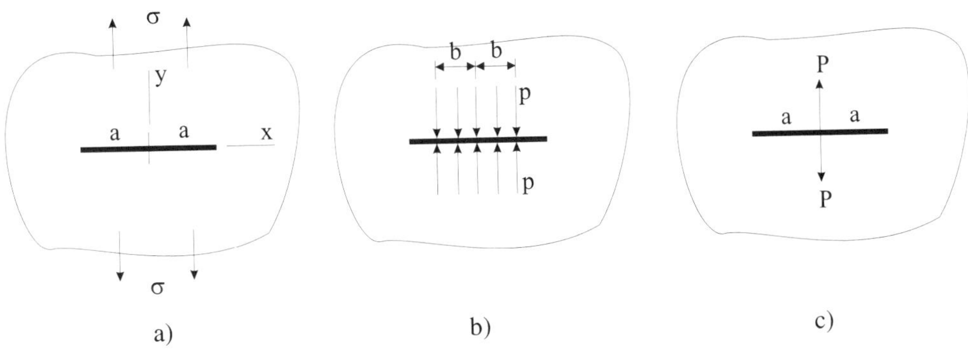

Figure 17.20: Example: A centre cracked plate. a) Subject to far-field loading σ. b) Subject to crack face distributed loading p over the interval $-b \leq x \leq +b$. c) Subject to crack face point loading P at $x = 0$.

$$\frac{\partial v}{\partial a} = \left(\frac{\kappa+1}{4\mu}\right)\frac{\sigma a}{\sqrt{a^2-x^2}}$$

From (17.60) the Beuckner weight function is therefore

$$m(x,a) = \frac{E'}{2K_I}\frac{\partial v(x,a)}{\partial a} = \frac{E'}{2\sigma\sqrt{\pi a}}\left(\frac{\kappa+1}{4\mu}\right)\frac{\sigma a}{\sqrt{a^2-x^2}} = \sqrt{\frac{a}{\pi}}\frac{1}{\sqrt{a^2-x^2}}$$

Let us now use this Bueckner weight function to determine an expression for the stress intensity factor for the configuration shown in Figure 17.20b) in which a uniformly distributed pressure, p, acts over the section $2b$ ($b \leq a$) on both the lower and upper crack faces. From (17.61) with $p^* = p$ then

$$K_I^*(p) = \int_a p^*(x)m(x,a)\,\mathrm{d}x = 4\int_0^b p\sqrt{\frac{a}{\pi}}\frac{1}{\sqrt{a^2-x^2}}\,\mathrm{d}x = 4p\sqrt{\frac{a}{\pi}}\sin^{-1}\left(\frac{b}{a}\right)$$

Finally, we can determine an expression for the stress intensity factor for the configuration of Figure 17.20c) by letting $P = 4bp$ and taking the limit $b \to 0$

$$K_I^*(P) = \lim_{b\to 0} 4\sqrt{\frac{a}{\pi}}\frac{P}{4b}\sin^{-1}\left(\frac{b}{a}\right) = \frac{P}{\sqrt{\pi a}}$$

and is seen to agree with (17.38).

Example 17.2 Determination of stress intensity factors using the compounding method

Figure 17.21a) illustrates a cracked geometry for which the stress intensity factors at tips A and B are required. By using the known stress intensity factors of the two sub-systems shown in cases b) and c) of Figure 17.21 we will use the compounding method to determine the stress intensity factors for case a). With the stress intensity factors for cases b) and c) given by

$$\left(\frac{K_b}{K_0}\right)_{tipA} = 1.34; \quad \left(\frac{K_b}{K_0}\right)_{tipB} = 1.34$$
$$\left(\frac{K_c}{K_0}\right)_{tipA} = 2.61; \quad \left(\frac{K_c}{K_0}\right)_{tipB} = 1.06$$

then the stress intensity factors for case a) are, (17.64)

$$\left(\frac{K_a}{K_0}\right)_{tipA} = 1 + (1.34 - 1) + (2.61 - 1) = 2.95$$
$$\left(\frac{K_a}{K_0}\right)_{tipB} = 1 + (1.34 - 1) + (1.06 - 1) = 1.42$$

neglecting interaction effects between boundaries.

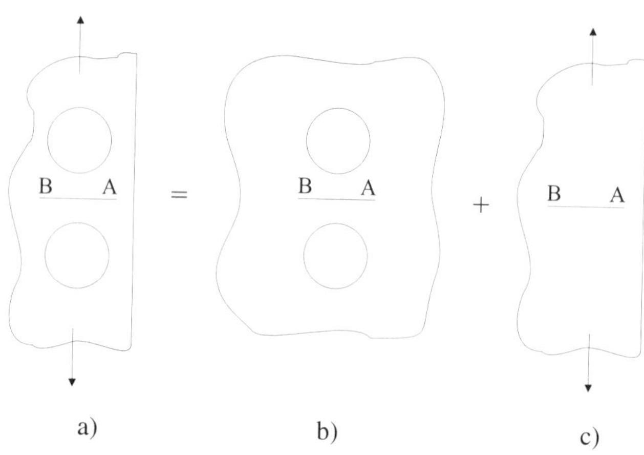

Figure 17.21: Example: Compounding method.

17.7 Crack Tip Plasticity

This section examines four popular small-scale crack tip plasticity models. The first model is Irwin's redistribution model. The second model makes use of the elastic crack tip stresses by re-arranging for plastic zone size for the Tresca and Huber-von Mises yield criteria and for both plane stress and plane strain. The final two models examine the Dugdale model with a variation due to Rice which accounts for the T-stress.

17.7.1 Irwin's Plastic Zone Model

Consider the variation of the σ_{yy} stress along $y = 0$ for a remotely loaded cracked plate in a state of plane stress, Figure 17.22a). Assume that the σ_{yy} stress exceeds the yield stress of the material, σ_Y, at some distance ahead of the crack tip, r^*. If we use r^* as the first estimate of the extent of the crack tip plastic zone then the force produced by the stress shown shaded in Figure 17.22a) acting over distance r^* will produce further yielding, that is stress redistribution. Thus, the whole curve must be shifted so as to obtain equilibrium which extends the plastic zone to a length of r_p, Figure 17.22b). Using the crack tip

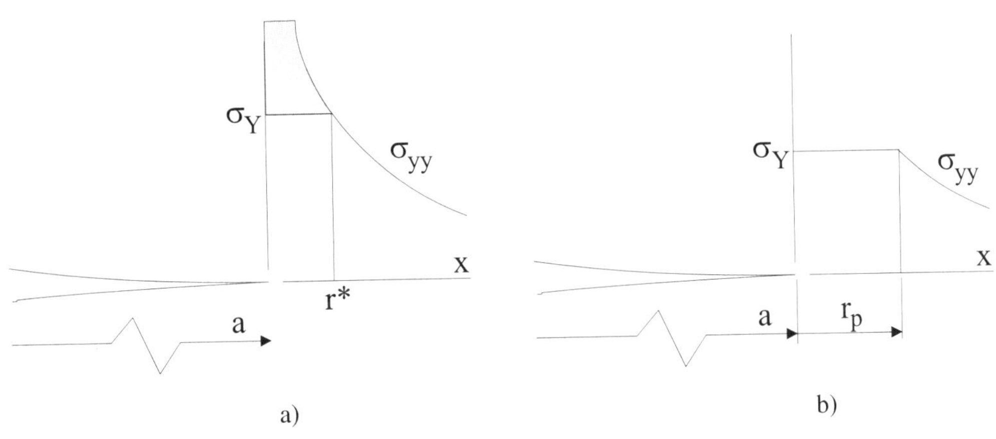

Figure 17.22: Irwin's crack tip plastic zone model. a) The stress σ_{yy} exceeds σ_Y for $x < r^*$. b) Redistribution of σ_{yy} for $x < r^*$ extending the crack tip plastic zone to r_p.

stress field (17.3) we have

$$\int_0^{r^*} \frac{K_I}{\sqrt{2\pi r}} \, \mathrm{d}r = r_p \sigma_Y \tag{17.65}$$

Performing the integration we find

$$\sqrt{\frac{2}{\pi}} K_I \sqrt{r^*} = r_P \sigma_Y \tag{17.66}$$

At $r = r^*$

$$\sigma_{yy} = \sigma_Y = \frac{K_I}{\sqrt{2\pi r^*}} \tag{17.67}$$

Re-arranging for r^* and substituting into (17.66) we arrive at the plastic zone size r_p

$$r_p = \frac{1}{\pi} \left(\frac{K_I}{\sigma_Y} \right)^2 \approx 0.32 \left(\frac{K_I}{\sigma_Y} \right)^2 \tag{17.68}$$

noting that the term $(K_I/\sigma_Y)^2$ is characteristic of all the crack tip plastic zone models we will discuss in the present section.

17.7.2 Crack Tip Plasticity based on the Crack Tip Stresses

For mode I loading the crack tip stresses are given by, neglecting T and higher order terms, (17.3)

$$\sigma_{ij} = \frac{K_I}{\sqrt{2\pi r}} \cos\frac{\theta}{2} f_{ij}(\theta) \tag{17.69}$$

Substituting these stresses into (7.19) we can determine the principal stresses σ_1 and σ_2

$$\sigma_{1,2} = \left(\frac{\sigma_{xx}+\sigma_{yy}}{2}\right) \pm \sqrt{\left(\frac{\sigma_{xx}+\sigma_{yy}}{2}\right)^2 + \tau_{xy}^2} = \frac{K_I}{\sqrt{2\pi r}} \cos\frac{\theta}{2}\left[1 \pm \sin\frac{\theta}{2}\right] \tag{17.70}$$

with the through-thickness stress, σ_{zz}, equal to zero for a state of plane stress and given by the following for a state of plane strain

$$\sigma_{zz} = \nu\left(\sigma_{xx}+\sigma_{yy}\right) = 2\nu\frac{K_I}{\sqrt{2\pi r}}\cos\frac{\theta}{2} \tag{17.71}$$

In the following we will consider both the Tresca and Huber-von Mises yield criteria separately.

17.7.2.1 Tresca Yield Criteria

The Tresca yield criterion predicts that yielding commences when the maximum shear stress reaches a critical value, (12.28)

$$|\sigma_1 - \sigma_2| = \sigma_Y \tag{17.72}$$

with $\sigma_3 = 0$ for plane deformation. Substituting σ_1 and σ_2 from (17.70) and re-arranging for $r = r_p$ then

$$\begin{aligned} r_p &= \left(\frac{K_I}{\sigma_Y}\right)^2 \frac{1}{2\pi}\cos^2\frac{\theta}{2}\left[1+\sin\frac{\theta}{2}\right]^2 && \text{for plane stress} \\ r_p &= \left(\frac{K_I}{\sigma_Y}\right)^2 \frac{1}{2\pi}\cos^2\frac{\theta}{2}\left[1-2\nu+\sin\frac{\theta}{2}\right]^2 && \text{for plane strain} \end{aligned} \tag{17.73}$$

17.7.2.2 Huber-von Mises Yield Criteria

The Huber-von Mises yield criterion postulates that yielding commences when the shear strain energy reaches a critical value, (12.46)

$$\sigma_1^2 - \sigma_1\sigma_2 + \sigma_2^2 = \sigma_Y^2 \tag{17.74}$$

with $\sigma_3 = 0$ for plane deformation. Substituting σ_1 and σ_2 from (17.70) and re-arranging for $r = r_p$ gives

$$\begin{aligned} r_p &= \left(\frac{K_I}{\sigma_Y}\right)^2 \frac{1}{4\pi}\left[1+\frac{3}{2}\sin^2\theta+\cos\theta\right] && \text{for plane stress} \\ r_p &= \left(\frac{K_I}{\sigma_Y}\right)^2 \frac{1}{4\pi}\left[\frac{3}{2}\sin^2\theta+(1-2\nu)^2(1+\cos\theta)\right] && \text{for plane strain} \end{aligned} \tag{17.75}$$

Both of the above estimates of r_p for the Tresca and Huber-von Mises criteria are illustrated in Figure 17.23 for both plane stress and plane strain. The dimensions of the plastic zone due to the Tresca yield criterion are greater than or equal to that of the Huber-von Mises criterion. Note also that the plastic zones are significantly greater for a state of plane stress than for plane strain. Finally, we will conclude by schematically illustrating the variation of crack tip plasticity through the thickness of a specimen, Figure 17.24. The plastic zone is greatest at the surface where plane stress conditions prevail due to the reduced constraint.

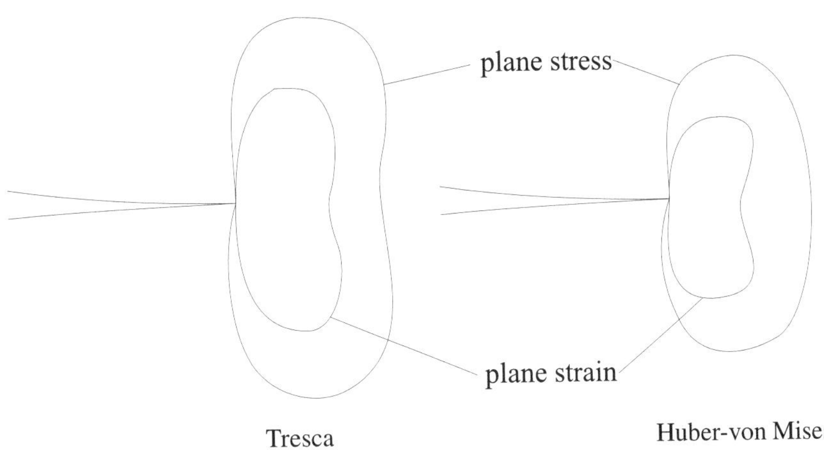

Figure 17.23: Variation of crack tip plasticity for the Tresca and Huber-von Mises yield criteria for plane stress and plane strain.

17.7.3 The Dugdale-Barenblatt Yield Model

D. S. Dugdale modelled crack tip plasticity ahead of a crack tip as a yielded strip in a non-hardening elastic-perfectly plastic material so that $\sigma_{yy} = \sigma_Y$ within the strip; where σ_Y is the tensile yield stress. Dugdale postulated that the effect of yielding is effectively to increase the crack length by the extent of the plastic zone to a total length of $2c$, Figure 17.25. Within the strip $a < |x| < c$ the opening of the crack faces is restrained by the stress $\sigma_{yy} = \sigma_Y$. The length of the strip is established from the condition that the stress field at the crack tip must be non-singular due to the presence of plasticity. The cohesive resisting force at a point b for an elemental strip of length δb within the interval $a \leq b \leq c$ is $\delta P = \sigma_Y \delta b$. To determine the stress intensity factor K_I we require the solution of a point loaded crack, Figure 17.26, for which

$$K_I = P\sqrt{\frac{a}{(a-b)(a+b)}} \tag{17.76}$$

Integrating over the interval $a \leq b \leq c$ we have

$$K_I = 2\sigma_Y \int_a^c \sqrt{\frac{c}{\pi(c-b)(c+b)}}\, db = 2\sigma_Y \sqrt{\frac{c}{\pi}} \left[\cos^{-1}\left(\frac{b}{c}\right)\right]_a^c = 2\sigma_Y \sqrt{\frac{c}{\pi}} \cos^{-1}\left(\frac{a}{c}\right) \tag{17.77}$$

where the multiplicative factor of 2 accounts for both the lower and upper crack faces. This stress intensity factor must be equivalent to the stress intensity factor for a crack of length $2c$

$$\sigma\sqrt{\pi c} = 2\sigma_Y \sqrt{\frac{c}{\pi}} \cos^{-1}\left(\frac{a}{c}\right) \tag{17.78}$$

which leads to

$$\frac{a}{c} = \cos\left(\frac{\pi\sigma}{2\sigma_Y}\right) \tag{17.79}$$

With $d = c - a$ and $c = a\sec(\pi\sigma/2\sigma_Y)$ from (17.79) then d is given by

$$d = c - a = a\left[\sec\left(\frac{\pi\sigma}{2\sigma_Y}\right) - 1\right] \tag{17.80}$$

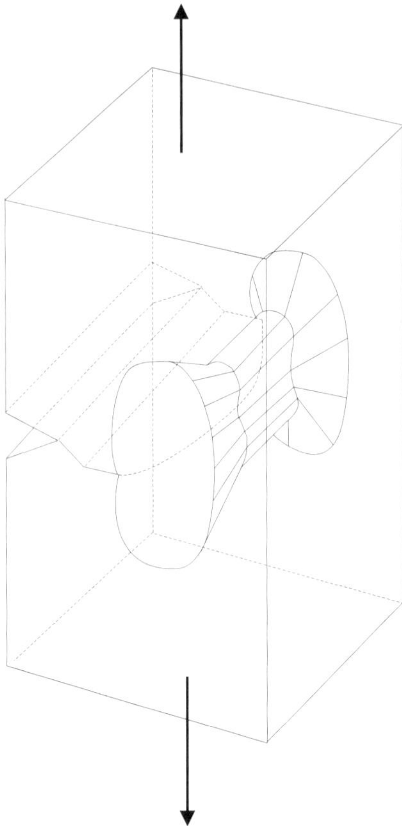

Figure 17.24: Variation of crack tip plasticity through the thickness of a specimen.

We can arrive at an expression for d in terms of K_I by performing a series expansion of the $\sec(\pi\sigma/2\sigma_Y)$ term. In general

$$\sec x = 1 + \frac{x^2}{2!} + 5\frac{x^4}{4!} + 61\frac{x^6}{6!} + \dots \tag{17.81}$$

so that

$$\sec\left(\frac{\pi\sigma}{2\sigma_Y}\right) = 1 + \frac{1}{2}\left(\frac{\pi\sigma}{2\sigma_Y}\right)^2 + \dots = 1 + \frac{\pi^2\sigma^2}{8\sigma_Y^2} + \dots \tag{17.82}$$

Considering only the first two terms

$$d = a\left[\left(1 + \frac{\pi^2\sigma^2}{8\sigma_Y^2}\right) - 1\right] = \frac{\sigma^2\pi^2 a}{8\sigma_Y^2} \tag{17.83}$$

With $K_I = \sigma\sqrt{\pi a}$ then

$$d = \frac{\pi}{8}\left(\frac{K_I}{\sigma_Y}\right)^2 \approx 0.39\left(\frac{K_I}{\sigma_Y}\right)^2 \tag{17.84}$$

and can be compared to Irwin's plastic zone model, (17.68).

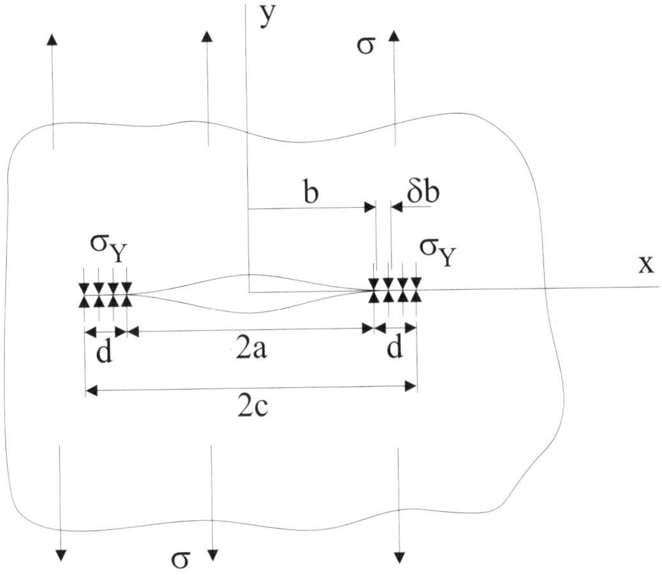

Figure 17.25: The Dugdale crack tip yield model.

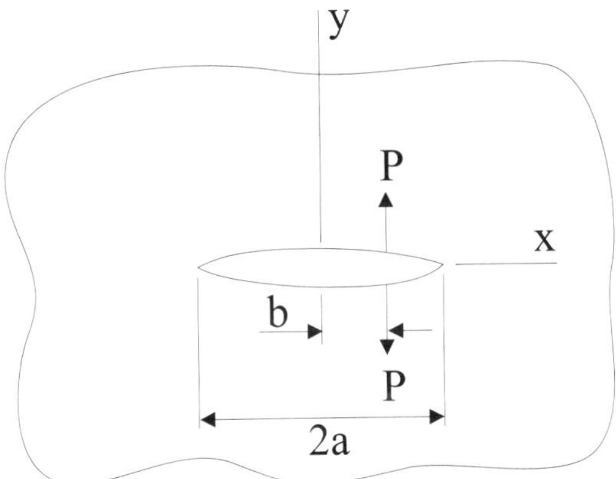

Figure 17.26: A point loaded crack.

The crack tip opening displacement, δ_{tip}, can be shown to be

$$\delta_{tip} = \frac{8a\sigma_Y}{\pi E} \ln\left[\sec\left(\frac{\pi\sigma}{2\sigma_Y}\right)\right] = \frac{K_I^2}{E\sigma_Y}\left[1 + \frac{\pi^2\sigma^2}{24\sigma_Y} + \dots\right] \tag{17.85}$$

and agrees with the linear elastic estimate of δ_{tip} for small scale yielding, $\sigma \leq \sigma_Y$.

17.7.4 K_I and T Characterisation of Crack Tip Plasticity

Rice (1974) developed a plane strain crack tip yield model which makes use of Dugdale's model resolved along two discrete shear bands inclined at an angle θ to the crack plane, Figure 17.27. By comparison of

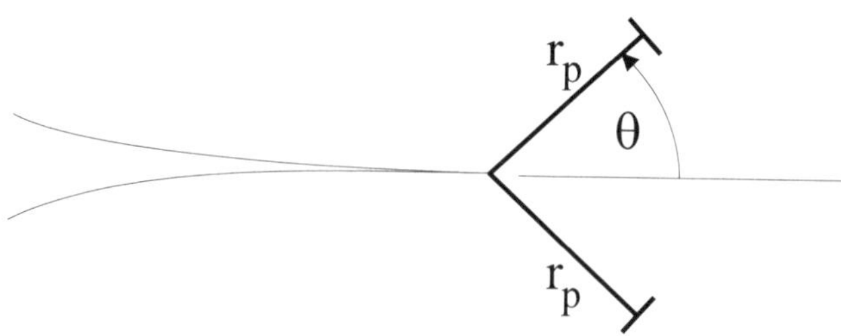

Figure 17.27: A crack tip with plastic zones along shear bands at angle θ to the crack plane.

the mode II shear stress intensity factor along the yield bands (of length r_p and yield stress in pure shear $k = \sigma_Y/\sqrt{3}$) with the analogous mode I stress intensity factor, Rice obtained the expression

$$r_p(\theta, K_I, T) = \frac{\pi}{64}\sin^2\theta\,[1+\cos\theta]\,\frac{K_I^2}{[k+T\sin\theta\cos\theta]^2} \tag{17.86}$$

Equation (17.86) illustrates that a negative T-stress increases the plastic zone size whereas a positive T-stress$z$ decreases the plastic-zone size.

Neglecting the T-stress then the value of θ which maximises r_p (i.e. θ^*) is obtained by differentiating (17.86). The solution is that $\cos\theta^* = 1/3$ or $\theta^* = 70.53°$ in which case (17.86) is given by

$$r_p(\theta^*, K_I) = \frac{\pi}{18}\left(\frac{K_I}{\sqrt{3}k}\right)^2 \approx 0.17\left(\frac{K_I}{\sqrt{3}k}\right)^2 \tag{17.87}$$

However, when T is considered θ^* varies with T/k for a given K_I and (17.86) requires numerical evaluation. As a comparison, assuming that θ^* remains constant at the value $\theta^* = 70.53°$ for $T = 0$ then we can compare the size of r_p for the three cases of uniaxial ($\lambda = 0$), equibiaxial ($\lambda = 1$) and pure shear ($\lambda = -1$) loading for a centrally cracked plate; where λ is the biaxiality parameter. The results of this comparison are shown in Table 17.2 and, as mentioned above, a negative T-stress is seen to increase crack tip plastic zone sizes.

λ	T/σ	$r_p(\theta^*, K_I, T)/r_p(\theta^*, K_I)$
0	-1	1.41
1	0	1
-1	-2	2.13

Table 17.2: Comparison of plastic zone sizes with and without the T-stress for uniaxial, equibiaxial and pure shear loadings.

17.8 Fracture Criteria and Plane Strain Fracture Toughness

In general engineering materials can be categorised as failing in either a brittle or ductile manner with a transitional type of failure in between. A highly ductile material will continuously neck whereas a highly brittle material will experience little deformation before failing and produce a macroscopically flat failure surface. The most common type of failure is one in which some necking and deformation is experienced with a brittle type of fracture. A similar behaviour is experienced with the failure of plates with through-thickness cracks since there is a pronounced dependence of the fracture toughness upon the degree of constraint at the crack tip. Constraint at the tips of cracks in through-thickness cracked plates primarily manifests itself via the plate thickness with typical experimentally observed variations shown schematically in Figure 17.28. The critical stress intensity factor, K_C, attains a minimum when the plate

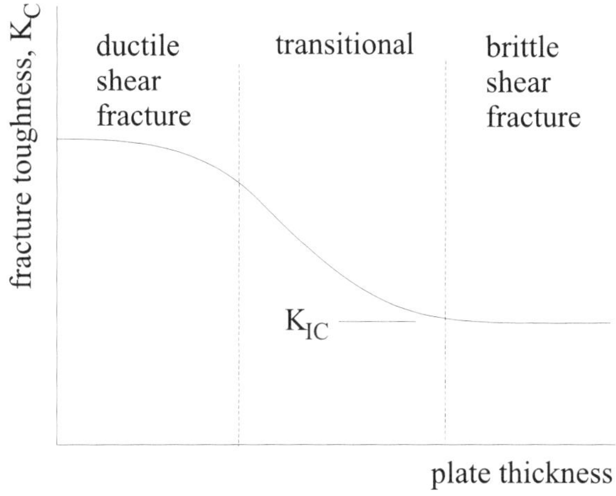

Figure 17.28: Variation of fracture toughness against plate thickness.

is sufficiently thick and is designated by K_{IC}. Thus, K_{IC} is geometry independent and is referred to as the *plane strain fracture toughness*. From a design perspective the plane strain fracture toughness should be used since it is the minimum fracture toughness of a given material. Figure 17.29 illustrates the propagation of a crack through a moderately thick plate and serves to illustrate that propagation is fastest in the centre of the wall thickness where plane strain conditions prevail.

Figure 17.30 illustrates typical ranges of fracture toughness for aluminium, titanium and steel alloys and demonstrates the typically observed inverse relationship between yield strength and fracture toughness.

When we discussed crack tip plasticity in §17.7 we obtained the following expression for small-scale crack tip plastic zone size, r_p

$$r_p = \alpha \left(\frac{K_I}{\sigma_Y} \right)^2 \tag{17.88}$$

where $\alpha = 1/\pi$ for Irwin's model, (17.68), and $\alpha = \pi/8$ for Dugdale's model, (17.84). Provided that $r_p \ll a$ then, as in Dugdale's model, an effective crack length of $2(a + r_p)$ can be used instead of a. At fracture $\sigma = \sigma_f$ and $K_I = K_{IC}$ where K_{IC} is given by, for a centre cracked plate

$$K_{IC} = \sigma_f \sqrt{\pi(a + r_p)} \tag{17.89}$$

with

$$r_p = \alpha \left(\frac{K_{IC}}{\sigma_Y} \right)^2 = \alpha \left(\frac{\sigma_f \sqrt{\pi a}}{\sigma_Y} \right)^2 = \alpha \pi a \left(\frac{\sigma_f}{\sigma_Y} \right)^2 \tag{17.90}$$

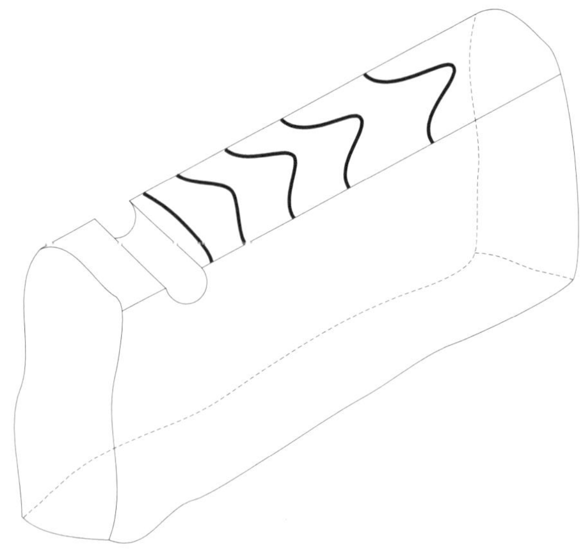

Figure 17.29: Crack propagation through the thickness of a plate.

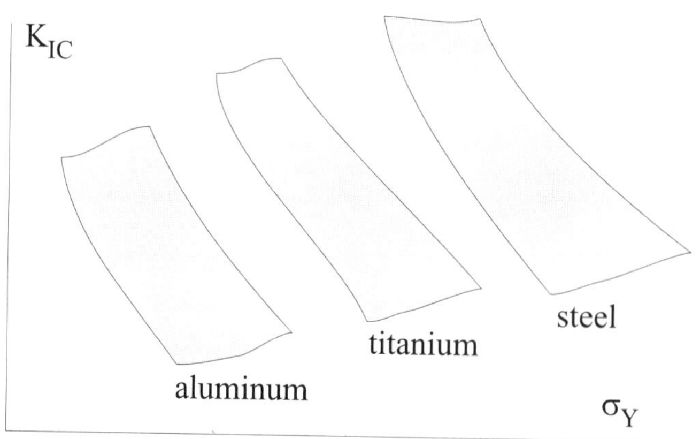

Figure 17.30: Plane strain fracture toughness against yield strength for aluminium, titanium and steel alloys.

Substituting r_p into (17.89) we have

$$K_{IC} = \sigma_f \sqrt{\pi a \left[1 + \alpha \pi \left(\frac{\sigma_f}{\sigma_Y}\right)^2\right]} \tag{17.91}$$

For brittle materials $\sigma_Y \gg \sigma_f$ and the influence of crack tip plasticity on K_{IC} is negligible. For $\sigma_f/\sigma_Y \leq 0.5$ the influence of crack tip plasticity is less than 10% on K_{IC}.

A convenient expression which relates plane strain fracture toughness, yield stress and plate thickness is the following semi-empirical result due to Irwin

$$K_C = K_{IC}\left[1 + \frac{1.4}{B^2}\left(\frac{K_{IC}}{\sigma_Y}\right)^4\right]^{1/2} \tag{17.92}$$

where B is the plate thickness and σ_Y is the yield stress of the material.

17.8.1 Elastic-Plastic Failure Criterion based on the Dugdale Model

Within the context of the Dugdale model an elastic-plastic fracture criterion can be expressed in terms of a critical value of the crack tip opening displacement, δ_{tip}. At initiation of crack growth

$$\delta_{tip} = \delta_c \tag{17.93}$$

where δ_c denotes the critical value of δ_{tip} and is considered to be a material property. It will be shown later in §17.12.3 that the J-integral for the Dugdale model is given by

$$J = \sigma_Y \delta_{tip} = \frac{8\sigma_Y^2 a}{\pi E} \ln\left[\sec\left(\frac{\pi\sigma}{2\sigma_Y}\right)\right] \tag{17.94}$$

An equivalent failure criterion can be expressed when J reaches a critical value, J_c

$$J = J_c \tag{17.95}$$

where J_c is assumed to be a material property. With $K_I = \sigma\sqrt{\pi a}$ for linear elastic materials or small-scale yielding then, for plane stress

$$J_{ssy} = \frac{K_I^2}{E} = \frac{\sigma^2 \pi a}{E} \tag{17.96}$$

and in view of (17.94)

$$\frac{J}{J_{ssy}} = \frac{8}{\pi^2}\left(\frac{\sigma_Y}{\sigma}\right)^2 \ln\left[\sec\left(\frac{\pi\sigma}{2\sigma_Y}\right)\right] \tag{17.97}$$

At fracture $\sigma = \sigma_f$ and $J = J_c$

$$\left(\frac{J_{ssy}}{J_c}\right)_f = \left\{\frac{8\sigma_Y^2}{\pi^2\sigma_f^2} \ln\left[\sec\left(\frac{\pi\sigma_f}{2\sigma_Y}\right)\right]\right\}^{-1} \tag{17.98}$$

For small-scale yielding we can also write $J_c = K_c^2/E$

$$\left(\frac{J_{ssy}}{J_c}\right)_f^{1/2} = \left(\frac{K_I}{K_c}\right)_f = \frac{\sigma_f}{\sigma_Y}\left\{\frac{8}{\pi^2} \ln\left[\sec\left(\frac{\pi\sigma_f}{2\sigma_Y}\right)\right]\right\}^{-1/2} \tag{17.99}$$

and can be used to estimate K_c if σ_f is measured.

17.8.2 Measurement of Plane Strain Fracture Toughness

Several different specimens can be used to measure K_{IC}, a few of which are shown in Figure 17.31. Both the centre cracked and edge cracked plates suffer from the property that a large plate thickness is required in order to achieve K_{IC} values. The two most popular types of specimen are the three-point bend and

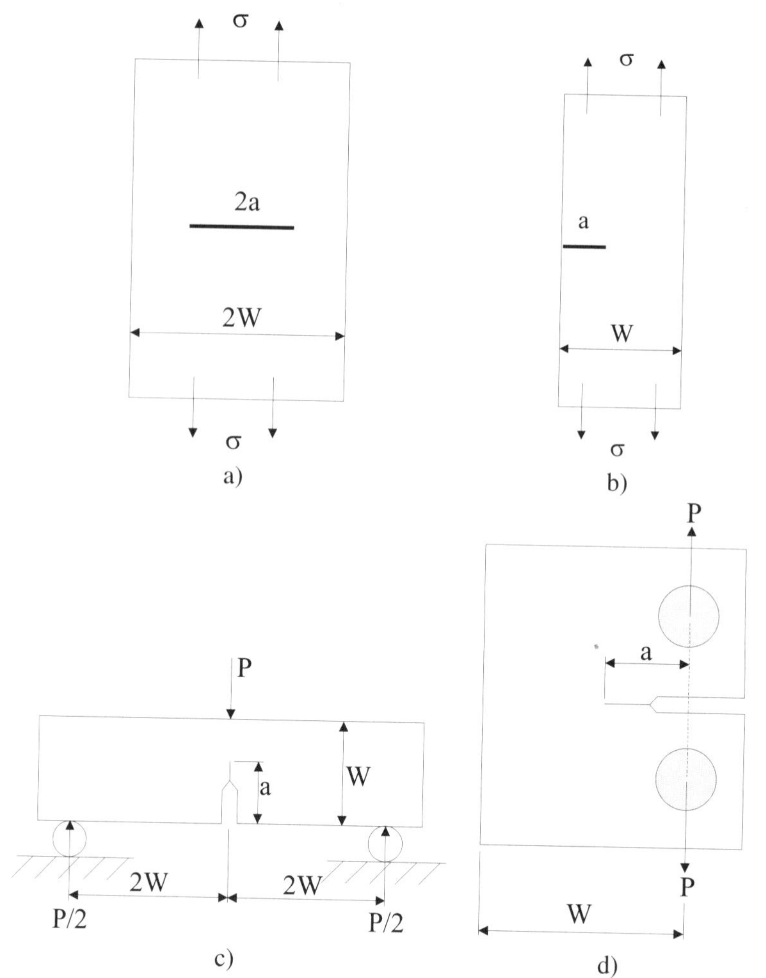

Figure 17.31: Fracture toughness test specimens. a) Centre cracked plate. b) Single edge cracked plate. c) Three-point bend specimen. d) Compact tension specimen.

compact tension specimens. Experimentally the crack length is measured using a travelling microscope and the crack tip opening displacement using strain gauge callipers. The mode I stress intensity factor for the three-point bend specimen is given by the following interpolating polynomial, BS 5447

$$K_I = \frac{3PL}{BW^{3/2}} \left[1.93 \left(\frac{a}{W} \right)^{1/2} - 3.07 \left(\frac{a}{W} \right)^{3/2} + 14.53 \left(\frac{a}{W} \right)^{5/2} - 25.11 \left(\frac{a}{W} \right)^{7/2} + 25.80 \left(\frac{a}{W} \right)^{9/2} \right] \tag{17.100}$$

where $L = 2W$ and B is the specimen thickness. The stress intensity factor for the compact tension specimen is similarly defined

$$K_I = \frac{P}{BW^{1/2}} \left[29.6 \left(\frac{a}{W} \right)^{1/2} - 185.5 \left(\frac{a}{W} \right)^{3/2} + 655.7 \left(\frac{a}{W} \right)^{5/2} - 1017 \left(\frac{a}{W} \right)^{7/2} + 638.9 \left(\frac{a}{W} \right)^{9/2} \right] \tag{17.101}$$

With reference to the load-displacement curve of Figure 17.32 the plane strain fracture toughness is measured by following the procedure:

1) Draw line OA through the initial elastic region.

2) Draw line OB having a slope of 5% less than OA. For example if line OA is at 60° to the CTOD axis then line OB is at $60^\circ - (60^\circ \times 5\%) = 57^\circ$.

3) Let OB intersect the load-displacement curve at point Q and let the horizontal distance between OA and OB be v.

4) Draw a line at 80% of PQ and record v_1.

5) The plane strain fracture toughness is given by

$$K_I = \frac{YP_Q}{BW^{1/2}} \tag{17.102}$$

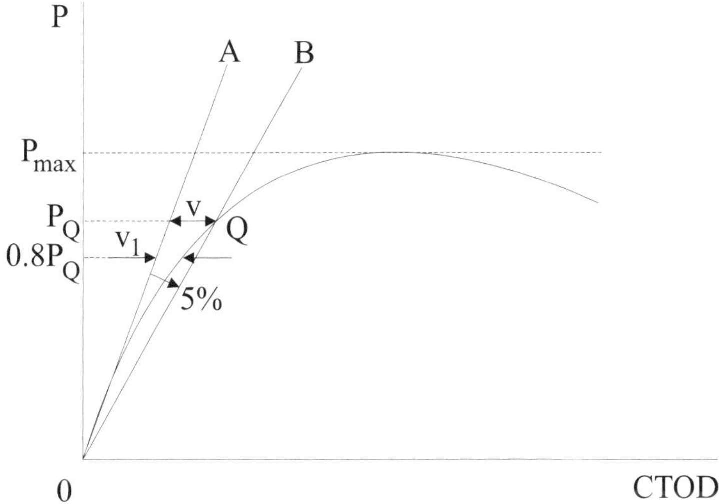

Figure 17.32: Measurement of plane strain fracture toughness from a load-displacement curve.

The following conditions must be satisfied so that the above test procedure is valid:

1) The fracture stress, σ_f, must satisfy

$$\sigma_f \leq 0.8\sigma_Y \tag{17.103}$$

where σ_Y is the 0.2% proof tensile yield stress.

2) The crack length and thickness must satisfy the following relation to ensure that the plane strain fracture toughness has been obtained with minimum crack tip plasticity

$$a, B \geq 2.5 \left(\frac{K_{IC}}{\sigma_Y}\right)^2 \tag{17.104}$$

3) The crack length must also satisfy the following so as to minimise specimen boundary effects

$$0.45 < \frac{a}{W} < 0.55 \tag{17.105}$$

and illustrates that the Y functions given in (17.100) and (17.101) are restricted to a small range of a/W.

4) $P_{\max}$ and v_1 must satisfy the following relations to ensure that $P_{\max}$ is sufficiently close to P_Q

$$v_1 \leq \frac{v}{4}; \quad P_{\max} \leq 1.1 P_Q \tag{17.106}$$

17.9 Mixed-Mode Fracture

Compared to single mode fracture, surprisingly little is known to date about the mixed-mode fracture of materials. Existing theories on mixed-mode fracture generally assume that an effective stress intensity factor is produced and the loci of failure are a combination of the individual modes. One of the more popular expressions for mixed-mode failure is that proposed by Wu (1967)

$$\left(\frac{K_I}{K_{IC}}\right)^2 + \left(\frac{K_{II}}{K_{IIC}}\right)^2 = 1 \tag{17.107}$$

and fits well with a variety of experimental data, Figure 17.33. Available experimental data suggests that

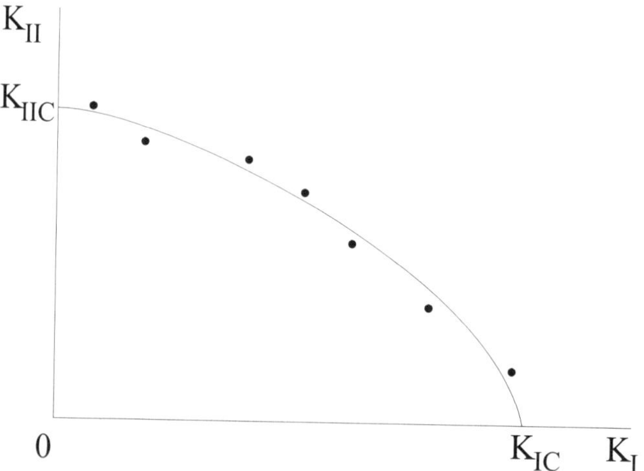

Figure 17.33: K_I against K_{II} for mixed-mode fracture.

$K_{IIC} \approx 0.75 K_{IC}$ from which (17.107) reduces to

$$K_I^2 + \frac{16}{9} K_{II}^2 = K_{IC}^2 \tag{17.108}$$

Several alternative theories have been proposed for mixed-mode fracture such as the *maximum tangential stress* criterion of Erdogan and Sih (1963) and the *strain energy density* criterion of Sih (1973). The maximum tangential stress criterion states that the crack commences extension in a radial direction and that the crack extends in a plane which is perpendicular to the direction of greatest tension. Let the crack extend in a direction α along which the tangential stress $\sigma_{\theta\theta}$ is a maximum and the shear stress $\tau_{r\theta}$ is zero. Hence, the direction of crack propagation, α, is given by

$$\frac{\mathrm{d}\sigma_{\theta\theta}}{\mathrm{d}\theta} = 0 \quad \text{and} \quad \frac{\mathrm{d}^2\sigma_{\theta\theta}}{\mathrm{d}\theta^2} < 0 \quad \text{at} \quad \theta = \alpha \tag{17.109}$$

Differentiating $\sigma_{\theta\theta}$ from (17.10) with respect to θ and from condition (17.109)

$$\frac{\mathrm{d}\sigma_{\theta\theta}}{\mathrm{d}\theta} = -\frac{3}{2}\frac{K_I}{\sqrt{2\pi r}}\cos^2\frac{\theta}{2}\sin\frac{\theta}{2} - \frac{3}{2}\frac{K_{II}}{\sqrt{2\pi r}}\left[\cos\frac{\theta}{2} - 3\sin^2\frac{\theta}{2}\cos\frac{\theta}{2}\right] = 0 \tag{17.110}$$

Setting $\theta = \alpha$ then the above equation simplifies to

$$K_I \sin\alpha + K_{II}(3\cos\alpha - 1) = 0 \tag{17.111}$$

As an example let us consider the specific case of an inclined crack in a centre cracked plate subject to a uniform tension, Figure 17.34a). With the crack angle, β, measured from the y-axis then the mixed-mode I and II stress intensity factors are, (17.37)

$$K_I = \sin^2\beta\sigma\sqrt{\pi a}; \quad K_{II} = \sin\beta\cos\beta\sigma\sqrt{\pi a} \tag{17.112}$$

Substituting (17.112) into (17.111) we have

$$\sin\alpha + (3\cos\alpha - 1)\cot\beta = 0 \tag{17.113}$$

Figure 17.34b) illustrates β against fracture angle, α, with the solid line indicating the maximum tangential stress criterion prediction.

Consider now the strain energy density criterion with the strain energy density, $\mathrm{d}W$, for an elemental area $\mathrm{d}A$ at the tip of a crack of the form

$$\frac{\mathrm{d}W}{\mathrm{d}A} = \frac{1}{r}\left(a_{11}K_I^2 + 2a_{12}K_IK_{II} + a_{22}K_{II}^2\right) + \text{non-singular terms} \tag{17.114}$$

with the strain energy density experiencing a $1/r$ singularity at the crack tip. Equation (17.114) can alternatively be written as

$$\frac{\mathrm{d}W}{\mathrm{d}A} = \frac{S}{r}; \quad S = a_{11}K_I^2 + 2a_{12}K_IK_{II} + a_{22}K_{II}^2 \tag{17.115}$$

with S referred to as the strain energy density factor and the constants a_{11}, a_{12} and a_{22} given by

$$\begin{aligned} a_{11} &= \frac{1}{16\pi\mu}(\kappa - \cos\theta)(1 + \cos\theta) \\ a_{12} &= \frac{2\sin\theta}{16\pi\mu}[\cos\theta - (\kappa - 1)] \\ a_{22} &= \frac{1}{16\pi\mu}[(\kappa + 1)(1 - \cos\theta) + (1 + \cos\theta)(3\cos\theta - 1)] \end{aligned} \tag{17.116}$$

with a crack tip polar coordinate system (r, θ). The strain energy density criterion postulates that crack extension occurs in the direction of minimum strain energy density factor, that is

$$\frac{\partial S}{\partial\theta} = 0 \quad \text{at some } \theta = \alpha \tag{17.117}$$

The criterion goes on to state that crack initiation commences when S attains a critical value, S_c when $\theta = \alpha$. For the example of an inclined crack in a centre cracked plate, Figure 17.34a), with K_I and K_{II} given by (17.37) then S can be shown to be given by

$$S = \sigma^2 \pi a \left(a_{11} \sin^2 \beta + 2a_{12} \sin \beta \cos \beta + a_{22} \cos^2 \beta \right) \sin^2 \beta \tag{17.118}$$

Differentiating with respect to θ and equating to zero then the fracture angle α is given by

$$(\kappa - 1) \sin(\alpha - 2\beta) - 2 \sin\left[2(\alpha - \beta)\right] - \sin 2\alpha = 0 \tag{17.119}$$

and is seen to be a function of Poisson's ratio. For the case of plane strain, $\kappa = 3 - 4\nu$, then α against β for several values of ν is shown in Figure 17.34c).

An alternative mixed mode failure criterion originally proposed by Hellen and Blackburn (1975) is based on the J_x and J_y integrals, §17.12, and assumes that the crack propagates along a path of maximum energy release rate. Figure 17.34d) illustrates the tip of a crack with components J_x and J_y acting parallel and perpendicular to the crack respectively. The maximum energy release rate is for a crack extending at an angle of

$$\alpha = \tan^{-1}\left(\frac{J_y}{J_x}\right) = \tan^{-1}\left(-\frac{2K_I K_{II}}{K_I^2 + K_{II}^2}\right) \tag{17.120}$$

to the crack plane and with magnitude

$$J(\theta) = J_x \cos\alpha + J_y \sin\alpha = \sqrt{J_x^2 + J_y^2} = \left(\frac{\kappa + 1}{8\mu}\right) \sqrt{K_I^4 + 6K_I^2 K_{II}^2 + K_{II}^4} \tag{17.121}$$

For the case of the inclined crack in a centre cracked plate shown in Figure 17.34a) then α is, substituting (17.112)

$$\alpha = \tan^{-1}(-2 \sin\beta \cos\beta) \tag{17.122}$$

and is graphically illustrated in Figure 17.34b) by the dashed line. It is observed that fracture angle predictions based on the maximum energy release rate (J_x and J_y integrals) are applicable when K_{II}/K_I is small, that is $\beta \to 90°$.

17.10 Pressure Vessels and Damage Analysis

Previous sections have considered general results for fracture mechanics and summarised several expressions for stress intensity factors. In the present section we specifically address the application of fracture mechanics to the important application of cylindrical pressure vessels. In Chapters 4 and 11 we examined both thin and thick walled pressure vessels assuming the vessels to be defect free. Of the three circumferential, $\sigma_{\theta\theta}$, axial, σ_{zz}, and radial, σ_{rr}, stress components the largest stress is $\sigma_{\theta\theta}$ and in the case of a thin-walled vessel $\sigma_{\theta\theta}$ is given by, (4.2)

$$\sigma_{\theta\theta} = \frac{pd}{2t} \tag{17.123}$$

where d is the mean diameter, t is the wall thickness and p is the internal pressure with zero external pressure. Equating $\sigma_{\theta\theta}$ to the yield stress, σ_Y, of the vessel material then a safe operating pressure, p_{op}, follows

$$p_{op} = \left(\frac{2t}{Sd}\right) \sigma_Y \tag{17.124}$$

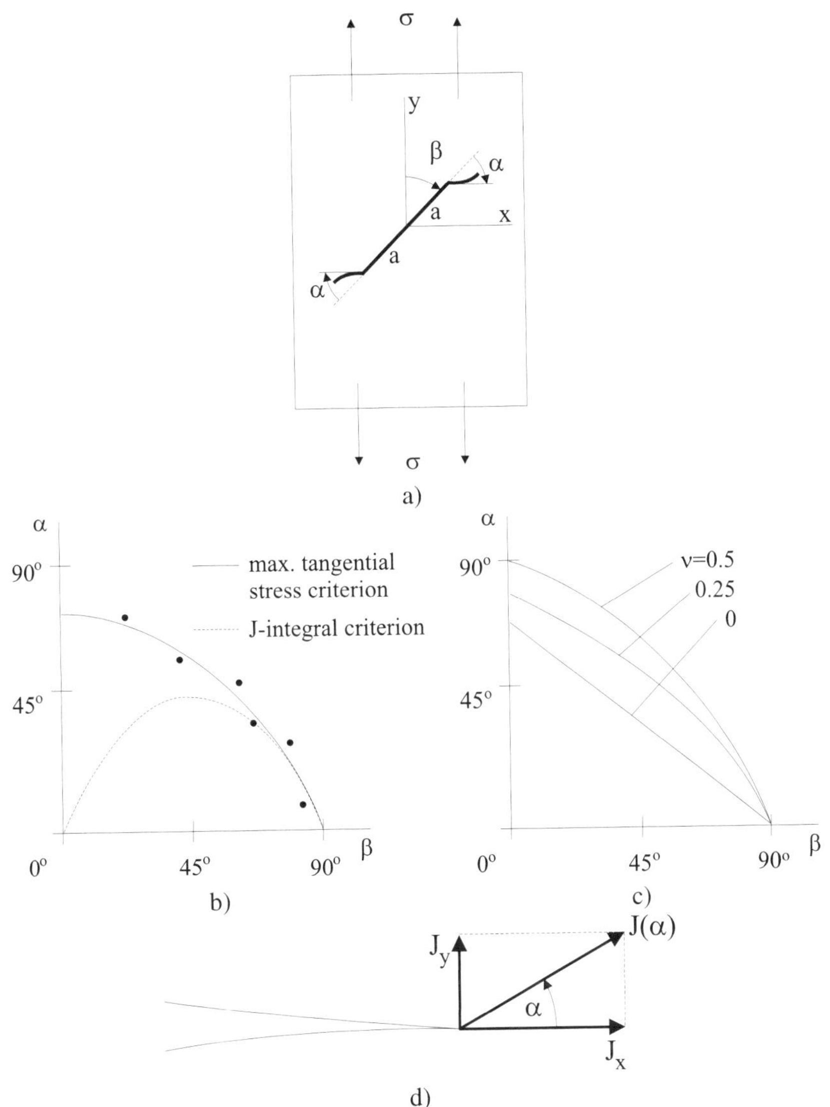

Figure 17.34: Mixed-mode fracture. a) An inclined crack in a centre cracked plate with crack inclination angle β and fracture angle α. b) Crack inclination angle, β, against fracture angle, α, for both the maximum tangential stress and maximum energy release rate criteria. c) Crack inclination angle, β, against fracture angle, α, for the strain energy density criterion as a function of Poisson's ratio. d) Parallel and perpendicular energy release rate components at the tip of a crack.

where S represents a factor of safety.

Now suppose that a shallow axial surface crack of depth a is present on the inner surface of the vessel. In general the defects are semi-elliptical in nature but, as a first approximation, we can make the conservative estimate that a long semi-elliptical surface flaw can be modelled as a semi-infinite edge-cracked plate in tension, (17.43). Equating this K_I to K_{IC}/S then we arrive at a new estimate for p_{op} which takes account for defect size

$$p_{op} = \left(\frac{2t}{Sd}\right)\frac{K_{IC}}{1.1215\sqrt{\pi a}} \tag{17.125}$$

In the case of a through-thickness axial crack then K_I can be approximated by, (17.47)

$$K_I = \frac{pd}{2t}\sqrt{\pi a}\left[1 + 3.22\frac{a^2}{dt}\right]^{1/2} \tag{17.126}$$

from which, equating K_I to K_{IC}/S, we find a third estimate of operating pressure

$$p_{op} = \left(\frac{2t}{Sd}\right)\left[1 + 3.22\frac{a^2}{dt}\right]^{1/2} K_{IC} \tag{17.127}$$

The above expressions (17.125) and (17.127) can be used to determine the maximum vessel pressure for a given defect size. Equally, we may require to know the size of defect which would lead to failure. The critical crack depth, a_c, can be determined by equating (17.124) and (17.125) to give

$$a_c = \frac{1}{(1.1215)^2\pi}\left(\frac{K_{IC}}{\sigma_Y}\right)^2 = \alpha\left(\frac{K_{IC}}{\sigma_Y}\right)^2; \quad \alpha \approx 0.25 \tag{17.128}$$

where α is a geometry-dependent constant.

Consider now the part-through surface flaw shown in Figure 17.35a). The depth of the flaw in the thickness direction is denoted by a, its length by $2c$ and the wall thickness by t. The wall is subject to a uniform tensile stress that acts remotely in the direction normal to the crack plane. The worst case scenario would be when $c \gg t > a$ and the crack front would approach a horizontal line with single crack dimension a. The results of both part-through and through-wall analyses can be coupled to give a conservative estimate of the critical dimensions of an initial flaw. Consider the three cases A, B and C shown in Figure 17.35b):

A This crack is benign even though its surface dimension is greater than the crack critical surface dimension because its depth dimension is less than $a_{critical}$.

B This crack is also benign. Although its depth dimension exceeds $a_{critical}$ its surface dimension is less than $2c_{critical}$. Note also that the through-wall crack resulting from flaw B will not be critical.

C This crack is critical. With both dimensions exceeding the corresponding critical dimensions this defect will lead to fracture.

In the case of pressure vessels flaw type B would give rise to the so-called *leak-before-break* in which the crack would grow through the vessel wall and consequently allow a contained fluid to escape at, presumably, a limited rate. Thus, a crack of type B could be detected in time to take remedial action whereas a crack of type C would not exhibit this desirable behaviour. The previous discussion of leak-before-break assumes that the applied stress does not alter during the crack growth process. In most instances this assumption would be conservative because a loss of fluid is generally accompanied with a decrease in pressure.

Rather than considering a part-through wall crack relative to the two conditions of through-thickness defect and an infinitely long horizontal defect we can analyse the exact crack shape. One of the first analyses of a part-through wall crack was that of Irwin (1962). Irwin's solution starts from an elliptical or penny-shaped crack embedded inside an infinitely large elastic body. By introducing approximate

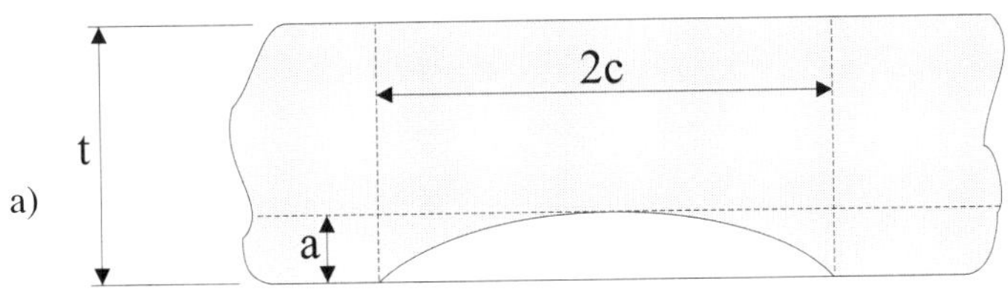

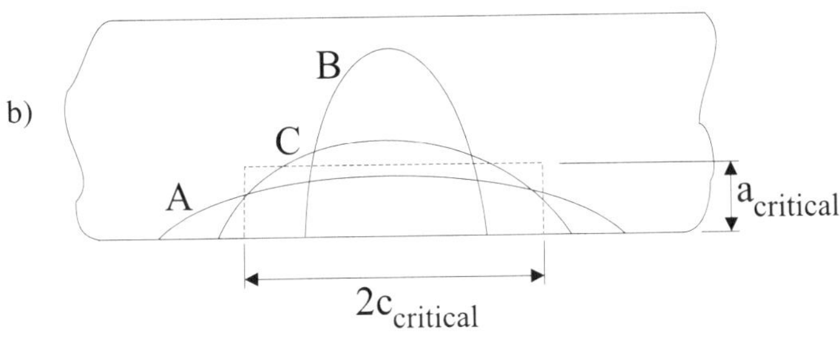

Figure 17.35: A part-through the wall surface flaw. a) Crack dimensions. b) Three cases A, B and C.

correction factors to take account of the free surface and crack tip plasticity Irwin arrived at the following expression for K_I

$$K_I = \frac{1.1215\sigma\sqrt{\pi a}}{\left[\Phi^2 - 0.212\left(\dfrac{\sigma}{\sigma_Y}\right)^2\right]^{1/2}}; \quad \Phi = \int_0^{\pi/2}\left(\sin^2\phi + \left(\frac{a}{c}\right)^2\cos^2\phi\right)\,d\phi \tag{17.129}$$

where $\Phi = \Phi(a, c)$ depends on the crack shape and is expressed in terms of the elliptical integral. Angle ϕ is measured as shown in Figure 17.36.

Extensive studies of part-through elliptical cracks have been performed by Newman and Raju (1981) and they developed the following expression for K_I

$$\begin{aligned} K_I = \sigma\sqrt{\pi a}\left[M_1 + M_2\left(\frac{a}{t}\right)^2 + M_3\left(\frac{a}{t}\right)^4\right]\left[1 + 1.1464\left(\frac{a}{c}\right)^{1.65}\right]^{-1/2} \times \\ \left[\left(\frac{a}{c}\right)^2\cos^2\phi + \sin^2\phi\right]^{1/4}\left\{1 + \left[0.1 + 0.35\left(\frac{a}{t}\right)^2\right](1-\sin\phi)^2\right\} \end{aligned} \tag{17.130}$$

where M_1, M_2 and M_3 are functions of (a/c). The Newman-Raju empirical equation is based on three-dimensional finite element analyses and is accurate to within $\pm 5\%$ provided that $0 < a/c < 1$ and $a/t \leq 0.8$. The Newman-Raju equation reveals the following key properties:

1) For small a/c the maximum value of K_I occurs at $\phi = \pi/2$, that is the deepest point in the wall.

2) For $a/c \approx 0.25$, K_I is approximately independent of ϕ.

3) For larger values of a/c then K_I exhibits a maximum on the plate surface.

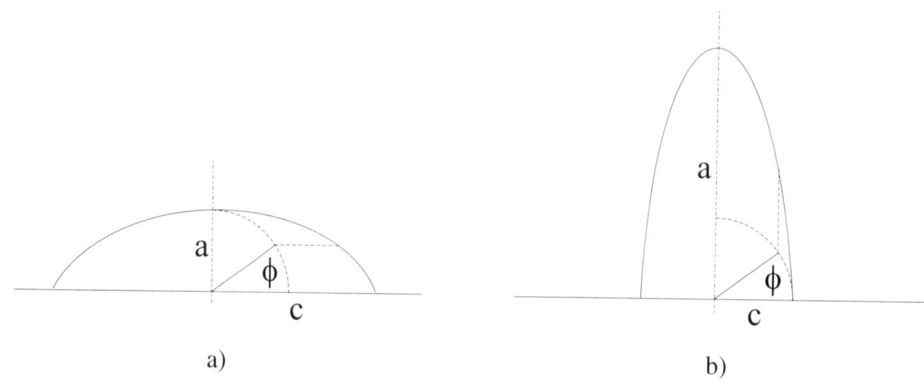

Figure 17.36: Part-through elliptical surface flaws. a) $a/c < 1$. b) $a/c > 1$.

4) Based on the above points then a surface flaw attempts to grow in such a manner as to optimise K_I around the entire crack profile so that the crack has uniform propagation for all ϕ.

The Newman-Raju equation reduces to the following for $a \ll t$ and $a \ll c$

$$\begin{aligned} K_I &= 1.13\sigma\sqrt{\pi a}\left[1 - 1.08\left(\frac{a}{c}\right)\right]\left[1 + 1.464\left(\frac{a}{c}\right)^{1.65}\right]^{-1/2} && \text{for } a \ll t \text{ (shallow cracks)} \\ K_I &= 1.13\sigma\sqrt{\pi a}\left[1 + 3.46\left(\frac{a}{c}\right)^2 + 11.5\left(\frac{a}{t}\right)^4\right] && \text{for } a \ll c \text{ (long cracks)} \end{aligned} \tag{17.131}$$

17.10.1 Leak-Before-Break

Leak-before-break is one of several possible outcomes:

1) Sub-critical crack growth by fatigue or corrosion to a critical depth.

2) Stable crack growth under operating loads.

3) Fracture instability and subsequent rapid growth through the wall.

4) Arrest of a through-thickness crack.

5) Re-initialisation of a through-thickness crack and subsequent stable growth along wall.

6) Fracture instability and rapid crack growth leading to failure.

The leak-before-break condition occurs when points 4) and 5) are well separated or when 5) and 6) are precluded. The factors affecting the leak-before-break condition are:

1) Orientation of the initial flaw.

2) Size and shape of crack at onset of stable growth.

3) Type, intensity and duration of applied loading.

4) Distribution of residual stresses, if present.

5) Mechanical and fracture properties of pressure vessel material.

Consider a crack which is much longer in length along the surface than in depth. Approximately, the initiation of unstable crack growth will occur in the radial direction when

$$1.1215\frac{pd}{2t}\left[\pi a \sec\left(\frac{\pi a}{2t}\right)\right]^{1/2} = K_{IC} \tag{17.132}$$

with a thin-walled pressure vessel assumed with $\sigma_{\theta\theta}$ being the dominant stress and given by (17.123) and K_I given by (17.45). Unstable crack growth will generally result in a through-wall crack of length equal to the surface length of the original part-through crack. Thus, the condition for re-initiation of crack growth to occur is, from (17.126)

$$\frac{p_{ins}\, d}{2t}\left[\pi c\left(1+3.22\frac{c^2}{dt}\right)\right]^{1/2} = K_C \tag{17.133}$$

where p_{ins} is the internal pressure at the onset of longitudinal crack instability and K_C is the fracture toughness corresponding to the given wall thickness in accordance with (17.92). From (17.132) and (17.133) we arrive at a relation which provides a bounding surface between leak and break

$$\frac{K_C}{K_{IC}} = \frac{\dfrac{p_{ins} d}{2t}\left[\pi c\left(1+3.22\dfrac{c^2}{dt}\right)\right]^{1/2}}{1.1215\dfrac{pd}{2t}\left[\pi a \sec\left(\dfrac{\pi a}{2t}\right)\right]^{1/2}} \tag{17.134}$$

Dividing by t and re-arranging we arrive at the following

$$\left(\frac{K_C}{K_{IC}}\right)^2\left(\frac{p}{p_{ins}}\right)^2 1.26\frac{a}{t}\sec\left(\frac{\pi a}{2t}\right) = \frac{c}{t}\left[1+3.22\frac{t}{d}\left(\frac{c}{t}\right)^2\right] \tag{17.135}$$

which provides a leak-before-break delimitation for axial cracks in pressurised vessels. Equation (17.135) requires a numerical solution using the Newton-Raphson method for c/t as a function of a/t. In the case of fluid leakage then the inequality $p_{ins} < p$ is expected. Figure 17.37 illustrates c/t against a/t for $d/t = 20$.

Example 17.3 Fracture mechanics assessment of a thin-walled pressure vessel

In this example we will perform a comparison between the two approaches of a defect-free *strength of materials* analysis and a *fracture mechanics* analysis of a thin-walled pressure vessel. The pressure vessel is of mean diameter d = 2m, wall-thickness t = 15mm and subject to an internal pressure p. Let us first assume that failure of the vessel is dominated by the circumferential stress which cannot exceed the tensile yield stress σ_Y = 350MPa with account for a factor of safety $S = 2$. To determine a safe operating pressure, p_{op}, we can equate the circumference stress, (17.123), $\sigma_{\theta\theta} = pd/2t$ to σ_Y/S

$$p_{op} = \left(\frac{2t}{Sd}\right)\sigma_Y \tag{17.136}$$

Now assume that a shallow crack of depth a exists on the inner surface of the vessel as shown in Figure 17.38. An exact expression for the stress intensity factor for such a defect could be used but as a first approximation let us assume that the crack approximates to an edge crack

$$K_I \approx 1.1215\sigma_{\theta\theta}\sqrt{\pi a}$$

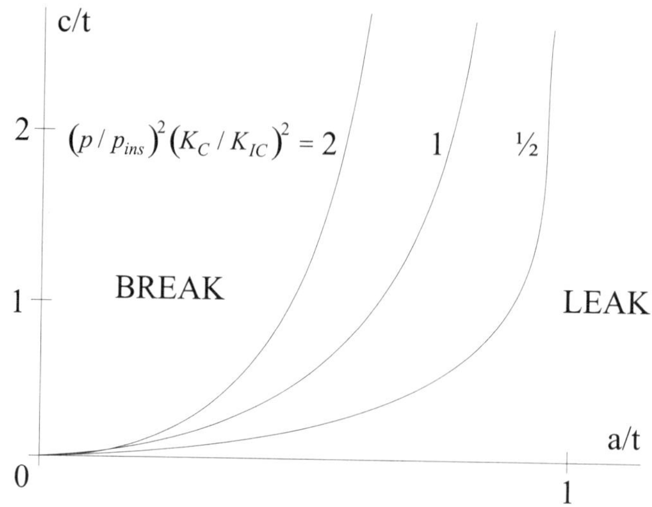

Figure 17.37: Leak-break loci given by (17.135) for $d/t = 20$.

Equating K_I to K_{IC}/S (with $K_{IC} = 35\text{MPa}\sqrt{\text{m}}$) to determine a safe operating pressure with $\sigma_{\theta\theta} = pd/2t$ as in (17.125) we have

$$p_{op} = \left(\frac{2t}{Sd}\right)\frac{K_{IC}}{1.1215\sqrt{\pi a}} \tag{17.137}$$

The structural geometry term $(2t/Sd)$ is common to both approaches with the material parameter, σ_Y, in (17.136) taking no account of defects whereas the material parameter in (17.137) is K_{IC} and defects are taken into account.

The critical crack depth at the operating pressure can be determined by equating (17.136) and (17.137)

$$a_c = \frac{1}{(1.1215)^2\pi}\left(\frac{K_{IC}}{\sigma_Y}\right)^2 = \alpha\left(\frac{K_{IC}}{\sigma_Y}\right)^2; \quad \alpha \approx 0.25 \tag{17.138}$$

where α is a geometry-dependent constant and noting the characteristic form of $\alpha(K_{IC}/\sigma_Y)^2$ for critical crack depth. Making use of (17.92) we can take account of the wall-thickness of the vessel since it is thin-walled

$$K_C = 35\left[1 + \frac{1.4}{(15\times 10^{-3})^2}\left(\frac{35}{350}\right)^4\right]^{1/2} = 44.58\text{MPa}\sqrt{\text{m}}$$

Using K_C instead of K_{IC} in (17.138) then the critical crack depth is

$$a_c = 0.25\left(\frac{44.58}{350}\right)^2 = 4.104\text{mm}$$

with the operating pressure given by (17.137) again using K_C instead of K_{IC}

$$p_{op} = \left(\frac{2\times 15\times 10^{-3}}{2\times 2}\right)\frac{44.58}{1.1215\sqrt{\pi\times 4.104\times 10^{-3}}} = 2.626\text{MPa} = \frac{2.626\times 10^6}{6,890}\text{psi} = 381\text{psi}$$

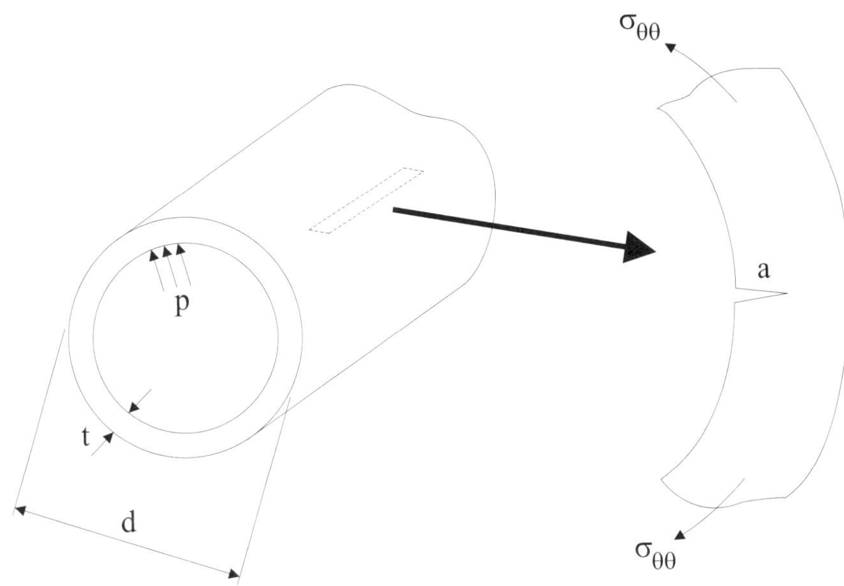

Figure 17.38: Example: A thin-walled pressure vessel with an axial crack on the inner wall.

17.11 Fatigue

In the following two sub-sections we will present an overview of fatigue crack growth considering separately long and short cracks. For an enjoyable overview of both long and short fatigue crack growth the reader is referred to Miller (1993). Firstly, however, we begin by examining fatigue from a non-fracture mechanics approach.

17.11.1 S-N Curve

The traditional strength of materials non-mechanistic approach to cyclic loading is the S-N curve with appropriate correction for notch effects, Figure 17.39. The maximum resistance to fatigue, *fatigue strength*, is obtained for smooth *hour glass* specimens and reduced for notched and cracked specimens. The fatigue strength is approximately equal to the tensile strength up to $10^2 \sim 10^3$ cycles. The most characteristic region for engineering alloys is between $10^2 \sim 10^3$ cycles and $10^6 \sim 10^8$ cycles where fatigue is highly dependent on the stress amplitude, $\Delta\sigma$. The relationship between $\Delta\sigma$ and the number of cycles to failure, N_f, in this range is given by

$$\Delta\sigma^n N = \text{constant} \tag{17.139}$$

where n is in the range $[8 : 15]$. Around $10^6 \sim 10^7$ cycles ferrous metals reach a fatigue limit whereas non-ferrous metals generally achieve a fatigue limit around 10^8 cycles.

Fatigue is frequently characterised as consisting of the following two regimes of cyclic loading: *low-cycle fatigue* and *high-cycle fatigue*. Low-cycle fatigue is characterised by a high stress level and by a low number of cycles to failure (thousands). High-cycle fatigue is characterised by a low stress level and a large number of cycles to failure (millions).

S-N curves tell us nothing of the mechanics of failure and subsequent sections will use a fracture mechanics approach to characterise fatigue crack growth. Adopting such an approach will enable us to

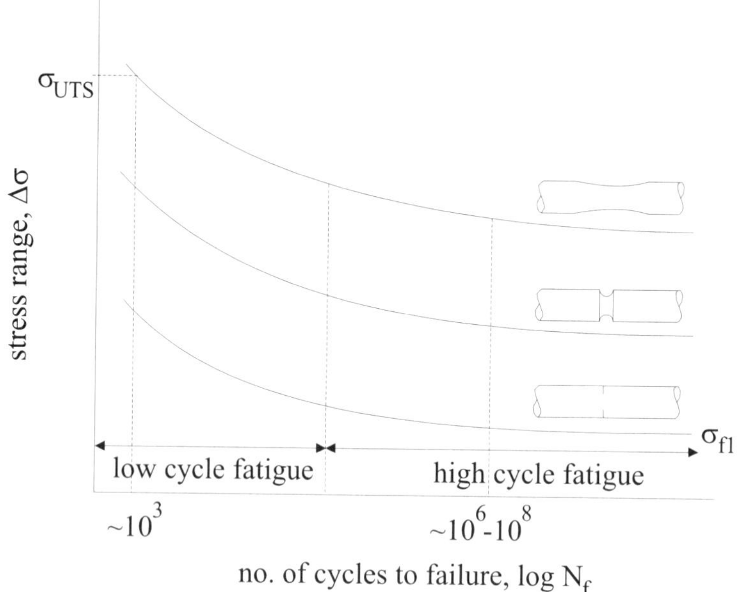

Figure 17.39: S-N curves for the same material but different geometries.

predict crack growth rates due to cyclic loading and to determine the number of cycles for a crack to grow to a predetermined length.

17.11.2 Effect of Mean Stress

The majority of engineering applications involve complex states of applied stress which may not be symmetric about the time axis and as a result have mean stress, $\sigma_m (= \sigma_{\max} - \sigma_{\min})$, as shown in Figure 17.40. Figure 17.41a) illustrates the typically observed variation of $\Delta\sigma$ against σ_m for a range of engineering alloys. For long fatigue life (low stress) the effect of mean stress is negligible whereas for short life (high stress) the effect of mean stress plays a more important role due to the increased specimen plasticity. An estimate of the effects of mean stress was proposed by Goodman and consists of the following two limiting conditions:

i) The allowable cyclic stress, $\Delta\sigma$, is constant until the mean stress is large enough so that the maximum stress is equal to the tensile strength, σ_{UTS}.

ii) There is a linear relationship between the cyclic stress and the mean stress.

These two conditions are satisfied by

$$\frac{\Delta\sigma}{\Delta\sigma_0} \leq 1 \quad \text{and} \quad \frac{\Delta\sigma + \sigma_m}{\sigma_{\text{UTS}}} \leq 1 \tag{17.140}$$

and combined to give the Goodman relation

$$\frac{\Delta\sigma}{\Delta\sigma_0} + \frac{\sigma_m}{\sigma_{\text{UTS}}} \leq 1 \tag{17.141}$$

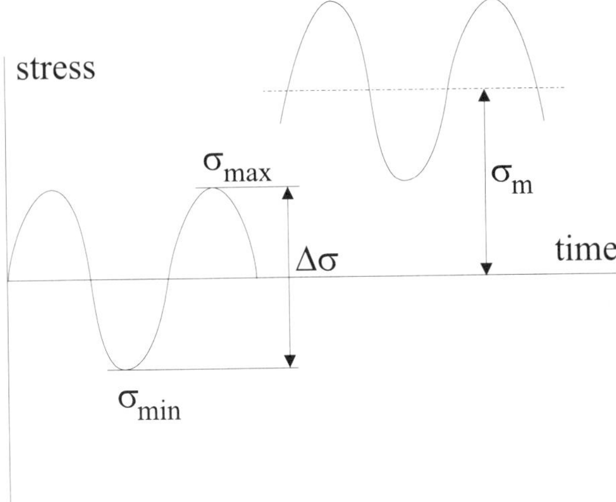

Figure 17.40: Stress cycles with different mean stress, σ_m.

and for a positive σ_m can be written

$$\Delta\sigma = \Delta\sigma_0 \left(1 - \frac{\sigma_m}{\sigma_{\mathrm{UTS}}}\right) \tag{17.142}$$

The Goodman relation provides a conservative estimate since the majority of experimental data falls above the curve with the exception of certain brittle materials.

17.11.3 Cumulative Damage

In practice the stress amplitude may vary with time and the analysis of fatigue is frequently based on the linear summation of damage. Consider the case of step-wise loading with $\Delta\sigma_k \quad (k = 1, 2, \ldots, s)$ denoting the amplitude of stress in the kth. group of cycles $\Delta N_k (= N_k - N_{k-1})$ and let N_k^f denote the number of cycles to failure for this amplitude. From the principle of linear summation the failure time is given by

$$\sum_{k=1}^{s} \frac{\Delta N_k}{N_k^f} = 1 \tag{17.143}$$

that is, the sum of the cycle ratios equals unity. This relation is generally referred to as the *Palmgren-Miner cumulative damage law*. If the stress amplitude changes continuously then (17.143) can be written

$$\int_0^{N_f} \frac{\mathrm{d}N}{N^f} = 1 \tag{17.144}$$

where N_f is the number of cycles to failure for the current amplitude $\mathrm{d}\sigma$ and N_f is the total number of cycles to failure.

Example 17.4 Estimation of component lifetime based on the Palmgren-Miner law

A component is subject to reversed cyclic loading of 800 cycles per day in a sequence consisting of the following: 1000 cycles at ±200MPa, 1000 cycles at ±150MPa, 2000 cycles at ±100MPa and 3000 cycles

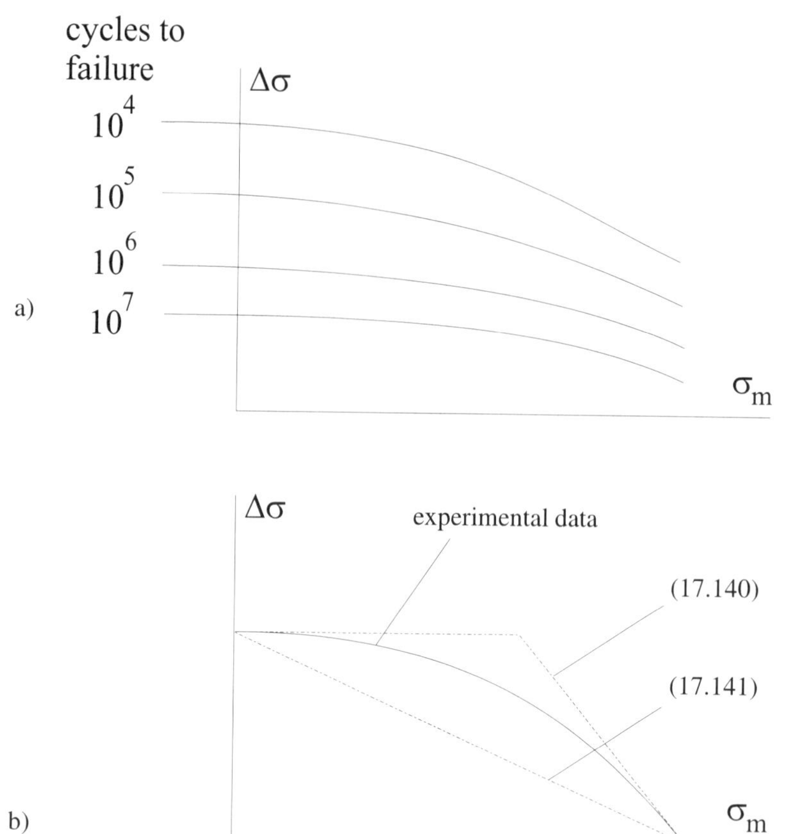

Figure 17.41: Effect of mean stress. a) Cyclic stress against mean stress as a function of number of cycles to failure. b) Cyclic stress versus mean stress and the Goodman relation.

at ±125MPa. The fatigue lives at these four stress levels are 1×10^4, 1×10^5, 2×10^6 and 1×10^6 cycles respectively. Determine the lifetime of the component using the Palmgren-Miner law.

Let the component fail due to S sequences of the applied cyclic loadings. From (17.143)

$$S\left(\frac{1000}{1 \times 10^4} + \frac{1000}{1 \times 10^5} + \frac{2000}{2 \times 10^6} + \frac{3000}{1 \times 10^6}\right) = 1$$

which gives $S = 8.77$. Therefore, for a total of 7000 cycles per sequence with 800 cycles per day leads to a component lifetime of $8.77 \times (7000/800) = 76$ days.

17.11.4 Long Cracks

Previous sections have demonstrated that the presence of cracks within a structure can result in failure when the applied loading causes the stress intensity factor to attain a critical value. However, it is also possible to induce failure as a result of sub-critical crack growth due to the cyclic loading of a component.

17.11.4.1 Crack Length Versus Number of Cycles

Figure 17.42 illustrates crack length against number of cycles for the three different stress levels $\Delta\sigma_1$, $\Delta\sigma_2$ and $\Delta\sigma_3$ in which $\Delta\sigma_1 > \Delta\sigma_2 > \Delta\sigma_3$. At a higher stress then a crack grows faster for a given number of cycles. Figure 17.42 can alternatively be plotted in terms of N/N_f (number of cycles/number

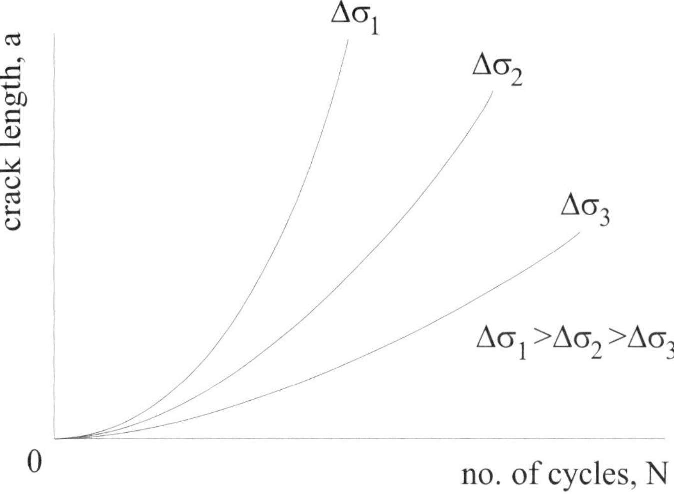

Figure 17.42: Crack growth as a function of cyclic stress.

of cycles to failure) for three different initial crack lengths, Figure 17.43. If the initial crack length is *long* ($\approx$10mm) then a low cyclic stress is required to propagate the crack. With such a low cyclic stress then the crack tip plasticity will be negligible and Linear Elastic Fracture Mechanics (LEFM) conditions will be applicable. If the initial crack length is of the order of 0.1mm and *physically short* then a higher value of cyclic stress will be required to propagate the crack. A higher applied cyclic stress will result in more extensive crack tip plasticity with an Elastic-Plastic Fracture Mechanics (EPFM) approach being more applicable. For initial *microstructurally short* cracks of the order 1μm in length then such cracks are dominated by the microstructure of the material and require high cyclic stresses in order to propagate. Such high cyclic stresses will result in both crack tip and bulk plasticity with crack propagation principally governed by microstructural barriers such as grain boundaries and inclusions.

17.11.4.2 Stress Intensity Factor Range

Consider the centre cracked plate shown in Figure 17.44 subject to constant-amplitude cyclic loading such that the loading varies between σ_{min} and σ_{max} $(\sigma_{\text{max}} > \sigma_{\text{min}} > 0)$. For mode I loading the minimum, K_{min}, and maximum, K_{max}, stress intensity factors are

$$K_{\text{min}} = Y\sigma_{\text{min}}\sqrt{\pi a}; \quad K_{\text{max}} = Y\sigma_{\text{max}}\sqrt{\pi a} \tag{17.145}$$

The stress intensity factor range, ΔK, is therefore

$$\Delta K = K_{\text{max}} - K_{\text{min}} = Y\Delta\sigma\sqrt{\pi a}; \quad \Delta\sigma = \sigma_{\text{max}} - \sigma_{\text{min}} \tag{17.146}$$

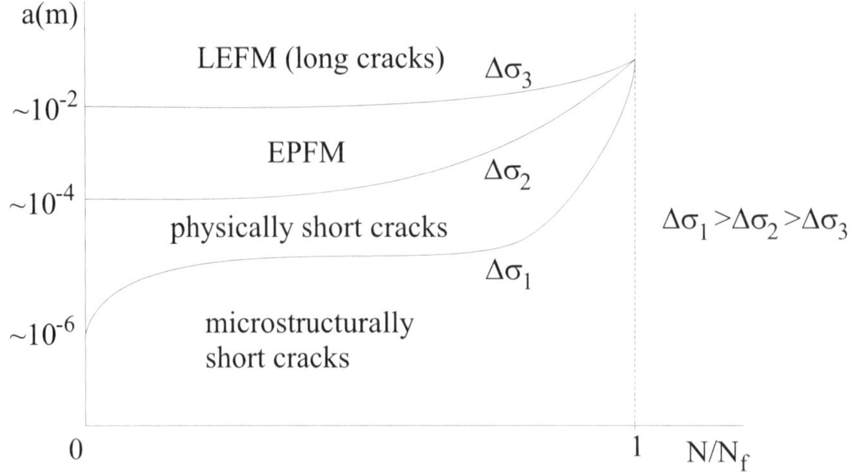

Figure 17.43: Crack growth as a function of cyclic stress for three different initial crack lengths.

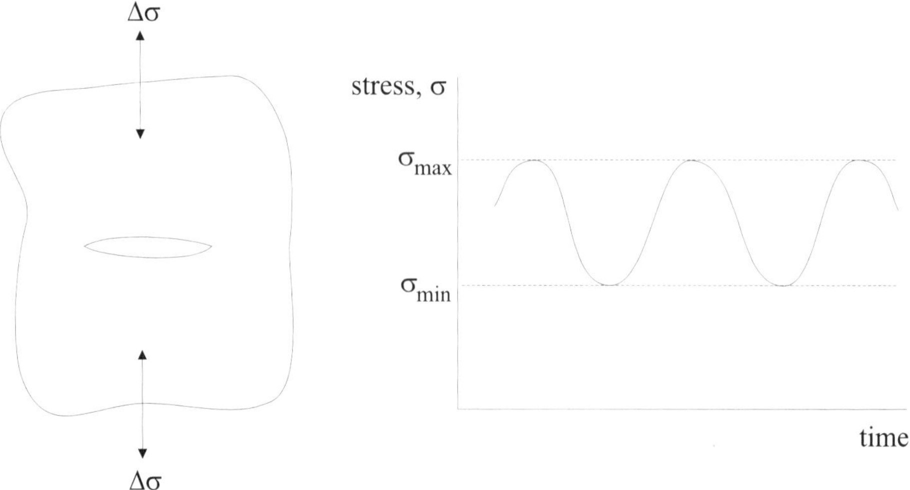

Figure 17.44: A centre cracked plate subject to an applied cyclic stress and the variation of the cyclic stress with time.

17.11.4.3 Crack Growth Laws

The growth curves of Figure 17.42 can be represented as a single curve if the slope $(\mathrm{d}a/\mathrm{d}N)$ is extracted for each crack length and plotted against the associated ΔK, Figure 17.45. The three key regimes of Figure 17.45 are as follows:

A This region extends from a threshold stress intensity factor, ΔK_{th}, below which it is assumed (for long cracks) that cracks do not propagate. This region extends to the constant slope regime of the $\mathrm{d}a/\mathrm{d}N$-ΔK curve.

B This effectively linear region can be represented by the following empirical growth law

$$\frac{\mathrm{d}a}{\mathrm{d}N} = C(\Delta K)^m \tag{17.147}$$

where C and m are experimentally determined material constants with $m \approx 3$ for steels and $m \approx$3-4 for aluminium alloys. Equation (17.147) is referred to as the *Paris growth law* after Paris and Erdogon (1963).

C This region exhibits a steep slope with $K_{\max}$ approaching the fracture toughness of the material, with failure being eminent.

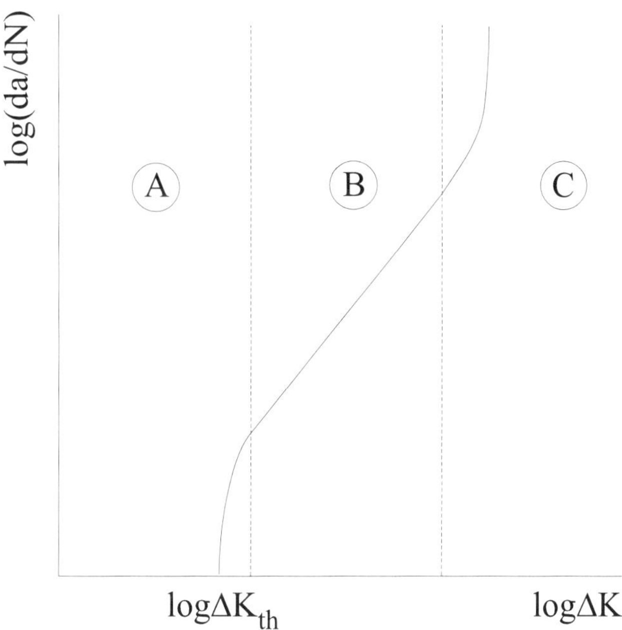

Figure 17.45: Crack growth in terms of $\mathrm{d}a/\mathrm{d}N$ versus ΔK.

There are several variants of Paris' law, a few of which we will now highlight. Forman *et al.* (1967) modified (17.147) to account for high values of ΔK

$$\frac{\mathrm{d}a}{\mathrm{d}N} = \frac{C(\Delta K)^m}{(1-R)K_C - \Delta K}; \quad R = \frac{K_{\min}}{K_{\max}} \tag{17.148}$$

Alternatively, Donahue *et al.* (1972) accounted for low values of ΔK as $\Delta K \to \Delta K_{th}$

$$\frac{\mathrm{d}a}{\mathrm{d}N} = C\left[\Delta K - \Delta K_{th}\right]^m \tag{17.149}$$

In §17.8 we noted the inverse relationship between plane strain fracture toughness, K_{IC}, and yield stress, σ_Y, for a variety of engineering materials. Similarly, there is an inverse relationship between crack growth rate and yield stress, Figure 17.46. McEvily and Groeger (1977) developed a growth law which takes account for σ_Y

$$\frac{\mathrm{d}a}{\mathrm{d}N} = \frac{A}{E\sigma_Y}\left[\Delta K - \Delta K_{th}\right]^2\left[1 + \frac{\Delta K}{K_{IC} - K_{\max}}\right] \tag{17.150}$$

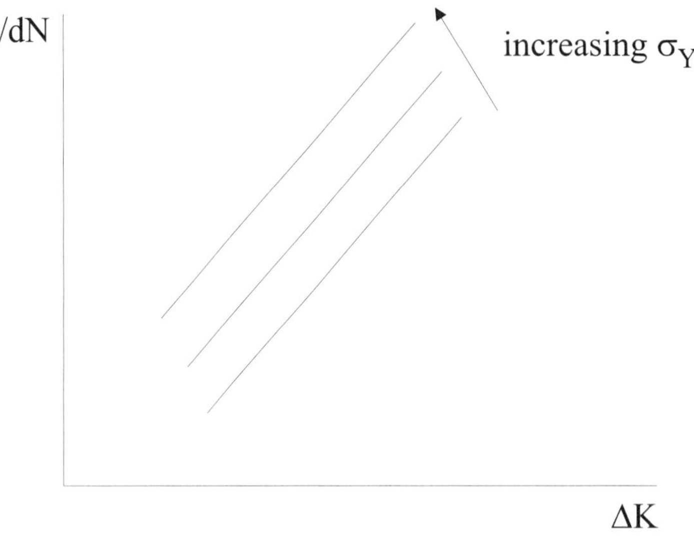

Figure 17.46: Crack growth rate, $\mathrm{d}a/\mathrm{d}N$, against stress intensity factor range, ΔK, as a function of yield stress, σ_Y.

where A is an environment sensitive material parameter.

Let us now examine the variation of crack length against number of cycles for Paris' law. With $\Delta K = Y\Delta\sigma\sqrt{\pi a}$ then substituting into (17.147)

$$\frac{\mathrm{d}a}{\mathrm{d}N} = C\left(Y\Delta\sigma\sqrt{\pi}\right)^m a^{m/2} \tag{17.151}$$

Separating variables and assuming that Y is not a function of a

$$\int \frac{\mathrm{d}a}{a^{m/2}} = \int C\left(Y\Delta\sigma\sqrt{\pi}\right)^m \mathrm{d}N \tag{17.152}$$

Performing the integration we have

$$\frac{1}{-m/2+1}a^{-m/2+1} = C\left(Y\Delta\sigma\sqrt{\pi}\right)^m N + \text{const} \tag{17.153}$$

With the initial condition of $a = a_i$ for $N = 0$ then we arrive at

$$\left(\frac{a}{a_i}\right)^{\kappa} = \frac{1}{1-\beta N} \tag{17.154}$$

where κ and β are given by

$$\kappa = \frac{m}{2} - 1; \quad \beta = \kappa a_i^{\kappa} C\left(Y\Delta\sigma\sqrt{\pi}\right)^m \tag{17.155}$$

Figure 17.47 illustrates the variation of a/a_i against N with fatigue crack growth accelerating as $N \to N_f$ where N_f is the number of cycles to failure and given by

$$N_f = \frac{1}{\beta} \tag{17.156}$$

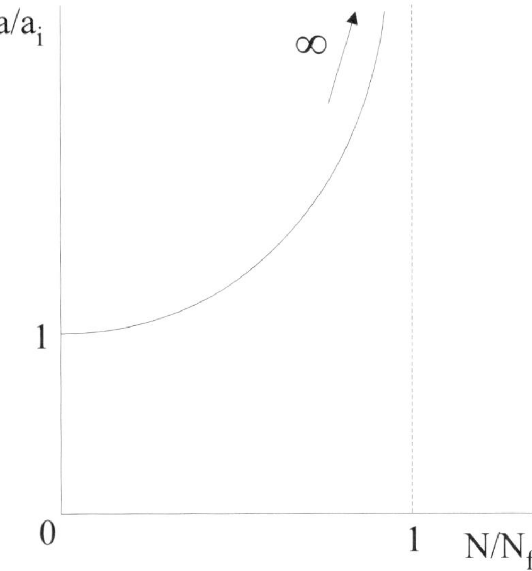

Figure 17.47: Normalised crack growth (a/a_i) as a function of normalised number of cycles (N/N_f) according to Paris' law.

with Figure 17.47 observed to be of an equivalent form as Figure 17.42 when $a_i = 0$.

Paris' law enables us to determine the number of cycles for a crack to grow from an initial length of a_i to a final length of a_f

$$\Delta N = \int_{a_i}^{a_f} \frac{1}{C(\Delta K)^m} \, \mathrm{d}a \tag{17.157}$$

Substituting for ΔK, (17.146)

$$\Delta N = \int_{a_i}^{a_f} \frac{1}{C\left(Y \Delta\sigma \sqrt{\pi a}\right)^m} \, \mathrm{d}a = \frac{1}{C(\Delta\sigma)^m \pi^{m/2}} \int_{a_i}^{a_f} \frac{1}{Y^m a^{m/2}} \, \mathrm{d}a \tag{17.158}$$

If the configuration correction factor, Y, is independent of a then

$$\Delta N = \frac{1}{C Y^m \pi^{m/2} (\Delta\sigma)^m} \frac{1}{1 - m/2} \left[a_f^{1-m/2} - a_i^{1-m/2} \right] \tag{17.159}$$

The following example will help illustrate the use of Paris' law to determine the number of cycles.

Example 17.5 Application of Paris' law to determine the number of fatigue cycles

A crack of length 25mm is detected at the toe of a fillet weld which is subject to cyclic loading that cycles from a minimum of 40MPa to a maximum of 50MPa. Let us determine the number of cycles for the crack to grow from 25mm to a length of 35mm assuming that Paris' law is representative of fatigue crack growth in the weld. The value of m will be assumed to be equal to 2 and the constant C is to be taken from the following which was fitted through available experimental data

$$C = \frac{1.315 \times 10^{-4}}{895.4^m}$$

It is to be assumed that Y is constant and equal to 1.15 throughout crack growth.

The cyclic stress range is $\Delta\sigma = \sigma_{\max} - \sigma_{\min} = 10$MPa and for m=2 then $C = 1.64 \times 10^{-10}$. From (17.158), re-arranging for N, we have for initial and final crack lengths a_i and a_f respectively

$$\Delta N = \frac{1}{CY^2\Delta\sigma^2\pi}\int_{a_i}^{a_f}\frac{\mathrm{d}a}{a}$$

Using the standard integral

$$\int\frac{\mathrm{d}x}{ax+b} = \frac{1}{a}\ln(ax+b)$$

then we find after integrating and substituting the relevant values

$$\Delta N = \frac{1}{CY^2\Delta\sigma^2\pi}\left[\ln a_f - \ln a_i\right] = \frac{1}{CY^2\Delta\sigma^2\pi}\ln\left(\frac{a_f}{a_i}\right) = 5\times 10^6\text{cycles}$$

17.11.4.4 The R-Ratio

Figure 17.48 illustrates equivalent ΔK values for two different constant amplitude cyclic loadings. Clearly, ΔK alone gives no indication of the proximity of $K_{\max 2}$ to the fracture toughness. To indicate

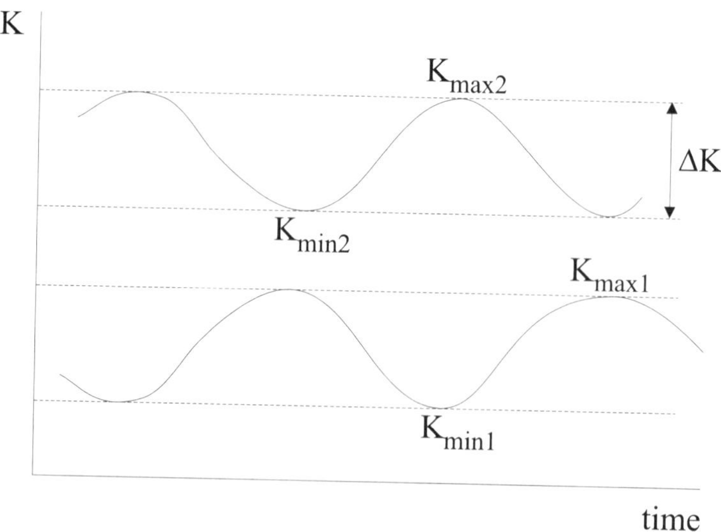

Figure 17.48: Same ΔK for different loading systems.

the amount by which the mean stress is removed from zero an R-ratio is generally considered

$$R = \frac{K_{\min}}{K_{\max}} \tag{17.160}$$

For the majority of steel alloys the variation of $R(R > 0)$ generally has negligible effect whereas aluminium alloys are more sensitive to R, Figure 17.49.

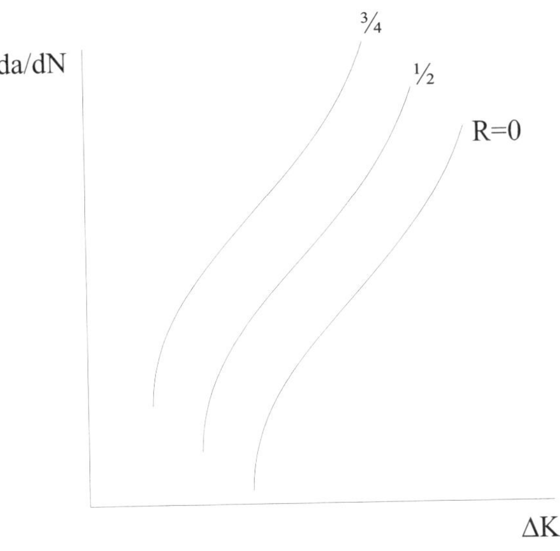

Figure 17.49: Influence of R for aluminium alloys.

17.11.4.5 Mixed-Mode Loading

Experimental investigations have shown that fatigue crack growth under combined mode I and mode II loadings is significantly increased when compared to pure mode I crack growth. Roberts and Kibler (1971) proposed the following mode II growth rate as a function of mode I

$$\frac{\mathrm{d}a}{\mathrm{d}N}(II, R_{II}) = \frac{\mathrm{d}a}{\mathrm{d}N}(I, R_I)\left(\frac{K_{IC}}{K_{IIC}}\right)^n; \quad R_i = \frac{K_{i\,\min}}{K_{i\,\max}}; \quad i = I, II \tag{17.161}$$

with n differing from Paris' exponent.

Investigations on inclined cracks in centre cracked plates which induce both mode I and mode II loadings have shown that higher growth rates are experienced when mode I is accompanied by a mode II component. Growth laws which use an effective stress intensity factor are of the form

$$\frac{\mathrm{d}a}{\mathrm{d}N} = C\,(\Delta K_e)^m \tag{17.162}$$

with one of the more popular forms of K_e given by

$$K_e = \left(K_I^4 + 8K_{II}^4\right)^{1/4} \tag{17.163}$$

17.11.5 Short Cracks

The previous section illustrated that the conventional way of representing fatigue crack growth data is in the form of $\mathrm{d}a/\mathrm{d}N$ versus ΔK in double logarithmic coordinates, Figure 17.45. When the cracks are physically and microstructurally small then fatigue crack growth experiences a decrease rather than an increase with respect to crack length, Figure 17.50. This is contrary to crack growth laws such as Paris' law in which fatigue crack growth is characterised by an increase in growth rate with crack length since $K \propto \sqrt{a}$. A physically short crack is one in which $a \leq 10/12$ grain sizes whereas a microstructurally short crack is one in which a is less than the characteristic grain size. When cracks are either

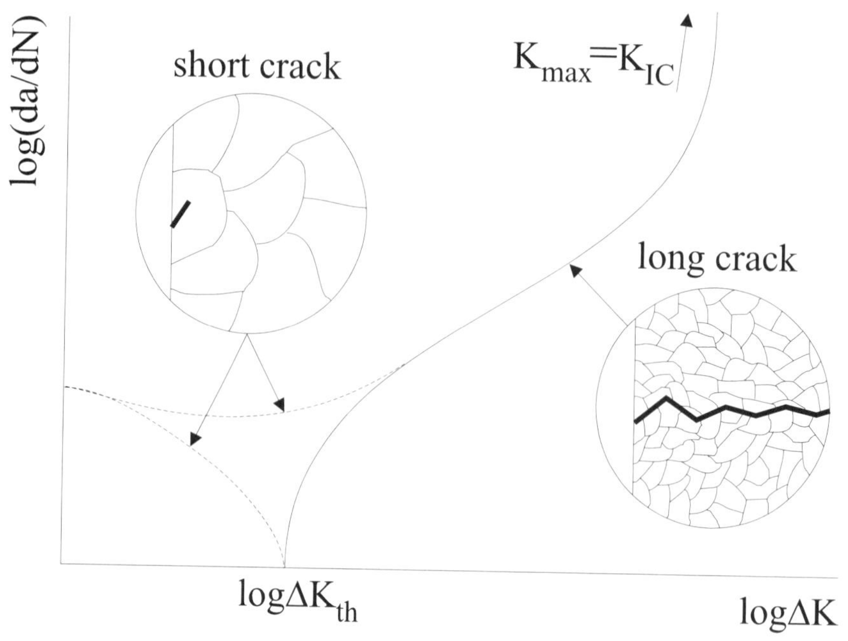

Figure 17.50: Crack growth rate against stress intensity factor range for both short and long cracks.

physically or microstructurally short then growth is highly influenced by the microstructure of the material. The grain size, shape and orientation of inclusions, microstructural barriers, material composition and local anisotropy all contribute to a highly complex fatigue crack growth mechanism. Figure 17.51 schematically illustrates typical experimentally observed crack growth rates against crack length. Case a) illustrates the case when the applied stress is below the fatigue limit, σ_{fl}, in which cracks propagate up to the microstructural threshold, d, but are unable to achieve the mechanical threshold crack length, a_{th}. At the fatigue limit, case b), then d corresponds with a_{th} and for applied stresses above the fatigue limit, case c), then the short and long crack regimes interact with the crack propagating to failure.

Hobson (1982) proposed one of the first quantitative models to describe short fatigue crack growth in terms of a microstructural parameter d which is assumed to be a material characteristic that is representative of the distance between obstacles to crack growth. From dimensional analysis Hobson obtained the following nonlinear crack growth equation

$$\frac{\mathrm{d}a}{\mathrm{d}N} = Ca^{\alpha}(d-a)^{1-\alpha}; \quad a \leq d \tag{17.164}$$

where C and α are empirical constants. The parameter C is assumed to be a function of both material and loading parameters such as Young's modulus, yield stress and the applied cyclic stress. Equation (17.164) models the observed behaviour of $\mathrm{d}a/\mathrm{d}N \to 0$ as $a \to d$.

Figure 17.52 illustrates the effect of α on $\mathrm{d}a/\mathrm{d}N$. From experimental observations Hobson argued that α must lie within the range $0 \leq \alpha \leq 1$ for (17.164) to be of practical use and proceeded to set $\alpha = 0$ which results in a linear decrease in $\mathrm{d}a/\mathrm{d}N$ with increasing a for $a < d$ according to (17.164). Hobson went further and equated d to the ferrite plate length of medium carbon steels and observed that C is a function of the applied cyclic stress, $\Delta\sigma$; that is $C = k\Delta\sigma^m$ where k and m are material constants.

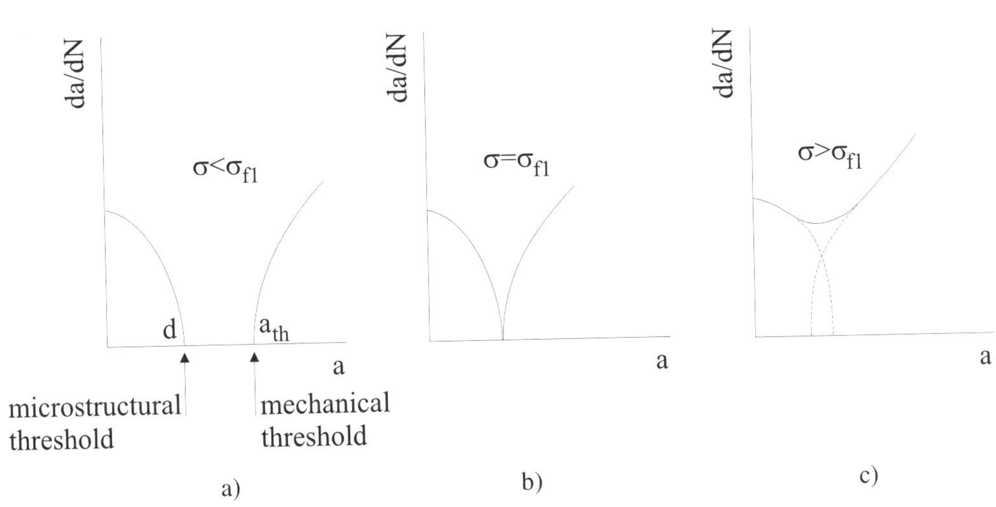

Figure 17.51: Schematic illustration of short fatigue crack growth. The short fatigue crack growth microstructural barrier size dimension is d and the linear elastic fatigue crack growth threshold is a_{th}. a) Applied stress is below the fatigue limit. b) Applied stress is equal to the fatigue limit. c) Applied stress is above the fatigue limit.

Consequently, short fatigue crack growth laws of the form follow from (17.164)

$$\frac{\mathrm{d}a}{\mathrm{d}N} = A\Delta\sigma^n(d-a) \tag{17.165}$$

where again A and n are empirically derived material constants.

When $\alpha = 0$ we can solve (17.164) exactly for a as a function of N. Separating variables we have

$$\int \frac{\mathrm{d}a}{d-a} = \int C\,\mathrm{d}N \tag{17.166}$$

Performing the integration with the initial condition $a = a_i$ at $N = 0$ and re-arranging we arrive at

$$a = d\left[1 - \left(1 - \frac{a_i}{d}\right)e^{-CN}\right] \tag{17.167}$$

and demonstrates the growth illustrated in Figure 17.43 for $a_i \leq a \leq d$.

17.11.5.1 The Kitagawa-Takahashi Curve

We will conclude our discussion of short fatigue cracks by examining the so-called *Kitagawa-Takahashi curve* which helps place short and long cracks into context, Figure 17.53. Examination of the Figure illustrates why cracks can grow at stress levels below ΔK_{th}. The line given by ΔK_{th} represents the threshold condition below which a crack should not grow according to LEFM. Linear elastic fracture mechanics is invalid when small-scale yielding conditions are exceeded and this approximately occurs when $\Delta\sigma \geq (2/3)\sigma_{cy}$ where σ_{cy} is the cyclic yield stress. The fatigue limit can be approximated to the cyclic yield stress range with LEFM not applicable in this regime. Several lines d_1, d_2, d_3 and so on can be shown to represent the size of microstructural units such as grain size, inclusion spacing and surface

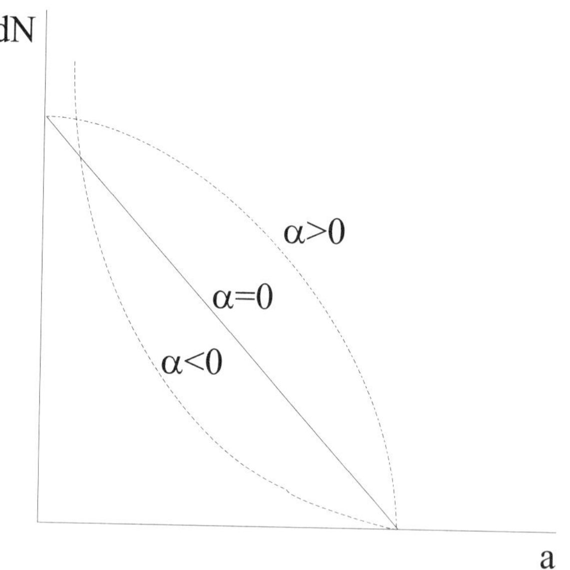

Figure 17.52: Influence of α on $\mathrm{d}a/\mathrm{d}N$.

finish which will all be expected to affect crack growth behaviour. However, at crack lengths $a > d_3$ a continuum mechanics approach is assumed to be applicable but may only be of sufficient accuracy for LEFM cracks in which $a > l$.

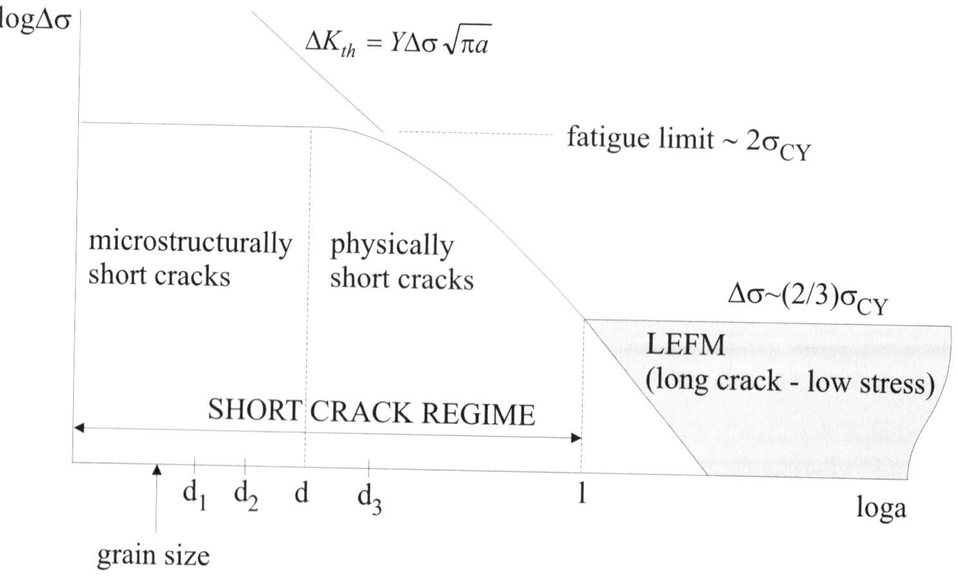

Figure 17.53: The Kitagawa-Takahashi curve.

17.12 The J-Integral

Rice (1968a,b) introduced an integral called J which can be shown to be path-independent and plays an important role in both elastic and elastic-plastic fracture mechanics. Consider a path of integration which extends from the lower face of a crack (in an anticlockwise sense) to the upper crack face and encompasses the crack tip, Figure 17.54

$$J = \int_\Gamma \left(W n_x - T_i \frac{\partial u_i}{\partial x} \right) \mathrm{d}s \tag{17.168}$$

where W denotes the strain energy density, n_i are the direction cosines of the outward unit normal vector, $T_i (= \sigma_{ij} n_j)$ is the traction or stress vector, u_i is the displacement vector and s is the arc length along the contour Γ.

The J-integral is in fact equivalent to the rate of decrease of potential energy, Π, with respect to the crack length, that is

$$J = -\frac{\partial \Pi}{\partial a} \tag{17.169}$$

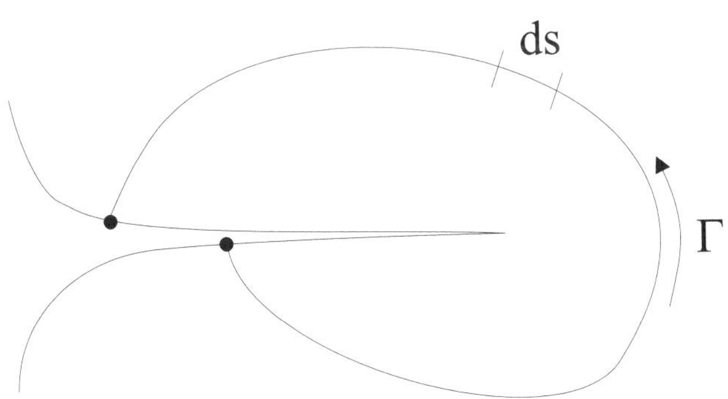

Figure 17.54: A contour enclosing a crack tip.

17.12.1 J is Equal to the Decrease in Potential Energy

To demonstrate that the J-integral is equivalent to the decrease in potential energy let us start by observing that the total potential energy of a two-dimensional body of a non-linear elastic material is given by, (10.37)

$$\Pi(a) = \int_A W \,\mathrm{d}A - \int_{\Gamma_T} T_i u_i \,\mathrm{d}s \tag{17.170}$$

where Γ_T denotes the section of the boundary Γ_0 which has prescribed tractions, Figure 17.55. Differentiating Π with respect to a

$$\frac{\mathrm{d}\Pi}{\mathrm{d}a} = \int_A \frac{\mathrm{d}W}{\mathrm{d}a} \mathrm{d}A - \int_{\Gamma_0} T_i \frac{\mathrm{d}u_i}{\mathrm{d}a} \mathrm{d}s \tag{17.171}$$

The boundary Γ_0 can be extended along the crack faces because $\mathrm{d}u_i/\mathrm{d}a = 0$ on Γ_u where the displacements are prescribed independently of a. In performing the integration let us introduce a translated

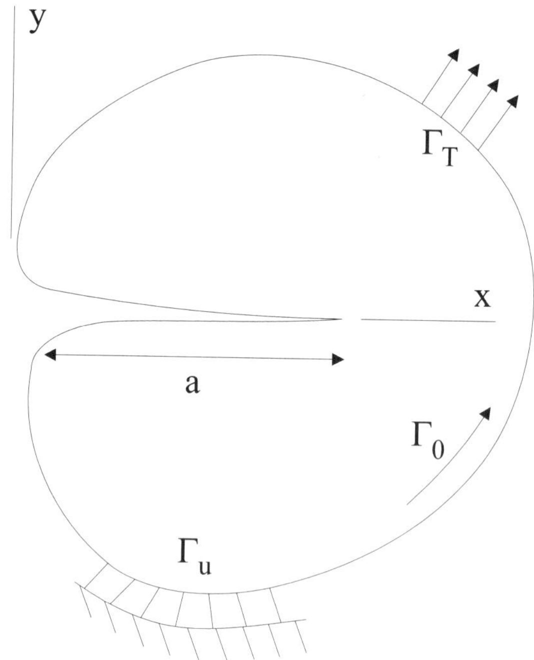

Figure 17.55: A cracked body with boundary Γ_0 and prescribed tractions and displacements on sections Γ_T and Γ_u respectively.

coordinate system

$$X_i = x_i - a\delta_{xx} \tag{17.172}$$

which is attached to the crack tip and where δ_{ij} denotes the Kronecker delta. It follows therefore that

$$\frac{\mathrm{d}}{\mathrm{d}a} = \frac{\partial}{\partial a} + \frac{\partial X_x}{\partial a}\frac{\partial}{\partial X_u} = \frac{\partial}{\partial a} - \frac{\partial}{\partial X_u} = \frac{\partial}{\partial a} - \frac{\partial}{\partial x} \tag{17.173}$$

since $\partial X_u/\partial a = -1$ and $\partial/\partial X_u = \partial/\partial x$. Thus, (17.171) can be written

$$\frac{\mathrm{d}\Pi}{\mathrm{d}a} = \int_A \left(\frac{\partial W}{\partial a} - \frac{\partial W}{\partial x}\right) \mathrm{d}A - \int_{\Gamma_0} T_i \left(\frac{\partial u_i}{\partial a} - \frac{\partial u_i}{\partial x}\right) \mathrm{d}s \tag{17.174}$$

From the definition of W, (9.2), and the constitutive relations (1.31)

$$\frac{\partial W}{\partial a} = \frac{\partial W}{\partial \varepsilon_{ij}}\frac{\partial \varepsilon_{ij}}{\partial a} = \sigma_{ij}\left(\frac{\partial u_i}{\partial a}\right)_{,j} \tag{17.175}$$

From the principle of virtual work

$$\int_A \frac{\partial W}{\partial a}\,\mathrm{d}A = \int_A \sigma_{ij}\left(\frac{\partial u_i}{\partial a}\right)_{,j} = \int_{\Gamma_0} T_i \frac{\partial u_i}{\partial x}\,\mathrm{d}s \tag{17.176}$$

Substituting (17.176) into (17.174) we have

$$-\frac{\mathrm{d}\Pi}{\mathrm{d}a} = \int_A \frac{\partial W}{\partial a}\,\mathrm{d}A - \int_{\Gamma_0} T_i \frac{\partial u_i}{\partial x}\,\mathrm{d}s \tag{17.177}$$

and an application of Gauss' divergence theorem to convert the area integral to a line integral gives

$$-\frac{\mathrm{d}\Pi}{\mathrm{d}a} = \int_{\Gamma_0} \left(W n_x - T_i \frac{\partial u_i}{\partial x} \right) \mathrm{d}s \tag{17.178}$$

Comparing (17.168) and (17.178) we observe that $J = -\partial\Pi/\partial a$ as required.

17.12.2 Path Independence

To demonstrate that the J-integral is path-independent consider the crack tip shown in Figure 17.56 with the two contours Γ and Γ_1 encompassing the crack tip. Consider the difference $J_1 - J$ evaluated on the

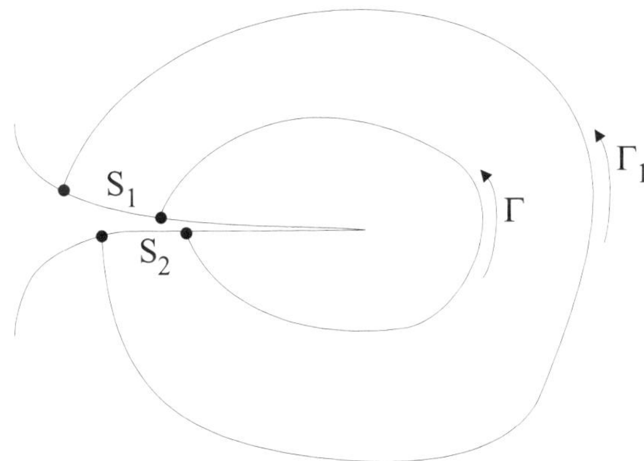

Figure 17.56: A crack tip enclosed by the two contours Γ and Γ_1.

paths Γ and Γ_1 respectively

$$J_1 - J = \int_{\Gamma_1+\Gamma+S_1+S_2} \left(W n_x - T_i \frac{\partial u_i}{\partial x} \right) \mathrm{d}s \tag{17.179}$$

where the contour is closed by including the crack face segments S_1 and S_2. With an application of the Gauss/Green divergence theorem we can convert the line integral to an area integral, where in general

$$\int_C (M\,\mathrm{d}x + N\,\mathrm{d}y) = \iint_A \left(\frac{\partial N}{\partial x} - \frac{\partial M}{\partial y} \right) \mathrm{d}x\,\mathrm{d}y \tag{17.180}$$

Referring to Figure 17.57 we observe that $n_x \mathrm{d}s = \mathrm{d}y$ and $n_y \mathrm{d}s = -\mathrm{d}x$ so that $W n_x \mathrm{d}s = W \mathrm{d}y$ and

$$T_i \frac{\partial u_i}{\partial x} \mathrm{d}s = \sigma_{ij} n_j \frac{\partial u_i}{\partial x} \mathrm{d}s = \sigma_{ix} \frac{\partial u_i}{\partial x} n_x\,\mathrm{d}s + \sigma_{iy} \frac{\partial u_i}{\partial x} n_y\,\mathrm{d}s = \sigma_{ix} \frac{\partial u_i}{\partial x} \mathrm{d}y - \sigma_{iy} \frac{\partial u_i}{\partial x} \mathrm{d}x \tag{17.181}$$

Applying these results to (17.179) we have

$$\begin{aligned} J_1 - J &= \iint_A \left[\frac{\partial W}{\partial x} - \frac{\partial}{\partial x_j} \left(\sigma_{ij} \frac{\partial u_i}{\partial x} \right) \right] \mathrm{d}A \\ &= \iint_A \left[\frac{\partial W}{\partial \varepsilon_{ij}} \frac{\partial \varepsilon_{ij}}{\partial x} - \sigma_{ij} \frac{\partial}{\partial x} \left(\frac{\partial u_i}{\partial x_j} \right) \right] \mathrm{d}A \\ &= \iint_A \left[\sigma_{ij} \frac{\partial \varepsilon_{ij}}{\partial x} - \sigma_{ij} \frac{\partial}{\partial x} \left(\frac{\partial u_i}{\partial x_j} \right) \right] \mathrm{d}A = 0 \end{aligned} \tag{17.182}$$

Hence, $J = J_1$ and J is therefore path-independent. The property that J is path-independent makes J a particularly attractive tool for analysing cracks since it has the same value for all paths of integration. This allows J to be evaluated on integration paths that are collapsed onto the near crack tip region or on far-field paths of integration.

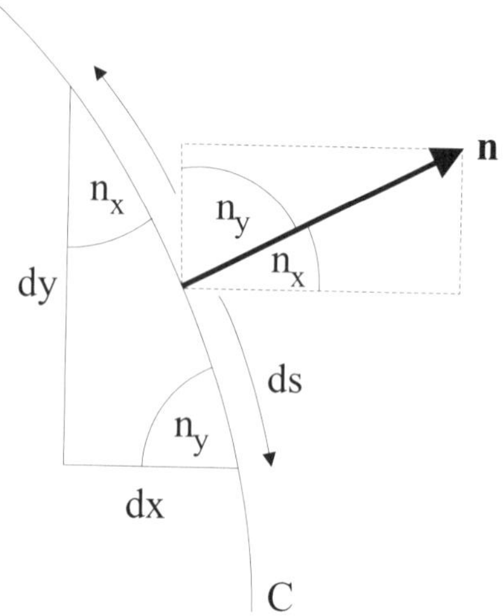

Figure 17.57: A contour C and outward normal **n**.

17.12.3 J as a Fracture Mechanics Parameter

The property that J is equivalent to the decrease in potential energy means that J and Griffith's energy release rate G are synonymous for linear elastic materials, (17.30)

$$J = G = \frac{1}{E'}\left(K_I^2 + K_{II}^2\right) + \frac{1}{2\mu}K_{III}^2 \tag{17.183}$$

where μ is the shear modulus and $E' = E$ for plane stress and $E' = (1-\nu^2)/E$ for plane strain.

The J-integral is equally applicable to non-linear materials and as an example let us consider evaluating J for the Dugdale cohesive zone model with yield zone d ahead of the crack tip, Figure 17.58. The extent of d is determined by the condition that the crack tip stresses are non-singular, §17.7.3. Let σ be the cohesive stress that can depend upon the separation $\delta = v^+ - v^-$ of the lower and upper crack faces. For a contour shrunk onto the boundary of the right hand side yielded zone then $n_x = T_x (= \sigma_{xj} n_j) = 0$ and J reduces to

$$J = -\int_0^d \sigma \left[\frac{\partial v^+}{\partial x} - \frac{\partial v^-}{\partial x}\right] \mathrm{d}x = -\int_0^d \sigma \frac{\partial \delta}{\partial x}\, \mathrm{d}x = \int_0^{\delta_{tip}} \sigma(\delta)\, \mathrm{d}\delta \tag{17.184}$$

where δ_{tip} is the crack tip opening displacement. If the cohesive stress is taken to be constant and equal to the tensile yield stress, σ_Y, then

$$J = \sigma_Y \delta_{tip} \tag{17.185}$$

Collecting the above results (17.183) and (17.185) we have, for mode I

$$J = G = \frac{K_I^2}{E'} = \sigma_Y \delta_{tip} \tag{17.186}$$

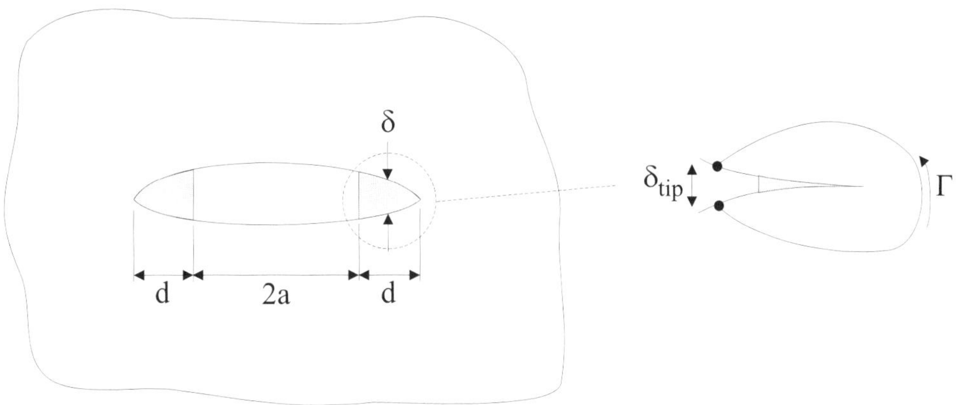

Figure 17.58: The Dugdale cohesive zone crack model.

Thus, the J-integral, Griffith's energy release rate, the stress intensity factor and the crack tip opening displacement are all equivalent fracture parameters. A fracture criterion can therefore be expressed in terms of any one of these four parameters and since the choice of parameter is academic then the generally adopted parameter is based on the particular application. This also helps to illustrate why the J-integral is so popular because it can be readily incorporated into finite element analyses with arbitrary paths of integration and eliminates the need for measuring small displacements or extrapolating stress intensity factors.

It can be shown for general three-dimensional deformation fields that J generalises to

$$J_k = \int_\Gamma \left(W n_k - T_i \frac{\partial u_i}{\partial x_k} \right) \mathrm{d}s \tag{17.187}$$

By evaluating J_k for crack extensions parallel and perpendicular to the crack it can be shown that

$$J_x = J = \left(\frac{\kappa + 1}{8\mu} \right) \left(K_I^2 + K_{II}^2 \right); \quad J_y = -\left(\frac{\kappa + 1}{8\mu} \right) 2K_I K_{II} \tag{17.188}$$

with only the singular (inverse square-root) contributions to the elastic crack tip stress field used. It is observed that J_y vanishes when either or both of K_I or K_{II} are zero.

17.13 The L and M Integrals and their Application

The previous section introduced the $J_k(k = x, y)$ integrals and the relationship between J, K_I, G and δ. We will now introduce two additional conservation integrals, L and M, and illustrate how the conservation laws $L(C) = 0$ and $M(C) = 0$, for any closed contour, can be used in the analysis of defects. The integrals J_k, L and M are given by, Knowles and Sternberg (1972)

$$\begin{aligned} J_k &= \oint_C (W n_k - T_i u_{i,k}) \, \mathrm{d}s \\ L &= \varepsilon_{zkj} \oint_C (W n_k - T_i u_{i,k}) \, x_j \, \mathrm{d}s + \varepsilon_{zkj} \oint_C T_k u_j \, \mathrm{d}s \\ M &= \oint_C (W n_k - T_i u_{i,k}) \, x_k \, \mathrm{d}s \end{aligned} \tag{17.189}$$

where $W(\varepsilon_{ij})$, $\varepsilon_{ij} = (u_{i,j} + u_{j,i})/2$, is the strain energy density, $T_i(= \sigma_{ij}n_j)$ is the traction along the contour C, n_j is the unit normal, ds is a differential element of contour C and ε_{ijk} is the Ricci alternating tensor. Generally, the second integral in L is neglected because it physically differs from the other integrals. In this case the similarity between J_k, L and M is observed. Evaluating J_k, L and M for an idealised centre cracked plate subject to a far-field loading of σ and τ it is found that, Figure 17.59a)

$$\begin{aligned} J_x &= \left(\frac{\kappa+1}{8\mu}\right)(K_I^2 + K_{II}^2); \quad J_y = -\left(\frac{\kappa+1}{8\mu}\right)2K_I K_{II} \\ M &= \left(\frac{\kappa+1}{8\mu}\right)(K_I^2 + K_{II}^2)\,a; \quad L = -\left(\frac{\kappa+1}{8\mu}\right)2K_I K_{II} a \end{aligned} \tag{17.190}$$

where an open contour of integration is taken for evaluating J_x and J_y with the origin at the crack tip, Figure 17.59b), whereas the origin of coordinates is taken at the centre of the crack for L and M, Figure 17.59c).

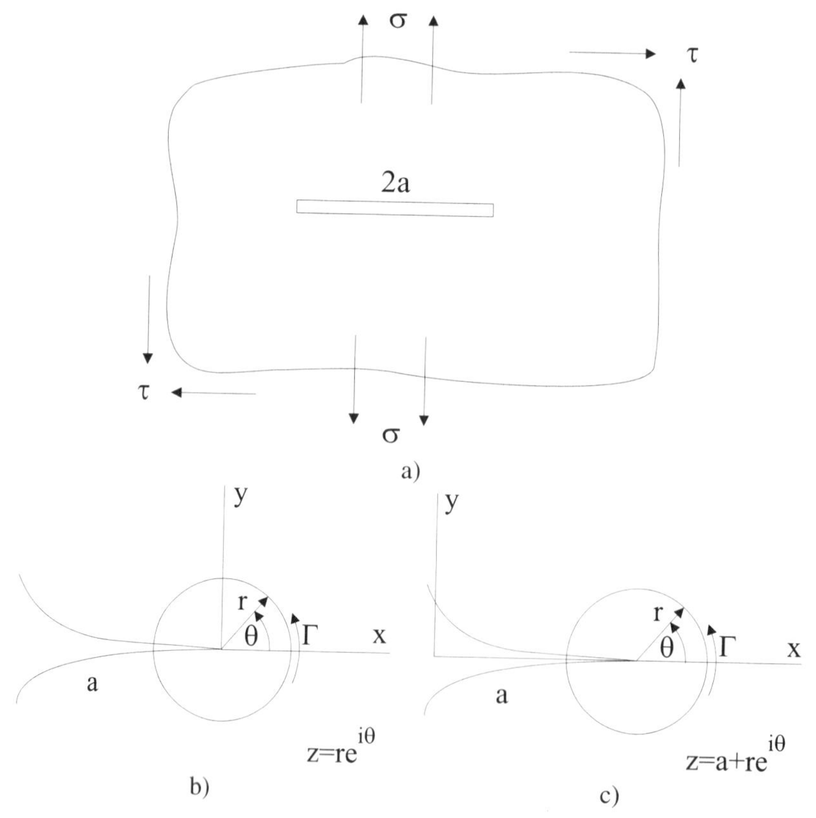

Figure 17.59: A centre cracked plate. a) A plate containing a centre crack of length $2a$ and subject to far-field loadings of σ and τ. b) A crack with coordinate system centred at the crack tip. c) A crack with coordinate system centred at the centre of the crack.

The physical meaning of the integrals is interpreted that J_k is the potential energy release rate of a crack translating in the x_k direction, while L and M are the potential energy release rates of a crack uniformly rotating and expanding about the x_k axis respectively. From (17.190) we note that

$$M = aJ_x; \quad L = -aJ_y \tag{17.191}$$

It is exactly these simple relationships between the integrals that allows us to determine stress intensity factors for various configurations. Later we will consider the determination of K_I for several geometries but first we require a result for M for the Boussinesq wedge subject to point loading at its apex, Figure 17.60. The stress field for this configuration is purely radial and given by, (14.218)

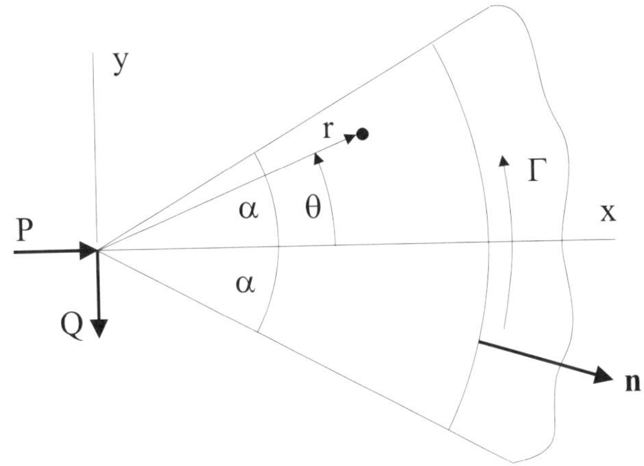

Figure 17.60: Wedge geometry of included angle α subject to point forces P and Q at the wedge apex.

$$\sigma_{rr} = -\frac{2P\cos\theta}{r(2\alpha+\sin 2\alpha)} + \frac{2Q\sin\theta}{r(2\alpha-\sin 2\alpha)}; \quad \sigma_{\theta\theta} = \tau_{r\theta} = 0 \tag{17.192}$$

It can be shown that the integrand of M is, Freund (1978)

$$Wx_in_i - T_ku_{k,i}x_i = -\frac{1}{2}\left(\frac{\kappa+1}{8\mu}\right) r\sigma_{rr}^2 \tag{17.193}$$

and evaluating M, (17.189), for the arc $\Gamma(-\alpha \leq \theta \leq \alpha)$ we have

$$M = -\left(\frac{\kappa+1}{8\mu}\right)\left[\frac{P^2}{2\alpha+\sin 2\alpha} + \frac{Q^2}{2\alpha-\sin 2\alpha}\right] \tag{17.194}$$

L can be similarly evaluated but we will only consider M in the present discussion. Note also that M is path-independent whereas J_x and J_y are path-dependent for the Boussinesq wedge and cannot therefore be used directly to determine K_I for cracked geometries via the Boussinesq wedge, Seed (1997).

As an application of the above results let us consider the following well-known geometries.

17.13.1 Semi-Infinite Crack with Symmetrical Point Loading

Consider the determination of K_I for the semi-infinite crack shown in Figure 17.61 subject to point loading on both faces of the crack. There is no contribution to M from the straight segments of Γ along the crack faces since they are radial segments in which there is zero traction. There is no contribution from the infinitely large contour Γ_∞ since the applied loads are self-equilibrating. It is assumed that the stress state in the vicinity of the applied loads approximates to that of a half-plane which is obtained by

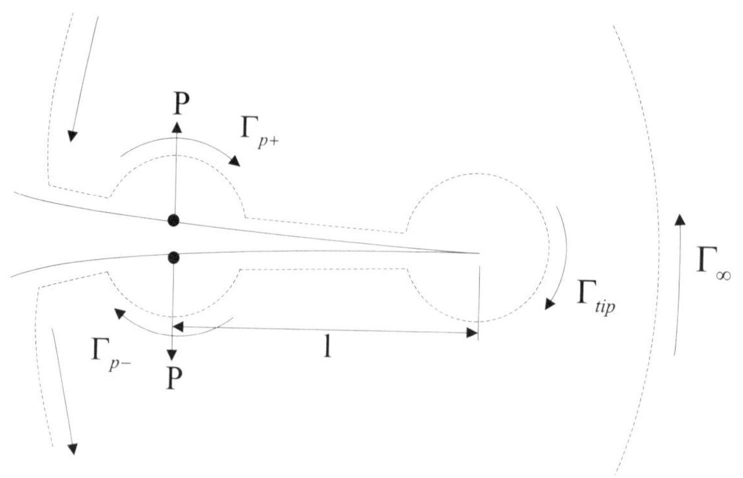

Figure 17.61: A semi-infinite crack subject to symmetrical crack face point loading.

letting $\alpha = \pi/2$ in the Boussinesq wedge. Thus, the value of $M(\Gamma_{p+,-})$ with $\alpha = \pi/2$, point loading of $-P$ and $Q = 0$ is, (17.194)

$$M\left(\Gamma_{p+,-}\right) = -\left(\frac{\kappa+1}{8\mu}\right)\frac{P^2}{\pi} \tag{17.195}$$

From (17.191) the value of M at the crack tip is

$$M\left(\Gamma_{tip}\right) = lJ \tag{17.196}$$

where l is the distance from the point loads to the crack tip. From the conservation law $M(C) = 0$

$$M\left(\Gamma_{p+}, \Gamma_{tip}, \Gamma_{p-}, -\Gamma_{\infty}\right) = 0 \tag{17.197}$$

we have

$$-2\left(\frac{\kappa+1}{8\mu}\right)\frac{P^2}{\pi} + lJ = 0 \tag{17.198}$$

Using the result $J = ((\kappa+1)8\mu)/K_I^2$ from (17.188) and solving for K_I we arrive at the well-known result

$$K_I = \sqrt{2}\frac{K_I}{\sqrt{\pi l}} \tag{17.199}$$

noting that K_I is independent of the crack length a and that $K_I \to \infty$ as $l \to 0$.

17.13.2 Symmetrically Point Loaded Finite Crack

Figure 17.62 illustrates a finite crack of length a subject to crack face point loading P. It follows that $M(\Gamma_{\infty}) = 0$ and the value of M at the points of loading is for $\alpha = \pi/2$, (17.194)

$$M\left(\Gamma_{p-}\right) = M\left(\Gamma_{p+}\right) = -\left(\frac{\kappa+1}{8\mu}\right)\frac{P^2}{\pi} \tag{17.200}$$

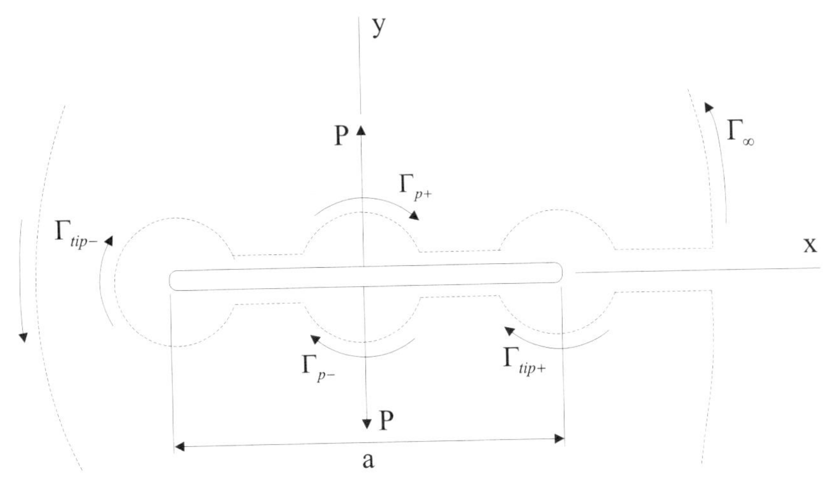

Figure 17.62: A finite crack of length a subject to symmetrical crack face point loading.

whereas the crack tip contributions are, (17.191)

$$M\left(\Gamma_{tip-}\right) = M\left(\Gamma_{tip+}\right) = \frac{a}{2}J = \frac{a}{2}\left(\frac{\kappa+1}{8\mu}\right)K_I^2 \tag{17.201}$$

Therefore, the conservation law

$$M\left(\Gamma_{tip-}, \Gamma_{p+}, \Gamma_{tip+}, \Gamma_{p-}, -\Gamma_{\infty}\right) = 0 \tag{17.202}$$

leads to

$$K_I = \sqrt{2}\frac{P}{\sqrt{\pi a}} \tag{17.203}$$

17.13.3 Corner-Loaded Surface Crack in a Semi-Infinite Half-Plane

Let us conclude our application of the $M = 0$ conservation law by considering the indentation problem of a half-plane loaded by a point force P a distance b from the mouth of a crack of length a which is lying normal to and breaking the surface, Figure 17.63a). When $b = 0$ then we essentially have two Boussinesq wedges; one point loaded (Γ_{0+}) and the other with no loading (Γ_{0-}), Figure 17.63b), with $M(\Gamma_{0-}) = 0$ and, (17.191)

$$M\left(\Gamma_{tip}\right) = aJ = a\left(\frac{\kappa+1}{8\mu}\right)K_I^2 \tag{17.204}$$

$M(\Gamma_{\infty})$ is found by approximating a far-field contour as detecting a half-plane solution with $\alpha = \pi/2$, (17.194)

$$M\left(\Gamma_{\infty}\right) = -\left(\frac{\kappa+1}{8\mu}\right)\frac{P^2}{\pi} \tag{17.205}$$

$M(\Gamma_{0+})$ is found by resolving P into normal and perpendicular components to the wedge apex; $P_r = P\cos\alpha$ and $Q_r = P\sin\alpha$ with $\alpha = \pi/4$, (17.194)

$$M\left(\Gamma_{0+}\right) = -\left(\frac{\kappa+1}{8\mu}\right)\left[\frac{P^2\cos^2\alpha}{2\alpha+\sin 2\alpha} + \frac{P^2\sin^2\alpha}{2\alpha-\sin 2\alpha}\right] = -\left(\frac{\kappa+1}{8\mu}\right)\left[\frac{2\pi P^2}{\pi^2-4}\right] \tag{17.206}$$

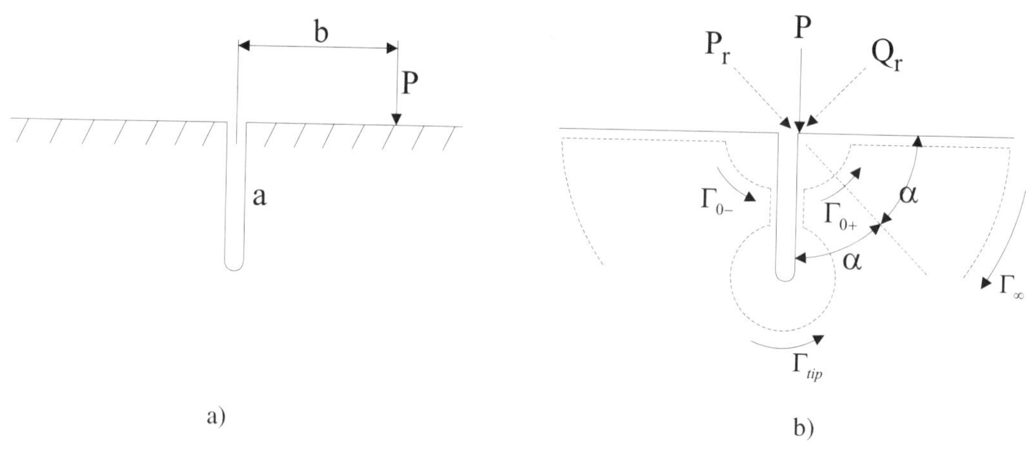

Figure 17.63: A semi-infinite half-plane containing a surface breaking crack that is lying normal to the free surface. a) The general case of $b \neq 0$. b) A corner-loaded crack with $b = 0$.

Using the conservation law

$$M\left(\Gamma_{0-}, \Gamma_{tip}, \Gamma_{0+}, \Gamma_{\infty}\right) = 0 \tag{17.207}$$

we arrive at the following value for K_I

$$K_I = -\frac{P}{\sqrt{\pi a}} \frac{2}{\sqrt{\pi^2 - 4}} \tag{17.208}$$

It is customary to normalise stress intensity factors resulting from point loading by $P/\sqrt{\pi a}$ so that

$$\frac{K_I \sqrt{\pi a}}{P} = -\frac{2}{\sqrt{\pi^2 - 4}} \approx -0.8255 \tag{17.209}$$

noting that K_I is negative.

Finally, Figure 17.64 illustrates the variation of both K_I and K_{II} for $b \neq 0$ with K_I being negative for all values of b/a and with K_{II} exhibiting a discontinuity at $b/a = 0$.

17.14 Conclusion

Hopefully the size of the present Chapter has helped emphasise the current importance attached to the discipline of fracture and fatigue in assessing the strength of engineering materials and components. The classical approach adopted when assessing the strength of a component was to assume the component to be defect free. Fracture mechanics allows us to take the alternative approach of assessing the worst case scenario by asking what if a defect of a given size and orientation is present. Throughout the present Chapter we have covered all of the key topics of both fracture and fatigue such as the stress intensity factor, the T-stress, crack tip plasticity, plane strain fracture toughness, methods for determining stress intensity factors, long and short fatigue cracks and the elastic-plastic J-integral.

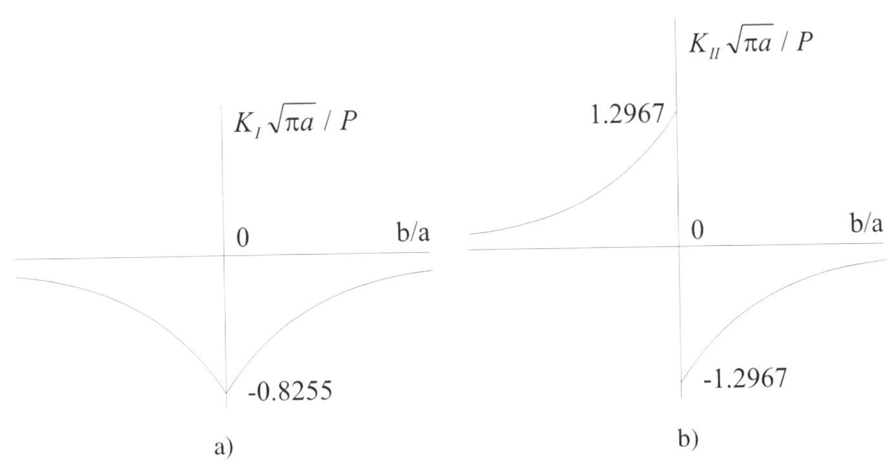

Figure 17.64: Variation of a) K_I and b) K_{II} for a surface breaking crack of length a subject to a point load P acting a distance b on the free surface.

17.15 References and Further Reading

○ BS 5447 (1977) *Methods of Test for Plane Strain Fracture of Metallic Materials*, British Standards Institute. Refer also to BS 7448:1991 and BS EN ISO 12737:1999 which have superseded BS 5447.

○ Cartwright, D. J. and Rooke, D. P. (1974) *Approximate Stress Intensity Factors Compounded from Known Solutions*, Engng. Fract. Mech., 6, 563-571.

○ Donahue, R. J., Clark, H. M., Atammi, P., Kumble, R. and McEvily, A. J. (1972) *Crack Opening Displacement and the Rate of Fatigue Crack Growth*, Int. J. Fract. Mech., 8, 209-219.

○ Erdogan, F. and Sih, G. C. (1963) *On Crack Extension in Plates under Plane Loading and Transverse Shear*, J. Basic Engng., 85, 519-527.

○ Forman, R. G., Kearney, V. E. and Engle, R. M. (1967) *Numerical Analysis of Crack Propagation in Cycli-Loaded Structures*, J. Basic Engng., 89, 459-464.

○ Freund, L. B. (1978) *Stress Intensity Factor Calculations Based on a conservation Integral*, Int. J. Solid Struct., 14, 241-250.

○ Gordon, J. E. (1976) *The New Science of Strong Materials or Why You Don't Fall Through the Floor*, Penguin, New York.

○ Griffith, A. A. (1921) *The Phenomena of Rupture and Flow in Solids*, Phil. Trans. Roy. Soc. Lond., A221, 163-197.

○ Hellen, T. K. and Blackburn, W. S. (1975) *The Calculation of Stress Intensity Factors for Combined Tensile and Shear Loading*, Int. J. Fract., 11, 605-617.

○ Hobson, P. D. (1982) *The Formation of a Crack Growth Equation for Short Cracks*, Fatigue Fract. Engng. Mater. Struct., 5, 323-327.

○ Hutchinson, J. W. (1968) *Singular Behaviour at the End of a Tensile Crack in Hardening Material*, J. Mech. Phys. Solids, 16, 13-31.

○ Inglis, C. E. (1913) *Stresses in a Plate due to the Presence of Cracks and Sharp* Corners, Trans. Institute Naval architects, London, 60, 219.

○ Irwin, G. R. (1957) *Analysis of Stresses and Strains Near the End of a Crack Traversing a Plate*, J. Appl. Mech., 24, 361-364.

∘ Irwin, G. R. (1962) *Crack Extension Force for a Part-Through Crack in a Plate*, J. Appl. Mech., 29, 651-654.

∘ Irwin, G. R. (1964) *Structural Aspects of Brittle Fracture*, Applied Materials Research, 3, 65-81.

∘ Knowles, J. K. and Sternberg, E. (1972) *On a Class of Conservation Laws in Linearised and Finite Elastostatics*, Arch. rat. Math. Anal., 44, 187-211.

∘ McEvily, A. J. and Groeger, J. (1977) *On the Threshold for Fatigue Crack Growth*, In *The Fourth Int. Conference on Fracture*, Vol. II, Canada, 1293-1298.

∘ Miller, K. J. (1993) *Materials Science Perspective of Metal Fatigue Resistance*, Mater. Sci. Technol., 9, 453-462.

∘ Newman, J. C. (1971) *An Improved Method of Collocation for the Stress Analysis of Cracked Plates with Various Shaped Boundaries*, NASA TND-6376.

∘ Newman, J. C. and Raju, I. S. (1981) *An Empirical Stress Intensity Factor Equation for Surface Cracks*, Engng. Fract. Mech., 15, 185-192.

∘ Paris, P. and Erdogon, F. (1963) *Critical Analysis of Crack Propagation Laws*, J. Basic Engng., 85, 528-534.

∘ Parker, A. P. (1981) *The Mechanics of Fracture and Fatigue: An Introduction*, E. & F. N. Spon, London.

∘ Rice, J. R. (1968a) *Mathematical Analysis in the Mechanics of Fracture*, In *Fracture: An Advanced Treatise*, Vol. II, ed. H. Liebowitz, Academic Press, New York, pp.191-311.

∘ Rice, J. R. (1968b) *A Path Independent Integral and the Approximate Analysis of Strain Concentration by Notches and Cracks*, J. Applied Mech., Transactions of the ASME, June, 379-386.

∘ Rice, J. R. (1974) *Limitations of the Small Scale Yieling Approximation for Crack Tip Plasticity*, J. Mech. Phys. Solids, 22, 17-26.

∘ Roberts, R. and Kibler, J. J. (1971) *Some Aspects of Fatigue Crack Propagation*, Engng. Fract. Mech., 2, 243-260.

∘ Seed, G. M. and Miller, K. J. (1991) *Short Cracks at Biaxially Stressed Notches: A Fine-Mesh Elastic-Plastic Finite Element Analysis*, Fatigue Fract. Engng. Mater. Struct., 14, 259-275.

∘ Seed, G. M. (1997) *The Boussinesq Wedge and the J_k, L and M Integrals*, Fatigue Fract. Engng. Mater. Struct., 20, 907-916.

∘ Sih, G. C. (1973) *Methods of Analysis and Solutions of Crack Problems*, Mechanics of Fracture, Vol. I, ed. G. C. Sih, Noordhoff, Leiden.

∘ Sneddon, I. N. (1946) *The Distribution of Stress in the Neighbourhood of a Crack in an Elastic Solid*, Proc. R. Soc. Lond., A, 187, 229-260.

∘ Timoshenko, S. P. (1953) *History of the Strength of Materials*, McGraw-Hill, New York.

∘ Williams, M. L. (1957) *On the Stress Distribution at the Base of a Stationary Crack*, J. Appl. Mech., 24, 109-114.

∘ Wu, E. M. (1967) *Application of Fracture Mechanics to Anisotropic Plates*, J. Appl. Mech., 34, 967-974.

17.16 Exercises

17.1 Describe what is meant by the stress intensity factor and the T-stress giving typical results for well known geometries such as centrally-cracked and edge-cracked plates subject to uniaxial, biaxial and shear loadings.

17.2 An approximate expression for the mode I stress intensity factor for a symmetrically cracked circular hole of radius R which is subject to a remote uniformly applied loading of σ is

$$K_I = \left[1 + 2.365\left(\frac{R}{R+a}\right)^{2.4}\right]\sigma\sqrt{\pi a}$$

The stress intensity factor and T-stress for a single-edge cracked semi-infinite plate subject to remote uniformly applied loading of σ are

$$K_I = 1.1215\sigma\sqrt{\pi a}; \quad T = -0.5258\sigma$$

Discuss the behaviour of the expressions for K_I and T for both configurations and their interaction with respect to an uncracked hole in a plate. Also discuss the above configurations with reference to a centrally cracked infinite plate with remote uniformly applied loading.

17.3 Discuss the effect of the thickness of a test specimen on the fracture toughness of a material. Describe how you would determine the plane strain fracture toughness of a material and state the requirements that must be satisfied before a test is considered valid.

17.4 A large plate made of aluminium alloy with yield stress $\sigma_Y = 425$MPa is tested with a central crack ($Y = 1$) of length 25mm and the fracture stress is observed to be $\sigma_f = 220$MPa. Calculate the fracture toughness for the following two cases:

i) using LEFM analyses
ii) using the small-scale plasticity correction for Irwin's model.

17.5 A circular rotor is of radius $R = 350$mm and rotates at a speed of 12×10^3 revolutions per minute. A defect is detected in the rotor to be of length $a = 12$mm as shown in Figure 17.65. Obtain a value for the critical crack length. Based on Paris' crack growth law how many times can the rotor be run up to speed before the crack attains the critical crack length? Assume that the rotor has the following properties: Poisson's ratio of $\nu = 0.33$, density of $\rho = 8 \times 10^3$kgm^{-3}, plane strain fracture toughness of $K_{IC} = 85$MNm$^{-3/2}$ and Paris law exponent $m = 3$ and constant $C = 4.1 \times 10^{-11}$ for growth in m/cycle.

The mode I stress intensity factor for the above geometry is to be taken as

$$K_I = Y\rho\frac{\omega^2 R^2}{8}\left(\frac{3-2\nu}{1-\nu}\right)\sqrt{\pi a}$$

where Y is the configuration correction factor and is a function of a/R and should be taken to be equal to 0.55 up to the critcal crack length and ω is the rotor rotational speed in radians/second.

17.6 Describe both long and short fatigue crack growth, emphasising the key differences between the two types of growth. In your discussion make use of the Kitagawa-Takahashi diagram.

Short fatigue cracks are observed to propagate in a medium carbon steel according to Hobson's law. From experimental investigations it is found that $\alpha = 0$ and C is given by the following empirical formula

$$C = 1.64 \times 10^{-34}(\Delta\sigma)^{11.14}$$

where $\Delta\sigma$ represents the cyclic stress range. The value of the microstructure barrier dimension, d, is taken to be equal to the long-crack threshold value assuming that $\Delta K_{th} = 6\text{MPa}\sqrt{\text{m}}$ and the configuration correction factor $Y = 2/\pi$. Determine the number of cycles of a short crack to propagate from an initial *total* length of 10μm to a final length which corresponds to 84% of the threshold crack length with the cyclic stress range $\Delta\sigma = 638$MPa.

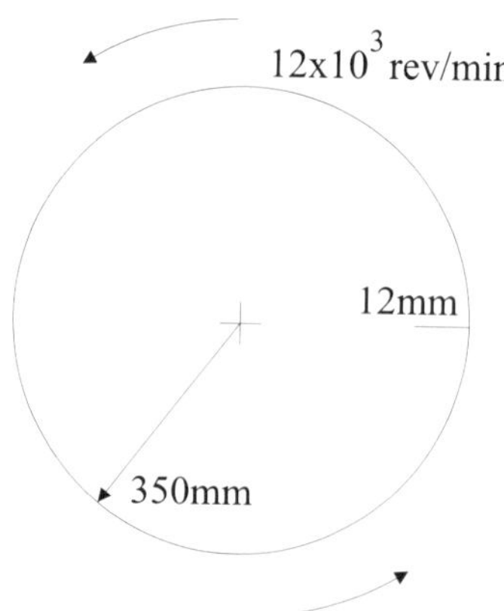

Figure 17.65: Exercise: A circular rotor with a radial crack of length 12mm.

17.7 In a purely mode I plane strain finite element analysis of a cracked plate the J-integral is found to equal 2.4×10^3N/m. If Young's modulus and Poisson's ratio of the plate are 210GPa and 0.3 respectively then determine the mode I stress intensity factor.

Chapter 18

Creep of Materials

18.1 Introduction

This Chapter examines the creep of materials and begins by defining the three key phases of creep. An overview of the driving mechanisms of creep is then presented. The following sections then examine both the separate and combined time, stress and temperature dependencies on the creep strain. The Chapter concludes by examining the three applications of the tension of a bar, the torsion of a circular tube and the bending of a rectangular beam. These are the same problems discussed in Chapter 12 when we examined the elastic-plastic behaviour of materials, and it is instructive to compare the two sets of problems.

18.2 Creep

Creep is the time-dependent inelastic deformation of a material subject to a given loading over a prolonged period of time. While some metallic materials such as lead creep at room temperature, creep is generally associated with loading at significantly higher temperatures. For crystalline materials, the mechanics of materials approach to creep focuses on a macroscopic description of the accumulation of creep strain, ε_c, with time t at a given stress, σ, and temperature, T. For constant σ and T the three-stage creep curve of Figure 18.1 is typically observed. The three key regions of Figure 18.1 are as follows:

A Primary Creep: period of work hardening in which the creep rate, $\mathrm{d}\varepsilon_c/\,\mathrm{d}t$, decreases with time.

B Secondary Creep: period where there is a balance between work-hardening due to plastic flow and thermal softening and the creep rate is constant.

C Tertiary Creep: period in which metallurgical instability occurs with an increasing creep rate until failure.

The secondary creep period is generally the key region of interest from an engineering design perspective. In the following sections we will examine the individual and combined dependencies of t, σ and T on the creep strain but first let us briefly discuss the mechanisms of creep.

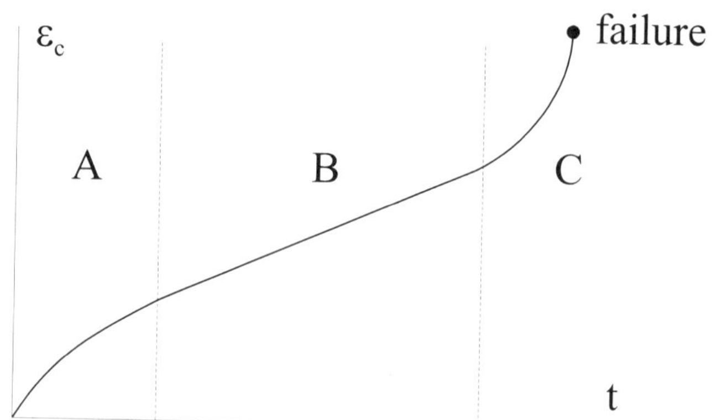

Figure 18.1: Creep strain against time for constant stress and temperature.

18.3 Mechanisms of Creep

18.3.1 Single Crystals

Of greatest engineering significance is secondary creep and as a result several models have been proposed. Weertman (1955) assumed the driving mechanism of secondary creep to be the climb of edge dislocations over obstacles which can be either precipitates or dislocation entanglements, Figure 18.2. Once a dislocation has climbed out of its slip plane and cleared an obstacle then it can further propagate and generate creep strain. The rate of creep strain then becomes a function of i) the strain resulting from the glide of the dislocation as it propagates from one obstacle to another, ii) the number of dislocations undergoing this climb-glide sequence and iii) the time required to overcome obstacles. Other models of secondary creep have been proposed such as the reduced propagation of screw dislocations due to the drag of vacancy-forming sessile jogs, Hirsch and Warrington (1961).

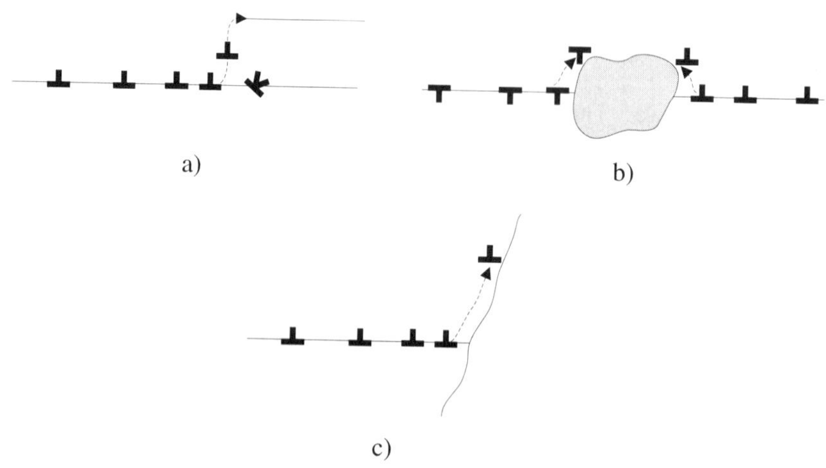

Figure 18.2: Climb of edge dislocations over a) a dislocation entanglement, b) a precipitate and c) along a grain boundary.

18.3.2 Polycrystalline Metals

In a polycrystalline material there are several additional mechanisms which contribute to creep, such as direct diffusion and the presence of grain boundaries. The energy of formation of vacancies and interstitials simultaneously inside a material is very high and it is therefore easier for vacancies to migrate in from grain boundaries and free surfaces. However, since vacancy migration rate varies linearly with stress then it is only important at low stresses combined with high temperatures in order to achieve a high diffusion coefficient. Of greater importance is the mechanism of sliding between two grains. At low temperatures grain-boundary sliding results in more rapid hardening whereas at high temperatures grain-boundary sliding accelerates the creep rate. The sliding of grain boundaries can be accompanied by diffusion and can lead to several experimentally observed mechanisms such as fold-formation at triple junctions of grains, opening up of a crack and the general deformation of the grains themselves. For further discussions on the mechanisms of creep refer to McClintock and Argon (1966) and Friedel (1964).

18.4 Time Dependence of Creep Strain

In this section we will consider the time dependence on creep. For the lower range of homologous[1] temperatures $T/T_m < 1/2$ primary creep dominates; where T_m is the absolute melt temperature. Figure 18.3 illustrates the creep strain, ε_c, against time, t, for a high stress σ. When σ is low then a limited amount of

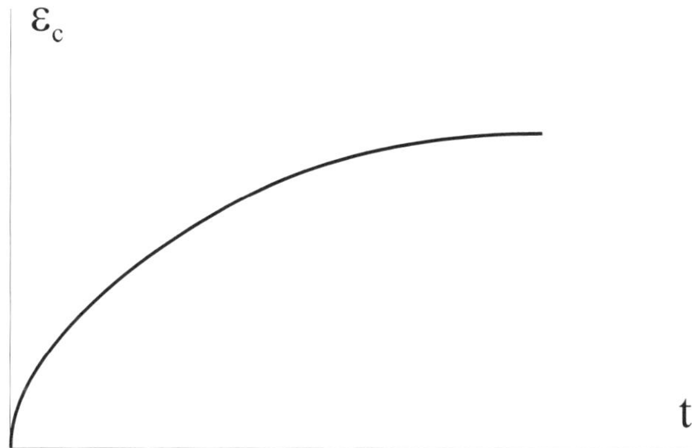

Figure 18.3: Creep strain against time for $T/T_m < 1/2$ and high stress.

logarithmic primary creep is present, Figure 18.4

$$\varepsilon_c(t) = \alpha \ln t; \quad \alpha = \alpha(\sigma, T) \tag{18.1}$$

For example, below room temperature both copper and aluminium exhibit logarithmic primary creep. When σ increases then primary creep is better described by Bailey's (1935) power law

$$\varepsilon_c(t) = \alpha t^m; \quad 0 < m < 1 \tag{18.2}$$

where α depends on plastic strain, stress and temperature. When $m = 1/3$ then Bailey's power law appears in Andrade's (1910) law

$$\varepsilon_c(t) = (1 + \alpha t^{1/3})e^{\beta t} - 1 \approx \alpha t^{1/3}; \quad \beta = \beta(\sigma, T) \tag{18.3}$$

[1]homologous - having the same relative value or position.

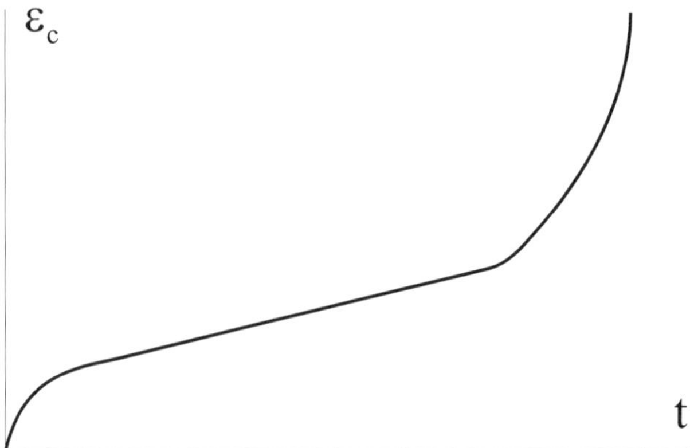

Figure 18.4: Creep strain against time for low stress.

since $e^{\beta t} \to 1$ as $t \to 0$ for primary creep. As we move from primary to secondary creep then the $e^{\beta t}$ term plays a greater role.

An alternative description of ε_c for high-temperature creep simply adds a linear term to Bailey's law

$$\varepsilon_c(t) = \alpha t^m + \beta t \tag{18.4}$$

Graham and Walles (1955,1958) for $m = 1/3$ added a cubic term to model the tertiary creep of a material

$$\varepsilon_c(t) = \alpha t^{1/3} + \beta t + \gamma t^3; \quad \gamma = \gamma(\sigma, T) \tag{18.5}$$

This expression models the complete creep curves of several engineering alloys subject to high temperatures.

18.5 Stress Dependence of Secondary Creep Rate

For a constant temperature, T, for the four stresses $\sigma_1 > \sigma_2 > \sigma_3 > \sigma_4$ then the creep strain curves are as shown in Figure 18.5. At low stresses primary creep may not be present whereas for high stresses the primary creep stage can be indistinguishable from plastic flow. At high stress, fracture is generally ductile whereas fracture is brittle when the stress is low. The transition between ductile and brittle fracture is apparent on the creep rupture curve $(\ln \sigma, \ln t^*)$ where t^* is the time to fracture, Figure 18.6. Accounting for both t and σ, $\varepsilon_c = \varepsilon_c(t, \sigma)$ is very complex and for the purposes of design we generally restrict ourselves to secondary creep and when the creep rate is a function of σ only. Neglecting the short primary region then for low σ

$$\dot{\varepsilon}_s(\sigma) = A\sigma; \quad A = A(T) \tag{18.6}$$

and we have a linearly viscous medium (Newtonian fluid). For medium σ then the Norton (1929) power law is generally used for modelling the creep of metals

$$\dot{\varepsilon}_s(\sigma) = A\sigma^n \tag{18.7}$$

where n is a material constant and typically in the range $3 \le n \le 8$. For high stresses n becomes dependent on σ and an exponential law is more applicable

$$\dot{\varepsilon}_s(\sigma) = Ae^{B\sigma} \tag{18.8}$$

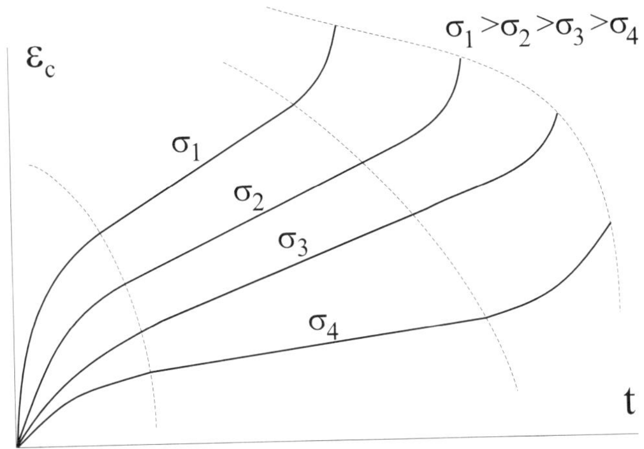

Figure 18.5: Creep strain against time for constant temperature at four different stress levels.

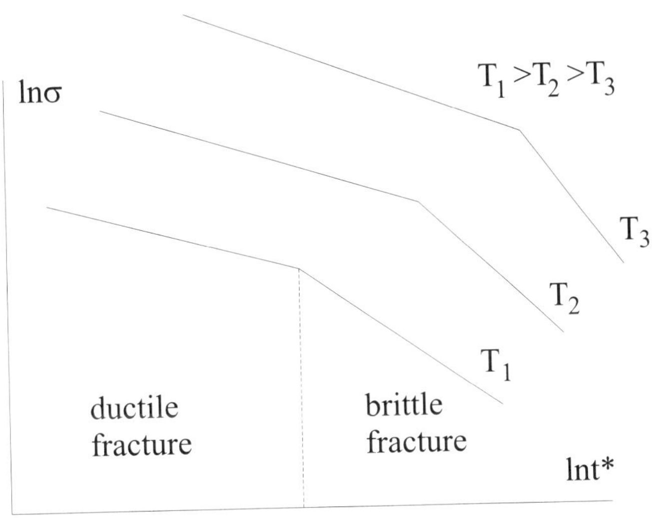

Figure 18.6: Creep rupture curves at three different temperatures.

where B is a material constant. Finally, Garofalo (1965) proposed the following law which is applicable over the entire range of σ

$$\dot{\varepsilon}_s(\sigma) = A[\sinh(C\sigma)]^p \tag{18.9}$$

where C and p are experimentally determined constants.

18.6 Temperature Dependence of Secondary Creep Rate

For a constant stress, σ, for the absolute temperatures $T_1 > T_2 > T_3 > T_4$ the creep strain curves are as shown in Figure 18.7. As in the case of the stress-dependence of secondary creep rate the time and

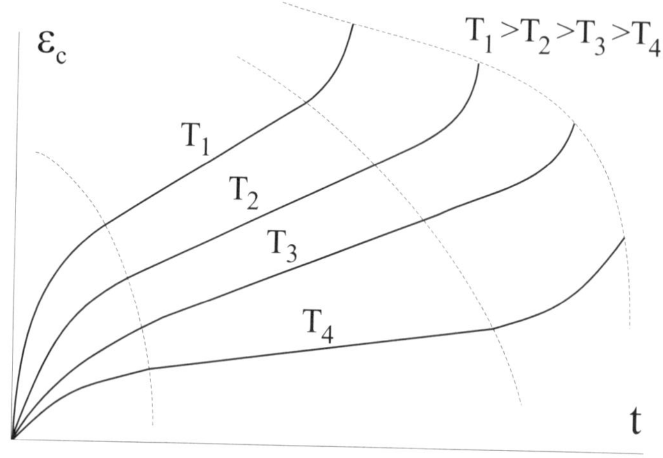

Figure 18.7: Creep strain against time for constant stress at four different temperatures.

temperature variation is very complex and generally we examine the secondary creep rate as a function of temperature only, $\dot{\varepsilon}_s = \dot{\varepsilon}_s(T)$. A widely used model is that of Dorn (1957) which is based on the thermal activation energy for dislocation recovery processes of combined climb and cross-slip and leads to the Arrhenius expression

$$\dot{\varepsilon}_s(T) = De^{-Q/RT} \tag{18.10}$$

where D is a constant, R is the gas constant ($R = 1.98$cal/mol $= 8.29$J/mol) and Q is the activation energy (e.g., $Q = 30$kcal/mol $= 125.6$kJ/mol for aluminium). The activation energy is found by plotting $\dot{\varepsilon}_s$ against $1/T$, Figure 18.8.

Example 18.1 Determination of the constants Q and D in $\dot{\varepsilon}_s(T)$

In a constant stress creep test the secondary creep rates of Table 18.1 were measured. Determine the constants Q and D for the Arrhenius expression (18.10).

Firstly, it is worth noting that to maintain a constant stress throughout a creep test the applied stress should gradually reduce in accordance with the loss of cross-sectional area. From (18.10) and taking logarithms we have

$$\ln \dot{\varepsilon}_s = \ln D - \frac{Q}{RT}$$

From a $\ln \dot{\varepsilon}_s$ versus $1/T$ plot with two points 1 ($T = 350$K) and 2 ($T = 430$K) then the slope is

$$-\frac{Q}{T} = \frac{\ln \dot{\varepsilon}_{s2} - \ln \dot{\varepsilon}_{s1}}{1/T_2 - 1/T_1} = \frac{\ln(120 \times 10^{-6}) - \ln(6 \times 10^{-6})}{1/430 - 1/350} = -22,627$$

from which Q is found to be

$$Q = 1.98 \times 22,627 = 45\text{kcal/mol}$$

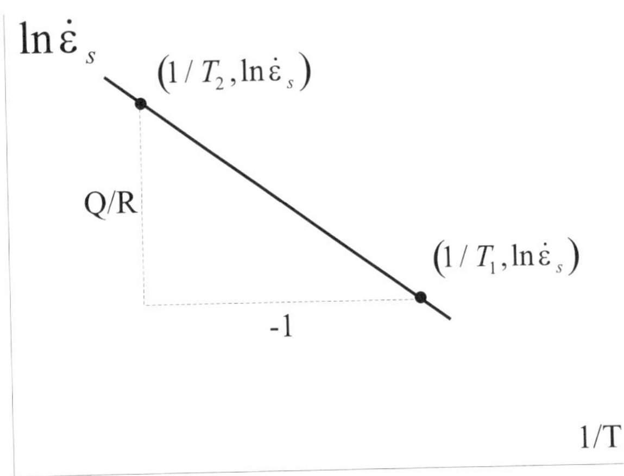

Figure 18.8: Creep strain rate against the reciprocal of temperature to determine the activation energy.

Finally, solving for D at $T = 350$K we have

$$\ln D = \ln \dot{\varepsilon}_s + \frac{Q}{RT} = \ln(6 \times 10^{-6}) + \frac{45 \times 10^3}{1.98(350)} \Rightarrow D = 952 \times 10^{20}$$

T(K)	350	370	390	410	430
$\dot{\varepsilon}_s \times \mathbf{10^{-6}(h^{-1})}$	6	12	30	48	120

Table 18.1: Example: Secondary creep rates against temperature.

18.7 Stress and Temperature Dependence of Secondary Creep Rate

The separate stress (Norton's power law) and temperature (Arrhenius expression) creep rates can be combined to give

$$\dot{\varepsilon}_s(\sigma, T) = D\sigma^n e^{-Q/RT} \tag{18.11}$$

for medium σ. For the entire range of σ Garofalo's expression, (18.9), can be combined with the Arrhenius expression, (18.10), to give

$$\dot{\varepsilon}_s(\sigma, T) = D[\sinh(C\sigma)]^p e^{-Q/RT} \tag{18.12}$$

18.8 Stress, Temperature and Time Dependence of Secondary Creep Rate

For primary creep the strain can be obtained by combining Bailey's power law, (18.2) and the Arrhenius stress-temperature expression, (18.10), to give

$$\dot{\varepsilon}_s(\sigma, T, t) = D\sigma^n t^m e^{-Q/RT} \tag{18.13}$$

For constant temperature T equation (18.13) can be generalised to

$$\dot{\varepsilon}_s(\sigma, t) = A\sigma^n f(t); \quad A = De^{-Q/RT} \tag{18.14}$$

and is analogous to Norton's power law; where $f(t)$ is an arbitrary function.

For both primary and secondary creep an expression of the form $\varepsilon_c(t) = \alpha t^m + \beta t$ can be used

$$\dot{\varepsilon}_s(\sigma, t) = D\sigma^n t^m e^{-Q/RT} + D'\sigma^n t e^{-Q/RT} \tag{18.15}$$

18.9 Tension of a Bar

Consider the creep of a bar under the action of a constant tensile load P; see §12.9 for the elongation of an elastic-plastic bar under constant load. Denote the initial length and cross-sectional area of the bar by L_0 and A_0 respectively and the current length and cross-sectional area by L and A. Then $\sigma_0 = P/A_0$ is the initial stress and $\sigma = P/A$ is the current stress. With incremental strain $\mathrm{d}\varepsilon = \mathrm{d}L/L$ then the strain and strain rate are

$$\varepsilon = \int_{L_0}^{L_1} \frac{\mathrm{d}L}{L} = \ln\left(\frac{L_1}{L_0}\right); \quad \frac{\mathrm{d}\varepsilon}{\mathrm{d}t} = \frac{1}{L}\frac{\mathrm{d}L}{\mathrm{d}t} \tag{18.16}$$

From the flow theory the secondary creep rate is given by

$$\dot{\varepsilon} = \frac{\mathrm{d}\varepsilon_s}{\mathrm{d}t} = \frac{1}{L}\frac{\mathrm{d}L}{\mathrm{d}t} = f(\sigma) \tag{18.17}$$

where $f(\sigma)$ is an arbitrary function. Letting $\lambda = L/L_0$ then the condition of incompressibility requires that

$$AL = A_0 L_0 \tag{18.18}$$

so that

$$\sigma = \frac{P}{A} = \left(\frac{L}{L_0}\right)\frac{P}{A_0} = \left(\frac{L}{L_0}\right)\sigma_0 = \lambda\sigma_0 \tag{18.19}$$

Thus, equation (18.17) can be re-written to give the differential equation

$$\frac{1}{\lambda}\frac{\mathrm{d}\lambda}{\mathrm{d}t} = f(\sigma_0\lambda) \tag{18.20}$$

subject to the initial condition $\lambda = 1$ at $t = 0$.

In the specific case of the Norton power law, (18.7)

$$f(\sigma) = A\sigma^n = A(\sigma_0\lambda)^n \tag{18.21}$$

Substituting into (18.20) and separating variables

$$\int \frac{\mathrm{d}\lambda}{\lambda A(\sigma_0 \lambda)^n} = \int \mathrm{d}t \tag{18.22}$$

Integrating we have

$$-\frac{\lambda^{-n}}{An\sigma_0^n} = t + C \tag{18.23}$$

where C is a constant of integration. In view of the initial condition $\lambda = 1$ at $t = 0$ then C is found to be

$$C = -\frac{1}{An\sigma_0^n} \tag{18.24}$$

and substituting into (18.23) and re-arranging for t

$$t = \frac{1}{An\sigma_0^n}(1 - \lambda^{-n}) = \frac{1}{n\dot{\varepsilon}_0}(1 - \lambda^{-n}) \tag{18.25}$$

where $\dot{\varepsilon}_0 = A\sigma_0^n$ is the initial creep strain rate. Letting $\lambda \to \infty$ with $t \to t_f$ then

$$t_f = \frac{1}{n\varepsilon_0} \tag{18.26}$$

where t_f is the time to viscous fracture and is found to be in good agreement with experimental observations. Substituting (18.26) into (18.25)

$$t = t_f(1 - \lambda^{-n}) \tag{18.27}$$

Re-arranging for λ

$$\lambda = \left(1 - \frac{t}{t_f}\right)^{-1/n} \tag{18.28}$$

and using the condition of incompressibility it follows that

$$\frac{A}{A_0} = \left(1 - \frac{t}{t_f}\right)^{1/n} \tag{18.29}$$

and is illustrated for $1 \leq n \leq 8$ in Figure 18.9. For $n > 1$

$$\frac{\mathrm{d}}{\mathrm{d}t}\left(\frac{A}{A_0}\right) \to \infty \quad \text{as} \quad t \to t_f \tag{18.30}$$

For $n \gg 1$ the cross-section decreases rapidly only in the latter stages of the specimen lifetime.

18.10 Torsion of a Circular Tube

A tube of inner and outer radius r_i and r_o is subject to a torque T with the creep shear strain γ at radius r after time t given by

$$\gamma = A\tau^n f(t) = \frac{r\theta}{l}; \quad \tau = \tau(r) \tag{18.31}$$

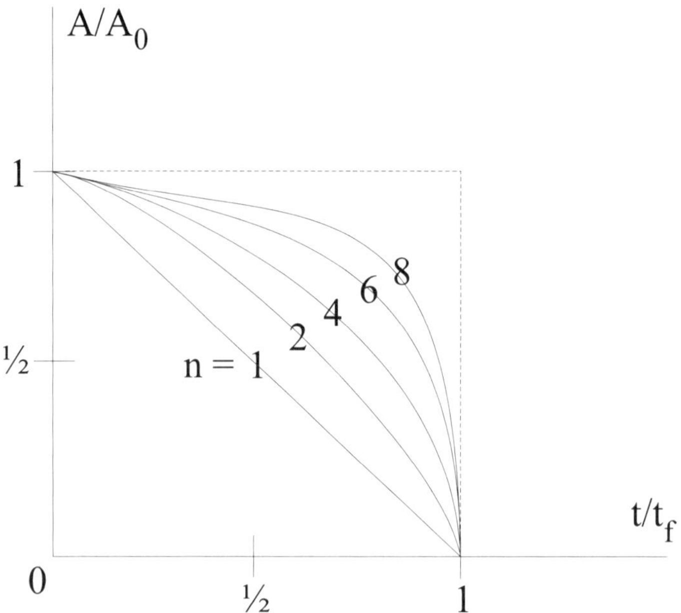

Figure 18.9: Variation of cross-sectional area against time to failure as a function of Norton's power law exponent, n.

from (18.14) for constant temperature; where θ is the angle of twist and l is the length of the bar. Rearranging for τ

$$\tau = \left[\frac{r\theta}{Alf(t)}\right]^{1/n} = \left[\frac{\theta}{Alf(t)}\right]^{1/n} r^{1/n} = Kr^{1/n} \tag{18.32}$$

Referring to Figure 18.10, from equilibrium for a circular ring of thickness $\mathrm{d}r$ then the elemental torque is $\mathrm{d}T = \mathrm{d}F.r$ where $\mathrm{d}F(= \tau.2\pi r\,\mathrm{d}r)$ is the elemental force and therefore the total torque for the tube is

$$T = 2\pi \int_{r_i}^{r_o} \tau r^2 \,\mathrm{d}r \tag{18.33}$$

Substituting τ from (18.32)

$$T = 2\pi \int_{r_i}^{r_o} K r^{1/n} r^2 \,\mathrm{d}r = 2\pi K \int_{r_i}^{r_o} r^{(1+2n)/n} \,\mathrm{d}r \tag{18.34}$$

Performing the integration

$$T = 2\pi K \frac{n}{1+3n}\left[r^{(1+3n)/n}\right]_{r_i}^{r_o} = \frac{2\pi K n}{1+3n}\left[r_o^{(1+3n)/n} - r_i^{(1+3n)/n}\right] \tag{18.35}$$

Substituting for K from (18.35) into (18.32) the shear stress is given by

$$\tau = Kr^{1/n} = \frac{(1+3n)Tr^{1/n}}{2\pi n\left[r_o^{(1+3n)/n} - r_i^{(1+3n)/n}\right]} \tag{18.36}$$

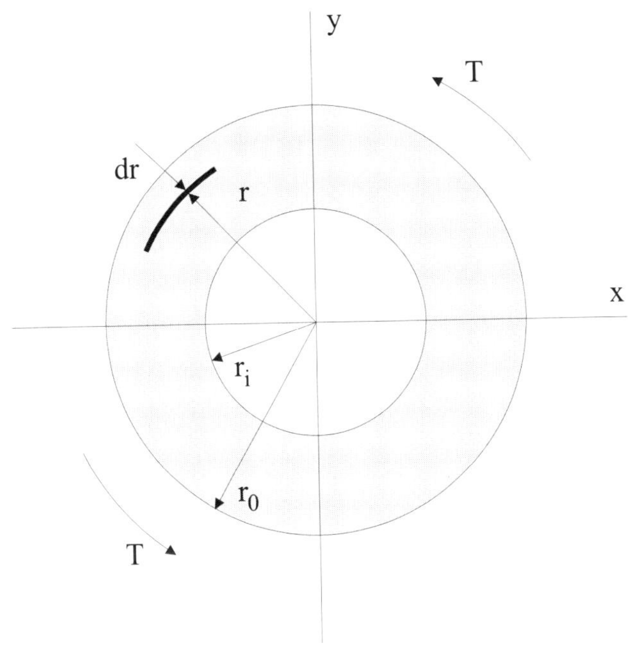

Figure 18.10: A circular tube of inner radius r_i and outer radius ro subject to a torque T.

Comparing this shear stress to the elastic torsional formula, (3.11)

$$\tau = \frac{Tr}{J}; \quad J = \frac{\pi}{2}(r_o^4 - r_i^4) \tag{18.37}$$

we observe that J for creep, J_c, is

$$\tau = \frac{Tr^{1/n}}{J_c}; \quad J_c = \frac{2\pi n \left[r_o^{(1+3n)/n} - r_i^{(1+3n)/n} \right]}{1+3n} \tag{18.38}$$

If $r_i = 0$ then the tube becomes a solid bar and J_c reduces to

$$J_c = \frac{2\pi n r_o^{(1+3n)/n}}{1+3n} \tag{18.39}$$

Noting that when $n = 1$ we have an elastic material and J_c is equal to J

$$J_c = J = \frac{\pi r_o^4}{2} \tag{18.40}$$

By comparing the elastic and creep shear stresses subject to the same torque, T, we have

$$\tau = \frac{Tr_{sk}}{J} = \frac{Tr_{sk}^{1/n}}{J_c} \tag{18.41}$$

where r_{sk} is referred to as the *skeletal radius* and is given by

$$r_{sk}^{1-1/n} = \frac{J}{J_c} \tag{18.42}$$

or

$$r_{sk} = \left(\frac{J}{J_c}\right)^{n/(n-1)} = \left\{\frac{(1+3n)(r_o^4 - r_i^4)}{4n\left[r_o^{(1+3n)/n} - r_i^{(1+3n)/n}\right]}\right\}^{n/(n-1)} \tag{18.43}$$

Figure 18.11 illustrates the variation of τ against r for both elastic and creep materials. For a prescribed temperature, the initially elastic stresses re-distribute during primary creep to attain the more uniform steady-state condition during secondary creep.

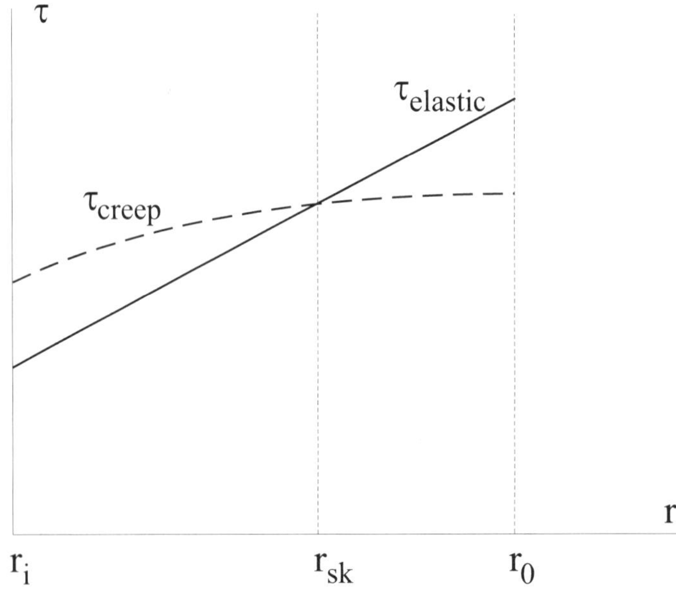

Figure 18.11: Variation of both elastic and creep shear stresses against radial distance across the tube thickness.

18.11 Bending of a Rectangular Beam

Figure 18.12 illustrates a section through a beam of rectangular cross-section $(b, 2h)$ subject to a pure bending moment M. For the elemental strip of thickness $\mathrm{d}y$ at a distance y from the neutral axis then from equilibrium we have that the elemental bending moment is $\mathrm{d}M = \mathrm{d}F.y$ where $\mathrm{d}F(= \sigma.b\,\mathrm{d}y)$ is the elemental force

$$M = \int_{-h}^{+h} \sigma b y \,\mathrm{d}y = 2b\int_0^h \sigma y \,\mathrm{d}y \tag{18.44}$$

As in the previous section we will assume a creep strain of the form $\varepsilon_c = A\sigma^n f(t)$ and with $\varepsilon_c = y/R$ from simple bending theory, (5.18)

$$\varepsilon_c = A\sigma^n f(t) = \frac{y}{R} \tag{18.45}$$

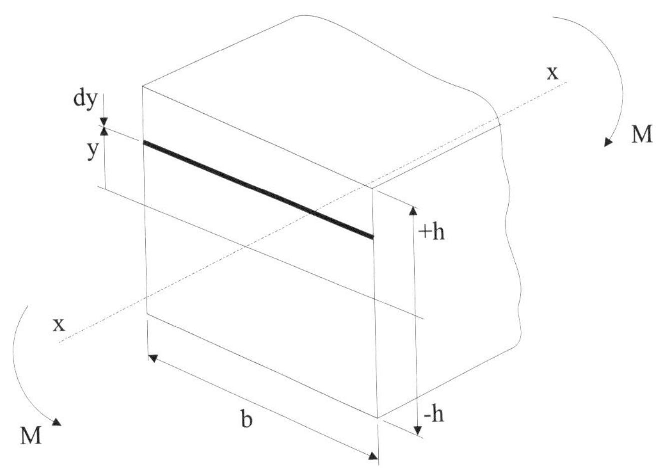

Figure 18.12: A rectangular beam subject to a bending moment.

where R is the radius of curvature. Substituting into (18.44)

$$M = 2b \int_0^h \left[\frac{y}{ARf(t)} \right]^{1/n} y \, \mathrm{d}y = 2b[ARf(t)]^{-1/n} \int_0^h y^{1+1/n} \, \mathrm{d}y = [ARf(t)]^{-1/n} I_c \quad (18.46)$$

where I_c is

$$I_c = 2b \int_0^h y^{1+1/n} \, \mathrm{d}y = \frac{2bn}{2n+1} h^{(2n+1)/n} \quad (18.47)$$

Comparing I_c to its elastic equivalent, I

$$I = \int_A y^2 \, \mathrm{d}A; \quad I_c = \int_A y^{1+1/n} \, \mathrm{d}A \quad (18.48)$$

with $I = I_c$ for $n = 1$.

Comparing the elastic and creep bending moments, we have for an elastic material, (5.30)

$$\frac{M}{I} = \frac{\sigma}{y} \quad (18.49)$$

and for the present creep material, (18.45)

$$\frac{M}{I_c} = \frac{1}{[ARf(t)]^{1/n}} = \frac{\sigma}{y^{1/n}} \quad (18.50)$$

Re-arranging for σ

$$\sigma = \frac{My_{sk}}{I} = \frac{My_{sk}^{1/n}}{I_c} \quad (18.51)$$

where y_{sk} is the *skeletal depth* and is given by

$$y_{sk}^{1-1/n} = \frac{I}{I_c} \quad \text{or} \quad y_{sk} = \left(\frac{I}{I_c} \right)^{n/(n-1)} \quad (18.52)$$

Figure 18.13 illustrates the variation of σ against y for both elastic and creep materials.

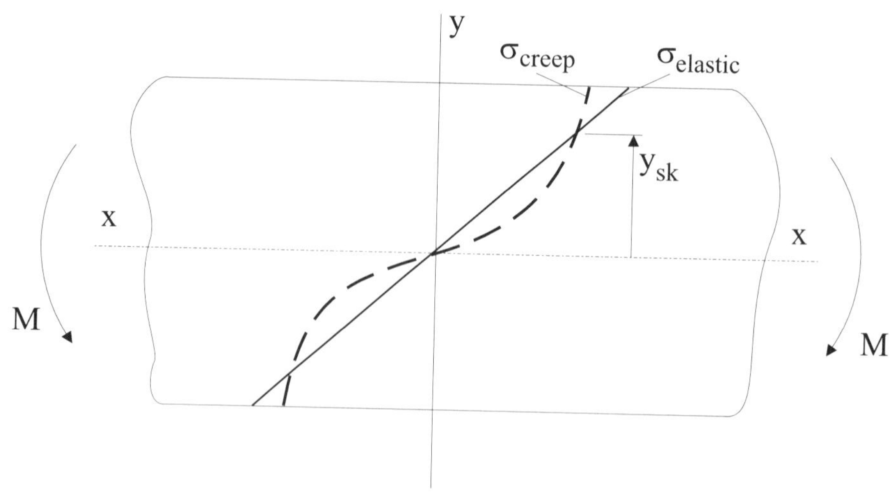

Figure 18.13: Variation of both elastic and creep bending stresses against depth across the beam thickness.

18.12 Conclusion

This Chapter has introduced the creep of materials. The key laws governing the separate and combined time, stress and temperature dependencies on the creep strain have been examined for engineering materials. The Chapter concluded by examining the three applications of the tension of a bar, the torsion of a circular tube and the bending of a rectangular beam.

18.13 References and Further Reading

∘ Bailey, R.W. (1935), Proc. Inst. Mech. Eng., 131, 131.

∘ Dorn, J. E. (1957) *Creep and Recovery*, American Society for Metals, Cleveland, Ohio.

∘ Friedel, J. (1964) *Dislocations*, Pergamon, London.

∘ Garofalo, F. (1965) *Fundamentals of Creep and Creep Rupture in Metals*, Macmillan, New York.

∘ Graham, A. and Walles, K. F. A. (1955), J. Iron Steel Inst., 179, 105.

∘ Graham, A. and Walles, K. F. A. (1958) *Regularities in Creep and Hot Fatigue Data*, Part I of Aero. Res. Cl. Report No. 379, HMSO, London.

∘ Hirsch, P. B. and Warrington, D. H. (1961) *The Flow Stress of Aluminium and Copper at High Temperatures*, Phil. Mag., 6, 735-768.

∘ Honeycombe, R. W. K. (1984) *The Plastic Deformation of Metals*, 2nd Ed, Arnold, London.

∘ McClintock, F. A. and Argon, A. S. (1966) *Mechanical Behaviour of Materials*, Addison-Wesley, MA.

∘ Norton, F. H. (1929) *Creep of Steel at High Temperatures*, McGraw-Hill, New York.

∘ Weertman, J. (1955) *Theory of Steady State Creep based on Dislocation Climb*, J. Appl. Phys., 26, 1213-1217.

18.14 Exercises

18.1 Describe the creep of engineering materials making reference to the time, stress and temperature dependencies on the creep strain.

18.2 In a constant temperature creep test the secondary creep rates of Table 18.2 were measured. Assuming that a Norton's power law is applicable then determine the constants A and n.

σ**(MPa)**	100	125	150	175	200
ε_s**(h^{-1})**	4.4×10^{-6}	9.4×10^{-6}	6.1×10^{-5}	1.2×10^{-4}	7.6×10^{-4}

Table 18.2: Exercise: Secondary creep rates against temperature.

18.3 In a constant stress creep test the secondary creep rates of Table 18.3 were measured. Determine the constants Q and D for a secondary creep rate given by

$$\varepsilon_s(T) = De^{-Q/RT}$$

T(K)	160	170	180	190	200
$\varepsilon_s \times 10^{-6}$**(h^{-1})**	7.24×10^{-6}	1.35×10^{-5}	9.42×10^{-5}	3.69×10^{-4}	8.19×10^{-3}

Table 18.3: Exercise: Secondary creep rates against temperature.

18.4 An experimental method for determining the activation energy, Q, of aluminium involves separate creep tests at the same stress but at different temperatures. The following power law is assumed to be applicable

$$\varepsilon = A[te^{-Q/RT}]^n$$

where A and n are constants. Show that if two tests are conducted at temperatures T_1 and T_2 and the times to reach the desired strains are t_1 and t_2 then Q is given by

$$Q = \frac{RT_1T_2}{T_2 - T_1}[\ln t_1 - \ln t_2]$$

18.5 A bar of constant cross-sectional area is subject to constant tensile load under conditions of creep. Determine the time taken, as a fraction of the total time to failure, for the cross-sectional area to equal 3/4 of the original cross-sectional area. Assume that the bar material conforms to Norton's power law with an exponent n equal to 3.

18.6 A solid bar of circular cross-section and outer diameter 100mm is subject to a pure torque of 10kNm. The creep shear strain of the bar material is of the form $\gamma_c = A\sigma^n f(t)$; where $A = 4.2 \times 10^{-4}$ and $n = 4$. Determine the skeletal radius of the bar.

18.7 A beam of rectangular cross-section of height 75mm and width 50mm is subject to a pure bending moment of 15kNm and is in a state of creep in which the creep strain response adopts a Norton power law $\varepsilon_c = A\sigma^n f(t)$; where $A = 3.34 \times 10^{-4}$ and $n = 6$. Determine the skeletal depth of the beam.

Chapter 19

Viscoelasticity

19.1 Introduction

In the present Chapter we examine viscoelasticity and the three most popular models of Maxwell, Voigt and Standard Linear Solid. Both creep and relaxation will be discussed for each model. In addition, we will examine linear viscoelastic materials and the Boltzmann linear superposition model for modelling viscoelastic materials that are subject to fluctuating loading.

19.2 Viscoelasticity

A viscoelastic material in one that possesses both elastic and viscous properties and exhibits a time-dependent stress-strain response. While the majority of materials are viscoelastic and sensitive to the rate of loading to some degree, materials such as polymers, adhesives and composites in particular are observed to exhibit viscoelastic behaviour. Viscoelastic behaviour is typically categorised in terms of either creep and recovery or stress relaxation, see Figure 19.1. In the case of a step loading σ_0 a viscoelastic material may exhibit an initial elastic response followed by a delayed elastic response with increasing time and with the strain given by $\varepsilon = \sigma_0 C(t)$ where $C(t)$ is the *creep compliance* and is a monotonically increasing function of t. If the stress σ_0 is subsequently removed then the material may at first experience an instantaneous partial recovery of elastic strain followed by a delayed elastic response with decreasing strain with time. The material is left with a permanent strain which is acquired through the action of creep. In the case of a constant strain ε_0 a viscoelastic material will exhibit a continuous decrease in stress or relaxation with increasing time and is given by $\sigma = \varepsilon_0 G(t)$ where the monotonically decreasing function $G(t)$ is the *relaxation modulus*.

Viscoelastic materials are typically modelled with the aid of rheological models. Numerous models have been proposed but we will examine three of the more well-known models, the simplest of which is the Maxwell model. The Voigt and Standard Linear Solid models are then examined. Finally, the Chapter concludes with the Boltzmann's linear superposition model for linear viscoelastic materials.

19.3 The Maxwell Model

The Maxwell model represents a viscoelastic material as a spring and dashpot in series, Figure 19.2. The

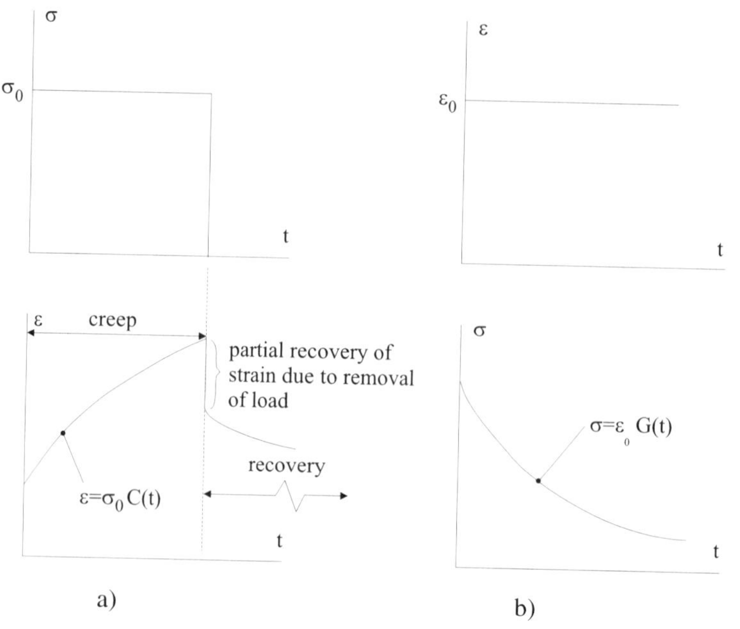

Figure 19.1: Viscoelastic response. a) Creep and recovery. b) Relaxation under constant strain ε_0.

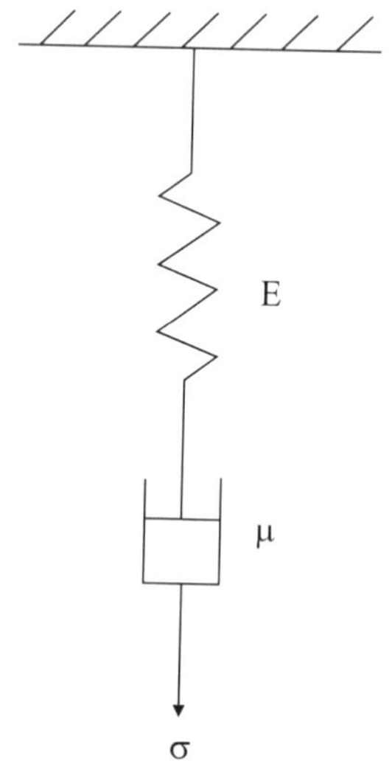

Figure 19.2: Maxwell model.

elastic component of the material is modelled by an elastic spring with Young's modulus E

$$\sigma = E\varepsilon_s \tag{19.1}$$

The viscous component of the material is modelled by a Newtonian dashpot with coefficient of viscosity μ

$$\sigma = \mu \frac{d\varepsilon_d}{dt} \tag{19.2}$$

It is the combination of these two components which enables us to model the viscoelastic response of a material. Note, however, that for the Maxwell model to be representative of a solid rather than a fluid then the viscosity must be large and comparable in magnitude with the modulus of elasticity.

In the Maxwell model under conditions of creep and recovery for a step stress σ the total strain, ε, is the sum of ε_s and ε_d

$$\varepsilon = \varepsilon_s + \varepsilon_d \tag{19.3}$$

which upon differentiating with respect to t gives

$$\frac{d\varepsilon}{dt} = \frac{d\varepsilon_s}{dt} + \frac{d\varepsilon_d}{dt}\frac{1}{E}\frac{d\sigma}{dt} + \frac{\sigma}{\mu} \tag{19.4}$$

For constant stress $\sigma = \sigma_0$ then

$$\frac{d\varepsilon}{dt} = \frac{\sigma_0}{\mu} \quad \text{or} \quad \varepsilon = \frac{\sigma_0}{\mu}t + A \tag{19.5}$$

where A is a constant of integration. If at $t = 0$ the instantaneous strain is elastic only then $\varepsilon = \sigma_0/E$ and $A = \sigma_0/E$ so that

$$\varepsilon = \left(\frac{t}{\mu} + \frac{1}{E}\right)\sigma_0 \tag{19.6}$$

The linear variation of ε against t according to (19.6) is shown in Figure 19.3a) with the strain continuously increasing with time. The Figure also illustrates the removal of σ_0 after time τ with an instantaneous recovery of the elastic strain σ_0/E but with no viscous recovery.

In the case of relaxation with a constant strain ε_0 then we have from (19.4)

$$\frac{1}{E}\frac{d\sigma}{dt} + \frac{\sigma}{\mu} = 0 \tag{19.7}$$

which upon separating variables gives

$$\frac{d\sigma}{\sigma} = -\frac{E}{\mu}dt \tag{19.8}$$

Integrating with respect to t

$$\ln \sigma = -\frac{E}{\mu}t + A \tag{19.9}$$

With $\sigma = \sigma_0$ at $t = 0$ then $A = \ln \sigma_0$ and

$$\sigma = \sigma_0 e^{-Et/\mu} \tag{19.10}$$

the variation of which is shown in Figure 19.3b).

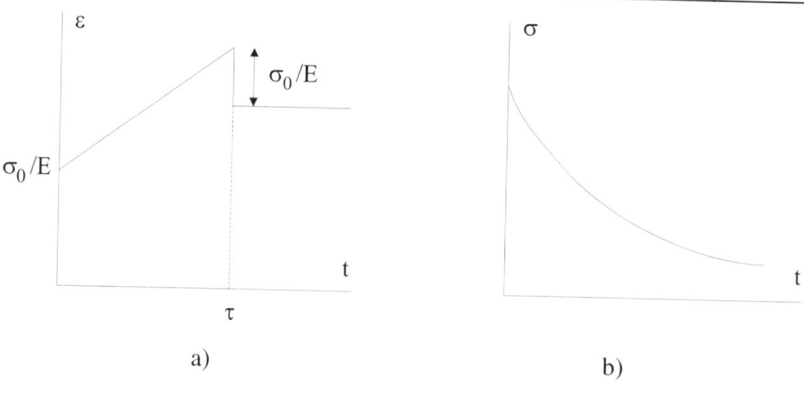

Figure 19.3: The Maxwell model. a) Creep and recovery. b) Relaxation.

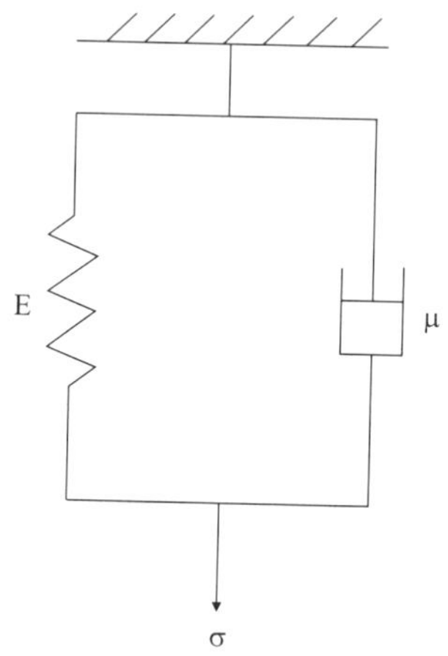

Figure 19.4: Voigt model.

19.4 The Voigt Model

The Voigt or Kelvin model represents a viscoelastic material as a spring and dashpot in parallel, Figure 19.4. The total stress σ is the sum of the spring and dashpot stresses σ_s and σ_d respectively

$$\sigma = \sigma_s + \sigma_d \tag{19.11}$$

Substituting $\sigma_s (= E\varepsilon_s)$ and $\sigma_d (= \mu \mathrm{d}\varepsilon_d / \mathrm{d}t)$ into (19.11)

$$\sigma = E\varepsilon + \mu \frac{\mathrm{d}\varepsilon}{\mathrm{d}t} \tag{19.12}$$

For constant stress $\sigma = \sigma_0$

$$\frac{\mathrm{d}\varepsilon}{\mathrm{d}t} = \frac{\sigma_0}{\mu} - \frac{E}{\mu}\varepsilon = \frac{1}{\mu}(\sigma_0 - E\varepsilon) \tag{19.13}$$

Separating variables

$$\int \frac{\mathrm{d}\varepsilon}{\sigma_0 - E\varepsilon} = \frac{1}{\mu}\int \mathrm{d}t \tag{19.14}$$

and integrating we have

$$-\frac{1}{E}\ln\left(\sigma_0 - E\varepsilon\right) = \frac{t}{\mu} + A \tag{19.15}$$

At time $t = 0$ then $\varepsilon = 0$ since the dashpot does not allow an instantaneous strain and therefore $A = -\ln \sigma_0/E$ so that

$$\varepsilon = \left(1 - e^{-Et/\mu}\right)\frac{\sigma_0}{E} \tag{19.16}$$

Figure 19.5 illustrates the variation of ε against t with a removal of stress σ_0 at $t = \tau$. The Voigt model is seen to exhibit delayed elasticity with the ultimate strain limited to a finite value, unlike the Maxwell model.

To determine the recovery strain for $t \geq \tau$ then from (19.13) with $\sigma_0 = 0$ and $t' = t - \tau$ then

$$\frac{\mathrm{d}\varepsilon}{\mathrm{d}t'} = -\frac{E}{\mu}\varepsilon \tag{19.17}$$

Separating variables

$$\int \frac{\mathrm{d}\varepsilon}{\varepsilon} = -\frac{E}{\mu}\int \mathrm{d}t' \tag{19.18}$$

and integrating with respect to t'

$$\ln \varepsilon = -\frac{E}{\mu}t' + A \tag{19.19}$$

With $\varepsilon = \varepsilon_\tau$ at $t = \tau$ then the constant of integration A is

$$A = \ln \varepsilon_\tau + \frac{E}{\mu}\tau \tag{19.20}$$

Substituting A into (19.19) gives

$$\varepsilon = \varepsilon_\tau e^{-\frac{E}{\mu}(t'-\tau)} \tag{19.21}$$

With ε_τ given by (19.16) for $t = \tau$ then the recovery strain is

$$\varepsilon = \varepsilon_\tau e^{-Et'/\mu} = \frac{\sigma_0}{E}\left(1 - e^{-E\tau/\mu}\right)e^{-Et'/\mu} \quad , \quad t \geq \tau \tag{19.22}$$

Unfortunately the Voigt model does not permit relaxation for constant strain ε_0 since from (19.12) $\sigma = E\varepsilon$.

19.5 The Standard Linear Solid Model

The *standard linear solid* or Zener model represents a viscoelastic material as a spring and Voigt model connected in series, Figure 19.6, and, as will be shown, models observed behaviour well. The stress between the spring and Voigt model elements must be equal

$$E_1\varepsilon_1 = E_2\varepsilon_2 + \mu\frac{\mathrm{d}\varepsilon_2}{\mathrm{d}t} \tag{19.23}$$

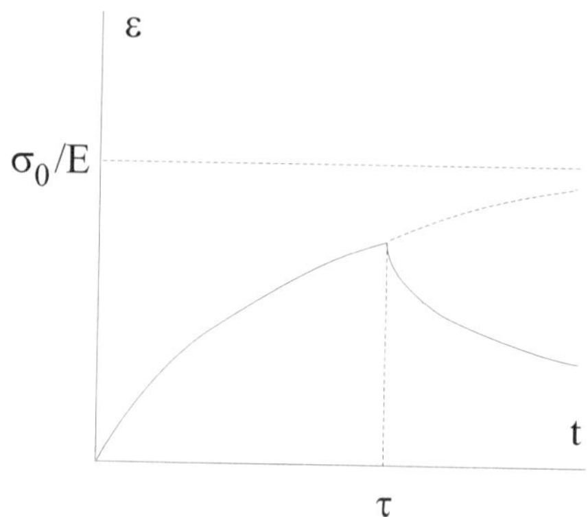

Figure 19.5: Creep and recovery for the Voigt model.

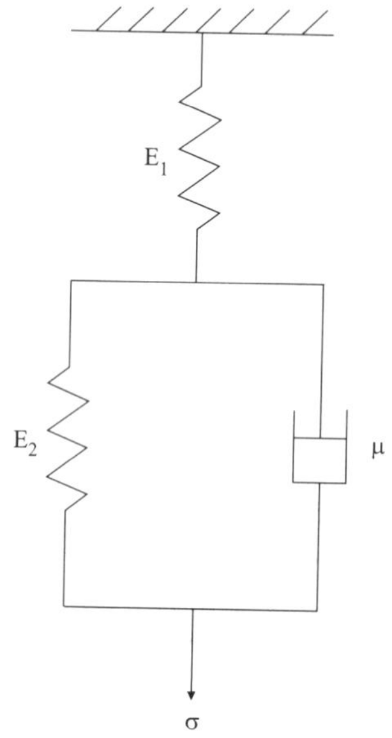

Figure 19.6: Standard linear solid model.

and the total strain, ε, is the sum of ε_1 and ε_2

$$\varepsilon = \varepsilon_1 + \varepsilon_2 \tag{19.24}$$

Substituting ε into (19.23)

$$E_1\varepsilon_1 = E_2\left(\varepsilon - \varepsilon_1\right) + \mu\frac{\mathrm{d}}{\mathrm{d}t}\left(\varepsilon - \varepsilon_1\right) \tag{19.25}$$

and with $\sigma = E_1\varepsilon_1$ gives the constitutive equation

$$\frac{\mathrm{d}\varepsilon}{\mathrm{d}t} + \frac{E_2}{\mu}\varepsilon = \frac{\sigma}{\mu}\left(1 + \frac{E_2}{E_1}\right) + \frac{1}{E_1}\frac{\mathrm{d}\sigma}{\mathrm{d}t} \tag{19.26}$$

For constant stress $\sigma = \sigma_0$

$$\frac{d\varepsilon}{dt} + \frac{E_2}{\mu}\varepsilon = \frac{\sigma_0}{\mu}\left(1 + \frac{E_2}{E_1}\right) \tag{19.27}$$

Separating variables

$$\frac{d\varepsilon}{dt} = \frac{\sigma_0}{\mu}\left[1 + \frac{E_2}{E_1} - \frac{E_2}{\sigma_0}\varepsilon\right] \tag{19.28}$$

and integrating

$$\mu \int \frac{d\varepsilon}{\sigma_0 + E_2(\sigma_0/E_1 - \varepsilon)} = \int dt \tag{19.29}$$

Performing the integration with the condition $\varepsilon = \sigma_0/E_1$ at $t = 0$ and re-arranging for ε we find

$$\varepsilon = \frac{\sigma_0}{E_1} + \frac{\sigma_0}{E_2}\left[1 - e^{-E_2 t/\mu}\right] \tag{19.30}$$

The variation of ε against t with a removal of stress σ_0 at time $t = \tau$ is shown in Figure 19.7a). The elastic strain σ_0/E_1 is instantaneously recovered. For $t \geq \tau$ then the recovery strain is given by, from the Voigt model (19.22)

$$\varepsilon = \frac{\sigma_0}{E_2}\left(1 - e^{-E_2\tau/\mu}\right) e^{-E_2 t'/\mu} \tag{19.31}$$

where $t' = t - \tau$. Thus, the standard linear solid model, like the Voigt model, exhibits delayed elasticity with the ultimate strain limited to a finite value.

In the case of relaxation with constant strain $\varepsilon = \varepsilon_0$ then from the constitutive equation (19.26), upon re-arranging

$$\frac{d\sigma}{dt} = \frac{E_1 E_2}{\mu}\varepsilon_0 - \frac{\sigma}{\mu}(E_1 + E_2) \tag{19.32}$$

After separating variables and with $\sigma = \sigma_0 = E_1\varepsilon_0$ at $t = 0$ then

$$\sigma = \frac{E_1\varepsilon_0}{E_1 + E_2}\left[E_2 + E_1 e^{-(E_1+E_2)t/\mu}\right] \tag{19.33}$$

and is illustrated in Figure 19.7b) noting that

$$\sigma \to \frac{E_1 E_2}{E_1 + E_2}\varepsilon_0 \quad \text{as} \quad t \to \infty \tag{19.34}$$

19.6 A Note on Temperature

It is generally assumed in viscoelastic models that the spring constant is insensitive to changes in temperature. However, dashpot viscosity shows a marked temperature dependence according to an Arrhenius type relationship

$$\mu_T = \mu_0 e^{Q/RT} \tag{19.35}$$

where μ_0 is the viscosity at a reference temperature, Q is the flow activation energy, R is the gas constant and T is the temperature (K).

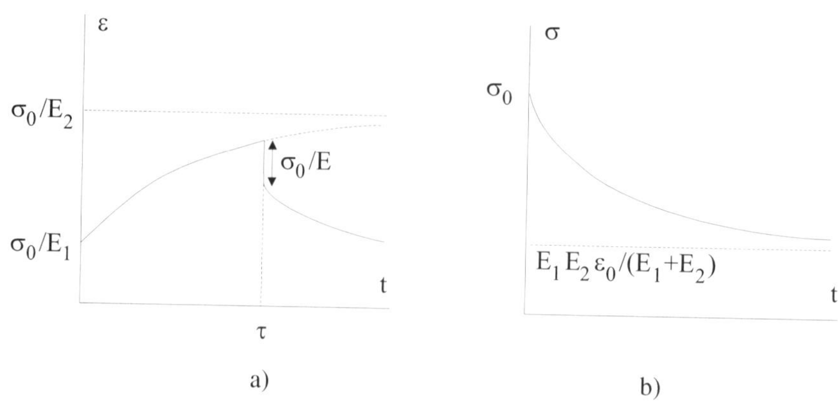

Figure 19.7: The standard linear solid model. a) Creep and recovery. b) Relaxation.

19.7 Linear Viscoelasticity

In practice, materials generally experience fluctuating loads at periodic or random intervals and of differing magnitudes. When dealing with applied stresses that are functions of time we can use Boltzmann's linear superposition principle. Consider the stress history

$$\sigma(t) = c_1\sigma_1(t) + c_2\sigma_2(t-\tau) \tag{19.36}$$

where c_1 and c_2 are constants and which is illustrated in 19.8a). If the net strain $\varepsilon[\sigma(t)]$ is the linear addition of the strains at stress levels σ_1 and σ_2, that is

$$\varepsilon = [c_1\sigma_1(t) + c_2\sigma_2(t-\tau)] = c_1\varepsilon\left[\sigma_1(t)\right] + c_2\varepsilon\left[\sigma_2(t-\tau)\right] \tag{19.37}$$

then the material is referred to as *linear viscoelastic*. For this requirement to be satisfied the strains must be small, as in the linear theory of elasticity.

For a linear viscoelastic material under creep and recovery conditions and stress relaxation we have seen in the previous sections that

$$\varepsilon(t) = C(t)\sigma \quad , \quad \sigma(t) = G(t)\varepsilon \tag{19.38}$$

In terms of the creep compliance, if the stress experiences a step change from σ_1 to $\sigma_1 + \sigma_2$ then the strain is

$$\varepsilon(t) = \sigma_1 C(t) + \sigma_2 C(t-\tau) \tag{19.39}$$

where, in general, $H(t)$ denotes the Heaviside unit step function defined as $H(t) = 1$ for $t \geq 0$ and $H(t) = 0$ for $t < 0$. If we now assume that the material experiences an infinitesimal change in stress $\mathrm{d}\sigma(\tau)H(t-\tau)$ at time τ then the associated incremental strain $\mathrm{d}\varepsilon$ is

$$\mathrm{d}\varepsilon = \mathrm{d}\sigma(\tau)C(t-\tau) \tag{19.40}$$

Integrating from $\tau = -\infty$ to t

$$\varepsilon(t) = \int_{-\infty}^{t} C(t-\tau)\frac{\partial\sigma(\tau)}{\partial\tau}\,\mathrm{d}\tau \tag{19.41}$$

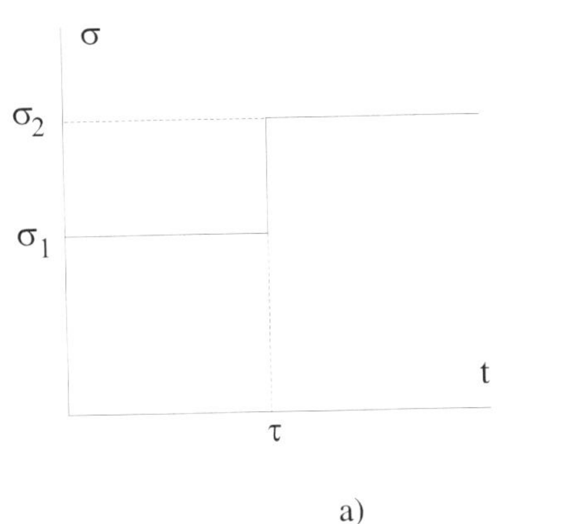

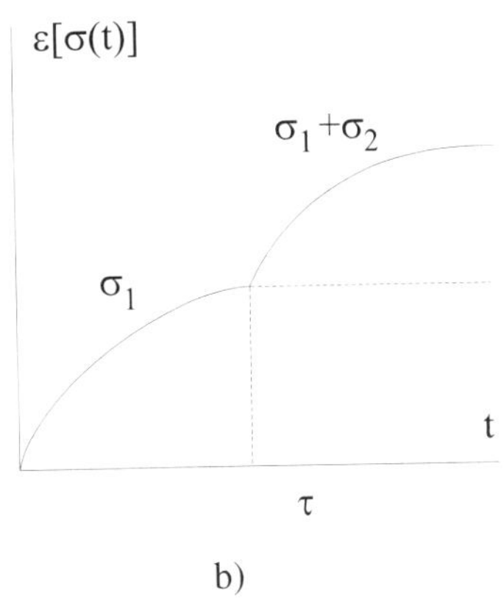

Figure 19.8: Linear superposition of creep strain b) subject to a step change in stress from σ_1 to $(\sigma_1 + \sigma_2)$, a).

with $\mathrm{d}\sigma = (\partial\sigma/\partial\tau)\mathrm{d}\tau$ since $\sigma(\tau)$ is assumed to be a continuous function.

As in the case of plasticity, $\varepsilon(t)$ is a function of the previous stress history and is accounted for by the lower limit of integration $(-\infty)$ for a stress history prior to $t = 0$. The lower limit of integration may be set to zero if no previous history exists in which case (19.41) is given by

$$\varepsilon(t) = \sigma_i(0) + \int_0^t C(t-\tau)\frac{\partial\sigma(\tau)}{\partial\tau}\,\mathrm{d}\tau \tag{19.42}$$

Similarly, the stress, $\sigma(t)$, is given by

$$\sigma(t) = \varepsilon_i(0) + \int_0^t G(t-\tau)\frac{\partial\varepsilon(\tau)}{\partial\tau}\,\mathrm{d}\tau \tag{19.43}$$

where $G(t)$ is the relaxation function which specifies the stress response to a unit step change of strain.

Example 19.1 Linear viscoelasticity for a material obeying secondary creep

In this example we illustrate the linear viscoelasticity model for a material which obeys the secondary creep law, (18.14), $\varepsilon(t) = At^m\sigma$. During creep the material experiences the following loading, Figure 19.9a):

1) $0 \le t \le t_1$ $\quad\sigma = kt$; k=constant

2) $t_1 \le t \le 3t_1$ $\quad\sigma = \sigma_1$=constant

3) at $t = 3t_1$ $\quad$ stress instantaneously removed

Let us consider these three phases separately:

1) $0 \le t \le t_1$
With $C(t) = At^m$, $\sigma = kt$, $\mathrm{d}\sigma/\mathrm{d}t = k$ then from (19.42) we have

$$\varepsilon(t) = \int_0^t A(t-\tau)^m k\,\mathrm{d}\tau = kA\int_0^t (t-\tau)^m\,\mathrm{d}\tau = \left[-\frac{kA}{m+1}(t-\tau)^{m+1}\right]_0^t = \frac{kAt^{m+1}}{m+1}$$

2) $t_1 \leq t \leq 3t_1$
From the creep law $\varepsilon(t)$ is given by, with $\sigma = kt_1$

$$\varepsilon(t) = A(t - t_1)^m \sigma = Akt_1(t - t_1)^m$$

Adding $\varepsilon(t)$ for both phases gives the net strain for the interval $t_1 \leq t \leq 3t_1$

$$\varepsilon(t) = \frac{kAt^{m+1}}{m+1} + kAt_1(t - t_1)^m$$

3) $t = 3t_1$
With the removal of stress at $t = 3t_1$ then, with $\tau = 3t_1$

$$\varepsilon(t) = \frac{kAt^{m+1}}{m+1} + kAt_1(t - t_1)^m - kAt_1(t - 3t_1)^m$$

Superimposing the strains in the above three phases we have the variation shown in Figure 19.9b).

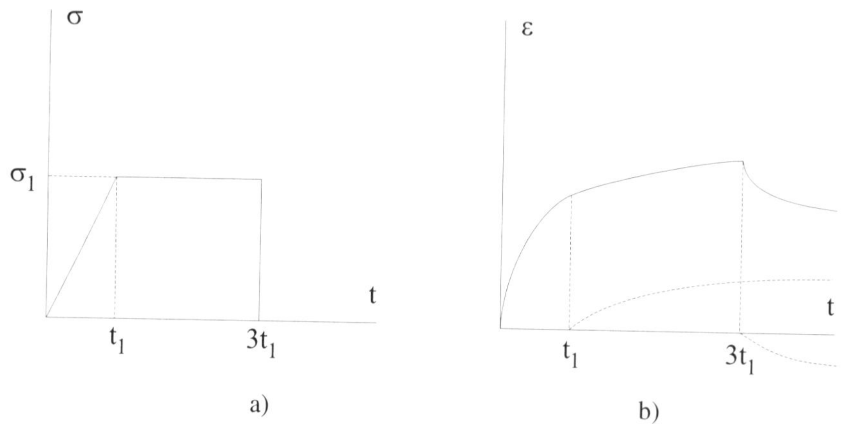

Figure 19.9: Example: Variation of a) stress and b) strain using the linear viscoelastic model.

Example 19.2 Determination of strain for a linear viscoelastic material

In this example we will determine the strain at a time of 3000s for the prescribed stress shown in Figure 19.10:

1) at $t = 0$ $\Delta\sigma_1 = 1$MPa

2) at $t = 1000$s $\Delta\sigma_2 = 1$MPa

3) at $t = 2000$s $\Delta\sigma_3 = -2$MPa

Assume that the compliance of the material is given by $C(t) = At^m\text{GPa}^{-1}$ where $A = 1.4 \times 10^{-9}$ and $m = 0.1$. From (19.42) the strain response due to i step changes in applied load is the sum of the responses to each change in load

$$\varepsilon(t) = \sum_{i=1}^{n} C(t - t_i)\Delta\sigma_i$$

In the present example the strain at time $t = 3000$s is

$$\begin{aligned}\varepsilon(3000) =& 1 \times 10^6(1.4 \times 10^{-9})(3000)^{0.1} + \\ & 1 \times 10^6(1.4 \times 10^{-9}(3000 - 2000)^{0.1} - \\ & 2 \times 10^6(1.4 \times 10^{-9})(3000 - 2000)^{0.1} = 5.249 \times 10^{-4}\end{aligned}$$

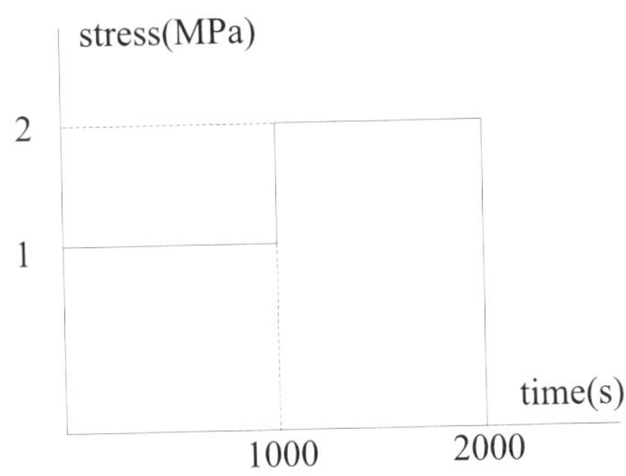

Figure 19.10: Example: Variation of applied stress against time.

19.8 Conclusion

This Chapter has examined the three most popular viscoelastic models of Maxwell, Voigt and Standard Linear Solid. Where possible both creep and recovery and relaxation have been examined for each model. The Chapter concluded by examining linear viscoelastic materials and the Boltzmann linear superposition model for modelling viscoelastic materials that are subject to fluctuating loading.

19.9 References and Further Reading

∘ Ferry, J. D. (1970) *Viscoelastic Properties of Polymers*, Wiley, New York.

19.10 Exercises

19.1 A polymer is modelled using both the Maxwell model ($E = 0.3$GPa, $\mu = 1000 \times 10^6$Ns/m^2) and Voigt model ($E = 100$kPa, $\mu = 10 \times 10^6$Ns/m^2). Determine the strain after 100s using both models if the polymer is subject to an initial stress of 5kPa.

19.2 Determine the creep response of a rheological model consisting of a Maxwell model and a Voigt model in series and sketch the strain-time curve.

19.3 For the model of Exercise 19.2 sketch the strain-time curve assuming that a stress σ_0 is instantaneously applied at time $t = 0$ and maintained for a time $t = \tau$ at which point σ_0 is instantaneously removed, Figure 19.11. What is the value of the strain as $t \to \infty$?

19.4 A polymer is modelled using the standard linear solid model with $E_1 = 0.8$GPa, $E_2 = 0.8$GPa and $\mu = 5$GPa. Determine the relaxation stress after 10s if an instantaneous stress of 8kPa is initially applied.

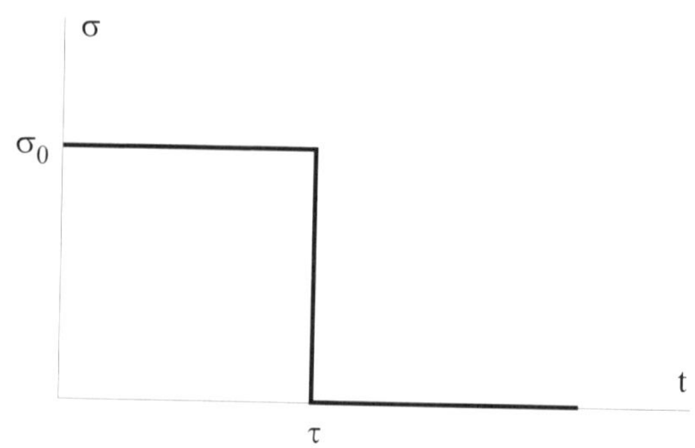

Figure 19.11: Exercise: Applied stress for the model of Exercise 19.3.

19.5 What are the creep compliance and relaxation moduli for the Maxwell and Voigt models?

19.6 Derive an expression for the creep strain for the Maxwell model but letting the dashpot behave according to the following non-linear strain rate

$$\dot{\varepsilon}_d = A\sigma^n$$

where A and n are material constants.

19.7 A viscoelastic material is subject to the following stress history:

1) at $t = 0$ $\Delta\sigma_1 = 0.5$MPa

2) at $t = 1000$s $\Delta\sigma_2 = 0.25$MPa

3) at $t = 2000$s $\Delta\sigma_3 = 0.25$MPa

4) at $t = 3000$s $\Delta\sigma_4 = -1$MPa

Assuming the compliance of the material is given by $C(t) = At^m$GPa^{-1} where $A = 1.52 \times 10^{-9}$ and $m = 0.15$ then determine the strain at time 3500s.

Chapter 20

Continuum Damage Mechanics

20.1 Introduction

In this Chapter we present an introduction to the subject of continuum damage mechanics. Firstly we review several different types of damage. We then introduce the damage and continuity damage parameters. The following sections illustrate the use of damage parameters and revisit fatigue crack growth and creep from a damage mechanics perspective.

20.2 Types of Damage

The following sections summarise several examples of damage.

20.2.1 Creep Damage

In ductile transgranular creep fracture the combination of applied stress, high temperatures and accumulation and growth of microcracks in metallic materials is one form of damage. Similarly, an accumulation and growth of microcracks at intergranular boundaries lead to brittle intergranular creep fracture.

20.2.2 Ductile Plastic Damage

The nucleation and growth of microvoids and microcracks due to large plastic strains can lead to plastic fracture.

20.2.3 Fatigue Damage

Under the action of cyclic loading a gradual deterioration of a material is experienced due to the accumulation of both micro and macro cracks.

20.2.4 Embrittlement of Steels

The action of atomic radiation alters the material properties of steels and decreases the plasticity of the material and leads to embrittlement. The contact of steel with free hydrogen results in the hydrogen atoms diffusing into the atomic structure of the steel and leads to hydrogen embrittlement.

20.2.5 Chemical-Mechanical Damage

The combined action of a cyclic stress and an aggressive environment such as sea water acting on a material can lead to intensive corrosion or stress corrosion cracking.

20.2.6 Concrete Damage

Concrete is a non-homogeneous material with zones of weak mechanical resistance and is subject to brittle fracture occurring at the matrix-stone interfaces. Similarly, continuum damage mechanics lends itself well to the analysis of fibre-reinforced composites, Chapter 21, and highly brittle materials such as ceramics.

20.2.7 Environmental Degradation

Materials such as geo-materials and polymers change their mechanical properties as a result of the environment even in the absence of stress. For example, the effects of humidity and temperature can result in the slippage and creep of soil.

20.3 An Isotropic Damage Parameter

Consider the section S through the arbitrary body shown in Figure 20.1. Let A_0 denote the initial area of

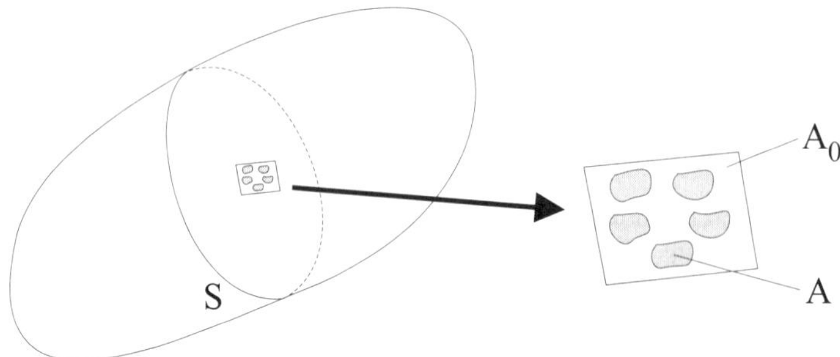

Figure 20.1: A section through a body.

the undamaged section. As a result of damage a certain part of the section is damaged or fractured and we will denote this damaged area by A. The values of A_0 and A are in the context of local damage but could equally be subject to appropriate averaging. The actual material area of the section is clearly $A_0 - A$.

Assuming now that the damage is isotropic and that cracks and voids are equally distributed in all directions then we can introduce a scalar *damage* parameter ω

$$\omega = \frac{A}{A_0} \tag{20.1}$$

and is a positive monotonically increasing function, $\mathrm{d}\omega/\,\mathrm{d}t > 0$, and restricted to the range $0 \leq \omega \leq 1$. Frequently, an additional term is introduced called the *continuity*, ψ, which is a positive monotonically decreasing parameter, $\mathrm{d}\psi/\,\mathrm{d}t < 0$,

$$\psi = 1 - \omega = \frac{A_0 - A}{A_0} \tag{20.2}$$

and is within the range $0 \leq \psi \leq 1$.

Denoting the nominal and actual stresses by σ_0 and σ_a for uniaxial tension with applied load P then we have for an elastic material

$$\sigma_a = \frac{P}{A_0 - A} = \frac{P}{A_0(1-\omega)} = \frac{P}{A_0\psi} = \frac{\sigma_0}{\psi} \tag{20.3}$$

illustrating that σ_a is related to the undamaged (actual area) of the section. Re-arranging (20.3) for ψ

$$\psi = \frac{\sigma_0}{\sigma_a} \tag{20.4}$$

Assuming that damage is mainly attributed to the actual stress then

$$\varepsilon = \frac{\sigma_a}{E} = \frac{1}{E}\frac{\sigma_0}{\psi} \quad \text{or} \quad \psi = \frac{1}{E}\frac{\sigma_0}{\varepsilon} \tag{20.5}$$

illustrating that Hooke's law is still applicable with Young's modulus E replaced by $E' = E\psi$.

In the case of elastic-plastic deformation then the damage is a result of large strains in which case it is assumed that damage does not depend on the elastic strain

$$\frac{\mathrm{d}\psi}{\mathrm{d}\varepsilon} = 0 \tag{20.6}$$

which leads to, from (20.5)

$$\frac{\mathrm{d}\psi}{\mathrm{d}\varepsilon} = -\frac{1}{\varepsilon^2}\frac{\sigma_0}{E} + \frac{1}{E_\varepsilon}\frac{\mathrm{d}\sigma_0}{\mathrm{d}\varepsilon} = 0 \tag{20.7}$$

from which

$$\frac{\sigma_0}{\varepsilon} = \frac{\mathrm{d}\sigma_0}{\mathrm{d}\varepsilon} \tag{20.8}$$

and substituting into (20.5) we find

$$\psi = \frac{1}{E}\frac{\sigma_0}{E} = \frac{1}{E}\frac{\mathrm{d}\sigma_0}{\mathrm{d}\varepsilon} = \frac{E'}{E} \tag{20.9}$$

Thus, damage may be estimated by measuring the elastic response with E' equivalent to the unloading modulus, Figure 20.2.

20.4 Kinetic Equation of Damage

The kinetic equation of damage assumes that the rate of damage growth is determined by the actual stress σ/ψ

$$\frac{\mathrm{d}\psi}{\mathrm{d}t} = f\left(\frac{\sigma}{\psi}\right) \tag{20.10}$$

In the case of uniaxial tension then $f(\sigma/\psi)$ can be assumed to be a power law

$$\frac{\mathrm{d}\psi}{\mathrm{d}t} = -A\left(\frac{\sigma}{\psi}\right)^n \tag{20.11}$$

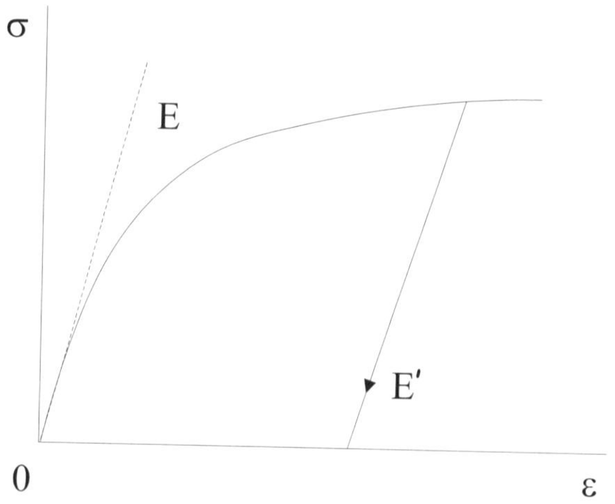

Figure 20.2: Young's modulus and the unloading modulus.

where $A(>0)$ and $n(\geq 1)$ are material constants. Equation (20.11) is found to agree with experimental observations for a range of different materials. The continuity ψ is equal to 1 and 0 at initiation $(t = 0)$ and fracture $(t = t_f)$ respectively. Under conditions of constant stress, $\sigma = \sigma_0$, then separating variables in (20.11) we have

$$\int \psi^n \, \mathrm{d}\psi = -A\sigma_0^n \int \mathrm{d}t \tag{20.12}$$

Integrating, we have

$$\frac{1}{n+1}\psi^{n+1} = -A\sigma_0^n t + C \tag{20.13}$$

where C is a constant of integration. With the initial condition $\psi = 1$ at $t = 0$ then the constant of integration C is found to equal unity. Therefore, re-arranging (20.13) for t

$$t = \frac{1 - \psi^{n+1}}{A(n+1)\sigma_0^n} \tag{20.14}$$

At the point of fracture $(t = t_f)$ then $\psi = 0$ and t_f is equal to

$$t_f = \frac{1}{A(n+1)\sigma_0^n} \tag{20.15}$$

and substituting (20.15) into (20.14) then t can be expressed in terms of t_f as

$$t_f = t_f \left[1 - \psi^{n+1}\right] \tag{20.16}$$

Alternatively, ψ can be expressed as

$$\psi = \left(1 - \frac{t}{t_f}\right)^{1/(n+1)} \tag{20.17}$$

20.5 Fatigue Crack Growth as a Damage Parameter

When we discussed fatigue in §17.11.4.3 we noted that Paris' law is found to be applicable for estimating the fatigue crack growth rate of long defects for a range of materials, (17.147)

$$\frac{da}{dN} = C(\Delta K)^m \tag{20.18}$$

where a is the crack length, N is the number of cycles, C and m are empirical constants and ΔK is the stress intensity factor range. The constant C depends on the frequency and mean stress of the cyclic loading. With $\Delta K = Y\Delta\sigma\sqrt{\pi a}$ then it was shown that the crack length relative to the initial crack length, a_i, can be expressed as, (17.154)

$$\frac{a}{a_i} = \left(\frac{1}{1-\beta N}\right)^{1/\kappa} \tag{20.19}$$

where κ and β are given by

$$\kappa = \frac{m}{2} - 1; \quad \beta = \kappa a_0^{\kappa} C \left(Y\Delta\sigma\sqrt{\pi}\right)^m \tag{20.20}$$

Figure 20.3 illustrates the variation of a/a_i against N/N_f with fatigue crack growth accelerating as $N \to N_f$ where N_f is the number of cycles to failure and given by

$$N_f = \frac{1}{\beta} \tag{20.21}$$

The ratio N/N_f can be considered as a damage parameter in an analogous manner to $\omega = A/A_0$.

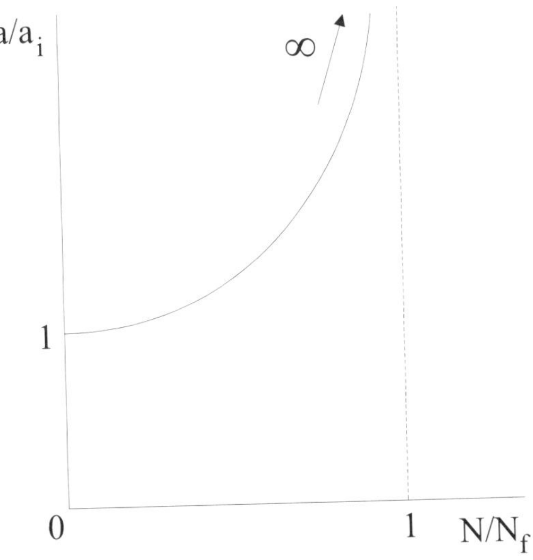

Figure 20.3: Normalised crack growth (a/a_i) as a function of normalised number of cycles (N/N_f).

20.6 Damage Analysis of Tertiary Creep

In §18.9 we analysed the uniaxial tension of a bar under conditions of secondary creep and arrived at the following relation, (18.29)

$$\frac{A}{A_0} = \left(1 - \frac{t}{t_f}\right)^{1/n} \tag{20.22}$$

which expresses the current cross-sectional area of the bar, A, relative to the original cross-sectional area, A_0, as a function of time t and the time to failure, t_f, and with n representing Norton's power law exponent. The variation of A/A_0 against t/t_f as a function of n is illustrated in Figure 18.9. Equally, A/A_0 can now be viewed as a continuity parameter since it monotonically decreases from 1 to 0 as t/t_f increases from 0 to 1.

Rabotnov (1969) and Kachanov (1986) proposed a damage model of tertiary creep under conditions of low stress and in which a dominant tertiary regime exists from grain-boundary cavitation and cracking. For a constant temperature then the tertiary creep strain rate and rate of change of damage parameter are assumed to be of the form

$$\dot{\varepsilon} = \frac{A\sigma^n}{(1-\omega)^q}; \qquad \dot{\omega} = \frac{B\sigma^k}{(1-\omega)^r} \tag{20.23}$$

and are based on Norton's power (18.7). At the initiation of tertiary creep ($t = 0$) then $\omega = 0$ and at fracture ($t = t_f$) then $\omega = 1$.

To determine an expression for the tertiary creep curve we begin by separating variables in the second equation of (20.23)

$$\int_0^{\omega} (1-\omega)^r \, \mathrm{d}\omega = B\sigma^k \int_0^t \mathrm{d}t \tag{20.24}$$

Performing the integrations we have

$$\frac{1}{1+r}\left[1 - (1-\omega)^{1+r}\right] = B\sigma^k t \tag{20.25}$$

At $t = t_f$ then $\omega = 1$ and from (20.25) we have

$$t_f = \frac{1}{(1+r)B\sigma^k} \tag{20.26}$$

and upon substitution back into (20.25) then t is given by

$$\frac{t}{t_f} = 1 - (1-\omega)^{1+r} \tag{20.27}$$

Now substitute (20.27) into the first equation of (20.23) to determine the tertiary creep strain

$$\frac{\mathrm{d}\varepsilon}{\mathrm{d}t} = \frac{A\sigma^n}{(1-\omega)^q} \Rightarrow \varepsilon = A\sigma^n \int_0^t \left(1 - \frac{t}{t_f}\right)^{-q/(1+r)} \mathrm{d}t = A\sigma^n t_f \lambda \left[1 - \left(1 - \frac{t}{t_f}\right)^{1/\lambda}\right] \tag{20.28}$$

where λ is

$$\lambda = \frac{1+r}{1+r-q} \tag{20.29}$$

At $t = t_f$ the failure strain is

$$\varepsilon_f = A\sigma^n t_f \lambda \tag{20.30}$$

Finally, substituting ε_f into (20.28) we have

$$\frac{\varepsilon}{\varepsilon_f} = 1 - \left(1 - \frac{t}{t_f}\right)^{1/\lambda} \tag{20.31}$$

which can be compared to (20.17), (20.19) and (20.22). The variation of ε is illustrated in Figure 20.4 and is seen to exhibit a continuously increasing slope as $t \to t_f$ which is consistent with an increasing creep strain rate in tertiary creep.

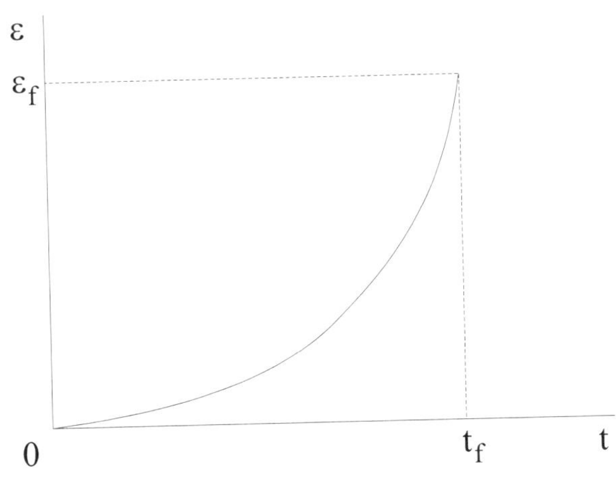

Figure 20.4: Tertiary creep curve based on a damage mechanics analysis.

20.7 Conclusion

In this Chapter we have been introduced to the damage and continuity parameters. Damage mechanics provides an alternative method of analysis for a range of problems. Throughout this Chapter we re-examined the problems of fatigue crack growth and tertiary creep considered in previous Chapters but this time from a continuum damage mechanics perspective.

20.8 References and Further Reading

- Kachanov, L. M. (1986) *Introduction to Continuum Damage Mechanics*, Martinus Nijhoff.
- Rabotnov, Y. N. (1969) *Creep Problems in Structural Members*, North Holland, Amsterdam.

20.9 Exercises

20.1 Name several different types of damage.

20.2 Examination of tensile test specimens upon failure reveals that the average damaged cross-sectional area due to the growth of micro-voids is equal to 826mm^2. If the original undamaged cross-sectional area is 1,018mm^2 then determine the average damage parameter at the point of failure.

20.3 For the tensile tests of Exercise 20.2 determine the average continuity parameter at the point of failure.

20.4 The Young's and unloading moduli of an elastic-plastic material are equal to 190GPa and 67GPa respectively. Estimate the level of damage.

20.5 Table 20.1 consists of tensile creep (σ_0, t_f) data. If the kinetic equation of damage is given by

$$\frac{\mathrm{d}\Psi}{\mathrm{d}t} = -A\left(\frac{\sigma}{\Psi}\right)^n$$

then determine the material constants A and n.

σ_0 **(MPa)**	14	13	12	11	10
t_f **(h)**	2.17	4.95	12.18	29.96	97.11

Table 20.1: Tensile creep (σ_0, t_f) data for Exercise 20.5.

20.6 For the experimental data of Exercise 20.5 determine the damage parameter at a time which is equal to 80% of the total time to failure.

20.7 Show that the ductility parameter λ for damage under conditions of tertiary creep is given by

$$\lambda = \frac{\varepsilon_f}{\dot{\varepsilon}_s t_f}$$

where ε_f is the failure strain at time $t = t_f$ and $\dot{\varepsilon}_s (= A\sigma^n)$ is the strain rate at the end of secondary creep and the start of tertiary creep.

Chapter 21

Composite Materials

21.1 Introduction

The aim of this Chapter is to present an overview of composite materials. We begin by classifying composite materials and examining the mechanical properties of several popular composites. Following brief discussions on thermoset and thermoplastic resins and chopped fibre composites the Chapter then focuses on unidirectional laminae. The elastic properties are then derived for unidirectional laminae with applied loading both parallel and perpendicular to the fibres. Furthermore, section 21.7 derives the elastic properties of unidirectional laminae with orthotropic material symmetry. The Chapter concludes by examining the strength of unidirectional laminae and the failure mechanism of fibre pull-out.

21.2 Classification of Composite Materials

Engineering materials can be classified broadly into six groups: metals, polymers, elastomers, ceramics, glasses and composites. A composite consists of two or more different material components to form a single material without the physical blending of the individual constituents. In other words, the constituents of a composite are still identifiable in the bulk material. The group of composites covers a wide range of different materials. Naturally occurring composites such as wood, bamboo and bone all exhibit high strength, low weight and great flexibility. These naturally occurring composites obtain their strength from the alignment of their fibres so as to provide maximum stiffness in the direction of loading. The majority of engineering materials can also be viewed as composites since they are combinations of two or more phases. For example BS 302-S25 (En 58A) austenitic stainless steel contains 0.08% carbon, 18% chromium and 9% nickel and possesses excellent ductility, good resistance to corrosion and good toughness at low temperature. Such metallic alloys achieve their strength by combining high strength and ductile phases into a single material. Both natural and engineered metallic alloys are referred to as micro-composites since the combined structure is very fine and cannot be viewed by the naked eye. At the opposite end of the scale are macro-composites. Examples of these are laminated bows and more modern engineering examples are reinforced concrete, skis and galvanised steel.

Modern composite materials are generally associated with non-metallic engineered materials. These date back to the 1910s with the use of doped fabric for aircraft surfaces. In the 1930s/40s fabric reinforced phenolic and glass fibre reinforced plastics were developed. The 1960s/70s saw the emergence of composites such as boron and carbon fibres and kevlar-49. Today there exists a vast library of composite materials such as alumina, boron nitride, silica, alumina/silica (kaowool) and polyolefin (spectra).

In general, composites consist of two separate phases: the *matrix* and the *fibres*. The matrix is the phase that bonds the fibres to form the bulk material. The fibres are the phase of the material that has been added to the matrix to give the composite its advantage (usually defined in terms of strength) compared solely to the matrix. Composites can be classified into the following three groups according to the type of fibres used in the composite: *particulate, short/long fibre* and *laminate*. Particulate composites consist of approximately spherical fillers. For example, concrete is a particulate composite consisting of a cement matrix and sand and stone filler. Short and long fibre composites are composites in which the length to diameter, l/d, ratio of the fibres is greater than unity. Short fibre composites are approximately categorised as having $l/d \leq 100$ whereas long fibres are for $l/d > 100$. Fibre glass is an example of a short fibre and carbon fibre is an example of a long fibre. Laminae are composites in which the filler is in the form of sheet such as a Formica work surface.

21.3 Material Properties

Table 21.1 lists the properties of a low alloy steel, nylon-6 and a carbon-fibre with an epoxy resin with unidirectional fibres. Compare the properties of the carbon-fibre composite for both parallel and perpendicular directions to the fibre direction. The Young's modulus and tensile strength parallel to the fibres are significantly greater than those perpendicular to the fibres. At first, these highly directional strength characteristics may be viewed as a significant disadvantage but in fact they allow a designer to provide increased strength and stiffness into a component where it is needed. Composites are non-homogeneous

Composite	**Density ($\mathbf{Mgm^{-3}}$)**	**Young's modulus (GPa)**	**Tensile strength (MPa)**	**Elongation to fracture (%)**	**Coefficient of thermal exp. ($\mathbf{10^{-6}\,{}^{\circ}C^{-1}}$)**
Quenched and tempered low alloy steel	7.86	207	600-2050	12-28	11
Nylon 6	1.14	1-2.5	70-85	60	90
Carbon fibre-epoxy ($V_f = 0.6$)					
i) $\parallel$ to fibres	1.62	220	1400	0.8	0.2
ii) $\perp$ to fibres	1.62	7	38	0.6	30

Table 21.1: Material properties of four composite materials.

and as a result they exhibit different responses when loaded in tension, compression or bending. Table 21.2 lists the uniaxial failure strength, σ^*, for glass-polyester and Kevlar 49-epoxy composites tested in tension (T) and compression (C) for both parallel and perpendicular to the fibre direction. Observe that the transverse strengths are significantly less than the parallel strengths.

Composite	$\sigma_{\parallel}^*(T)$ **(MPa)**	$\sigma_{\parallel}^*(C)$ **(MPa)**	$\sigma_{\perp}^*(T)$ **(MPa)**	$\sigma_{\perp}^*(C)$ **(MPa)**
glass-polyester	650-750	600-900	20-25	90-120
Kevlar 49-epoxy	1100-1250	240-290	20-30	110-140

Table 21.2: Uniaxial failure strength for glass-polyester and Kevlar-epoxy tested in tension and compression both parallel and perpendicular to the fibre direction.

21.3.1 Specific Properties

Perhaps the most important factor leading to the rapid development and use of composite materials is the savings in mass that can be achieved from the use of low density fibres with high modulus and strength. The *specific Young's modulus* and *specific tensile strength* are defined as the measured modulus and strength divided by the density with some typical values shown in Table 21.3 for carbon-fibre, E glass and Kevlar 49. The specific Young's modulus of carbon fibres are far superior to those of E glass which in

Specific Young's modulus/Tensile strength	**Carbon PAN-based (Type I)**	**E glass**	**Kevlar 49**
specific Young's modulus (Nm^{-2}/kgm^{-3})	200	30	86
specific tensile strength (Nm^{-2}/kgm^{-3})	1.1	1.4	2.2

Table 21.3: Specific properties for carbon fibre, E glass and Kevlar composites.

turn are approximately 50% less than Kevlar. The specific Young's moduli of Table 21.3 can be compared to those of quenched and tempered low alloy steel ($207/7.85 \approx 26$) and high strength steel ($72/2.8 \approx 25$).

21.3.2 Thermosetting and Thermoplastic Resins

A composite matrix is generally categorised as either a thermoset or thermoplastic resin and these are discussed separately in the following sub-sections.

21.3.2.1 Thermosetting Resins

Thermosetting polymers are cross-linked which leads to hard brittle solids consisting of tightly bound networks of polymer chains. As a result, the material properties of themosetting polymers are dependent on both the length and density of the cross-links and the molecular components making up the polymer. Thermosetting resins are isotropic and do not melt on heating. Epoxy resins have high distortion temperatures although their mechanical properties do suffer as the temperature approaches that of the distortion temperature. Furthermore, epoxy resins have higher tensile strength and Young's modulus than polyester resins although they have a higher viscosity before curing and are more expensive. Examples of thermosetting resins for composites are unsaturated polyester, vinyl ester, epoxy, urethane and phenolic. Table 21.4 lists several properties for epoxy, polyester and acetals thermosetting resins.

Material/Property	**Epoxy**	**Polyester**	**Acetals (glass filled)**
Density (Mgm^{-3})	1.1-1.4	1.2-1.5	1.6
Young's modulus (GPa)	3-6	2-4.5	7
Poisson's ratio	0.4	0.38	
Tensile strength (MPa)	35-100	40-90	58-75
Elongation at break (%)	1-6	2	2-7
Heat distortion temperature (°C)	50-300	50-110	

Table 21.4: Properties of some thermosetting plastics.

21.3.2.2 Thermoplastic Resins

Thermoplastics are not cross-linked and derive their strength from the high molecular weight of the polymers. Thermoplastics are generally combined with short fibres and used for products manufactured by injection moulding. In amorphous thermoplastics there is a high concentration of molecular entanglements which act like cross-links. In crystalline thermoplastics there is a high degree of molecular alignment. Unlike thermosetting resins, thermoplastics melt with amorphous thermoplastics disentangling when subjected to heat and becoming a viscous liquid while crystalline thermoplastics become an amorphous viscous liquid when subject to heat. Thermoplastics, unlike thermosets, experience yielding and undergo extensive deformation before final failure. For example, Nylon is a tough matrix with fracture energy of the order $10^3 \sim 10^5 \text{Jm}^{-2}$. As a result, it is difficult to propagate brittle cracks unless they are cold-drawn, fatigued or in thick sections. Another important property of thermoplastics is that they creep under constant load. Examples of thermoplastic resins for composites are polypropylene, polyethylene, polystyrene, nylon and PEEK (poly-ether-ether-ketone). Table 21.5 lists several properties for polypropylene, nylon and PTFE thermoplastics.

Material/Property	**Polypropylene**	**Nylon**	**PTFE**
Density (Mgm^{-3})	0.9	1.16	2.1
Young's modulus (GPa)	1-1.4	2.4	0.3
Posson's ratio	0.3	0.3	
Tensile strength (MPa)	25-38	60	13
Elongation at break (%)	200-700	90	100
Melting temperature (°C)	175	264	

Table 21.5: Properties of some thermoplastics.

21.4 Chopped Fibre Composites

In the previous section it was noted that a common use of thermoplastic resins is with short fibre reinforcement. In the next two sub-sections we briefly examine the two most important properties of chopped fibre composites: orientation and length.

21.4.1 Fibre Orientation

The most important property of chopped fibre composites is the orientation of the fibres since fibre orientation defines the directional strength of a composite. Assuming the fibres to lie within a plane then a typical variation of fibre orientation is shown in Figure 21.1 for both random and non-random distributed fibres. The inclination of each fibre with respect to the coordinates axes (x, y) is plotted as a normalised percentage for all fibres having the same orientation. For a fully random distribution of fibres then the orientation will be independent of angular position. However, a non-random distribution will exhibit preferential fibre alignment for certain angular positions.

21.4.2 Fibre Length Distribution

Chopped fibres are usually chopped to a predetermined length before they are mixed with the matrix resin. In the case of thermosetting resins the resin is added to the fibres as a low viscosity liquid so that

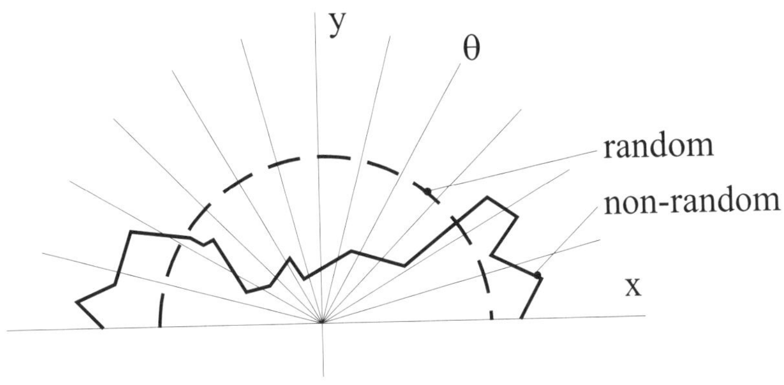

Figure 21.1: In-plane chopped fibre orientation for random and non-random distributions.

the original fibres remain unbroken. On the other hand, the addition of chopped fibres to a thermoplastic resin is usually performed in an extruder with extensive shearing leading to fibre breakage. Subsequent use of injection moulding leads to further fibre breakages. Figure 21.2 illustrates a frequency spectrum of fibre length and serves to illustrate that it is not uncommon that no fibres remain of the original length.

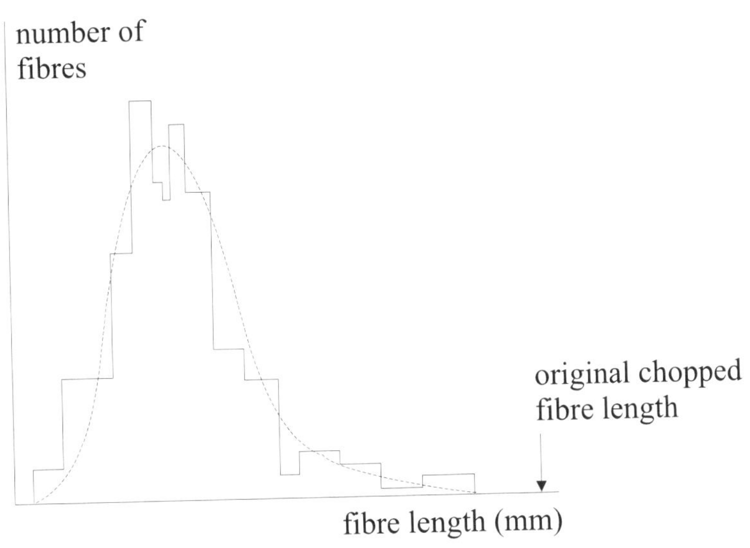

Figure 21.2: Fibre length distribution.

21.5 Ideal Unidirectional Laminae

In an ideal configuration the fibres can be considered to be arranged either on a hexagonal or square lattice, Figure 21.3. For both cases all fibres are assumed to have the same diameter $2r$, distance between fibre centres $2R$ and separation distance s. To determine the volume fraction of fibres V_f, for the square

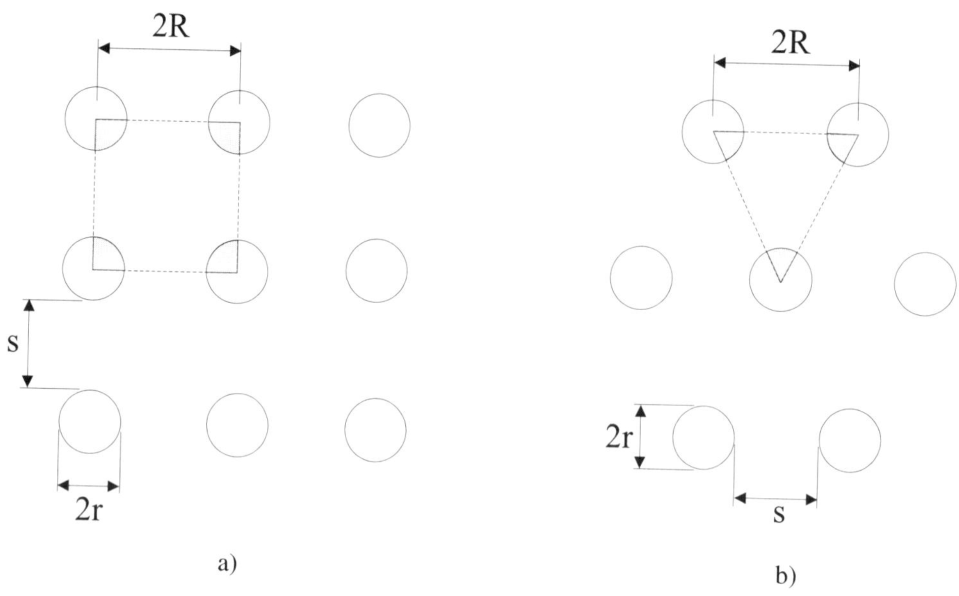

Figure 21.3: Idealised a) square and b) hexagonal fibre arrangements for unidirectional laminae.

configuration consider the square of side $2R$ in Figure 21.3a). The volume fraction is

$$V_f = \frac{A_f}{A} = \frac{4\left(\pi r^2/4\right)}{4R^2} = \frac{\pi}{4}\left(\frac{r}{R}\right)^2 \tag{21.1}$$

where A_f and A are the areas of the fibres and square respectively. The separation distance between the fibres is, with $R = (\pi r^2/4V_f)^{1/2}$ from (21.1)

$$s = 2(R-r) = 2\left[\left(\frac{\pi r^2}{4V_f}\right)^{1/2} - r\right] = 2r\left[\left(\frac{\pi}{4V_f}\right)^{1/2} - 1\right] \tag{21.2}$$

To determine the volume fraction for the hexagonal configuration consider the equilateral triangle of side length $2R$ in Figure 21.3b)

$$V_f = \frac{A_f}{A} = \frac{3\left(\pi r^2/6\right)}{\sqrt{3}R^2} = \frac{\pi}{2\sqrt{3}}\left(\frac{r}{R}\right)^2 \tag{21.3}$$

With $R = (\pi r^2/2\sqrt{3}\,V_f)^{1/2}$ then the separation distance is

$$s = 2(R-r) = 2\left[\left(\frac{\pi r^2}{2\sqrt{3}\,V_f}\right)^{1/2} - r\right] = 2r\left[\left(\frac{\pi}{2\sqrt{3}\,V_f}\right)^{1/2} - 1\right] \tag{21.4}$$

The maximum value of V_f occurs when the fibres touch ($s = 0$) and leads to $V_{f,\max} = \pi/4 \approx 0.785$ for the square case and $V_{f,\max} = \pi/2\sqrt{3} \approx 0.907$ for the hexagonal case. The separation distances for both cases are shown in Figure 21.4. The separation distance is equal to the fibre diameter ($s = 2r$) when $V_f = \pi/16 \approx 0.196$ for the square case and when $V_f = \pi/8\sqrt{3} \approx 0.227$ for the hexagonal case. Thus, even for low volume fractions the separation distance of the fibres is less than the fibre diameter and as a result will lead to strain magnification.

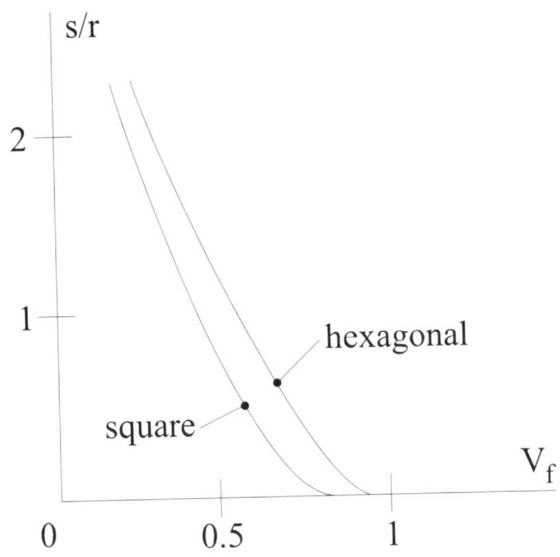

Figure 21.4: Separation distance against volume fraction.

21.6 Elastic Properties of Unidirectional Laminae

In this section we take a more quantitative look at composites by examining the distribution of stress and strain in a composite material in terms of geometry, distribution of volume fraction of the fibres and elastic properties of the fibres and matrix. When either a tensile or compressive load is applied in a direction parallel to the fibres in a unidirectional laminae then the strain ε_{xx} in the matrix will be the same as the strain in the fibres if the bond between the fibres and matrix is perfect. If both the fibres and matrix behave elastically then the stresses are, see Figure 21.5

$$\sigma_f = E_f \varepsilon_{xx} \quad , \quad \sigma_m = E_m \varepsilon_{xx} \tag{21.5}$$

where the subscripts f and m refer to the fibres and matrix respectively. Equations (21.5) assume that $\varepsilon_{xx} = \varepsilon_f = \varepsilon_m$ with a strong matrix-fibre bond.

The underlying design consideration for fibre reinforced composites is that the Young's modulus, E, of the fibres is significantly greater for the fibres than the matrix

$$E_f > E_m \tag{21.6}$$

This ensures that the stress in the fibres is greater than in the matrix so that the fibres carry the majority of the applied loading.

For a composite with a total cross-sectional area A then the average stress is

$$\sigma_{xx} = \frac{P}{A} \tag{21.7}$$

where

$$P = P_f + P_m = \sigma_f A_f + \sigma_m A_m \tag{21.8}$$

Substituting $\sigma_f = E_f \varepsilon_{xx}, \sigma_m = E_m \varepsilon_{xx}$ then P is given by

$$P = E_f \varepsilon_{xx} A_f + E_m \varepsilon_{xx} A_m \tag{21.9}$$

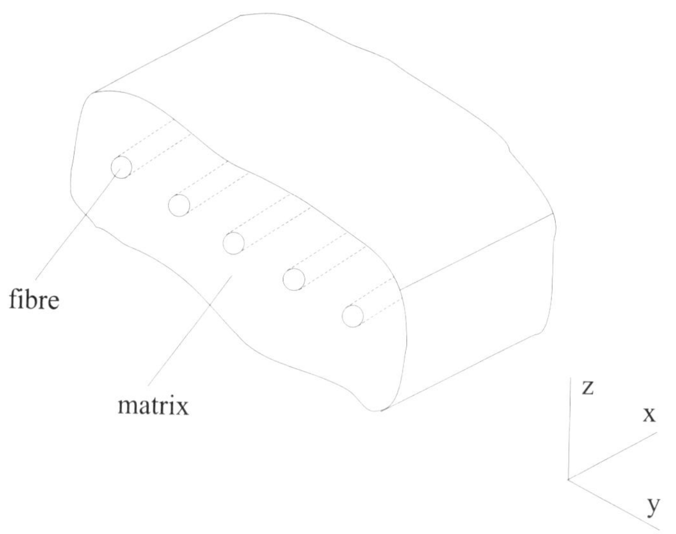

Figure 21.5: Matrix-fibre composite.

and substituting into (21.7) and noting $\sigma_{xx} = E_{xx}\varepsilon_{xx}$

$$\sigma_{xx} = \frac{P}{A} = E_{xx}\varepsilon_{xx} = E_f\varepsilon_{xx}\frac{A_f}{A} + E_m\varepsilon_{xx}\frac{A_m}{A} \tag{21.10}$$

from which we find E_{xx}

$$E_{xx} = E_f\frac{A_f}{A} + E_m\frac{A_m}{A} \tag{21.11}$$

Defining the fibre and matrix volume fractions as $V_f = A_f/A$ and $V_m = A_m/A$ respectively then

$$E_{xx} = E_f V_f + E_m V_m \tag{21.12}$$

With $V_m = 1 - V_f$ then

$$E_{xx} = E_f V_f + E_m (1 - V_f) \tag{21.13}$$

and is referred to as the *rule of mixtures*. The rule of mixtures is based on (21.5) which takes no account of the Poisson contractions when $\nu_m \neq \nu_f$. However, experimental evidence informs us that accounting for differing Poisson contractions leads to approximately 2% difference in E_{xx} and can therefore be neglected.

For unidirectional laminae the notation $E_{\|}$, $E_{\perp}$ and $E_{\|\perp}$ is frequently used to indicate E_{xx} parallel to fibres, E_{yy} perpendicular to fibres and the coupled E_{xy}. Figure 21.6a) illustrates the typical variation of $E_{\|}$ for a unidirectional laminae of glass fibres in polyester resin against volume fraction. The majority of the available experimental data is in the range $0.2 \leq V_f \leq 0.6$ and is found to be in good agreement with the rule of mixtures (21.13).

A similar procedure can be performed to determine the transverse modulus of a unidirectional laminae, $E_{yy} = E_{\perp}$. Assuming that the applied load transverse to the fibres acts equally on both the fibres and matrix then we can assume that the following holds

$$\sigma_f = \sigma_m \tag{21.14}$$

The corresponding strains are

$$\varepsilon_f = \frac{\sigma_{yy}}{E_f} \quad , \quad \varepsilon_m = \frac{\sigma_{yy}}{E_m} \tag{21.15}$$

for an applied stress, σ_{yy}, in the y-direction. The strain in the y-direction is

$$\varepsilon_{yy} = V_f \varepsilon_f + V_m \varepsilon_m \tag{21.16}$$

and substituting the strains (21.15)

$$\varepsilon_{yy} = V_f \frac{\sigma_{yy}}{E_f} + V_m \frac{\sigma_{yy}}{E_m} \tag{21.17}$$

Substituting $\sigma_{yy} = E_{yy}\varepsilon_{yy}$ into (21.17) and re-arranging we have

$$E_{\perp} = \frac{E_f E_m}{E_f\left(1 - V_f\right) + E_m V_f} \tag{21.18}$$

and is illustrated in Figure 21.6b). Accounting for Poisson's ratio of the matrix through the use of the plane stress-plane strain effective Young's modulus $E' = E/(1-\nu^2)$ then (21.18) can be written

$$E_{\perp} = \frac{E_f E'_m}{E_f\left(1 - V_f\right) + E'_m V_f}; \quad E'_m = \frac{E_m}{1 - \nu_m^2} \tag{21.19}$$

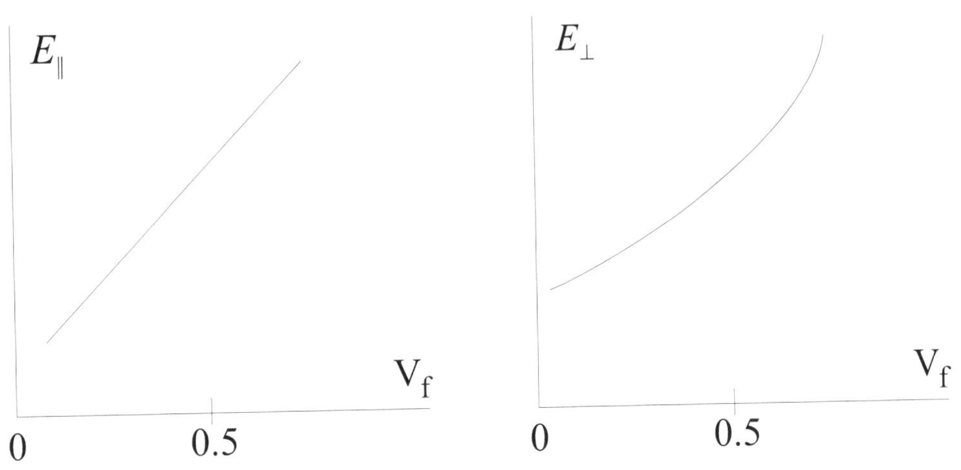

Figure 21.6: Variation of elastic moduli against volume fraction for a unidirectional laminae of glass fibres in polyester resin. a) $E_{\|}$ against V_f. b) $E_{\perp}$ against V_f.

The key assumption behind the derivation of (21.13) and (21.18) is that within both the fibres and matrix the strain is uniform. The strain is far from uniform throughout a cross-section of a unidirectional laminae due to the strain magnification in the matrix between fibres. Numerous studies have been undertaken to modify $E_{\|}$ and $E_{\perp}$ to take account of strain magnification. One of the first examinations was due to Kies (1962) who arrived at the following strain magnification for a square array of fibres subject to tensile loading

$$\frac{\varepsilon_{xx}}{\varepsilon_{xx}^0} = \frac{2 + \dfrac{s}{r}}{\dfrac{s}{r} + 2\left(\dfrac{E_m}{E_f}\right)} \tag{21.20}$$

where $2r$ is the fibre diameter, s is the fibre separation distance and ε_{xx}^0 is the far-field applied tensile strain. Figure 21.7 illustrates the variation of the strain magnification against V_f with s given by (21.2) for the square configuration of fibres in which $E_f/E_m = 10$. Note that $\varepsilon_{xx}/\varepsilon_{xx}^0 \to E_f/E_m$ as $s \to 0$.

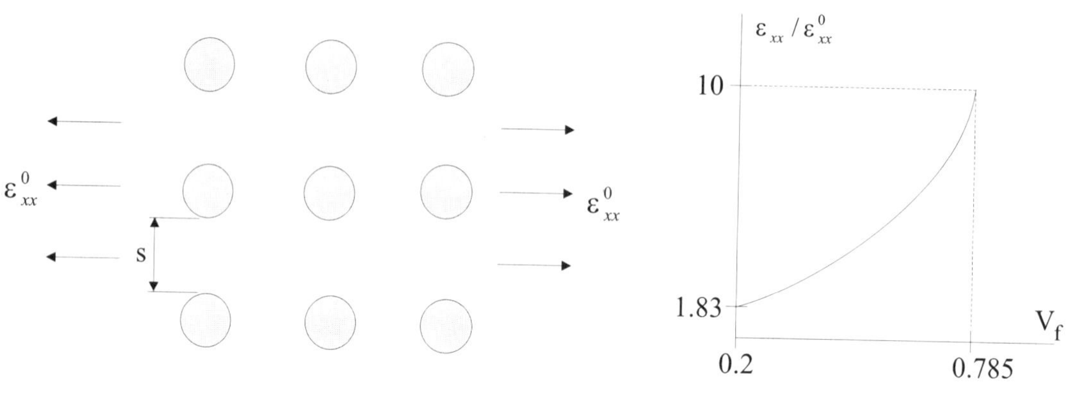

Figure 21.7: Strain magnification $\varepsilon_{xx}/\varepsilon_{xx}^0$ against volume fraction V_f for a square array of fibres.

21.7 Elastic Properties of Unidirectional Laminae with Orthotropic Material Symmetry

In the previous section we derived the rules of mixtures for unidirectional laminae and obtained expressions for the Young's moduli measured both parallel and perpendicular to the fibre direction. The present section derives expressions for Young's modulus, shear modulus and Poisson's ratio by modelling the laminae as an orthotropic body with 3 planes of material symmetry, Figure 21.5. Assuming that the laminae is sufficiently thin then a state of plane stress may be assumed with the stress-strain relations given by (14.99)

$$\begin{Bmatrix} \varepsilon_{xx} \\ \varepsilon_{yy} \\ \gamma_{xy} \end{Bmatrix} = \begin{bmatrix} 1/E & -\nu/E & 0 \\ -\nu/E & 1/E & 0 \\ 0 & 0 & 1/G \end{bmatrix} \begin{Bmatrix} \sigma_{xx} \\ \sigma_{yy} \\ \tau_{xy} \end{Bmatrix}$$
$$\begin{Bmatrix} \sigma_{xx} \\ \sigma_{yy} \\ \tau_{xy} \end{Bmatrix} = \begin{bmatrix} E/(1-\nu^2) & \nu E/(1-\nu^2) & 0 \\ \nu E/(1-\nu^2) & E/(1-\nu^2) & 0 \\ 0 & 0 & G \end{bmatrix} \begin{Bmatrix} \varepsilon_{xx} \\ \varepsilon_{yy} \\ \gamma_{xy} \end{Bmatrix} \tag{21.21}$$

Since the lamina properties are independent of direction in the plane normal to the x-axis then the lamina can be assumed to be isotropic in that plane. Assuming plane stress conditions apply then

$$\begin{Bmatrix} \varepsilon_{xx} \\ \varepsilon_{yy} \\ \gamma_{xy} \end{Bmatrix} = \begin{bmatrix} 1/E_1 & -\nu_{12}/E_2 & 0 \\ -\nu_{12}/E_2 & 1/E_2 & 0 \\ 0 & 0 & 1/G_{12} \end{bmatrix} \begin{Bmatrix} \sigma_{xx} \\ \sigma_{yy} \\ \tau_{xy} \end{Bmatrix} \tag{21.22}$$

with directions (1,2) corresponding to the (x, y) axes. The Poisson's ratio $\nu_{12}(= -\varepsilon_{yy}/\varepsilon_{xx})$ denotes the strains produced in the 2-direction when a body is loaded in the 1-direction. Similarly, $\nu_{21} = -\varepsilon_{xx}/\varepsilon_{yy}$ with $-\nu_{12}/E_1 = -\nu_{21}/E_2$.

Inverting (21.22) then the stress-strain stiffness matrix is given by

$$\begin{Bmatrix} \sigma_{xx} \\ \sigma_{yy} \\ \tau_{xy} \end{Bmatrix} = \begin{bmatrix} Q_{11} & Q_{12} & 0 \\ Q_{12} & Q_{22} & 0 \\ 0 & 0 & Q_{66} \end{bmatrix} \begin{Bmatrix} \varepsilon_{xx} \\ \varepsilon_{yy} \\ \gamma_{xy} \end{Bmatrix} \tag{21.23}$$

where

$$\begin{aligned} Q_{11} = C_{11} &= \frac{E_1}{1-\nu_{12}\nu_{21}} \quad , \quad Q_{12} = C_{12} = \frac{\nu E_2}{1-\nu_{12}\nu_{21}} = \frac{\nu_{21}E_1}{1-\nu_{12}\nu_{21}} \\ Q_{22} = C_{22} &= \frac{E_2}{1-\nu_{12}\nu_{21}} \quad , \quad Q_{66} = \frac{1}{2}\left(C_{11} - C_{12}\right) = G_{12} \end{aligned} \tag{21.24}$$

The state of stress with respect to a set of local coordinates axes (x', y') rotated through an angle θ to the (x, y) axes can be determined by aid of the transformation equations, §7.2. For example, consider the evaluation of $\sigma_{x'x'}$ from (7.8)

$$\sigma_{x'x'} = \sigma_{xx}\cos^2\theta + \sigma_{yy}\sin^2\theta + 2\tau_{xy}\sin\theta\cos\theta \tag{21.25}$$

and substituting σ_{xx}, σ_{yy} and τ_{xy} from (21.23)

$$\begin{aligned} \sigma_{xx} &= Q_{11}\varepsilon_{xx} + Q_{12}\varepsilon_{yy} \\ \sigma_{yy} &= Q_{12}\varepsilon_{xx} + Q_{22}\varepsilon_{yy} \\ \tau_{xy} &= Q_{66}\gamma_{xy} \end{aligned} \tag{21.26}$$

then $\sigma_{x'x'}$ is given by

$$\sigma_{x'x'} = \left(Q_{11}c^2 + Q_{12}s^2\right)\varepsilon_{xx} + \left(Q_{12}c^2 + Q_{22}s^2\right)\varepsilon_{yy} + 2Q_{66}sc\gamma_{xy} \tag{21.27}$$

where s and c denote $\sin\theta$ and $\cos\theta$ respectively. Similarly, from the strain transformation equations (8.10) then ε_{xx}, ε_{yy} and γ_{xy} are given by

$$\begin{aligned} \varepsilon_{xx} &= \varepsilon_{x'x'}\cos^2\theta + \varepsilon_{y'y'}\sin^2\theta - \gamma_{x'y'}\sin\theta\cos\theta \\ \varepsilon_{yy} &= \varepsilon_{x'x'}\sin^2\theta + \varepsilon_{y'y'}\cos^2\theta + \gamma_{x'y'}\sin\theta\cos\theta \\ \frac{\gamma_{xy}}{2} &= \varepsilon_{x'x'}\sin\theta\cos\theta - \varepsilon_{y'y'}\sin\theta\cos\theta + \frac{\gamma_{x'y'}}{2}\left(\cos^2\theta - \sin^2\theta\right) \end{aligned} \tag{21.28}$$

substituting these into (21.27) we arrive at

$$\sigma_{x'x'} = \bar{Q}_{11}\varepsilon_{x'x'} + \bar{Q}_{12}\varepsilon_{y'y'} + \bar{Q}_{16}\gamma_{x'y'} \tag{21.29}$$

where $\bar{Q}_{11}$, $\bar{Q}_{12}$ and $\bar{Q}_{16}$ are given below. Following a similar procedure for $\sigma_{y'y'}$ and $\tau_{x'y'}$ we arrive at the following stress-strain relation, Hull (1981)

$$\begin{Bmatrix} \sigma_{x'x'} \\ \sigma_{y'y'} \\ \tau_{x'y'} \end{Bmatrix} = \begin{bmatrix} \bar{Q}_{11} & \bar{Q}_{12} & \bar{Q}_{16} \\ \bar{Q}_{12} & \bar{Q}_{22} & \bar{Q}_{26} \\ \bar{Q}_{16} & \bar{Q}_{26} & \bar{Q}_{66} \end{bmatrix} \begin{bmatrix} \varepsilon_{x'x'} \\ \varepsilon_{y'y'} \\ \gamma_{x'y'} \end{bmatrix} \tag{21.30}$$

where

$$
\begin{aligned}
\bar{Q}_{11} &= Q_{11}c^4 + 2\left(Q_{12} + 2Q_{66}\right)s^2c^2 + Q_{22}s^4 \\
\bar{Q}_{12} &= \left(Q_{11} + Q_{22} - 4Q_{66}\right)s^2c^2 + Q_{12}\left(s^4 + c^4\right) \\
\bar{Q}_{22} &= Q_{11}s^4 + 2\left(Q_{12} + 2Q_{66}\right)s^2c^2 + Q_{22}c^4 \\
\bar{Q}_{16} &= \left(Q_{11} - Q_{12} - 2Q_{66}\right) + \left(Q_{12} - Q_{22} + 2Q_{66}\right)s^3c \\
\bar{Q}_{26} &= \left(Q_{11} - Q_{12} - 2Q_{66}\right)s^3c + \left(Q_{12} - Q_{22} + 2Q_{66}\right)sc^3 \\
\bar{Q}_{66} &= \left(Q_{11} + Q_{22} - 2Q_{12} - 2Q_{66}\right)s^2c^2 + Q_{66}\left(s^4 + c^4\right)
\end{aligned}
\tag{21.31}
$$

From the above procedure the local E, G, ν are found to be

$$
\begin{aligned}
\frac{1}{E_{x'x'}} &= \frac{c^4}{E_1} + \left(\frac{1}{G_{12}} - \frac{2\nu_{12}}{E_1}\right)s^2c^2 + \frac{s^4}{E_2} \\
\frac{1}{E_{y'y'}} &= \frac{s^4}{E_1} + \left(\frac{1}{G_{12}} - \frac{2\nu_{12}}{E_1}\right)s^2c^2 + \frac{c^4}{E_2} \\
\frac{1}{G_{x'y'}} &= 2\left(\frac{2}{E_1} + \frac{2}{E_2} + \frac{4\nu_{12}}{E_1} - \frac{1}{G_{12}}\right)s^2c^2 + \frac{s^4 + c^4}{G_{12}} \\
\nu_{x'y'} &= \left[\frac{\nu_{12}\left(s^4 + c^4\right)}{E_1} - \left(\frac{1}{E_1} + \frac{1}{E_2} - \frac{1}{G_{12}}\right)s^2c^2\right]E_{x'y'}
\end{aligned}
\tag{21.32}
$$

and are solely in terms of E_1, E_2, G_{12}, ν_{12} and θ. The variation of $E_{x'x'}$, $E_{y'y'}$, $G_{x'y'}$ and $\nu_{x'y'}$ are shown in Figure 21.8 for a unidirectional laminae of glass fibre and polyester resin with $V_f = 0.45$, $E_1 = 40$GPa, $E_2 = 10$GPa, $G_{12} = 4G$Pa and $\nu_{12} = 0.25$.

Neglecting end effects the above equations (21.32) can be used to provide an approximate expression for the Young's modulus, $\bar{E}$, for a randomly distributed long fibre laminae, Nielsen and Chen (1968)

$$
E = 2\pi \int_0^{\pi/2} E(\theta)\, \mathrm{d}\theta \tag{21.33}
$$

where $E(\theta)$ is either $E_{x'x'}$ or $E_{y'y'}$. Performing the integration in (21.33) is a lengthy procedure and alternatively the following simplified equations may be used

$$
\bar{E} = \frac{3}{8}E_1 + \frac{5}{8}E_2 \quad , \quad \bar{G} = \frac{1}{8}E_1 + \frac{1}{4}E_2 \quad , \quad \bar{\nu} = \frac{\bar{E}}{2\bar{G}} - 1 \tag{21.34}
$$

21.8 Strength of Unidirectional Laminae

The following two sub-sections examine both the longitudinal and transverse strength of unidirectional laminae.

21.8.1 Longitudinal Tensile Strength

When we discussed material properties in §21.3 it was noted that the transverse failure strength was significantly less than the parallel failure strength for unidirectional laminae subject to both tensile and compressive loading. This is because the transverse strength is governed by the matrix properties while the parallel strength is governed by the fibre properties with $E_f \gg E_m$. If the matrix-fibre bond is perfect

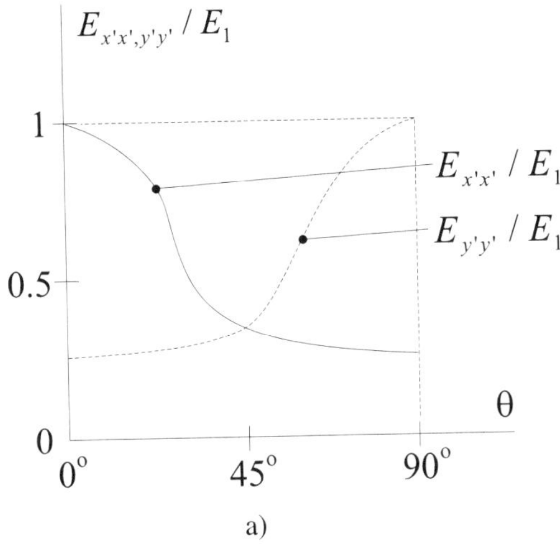

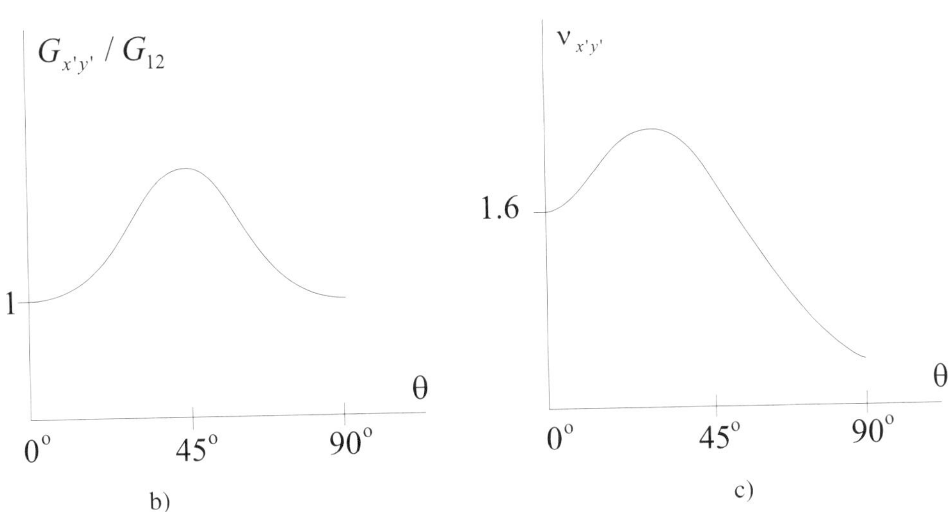

Figure 21.8: Variation of material properties for an orthotropic unidirectional laminae of glass fibre and polyester resin. a) $E_{x'x'}$ and $E_{y'y'}$ against θ. b) $G_{x'y'}$ against θ. c) $\nu_{x'y'}$ against θ.

then $\varepsilon_{||} = \varepsilon_m + \varepsilon_f$ for a unidirectional laminae loaded in tension and from the rule of mixtures, (21.13), we have

$$\sigma_{||} = E_f \varepsilon_{||} V_f + E_m \varepsilon_{||} (1 - V_f) = \sigma_f V_f + \sigma_m (1 - V_f) \tag{21.35}$$

There are two possible cases to consider: i) $\varepsilon_f^* > \varepsilon_m^*$ and ii) $\varepsilon_m^* > \varepsilon_f^*$ where ε_f^* and ε_m^* are the failure strains in uniaxial tension of the fibres and matrix respectively. The following sub-sections consider these two cases separately.

i) $\varepsilon_f^* > \varepsilon_m^*$

When V_f is low then the strength of the laminae depends primarily on σ_m^*. The matrix will fail before the fibres and then all of the applied load will be transferred to the fibres, Figure 21.9. When V_f is low then

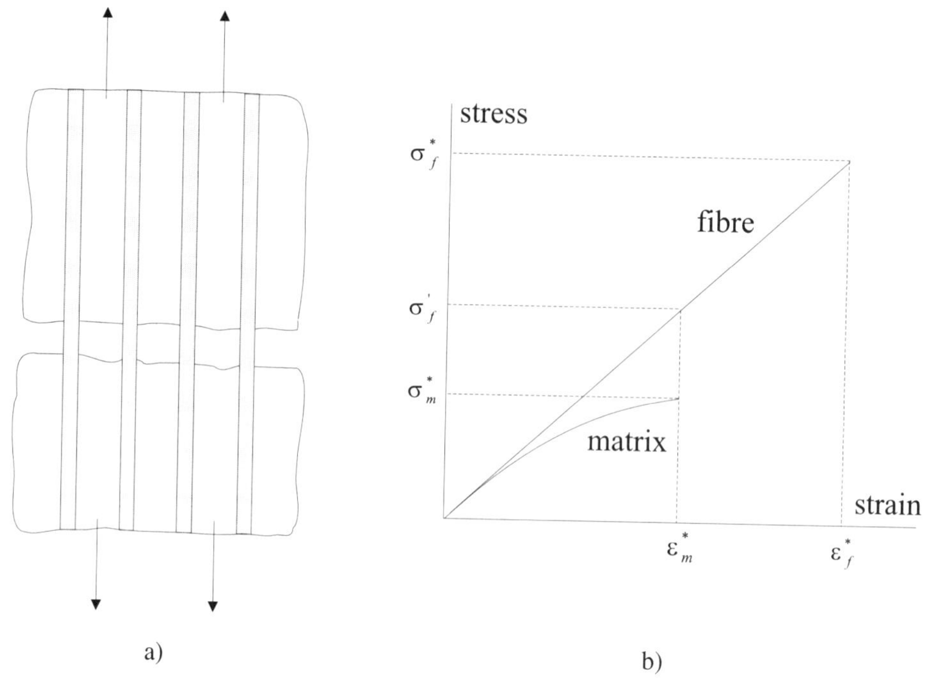

Figure 21.9: Failure of a unidirectional laminae when $\varepsilon^*_f > \varepsilon^*_m$. a) Failure of the matrix before failure of the fibres. b) Stress against strain for the composite.

the fibres will be unable to support the load following the failure of the matrix, from (21.35)

$$\sigma^*_{\parallel} = \sigma^*_f V_f + \sigma^*_m (1 - V_f) \qquad (21.36)$$

However, when V_f is large then the transfer of load to the fibres will be insufficient to cause failure of the fibres. With the load transferred to the fibres then the failure strength is increased to the failure strength of the fibres. The variation of $\sigma^*_{\parallel}$ with V_f is illustrated in Figure 21.10. The cross-over point (indicated by a dash) is obtained from (21.36) when the matrix has failed

$$\sigma^*_{\parallel} = \sigma'_f V'_f + \sigma^*_m (1 - V'_f) = \sigma^*_f V'_f \qquad (21.37)$$

and re-arranging for V'_f

$$V'_f = \frac{\sigma^*_m}{\sigma^*_f - \sigma'_f + \sigma^*_m} \qquad (21.38)$$

When $V_f > V'_f$ then the fibre strength dominates.

ii) $\varepsilon^*_m > \varepsilon^*_f$
A similar procedure can be adopted as in the previous section for the case $\varepsilon^*_f > \varepsilon^*_m$. Following the fracture of the fibres, Figure 21.11, when V_f is low then the additional load on the matrix is insufficient to fracture the matrix. However, the effective cross-sectional area of the matrix will be reduced due to the presence of holes at the fibre ends and as a result the failure strength of the matrix will be less than σ^*_m. Thus, from (21.36)

$$\sigma^*_{\parallel} = \sigma^*_m V_m = \sigma^*_m (1 - V_f) \qquad (21.39)$$

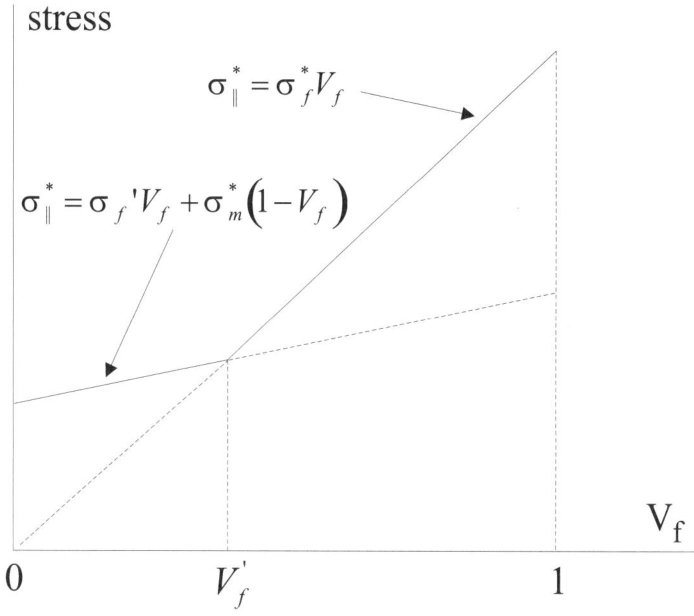

Figure 21.10: Stress against volume fraction when $\varepsilon_f^* > \varepsilon_m^*$.

When V_f is large then the load transferred to the matrix when the fibres fail will be sufficient to cause the matrix to fail when the fibres fail. The variation of $\sigma_{\|}^*$ with V_f is shown in Figure 21.12 with a cross-over point of

$$V_f' = \frac{\sigma_m^* - \sigma_m'}{\sigma_f^* + \sigma_m^* - \sigma_m'} \tag{21.40}$$

21.8.1.1 Failure due to Debonding

Throughout deformation microstructural damage can be widespread throughout a composite although composites can sustain significant amounts of damage before their load-bearing capability is impaired and final failure occurs. A common form of failure is that of debonding between the matrix and fibres, Figure 21.13. Case a) illustrates a strong bond between the matrix and a single fibre. In case b) a crack has propagated through the matrix and has been halted by the fibre due to the following two reasons: i) the high stiffness of the fibre inhibits the crack tip opening displacement of the matrix crack and ii) the strength of the fibre is too high for it to be broken by the current level of applied load. If the applied load is increased further then the matrix and fibre will deform differently and a large local stress will build up in the fibre. This will result in a local Poisson contraction which, if sufficient, will overcome the interface bond between the matrix and fibre. The interface bond is due to interfacial shear strength due to chemical bonding and the frictional resistance due to mechanical keying. At some critical applied load then the shear stress at the interface will exceed the interfacial shear strength of the bond and local matrix-fibre debonding will occur at the crack tip. This debonding can travel along the fibre in both directions from the crack tip, case c). Further matrix crack tip opening displacement will subsequently follow (possibly combined with localised fibre failure, case d)) resulting in propagation of the crack beyond the fibre. The

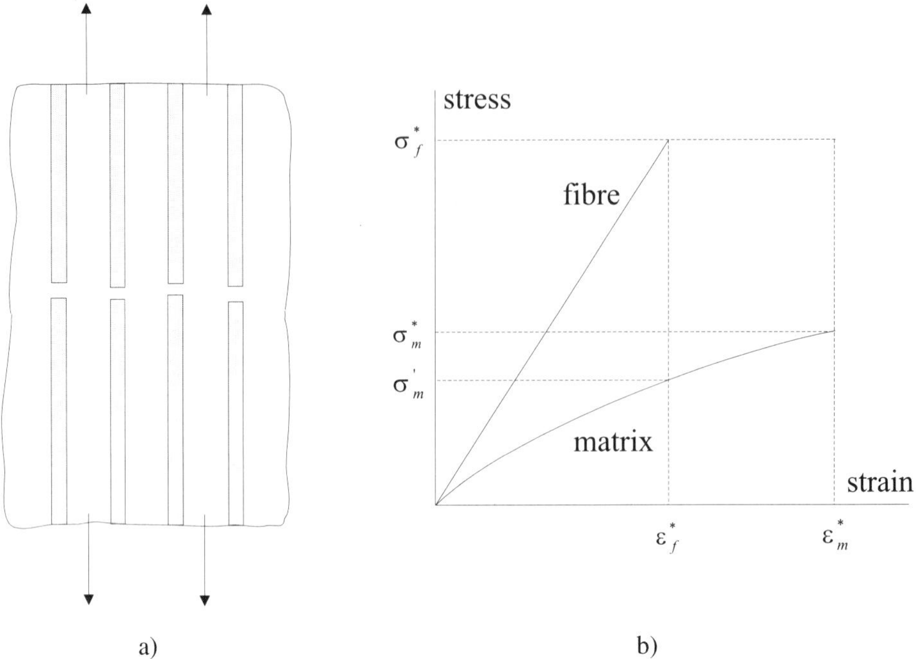

Figure 21.11: Failure of a unidirectional laminae when $\varepsilon^*_m > \varepsilon^*_f$. a) Failure of the matrix before failure of the fibres. b) Stress against strain for the composite.

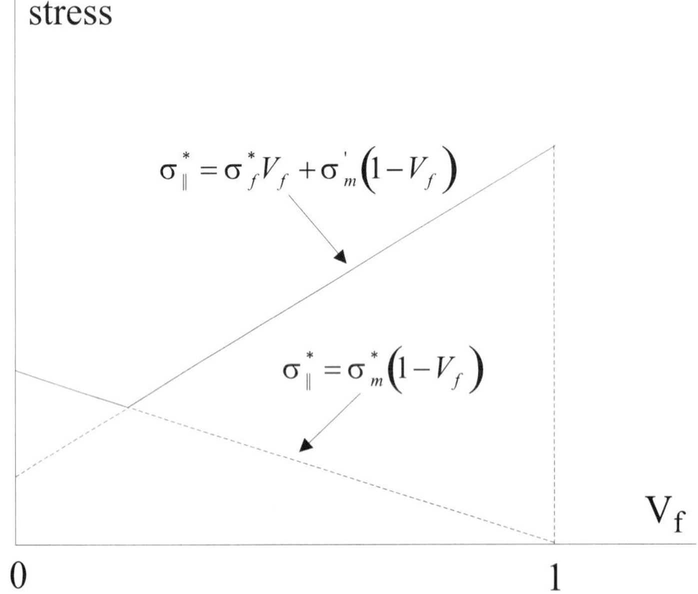

Figure 21.12: Stress against volume fraction when $\varepsilon^*_m > \varepsilon^*_f$.

composite finally fails following complete matrix failure due to crack propagation and fibre pull-out, case e).

Failure due to debonding acts as a main contributor to the fracture energy of several composites, particularly in the case of GRP. Debonding permits a substantial increase in the volume of fibre that becomes highly stressed, thus transferring the applied load from the weak matrix to the stronger fibres as in the case $\varepsilon_f^* > \varepsilon_m^*$ discussed above.

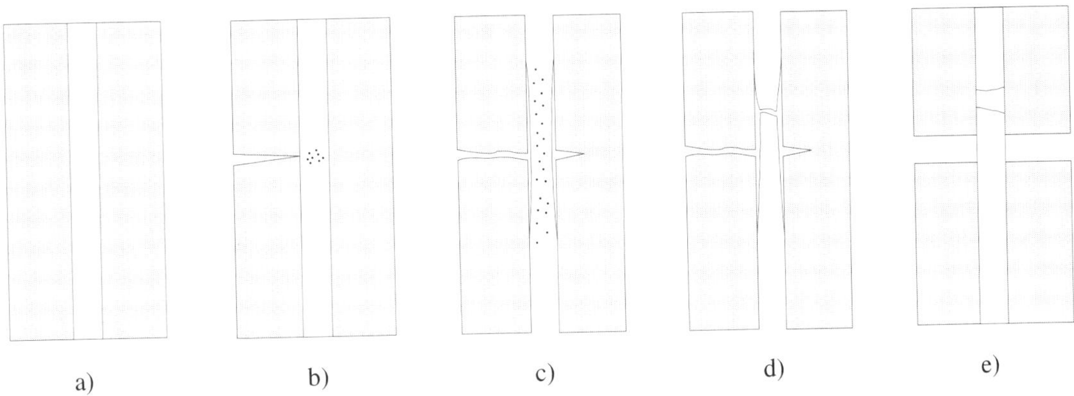

Figure 21.13: Stages of composite failure due to debonding. a) Strong bond exists between matrix and fibre. b) Matrix crack is halted by a fibre. c) Interfacial shearing and lateral contraction of fibre result in debonding and further crack growth. d) Fibre breakage with further crack growth. e) Broken fibre pull-out followed by matrix failure.

21.8.2 Transverse Tensile Strength

The transverse tensile strength of unidirectional laminae is significantly lower than the longitudinal tensile strength. For example, Table 21.2 illustrates that the transverse tensile strength of a unidirectional glass-polyester composite is approximately 33 times less than its longitudinal strength. The reasons for the reduced strength in the transverse direction are complex and a function of the matrix-fibre interface bond strength, presence of voids and interaction effects between fibres. Essentially, under transverse loading the matrix is having to bear the majority of the applied loading rather than the fibres. The overall composite transverse strength is generally less than the inherent strength of the matrix with the fibres thus having a negative reinforcing effect. This is partly because of the low fracture energy of the fibres and due to the lower effective toughness of the plastically constrained matrix between the fibres, Harris (1980).

A key factor governing the transverse strength of a lamina is interfacial debonding between the matrix and fibres. As a first approximation assume that the matrix-fibre interface has fully debonded in which case the composite can be modelled as a matrix with cylindrical holes. We can determine the percentage reduction of the cross-sectional area by considering the idealised square configuration of fibres as discussed in §21.5 and Figure 21.3. Letting $a = (\pi/4V_f)^{1/2}$ then from (21.1) and (21.2) we have

$$R = ra \quad , \quad s = 2r(a-1) \tag{21.41}$$

where $2r$ is the fibre diameter, $2R$ is the fibre centre separation and s is the fibre separation. Therefore, the ratio of separation distance to fibre centre separation is

$$\frac{s}{2R} = 1 - \frac{1}{a} = 1 - \left(\frac{\pi}{4V_f}\right)^{-1/2} = 1 - 2\sqrt{\frac{V_f}{\pi}} \tag{21.42}$$

and hence the transverse failure strength is

$$\sigma_\perp^* = \sigma_m^* \left(1 - 2\sqrt{\frac{V_f}{\pi}}\right) \tag{21.43}$$

where σ_m^* is the failure strength of the matrix. The variation of $\sigma_\perp^*$ against volume fraction is illustrated in Figure 21.14.

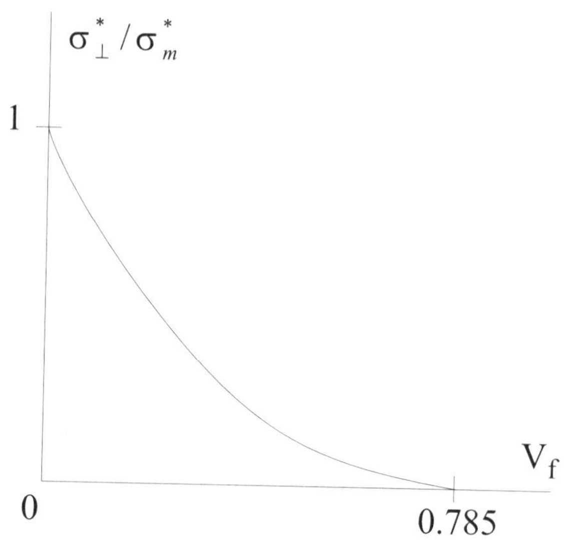

Figure 21.14: Reduction in transverse strength due to full interfacial debonding as a function of volume fraction.

21.9 Fibre Pull-Out

Composites can fail due to the fibres being pulled out of the matrix, Figure 21.15. The fibre is embedded in the matrix by length l_e. If the shear strength of the interface is τ then the tensile stress applied to the fibre which is required to break the bond is, to a first approximation, given by

$$\pi r^2 \sigma = (2\pi r l_e)\tau \quad \text{or} \quad \frac{l_e}{r} = \frac{\sigma}{2\tau} \tag{21.44}$$

When the tensile stress required for breaking the bond is plotted against the embedded length then there is a sharp cut off due to fibre failure before debonding, Figure 21.16. The critical fibre length, l_{ce}, depends on the strength of the bond and from (21.44) is given by

$$l_{ce} = \frac{\sigma_f^* r}{2\tau} \tag{21.45}$$

where σ_f^* is the failure strength of the fibres.

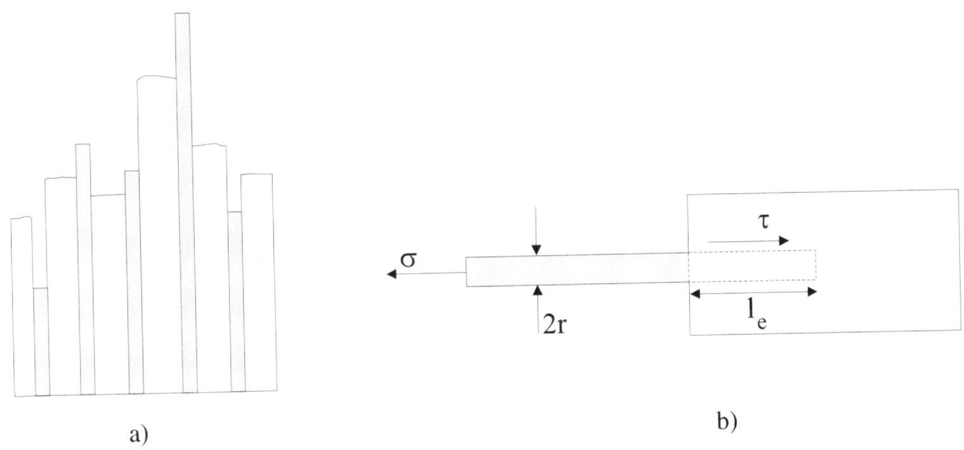

Figure 21.15: Fibre pull-out. a) Schematic illustration of the failure of a composite due to fibre pull-out. b) Pull-out of a single fibre from a matrix.

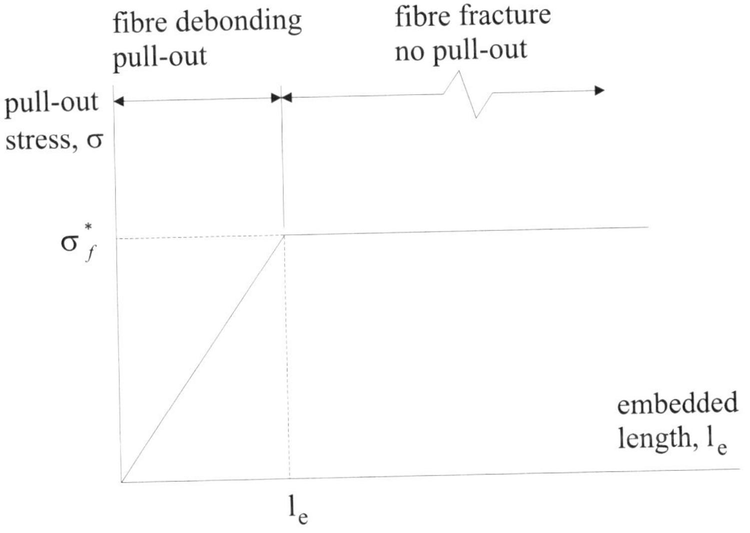

Figure 21.16: Variation of pull-out stress against embedded length.

Example 21.1 Determination of the critical embedded fibre length

In a fibre pull-out test of glass-polyester fibres the fibre radius is $r = 0.25$mm, the shear strength of the fibre-matrix interface is $\tau = 50$MPa and the fracture strength of the fibre is $\sigma^*_f = 680$MPa. Determine the critical embedded fibre length.

From (21.45) the critical embedded fibre length, l_{ce}, is

$$l_{ce} = \frac{\sigma^*_f r}{2\tau} = \frac{680 \times 10^6 \times 0.25 \times 10^{-3}}{2 \times 50 \times 10^6} = 1.7\text{mm}$$

which is approximately 7 times the fibre radius.

21.10 Conclusion

This Chapter has presented an introduction to composite materials. Although there are numerous different types of engineered composites this Chapter has focused on unidirectional laminae due to their ease of modelling and practical application. The elastic properties of unidirectional laminae with applied loading both parallel and perpendicular to the fibres have been derived, as well as the elastic properties of unidirectional laminae with orthotropic material symmetry. Both the longitudinal and transverse strength of unidirectional laminae have been examined and the failure mechanism of fibre pull-out.

An excellent source of information on polymer-, metal- and ceramic-matrix composites can be found in the Composite Materials Handbook (2000).

21.11 References and Further Reading

- Composite Materials Handbook (2000) http://mil-17.udel.edu/index.html.
- Harris, B. (1980) *Micromechanisms of Crack Extension in Composites*, Metal Science, August-September, 351-362.
- Hull, D. (1981) *An Introduction to Composite Materials*, Cambridge University Press, Cambridge.
- Jones, R. M. (1975) *Mechanics of Composite Materials*, Hemisphere Publishers, New York.
- Kies, J. A. (1962) *Maximum Strains in the Resin of Fibre Glass Composites*, US Naval Research Lab. Report NRL 5762.
- Neilsen, L. E. and Chen, P. E. (1968) *Young's Modulus of Composites Filled with Randomly Oriented Fibres*, J. Mater., 3, 352-358.

21.12 Exercises

21.1 Compare the values of $E_{||}$ and $E_{\perp}$ for idealised square and hexagonal arrangements of fibres of circular cross-section. In your comparison assume that both the square and hexagonal fibre arrangements have their respective maximum values of volume fraction.

21.2 Determine the elastic modulus measured parallel to the fibres of a unidirectional laminae of glass fibre-polyester resin in which $E_m = 4$GPa, $E_f = 75$GPa and $V_f = 0.45$.

21.3 Determine the elastic modulus measured perpendicular to the fibres for the unidirectional laminae of Exercise 21.2.

21.4 For a randomly distributed long fibre laminae of glass fibre-polyester resin determine approximate values for Young's modulus, shear modulus and Poisson's ratio. Assume that $E_{||}$ and $E_{\perp}$ are those determined in Exercises 21.2 and 21.3.

21.5 Determine the volume fraction of fibres for which the fibre strain dominates for a unidirectional laminae loaded longitudinally. Assume that $E_f = 75$GPa, the failure stresses of the matrix and fibres are $\sigma_m^* = 75$MPa and $\sigma_f^* = 2,000$MPa and the cross-over failure stress between low and high V_f is $\sigma_f' = 1,500$MPa.

21.6 Estimate the transverse failure strength of a glass fibre-polyester resin in which $\sigma_m^* = 65$MPa and $V_f = 0.3$.

21.7 If the interface strength of a glass fibre-polyester resin composite is 45MPa and the fracture strength of the fibres, of radius 0.2mm, is 750MPa then determine the critical embedded fibre length.

Chapter 22

Contact Mechanics

22.1 Introduction

This Chapter is concerned with the displacements and stresses arising from the contact of two bodies. The Chapter begins by outlining the problem of contact under conditions of linear elastic bodies assuming small displacement theory. The two cases of concentrated normal and tangential forces acting on the surface of a half-plane are then examined. These concentrated forces form the basis of the more general problem of distributed normal and tangential pressures with the specific cases of uniform normal and tangential distributed pressures examined. The indentation of a rigid flat punch is then examined and is followed by a discussion of Hertzian contact for both linear elastic and viscoelastic materials. The Chapter concludes with an overview of indentation fracture.

22.2 The Elastic Half-Plane Subject to Surface Tractions

Figure 22.1 illustrates a half-plane with origin, O, of coordinates on the free surface with x measured along the free-surface, y measured normal to the free surface and the z-axis measured into the half-plane. Along the strip $-b \leq x \leq a$, $y = 0$ the surface is subject to both normal, $p(x)$, and tangential, $q(x)$, tractions. The half-plane is assumed to be linear elastic and deformations are assumed to be sufficiently small so that small displacement theory is applicable. The interval over which the surface tractions are prescribed is assumed to be small compared to the unbounded half-plane and the radius of curvature of the interacting bodies. It will be shown that the stresses induced by the contact of two bodies are highly localised to the region of contact and vanish at large distances from the region of contact. This ensures that the local stress-displacement distribution is governed by the geometry of the contact bodies in the vicinity of the contact region and not governed by the geometry of the contact bodies at large distances from the contact region.

The through-thickness direction is assumed to be sufficiently thick so that the through-thickness strain, ε_{zz}, is zero and conditions of plane strain prevail. In addition to the equations of elasticity discussed in Chapter 14 the boundary conditions on $y = 0$ within the loaded region are

$$\bar{\sigma}_{yy} = -p(x); \quad \bar{\tau}_{xy} = -q(x) \qquad -b \leq x \leq a \tag{22.1}$$

where an over bar denotes a value on the surface $y = 0$. Outside the loaded region the surface is assumed to be traction free

$$\bar{\sigma}_{yy} = \bar{\tau}_{xy} = 0 \qquad x < -b\,,\; x > a \tag{22.2}$$

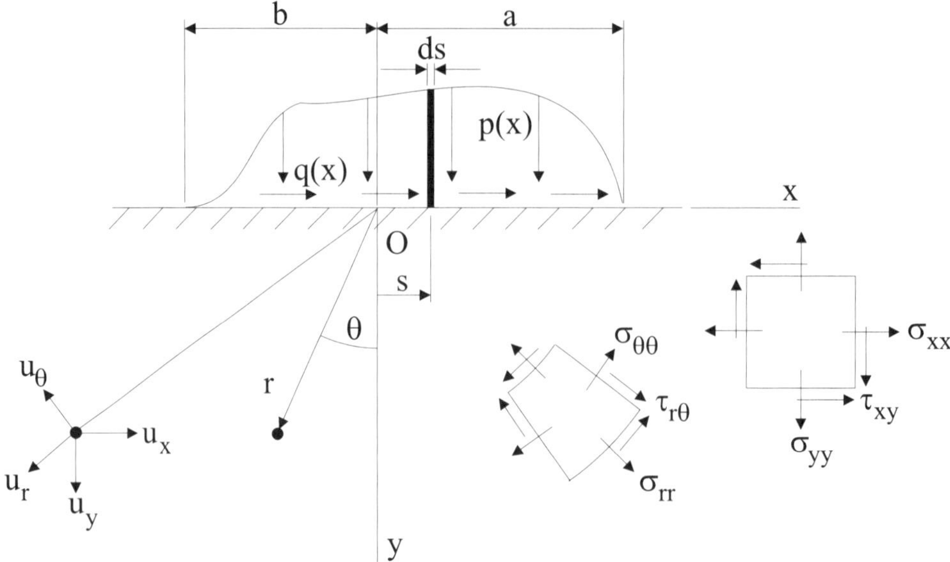

Figure 22.1: A half-plane subject to normal and tangential surface tractions.

At large distances from the loaded region the in-plane stresses (22.1) are all assumed to vanish. Alternatively, the normal, $\bar{u}_y(x)$, and tangential, $\bar{u}_x(x)$, surface displacements can be prescribed which again are assumed to vanish as $x, y \to \infty$.

22.3 Concentrated Normal and Tangential Forces

The following two sub-sections examine the problems of normal and tangential point forces acting on the free surface of a half-plane.

22.3.1 Concentrated Normal Force

The first problem that we will investigate is that of a half-plane subject to a point force P acting at the origin of coordinates in a normal direction to the surface; note that P is assumed to be the force per unit length along the through-thickness z-axis, Figure 22.2. We have previously examined this problem when we considered the Boussinesq wedge in §14.15 for the particular case when the wedge apex angle is equal to 180° and the wedge degenerates to a half-plane. It was shown that the Airy stress function in polar coordinates is given by

$$\phi = Cr\theta \sin\theta \tag{22.3}$$

where θ is measured from the line of action of P and where C is a constant to be determined from the prescribed boundary conditions. From the Airy stresses, (14.55), the polar stress components are

$$\sigma_{rr} = 2C\frac{\cos\theta}{r}; \quad \sigma_{\theta\theta} = \tau_{r\theta} = 0 \tag{22.4}$$

and illustrates a purely radial stress distribution with $\sigma_{rr} \to \infty$ as $r \to 0$ and $\sigma_{rr} \to 0$ as $r \to \infty$. The constant C can be determined from equilibrating the applied point force P to the resolved elemental

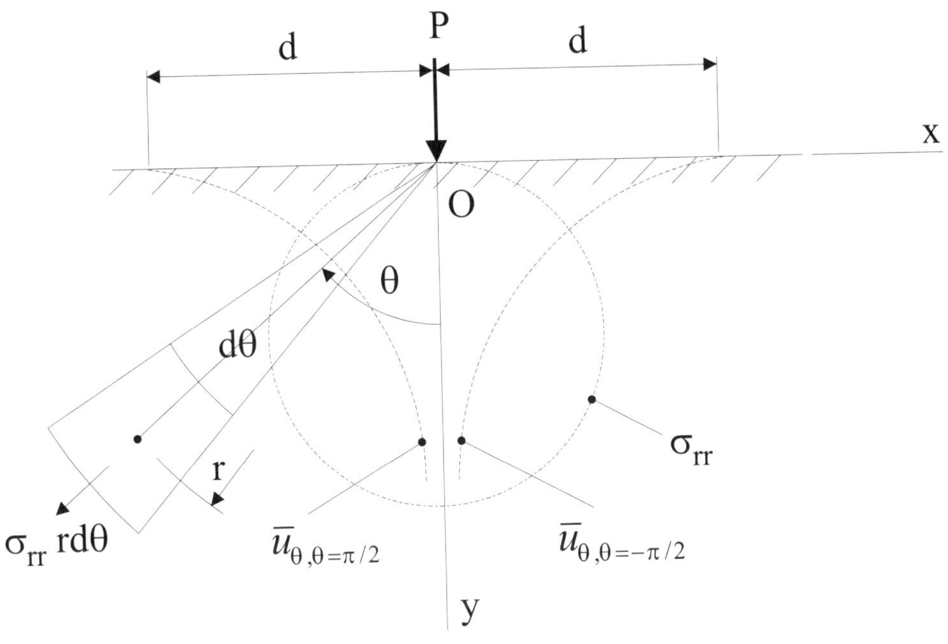

Figure 22.2: A half-plane subject to a concentrated normal force P.

forces on a semi-circle at distance r

$$\int_{-\pi/2}^{\pi/2} \sigma_{rr} \cos\theta r \, \mathrm{d}\theta = -P \tag{22.5}$$

Substituting σ_{rr} from (22.4) and performing the integration we find that $C = -P/\pi$ so that σ_{rr} is given by

$$\sigma_{rr} = -\frac{2P\cos\theta}{\pi r} \tag{22.6}$$

The in-plane Cartesian stresses can be found by an application of the transformation equations, (14.225)

$$\begin{aligned} \sigma_{xx} &= \sigma_{rr}\sin^2\theta = -\frac{2Px^2y}{\pi\left(x^2+y^2\right)^2} \\ \sigma_{yy} &= \sigma_{rr}\cos^2\theta = -\frac{2Py^3}{\pi\left(x^2+y^2\right)^2} \\ \tau_{xy} &= \sigma_{rr}\sin\theta\cos\theta = -\frac{2Pxy^2}{\pi\left(x^2+y^2\right)^2} \end{aligned} \tag{22.7}$$

with $\cos\theta = y/r$ and $\sin\theta = x/r$.

The polar strains are found from the Hookian equations, (1.16), with $\sigma_{zz} = \nu(\sigma_{rr} + \sigma_{\theta\theta})$ for plane strain

$$\begin{aligned}\varepsilon_{rr} &= \frac{1}{E}\left[\sigma_{rr} - \nu\left(\sigma_{\theta\theta} + \sigma_{zz}\right)\right] = \frac{1-\nu^2}{E}\sigma_{rr} = -\frac{1-\nu^2}{E}\frac{2P\cos\theta}{\pi r} \\ \varepsilon_{\theta\theta} &= \frac{1}{E}\left[\sigma_{\theta\theta} - \nu\left(\sigma_{rr} + \sigma_{zz}\right)\right] = -\frac{\nu(1+\nu)}{E}\sigma_{rr} = \frac{\nu(1+\nu)}{E}\frac{2P\cos\theta}{r} \\ \gamma_{r\theta} &= \frac{\tau_{r\theta}}{G} = 0\end{aligned} \tag{22.8}$$

The displacements can be expressed in terms of the strains as; (14.6), (14.7) and (14.8)

$$\begin{aligned}\frac{\partial u_r}{\partial r} &= \varepsilon_{rr} \\ \frac{u_r}{r} + \frac{1}{r}\frac{\partial u_\theta}{\partial \theta} &= \varepsilon_{\theta\theta} \\ \frac{1}{r}\frac{\partial u_r}{\partial \theta} + \frac{\partial u_\theta}{\partial r} - \frac{u_\theta}{r} &= \gamma_{r\theta}\end{aligned} \tag{22.9}$$

Therefore, from the first of (22.9) we find for u_r

$$u_r = -\frac{1-\nu^2}{E\pi}2P\cos\theta\ln r + f(\theta) \tag{22.10}$$

From the second of (22.9) we find for u_θ

$$u_\theta = \frac{\nu(1+\nu)}{E\pi}2P\sin\theta + \frac{1-\nu^2}{E\pi}2P\sin\theta\ln r - \int f(\theta)\,\mathrm{d}\theta + g(r) \tag{22.11}$$

From the third of (22.9) we find for f and g

$$f(\theta) = -\frac{(1+\nu)(1-2\nu)}{E\pi}P\theta\sin\theta + A\sin\theta + B\cos\theta; \quad g(r) = Cr \tag{22.12}$$

where A, B and C are constants. Substituting f and g then the radial and circumferential displacements are

$$\begin{aligned}u_r &= -\frac{1-\nu^2}{E}\frac{2P}{\pi}\cos\theta\ln r - \frac{(1+\nu)(1-2\nu)}{E}\frac{P}{\pi}\theta\sin\theta + A\sin\theta + B\cos\theta \\ u_\theta &= \frac{\nu(1+\nu)}{E}\frac{2P}{\pi}\sin\theta + \frac{(1-\nu^2)}{E}\frac{2P}{\pi}\sin\theta\ln r + \\ &+ \frac{(1+\nu)(1-2\nu)}{E}\frac{P}{\pi}\sin\theta - \frac{(1+\nu)(1-2\nu)}{E}\frac{P}{\pi}\theta\cos\theta + A\cos\theta - B\sin\theta + Cr\end{aligned} \tag{22.13}$$

Assuming that points on the y-axis ($\theta = 0°$) have no lateral displacement ($u_\theta = 0$) then $A = C = 0$. On the surface, $\theta = \pm\pi/2$, we have

$$\begin{aligned}\bar{u}_{r,\theta=\pi/2} &= \bar{u}_{r,\theta=-\pi/2} = -\frac{(1+\nu)(1-2\nu)}{2E}P \\ \bar{u}_{\theta,\theta=\pi/2} &= -\bar{u}_{\theta,\theta=-\pi/2} = \frac{1-\nu^2}{E\pi}2P\ln r + C\end{aligned} \tag{22.14}$$

noting that $\bar{u}_r$ is constant along the entire surface. The constant C is determined by choosing a point on the surface at a distance d from the y-axis at which the u_θ displacement is zero. Selecting the point $r = d$ on $\theta = -\pi/2$ then C is

$$C = \frac{1-\nu^2}{E\pi}2P\ln d \tag{22.15}$$

so that u_θ is given by

$$\bar{u}_{\theta,\theta=\pi/2} = -\bar{u}_{\theta,\theta=-\pi/2} = -\frac{1-\nu^2}{E\pi}2P\ln\left(\frac{d}{r}\right) \tag{22.16}$$

The u_θ surface displacement is illustrated in Figure 22.2. The logarithmic singularity informs us that u_θ directly underneath P becomes unbounded although $\ln(1/r)$ is a weak singularity; for example $\ln(1 \times 10^6) \approx 13.82$.

22.3.2 Concentrated Tangential Force

The case of a concentrated force Q (per unit length) acting at the origin O is shown in Figure 22.3. If we

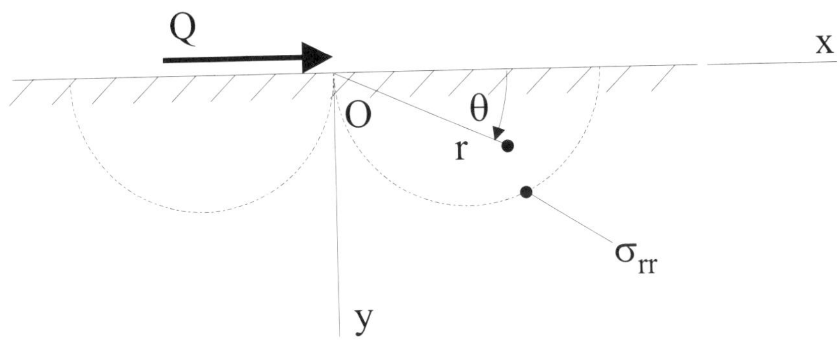

Figure 22.3: A half-plane subject to a concentrated tangential force Q.

again measure θ from the line of action Q then the stresses are equivalent to those of the normal force, (22.6)

$$\sigma_{rr} = -\frac{2Q\cos\theta}{\pi r}; \quad \sigma_{\theta\theta} = \tau_{r\theta} = 0 \tag{22.17}$$

The stress distribution is again purely radial with σ_{rr} being compressive for positive x and tensile for negative x. The Cartesian stresses are obtained in an analogous manner to (22.7) but with $\sigma_{xx}(Q) = \sigma_{yy}(P)$, $\sigma_{yy}(Q) = \sigma_{xx}(P)$, $\cos\theta = x/r$ and $\sin\theta = y/r$ due to the line of action of Q being at 90° to that of P

$$\begin{aligned}
\sigma_{xx} &= \sigma_{rr}\cos^2\theta = -\frac{2Qx^3}{\pi(x^2+y^2)} \\
\sigma_{yy} &= \sigma_{rr}\sin^2\theta = -\frac{2Qxy^2}{\pi\,(x^2+y^2)} \\
\tau_{xy} &= \sigma_{rr}\sin\theta\cos\theta = -\frac{2Qx^2y}{\pi\,(x^2+y^2)}
\end{aligned} \tag{22.18}$$

The displacements are analogous to those of (22.13) and on the surface are given by

$$\begin{aligned}
\bar{u}_{r,\theta=0} &= -\bar{u}_{r,\theta=\pi} = -\frac{1-\nu^2}{\pi E}2Q\ln\left(\frac{d}{r}\right) \\
\bar{u}_{\theta,\theta=0} &= \bar{u}_{\theta,\theta=\pi} = \frac{(1+\nu)(1-2\nu)}{2E}Q
\end{aligned} \tag{22.19}$$

22.4 Distributed Normal and Tangential Pressure

Returning now to Figure 22.1 we address the problem of a surface subject to distributed normal, $p(x)$, and tangential, $q(x)$ surface tractions. The stresses in the half-plane can be determined by considering an elemental strip ds at a distance s from the origin O. The concentrated normal and tangential forces are pds and qds respectively. It follows that the in-plane Cartesian stresses can be derived from (22.7) and (22.18) with x replaced by $(x - s)$ and integrating over the interval $-b \leq x \leq a$

$$\begin{aligned}
\sigma_{xx} &= -\frac{2y}{\pi}\int_{-b}^{a}\frac{p(s)\,(x-s)^2\,\mathrm{d}s}{\left[(x-s)^2+y^2\right]^2} - \frac{2}{\pi}\int_{-b}^{a}\frac{q(s)\,(x-s)^3\,\mathrm{d}s}{\left[(x-s)^2+y^2\right]^2} \\
\sigma_{yy} &= -\frac{2y^3}{\pi}\int_{-b}^{a}\frac{p(s)\,\mathrm{d}s}{\left[(x-s)^2+y^2\right]^2} - \frac{2y^2}{\pi}\int_{-b}^{a}\frac{q(s)\,(x-s)\,\mathrm{d}s}{\left[(x-s)^2+y^2\right]^2} \\
\tau_{xy} &= -\frac{2y^3}{\pi}\int_{-b}^{a}\frac{p(s)\,(x-s)\,\mathrm{d}s}{\left[(x-s)^2+y^2\right]^2} - \frac{2y}{\pi}\int_{-b}^{a}\frac{q(s)\,(x-s)^2\,\mathrm{d}s}{\left[(x-s)^2+y^2\right]^2}
\end{aligned} \tag{22.20}$$

Similarly, the $\bar{u}_x$ and $\bar{u}_y$ displacements on the surface are, from (22.14) and (22.19)

$$\begin{aligned}
\bar{u}_x &= -\frac{(1+\nu)(1-2\nu)}{2E}\left\{\int_{-b}^{x}p(s)\,\mathrm{d}s - \int_{x}^{a}p(s)\,\mathrm{d}s\right\} - \frac{2\,(1-\nu)^2}{E\pi}\int_{-b}^{a}q(s)\ln|x-s|\,\mathrm{d}s + C_1 \\
\bar{u}_y &= \frac{(1+\nu)(1-2\nu)}{2E}\left\{\int_{-b}^{x}q(s)\,\mathrm{d}s - \int_{x}^{a}q(s)\,\mathrm{d}s\right\} - \frac{2\,(1-\nu)^2}{E\pi}\int_{-b}^{a}p(s)\ln|x-s|\,\mathrm{d}s + C_2
\end{aligned} \tag{22.21}$$

with the constants C_1 and C_2 determined by specifying zero displacements at a given distance d from the origin. The interval of integration $-b \leq x \leq a$ is divided into the two sub-intervals $-b \leq s \leq x$ and $x \leq s \leq a$ due to the discontinuity in displacement at the origin. By prescribing displacement gradients ($\partial\bar{u}_x/\partial x$ and $\partial\bar{u}_y/\partial x$) on the surface rather than $\bar{u}_x$ and $\bar{u}_y$ then (22.21) take on a simpler form and eliminates the need to determine C_1 and C_2

$$\begin{aligned}
\frac{\partial\bar{u}_x}{\partial x} &= -\frac{(1+\nu)(1-2\nu)}{E}p(x) - \frac{2\,(1-\nu)^2}{E\pi}\int_{-b}^{a}\frac{q(s)\,\mathrm{d}s}{x-s} \\
\frac{\partial\bar{u}_y}{\partial x} &= -\frac{(1+\nu)(1-2\nu)}{E}q(x) - \frac{2\,(1-\nu)^2}{E\pi}\int_{-b}^{a}\frac{p(s)\,\mathrm{d}s}{x-s}
\end{aligned} \tag{22.22}$$

Equations (22.20) and (22.22) provide the necessary stress and displacement solutions although performing the necessary integrations can be difficult. In the following two sub-sections we will examine the two specific cases of uniformly distributed normal and tangential tractions.

22.4.1 Uniformly Distributed Normal Pressure

For the case of a uniform normal pressure applied over the interval $-a \leq x \leq a$ then $p(s) = p$ and $q(s) = 0$ and from (22.20)

$$\begin{aligned}\sigma_{xx} &= -\frac{2yp}{\pi}\int_{-a}^{a}\frac{(x-s)^2\,\mathrm{d}s}{\left[(x-s)^2+y^2\right]^2}\\ \sigma_{yy} &= -\frac{2y^3p}{\pi}\int_{-a}^{a}\frac{\mathrm{d}s}{\left[(x-s)^2+y^2\right]^2}\\ \tau_{xy} &= -\frac{2y^2p}{\pi}\int_{-a}^{a}\frac{\mathrm{d}s}{\left[(x-s)^2+y^2\right]^2}\end{aligned} \tag{22.23}$$

Performing the integrations we arrive at, see Figure 22.4.

$$\begin{aligned}\sigma_{xx} &= -\frac{p}{2\pi}\left[2\alpha + (\sin 2\theta_1 - \sin 2\theta_2)\right]\\ \sigma_{yy} &= -\frac{p}{2\pi}\left[2\alpha - (\sin 2\theta_1 - \sin 2\theta_2)\right]\\ \tau_{xy} &= \frac{p}{2\pi}\left[\cos 2\theta_1 - \cos 2\theta_2\right]\end{aligned} \tag{22.24}$$

where $\alpha = \theta_1 - \theta_2$, $\tan\theta_{1,2} = y/(x \mp a)$ and $r_{1,2} = [(x \mp a) + y^2]^{1/2}$. The maximum shear stress,

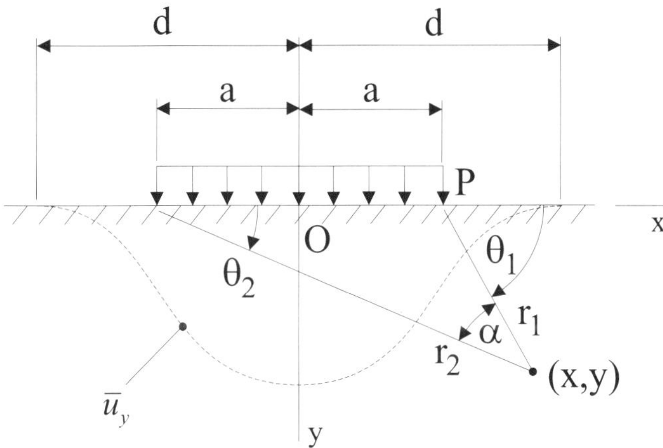

Figure 22.4: A half-plane subject to a uniform normal pressure p.

$\tau_{\max}$, can be found from (7.25)

$$\tau_{\max} = \sqrt{\left(\frac{\sigma_{xx}-\sigma_{yy}}{2}\right)^2 + \tau_{xy}^2} \tag{22.25}$$

with

$$\begin{aligned}\left(\frac{\sigma_{xx}-\sigma_{yy}}{2}\right)^2 &= \frac{p^2}{4\pi^2}\left(\sin^2 2\theta_1 + \sin^2 2\theta_2 - 2\sin 2\theta_1 \sin 2\theta_2\right)\\ \tau_{xy}^2 &= \frac{p^2}{4\pi^2}\left(\cos^2 2\theta_1 + \cos^2 2\theta_2 - 2\cos 2\theta_1 \cos 2\theta_2\right)\end{aligned} \tag{22.26}$$

so that $\tau_{\max}$ is

$$\tau_{\max} = \left\{ \frac{p^2}{\pi^2} \frac{1}{2} \left[1 - (\sin 2\theta_1 \sin 2\theta_2 + \cos 2\theta_1 \cos 2\theta_2)\right] \right\}^{1/2} \tag{22.27}$$

Noting the trigonometric identities

$$\begin{aligned} \cos(a-b) &= \sin a \sin b + \cos a \cos b \\ \sin^2 a &= \frac{1}{2}(1 - \cos 2a) \end{aligned} \tag{22.28}$$

then $\tau_{\max}$ is

$$\tau_{\max} = \frac{p}{\pi} \sin \alpha \tag{22.29}$$

and attains a maximum value, p/π, when $\alpha = 90°$, that is a semi-circle of centre O, radius a and passing through $x = \pm a$ as shown in Figure 22.5a). The principal stresses, σ_1 and σ_2, are, (7.19)

$$\sigma_{1,2} = \left(\frac{\sigma_{xx} + \sigma_{yy}}{2} \right) \pm \tau_{\max} = -\frac{p}{\pi}\alpha \pm \frac{p}{\pi} \sin \alpha = -\frac{p}{\pi}(\alpha \mp \sin \alpha) \tag{22.30}$$

and act on the principal planes, (7.15)

$$\tan 2\theta_p = \frac{\tau_{xy}}{(\sigma_{xx} - \sigma_{yy})/2} = -\frac{(\cos 2\theta_1 - \cos 2\theta_2)}{(\sin 2\theta_1 - \sin 2\theta_2)} \tag{22.31}$$

Making use of the identities

$$\begin{aligned} \cos a - \cos b &= 2 \sin \frac{1}{2}(a+b) \sin \frac{1}{2}(b-a) \\ \sin a - \sin b &= 2 \cos \frac{1}{2}(a+b) \sin \frac{1}{2}(a-b) \\ \sin(-a) &= -\sin(a) \end{aligned} \tag{22.32}$$

then we arrive at

$$\theta_p = \frac{\theta_1 + \theta_2}{2} \tag{22.33}$$

with the planes of the maximum shear stress at $\theta_s = \theta_p \pm 45°$. The principal stresses are therefore seen to be a set of confocal ellipses and hyperbola with foci at $x = \pm a$, $y = 0$ as shown in Figure 22.5b).

We will now determine the $\bar{u}_x$ and $\bar{u}_y$ displacements from (22.22) with $q(s) = 0$ and $p(x) = p$. Firstly, $\bar{u}_x$ is

$$\frac{\partial \bar{u}_x}{\partial x} = -\frac{(1+\nu)(1-2\nu)}{E} p; \quad \bar{u}_x = -\frac{(1+\nu)(1-2\nu)}{E} px + C \qquad -a \le x \le a \tag{22.34}$$

Assuming that the origin remains fixed laterally ($\bar{u}_x = 0$ at $x = 0$) then $C = 0$ and $\bar{u}_x$ is given by

$$\bar{u}_x = -\frac{(1+\nu)(1-2\nu)}{E} px \tag{22.35}$$

From (22.22) $\bar{u}_y$ is

$$\frac{\partial \bar{u}_y}{\partial x} = -\frac{2(1-\nu^2)}{E} \frac{p}{\pi} \int_{-a}^{a} \frac{\mathrm{d}s}{x-s} \tag{22.36}$$

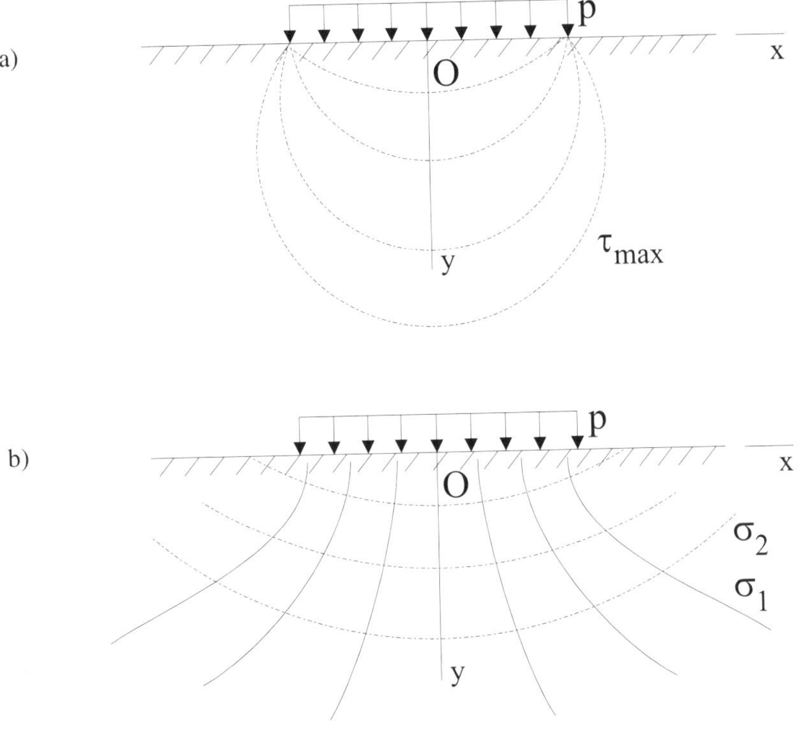

Figure 22.5: Variation of a) maximum shear stress, τ_{max}, and b) principal stresses, σ_1 and σ_2, for a uniformly distributed normal pressure p.

The integrand has a singularity at $s = x$ within the interval of integration, so before proceeding let us take a brief look at the following general improper definite integral, Kreyszig (1983)

$$\int_A^B f(x)\,\mathrm{d}x \tag{22.37}$$

whose integrand becomes infinite at a point a within the interval of integration, that is $\lim_{x\to a} |f(x)| = \infty$. Performing the integration in two parts

$$\int_A^B f(x)\,\mathrm{d}x = \lim_{\varepsilon\to 0}\int_A^{a-\varepsilon} f(x)\,\mathrm{d}x + \lim_{\eta\to 0}\int_{a+\eta}^B f(x)\,\mathrm{d}x \tag{22.38}$$

with both ε and η approaching zero independently. Neither of the two limits may exist if $\varepsilon \to 0$ and $\eta \to 0$ independently but the following exists and is referred to as the *Cauchy principal value*

$$\lim_{\varepsilon\to 0}\left[\int_A^B f(x)\,\mathrm{d}x + \int_{a+\varepsilon}^B f(x)\,\mathrm{d}x\right] \quad \text{or pr.v.} \int_A^B f(x)\,\mathrm{d}x \tag{22.39}$$

Returning to the improper integral of (22.36) then the Cauchy principal value is

$$\text{pr.v.} \int_a^b \frac{\mathrm{d}s}{x-s} = \ln\left(\frac{x-a}{b-x}\right) \tag{22.40}$$

and hence from (22.36) for the interval $-a \leq x \leq a$

$$\frac{\partial \bar{u}_y}{\partial x} = -\frac{2\left(1-\nu^2\right)}{E}\frac{p}{\pi}\ln\left(\frac{x-a}{a-x}\right) \tag{22.41}$$

Integrating with respect to x

$$\bar{u}_y = -\frac{\left(1-\nu^2\right)}{E}\frac{p}{\pi}\left[(a+x)\ln\left(\frac{a+x}{a}\right)^2 + (a-x)\ln\left(\frac{a-x}{a}\right)^2\right] + C \tag{22.42}$$

where use of the following standard indefinite integral has been made

$$\int \ln(ax+b)\,\mathrm{d}x = \frac{ax+b}{a}\left[\ln(ax+b)-1\right] \tag{22.43}$$

The constant of integration C is found by requiring that $\bar{u}_y = 0$ at a fixed distance along the x-axis, at $x = \pm d$.

22.4.2 Uniformly Distributed Tangential Pressure

Now consider the case of a uniform tangential pressure, q, applied over the interval $-a \leq x \leq a$, Figure 22.6. With $p(s) = 0$ and $q(s) = q$ then from (22.20) we have

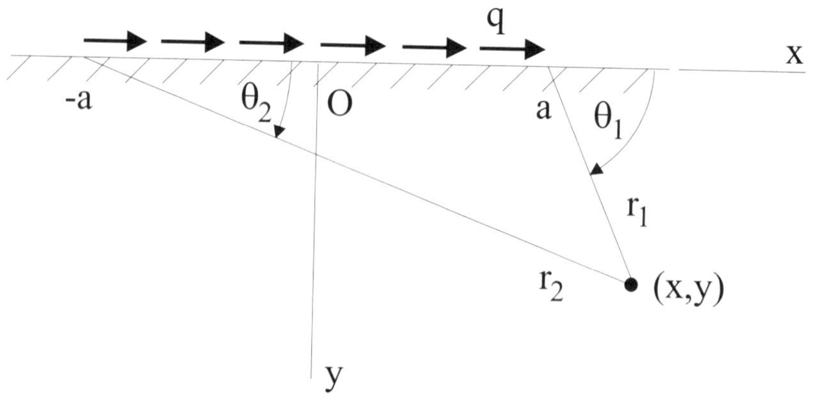

Figure 22.6: A half-plane subject to a uniform tangential pressure q.

$$\begin{aligned}
\sigma_{xx} &= -\frac{2}{\pi}q\int_{-a}^{a}\frac{(x-s)^3\,\mathrm{d}s}{\left[(x-s)^2+y^2\right]^2} \\
\sigma_{yy} &= -\frac{2y^2}{\pi}q\int_{-a}^{a}\frac{(x-s)\,\mathrm{d}s}{\left[(x-s)^2+y^2\right]^2} \\
\tau_{xy} &= -\frac{2y}{\pi}q\int_{-a}^{a}\frac{(x-s)^2\,\mathrm{d}s}{\left[(x-s)^2+y^2\right]^2}
\end{aligned} \tag{22.44}$$

Peforming the necessary integrations we obtain

$$\begin{aligned}\sigma_{xx} &= \frac{q}{2\pi}\left[4\ln\left(\frac{r_1}{r_2}\right) - (\cos 2\theta_1 - \cos 2\theta_2)\right] \\ \sigma_{yy} &= \frac{q}{2\pi}\left[\cos 2\theta_1 - \cos 2\theta_2\right] \\ \tau_{xy} &= -\frac{q}{2\pi}\left[2\alpha + (\sin 2\theta_1 - \sin 2\theta_2)\right]\end{aligned} \tag{22.45}$$

with α, $\theta_{1,2}$ and $r_{1,2}$ as before. The variation of $\bar{\sigma}_{xx}$ is shown in Figure 22.7. Examining (22.21) we ob-

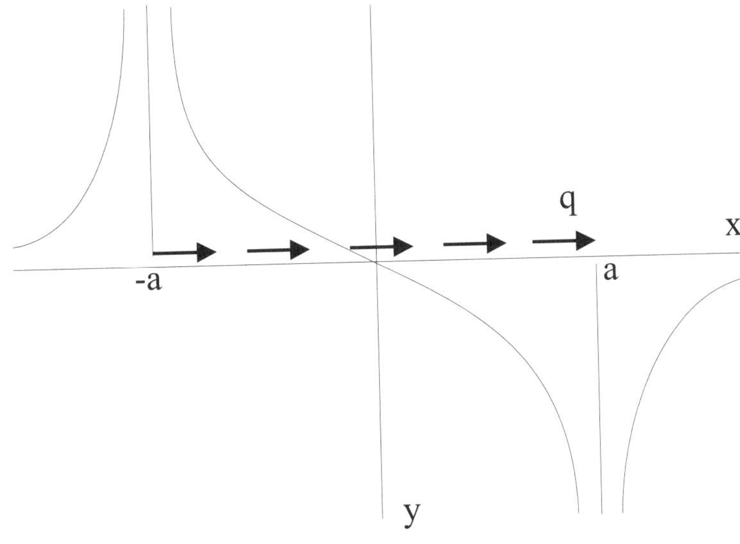

Figure 22.7: Variation of $\bar{\sigma}_{xx}$ for a uniformly distributed tangential pressure q.

serve the following equivalencies between the surface displacements $\bar{u}_x$ and $\bar{u}_y$ for uniformly distributed normal and tangential loadings

$$\bar{u}_x(q) = \bar{u}_y(p); \quad \bar{u}_y(q) = -\bar{u}_x(p) \tag{22.46}$$

22.5 Indentation by a Rigid Flat Punch

Figure 22.8 illustrates a half-plane subject to a rigid flat punch of width $2a$ being pressed into the surface by an applied force P. Since the punch is rigid (unbounded modulus of elasticity) it will not deform upon increasing applied load. In practice, no punch can be fully rigid but the relative moduli of punch and half-plane may be such that the punch can be effectively considered as rigid; for example, a metal punch indenting a polymer half-plane.

This problem differs from the previously examined problems since the displacement is specified within the loaded region; namely

$$\bar{u}_y(x) = \delta_y \qquad -a \leq x \leq a \tag{22.47}$$

where δ_y is a constant. In addition it will be assumed that the punch does not tilt throughout loading and remains parallel to the initial undeformed half-plane surface. Furthermore, consider the additional

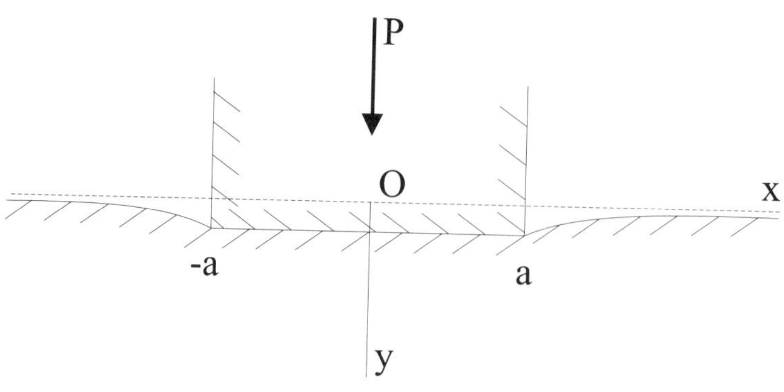

Figure 22.8: A half-plane subject to a rigid flat punch.

boundary condition of a frictionless punch such that $q(x) = 0$. From (22.22) re-arranging for $p(s)$ and $q(s)$

$$\begin{aligned}\int_{-b}^{a} \frac{q(s)\,\mathrm{d}s}{x-s} &= -\frac{\pi(1-2\nu)}{2(1-\nu)}p(x) - \frac{\pi E}{2\,(1-\nu^2)}\frac{\partial \bar{u}_x}{\partial x} \\ \int_{-b}^{a} \frac{p(s)\,\mathrm{d}s}{x-s} &= -\frac{\pi(1-2\nu)}{2(1-\nu)}q(x) - \frac{\pi E}{2\,(1-\nu^2)}\frac{\partial \bar{u}_y}{\partial x}\end{aligned} \tag{22.48}$$

With $\partial \bar{u}_y/\partial x = 0$ and $q(x) = 0$ then $(22.48)_2$ reduces to

$$\int_{-b}^{a} \frac{p(s)\,\mathrm{d}s}{x-s} = 0 \tag{22.49}$$

Let us briefly consider the general singular integral

$$\frac{1}{\pi}\int_{-a}^{a} \frac{f(t)\,\mathrm{d}t}{t-x} = g(x) \qquad |x| < a \tag{22.50}$$

which has the following solution for $f(x)$ with integrable singularities at $\pm a$, Gladwell (1980)

$$f(t) = \frac{\left(a^2-t^2\right)^{-1/2}}{\pi}\int_{-a}^{a} \frac{\left(a^2-x^2\right)^{1/2}}{x-t} g(x)\,\mathrm{d}x + C\left(a^2-t^2\right)^{-1/2} \tag{22.51}$$

where C is an arbitrary constant and is found from the condition

$$C = \frac{1}{\pi}\int_{-a}^{a} f(t)\,\mathrm{d}t \tag{22.52}$$

Alternatively, if $f(t)$ is to be finite at $\pm a$ then it will in fact be zero there and

$$f(t) = \frac{\left(a^2-t^2\right)^{1/2}}{\pi}\int_{-a}^{a} \frac{\left(a^2-x^2\right)^{-1/2}}{x-t} g(x)\,\mathrm{d}x \tag{22.53}$$

provided that $g(x)$ satisfies

$$\int_{-a}^{a} \frac{g(x)\,\mathrm{d}x}{\left(a^2-x^2\right)^{1/2}} = 0 \tag{22.54}$$

Returning to the integral in (22.49) we have $g(s) = 0$ so that the solution is

$$p(x) = \frac{P}{\pi\left(a^2 - x^2\right)^{1/2}} \tag{22.55}$$

with C equal to P/π from (22.52) since

$$P = \int_{-a}^{a} p(s)\,\mathrm{d}s \tag{22.56}$$

The variation of $p(x)$ is shown in Figure 22.9 and illustrates that $p(0) = P/\pi a$ and $p(\pm a) = \infty$. From

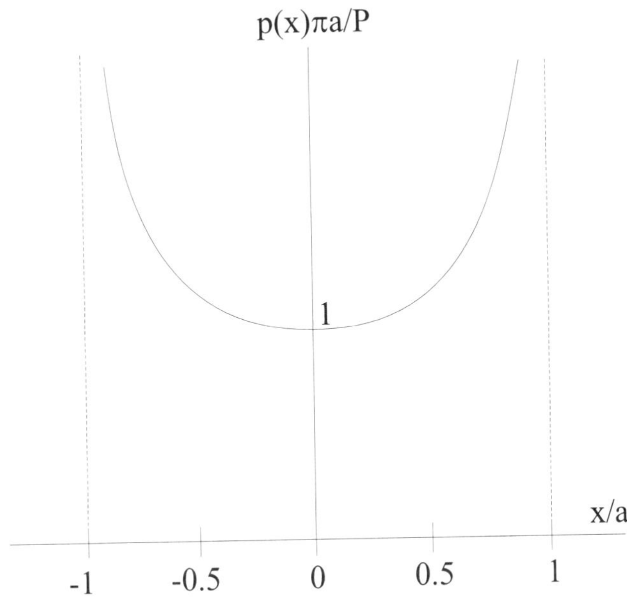

Figure 22.9: Variation of pressure directly underneath a flat rigid frictionless punch.

(22.21) the surface displacements are

$$\begin{aligned}\bar{u}_x &= -\frac{(1+\nu)(1-2\nu)P}{E\pi}\sin^{-1}\left(\frac{x}{a}\right)\\ \bar{u}_y &= \delta_y - \frac{2\left(1-\nu^2\right)P}{E\pi}\ln\left[\frac{x}{a} + \left(\frac{x^2}{a^2} - 1\right)^{1/2}\right]\end{aligned} \tag{22.57}$$

noting that $\partial\bar{u}_y/\partial x = \infty$ at $x = \pm a$.

The case of a frictionless punch is the simplest of prescribed boundary conditions. In addition we could consider the cases of i) no slip between the punch and half-plane, ii) partial slip in which no slip occurs in the central region $-c \leq x \leq c$ with slip zones $c \leq |x| \leq a$ in which $q(x) = \pm\mu p(x)$; where $c < a$ and μ is the coefficient of friction and iii) a sliding punch in which $q(x) = \mu p(x)$ over the entire length of the punch. These cases are examined by Johnson (1985) and it is found that for all these cases the normal pressure $p(x)$ is approximately equivalent to the frictionless case, (22.55).

22.6 Hertzian Contact

Previous sections have examined the cases of concentrated and distributed surface tractions applied to a half-plane. In the present section we present an overview of the problem of contact of two elastic bodies. Figure 22.10 illustrates two three-dimensional elastic bodies S_1 and S_2 in contact along the segment $2a$. This general problem was solved by Hertz subject to the following conditions i) the area of contact is

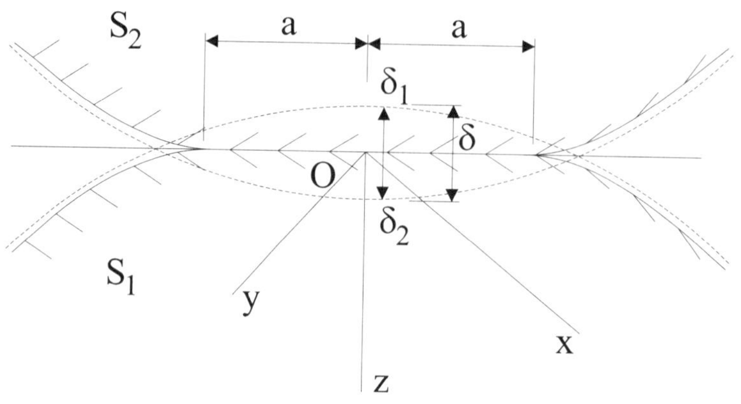

Figure 22.10: Contact of two elastic bodies.

small compared to the dimensions of the bodies, ii) the surfaces of contact are topographically smooth and iii) the equations of the undeformed surfaces in the region of contact can be approximated by expressions of the form

$$z = Ax^2 + By^2 + Cxy \tag{22.58}$$

By orienting the x and y axes so that the term in xy vanishes then

$$z_1 = \frac{1}{2R_1'}x_1^2 + \frac{1}{2R_1''}y_1^2; \quad z_2 = -\left(\frac{1}{2R_2'}x_2^2 + \frac{1}{2R_2''}y_2^2\right) \tag{22.59}$$

where $R_{1,2}'$ and $R_{1,2}''$ are the principal radii of curvature of the surfaces of S_1 and S_2 at the origin O, that is the maximum and minimum radii of curvature for all cross-sections through S_1 and S_2. The separation distance between S_1 and S_2 is denoted by $h = z_1 - z_2$ and is of a similar form to (22.58)

$$h = Ax^2 + By^2 = \frac{1}{2R'}x^2 + \frac{1}{2R''}y^2 \tag{22.60}$$

where R' and R'' are the principal relative radii of curvature for S_1 and S_2. If S_1 and S_2 are solids of revolution then $R_1 = R_1' = R_1''$ and $R_2 = R_2' = R_2''$ and $A = B = (1/R_1 + 1/R_2)/2$ and the relative curvature is $1/R = (1/R_1 + 1/R_2)$. If, after deformation, bodies S_1 and S_2 are displaced by $\bar{u}_{z1}$ and $\bar{u}_{z2}$ in the contact region then

$$\bar{u}_{z1} + \bar{u}_{z2} + h = \delta_1 + \delta_2 \tag{22.61}$$

where $\delta = \delta_1 + \delta_2$ is the total displacement of the two bodies along the z-axis. Substituting (22.60) into (22.61)

$$\bar{u}_{z1} + \bar{u}_{z2} = \delta - Ax^2 - By^2 \tag{22.62}$$

With $A = B = (1/R_1 + 1/R_2)/2$ then (22.62) can be written

$$\bar{u}_{z1} + \bar{u}_{z2} = \delta - \left(\frac{1}{2R}\right) r^2 \tag{22.63}$$

From experimental observations of studying Newton's optical interference fringes Hertz proposed that the contact area is, in general, elliptical with a pressure distribution given by

$$p = \frac{p_0}{a}\left(a^2 - r^2\right)^{1/2} = p_0\left[1 - \left(\frac{r}{a}\right)^2\right]^{1/2} \tag{22.64}$$

where p_0 is the maximum pressure, see Figure 22.11. The normal displacement can be shown to be,

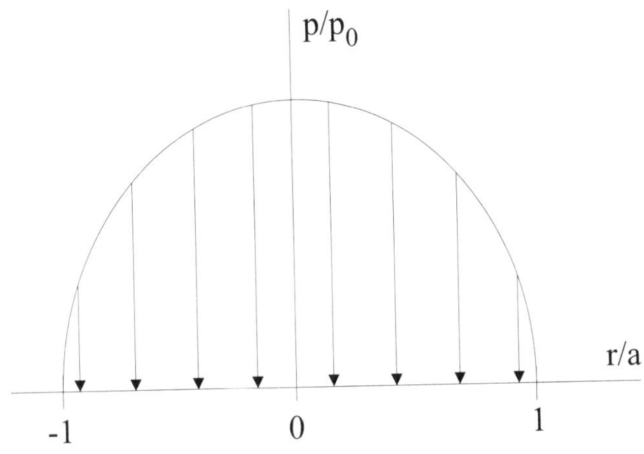

Figure 22.11: Variation of applied pressure for Hertzian contact.

Johnson (1985)

$$\bar{u}_z = \frac{1-\nu^2}{E}\frac{\pi p_0}{4a}\left(2a^2 - r^2\right) \tag{22.65}$$

The pressure acting on S_1 due to S_2 is the same as the pressure acting on S_2 due to S_1, so that by letting

$$\frac{1}{E_{12}} = \frac{1-\nu_1^2}{E_1} + \frac{1-\nu_2^2}{E_2} \tag{22.66}$$

and substituting $\bar{u}_{z1}$ and $\bar{u}_{z2}$ into (22.63) we find

$$\frac{\pi p_0}{4aE_{12}}\left(2a^2 - r^2\right) = \delta - \left(\frac{1}{2R}\right) r^2 \tag{22.67}$$

from which a and δ are given by

$$a = \frac{\pi p_0 R}{2E_{12}}; \quad \delta = \frac{\pi a p_0}{2E_{12}} \tag{22.68}$$

Assuming the contact region to be a circle of radius a then the total load, P, compressing S_1 and S_2 is

$$P = \int_0^a p 2\pi r \, dr = \frac{2\pi p_0}{a}\int_0^a r\sqrt{a^2 - r^2}\, dr = -\frac{2\pi p_0}{3a}\left[\left(a^2 - r^2\right)^{3/2}\right]_0^a = \frac{2}{3}p_0\pi a^2 \tag{22.69}$$

In practice, the applied load P is generally specified, so re-arranging a, δ and p_0 in terms of P we have

$$a = \left(\frac{3PR}{4E_{12}}\right)^{1/3}; \quad \delta = \frac{a^2}{R} = \left(\frac{9P^2}{16RE_{12}^2}\right)^{1/3}; \quad p_0 = \frac{3P}{2\pi a^2} = \left(\frac{6PE_{12}^2}{\pi^3 R^2}\right)^{1/3} \tag{22.70}$$

22.6.1 Hertzian Contact with a Viscoelastic Material

In Chapter 19 we examined viscoelastic materials and derived the following relationships for a linear viscoelastic material subject to fluctuating loads, (19.42) and (19.43)

$$\begin{aligned}\varepsilon(t) &= \int_0^t C(t-\tau)\,\frac{\partial\sigma(\tau)}{\partial\tau}\,\mathrm{d}\tau \\ \sigma(t) &= \int_0^t G(t-\tau)\,\frac{\partial\varepsilon(\tau)}{\partial\tau}\,\mathrm{d}\tau\end{aligned} \tag{22.71}$$

assuming no stress-strain history prior to $t = 0$ and where $C(t)$ and $G(t)$ are the compliance and relaxation functions respectively. In this section we will examine the indentation of a rigid spherical indenter with a linear viscoelastic material. Rather than the linear elastic Hertzian contact discussed above the contact area will increase with time under the action of a constant applied normal load. When the solution is known for a linearly elastic material then Radok (1957) and Lee and Radok (1960) proposed a simple solution to the same problem but for a viscoelastic material which consists of replacing the elastic constants $2G$ and $1/2G$ by the corresponding integral operator for the linear viscoelastic stress-strain relations, (22.71). Provided that the contact area is always increasing then this approximation is found to be acceptable.

Let the spherical indenter 1 be fully rigid ($E_1 = \infty$) and spherical indenter 2 be incompressible ($\nu_2 = 1/2, E = E_2, G = G_2$). From (22.66) E_{12} is

$$\frac{1}{E_{12}} = \frac{1-\nu_2^2}{E_2} = \frac{3}{4E} = \frac{1}{4G} \tag{22.72}$$

with $E_2 = 2G_2(1+\nu_2) = 3G_2$. From (22.70) the contact area is

$$a^3 = \frac{3PR}{4E_{12}} = \frac{3}{8}\left(\frac{1}{2G}\right)RP \tag{22.73}$$

From (22.70) and substituting a^3 from (22.73)

$$\frac{p_0}{a} = \frac{3P}{2\pi a^3} = \frac{4}{\pi R}2G \tag{22.74}$$

so that the Hertzian contact pressure is given by, (22.64)

$$p = \frac{p_0}{a}\left(a^2-r^2\right)^{1/2} = \frac{4}{\pi R}2G\left(a^2-r^2\right)^{1/2} \tag{22.75}$$

Replacing $1/2G$ by the compliance operator and $2G$ by the relaxation operator the contact radius and pressure distribution as functions of time are now given by

$$\begin{aligned}a^3(t) &= \frac{3R}{8}\int_0^t C(t-\tau)\,\frac{\partial P(\tau)}{\partial\tau}\,\mathrm{d}\tau \\ p(r,t) &= \frac{4}{\pi R}\int_0^t G(t-\tau)\,\frac{\partial}{\partial\tau}\left[a^2(\tau)-r^2\right]^{1/2}\mathrm{d}\tau\end{aligned} \tag{22.76}$$

As an illustration consider the case of a constant applied load $P(t) = P_0$ for $t \geq 0$ applied instantaneously at $t = 0$. From (22.76) the contact radius is

$$a^3(t) = \frac{3}{8}RP_0C(t) \tag{22.77}$$

Choosing the standard linear solid viscoelastic material, §19.5, the strain response to a step change in stress, σ_0, is, (19.30)

$$\varepsilon(t) = \left[\frac{1}{E_1} + \frac{1}{E_2}\left(1 - e^{-E_2 t/\mu}\right)\right]\sigma_0 = C(t)\sigma_0 \tag{22.78}$$

where E_1, E_2 and μ are the spring and dashpot material constants. Substituting (22.78) into (22.77) the contact radius is

$$a^3(t) = \frac{3}{8}RP_0\left[\frac{1}{E_1} + \frac{1}{E_2}\left(1 - e^{-E_2 t/\mu}\right)\right] \tag{22.79}$$

At $t = 0$ the contact radius is initially $a_0 = (3RP_0/8E_1)^{1/3}$. The variation of $a(t)$ with increasing time is shown in Figure 22.12a) and is seen to approach the following as $t \to \infty$

$$a(\infty) = \left\{\frac{3}{8}RP_0\left[\frac{1}{E_1} + \frac{1}{E_2}\right]\right\}^{1/3} = a_0\left[\frac{E_1 + E_2}{E_2}\right]^{1/3} \tag{22.80}$$

since the standard linear solid viscoelastic material model exhibits delayed elasticity with the final strain limited to a finite value.

Evaluating the variation of the pressure distribution with increasing time from (22.76) requires numerical integration with $a(t)$ given by (22.79) and $G(t)$ given by (19.33)

$$\sigma(t) = \frac{E_1}{E_1 + E_2}\left[E_2 + E_1 e^{-(E_1+E_2)t/\mu}\right]\varepsilon_0 = G(t)\varepsilon_0 \tag{22.81}$$

Variation of $p(r,t)$ is shown in Figure 22.12b) as a function of t/T where $T = \mu/(E_1+E_2)$ and illustrates the growth of the contact area and redistribution of pressure with increasing time.

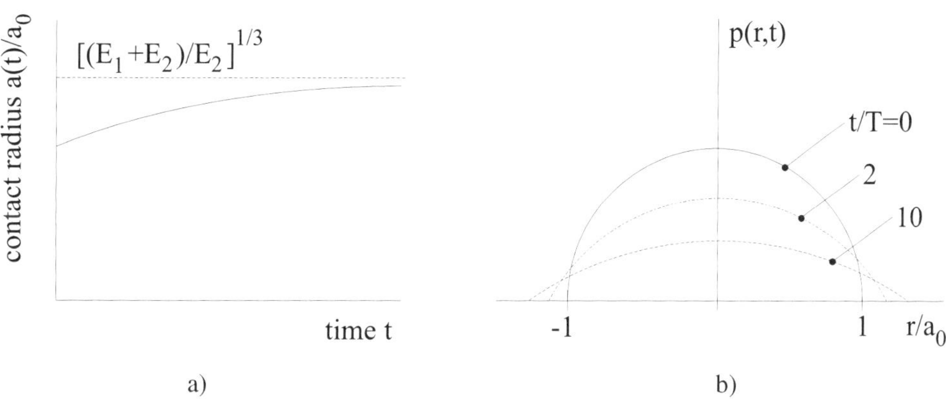

Figure 22.12: Figure 22.12. Indentation of a rigid spherical indenter into a spherical indenter of linear viscoelastic material. a) Growth of the contact radius with time. b) Variation of the pressure distribution with time.

22.7 Indentation Fracture

When surfaces come into contact and repeatedly rub together cracks are frequently formed at the mating surfaces. The growth of such cracks is referred to as *fretting fatigue*. When we discussed fracture mechanics in Chapter 17 we saw that under elastic conditions the stress field at the tip of a crack is governed by the stress intensity factor. The stress intensity factor informs us how crack size, geometry and loading conditions alter the fracture process and can also be used to determine the rate of growth of a crack with the aid of empirical crack growth laws such as Paris' law. As a result, the variation of the modes I and II stress intensity factors for a variety of different geometric and loading configurations provides an insight into the fretting fatigue process. In this section we will discuss various modes I and II stress intensity factors, K_I and K_{II}, for a crack breaking the surface of a half-plane subject to various forms of contact loading. The method used to determine K_I and K_{II} is that of replacing the crack by a continuous distribution of edge dislocations and assumes the crack to be traction-free over its entire length, Nowell and Hills (1987). A traction-free crack is achieved by cancelling the tractions along the crack site that would be present if the half-plane was uncracked. The stress distribution for an elastic uncracked half-plane is determined using the methods outlined in the previous sections by way of distributed surface tractions.

The first case we will consider is that of a half-plane subject to concentrated point forces P and $Q(= fP)$ acting normally and tangentially to the surface at a distance b from the crack mouth, Figure 22.13, where f is the coefficient of friction. Figure 22.14 illustrates the variation of K_I and K_{II} for a normal

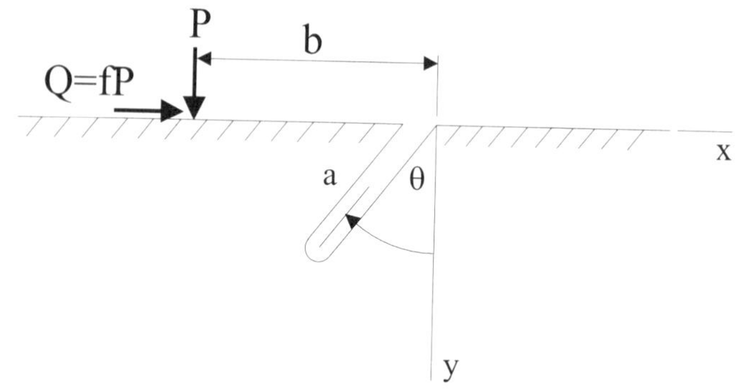

Figure 22.13: Concentrated point forces in contact with a half-plane containing a surface-breaking crack.

crack ($\theta = 0°$) and the coefficients of friction $f = 0$, 0.5 and 1. Both K_I and K_{II} are normalised with respect to $P/\sqrt{\pi a}$, that is the mode I stress intensity factor for a point loaded crack in an infinite plate. For $f = 0$ then K_I is symmetrical and continuous about $b/a = 0$ whereas K_{II} is asymmetric and discontinuous about $b/a = 0$. For $f \neq 0$ both K_I and K_{II} are discontinuous about $b/a = 0$. The effect of friction on K_I is to increase closure of the crack for $b/a < 0$ and to decrease closure for $b/a > 0$. For the case $b/a = 0$ the value of $K_I\sqrt{\pi a}/P = -0.8255$ is seen to agree with (17.209) determined by an application of the M-integral. The effect of varying crack angle, θ, on K_I and K_{II} for $f = 0$ is shown in Figure 22.15. Except for $K_I(\theta = 0°)$ both K_I and K_{II} are discontinuous about $b/a = 0$ and both K_I and K_{II} increase and decrease in absolute value for $b/a < 0$ and $b/a > 0$, respectively, as the crack angle increases.

The next case we will consider is that of uniformly distributed normal, P, and tangential, Q, tractions acting over an area $2c$ at a distance b from the mouth of the crack, Figure 22.16. In the case of a uniform pressure only the variation of K_I and K_{II} are shown in Figure 22.17 for $a/c = 1$ and $\theta = 0°$, $15°$ and $30°$. Both K_I and K_{II} are seen to be continuous throughout the shown ranges of b/a and θ noting that $K_{II} = 0$ when $\theta = 0°$ and $b/a = 0$. Figure 22.18 compares the variation of K_{II} with b/a for a concentrated force and a uniformly distributed normal pressure for the case of a crack normal to the surface. For $|b/a| > 1$ the point force is seen to be in close agreement with the uniformly distributed load whereas for the region $-1 < b/a < 1$ the distributed load passes over the crack and K_{II} is seen to be continuous and equal to zero for $b/a = 0$.

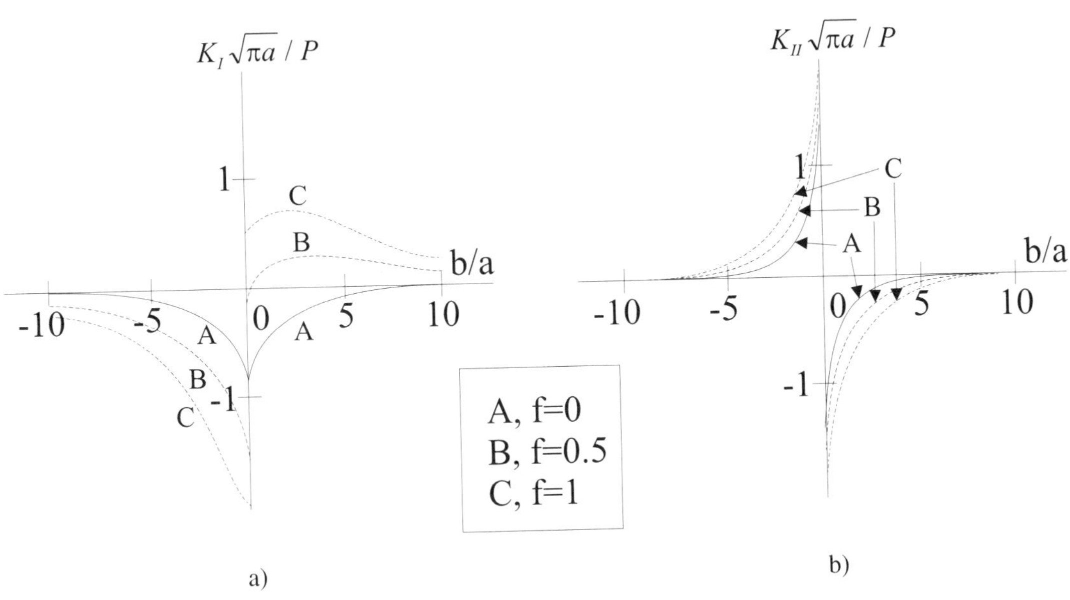

Figure 22.14: Variation of a) K_I and b) K_{II} with b/a for $f = 0, 0.5$ and 1 for a normal crack, $\theta = 0°$, subject to point forces.

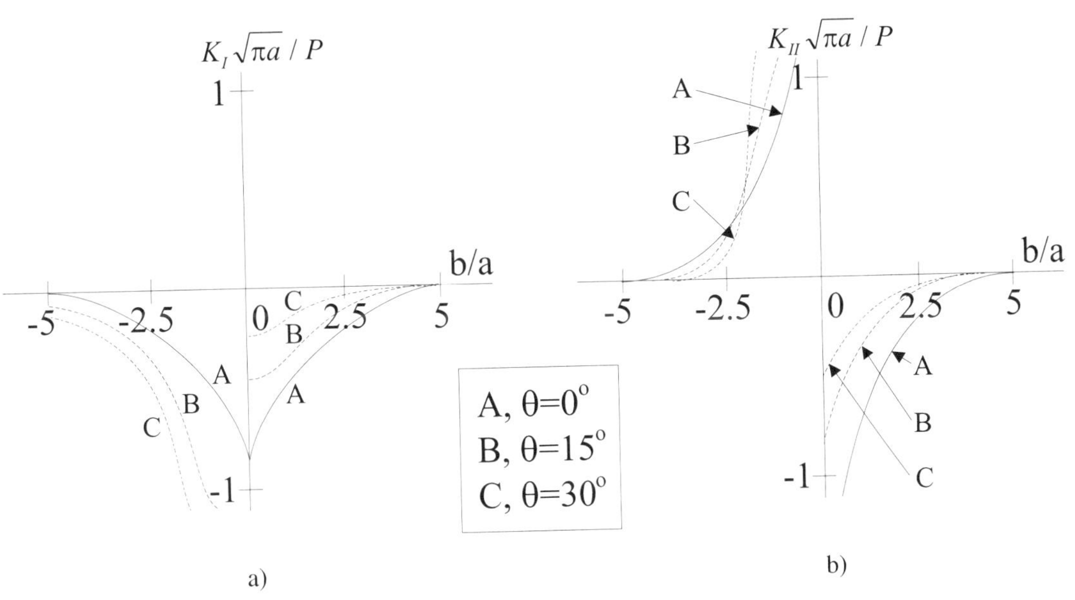

Figure 22.15: Figure 22.15. Variation of a) K_I and b) K_{II} with b/a for $\theta = 0°, 15°$ and $30°$ when $f = 0$ and subject to a normal point force.

When the indenter is that of a flat rigid punch of length $2c$, Figure 22.19, then the stresses at the corners of the indenter are infinite. Thus, we expect these large, but highly localised, stresses at the sharp corners of the indenter to have a significant influence on K_I and K_{II}. Various coefficients of friction K_I and K_{II}, against b/a, are given in Figure 22.20. As the indenter passes over the crack from left to right the

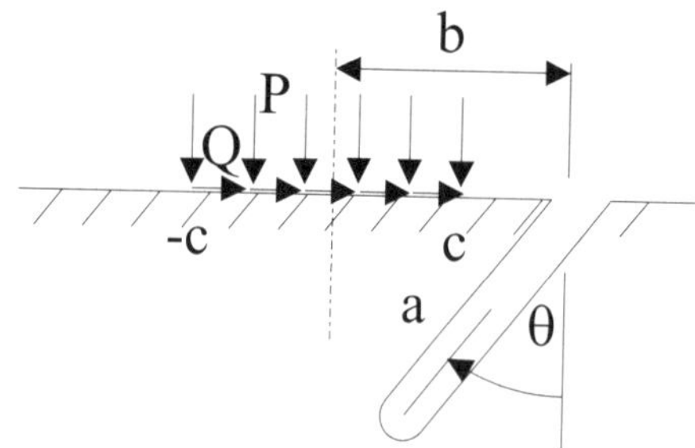

Figure 22.16: Uniformly distributed normal and tangential tractions acting on the surface of a half-plane containing a surface-breaking crack.

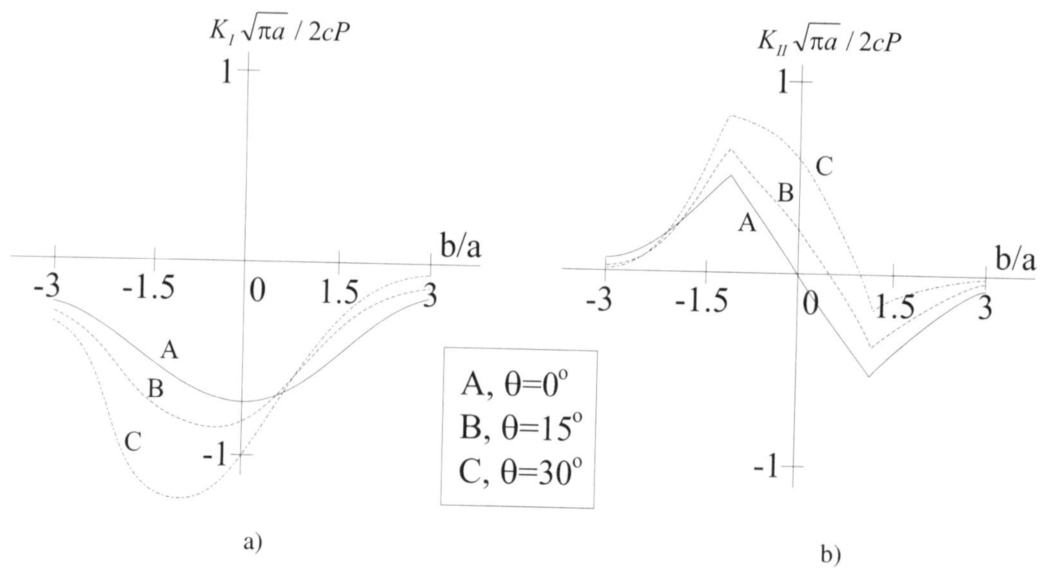

Figure 22.17: Variation of a) K_I and b) K_{II} with b/a for a uniformly distributed normal pressure when $a/c = 1$ and $\theta = 0°$, $15°$ and $30°$.

leading corner of the indenter is seen to have a significant influence on K_I and K_{II} when friction is non-zero, while the rear corner of the indenter does not play such an important role. The effect of friction is observed to increase closure of the crack faces for $b/a < -1$ and open the crack faces for $b/a > -1$. When $f \neq 0$ then both K_I and K_{II} are discontinuous about $a/b = -1$.

Finally, the case of a rigid circular indenter with sliding friction in contact with a cracked half-plane is considered. The general geometry is shown in Figure 22.21 with the radius of curvature of the circular indenter given by $R(R \gg c, a)$ where c is the contact radius. The problem of a moving Hertzian pressure is considered in an attempt to model the contact between a circular railway wheel and track which contains an inclined surface crack. The influence of friction on K_I and K_{II} is shown in Figure 22.22 for $a/c = 1$ and $\theta = 0°$. We note that for $f = 0$ then K_I is symmetrical about $b/a = 0$ while K_{II} is asymmetrical about $b/a = 0$ but is continuous for all b/a and passes through zero for $b/a = 0$. For large values of friction the crack is seen to open for $b/a > 0$. For the case $a/c = 1$ and $f = 0.2$ then the variation of K_I

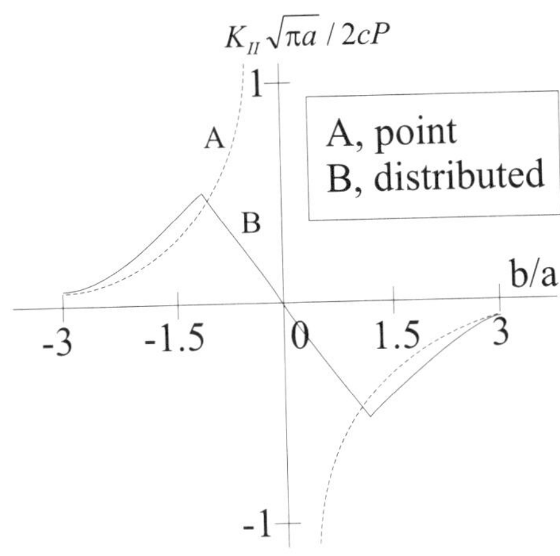

Figure 22.18: Comparison of K_{II} for a point force and a uniformly distributed normal pressure with b/a for $a/c = 1$.

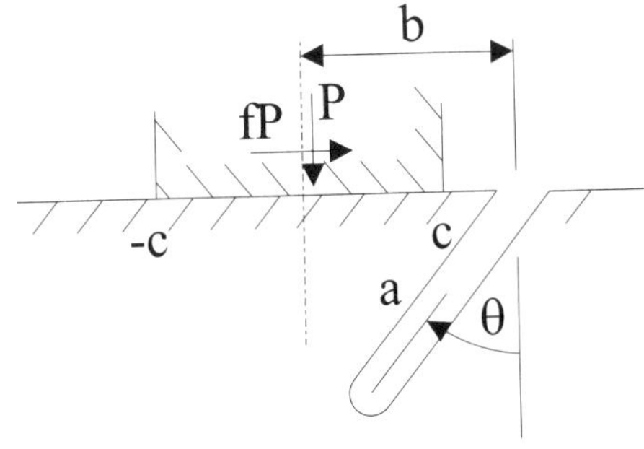

Figure 22.19: A flat rigid indenter in contact with a cracked half-plane.

and K_{II} for various crack angles is shown in Figure 22.23. Increasing θ is seen to generally increase, in absolute value, both K_I and K_{II} throughout the range of b/a shown.

22.8 Conclusion

This Chapter has addressed the contact of two bodies. We began by examining the simplest case of concentrated normal and tangential forces acting on the surface of a half-plane. The more general solutions of distributed normal and tangential pressure were then formulated by integrating the concentrated force formulae over a finite contact area. We also examined the indentation of a rigid flat punch and Hertzian contact for both linear elastic and viscoelastic materials. The Chapter concluded by presenting an overview of modes I and II stress intensity factors for surface-breaking cracks subject to a variety of contact loadings.

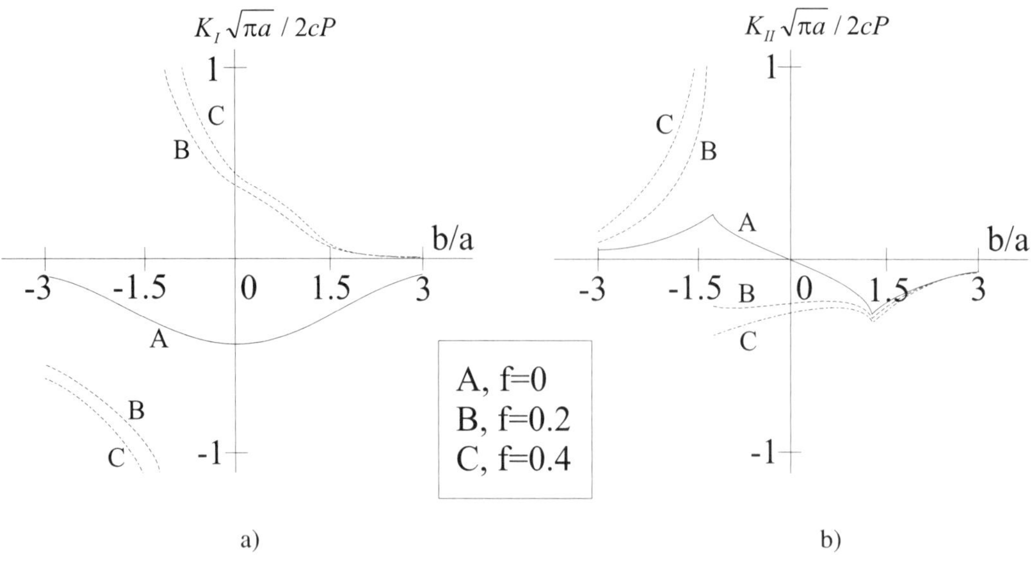

Figure 22.20: Variation of a) K_I and b) K_{II} with b/a for a flat rigid indenter in which $q = 0°$, $a/c = 1$ and f =0, 0.2, 0.4 and 0.6.

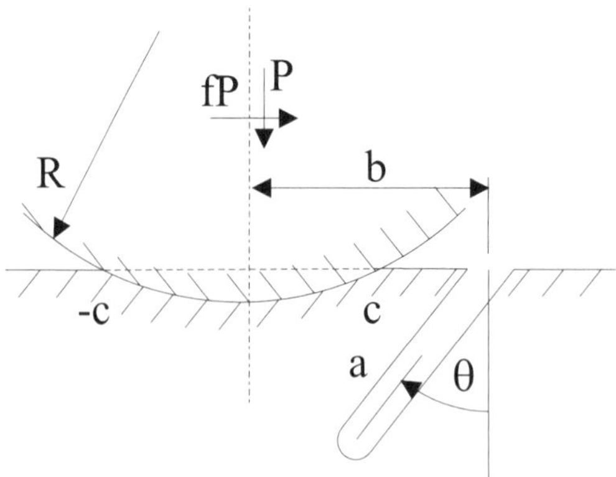

Figure 22.21: A rigid circular indenter in contact with a cracked half-plane.

22.9 References and Further Reading

∘ Gladwell, G. M. L. (1980) *Contact Problems in the Classical Theory of Elasticity*, Sijthoff and Noordhoff, The Netherlands.

∘ Johnson, K. L. (1985) *Contact Mechanics*, Cambridge University Press, Cambridge.

∘ Kreyszig, E. (1983) *Advanced Engineering Mathematics*, 5th Ed., Wiley & Sons, Canada.

∘ Lee, E. H. and Radok, J. R. M. (1960) *The Contact Problem for Viscoelastic Bodies*, Trans. ASME, Series E, J. Appl. Mech., 27, 438.

∘ Muskhelishvili, N. I. (1953) *Some Basic Problems of the Mathematical Theory of Elasticity*, 3rd Ed.,

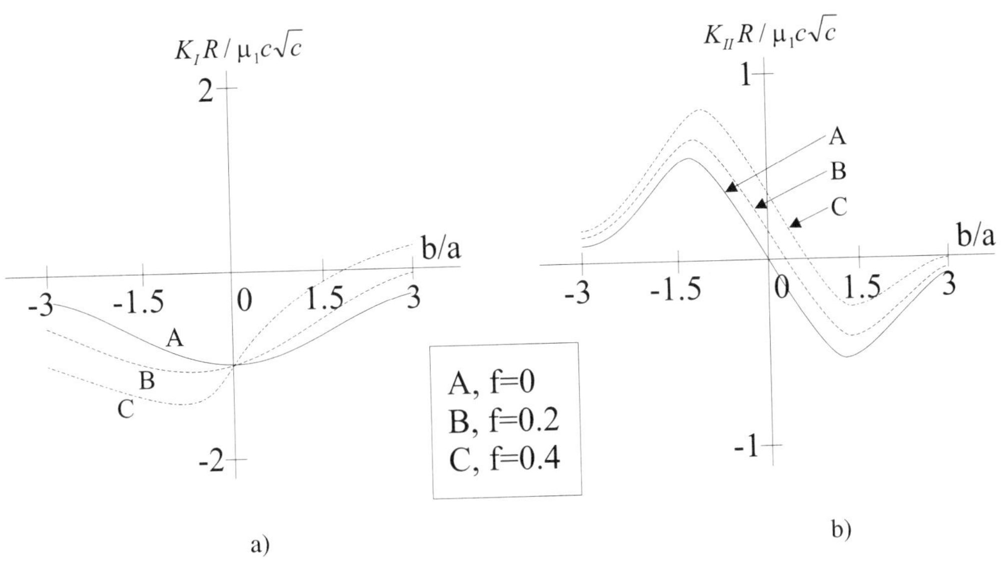

Figure 22.22: Variation of a) K_I and b) K_{II} with b/a for a rigid circular indenter in which $\theta = 0°$, $a/c = 1$ and $f = 0, 0.2$ and 0.4.

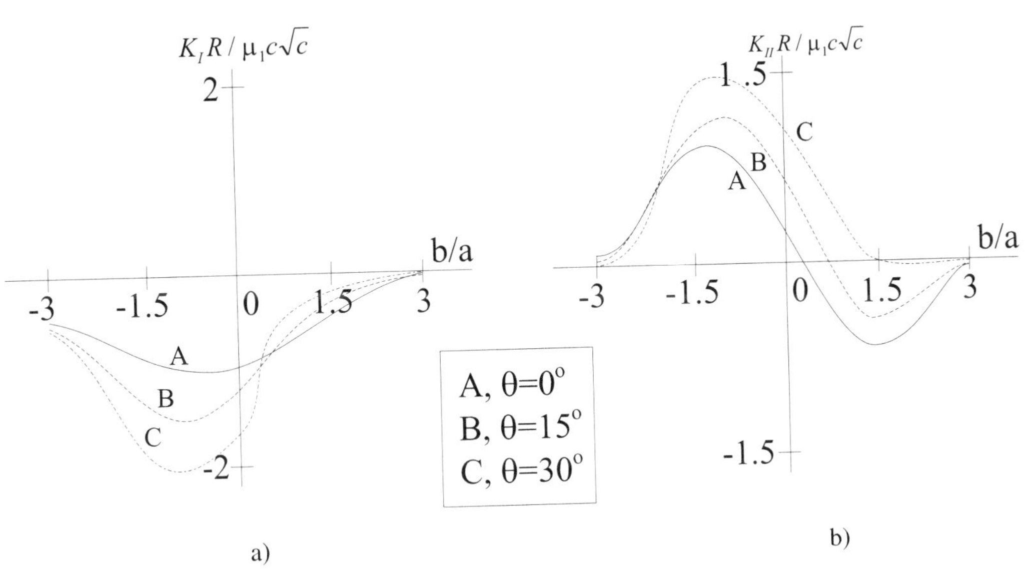

Figure 22.23: Variation of a) K_I and b) K_{II} with b/a for a rigid circular indenter in which $a/c = 1$, $f = 0.2$ and $\theta = 0°, 15°$ and $30°$.

(Translation by J. R. M. Radok), Noordhoff, Groningen, Holland.

∘ Nowell, D. and Hills, D. A. (1987) *Open Cracks at or Near Free Edges*, J. Strain Analysis, 22, 177-185.

∘ Radok, J. R. M. (1957) *Viscoelastic Stress Analysis*, Q. App. Math., 15, 198.

22.10 Exercises

22.1 Determine the stresses in the half-plane for the inclined point force P_r shown in Figure 22.24.

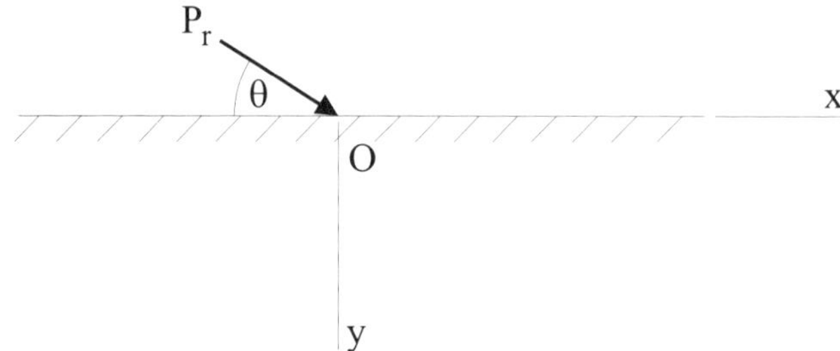

Figure 22.24: Exercise: A half-plane subject to an inclined concentrated normal force P_r.

22.2 Determine the stresses in the half-plane when a normal point force P acts at a distance b from the origin of coordinates O, as shown in Figure 22.25a). Also, show that the shear stress is zero along the line $x = b/2$ for the case of Figure 22.25b) in which point forces are applied at $x = 0$ and $x = b$.

22.3 Determine the in-plane stresses for the triangular normal pressure distribution shown in Figure 22.26.

22.4 The complex stress function $\Phi(z)$ and $\Psi(z)$ for a half-plane subject to both normal and tangential stresses ($\sigma_{yy} = N(t)$ and $\tau_{xy} = T(t)$ on Ox) on a segment $(-a \le t \le a)$ are given by, Muskhelishvili (1953)

$$\Phi(z) = -\frac{1}{2\pi i}\int_{-a}^{a}\frac{(N-iT)\,\mathrm{d}t}{t-z}; \quad \Psi(z) = -\frac{1}{\pi}\int_{-a}^{a}\frac{T\,\mathrm{d}t}{t-z} + \frac{z}{2\pi i}\int_{-a}^{a}\frac{(N-iT)\,t\,\mathrm{d}t}{(t-z)^2}$$

For a uniform pressure p show that $\Phi(z)$ and $\Psi(z)$ are given by

$$\Phi(z) = \frac{p}{2\pi i}\ln\left(\frac{z-a}{z+a}\right); \quad \Psi(z) = -\frac{paz}{\pi i\,(z^2-a^2)}$$

Also, determine expressions for the stresses in the half-plane.

22.5 A flat rigid punch of width 25mm is pressed into a half-plane by a force of 5kN. Determine the pressure in the half-plane at the centre of the punch.

22.6 Show that the radius of the contact region, a, and total displacement, δ, for a circular indenter of radius R pressed into a half-plane by a force P are given by

$$a = \left[\frac{3PR}{2}\left(\frac{1-\nu^2}{E}\right)\right]^{1/3}; \quad \delta = \left[\frac{9P^2}{16R}\left(\frac{1-\nu^2}{E}\right)^2\right]^{1/3}$$

where E and ν are Young's modulus and Poisson's ratio for both the indenter and half-plane.

22.7 Explain how you would determine expressions for the contact region, total displacement and maximum pressure for the circular ball in a circular seat shown in Figure 22.27.

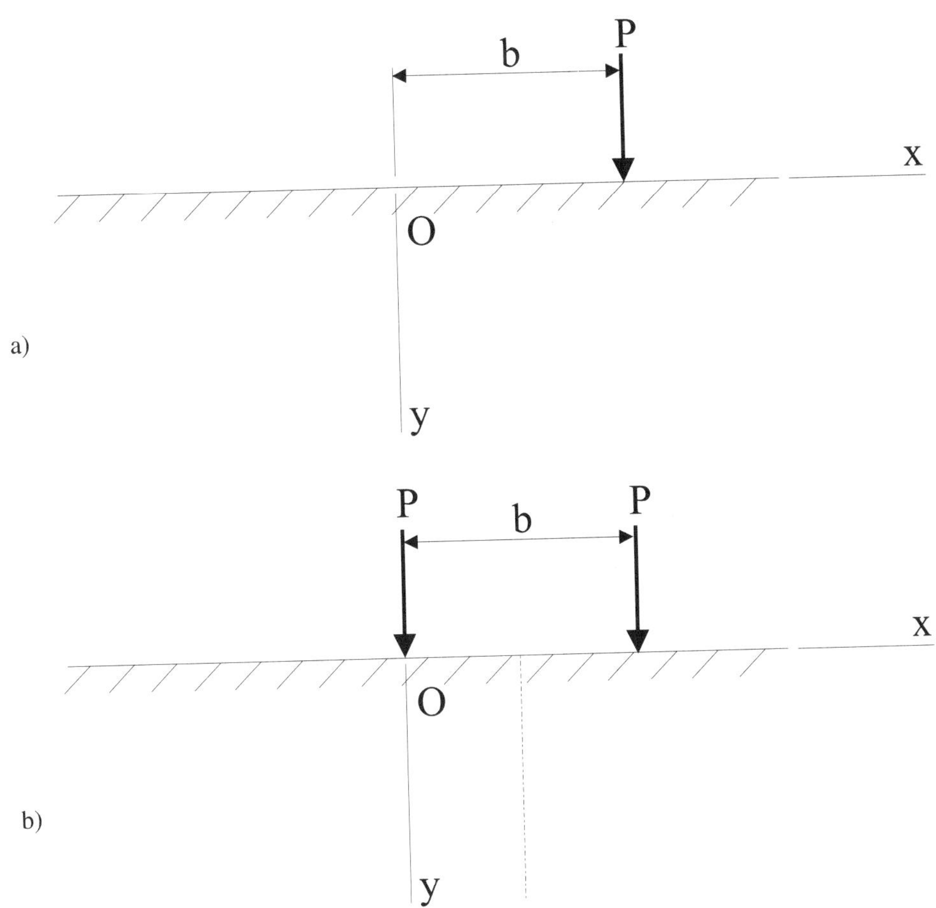

Figure 22.25: Exercise: A half-plane subject to concentrated normal forces. a) Point force P acting at a distance b from the origin O. b) Two point forces P acting at the origin O and at a distance b from O.

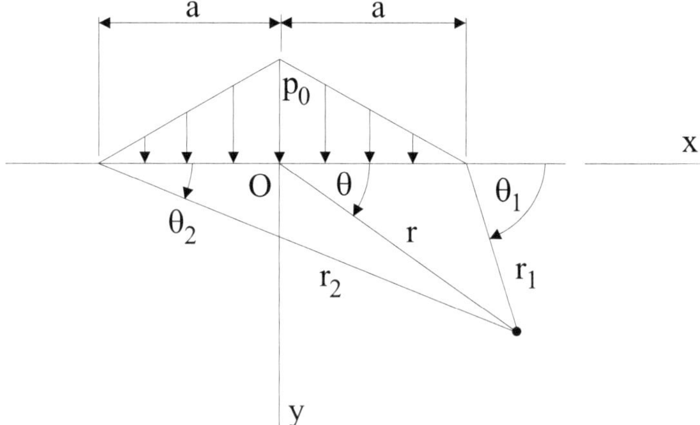

Figure 22.26: Exercise: A half-plane subject to a triangular normal pressure of maximum value p_0.

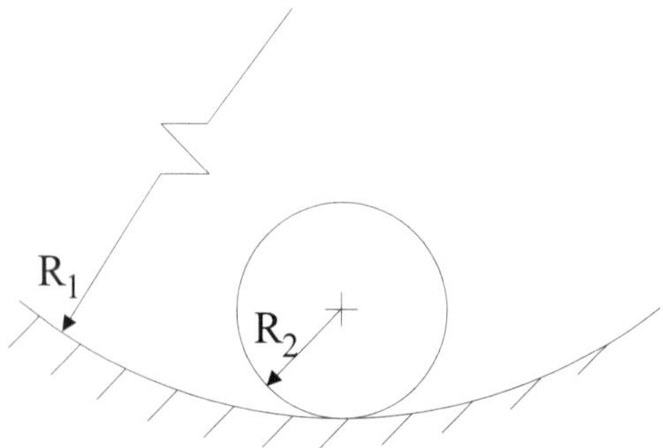

Figure 22.27: Exercise: Circular ball in a circular seat.

Chapter 23

Statistical Analysis of the Strength of Brittle Materials

23.1 Introduction

The variability of strength in identical specimens for ductile materials is typically not greater than 10% of the material's mean strength. In the case of brittle materials, however, the scatter of failure strengths is far more pronounced and can be as much as 100% of the mean strength. Therefore, from a design point of view, the use of the mean strength is an inadequate measure of the strength of a brittle material, and a statistical analysis of strength is required. This is particularly relevant to the failure of brittle ceramics. In Chapter 17 we examined the brittle fracture of a material containing a line crack. Real materials contain numerous flaws of different sizes and orientations. In addition, different types of flaws may be present such as inclusions, grain boundaries and dislocations. It follows that the greater the volume of material then the greater the number of flaws and hence the greater the probability of failure. This argument is in agreement with experimental observations that the strength of a brittle material decreases with increasing specimen volume. The failure of brittle materials is also a function of the loading system. The scatter in strength of brittle materials subject to compressive loading differs from that of tensile loading. Furthermore, different strengths are observed from the same specimen subject to three-point and four-point bending.

The Chapter begins by providing some necessary statistical background such as the representation of sample data and the relative and cumulative frequencies. The mean, variance and standard deviation are then discussed. Sample space, events and probability are covered and followed by a re-examination of the mean, variance and standard deviation in terms of the probability density function. The normal and Weibull distributions are then examined. Using the Weibull distribution estimates for the mean strength of both uniaxial tension and three-point bend specimens are determined. The Chapter concludes with a discussion of the size effect on the mean strength of brittle materials.

23.2 Representation of Data

As an illustration of the statistical analysis of experimental observations consider the example of testing specimens of glass in tension. Tests on 30 specimens produced the following failure strengths (MPa)

$$\{61, 67, 72, 54, 49, 30, 57, 57, 49, 49, 76, 44, 46, 56, 55, 57, 90, 56, 66, 57, 42, 67, 51, 59, 53, 55, 59, 55, 59, 58\}$$

The sample size is denoted by N and the failure strengths are the 30 sample points. Figure 23.1 illustrates the failure strength for each of the 30 sample points. In column 1 of Table 23.1 the sample points have

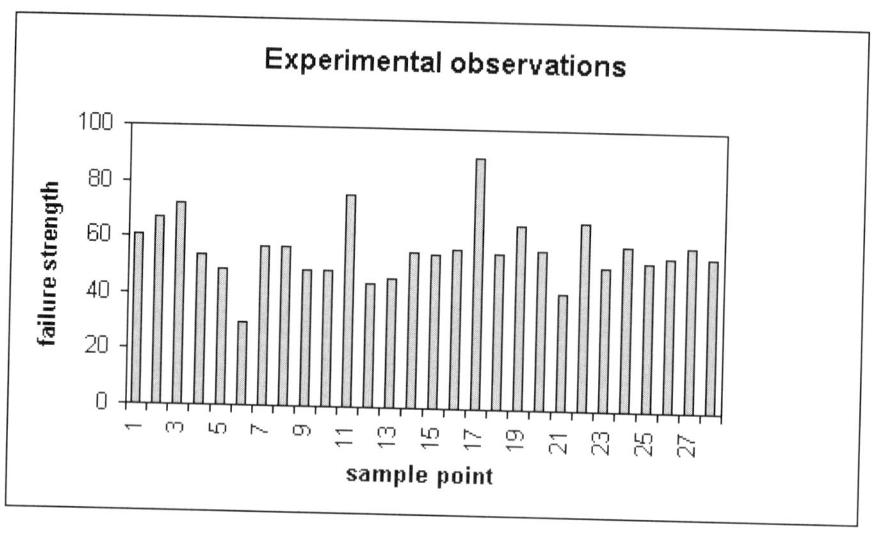

Figure 23.1: Failure strength (MPa) for 30 glass specimens tested in uniaxial tension.

been ordered from the smallest value (30MPa) to the largest value (90MPa). The number of occurrences of each sample point is listed in column 2 of Table 23.1 and is referred to as the *absolute frequency* and is illustrated in Figure 23.2. Dividing the absolute frequency by the sample size then we obtain the *relative frequency* and these are listed in column 3 of Table 23.1. For example, there are 3 occurrences of the failure strength 59MPa so that it has an absolute frequency of 3 and a relative frequency of 0.1. If we sum all of the absolute values which are smaller than or equal to a sample value x then we obtain the *cumulative absolute frequency* corresponding to that x. Similarly, dividing by the sample size we have the *cumulative relative frequency*. The cumulative frequencies are listed in columns 4 and 5 of Table 23.1 and graphically illustrated in Figure 23.3 and show that 80% of all sample values are less than or equal to 61MPa. The cumulative distribution curve has an inflexion in the vicinity of the median (50%) and exactly at this point only when the distribution is symmetrical. Note that if a certain value does not appear in the sample values then its absolute frequency is zero; for example no sample failed at 41MPa. Alternatively, if all N samples have the same sample value then the absolute frequency is N so that the relative frequency is equal to unity. Thus, the relative frequency lies in the range $[0 : 1]$.

23.3 Relative Frequency

Let there be m numerically different sample values in a sample of size N $(N \geq m)$ $x_1, x_2, \ldots, x_m$ with corresponding relative frequencies $\tilde{f}_1, \tilde{f}_2, \ldots, \tilde{f}_m$. We now introduce the following frequency function of the sample

$$\tilde{f}(x) = \begin{cases} \tilde{f}_i & \text{when } x = x_i \ (i = 1, 2, \ldots, m) \\ 0 & \text{for a value of } x \text{ not appearing in the sample} \end{cases} \tag{23.1}$$

failure strength	ab. frequency	rel. frequency	cum. ab. frequency	cum. rel. frequency
30	1	0.03	1	0.03
42	1	0.03	2	0.06
44	1	0.03	3	0.1
46	1	0.03	4	0.13
49	3	0.1	7	0.23
51	1	0.03	8	0.26
53	1	0.03	9	0.3
54	1	0.03	10	0.33
55	3	0.1	13	0.43
56	2	0.06	15	0.5
57	4	0.13	19	0.63
58	1	0.03	20	0.66
59	3	0.1	23	0.76
61	1	0.03	24	0.8
66	1	0.03	25	0.83
67	2	0.06	27	0.9
72	1	0.03	28	0.93
76	1	0.03	29	0.96
90	1	0.03	30	1

Table 23.1: Sample of 30 values of the tensile failure strength (MPa) of glass specimens.

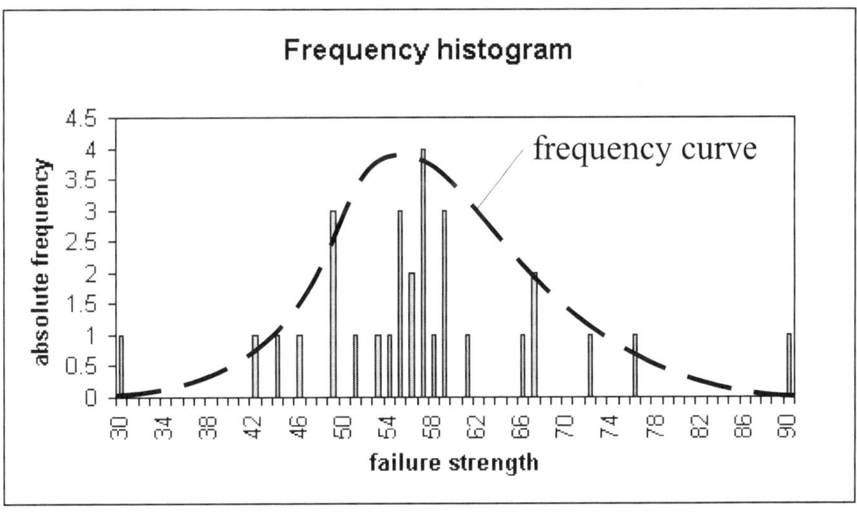

Figure 23.2: Frequency histogram of failure strength of the glass specimens.

For example, for the sample data in Table 23.1 $\tilde{f}(58) = 0.03$ and $\tilde{f}(82) = 0$. Clearly, the sum of all the relative frequencies is equal to unity

$$\sum_{i=1}^{m} \tilde{f}(x_i) = \tilde{f}(x_1) + \tilde{f}(x_2) + \cdots + \tilde{f}(x_m) = 1 \tag{23.2}$$

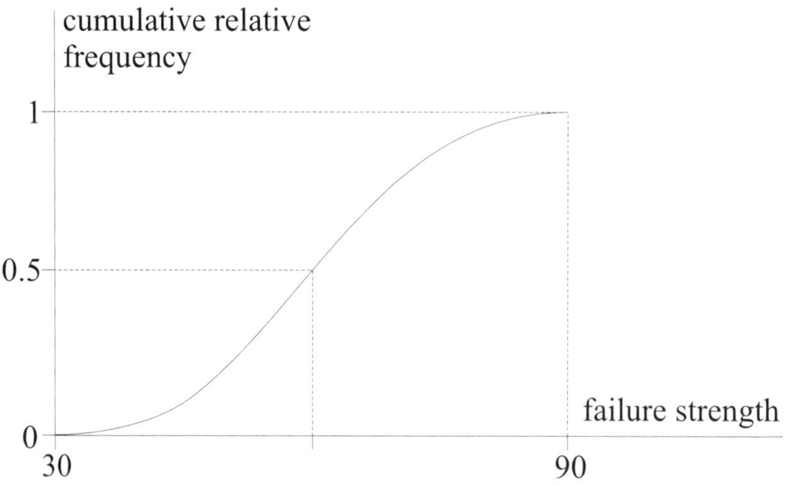

Figure 23.3: Cumulative relative frequency of the failure strength of the glass specimens.

23.4 Cumulative Frequency

We can also introduce the cumulative frequency function, $\tilde{F}(x)$, which is the sum of all relative frequencies of the sample values that are less than or equal to x

$$\tilde{F}(x) = \sum_{t \leq x} \tilde{f}(t) \tag{23.3}$$

where $t \leq x$ means that for a given x we sum all $\tilde{f}(t)$ for which t is less than or equal to x. The range of $\tilde{F}(x)$ is $[0:1]$.

23.5 Mean, Variance and Standard Deviation

The *mean* value of a sample $x_1, x_2, \ldots, x_N$ is denoted by $\bar{x}$ and defined by

$$\bar{x} = \frac{1}{N} \sum_{i=1}^{N} x_i = \frac{1}{N} (x_1 + x_2 + \cdots + x_N) \tag{23.4}$$

and measures the average value of the sample values. The *variance* of the sample is denoted by s^2 and defined by

$$s^2 = \frac{1}{N-1} \sum_{i=1}^{N} (x_i - \bar{x})^2 = \frac{1}{N-1} \left[(x_1 - \bar{x})^2 + (x_2 - \bar{x})^2 + \cdots + (x_N - \bar{x})^2 \right] \tag{23.5}$$

and measures the spread of the sample from the mean. The square root of the variance is the *standard deviation* and is denoted by s. From (23.2) the sum of the relative frequencies for all $x_i (i = 1, 2, \ldots, m)$ is equal to unity for m numerically different sample values, so it follows that $\bar{x}$ can alternatively be written as

$$\bar{x} = \frac{1}{N}\sum_{i=1}^{m} x_i N\tilde{f}(x_i) = \frac{1}{N}\sum_{i=1}^{m} x_i a(x_i); \quad a(x_i) = N\tilde{f}(x_i) \tag{23.6}$$

where $a(x_i)$ denotes the absolute frequencies. Similarly, the variance can be written

$$s^2 = \frac{1}{N-1}\sum_{i=1}^{m} (x_i - \bar{x})^2 \, a(x_i) \tag{23.7}$$

23.6 Sample Space and Events

When performing experiments we are interested in the *outcome* of a particular event. For example, the rolling of a die can result in any one of six possible outcomes. The set of all possible outcomes is referred to as the *sample space* and denoted by S and can consist of either a finite or infinite number of outcomes. Each outcome is an element of S. The sample space for the rolling of a die is finite and consists of six elements; $S = \{1, 2, 3, 4, 5, 6\}$. However, the sample space of the tensile failure strength of a material is infinite because the outcome may be any value within a given range.

The total space can consist of subset sample spaces referred to as *events*. Frequently, we are interested in the subset sample space to which a particular outcome belongs. Let $E(\in S)$ denote the elements of S associated with a particular event. The complement of E is denoted by E^c and refers to all elements in S but not in E. If an event contains no element then it is referred to as an *impossible event*. Let A and B be two events. The union of A and B, $A \cup B$, is the event consisting of elements that belong to both A and B. The intersection of A and B, $A \cap B$, is the event consisting of elements that belong to both A and B. If all elements of A belong to B then A is a subset or subevent of B, $A \subset B$. If A and B have no elements in common, $A \cap B = \emptyset$, then A and B are said to be mutually exclusive events.

Let us conclude with a result that will prove useful in subsequent sections. Consider an experiment in which the sample size is N and events A and B have relative frequencies of $\tilde{f}(A)$ and $\tilde{f}(B)$ respectively. The relative frequency of $A \cup B$ is

$$\tilde{f}\,(A \cup B) = \tilde{f}\,(A) + \tilde{f}\,(B) - \tilde{f}\,(A \cap B) \tag{23.8}$$

with the subtraction of $\tilde{f}(A \cap B)$ preventing a repetition of frequencies. If A and B are mutually exclusive then $\tilde{f}(A \cap B) = 0$ and (23.8) reduces to

$$\tilde{f}\,(A \cup B) = \tilde{f}\,(A) + \tilde{f}\,(B) \tag{23.9}$$

23.7 Probability

Earlier in §23.2 we considered the failure strength of glass specimens with a sample size $N = 30$. Let $l_1, l_2, \ldots, l_N$ denote the number of events or specimens whose failure strengths lie in the intervals (σ_0, σ_1), (σ_1, σ_2) and so on. The *probability* of interval i is

$$P_i = \frac{l_i}{N}; \quad i = 1, 2, \ldots, N \tag{23.10}$$

and is seen to be equal to the relative frequency $\tilde{f}(\sigma_i - \sigma_{i-1})$. Thus, when we refer to the probability of an event E, $P(E)$, we mean that if we perform a significantly large number of experiments then $P(E)$ is approximately equal to $\tilde{f}(E)$. It follows that the sum of probabilities for all events is equal to 1

$$\sum_{i=1}^{N} P_i = \sum_{i=1}^{N} \frac{l_i}{N} = \frac{1}{N}\sum_{i=1}^{N} l_i = 1 \tag{23.11}$$

If E is an event of sample space S then the probability of E, $P(E)$, is within the range

$$0 \leq P(E) \leq 1 \tag{23.12}$$

The probability of a certain event is 1 and the probability of an impossible event is 0. Since S is the entire sample space then the probability of S is certain, that is $P(S) = 1$. From (23.8), if A and B are arbitrary events in S then

$$P(A \cup B) = P(A) + P(B) - P(A \cap B) \tag{23.13}$$

whereas if A and B are mutually exclusive events then, (23.9)

$$P(A \cup B) = P(A) + P(B) \tag{23.14}$$

If the complement of an event is E^c then it follows that

$$P(E \cup E^c) = P(E) + P(E^c) = 1 \tag{23.15}$$

since E and E^c are mutually exclusive and $E \cup E^c = S$. If events A and B are independent such that A does not depend on B and vice versa then

$$P(A \cap B) = P(A)P(B) \tag{23.16}$$

that is, the probability of the joint occurrence of any number of mutually independent events is equal to the product of the probabilities of these events.

Let us now introduce the random variable X which lies in the interval $a < X < b$. The event corresponding to the number a will be denoted by $X = a$ and corresponding probability $P(X = a)$. The probability of X being any value in the interval $a < X < b$ is denoted by $P(a < X < b)$. The probability of the event $X \leq c$ (X has any value less than or equal to c) is denoted by $P(X \leq c)$ and the probability of the event $X > c$ is denoted by $P(X > c)$. Since $P(X \leq c) + P(X > c) = P(-\infty < X < \infty) = 1$ then it follows that $P(X > c) = 1 - P(X \leq c)$.

23.7.1 Discrete Distribution and Random Variable

Consider the case of a discrete variable X and corresponding discrete distribution with values of X: x_1, $x_2, \ldots$ and corresponding probabilities $p_1, p_2, \ldots$. Letting $P(X = x_1) = p_1$ etc. then the *probability density function* of X is defined as

$$f(x) = \begin{cases} p_i \text{ when } x = x_i \ (i = 1, 2, \ldots) \\ 0 \text{ for a value of } x \text{ not apparing in the sample} \end{cases} \tag{23.17}$$

Note the similarity between $f(x)$ and the relative frequencies function $\tilde{f}(x)$ in §23.3 with $f(x)$ tailored for probabilities. With $P(S) = 1$ then

$$\sum_{i=1}^{\infty} f(x_i) = 1 \tag{23.18}$$

If $f(x)$ is known then the probability $P(a < X \leq b)$ can be found

$$P(a < X \leq b) = \sum_{a<x_i\leq b} f(x_i) \tag{23.19}$$

for all x_i within the interval $a < x_i \leq b$. The probability $P(X \leq x)$ for X having a value less than or equal to x is given by the cumulative distribution function $F(x)$

$$F(x) = P(X \leq x) \tag{23.20}$$

which is analogous to the cumulative frequency function, $\tilde{F}(x)$, discussed in §23.4 and is equal to

$$F(x) = \sum_{x_i \leq x} f(x_i) \tag{23.21}$$

With $P(a < X \leq b) = P(X \leq b) - P(X \leq a)$ then

$$P(a < X \leq b) = F(b) - F(a) \tag{23.22}$$

$F(x)$ is a step function when $f(x)$ is a discrete function.

23.7.2 Continuous Distribution and Random Variable

A random variable X and corresponding distribution are said to be continuous when the cumulative distribution function, $F(x)$, can be represented by an integral form

$$F(x) = \int_{-\infty}^{x} f(\nu)\,\mathrm{d}\nu \tag{23.23}$$

where the integrand is continuous. Differentiating (23.23) then $f(x)$ is given by

$$f(x) = F'(x) \tag{23.24}$$

With $P(S) = 1$ then

$$\int_{-\infty}^{\infty} f(\nu)\,\mathrm{d}\nu = 1 \tag{23.25}$$

and from (23.22)

$$P(a < x \leq b) = F(b) - F(a) = \int_{a}^{b} f(\nu)\,\mathrm{d}\nu \tag{23.26}$$

which informs us that the probability is equal to the area under the curve of the probability density function $f(x)$ in the interval $[a : b]$, as shown in Figure 23.4.

23.8 Mean, Variance and Standard Deviation in Terms of the Probability Density Function

If a sample consists of the N values $x_1, x_2, \ldots, x_N$ then from (23.4) the mean is equal to $\bar{x} = (x_1, x_2, \ldots, x_N)$. If $x_1, x_2, \ldots, x_N$ is represented by $f(x)$ then $xf(x)\delta x$ gives the sum of all values in the interval x to $x + \delta x$. Therefore, the mean is given by

$$\bar{x} = \frac{\int_{-\infty}^{\infty} xf(x)\,\mathrm{d}x}{\int_{-\infty}^{\infty} f(x)\,\mathrm{d}x} \tag{23.27}$$

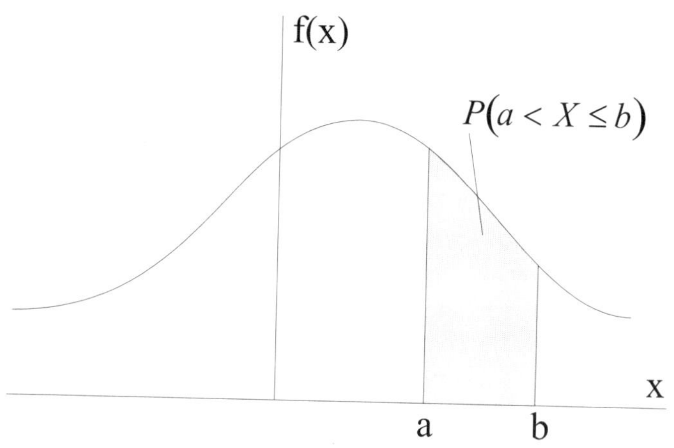

Figure 23.4: The probability $P(a < X \leq b)$ is equal to the area under the curve $f(x)$ in the interval $[a : b]$.

but in view of (23.25) then the mean for a continuous distribution is

$$\bar{x} = \int_{-\infty}^{\infty} x f(x)\, \mathrm{d}x \tag{23.28}$$

If $f(x)$ is a discrete distribution then $\bar{x}$ is given by

$$\bar{x} = \sum_i x_i f(x_i) \tag{23.29}$$

The variance for both continuous and discrete distributions are defined as, (23.5)

$$\sigma^2 = \int_{-\infty}^{\infty} (x - \bar{x})^2 f(x)\, \mathrm{d}x; \quad \sigma^2 = \sum_i (x_i - \bar{x})^2 f(x) \tag{23.30}$$

with the standard deviation, σ, equal to the square root of the variance.

Example 23.1 Mean and variance of a rectangular probability density function

Figure 23.6 illustrates a rectangular distribution with $f(x) = 1/(b - a)$ in the interval $a < x < b$ and $f(x) = 0$ elsewhere; also shown is $F(x)$. From (23.28) the mean is equal to

$$\bar{x} = \int_a^b \frac{x\, \mathrm{d}x}{b - a} = \frac{1}{2(b - a)} \left[x^2\right]_a^b = \frac{1}{2(b - a)} \left[b^2 - a^2\right] = \frac{b + a}{2}$$

The variance is given by, (23.30)

$$\sigma^2 = \int_a^b \left[x - \frac{b + a}{2}\right]^2 \frac{\mathrm{d}x}{b - a} = \frac{(b - a)^2}{12}$$

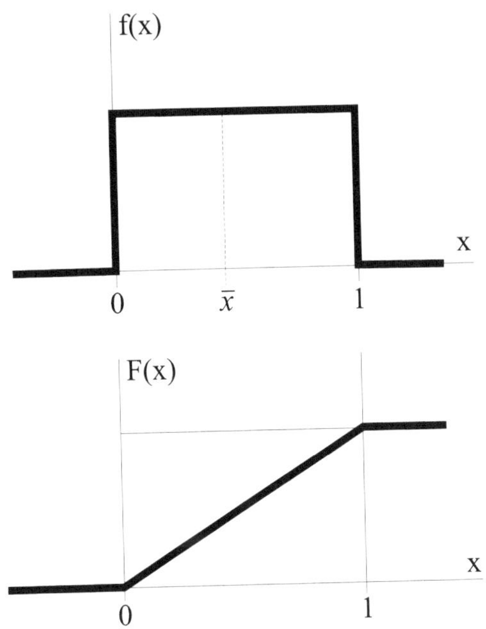

Figure 23.5: Example: Variation of $f(x)$ and $F(x)$ against x for a rectangular distribution.

23.9 Normal Distribution

A popular continuous distribution is the Gaussian or normal distribution

$$f(x) = \frac{1}{\sigma\sqrt{2\pi}} e^{-\frac{1}{2}\left(\frac{x-\bar{x}}{\sigma}\right)^2} \tag{23.31}$$

and produces a bell-shaped curve as shown in Figure 23.6 for $\bar{x} = 0$. The smaller the value of σ^2 then

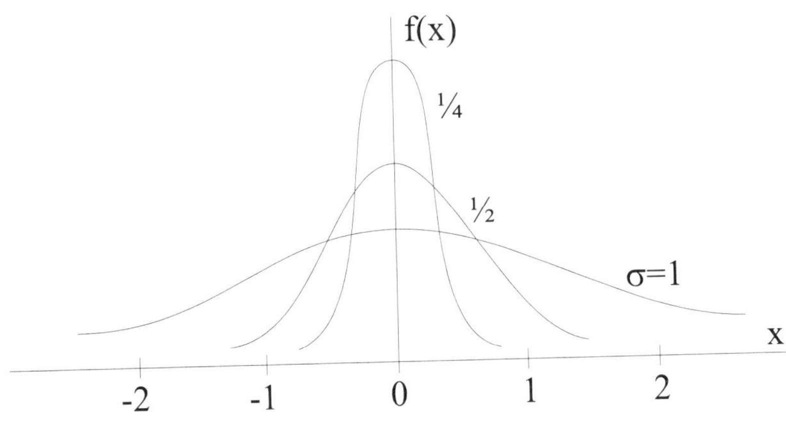

Figure 23.6: Variation of the normal probability density function as a function of σ with $\bar{x} = 0$.

the higher the peak at $x = 0$. If $\bar{x} > 0$ ($\bar{x} < 0$) then the normal distribution is the same shape but shifted $|\bar{x}|$ to the right (left). From (23.23) the cumulative distribution function is

$$F(x) = \int_{-\infty}^{x} f(\nu)\,\mathrm{d}\nu = \frac{1}{\sigma\sqrt{2\pi}} \int_{-\infty}^{x} e^{-\frac{1}{2}\left(\frac{\nu-\bar{x}}{\sigma}\right)^2} \mathrm{d}\nu \tag{23.32}$$

From (23.22) the probability that the random variable X lies in the interval $a < x \leq b$ is

$$P(a < X \leq b) = F(b) - F(a) = \frac{1}{\sigma\sqrt{2\pi}} \int_{a}^{b} e^{-\frac{1}{2}\left(\frac{\nu-\bar{x}}{\sigma}\right)^2} \mathrm{d}\nu \tag{23.33}$$

The integral in (23.32) cannot be easily evaluated and the use of a transformation assists the evaluation. Let $u = (\nu - \bar{x})/\sigma$ then $\mathrm{d}u/\,\mathrm{d}\nu = 1/\sigma$ and substituting into (23.32) we have

$$F(x) = \frac{1}{\sqrt{2\pi}} \int_{-\infty}^{(x-\bar{x})/\sigma} e^{-u^2/2}\,\mathrm{d}u \tag{23.34}$$

and letting $z = (x - \bar{x})/\sigma$ then $F(x)$ is seen to be equivalent to the following integral [1]

$$\Phi(z) = \frac{1}{\sqrt{2\pi}} \int_{-\infty}^{z} e^{-u^2/2}\,\mathrm{d}u \tag{23.35}$$

such that

$$F(x) = \Phi\left(\frac{x - \bar{x}}{\sigma}\right) \tag{23.36}$$

From (23.33) then

$$P(a < X \leq b) = F(b) - F(a) = \Phi\left(\frac{b - \bar{x}}{\sigma}\right) - \Phi\left(\frac{a - \bar{x}}{\sigma}\right) \tag{23.37}$$

The distribution function $\Phi(z)$ is tabulated in several mathematical tables and texts (see for example Kreyszig (1983)); values of $\Phi(z)$ for z equal 0 to 2 are listed in Table 23.2.

z	0	0.2	0.4	0.6	0.8	1	1.2	1.4	1.6	1.8	2
$\Phi(z)$	0.5	0.5793	0.6554	0.7257	0.7881	0.8413	0.8849	0.9192	0.9452	0.9641	0.9772

Table 23.2: Values of the distribution function $\Phi(z)$ for $0 \leq z \leq 2$, noting that $\Phi(-z) = 1 - \Phi(z)$.

For example, when $a = \bar{x} - \sigma$ and $b = \bar{x} + \sigma$ then $P = \Phi(1) - \Phi(-1)$ and similarly for $a = \bar{x} - 2\sigma$, $b = \bar{x} + 2\sigma$ and $a = \bar{x} - 3\sigma$, $b = \bar{x} + 3\sigma$ using the tabulated values of $\Phi(z)$ we have

$$\begin{aligned} P(\bar{x} - \sigma < X \leq \bar{x} + \sigma) &\approx 68\% \\ P(\bar{x} - 2\sigma < X \leq \bar{x} + 2\sigma) &\approx 95.5\% \\ P(\bar{x} - 3\sigma < X \leq \bar{x} + 3\sigma) &\approx 99.7\% \end{aligned} \tag{23.38}$$

These results illustrate that 68% of values lie in the interval $[\bar{x} - \sigma, \bar{x} + \sigma]$, 95% of values lie in the interval $[\bar{x} - 2\sigma, \bar{x} + 2\sigma]$ and practically all values lie in the interval $[\bar{x} - 3\sigma, \bar{x} + 3\sigma]$. The first two of these intervals are illustrated in Figure 23.7 and recalling that the probability is equal to the area under the curve of the probability density function $f(x)$ in the interval $[a, b]$.

[1] z is called the *standardised variable* corresponding to x. If x has a mean $\bar{x}$ and variance σ^2 then $z = (x - \bar{x})/\sigma$ has the mean 0 and variance 1.

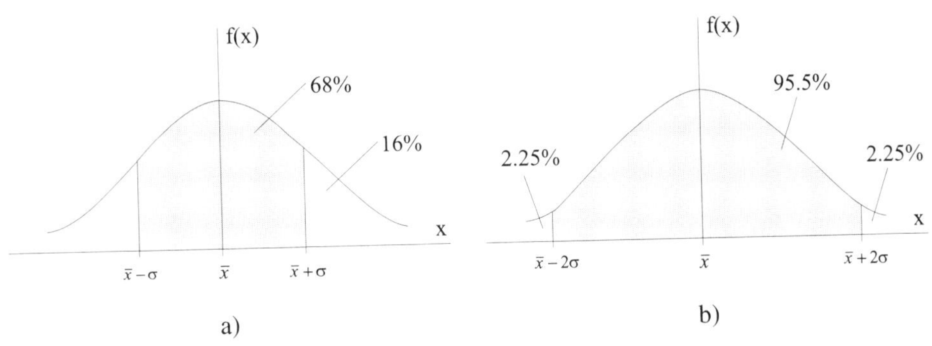

Figure 23.7: The normal distribution for the intervals a) $[\bar{x} - \sigma < X \leq \bar{x} + \sigma]$ and b) $[\bar{x} - 2\sigma < X \leq \bar{x} + 2\sigma]$.

23.10 Weibull Distribution

A statistical method that is frequently used to determine the strength of materials is that of Weibull (1951). The method assumes that the material is both isotropic and statistically homogenous, that is the probability of finding a flaw of a certain severity within an arbitrary small volume is the same throughout the entire material. The method also assumes that the 'weakest link of a chain' argument applies so that failure at the most critical flaw results in complete failure.

Consider a material with N flaws or links and let $F_1(\sigma)$ be the cumulative distribution or probability of failure by stress σ for one link. The probability of survival at σ for one link is therefore equal to $1 - F_1(\sigma)$. For N links the probability of failure by σ is $F_N(\sigma) = 1 -$(probabilities of survival of N links) and from (23.16)

$$F_N(\sigma) = 1 - [1 - F_1(\sigma)]^N \tag{23.39}$$

Noting the following limit

$$\lim_{n\to\infty} \left(1 - \frac{x}{n}\right)^n = e^{-x} \tag{23.40}$$

then (23.39) can be written

$$1 - F_N(\sigma) = [1 - F_1(\sigma)]^N = e^{-NF_1(\sigma)} \tag{23.41}$$

The number of links, N, is proportional to the volume of the material, V, so that (23.41) can be written; with $V\phi(\sigma) = NF_1(\sigma)$

$$P_f = F_N(\sigma) = 1 - e^{-V\phi(\sigma)} \tag{23.42}$$

with $P_f = F_N(\sigma)$ denoting the probability of failure by stress σ. The probability of survival is therefore $P_s = 1 - P_f$.

The function $\phi(\sigma)$ is unknown and Weibull assumed an empirical form given by

$$\begin{aligned} \phi(\sigma) &= \left(\frac{\sigma - \sigma_u}{\sigma_0}\right)^m \quad \text{for} \quad \sigma > \sigma_u \\ \phi(\sigma) &= 0 \quad \text{for} \quad \sigma \leq \sigma_u \end{aligned} \tag{23.43}$$

where σ_u is the threshold stress at which there is zero probability of failure, σ_0 is a normalising factor and m is the Weibull modulus and is a material parameter. Substituting $\phi(\sigma)$ into (23.42) we arrive at the Weibull distribution

$$\begin{aligned} P_f &= 1 - e^{-V\left(\frac{\sigma-\sigma_u}{\sigma_0}\right)^m} \quad \text{for} \quad \sigma > \sigma_u \\ P_f &= 0 \quad \text{for} \quad \sigma \leq \sigma_0 \end{aligned} \tag{23.44}$$

and is applicable for materials subject to a uniaxial tensile stress only. A non-uniform stress state (such as bending of beams) can be accounted for by the following

$$P_f = 1 - e^{-\int_V \left(\frac{\sigma-\sigma_u}{\sigma_0}\right)^m \mathrm{d}V} \tag{23.45}$$

The determination of σ_0 and m will be discussed later in §23.10.5 but we note at this stage that in general m increases as the material becomes less brittle and that a high variability of strength for a batch of specimens is characterised by a small value for m. Typical values of m for glass and cast iron are 2 and 38 respectively.

23.10.1 Mean Strength

From (23.28) the mean strength, $\bar{\sigma}$, of a sample of strength values $\sigma_1, \sigma_2, \ldots, \sigma_N$ is

$$\bar{\sigma} = \int_{-\infty}^{\infty} \sigma g(\sigma)\, \mathrm{d}\sigma = \int_0^{\infty} \sigma g(\sigma)\, \mathrm{d}\sigma \tag{23.46}$$

where $g(\sigma)$ is the probability density function and in terms of P_f is given by, (23.24)

$$g(\sigma) = \frac{\mathrm{d}P_f}{\mathrm{d}\sigma} \tag{23.47}$$

Substituting $g(\sigma)$ into (23.46)

$$\bar{\sigma} = \int_0^{\infty} \sigma \frac{\mathrm{d}P_f}{\mathrm{d}\sigma}\, \mathrm{d}\sigma = \int_0^1 \sigma\, \mathrm{d}P_f = \int_0^{\infty} P_s\, \mathrm{d}\sigma \tag{23.48}$$

with $P_s = 1 - P_f$. When $\sigma_u \neq 0$ then

$$\sigma = \sigma_u + \int_0^1 \sigma\, \mathrm{d}P_f \tag{23.49}$$

where the integral is derived when $\sigma_u = 0$. In the following sub-sections we will consider the cases of uniaxial tension and bending separately.

23.10.2 Uniaxial Tension

When σ_u=0 then the Weibull distribution for uniaxial tension is given by, (23.44)

$$P_f = 1 - e^{-V(\sigma/\sigma_0)^m} \tag{23.50}$$

From (23.48) the mean strength is

$$\bar{\sigma} = \int_0^{\infty} e^{-V(\sigma/\sigma_0)^m}\, \mathrm{d}\sigma \tag{23.51}$$

Making the substitution $t = (\sigma/\sigma_0)^m V$ and using the gamma function

$$\Gamma(z) = \int_0^\infty t^{z-1} e^{-t}\,\mathrm{d}t \qquad (z > 0) \tag{23.52}$$

then $\bar{\sigma}$ is

$$\bar{\sigma} = \frac{\sigma_0}{V^{1/m}} \Gamma\left(1 + \frac{1}{m}\right) \tag{23.53}$$

Table 23.3 lists several values of $\Gamma(z)$ for $1 \leq z \leq 2$.

z	1	1.1	1.2	1.3	1.4	1.5	1.6	1.7	1.8	1.9	2
$\Gamma(z)$	1	0.9514	0.9182	0.8975	0.8873	0.8862	0.8935	0.9086	0.9314	0.9618	1

Table 23.3: Values of the gamma function, $\Gamma(z)$, for $1 \leq z \leq 2$.

23.10.3 Three-Point Bending

Consider now the case of three-point bending. From (23.45) with $\sigma_u = 0$ and $\sigma = k\sigma_f$ where k is a constant that depends on the geometry and loading of the specimen and σ_f is the failure stress then

$$P_s = 1 - P_f = \exp\left[-\sigma_f^m \int_V \left(\frac{k}{\sigma_0}\right)^m \mathrm{d}V\right] = \exp\left[-\sigma_f^m h\right] \tag{23.54}$$

where

$$h = \int_V \left(\frac{k}{\sigma_0}\right)^m \mathrm{d}V = \frac{1}{\sigma_0^m} \int_V k^m\,\mathrm{d}V \tag{23.55}$$

From (23.48)

$$\bar{\sigma}_f = \int_0^\infty P_s\,\mathrm{d}\sigma_f = \int_0^\infty \exp\left[-\sigma_f^m h\right] \mathrm{d}\sigma_f = \frac{1}{h^{1/m}} \Gamma\left(1 + \frac{1}{m}\right) \tag{23.56}$$

Figure 23.8 illustrates a beam of rectangular cross-section $(b \times d)$, length L and subject to three-point bending with centre-span loading of W. From simple beam theory the bending stress throughout the section of the beam is, (5.30)

$$\sigma = \frac{My}{I} = \frac{Wxy}{2I} \tag{23.57}$$

where $M(= Wx/2)$ is the bending moment, y is measured from the neutral axis and I is the second moment of area. At $x = L/2$, $y = d/2$ then $\sigma = \sigma_f$ and therefore

$$\sigma_f = \frac{WLd}{8I} \tag{23.58}$$

Comparing (23.57) and (23.58) we have

$$\sigma = \left(\frac{4xy}{Ld}\right) \sigma_f \tag{23.59}$$

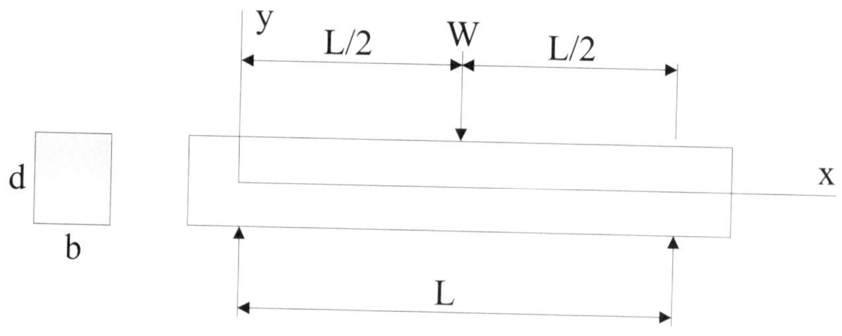

Figure 23.8: Three-point bending of a beam.

and with $\sigma = k\sigma_f$ then we find that $k = 4xy/Ld$. Substituting k into (23.55)

$$h = \frac{1}{\sigma_0^m}\int_V \left(\frac{4xy}{Ld}\right)^m \mathrm{d}V = \frac{1}{\sigma_0^m}\int_0^{L/2}\int_0^{d/2}\left(\frac{4xy}{Ld}\right)^m 2b\,\mathrm{d}x\,\mathrm{d}y = \frac{V}{2\,(m+1)^2\,\sigma_0^m} \tag{23.60}$$

and from (23.56) then the mean strength is

$$\bar{\sigma} = \frac{1}{h^{1/m}}\Gamma\left(1+\frac{1}{m}\right) = \frac{\sigma_0}{V^{1/m}}\Gamma\left(1+\frac{1}{m}\right)\left[2\,(m+1)^2\right]^{1/m} \tag{23.61}$$

Comparing (23.53) and (23.61) we see that $\bar{\sigma}$ is given by

$$\bar{\sigma} = \frac{\sigma_0}{V^{1/m}}\Gamma\left(1+\frac{1}{m}\right) f(m) \tag{23.62}$$

where $f(m) = 1$ for uniaxial tension and $f(m) = [2(m+1)^2]^{1/m}$ for three-point bending. The fact that $f(m)$ is different for different loadings illustrates that the mean strength is a function of the loading system. Table 23.4 lists $f(m)$ for three-point bending for several values of m and illustrates that the effect of the loading system, relative to uniaxial tension, reduces as m increases and the material becomes less brittle. Finally, if $\sigma_u \neq 0$ then

$$\bar{\sigma} = \sigma_u + \frac{\sigma_0}{V^{1/m}}\Gamma\left(1+\frac{1}{m}\right) f(m) \tag{23.63}$$

m	2	4	6	8	10	12
$f(m) = [3(m+1)^2]^{1/m}$	4.24	2.66	2.15	1.89	1.73	1.62

Table 23.4: Variation of $f(m)$ for three-point bending.

23.10.4 The Size Effect

In the previous section we derived (23.62) for the mean strength of a specimen subject to either tension or bending for the case $\sigma_u = 0$. The mean strength is a function of $f(m)$ which is observed to be dependent on the loading system only. This is a general result and if we therefore test (in a similar manner) two similar specimens of different volumes V_1 and V_2 then their respective mean strengths $\bar{\sigma}_1$ and $\bar{\sigma}_2$ are, (23.62)

$$\bar{\sigma}_1 = \frac{\sigma_0}{V_1^{1/m}}\Gamma\left(1+\frac{1}{m}\right) f(m); \quad \bar{\sigma}_2 = \frac{\sigma_0}{V_2^{1/m}}\Gamma\left(1+\frac{1}{m}\right) f(m) \tag{23.64}$$

so that

$$\frac{\bar{\sigma}_1}{\bar{\sigma}_2} = \left(\frac{V_2}{V_1}\right)^{1/m} \tag{23.65}$$

Equation (23.65) is useful for comparing the strengths of different sized test specimens and for determining the strength of large components from small test specimens. Table 23.5 lists the values of (23.65) for several values of V_2/V_1 and m and illustrates that the size effect reduces as m increases and the material becomes less brittle.

	$\bar{\sigma}_1/\bar{\sigma}_2$					
	m=2	4	6	8	10	12
V_2/V_1=2	1.41	1.19	1.12	1.09	1.07	1.06
4	2	1.41	1.26	1.19	1.15	1.12
6	2.45	1.57	1.35	1.25	1.2	1.16
8	2.83	1.68	1.41	1.3	1.23	1.19

Table 23.5: The size effect on mean strength.

23.10.5 Determination of σ_0 and m

For $\sigma_u = 0$ equation (23.44) can be transformed to the following

$$y = \ln\ln\left(\frac{1}{1-P_f}\right) = m\ln\sigma + \ln\left(\frac{V}{\sigma_0^m}\right) \tag{23.66}$$

so that the graph of y against σ is a straight line with slope m, Figure 23.9. By performing a series of experiments (such as three-point bending) and performing a least squares regression leads to estimates for m and σ_0. A least squares fit results in the following equations, Muller *et al.* (1995).

$$\begin{aligned} &P_i = \frac{i-0.3}{N+0.4}; \quad y_i = \ln\ln\left(\frac{1}{1-P_i}\right); \quad x_i = \ln\sigma_i; \quad A = -m\ln\sigma_0 \\ &m = \frac{N\sum_{i=1}^N x_i y_i - \sum_{i=1}^N x_i \sum_{i=1}^N y_i}{N\sum_{i=1}^N x_i^2 - \sum_{i=1}^N x_i \sum_{i=1}^N x_i}; \quad A = \frac{\sum_{i=1}^N y_i \sum_{i=1}^N x_i^2 - \sum_{i=1}^N x_i \sum_{i=1}^N x_i y_i}{N\sum_{i=1}^N x_i^2 - \sum_{i=1}^N x_i \sum_{i=1}^N x_i} \end{aligned} \tag{23.67}$$

The case of $\sigma_u \neq 0$ is more difficult and in practice the two Weibull parameters (σ_0, m) suffice with $\sigma_u \approx 0$ for brittle materials.

23.11 Conclusion

This Chapter has examined the statistical strength of brittle materials. Firstly we provided an introduction to various statistical measures such as the representation of sample data, relative and cumulative frequencies, mean, variance, standard deviation, probability and the probability density function. The normal

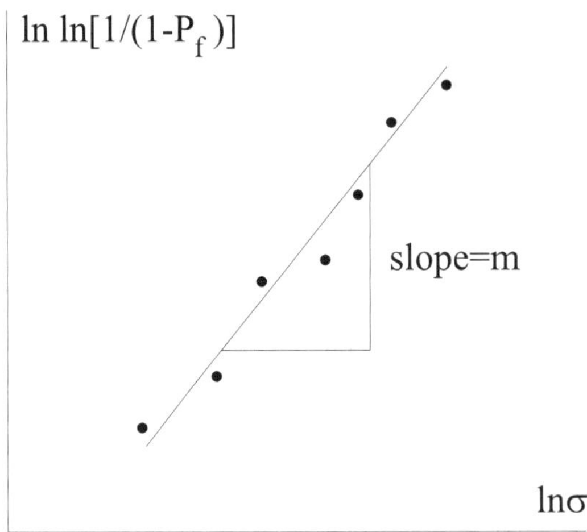

Figure 23.9: Determination of the Weibull modulus m.

and Weibull distributions were then examined. The Weibull distribution is frequently used to determine the strength of brittle materials and the Chapter concluded by deriving estimates for the mean strength of both uniaxial tension and three-point bend specimens.

23.12 References and Further Reading

∘ Jayatilaka, A. de S., (1979) *Fracture of Engineering Brittle Materials*, Applied Science Pub, London.

∘ Kreyszig, E. (1983) *Advanced Engineering Mathematics*, 5th Ed., Wiley & Sons, Canada.

∘ Muller, W. H., Romme, R. and Bornhauser, A. C. (1995) *Applications in Ceramic Structures, In Probablistic Structural Mechanics Handbook: Theory and Industrial Applications*, (Ed. C. R. Sundararajan), Chapter 30, Chapman and Hall.

∘ Weibull, W. J. (1951) J. Appl. Mech., 18, 293.

23.13 Exercises

23.1 Ten specimens of a glass-fibre-polyester resin unidirectional laminae composite were tested in tension in the direction of the fibres and the following failure strength (MPa) were measured

$$\{710, 650, 680, 700, 680, 710, 700, 750, 710, 740\}$$

Generate a table similar to Table 23.1 which tabulates for each sample point the failure strength, absolute frequency, relative frequency, cumulative absolute frequency and cumulative relative frequency.

23.2 Calculate the mean, variance and standard deviation of the failure strengths of Exercise 23.1.

23.3 Find the probability that one throw of a die is either an even number or a multiple of 3.

23.4 Determine the mean and variance for the probability density function $f(x) = x/2$ in the interval $0 \leq x \leq 3$.

23.5 Determine the probability $P(X \leq 2)$ where X is assumed to be of a normal distribution for the two cases: i) mean 0 and variance 1 and ii) mean 0.4 and variance 4.

23.6 A specimen of brittle material is tested in uniaxial tension. Determine the mean strength of the specimen assuming that the specimen material can be described by the Weibull distribution. The volume of the specimen is 0.225m^3 and the Weibull parameters are $\sigma_0 = 650$MPa and $m = 10$.

23.7 A specimen is made of the same material but of twice the volume as the specimen of Exercise 23.6. Determine the expected mean strength of this new test specimen.

Index

About the Author

Graham M. Seed completed a B.Eng. degree at Leeds University, followed by an M.Phil. and Ph.D. at Sheffield University in Mechanical Engineering. He has lectured in the disciplines of strength of materials and computer programming techniques for several years. His primary research interests involve computational mechanics and object-oriented methods. He is presently a lecturer at the Department of Mechanical and Chemical Engineering at Heriot-Watt University. Dr. Seed is also the author of *An Introduction to Object-Oriented Programming in C++: With Applications in Computer Graphics* published by Springer-Verlag.

SAXE-COBURG
PUBLICATIONS
mm